NON SEQUITUR

by Wiley

"TALK ABOUT LUCK! I RAN INTO A TOW TRUCK!"

LLAR & BILL HINDS

JERSEYS ARE
THE COURT IN
PROTESTING!

HOW CAN A TEAM PLAY WITH DIGNITY LOOKING LIKE--

The Origin of the Solar System

NATO ADVANCED STUDY INSTITUTE
'THE ORIGIN OF THE SOLAR SYSTEM'

Director of Institute **Professor S.K. Runcorn,** F.R.S.
Scientific Director **Professor W. H. McCrea,** F.R.S.
Scientific Secretary **Dr. S. F. Dermott**

Local Organizing Committee **Mr. W. F. Mavor**
Miss A. Codling
Mrs. J. Roberts

Held in the School of Physics at the
University of Newcastle upon Tyne—
29 March–9 April 1976

The Origin of the Solar System

Edited by

S. F. Dermott

Institute of Lunar and Planetary Sciences,
School of Physics,
University of Newcastle upon Tyne

A Wiley–Interscience Publication

JOHN WILEY & SONS

New York · Chichester · Brisbane · Toronto

Library of Congress Cataloging in Publication Data:

NATO Advanced Study Institute, University of
Newcastle upon Tyne, 1976.
The origin of the solar system.

'A Wiley–Interscience publication.'
Papers present at the NATO Advanced Study Institute held at the University of Newcastle upon Tyne, March 29–April 9, 1976.
Includes indexes.
1. Solar system—Origin—Congresses.
I. Dermott, S. F.
QB501.N23 1976 521'.54 77–7547

ISBN 0 471 27585 9

Printed in the United States of America

CONTRIBUTORS

ALFVÉN, H., *Department of Applied Physics and Information Science, University of California at San Diego, La Jolla, California 92093, U.S.A.*

ARRHENIUS, G., *Mail Code A-020, Scripps Institution of Oceanography, La Jolla, California 92093, U.S.A.*

BARSHAY, S. S., *Department of Chemistry and Department of Earth and Planetary Sciences, Room 54-1220, M.I.T., Cambridge, Massachusetts 02139, U.S.A.*

BLACK, D. C., *Theoretical Studies Branch, Space Science Division, NASA—Ames Research Center, Moffet Field, California 94035, U.S.A.*

BROWN, G. M., *Department of Geological Sciences, Science Laboratories, University of Durham, South Road, Durham DH1 3LE, U.K.*

CAMERON, A. G. W., *Center for Astrophysics, 60 Garden Street, Cambridge, Massachusetts 02138, U.S.A.*

DELSEMME, A. H., *Department of Physics and Astronomy, University of Toledo, 2801 West Bancroft Street, Toledo, Ohio 43606, U.S.A.*

FREEMAN, J. W., *Department of Space Physics and Astronomy, Rice University, P.O. Box 1892, Houston, Texas 77001, U.S.A.*

GOETTEL, K. A., *Department of Earth and Planetary Sciences, Box 1169, Washington University, St. Louis, Missouri, 63130, U.S.A.*

GOUDAS, C. L., *Department of Mechanics, University of Patras, Patras, Greece.*

HALIOULIAS, A. A., *Department of Geodesy, Technical University of Athens, Athens, Greece.*

HARRIS, A. W., *Jet Propulsion Laboratory, California Institute of Technology, 4800 Oak Grove Drive, Pasadena, California 91103, U.S.A.*

HEPPENHEIMER, T. A., *Max-Planck-Institut für Kernphysik, 69 Heidelberg 1, Postfach 103980, West Germany.*

HERBIG, G. H., *Lick Observatory, University of California, Santa Cruz, California 95064, U.S.A.*

HERNDON, J. M., *University of California, San Diego, Department of Chemistry, B-017, La Jolla, California 92093, U.S.A.*

KATSIARIS, G. A., *Department of Mechanics, University of Patras, Patras, Greece.*

KERRIDGE, J. F., *Institute of Geophysics and Planetary Physics, University of California, Los Angeles, California 90024, U.S.A.*

KIRSTEN, T., *Max-Planck-Institut für Kernphysik, 69 Heidelberg 1, Postfach 103980, West Germany.*

LARIMER, J. W., *Department of Geology, Arizona State University, Tempe, Arizona 85281, U.S.A.*

LARSON, R. B., *Yale University Observatory, New Haven, Connecticut, 06520, U.S.A.*

MCCREA, W. H., *87 Houndean Rise, Lewes, East Sussex BN7 1EJ, U.K.*

PRENTICE, A. J. R., *Department of Mathematics, Monash University, Clayton, Victoria, Australia 3168.*

REEVES, H., *SEP/SES/Bâtiments 28, Centre d'Etudes Nucleaires de Saclay, Boite Postale No. 2, 91190 Gif-sur-Yvette, Saclay, France.*

RUNCORN, S. K., *Institute of Lunar and Planetary Sciences, Department of Geophysics and Planetary Physics, School of Physics, The University, Newcastle upon Tyne, NE1 7RU, U.K.*

STEVENSON, D. J., *Research School of Earth Sciences, The Australian National University, Box 4, P.O., Canberra, A.C.T., Australia 2600.*

TOZER, D. C., *Department of Geophysics and Planetary Physics, School of Physics, The University, Newcastle upon Tyne, NE1 7RU, U.K.*

WILLIAMS, I. P., *Department of Applied Mathematics, Queen Mary College, Mile End Road, London, E1 4NS, U.K.*

WOOLFSON, M. M., *Department of Physics, University of York, Heslington, York YO1 5DD, U.K.*

CONTENTS

FOREWORD BY THE DIRECTOR OF THE NATO INSTITUTE

The suggestion to hold a meeting on the origin of the solar system in the United Kingdom was first suggested by Professor M. M. Woolfson of the University of York at a meeting of the Planetary Sciences Sub-Committee of the National Committee for Space Research. I was later asked to hold it as one of a series (the tenth) of the NATO Advanced Study Institutes of the School of Physics of the University of Newcastle upon Tyne. I always hoped that this series of meetings would have a role which would not unduly overlap that of other international meetings in geophysics. I believed that three features of the NATO scheme were of especial relevance to this field. These were (1) the emphasis on giving postgraduates and young research workers access to a wide group of prominent scientists in some fundamental field of geophysics and (2) the emphasis on international collaboration, especially between Europe and N. America, but not excluding other countries, so necessary in geophysics, and (3) the emphasis on the bridging of disciplines again so necessary in understanding the complex processes of Earth and planets formation and evolution. Many international conferences contribute to one or other of these aims but the NATO Advanced Study Institute uniquely, I believe, covers them all.

Few subjects could be as suitable as the origin of the solar system for a NATO Advanced Study Institute. It is a subject *par excellence* for the young, who can assimilate all the new data from space research, and for the revival of older subjects, celestial mechanics, meteorite research, etc., which space exploration has brought about. It is for them to inject the new ideas which may make sense of the new data and throw new light on the observations. It is a field in which Europe and N. America have presently a special role to fill. We have been lately reminded by the centenaries of Kepler, Copernicus, and Galileo and now this year Gauss, of the historic role Europe has played in the development of our understanding of the Earth and solar system and that a contribution worthy of its past one to the present space exploration of the solar system being carried out by the USA and USSR must be an aim of the European community.

It is perhaps in the study of the origin of the solar system that scientific endeavour and humanism make most contact. The remarkable advances in lunar and planetary exploration may well be mankind's most lasting achievement of the second half of the twentieth century. Just as the development of the technology of Gothic architecture in the eleventh and twelfth centuries resulted in a cultural achievement which has been understood and appreciated throughout history, so the development of space technology and its use for the peaceful purpose of understanding the immediate astronomical environment of the Earth will probably be regarded in future centuries as being the most significant contribution of our times to human culture. We can also be confident that, like other examples of pure, fundamental research, development of lunar and planetary exploration, apparently satisfying only scientific curiosity, is bound to have practical relevance for the human race. The better understanding of the physical and chemical processes in the solar system cannot but have direct relevance to the solution of the environmental problems that are now so much debated.

Finally no subject is in more need of an interdisciplinary approach: celestial mechanics and classical astronomy have made their important contribution to one of the most insistent questions occurring to the mind of Man 'How did the solar system form?' That they have not given the key probably means that essential facts and fundamental ideas elude us. Geochemistry and geophysics have an essential role to play. The meeting and the proceedings will have contributed to these objectives but perhaps particularly to the last: the chemists, physicists, and astronomers talked to each other.

As in such meeting the scientific and administrative planning was successful in so far as it was a cooperation: many ideas were fed in and those whose names appear on the title page were the willing instruments of the wider community of scientists and University staff whose contributions are gratefully acknowledged.

S. K. Runcorn F.R.S.

FOREWORD BY THE SCIENTIFIC DIRECTOR

The cosmogonic problem to which modern astronomy has given by far the most attention is the problem of the origin of the solar system. There is still no agreed solution, but astronomers are continually learning to formulate the problem in more meaningful terms as well as learning more about the basic physical processes that have been operative. They think they know most of the picture that the jig-saw has to reproduce and they believe they possess most of the pieces of the jig-saw. But they are still struggling to get the pieces to fit. This book presents the picture, or at any rate it describes parts of the picture and how they have been acquired. And it looks at pieces of the jig-saw and the manner in which some astronomers propose to fit some of them together. Not enough of the picture has been reproduced to convince everybody that any considerable part has been got correctly, but most workers hope that they have put together a few pieces that will not have to be pulled apart any more.

In regard to the origin of any physical body, the first question concerns the provenance of its raw material. When astronomers first considered in particular the case of the planets, it seemed to them obvious either that the necessary material must have been present when the Sun was formed and have been left over by whatever process produced the Sun as an isolated heavenly body, or else that the material was pulled out of the Sun, or maybe another star closely associated with it at the time—by bodily collision, tidal action or a nova outburst—after the Sun had been formed. The first possibility led to the hypothesis of a solar nebula; the second led to the various catastrophic theories. It was a long time before the general prevalence of interstellar matter (ISM) gained acceptance, and longer still before ISM came to be regarded (a) as the most natural raw material for the planets after some had been captured into the Sun's gravitational field, or (b) as going to produce numerous stars and planets all at about the same time, or (c) as going to produce stars in such conditions that catastrophic processes are likely actually to occur. At the present time, most studies of the formation of the solar system do treat ISM, of much the sort now observed, as the required raw material. Nevertheless the earlier ideas still colour much of the thinking about the problem.

In quite recent years, much new knowledge has been gained about ISM, its composition as regards abundances of the elements, its state as regards ionic, atomic, molecular, and grain components, its density, temperature, and state of turbulence, about the way in which these properties depend upon the progress of the material through the structural features of the Galaxy (density waves, shock fronts, etc.), and even about secular changes in the properties as some of the material is reprocessed through new generations of stars.

During the same time, there has been much study of the physical processes that are generally agreed to play essential parts in the formation of planets and satellites, the want of agreement being largely as to the manner in which these processes combine to produce the desired outcome. They include the differentiation of material, gas and dust, under gravity (the role of mutual adhesion of grains being still subject to uncertainty), the operation of gravitational instability to initiate condensations on various scales (although the case of gas–dust mixtures is not yet fully elucidated), various means for the transfer of angular momentum by various sorts of viscosity or by electromagnetic action, and chemical processes in interstellar and circumstellar material.

Also in recent times, partly as direct results of space missions and observations from outside the Earth's atmosphere, and partly in consequence of the stimulus given by such results to new ground-based studies in various parts of the electromagnetic spectrum, there has been amassed much fresh knowledge of the boundary conditions—or constraints, as it is fashionable to say—of the problem. Besides the already mentioned new knowledge about chemical abundances in ISM generally, there is new knowledge of element and isotope abundances in various parts of the solar system itself. There is new knowledge of the mechanical and magnetic properties of planets, of the gas-dynamical and thermodynamic properties of their outer layers, and of conditions in the circumsolar space in which they exist, including properties of the solar wind. Above all, there are greatly improved models of the individual planets and satellites, so that cosmogonists now know rather well the sort of bodies whose existence they are called upon to explain.

Here, however, a clear distinction must be kept in mind between two sorts of problem. On the one hand there is the problem of constructing a model of some particular body, being given an exact mass, an exact chemical composition and exact equations of state; here a highly precise result is demanded. On the other hand, there is the problem of *predicting* that so many bodies having such-and-such masses will be formed, being given a minimum of parameters of the raw material; here success as to order of magnitude would be beyond anything yet achieved. The comparison is rather like that between accountancy, where exactitude is of the essence, and budgeting, where overall reliability is the aim, but exactitude is meaningless.

Finally, it is well to recall how the cosmogony of the solar system is part of a larger problem of *evolution*. The formation of this system is a tiny incident in the evolution of the Galaxy, and an even tinier one in the evolution of the Universe. It follows that in whatever way we treat the problem, the stage from which we start is arbitrary, while the only natural place to end is with what we actually observe now. So a proper treatment of the formation of the solar system should include its evolution into its present state under the dynamical interaction of its parts, the action of the solar wind, and so on.

Although planets are such tiny bodies in the cosmos, as a species they are, so far as we can tell, the culmination of all material evolution. For if we consider any other species at any epoch, we infer that some of the matter of at any rate some of its bodies subsequently goes to form other bodies of the same or other species. But once any matter has helped to form a planet, it will never help to form anything else. In this way, our subject possesses unique significance.

Such in brief seems to be the setting of the topics of this book and of the proceedings of the NATO Advanced Study Institute upon which it has been based. Clearly, only a selection of the particular topics covered by this conspectus could be treated. Those actually treated were chosen because they are central to the subject, because they are some in which particular progress is currently being made, because it was desired that the proceedings should be not unduly disjointed, and for a variety of other reasons. I hope that this attempt to offer a broad survey may assist the reader to see how the topics fit into the subject as a whole.

As the one invited to serve as Scientific Director of the Institute and to contribute an introduction to the book, I should like to record the gratitude of all participants to the scientists who advised the organizers and to all the other contributors to the proceedings. Special thanks are due also to the University of Newcastle upon Tyne and to Professor S. K. Runcorn who directed the organization with the expert help of Mr W. F. Mavor and his staff.

All concerned as participants, speakers or contributors to this book owe a special debt to Dr S. F. Dermott for his indefatigable efforts as Scientific Secretary of the Institute and as Editor of this resulting record of its proceedings.

WILLIAM H. MCCREA F.R.S.
1976, September 17

PREFACE

The NATO Advanced Study Institute on 'The Origin of the Solar System' was held in the School of Physics of the University of Newcastle upon Tyne from 19 March–9 April, 1977, and was the tenth in a series of interdisciplinary meetings on Earth and Planetary Sciences inspired and directed by Professor S. K. Runcorn. The aim of the meeting was to appraise the constraints placed on models of the origin of the solar system by our knowledge of present and past states of stars, planets, satellites, comets, asteroids, and meteorites.

The format differed markedly from the others in the series in that the contributions were confined to reviews. The number of lecturers was comparatively small (less than a quarter of those attending read a paper) and a majority were given between two and three hours to present their views, with a further thirty minutes being allowed for discussion after each hour-long talk. This format proved to be remarkably fruitful, particularly for the younger students, even though it meant that a number of the scientists present who had made important contributions to the field were not able to present a paper. Everybody however, was able to contribute to the extensive discussion sessions and this was, perhaps, the most enjoyable and rewarding aspect of the meeting. The faculty deserves our thanks for respecting the educational aims of the Institute and the multidisciplinary background of their audience.

All of the talks given at the ASI are, without exception, published in this volume. Two other papers were not actually read at the meeting but are contributions invited by the editor to cover topics raised in the discussion periods. A further two papers have been published previously but these are reproduced here to complete the record. The student will find that most of the subjects which contribute to our understanding of the origin of the solar system are discussed in this book.

As most of the participants could not present papers, it was even more difficult than usual for them to obtain funds from their own institutions, and so we are particularly grateful to the Scientific Affairs Division of NATO whose generous grant made this meeting possible.

As Scientific Secretary of the Institute, I would like to thank Professor Runcorn, whose influence and actions ensured that the Institute would meet the high standards set by previous meetings, and Professor McCrea for his constant counsel in scientific matters. Thanks are also due to those other scientists, particularly Professor Woolfson, who advised us on the choice of faculty and other matters.

The domestic arrangements were handled, with tact and skill, by Mr Mavor, Mrs Roberts, and Miss Codling. I also thank Mrs Roberts for her considerable help with the editorial work and Mrs Dermott for preparing the index.

S. F. Dermott
Cornell University

THE ORIGIN OF THE SOLAR SYSTEM

H. REEVES
Centre d'Etudes Nucleaires de Saclay, Saclay, France

The first studies concerned with the origin of the solar system were those of Descartes (1644). During the last three hundred years many famous scientists, including Buffon, Kant, Laplace, Jeans, and Hoyle, have looked at this problem and have put forward various widely differing theories.

In 1975 where do we stand? Have we progressed since the time of Descartes? By synthesizing the contributions from astronomy, space physics, and mineralogy it is now possible to eliminate some theories and to assess the plausibility of others.

The 'best fit' explanation seems to be that of the Laplacian protosolar nebula: the Sun and its associated planets were formed at the same time from a cloud of interstellar matter which condensed about 4·6 billion years ago. The rapidly rotating protosolar nebula formed from this cloud as a flattened, gaseous disk. The Sun was formed in the central portion of this disk, the planets from the outer portions.

How much confidence can we place in this 'best fit' theory? Not very much, and it is this which is discussed here.

Paradoxically, the subjects least well understood in astronomy are those which concern us most. We know more about the internal structure of the Sun than we do of that of the Earth, and more about the evolution of stars than the origin of the solar system.

Astronomical observations can be made on thousands of stars but on only one solar system and it is this which causes all the difficulties. A biologist who was able to examine only one sort of tree would find the study of evolution very difficult.

Even the stars nearest to us are very far away. We do not know whether they have planetary systems analogous to our solar system. One or two stars, for example Barnard and Epsilon Eridani, do show evidence of having some planets associated with them but we are not able to determine the orbits of these planets.

Where should we begin our study of the origin of the solar system? How may we reconstruct the sequence of events, their causes and effects, which resulted in the formation of the Sun with its encircling planets which themselves possess orbiting satellites?

The steps in this investigation are very like those of a Sherlock Holmes, called in to discover the perpetrators of a particularly well-planned crime. At first he searches for useful clues: ash, footprints, etc; then he reconstructs the crime with the help of all these clues. If this reconstruction lacks psychological probability or inherent consistency or if new clues are discovered, the detective may be obliged to alter it to a greater or lesser degree, or even to reject it completely. If the crime is seemingly 'perfect' and the clues few or uninformative, many different reconstructions would be able to account for the facts. The detective would naturally have his preferred version but he would not be able to reject the others without new information. At the same time he continues his enquiries and examines new clues for inherent consistency with each of his theories. By feeling his way and using his intuition, the detective builds up his case until he understands exactly what has happened and can arrest the criminals.

For the last three centuries, numerous astronomers have tried to understand the origin of the solar system. They have used contemporary methods of observation to obtain data from which they have proposed cosmogonies or constructed 'models'. Many investigators have pursued these steps for a number of years, adding to or altering their existing models, or even adopting new ones as a result of new observations. For example Fred Hoyle, in the course of his career, has constructed many models which have very little in common with one another.

As a result, we are today confronted by a large number of models. None is universally accepted and very few can be definitely eliminated. The cause of this confusion is the difficulty in finding discriminatory criteria, that is to say observations which permit the preference of one model over another. The models are very flexible, they are relatively easily adapted to accommodate new observations and hence their ability to predict phenomena is relatively poor. This results from the great complexity of a subject which uses nearly all branches of the physical sciences at one time or another, but in conditions only remotely akin to those met in the laboratory. It is understandable that the subject has evolved slowly.

In this article I propose to give firstly a general discussion on a number of models of the solar system. I shall attempt to evaluate their respective acceptability and I will give the reasons for our belief in the 'best fit' model. Secondly, I will reconstruct the sequence of events leading to the formation of the solar system, choosing the chronological stages which seem to me to be the most likely. Even so the argument is sometimes highly speculative and some of it borders on science fiction.

The Various Theories

I would like to begin by dividing the different models into groups which are defined by the types of initial hypotheses, then I will test these groups with some data, partly observational and partly theoretical, in order to assess the degree of realism of each group. The groups are defined by the response of each model within them to the following two questions:

Were the Sun and the planets formed at the same time, in other words are they co-genetic?

Were the planets formed from interstellar material having a 'cosmic' chemical composition (see Table 1), or from stellar material, that is interstellar material which has been modified by previous passage in a star.

We define four groups of models according to their responses to these questions (see Table 2).

The first group supposes that the Sun and the planets were formed at the same time from the same cloud of interstellar material. This approach was first suggested by Kant and subsequently developed by Laplace; the hypothetical celestial object corresponds to the 'protosolar nebula' of Laplace. We believe in this case that after fragmentation of an interstellar cloud a rapidly rotating mass of gas was formed (we shall see the reason for this later), in the shape of a flattened disk. In this cloud, non-volatile elements condensed and accreted to form the planets in the relatively cold regions, relatively far from the rotational centre. The central portion contracted to form the Sun. In this group of models we have the recent work of Kuiper, Cameron, Schatzman, Levin and some theories of Hoyle.

The second group of models similarly requires the formation of the planets from a cloud of interstellar material but supposes that this cloud had been captured by the previously formed Sun. Hoyle and Lyttleton, for example have proposed the following sequence: at some stage the Sun passed through one of the many dense clouds which occur throughout the galaxy. Having a gravitational field, the Sun was able to capture a portion of this gas, leaving in its wake

Table 1. Abundances of chemical elements in the interstellar material (gas and dust). These are known as the 'cosmic' abundances in number of atoms relative to silicon

Hydrogen	$2{\cdot}5\times10^{10}$	Beryllium	0·45	Neon	3×10^{6}	Sulphur	4×10^{5}
Deuterium	4×10^{5}	Boron	4	Sodium	5×10^{4}	Calcium	6×10^{4}
Helium-3	4×10^{5}	Carbon	10^{7}	Magnesium	10^{6}	Iron	8×10^{5}
Helium-4	2×10^{9}	Nitrogen	3×10^{6}	Aluminium	8×10^{4}		
Lithium	25	Oxygen	2×10^{7}	Silicon	10^{6}		

Table 2. Groups of models for the origin of the solar system, divided in accordance to their response to the two questions:
1. Are the Sun and the planets co-genetic?
2. Was the planetary material altered or unaltered interstellar material?

	Planets formed from unaltered interstellar material	Planets formed from stellar material (material which has been at stellar temperatures)
Sun and planets are co-genetic	Kant, 1755 Laplace, 1796 Hoyle, 1960 Edgeworth, 1949 McCrea, 1960 Whipple, 1948 Urey, 1946 Von Weizsäcker, 1944 Kuiper, 1951 Ter Haar, 1950 Gurevich and Lebedinsky, 1950 Schmidt, 1959 Levin, 1972 Safronov, 1972 Cameron, 1962 Schatzman, 1963	Gunn, 1932 Lyttleton, 1940 Hoyle, 1944 Egyed, 1960
Sun and planets are not co-genetic	Berkeland, 1912 Berlage, 1927 Lyttleton, 1961 Hoyle, 1956 Sekiguchi, 1961 Schmidt, 1944 Alfvén, 1942 Woolfson, 1964	Buffon, 1745 Bickerton, 1818 Arrhenius, 1913 See, 1910 Jeans, 1916 Chamberlin, 1901 Moulton, 1905 Jeffreys, 1929 Lyttleton, 1937 Russell, 1935 Banerji and Srivastras, 1963

a long empty tunnel. The captured material subsequently encircled the Sun forming a nebula similar to that of Laplace.

This group includes the hypotheses of Alfvén and Arrhenius. They suppose that the young Sun was formed leaving no residual gas cloud, but having a strong magnetic field (as we shall see later, this is not unreasonable). The Sun's gravitational field attracted to it the surrounding interstellar material. The neutral atoms in the galactic gas moved towards the Sun and were ionized by the solar photon flux. When this happened they were captured by the Sun's magnetic field and became part of the rotating solar mass. This process formed, around the Sun, an envelope in which the non-volatile elements were gradually condensed and subsequently formed the planets. In contrast to the preceding models, in this case there is no massive cloud but a gradual addition of interstellar material. The density of this envelope was very low and the thermodynamic conditions in it very far from equilibrium.

The third group of models supposes that the Sun, at formation, was part of a binary system; in other words it was associated with a twin sister with which it formed a double star. This sister star, for some reason, disintegrated and its material was dissipated into space. A portion of this gas was captured by the Sun and formed a cloud similar to that discussed in the previous theories. This theory is often met in scientific literature and has been defended by Lyttleton, particularly, in the 1940's. In this case the material of the cloud is 'stellar' (since it has been modified by nuclear reactions occurring in the interior of the sister star), and there is a co-genetic relationship since the two stars were formed at the same time.

The fourth group includes models which involve stellar collisions. It is believed that a star could come close enough to, and induce violent tidal effects on the surface of, the Sun. These tides caused material to be ejected from the outermost layers and this material spread out and subsequently orbited the Sun. This model has been supported by Jeans and Jeffreys in particular early in this century and has been revived recently, in a somewhat modified form, by Woolfson. It requires stellar material for the cloud but does not imply a genetic relationship between the Sun and the planets.

Each of these groups of models will now be tested by confronting them with some observed facts. These facts will serve as 'criteria', allowing us in principle to accept one group of models in preference to another. Five criteria will be used for this and the degree to which each group of models is able to accommodate these criteria will be examined.

The Isotopic Abundances of the Elements

One of the more important astrophysical observations of the last two years has been the detection and measurement of the ratio of deuterium (heavy hydrogen) to hydrogen (D/H). For example, we know that this ratio is essentially the same in the atmosphere of Jupiter and in interstellar space whereas in the solar photosphere this ratio is much smaller (in Table 3 only an upper limit is given).

The explanation is simple: planetary material is typically interstellar in

Table 3. Comparison of the abundances of some elements in the interstellar material (gas and dust), in the Sun and in the planets

	Interstellar material	Sun	Planets
Deuterium/hydrogen	2×10^{-5}	$< 3 \times 10^{-7}$	$\sim 2 \times 10^{-5}$
Lithium/silicon	3×10^{-5}	3×10^{-7}	3×10^{-5}
Iron/silicon	~ 1	~ 1	~ 1

composition. Deuterium is an unstable isotope in thermonuclear reactions so that its abundance, already small, is reduced to zero under the conditions found in stellar interiors; this is the reason for its absence in the solar photosphere. This criterion eliminates the two groups of models which require a stellar composition for the planets. These models have previously encountered serious objections which have made them less plausible and it appears that consideration of the D/H ratio definitely rules them out. This assessment is confirmed if we include here another element, lithium, similarly unstable in stellar thermonuclear reactions though somewhat less so than deuterium. The planets and the interstellar medium have similar relative abundances of lithium, which are distinctly different from that of the solar photosphere in which it is lower by a factor of 100. This confirms that the planetary matter has retained its original character whereas the photospheric material has been altered by nuclear reactions occurring within the Sun.

The Sun and the Planets: Their Quasi-contemporary Formation

We are able to determine the age of the planets fairly precisely by measuring the abundances of radioactive isotopes with very long half-lives (rubidium-87, half-life 5×10^{10} years; thorium-232, half-life 2×10^{10} years; uranium-238, half-life $4{\cdot}5 \times 10^9$ years) in meteorites and in terrestrial and lunar rocks. These measurements indicate that the planets solidified about $4{\cdot}6 \times 10^9$ years ago with an uncertainty in this age of about 100 million years.

We are not able to measure the age of the Sun directly. It is possible, however, to obtain an indirect measurement by combining the measurement of the age of the Earth already described with determinations of the abundances of two isotopes with relatively short half-lives, (plutonium-244, half-life 8×10^7 years; iodine-129, half-life 16×10^6 years). These isotopes have decayed long ago and are now undetectable in planetary material. They were present, however, at the time of solidification of this material and have left behind traces of their existence, particularly in the distribution of certain xenon isotopes; this is termed 'fossil radioactivity'.

Why does this observation interest us? It should be stated here that these radioactive atoms, like many other heavy atoms, are formed during the final explosion which characterizes the death of a massive star, and these atoms are ejected into interstellar space. The galactic gas is continually fed with new atoms by successive generations of stars and these contributions maintain a certain degree of radioactivity, not only from the previously mentioned isotopes but also from some others. We find that the abundance of plutonium-244 and iodine-129 at the time of solidification of planetary material shows that only a short period of time separated the isolation of the protosolar material from the interstellar medium and the formation of the planets. Calculations indicate this period to have been about 100 million years. (Recent detection of isotopic anomalies have somewhat complicated the picture.) These observations fit

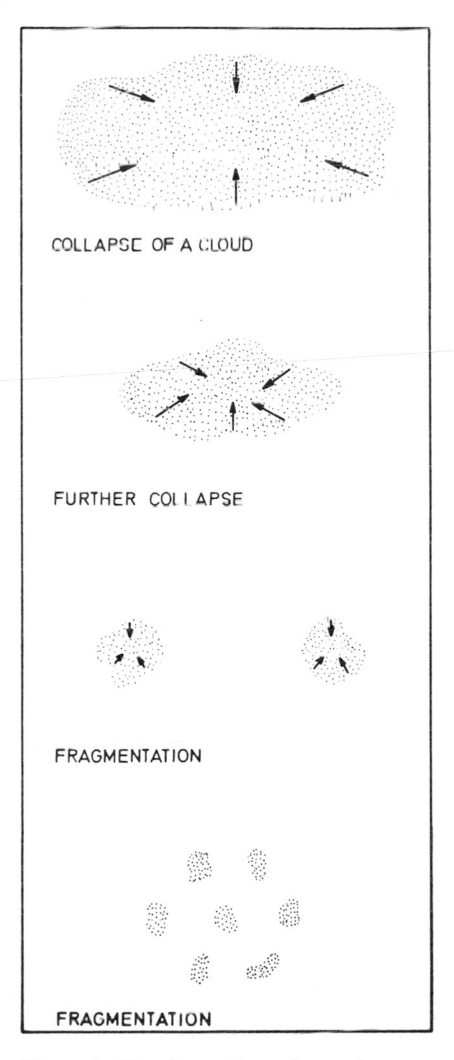

Figure 1. The formation of stars from an interstellar cloud. As it passes into a spiral arm of the galaxy, the massive cloud (many thousands of solar masses) suffers compressive deceleration. If this compression is great enough the cloud collapses and fragments, forming many protostellar nebulae. The final result being a multitude of stars. A small proportion of these may have planetary systems like ours which could support life as we know it

easily within the framework of the protosolar nebular theory of Laplace. The event which isolated the solar system (the Sun and the planets) from the interstellar gas was the collapse of a particularly massive interstellar cloud (Figure 1). The sequence of events—contraction, fragmentation into thousands of protosolar nebulae, re-condensation of fragments, solidification and accretion of non-volatile elements—must have occurred in less than 100 million years.

Our understanding of the structure and dynamics of the galaxy enables us to go further. The clouds of interstellar material, the future protostellar nebulae, are not stationary with respect to the galaxy; they revolve around its centre like the planets revolving around the Sun. As they orbit they pass through the spiral arms and the inter-arm regions of the galaxy (see Figure 2). It seems well-established that the initiation of the contraction of the clouds into stars

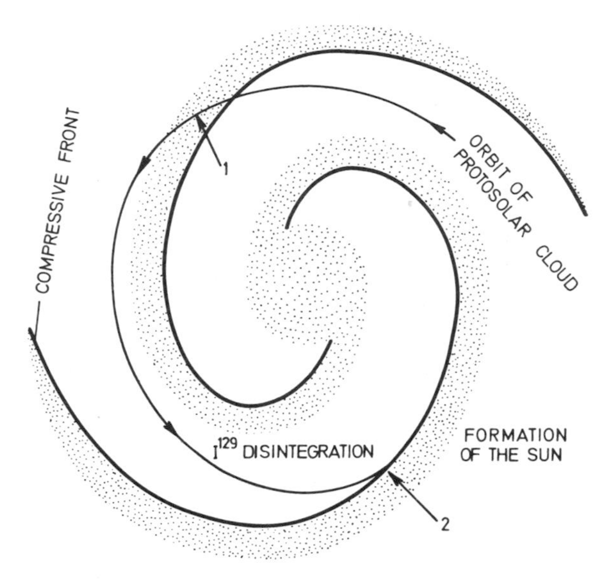

Figure 2. Diagram showing the path of a cloud through the galaxy and the stages preceding the formation of the Sun. The orbital period of the cloud is 200 million years. During each transit from an inter-arm region to an arm region the cloud is capable of being transformed into a star, but the probability of this happening is small. At the first transit illustrated here (1 on the diagram), the cloud does not collapse but is affected by the nucleosynthesis which occurs in the spiral arm and receives a fresh supply of newly-formed atoms. During the second transit (2 on the diagram) it collapses and the Sun is formed. About 100 million years separate these two transits. The ^{129}I atoms, (with a half-life of about 16 million years) which the cloud acquired in the region of active nucleosynthesis, have almost entirely disintegrated

only occurs in certain regions of space. In passing from an inter-arm region into an arm, a cloud suffers a considerable deceleration which strongly compresses it and initiates the self-contraction which forms the young stars within the spiral arms. Once formed, the stars continue in their circum-galactic path accompanied by their parent clouds. Massive stars are short-lived however (some tens of millions of years) and they explode and die before they have even left the galactic arms in which they were born. This suggestion is supported by observations on nearby galaxies; there is a good correlation between the position of the supernovae and the spiral structure.

In the galaxy, there are regions—the arms— where stellar formation and nucleosynthesis are particularly intense and where the level of radioactivity in the gas should be very high. When the gas leaves the arm and travels into the inter-arm regions its radioactivity begins to decline. Since the transit time through the inter-arm regions is about 100 million years the isotopes which have half-lives much shorter than this period will have largely decayed before the cloud enters another arm.

Using these considerations we are now able to reinterpret the difference in age of 100 million years obtained from the iodine-129 and plutonium-244 data. It seems to measure the time which elapsed between the last addition of radioactive atoms to the protosolar cloud and the time of solidification of the planetary system, that is to say the time between passing through an arm in which no sun was formed and passing through a second arm in which it was formed. In other words the greater part of this 100 million years has nothing to do with the condensation itself and cannot be assumed to be the difference in age between the Sun and the planets. This argument strengthens the concept of the quasi-simultaneous formation of the Sun and the planets.

How was the Sun Formed?

It is not easy to transform an interstellar cloud into stars and there are at least three difficulties to be overcome. The clouds are too hot, too magnetic and they rotate too rapidly.

The thermal problem seems to be overcome when the clouds meet the spiral arms, as described earlier.

The magnetic problem appears to be this; the interstellar material is slightly magnetized and we observe in space a field of about 3 microgauss. This field is closely associated with the gas and contracts with it. This contraction causes a progressive increase in the field strength which opposes and slows down further contraction; it is as though the gas has become elastic. In fact calculations indicate that the interstellar clouds are much too magnetic to condense into stars.

The rotation problem acts in a similar fashion; entrained by the differential rotation of the galaxy, the interstellar clouds themselves slowly rotate. During contraction this rotation speeds up; it is the familiar experience of a small child turning on a stool who extends his arms and then clasps them to himself. When the centrifugal force becomes equal to the gravitational force the contraction can proceed no further. Again we see that, quantitatively, the interstellar clouds have too much energy, this time rotational, for them to contract into stars. There is a mechanism, however, by which the excess magnetic field and rotational energy can be dissipated; it involves the galactic magnetic field which induces a coupling of the contracting cloud with the surrounding material. The rotation of the cloud, when accelerated by the contraction, has the effect of dragging along the lines of force which stretch and resist more and more, like an elastic band which is pulled out. This resistance of the surrounding material has the required double effect; it decreases the rotational energy and reduces the magnetic field of the rotating mass, (by a mechanism too involved to develop here).

We can now visualize these events as follows (Figure 3); gravitational energy which is liberated during the contraction of the cloud is partially transformed into rotational energy. When these two energies are approximately equal the cloud has the form of a very flattened disk which is strongly magnetized. The

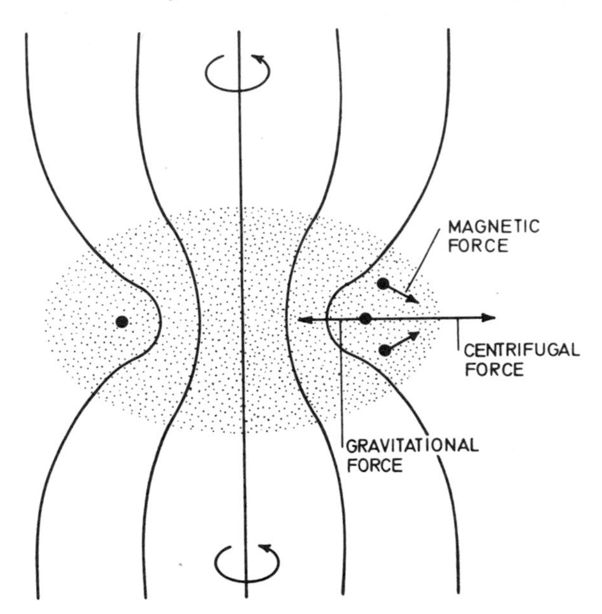

Figure 3. The interplay of the magnetic, gravitational, and centrifugal forces in the protosolar nebula. In contracting, the nebula deforms the surrounding interstellar magnetic field. The lines of force are deflected and a magnetic gradient created which plays an important role in the formation of the nebula by coupling the nebula to the surrounding interstellar gas and allowing a gradual decrease in the magnetic and rotational energy

lines of magnetic force, fixed in part to the interstellar medium and in part incorporated within the rotating cloud, apply a strong restraint which is particularly effective at the periphery of the disk where the stretching is greatest. As the excess energy is dissipated, the cloud contracts and approaches the size of the Sun. These phenomena are in accordance with the group of theories postulating co-genesis; we have produced a flattened disk from which it is easy to form the planets. On the other hand, the non co-genetic models are now even less likely on two grounds: they are required to explain—taking account of the above constraints—how the Sun was formed; they must also introduce a mechanism causing the deceleration and demagnetization of the subsequently captured clouds.

What is the Composition of the Planets?

In this section we study in some detail the chemical and mineralogical composition of different objects in the solar system. The meteorites which fall to Earth certainly make up the most varied sample and are particularly relevant to this study. They can generally be classified into a small number of chemical groups and are then divided up and distributed among chemists, physicists, mineralogists, geologists, etc. These scientists try to understand how, when, and under what thermodynamic conditions the meteorites were formed. For example, one group of meteorites, known as 'ordinary chondrites' contains iron sulphide (FeS) but not iron oxide (Fe_3O_4). If these meteorites were formed from non-volatile elements which had condensed from a nebular gas of interstellar composition, the presence of iron sulphide rather than iron oxide implies a condensation temperature of between ~ 400° and ~ 600° Kelvin (or ~ 125°C and ~ 325°C).

Similar information can be obtained from some of the bulk properties of the planets, in particular their densities. As the distance that they are from the Sun increases, the planets are less and less dense. This is explained quite simply if we accept that their formation temperature decreases with their distance from the Sun. Hence Mercury was formed at a very high temperature (~ 1500 K) at which only metallic iron and some refractory silicates were able to solidify. Venus, formed at about 1000 K, has a lower density and might contain in addition alkalis and alumino-silicates. The Earth, condensed at about 550 K, also containes FeS, etc. At even lower temperatures solids of lower density are formed (water-bearing tremolite, talc, ...). Finally, in the environment of the giant planets, ices are formed (of water, ammonia, and methane).

It is not surprising that the formation temperatures of the planets decreases as their distance from the Sun increases. Indeed this is still so but their formation temperatures were much higher than their present temperatures (fortunately it is no longer 550 K at the Earth's orbital distance). The initial heating could have been supplied either by a much brighter Sun or from within the opaque nebula by a normal Sun and invoking the 'greenhouse effect'. The degree

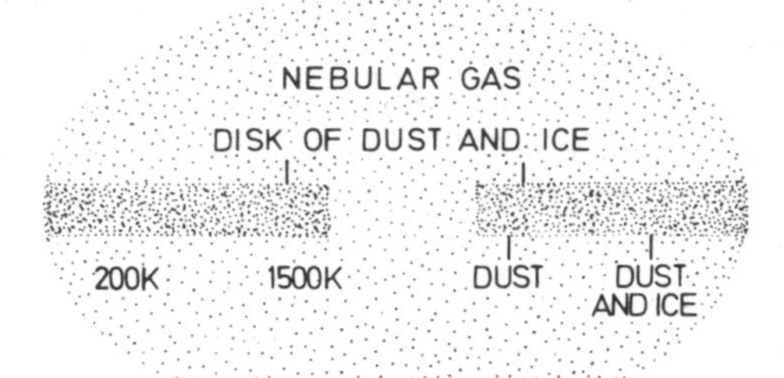

Figure 4. The protosolar nebula. The condensation of the chemical elements has produced a thin disk of dust and ices in the equatorial plane. The planets are formed within this disk

of chemical reaction implied by these observations depends primarily on the temperature but also, to a lesser extent, on the pressure and density within the nebula (Figure 4). If the pressures under which the solids condensed can be estimated, within limits, we can derive an approximate value for the total mass of the protosolar nebula. According to recent work, particularly by the group led by Professor Anders at the University of Chicago, the required pressures are between 10^{-3} and 10^{-6} atmospheres. This gives a nebular mass (without the Sun) of between some hundredths and a few solar masses.

It is interesting to compare this physico-chemical assessment with another derived from astronomy. The present total mass of the planets is about one-thousandth of a solar mass. This is only a lower limit to the nebular mass since it is certain that the nebula has lost a large portion of its initial complement of volatiles. In fact the minor planets have retained only the refractory components of the interstellar material from which the nebula was formed. This refractory component (Table 2 above) is estimated to be a thousandth of the initial mass. The volatile component, composed principally of hydrogen and helium, was dissipated into space, apparently blown away by the solar wind emitted by the young Sun. These considerations give us a new lower limit of about a hundreth of a solar mass. Other arguments, some more convincing than others, suggest a lower limit that is somewhat higher, about a twentieth of a solar mass.

Is it possible to establish an upper limit for the mass of the nebula? Not exactly, but it must be remembered that if we initially postulate a very large mass we also need a mechanism to reduce this. Astronomical observations on young stars (of the T-Tauri type) show that these eject a large proportion of their mass in the form of a very intense stellar wind. The excess mass could be dissipated in this way but the process has its limitations and is unlikely to be capable of reducing the nebula by more than a few solar masses. To summarise: the limits of mass of the Laplace type nebula agree quite well with the range of pressures needed for the different chemical processes responsible for the composition of the planets and meteorites.

On the other hand, hypotheses involving the slow capture of interstellar gas lead inevitably to much lower pressures and are slightly more difficult to reconcile with the observed physico-chemical phenomena. Unfortunately the situation is not as simple as this. Until recently, certain phenomena in meteorites could be explained in terms of physico-chemical processes occurring in thermodynamic equilibrium within a massive nebula and this was relatively well accepted. Different methods of evaluating pressure and temperature had been proposed which gave very satisfactory concordant results. New data, such as the identification of two isotopically distinct components of oxygen in certain meteorites (carbonaceous chondrites) have placed doubt on the most precise of these thermometers. Other thermometric methods, such as that using the abundances of bismuth, indium, and thallium, are equally in doubt. The support given by the physico-chemical criteria to the Laplace-type nebula theory has been similarly weakened but far from negated.

In contrast, the hypothesis involving the progressive capture of the nebular gas implies a large degree of thermodynamic disequilibrium between the gas, which is at high temperature and the solid matter which is at a low temperature. The reactions causing accretion occur within a partially ionized gas and the resulting phases should have certain characteristics. It is possible from this to test this theory: are we able to explain the chemistry of meteorites in terms of thermodynamic equilibria within a massive nebula where the degree of ionization is very small, or do we have definite proof that an ionized component was present? In spite of the work of Alfvén and his group the evidence put forward so far in favour of such a component is not convincing but further studies may shed more light on this possibility.

It should be mentioned here that the study of lunar rocks has shown that the Moon, early in its life, was associated with an intense magnetic field. Was this an after-effect of the initially strong solar magnetic field postulated by Alfvén? Indeed, the discussion of the previous criterion presented us with the idea of an equally strongly magnetized Laplace-type nebula. This shows how difficult it is to establish criteria which discriminate between the various hypotheses.

What do Astronomical Observations tell us?

It should be remembered that many models of the origin of the solar system were put forward before astronomical observations had shown the presence of diffuse clouds in interstellar space. The discovery of these clouds, the study of their properties and the identification within them of a large number of young stars, has shown that an intimate relationship exists between the concentrations of interstellar material and the places where stars are formed. The spiral arms have already been shown to be the best sites for stellar formation. In fact it is this continuous sequence of young stars (accompanied by their clouds of ionised hydrogen) and murky areas which mark out the spiral form of neighbouring galaxies. Seen from Earth, the spiral arms of our galaxy form the Milky Way. Astronomical observations using various parts of the electromagnetic spectrum (visual, infrared and millimetre wavelengths) enable us to identify a large number of clouds and to extend further the study of stellar formation, in particular by the detection of molecules very likely to be forming in condensed phases of the galactic gas. The association of clouds with young stars strongly suggests a co-genetic relationship between the Sun and the planets and supports hypotheses involving this concept. Clouds of several solar masses are also observed emitting in the infrared at the temperatures of around 1000 K; these are believed to be protostellar nebulae.

Here, as often in this subject, we are confronted once more by the following situation. The criteria examined support one group of models but do not eliminate the others. Hence it is easy for those who support less satisfactory models to defend their theories. Their arguments, apparently convincing, seem less natural, less likely, and perhaps this discussion is more a case of intuition. Certainly intuition is often unreliable but in the real world it is probably our best guide.

From this discussion a group of models stands out as particularly favoured; that implying the formation of the Sun and the planets from the same interstellar cloud. It is therefore not remarkable to find that the majority of workers in this field today adopt this point of view. Nevertheless it would be extremely dangerous to use this majority consensus as an objective criterion. Scientists well know the influence certain 'big names' are able to have on a discipline. The weight of authority is always there, in all branches of science and particularly in subjects as difficult and indeterminate as cosmology and cosmogony.

A Plausible Theory

Our galaxy was formed about 15 billion (10^9) years ago as a gigantic gaseous mass, equivalent to 100 billion solar masses, without stars and heavy atoms and containing only hydrogen and helium. The first generation of stars appeared very soon. After a few million years, the more massive stars began to eject the products of their stellar reactions into space, initiating the progressive enrichment of the interstellar gas in heavy atoms. At the same time as these atoms were formed, the initially spherical galaxy collapsed slowly into the flattened shape we now see and extended long tentacles, which subsequently formed the spiral arms. Under the influence of gravity, the interstellar gas separated into irregular clouds, more or less massive, which collided and exchanged thermal and kinetic energy. Orbiting the galactic centre, these clouds passed from one arm to another. During each passage through an arm, a small portion of the cloud—the most dense and the coldest—was violently compressed by the deceleration, passed the point of no return and collapsed. This was accompanied by a massive fragmentation into thousands of sub-systems. The result is the large number of stars scattered throughout the galaxy; the death of the more massive of these continues the process of nucleosynthesis.

After about 10 billion years the proportion of heavy atoms, mainly metallic, in the galactic gas was about 2 to 3 per cent by weight. At this time a cloud, unremarkable in itself but important for us, gave birth to a large number of stars, one of which was the Sun. During collapse, the rotational energy of the cloud was divided between the various masses which the fragmentation produced. These protostellar 'nebulae', flattened by rotation and crossed by the lines of force of the galactic magnetic field, existed in a semi-equilibrium state governed by the interplay of gravitational, magnetic and rotational forces. The slow contraction of the system was accompanied by a gradual increase in temperature and density, particularly in the region of the rotational axis. Collisions between atoms became more and more frequent, molecules were formed and the non-volatile elements condensed rapidly. The interstellar dust, already present in the initial cloud, provided nucleation centres. Gravity caused this dust to fall rapidly towards the equatorial plane of the nebular disk, forming what appeared to be a new dusty disk, much thinner than the nebular disk itself.

In the vicinity of the major planets, the temperature was relatively low and ices of water, ammonia and methane condensed on the surfaces of the dust grains. The condensed fraction of the nebular material was many times greater here than in the vicinity of the minor planets where only silicates and iron (and iron compounds) were able to condense.

What made the disk of dust divide into rings and what caused each ring to transform into a planet? This is the most obscure part of the story. The orbits of the dust grains might have been relatively irregular, and turbulent. Statistically these movements could produce not only changes in density but also the formation of 'clouds' of dust, in a similar way that clouds formed from interstellar material.

It follows, by analogy, that these clouds would collapse under their own weight when they reached a critical mass to form a large number of planetary nuclei. The nuclei in the outer regions, furthest from the rotational axis, where temperatures were low, contained a large proportion of ices and were able to retain, almost completely, their surrounding hydrogen and helium. It is this which has caused the disparity between the major and the terrestrial planets, which totally lost their hydrogen and nearly all their ices. If the composition of the terrestrial planets is subtracted from the solar composition, the remainder is close to the mass of the major planets. The formation of Jupiter and Saturn took place in stages very similar to those forming the solar system. The mass of gas, isolated from the rest of the nebula by the gravitational attraction of the planetary nucleus, possessed an appreciable magnetic field and kinetic energy compared with those of the central nucleus. The same mechanism for reducing rotational energy by coupling the magnetic fields was apparently capable of functioning on a planetary scale, in the same way as that forming the planetary system as a whole, and produced the satellites of the major planets. The minor planets were unable to retain their gas and rapidly decoupled from the external magnetic field. This is perhaps the reason why they have not developed satellite systems like those of Saturn, Jupiter, and the Sun. If this is the case how can the Moon and the two satellites of Mars, Phobos, and Deimos, be explained? There is at present no good explanation for this but there are many hypotheses, all very debatable. The chemical composition of the Moon suggests that it was formed very far from the Earth; later disturbances caused it to be captured by the gravitational field of this planet.

To return to our reconstruction of events. While the dust grains accreted and formed the planetary nuclei, the diameter of the protosolar disk became smaller. The nebular material withdrew towards the rotational centre leaving behind a large number of celestial bodies, large and small; it is as if it were low tide at the seaside. The volatile elements were carried towards the rotational axis where they subsequently formed the Sun. This recession of the nebula was accompanied by progressive re-heating. The heat was transported by convection to the surface of the nebula which appeared to be 'boiling'. The solar wind increased in intensity and as it travelled past the planets it carried away and removed from the solar system all the uncombined atoms. Is it possible to extend this further and suppose that the wind, strongly ionised, has caused reheating of the small bodies in the solar system? Certain phenomena observed in meteorites suggest that a reheating has occurred but there is no conclusive proof of how it came about.

A strong flux of fast particles (of energy exceeding a few MeV) accompanied this wind, apparently accelerated by very intense solar flares. Some primitive dust particles, now incorporated into meteorites, show effects produced by this radiation. Their surfaces are saturated with the solar wind and have been heavily bombarded by the fast-moving particles; about 100 billion impacts per square centimetre have been observed. In some instances the irradiation has produced a superficial amorphous skin on the surface of the dust particles and it is the adhesive properties of this skin that may well have dominated the agglomeration process. The nebula continued to contract, slowly forming the Sun as it is today. The increase in temperature combined with the change in the centrifugal force to counterbalance the increasing gravitational energy. The configuration of the gas at this stage was nearly spherical and in the terminology of astronomers, this was the beginning of the 'main sequence'. The rotation of the Sun continued to slow down gradually. This process was no longer caused by the coupling with the external magnetic field but by the solar wind. During the billion years following the formation of the Sun, its velocity decreased from a few hundred kilometres to a few kilometres per second. This last stage of slowing-down occurred much later than, and had nothing to do with, the formation of the planets. Contrary to our long-established belief, the Sun has not transferred its kinetic energy to the major planets but the ejected solar wind has removed this energy to the extremities of the solar system.

The circulation within the planetary system at this time was highly erratic and somewhat akin to a stock car race. A large number of small bodies travelled in very irregular, elliptical and inclined orbits crossing and colliding with each other, smashing themselves into fragments. As a result of their large surface area, the more massive planets progressively swept up the vast majority of these small objects. This was when the Moon trapped the complement of bolides which formed the characteristic surface which we see on her today. An even larger number collided with our Earth but intense magmatic activity and atmospheric erosion have removed nearly all trace of them. There was also great activity on the surface of the Sun which produced large variations in luminosity and enormous solar flares.

As time progressed the activity calmed down. The Sun has maintained part of its initial activity in an eleven-year cycle but this is of much lower intensity. The planets and asteroids which have survived the early period of major collisions now silently follow their prescribed orbits, at the very most only weakly disturbed by their respective gravitational fields. We are reminded of the earlier period of activity only by some shooting stars and a few meteorites, apparently from the asteroid belt, and by a few comets coming from the edges of the solar system. Further reminders are the photographs of the surfaces of the Moon, of Mercury and Mars, which show innumerable meteorite craters of all sizes. More impressive still is Phobos which has also suffered, small though it is; its surface is one big scar.

Acknowledgements

This paper was first published in *La Recherche*, **6**, No. 60, 807–818 (1975) [translated 1976 in *Mercury*] and is reproduced by permission of the publishers.

References

The Origin of the Solar System, proceedings of a symposium held at Nice, 1972, edited by C.N.R.S.

Ter Haar, D. (1967). *Ann. Rev. Astronom. Astrophys.*, **5**, 267.

Williams, I. P., and Cremin, A. W. (1968). *Quarter. J. Roy. astron. Soc.*, **9**, 40.

ORIGIN OF THE SOLAR SYSTEM

H. ALFVÉN
Department of Applied Physics and Information Science,
University of California, San Diego, California, U.S.A.

Creation Myths

Old Myths of Creation

Speculation about the origin and evolution of the Earth and the celestial bodies is probably as old as human thinking. During the millenia which is covered by the history of science, philosophy and religion we can distinguish three types of approach to this problem.

The first one is the theocratic-myth approach, according to which the evolution of the world was governed by gods who once upon a time *created* it. However, we must remember that the meaning of 'creation' has changed. The earliest meaning of this term seems to have been that the gods brought order into a pre-existing chaos. The world was 'ungenerated and indestructable'—as Aristotle puts it—and the gods were part of this world and also eternal. According to Indian mythology the 'creation' took place when Brahma woke up in the morning and finding the world in a chaotic state brought order into it, transforming chaos to cosmos. And when Brahma goes to sleep after a billion-year long Kalpa, chaos will again prevail. But the world is eternal, just as Brahma and the other gods.

The rise of the monotheistic religions changed this view. When one of the gods got a higher status than others (who in some cases became demons or devils), he continued to increase in prestige and power until he became the Supreme Lord, the undisputed ruler of the whole world. Then it was not enough that he had created the world in the sense of organizing a pre-existing chaos, he had created it all from nothing ('*ex nihilo*') by pronouncing a magic word or by his will power. This is the meaning of 'creation' when we speak of it today, but this is a relatively new concept. It was generally accepted in Christianity in the second century A.D. but the Genesis description of the Creation seems to have either meaning. The creation *ex nihilo* was not generally accepted by the philosophical–scientific community until the Saint Thomas synthesis of Christian dogma and Aristotelian philosophy.

In the theocratic mythologies the gods created the world and ruled its evolution according to their whims. We read in the Odyssey how Neptune was angry with Odysseus and generated storms to destroy him but how Pallas Athene saved him by producing other natural phenomena. In a similar way the actions of their parents or grandparents, (Zeus–Jupiter and Chronos–Saturn) had led to the creation of the world. There was no obvious reason why the world should be as it is. It was merely an accidental result of the activities of the gods. In the monotheistic religions God was sometimes thought to be a despot, who did whatever he liked, and it was not allowed to question or analyse his acts.

The Mathematical Myths

With the rise of philosophy and early science, the gods became less despotic and increasingly philosophically and scientifically minded. The creation of the world and its evolution were parts of a master plan, and it was not unreasonable that man should be able to understand this plan. A breakthrough in this thinking came with the Pythagorean philosophy.

The Pythagoreans had discovered how beautiful and powerful mathematics was. They had found that musical harmonies could be explained as ratios between integers, and they had demonstrated that there were five and only five regular polyhedra. I think there are few if any scientific discoveries which surpass these in beauty.

With such achievements it was quite natural that the Pythagoreans applied the same methods to other scientific and philosophical problems, one of them being the macroscopic structure of the world. They tried to explain this in terms of simple numerical relations and in terms of logically and mathematically beautiful concepts. They considered the sphere to be the most 'perfect' of all bodies and the uniform motion to be the simplest and most beautiful type of motion. *Ergo* the stars and the planets must be located on crystal spheres which revolved around the Earth with a uniform motion. The basic idea was that the macroscopic world must be structured according to simple mathematical laws—just like musical harmonies and geometrical figures.

Such views were not necessarily in conflict with religion. It was not necessary —although certainly possible—to question that the gods had created the world, because the gods no doubt understood the beauty of mathematics. Indeed, no one who studies mathematics can avoid the impression that the theorems

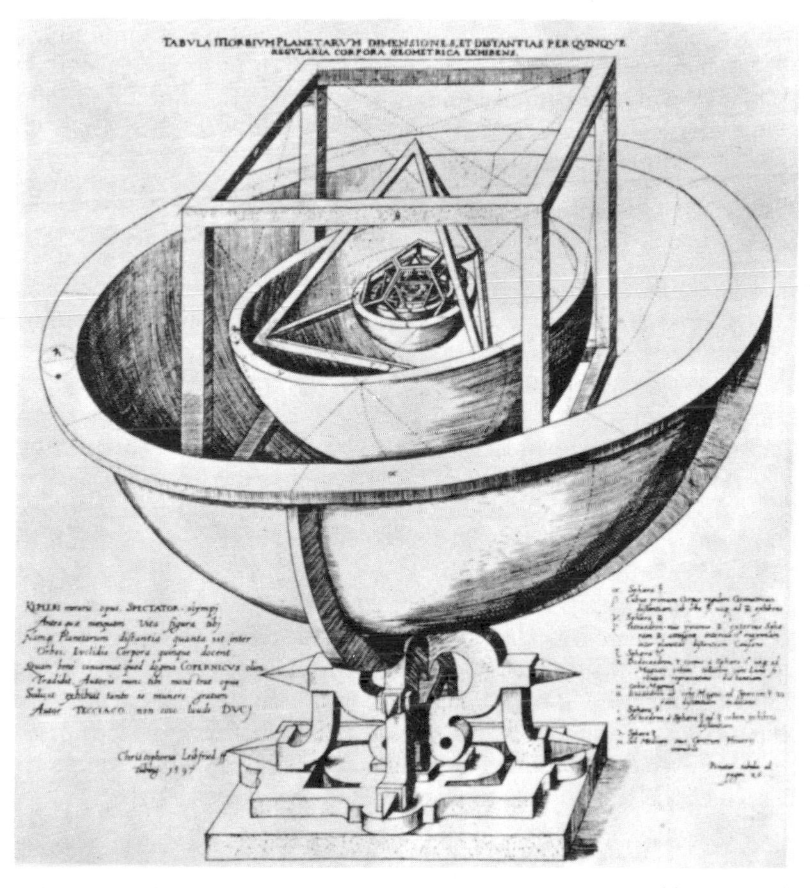

Figure 1. Relation between regular polyhedra and planets. In this model of the universe, the outermost sphere is Saturn's (From *Mysterium Cosmographicum* (1597, edition of 1621))

have a beauty which may be called divine. Hence one could expect the gods to structure the world according to some mathematically and logically beautiful principles. It was the task of philosophers and scientists to find what these cosmologic principles were. When they were found the cosmological problem was solved. We need only one principle, one formula, in order to understand the whole world.

This approach which may be called the *mathematical myth* developed during the centuries into the Ptolemaic cosmology. It is impressive by its logical reasoning and mathematical beauty. For example, it was demonstrated that there should be seven planets—including the sun and the moon—revolving around the Earth—seven was a holy number—seven days in a week, and seven tones in the scale, etc. Excluding the sun and the moon there were just as many planets as regular polyhedra (Figure 1).

However, a comparison between this cosmology and observations led to a number of discrepancies. In order to account for the observed motions of the celestial bodies it was necessary to introduce a series of epicycles etc. which made the system increasingly complicated. This was not diminishing the credibility of the theory—it just demonstrated that the material world is imperfect.

Empirical Approach

In the sixteenth and seventeenth century the Ptolemaic system broke down, and a new celestial mechanics was introduced. This represents the third type of approach, the *empirical approach.* This was based especially on the investigations of falling bodies by Galileo and the very accurate astronomical observations by Tycho Brahe. It is generally believed that it was the injection of this new empirical material which was fatal to the Ptolemaic system. This is of course partially true, but there is another factor which seems to have been at least equally important. A prerequisite for the breakthrough of the new approach was the collapse of the peer review system, which up to this time had been powerful enough to prevent the rise of new ideas. When Galileo claimed that the Earth moved, his peers in Italy reviewed these ideas and almost unanimously agreed that they were wrong, and Galileo had to recant them publicly. But the scientific establishment in Italy was not powerful enough to prevent German, Dutch and English scientists from accepting them and developing them further. The birth of modern science was possible because of a decay in the power of the philosophical–scientific establishment and a breakdown of their peer review system which for centuries had preserved the dark ages in Europe.

The Triumph of Science

With this breakthrough the scientific age started. The old myths, both the theocratic myths and the mathematical myths, are dead forever. We live in the scientific age, the age of reason. This is at least how we generally depict our own time. But is this really true?

Modern Myths

If we read a daily newspaper what do we find? There is normally a column devoted to an analysis of how the planets influence our life. But it is not the planets which the astronomers observe, those which are the targets of space research, it is the planets of the old Greek–Roman mythology. Venus is not the planet with a thick atmosphere of carbon dioxide, it is the goddess of love, Mars is not the sandstorm-ridden sphere of rock, it is the old god of war. And these old gods are believed by the newspaper readers to rule our lives in the same way as they once ruled the voyage of Odysseus. The theocratical myths of 2,000 years ago flourish today more than ever.

Of course this is all outside of the academic world. I do not believe that there is any respectable university in the world which has astrology as part of its curriculum. The theocratic myth approach to cosmology is dead in the academic community.

The Cosmological Formula

But what about the mathematical myths? Does the scientific community *intra muros* still subscribe to the Pythagorean belief that the structure of the universe could be solved by one simple mathematical formula? I am afraid that the answer is yes.

Although it is always dangerous to compare different cultures and different epochs, I think that there is an analogy between the special theory of relativity and the early Pythagorean results. In both cases simple and beautiful reasoning led to an important breakthrough. In both cases the success stimulated cosmological speculations. When I was a young student I was very impressed when Eddington, no doubt one of the leading astronomers of his time, claimed that the number 137 contained **the** solution of the cosmological problem. In his fascinating book *The Philosophy of Physical Science* he claims that sitting in his arm-chair he had counted the number of protons in the universe and found it to be $1{\cdot}57477 \times 10^{79}$ or more exactly $136 \times 2^{256} = 15, 747, 724, 136, 275, 002, 577, 605, 653, 961, 181, 555, 468, 044, 717, 914, 527, 116, 709, 366, 231, 425, 076, 185, 631, 031, 296$. Considered as a myth this is beautiful, but considered as science, it is nonsense, and is nowadays generally recognized to be so.

However, the collapse of Eddington's cosmology has not discredited mathematical myths in general. On the contrary it seems rather to have acted as a fertilizer for a rich flora of mathematical myths of which some no doubt are attractive from an aesthetic point of view but none from a scientific point of view. One of them, the 'big bang' cosmology, is at present 'generally accepted' by the scientific community. This is mainly because it was propagated by Gamow with his irresistible charm and vitality. The observational support for it, which he and others claimed, is totally obliterated but the less there is of scientific support, the more fanatical is the belief in it. As you know this cosmology is utterly absurd—it claims that the whole of the universe was created at a certain instant as an exploding atomic bomb with a size of much less than the head of a pin. It seems that in the present intellectual climate it is a great asset of the big bang cosmology that it offends common sense: *credo quia absurdum* (I believe because it is absurd!). When scientists attack the astrological nonsense *extra muros* it is wise to remember that there is still worse nonsense propagated *intra muros.*

Big Creation–Small Creation

The old problem of how the world was 'created'—if it was created—is today divided into two problems. One is the 'big creation' or how the universe as a whole has originated and developed—which we have discussed to some extent. The other is the 'small creation' or how in a small part of a small part of a small part of the universe the solar system originated. We shall devote the rest of this lecture exclusively to this restricted problem.

Modern Astrophysics: Myth or Science?

Like in many other parts of astrophysics, there is today a confrontation between a mythological approach and an empirical approach and like in many other parts of astrophysics it is the mythological approach which is 'generally accepted' by the scientific community. To those who believe that the structure and evolution of the whole universe can be solved by a single formula, all phenomena in the universe should in principle be derivable more or less directly from this. The formation of the solar system would be found to be a result of, for example, the big bang when all the consequences are drawn from this theory. There are few people who are so bold to try to do this. Usually one does not go back further than to the formation of stars. In fact, the 'generally accepted' theories start from a treatment of how stars are formed, and try to derive the formation of the solar system as a by-product of stellar formation.

The Formation of Stars

By this approach the theory of the formation of the solar system becomes critically dependent on the mechanism for star formation. What do we know about this?

What we really know is not very much. It is likely that stars are formed in dark interstellar clouds, and during the last few years infrared and radio astronomy have given us a wealth of information about such clouds. It has been demonstrated that they contain both dust and gas and that they contain rather complex molecules. As far as we know such molecules can be formed at a sufficient rate only in a plasma, so their presence gives a strong indication of the existence of electromagnetic phenomena. Observations of the Zeeman effect give further support for this. Lyman Spitzer, one of the pioneers in cosmic plasma physics, has devoted much interest to the formation of stars from an interstellar cloud and stressed the importance of hydromagnetic effects for this process. In spite of this there is a whole literature about the formation of stars and of solar systems in which hydromagnetic processes are neglected or treated in an erroneous way.

The Laplacian Theory

Speculations about the formation of solar systems from interstellar clouds were actually initiated by Laplace. He was inspired by the great interest for the origin of the solar system which resulted from speculations by Descartes, Kant and other leading philosophers and scientists 200–300 years ago. At this time the astronomers had discovered that besides the stars there were also an abundance of small nebular objects in the sky. Laplace understood that many of these consisted of a great multitude of stars—were galaxies with modern terminology—but thought that some of them, and also 'planetary nebulae', were solar systems in formation. With this as a background he developed a theory of the formation of the solar system (Figure 2). When later the advanced observational technique showed that the disk-like objects which were observed, were not solar systems in formation, the theory lost its observational foundation. But

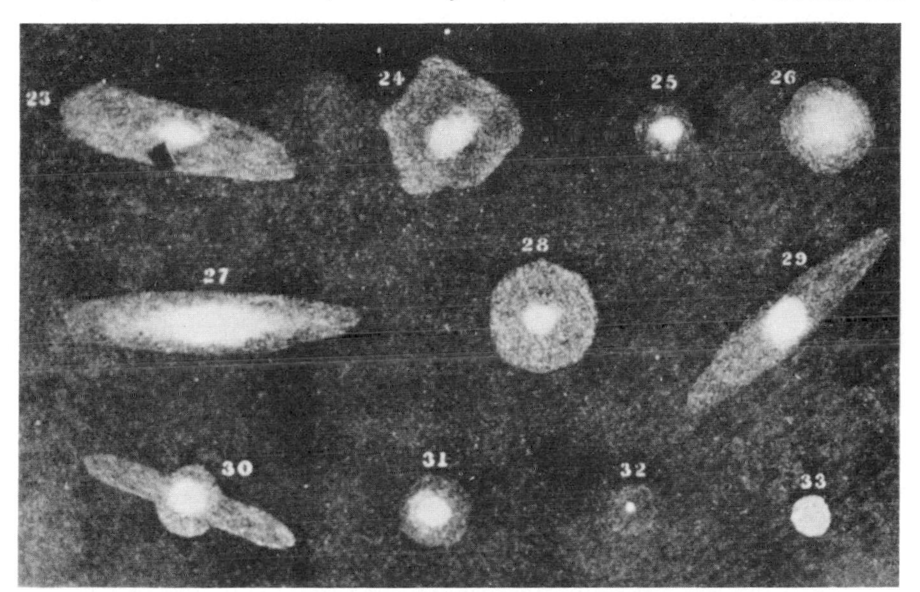

Figure 2. Herschel's nebulae, interpreted by Laplace as solar systems in formation

when the observational support for the 'nebular theory' disappeared, the theory itself continued to live a life of its own and has over the centuries become a sacrosanct myth.

The Laplacian theory has been supplemented with the theory of gravitational collapse as a mechanism for the formation of stars and solar systems. The history of this concept is the following:

If we consider a gravitating sphere of gas in which the variables (pressure, temperature, etc.) are functions of r and t alone, the gas pressure gradient will balance the gravitation and prevent the sphere from contracting. If the temperature decreases below a certain critical value, gravitation will dominate, and the sphere will begin to contract. When it does so, both the gravitation and the pressure gradient will increase, but the latter not enough to compensate the former. The result is a collapse which takes only some thousand years. It is generally believed that stars and solar systems are formed by this process.

A process of this kind has never been observed. From a theoretical point of view it depends critically on the *assumption* that the variables are functions of only r and t. This assumption is introduced only to make the problem mathematically easy to solve, and if it is dropped, it is obvious that the state from which the collapse starts can never be established (it is unstable!). In other words, in order to obtain a mathematically elegant solution, assumptions are introduced which make this solution uninteresting from a scientific point of view. This is a typical example of how a mathematical myth originates.

To this is added the assumption that the condensation from the nebula takes place in thermal equilibrium—another assumption of the same character.

The myth which is developed on the basis of the Laplacian mistake supplemented with these three erroneous physical concepts has become sacrosanct, and is the basis of most of the papers of today about the evolution of the solar system. It is defended by a strongly entrenched community, who seldom admits that there exists any objection to their myths. The peer review system will probably give this myth the same eternal life *intra muros* as astrology enjoys *extra muros*.

Empirical Approach

Methodology

After this brief review of some of the most interesting myths—old and new—we shall approach the origin and evolution of the solar system in an empirical way. As has been pointed out by Gustaf Arrhenius, the construction of models is not so important as an analysis of the methodology which is applicable. (Figure 3).

A realistic attempt to reconstruct the early history of the solar system must necessarily choose a procedure which reduces speculation as much as possible and connects the evolutionary models as closely as possible to experiment

Origin of the Solar System

General Principles

1. Reduce speculation as far as possible by relating all processes to laboratory experiments or space observations.

2. Approach the problem not by making more or less arbitrary assumptions about the primitive sun but by starting from present state of the solar system and systematically reconstructing increasingly older states.

3. We should not try to make a theory of the origin of planets around the sun but a general theory of the formation of secondary bodies around a central body. This theory should be applicable both to the formation of satellites and the formation of planets.

4. The aim is not primarily detailed theories but more a general framework into which the rich empirical material could be fitted. This framework must be acceptable from the point of view of celestial mechanics, plasma physics, plasma chemistry, geology, theory of hypersonic collisions, etc.

Figure 3. Methodology of empirical approach

and observation. Because no one can know *a priori* what happened four to five billion years ago we must start from the present state of the solar system and step by step reconstruct increasingly older periods. This '*actualistic principle*', which emphasizes reliance on observed phenomena, is the basis for the modern approach to the geological evolution of the Earth; 'the present is the key to the past.' This principle should also be used in the study of the solar system, especially now when NASA is supplying us with most valuable geological specimens from the space missions.

Hence we should proceed by establishing which experimentally verified laws are of controlling significance in the space environment. For this purpose laboratory studies of processes which are likely to be important in space are essential, but applying the results of laboratory investigations to cosmic conditions requires a thorough study of the scaling laws. During the last few years the rapidly increasing information on extra-terrestrial processes that modern space research is providing, has increased the reliability of this method. If the large body of available empirical knowledge is interpreted strictly in terms of these laws the speculative ingredient of cosmogonic theories can be significantly reduced.

When analysing the origin and evolution of the solar system we should recognize that its present structure is a result of a long series of complicated processes. The final aim is to construct theoretical partial models of all these processes. However, there is often a choice between different partial models, which *a priori* may appear equally acceptable. Before the correct choice can be made it is necessary to define a framework of boundary conditions which these models must satisfy.

Planetary System–Satellite System

Theories of the formation of the solar system must also account for the satellite systems in a manner consistent with the way in which the planetary system is treated. In certain respects the satellite systems provide even more significant information about evolutionary processes than does the planetary system, partly because of the uncertainty about the state of the early sun.

Observing that the highly regular systems of Jupiter, Saturn, and Uranus are in essential respects similar to the planetary system, we should aim at a *general theory of the formation of secondary bodies* around a primary body. This principle was stated already by Laplace, but seems to be forgotten by those who today work on the development of Laplacian-type theories.

The theoretical framework we try to construct should, consequently, be applicable both to the formation of satellite systems around a planet and to the formation of planets around the sun. Through this requirement we introduce the postulate that these processes are essentially analogous. Our analysis supports this postulate as reasonable. Indeed, we find evidence that the formation of the regular systems of secondary bodies around a primary body—either the sun or a planet— depends in a unique way on only two parameters of the primary body, its mass and spin. It is also necessary to assume that the central bodies were magnetized, but the strength of the magnetic field does not appear explicitly; it must only surpass a certain limit.

Five Stages in the Evolution

Applying these principles we find that the evolutionary history of the solar system can be understood in terms of five stages, in part overlapping in time (Figure 4):

(1) Most recently—during the last four billion years—a *slow evolution* of the primeval planets, satellites and asteroids which produced the present state of the bodies in the solar system. By studying this latest phase of the evolution (post-accretional evolution) we prepare a basis for reconstructing the state established by earlier processes.

(2) Preceding this stage, an *accretional evolution* of condensed grains, moving in Kepler orbits to form planetesimals, which by continuing accretion, grow in size. These planetesimals are the embryonic precursors of the bodies found today in the solar system. By clarifying the accretional processes we attempt to reconstruct the chemical and dynamic properties of the early population of grains.

(3) To account for grains moving in Kepler orbits around the sun and the protoplanets, *transfer of angular momentum* from these primary bodies to the

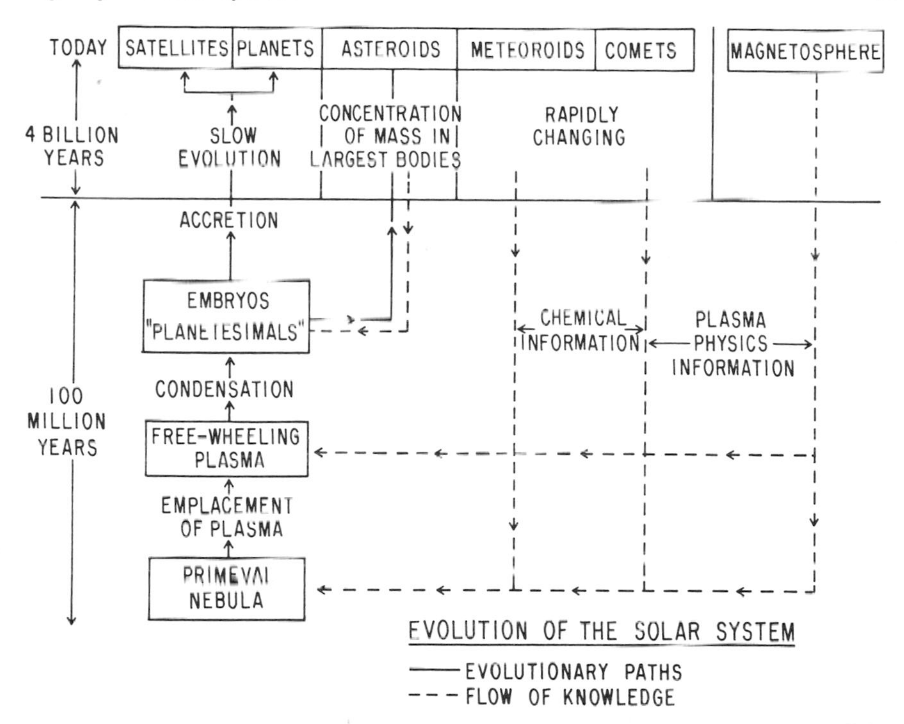

Figure 4. Diagram of the evolution of the solar system

surrounding medium must have occurred in the stage of evolution preceding accretion.

(4) *Emplacement of gas and dust* to form a medium around the magnetized central bodies in the regions where the planet and satellite groups later accreted.

(5) *Formation of the sun* as the first primary body to accrete from the source cloud of the solar system.

Extrapolation from Present Day Space Conditions

The next phase in our analysis is to try to find what processes have been active during the different phases of the evolution, or at least to give examples of what type of processes deserve to be more closely analysed. In doing so we must take warning of much of the earlier work which has been so speculative that it has lost contact with reality. In this field as in all other fields of science we can never avoid speculations, but when speculating we must all the time keep close contact with reality. If we forget this we will at the best substitute an old myth by a new one—and this is what necessarily must be avoided.

First of all we should realize that when the solar system was formed the conditions in our part of space were different in many respects from what they are today, but that *the same general laws of physics were working.* Solid bodies, including grains, moved at that time in Kepler orbits similar to the present ones, although viscosity effects and mutual collisions between the grains introduced perturbations. Space contained a plasma the parameters of which certainly differed from the present parameters but the differences were not so drastic. The conclusion is that in important respects we can regard the cosmogonic state to be *an extrapolation of present day conditions.*

In fact if we compare the present plasma in interplanetary space and the magnetospheres with the cosmogonic plasma out of which the planets, asteroids, and satellites once condensed, we find that the latter no doubt was much denser. But we have enough knowledge of the behaviour of dense plasmas from studies of the ionosphere, the solar corona, chromosphere, and photosphere to be able to make reasonable extrapolations. By choosing such an approach we can to a large extent avoid the introduction of blackboard mechanisms, which are pests in modern astrophysics.

The Latest Period

According to these principles we may try to reconstruct the early history of the planets-satellites in the following way:

We have good reasons to believe that during the last four billion years neither the chemical composition nor the orbital elements of the planets and satellites have changed very much. There has been a slow geological development at the surface of the Earth and some other bodies. The orbital elements of the bodies have been subject to what is called 'secular changes' of the semi-major axis (a), eccentricity (e), and inclination (i), but these are periodic variations within rather small limits. There are two exceptions: tidal effects have changed the orbits of the Moon and of the Neptunian satellite Triton. In almost all other respects, the solar system looked pretty much the same four billion years ago as it does today.

How the Earth Accreted from Planetesimals

Radioactive dating has demonstrated that this long and stable period was preceded by a period—perhaps some ten or a hundred million years long—during which the solar system was formed. The matter which now composes the planets and satellites aggregated from an earlier embryonic or 'planetesimal' state, in which it was dispersed as a number of small bodies. These moved in Kepler orbits around the sun, but collided mutually, with the end result that they accreted to the present celestial bodies. The craters we see on the moon and other bodies bear witness of the rain of planetesimals, which made the bodies grow to their present size.

In fact, by comparing the different space mission photographs of the Moon, Mercury, Mars and Phobos, we find that their surfaces look so similar with respect to cratering that we may conclude that all of these rocky bodies have developed in a similar way, and in some respects just represent different phases of an evolution. This makes it possible to reconstruct the history of the Earth (Figure 5).

The Earth must have passed through a stage as a very *small body*, similar in size to—say—*Phobos*, the smallest body yet observed. We see that Phobos has a number of craters which after all have been produced by impacting planetesimals. When Phobos reached its present state it had exhausted all the planetesimals in its surroundings. For the Earth, however, this state was only

Figure 5. Photographs of Phobos, Moon, and Mercury showing cratering presumably from planetesimal impact. It is highly desirable to obtain similar photographs from asteroids

transitory. The rain of planetesimals continued and the Earth grew bigger and bigger. By looking at the moon we get a snapshot of the Earth when it had accreted one percent of its present mass. Mercury and Mars show later phases of its childhood when its mass was 4 per cent and 10 per cent of the present mass. From these photographs we conclude that the early history of the Earth was rather monotonous, consisting of a perpetual rain of planetesimals. We further conclude that when a body reaches the size of Mars it begins to retain—or accrete—an atmosphere; the craters at its surface are weathered, and also modified by other geological effects. Such effects become more pronounced when the body grows and when it reaches the size of the Earth or Venus the geological evolution has obliterated most of the surface evidence of its planetesimal accretion.

Reconstruction of the Planetesimal State

From a study of the impact craters we can draw some conclusions about the planetesimal state. For example, we can derive the size distribution among the planetesimals. But this information is not enough to give us a very clear picture of what the planetesimal state was like. In order to clarify this it is important to observe that in the asteroidal region between Mars and Jupiter we have at present a state which in essential respects must be similar to the planetesimal state out of which, for example, the Earth was formed. Hence we need not make a speculative armchair model of the *planetesimal state.* We can derive it as an *extrapolation of the present state in the asteroidal region.*

In the main belt of the asteroidal region there is a large number of small bodies moving in orbits with rather high eccentricities (up to 0·30–0·35) and inclinations (up to 30° or more). A few thousand of these have been observed, but their total number is likely to be some orders of magnitude larger. They necessarily collide with each other. There has been a controversy whether these collisions result in fragmentation or in accretion. The answer no doubt is: both. There are good reasons to believe that the end result will be that most of the matter contained in the asteroidal belt will be concentrated into one or a few bodies. Already at the present state the three biggest bodies contain 80 per cent of the total mass. This concentration of mass will continue, and the eventual result of the evolution will be the formation of one or perhaps a few planets. In other words, in the asteroid belt we see something like a photograph of the Earth at an embryonic stage, before it had accreted.

The main difference between the early planetesimal state in the region of the terrestrial planets and the present one in the asteroidal belt is that the mass density in the former was a ten or thousand times larger, with the result that the Earth was accreted relatively rapidly—perhaps in 10 or a few 100 million years—whereas the similar evolutions in the asteroidal belt will take 10^{11} years or more.

The picture of the planetesimal state we get in this way is drastically different from the Laplacian disk. The planetesimals actually move in highly eccentric and inclined orbits and not in the circular orbits of a Laplacian disk (which a recent myth even claims to be an *extremely* thin Saturnian-like sheet of grains). These differences are essential for the understanding of the accretion of planets and satellites. They are equally essential for our next step backwards in time —the reconstruction of how planetesimals accreted from grains which were formed in a plasma or captured by it.

The Plasma Phase

One of the central problems in all attempts to reconstruct the origin of the planetesimal state is how the grains were put in orbit. This must have been associated with a transfer of angular momentum from a spinning central body—the sun or a planet—to the surrounding planetesimals. As there is no known mechanism for the transfer of momentum to a solid body, it is likely that the transfer has taken place when the matter was in a dispersed state; i.e. formed a plasma, more specifically *a dusty plasma* containing a large amount of dust grains. If we speculate about what mechanism may have produced this transfer, we find that a likely mechanism is a hydromagnetic transfer by the means of electric currents flowing in the way as depicted in Figure 6.

This is a nice model and we can demonstrate that it produces the effect which is necessary in order to understand how the matter which the secondary bodies now consists of once was put in orbit. However, the model is of a speculative character. Do we have any evidence that processes of this kind really occur in space? Only a few years ago the answer would have been no. Today it is yes, and the change in the situation is largely due to work by Zmuda, Armstrong, and their colleagues at the Johns Hopkins University. They have measured the so-called Birkeland currents flowing along the magnetic field lines to the auroral zone, and found that there are sheet currents flowing in opposite directions (Figure 7). From this follows that the current system observed in the magnetosphere actually transfers angular momentum from the earth to a

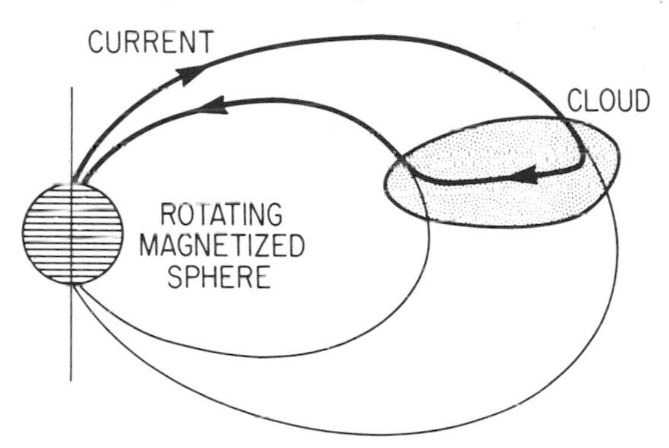

Figure 6. Transfer of angular momentum

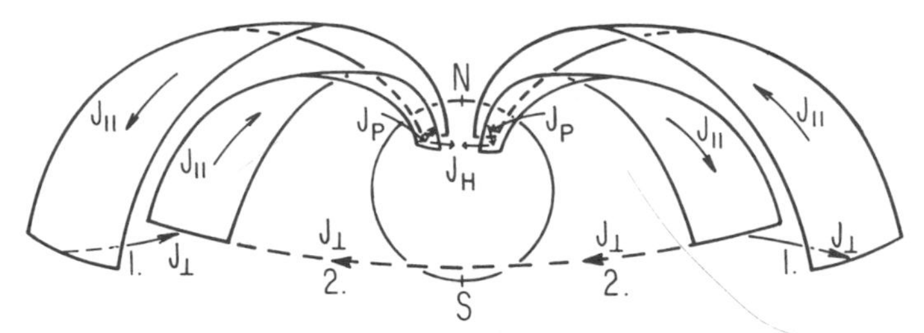

Figure 7. Zmuda–Armstrong current system. Large-scale Birkeland current sheets shown schematically for a dipolar field geometry with alternative closure paths for the lower latitude sheet currents

surrounding plasma. Hence the mechanism which is necessary in order to understand how natural satellites were put in orbit is no longer founded on armchair speculations but from actual observations from spacecrafts.

The Freewheeling Plasma

Hence we see that under present conditions there exists a plasma mechanism which transfers angular momentum from a central body to a surrounding plasma. From a purely hydromagnetic point of view we expect such a process to continue until the angular velocity of the plasma is the same as that of a central body, a state called Ferraro corotation. However, we know today that such a state is not necessarily reached because other spacecrafts observations have demonstrated that the Birkeland currents which tend to establish this are producing field-aligned electric fields and electrostatic double layers which decouple the plasma from the ionosphere. In this way only a *partial* corotation is attained. This means that when a certain quantity of angular momentum is transferred the surrounding plasma becomes essentially *freewheeling*.

In a freewheeling plasma an equilibrium is established between the main forces, acting on the plasma, viz. gravitation, centrifugal force, and the electromagnetic forces. Figure 8 shows that these balance each other in such a way that the plasma is supported against gravitation partially by the centrifugal force and partially by hydromagnetic forces. As an elementary calculation shows the kinetic energy of the freewheeling plasma is 2/3 of the kinetic energy of a body in Kepler motion. The factor 2/3 derives from the geometry of a dipole field.

What will happen to grains produced by condensation or capture in such a freewheeling plasma? We find that when the grains are large enough to move independently of the magnetic field, these will form bodies orbiting in Kepler ellipses with eccentricity $e = 1/3$ (again a factor deriving from the geometry of a dipole field). If a number of such bodies are produced in the same region of space they will interact, for example, by collisions, with the result that both e and i will diminish. The end result of this process is that the condensed bodies

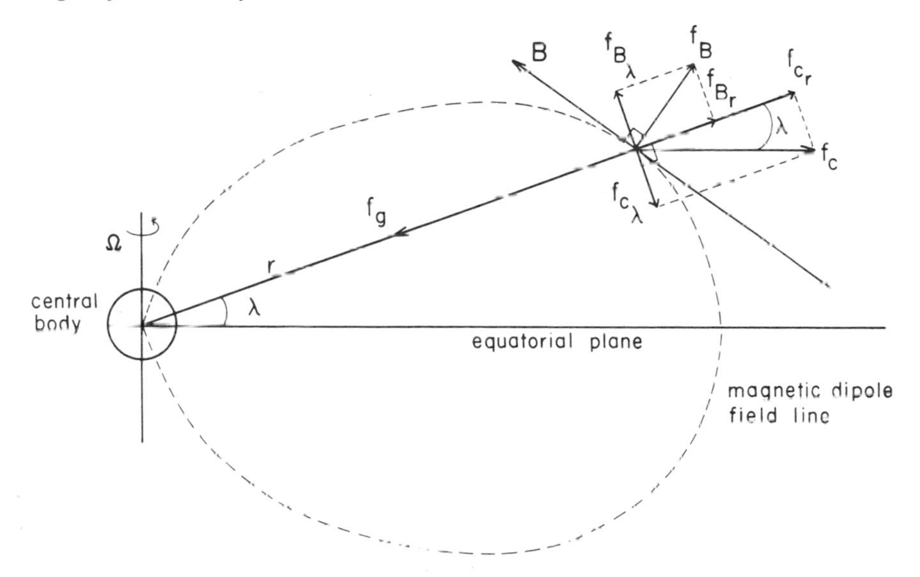

Figure 8. Equilibrium of freewheeling plasma

will move in circular orbits at 2/3 of the distance where the freewheeling plasma condensed.

Hence we find the important *laws of transition* from a state of *freewheeling plasma* to a state of Kepler motion:

1. The first result is solid bodies orbiting with $e = 1/3$
2. The end result is less eccentric orbits
3. There is a general contraction by a factor 2/3.

We have started from the conditions in the magnetosphere of today and made a fairly straightforward extrapolation to a cosmogonic plasma, which has a much higher density so that we can expect a condensation to take place. As all extrapolations this is necessarily dangerous, and unfortunately we cannot check the results by the study of present-day processes, because no similar condensation can be expected to occur in our solar system under present conditions.

However we can check our results by studying whether the structure of the asteroidal belt—being a present-day representation of the planetesimal state—can be explained by this process. We should also observe that in the Saturnian rings we have another example of matter in a dispersed state which should have been generated by condensation from a freewheeling plasma.

Dynamics of the Asteroidal Belt

The asteroidal belt is usually represented by the (n, a) diagram (n = number, a = semi-major axis). This gives the impression of a rather chaotic state, the only regular feature being the Kirkwood gaps which are a resonance phenomenon produced by Jupiter. However, if we instead plot the cosmogonically more relevant (m, a) diagram (m = mass, calculated under the assumption that the density and albedo are constant) we find that the belt has a much more regular structure, with a sharp cut-off both at the inner and outer edge (Figure 9). In fact, outside $2{\cdot}2 < a < 3{\cdot}5$ there is no appreciable mass, except the Hilda group at $a = 3{\cdot}95$. This is produced by a resonance with Jupiter, which we shall not discuss here.

There is no known process acting today which can account for the sharp cut-offs of the main belt. Hence there are reasons to suppose that they are of cosmogonic origin. An objection to this is that there are frequent collisions between asteroids and one would suppose that even if the asteroids originally were formed in a well-defined belt, the collisions would cause a diffusion so that they spread to adjacent parts of space. However, this picture is not correct, because inelastic collisions between bodies in Kepler orbits will produce a *negative* diffusion. This means that if the asteroidal belt originally had sharp borders, the diffusion will tend to make mass move *away* from the borders and concentrate the mass in those regions where the mass density already is high.

Moreover, the largest asteroids ($R > 100$ km) are so few that for them the chance for a disruptive or orbit-changing collision is statistically not very large. In fact, the largest asteroids probably represent a sample of the original condensed material, rather unaffected by planetary differentiation, and modified mainly by the sequence of collisions that they have undergone during accretion.

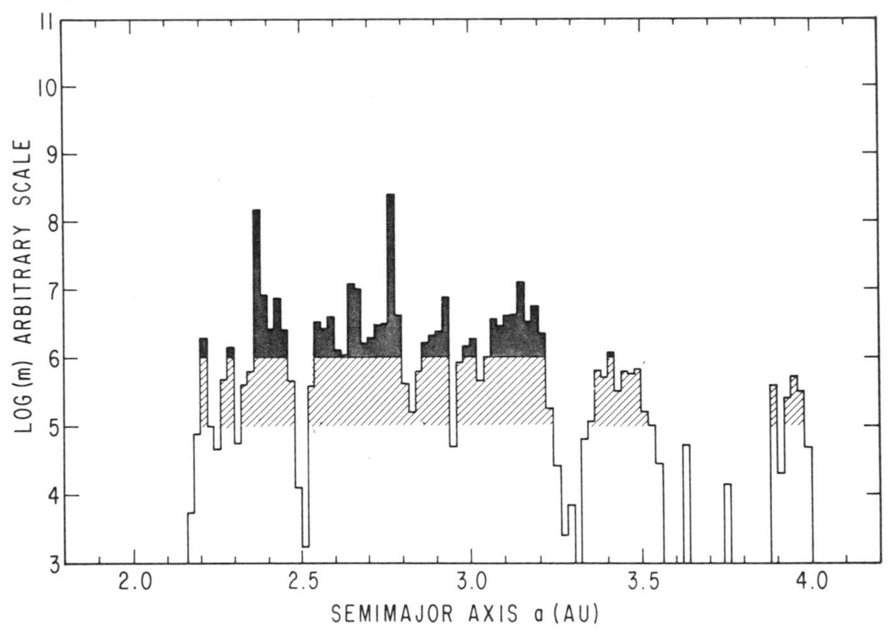

Figure 9. Mass distribution in asteroidal belt

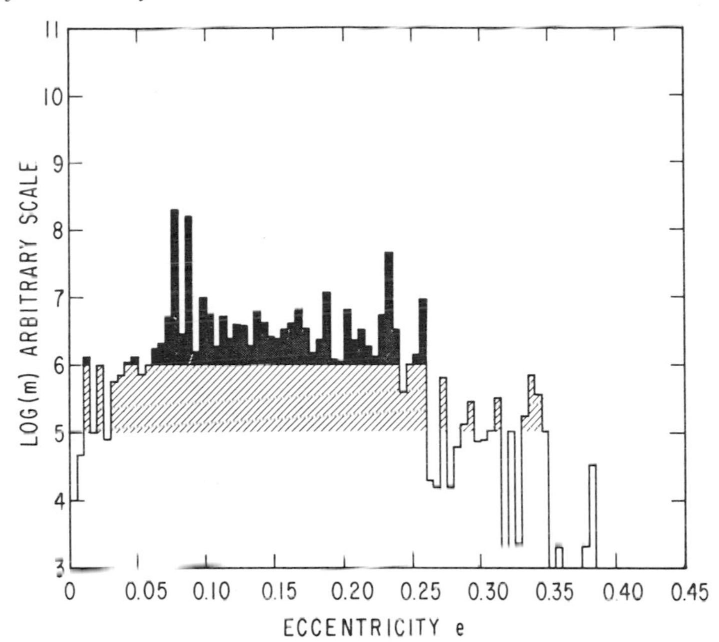

Figure 10. Eccentricities of asteroidal orbits

The Asteroidal Belt as Derived from a Grain Assemblage in a Freewheeling Plasma

With this as a background we can test the hypothesis that the primordial grains are derived from, or captured by, a freewheeling plasma. The results of a detailed analysis can be summarized in the following way:

1. The eccentricities of the main belt asteroids never exceed $e = 1/3$. This is what we should expect. Of course most of the asteroids have lower e-value, which is a natural result of collisions (Figure 10).

2. Asteroids containing a considerable part of the total mass orbit with inclinations as high as 30°. This is a natural result of a condensation if they derive from a freewheeling plasma, but it is impossible to reconcile with a formation from a flat Laplacian disk (Figure 11).

3. The fall down ratio of 2 : 3 explains the outer limit of the asteroidal belt as due to the 'shadow' of Jupiter. Because grains condensed outside Jupiter's orbit are perturbed or captured by Jupiter, the asteroidal region derives from a condensation and plasma capture of grains inside Jupiter. This explains why its outermost limit is almost exactly 2/3 the orbital radius of Jupiter.

4. Because the material in the asteroidal belt itself will sweep up plasma, the density will fall at 2/3 of the outer limit, and become negligible at 2/3 of the a-value, where the density has increased sufficiently. This means that the inner limit to the asteroidal belt is given by its 'own shadow'. See Figure 12.

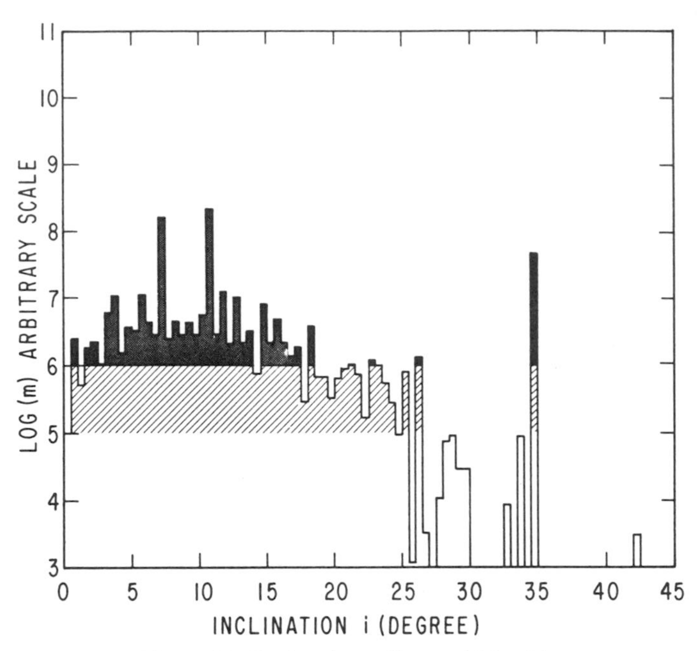

Figure 11. Inclinations of asteroidal orbits

Hence the dynamic structure of the asteroidal belt supports the view that it has been formed from grains in a freewheeling plasma. We can also understand how the excess energy associated with high eccentricities and inclinations is dissipated by collisions. This process leads slowly to the accretion of all the mass into one or a few planets.

These conclusions are very important. They mean that *we can study the basic process of planetesimal accretion under present conditions in the asteroidal belt.*

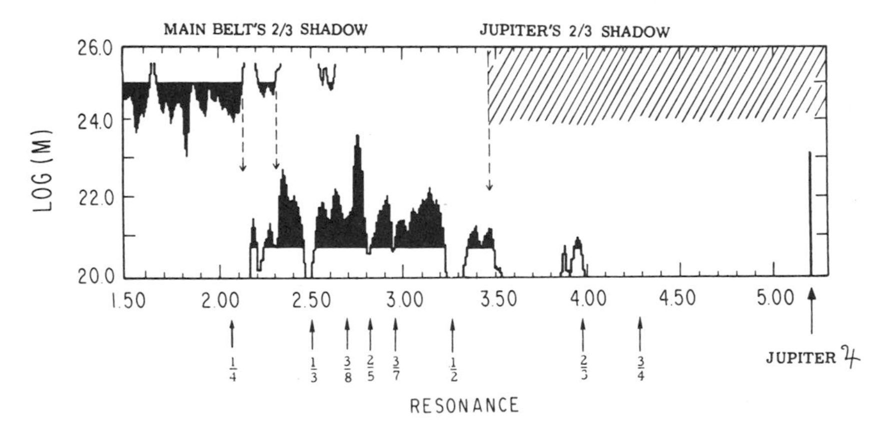

Figure 12. 'Shadow' effects in the asteroidal region

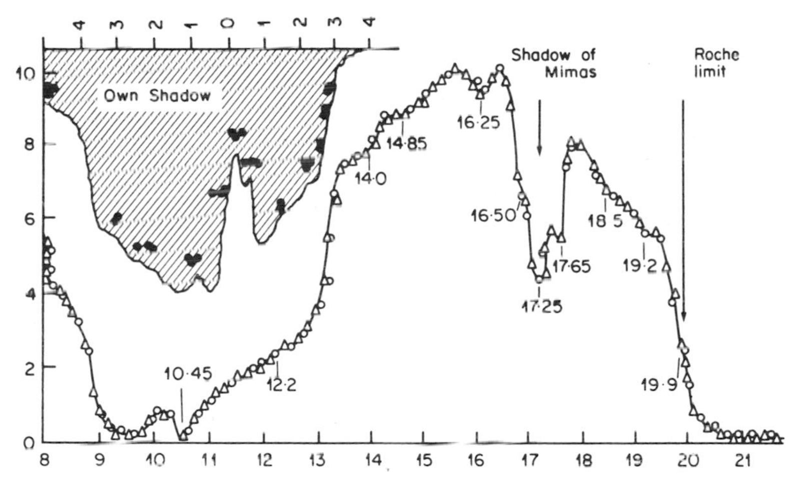

Figure 13. 'Shadow' effects in the Saturnian ring system

The Saturnian Rings

The Saturnian ring system gives us a second possibility to study the condensation from a freewheeling plasma. In this case the final accretion to planets or satellites is prohibited because the rings are located inside the Roche limit. Hence they still contain information which necessarily is lost at the accretion of large bodies.

The fine structure of the Saturnian ring system, for example the Cassini division, has been thought to be due to resonances produced by Mimas. Modern observational data rule out this possibility. Also, a theoretical study demonstrates both qualitatively and quantitatively that the observed structure cannot be explained by resonance effects. On the other hand, as shown by Figure 13 the structure can be understood rather much in detail as a result of condensation from a freewheeling plasma.

This means that the Saturnian rings should be considered as a beautiful time capsule, telling the physicists of today about the state of the plasma from which it condensed some billion years ago.

The Emplacement of Plasma

There are two more steps in our progress along the negative time-axis, which should only be mentioned briefly here.

One is the problem of how the plasma was emplaced in different regions of the solar system. This will explain the differences between the systems of secondary bodies around the difference primary bodies, and also account for the *chemical differences* between the bodies in the solar system. The key to this seems to be a plasma phenomenon called the 'critical velocity' which has been explored extensively in the laboratory and theoretically.

Formation of the Sun

In an empirical approach the formation of the sun should be the last problem we discuss. By first studying the formation of planets and the formation of satellites around them we have got a valuable insight in the general character of the formation of secondary bodies around a primary body. This knowledge allows us to define the constraints on theories of star formation. This information should be combined with the rapidly increasing observational data about the dark interstellar clouds in which stars probably are formed.

It is premature to draw very definite conclusions about the formation of stars. Only one thing can be stated with a high degree of confidence: they were *not* born by what is usually meant by a 'gravitational collapse'.

Acknowledgement

I wish to thank Mr. Albin Rhomberg and Dr. Wing-Huen Ip very much for their kind help in preparing the figures.

References

Alfvén, H. (1977). 'Cosmology: Myth or Science?' *Cosmology, History and Theology* (Eds. W. Yourgrau and A. D. Breck). Plenum Press, Chapter 1.

Alfvén, H. and Arrhenius, G. (1975). *Structure and Evolutionary History of the Solar System*. D. Reidel, Dordrecht, Holland.

Alfvén, H. and Arrhenius, G. (1976). *Origin and Evolution of the Solar System*. NASA SP-345.

Alfvén, H. and Carlqvist, P. (1978). 'Interstellar Clouds and the Formation of Stars'. *Astrophys. and Space Sci.*

THE BAND-STRUCTURE OF THE SOLAR SYSTEM

H. ALFVÉN

Department of Applied Physics and Information Science,
University of California, San Diego, California, U.S.A.

Criticism of 'Laplacian' Theories

Laplacian Type Theory as a Two-step Process

Laplace interpreted Herschel's observations of what we now know to be galaxies as solar systems in formation. Hence he concluded that in an early phase a forming solar system should be a rather *uniform gaseous disk*. When it became obvious that the objects Herschel observed had nothing to do with the formation of the solar system, it would have been logical to take a fresh look at the problem, and, among other things, question whether a cosmogonic theory should aim at the formation of a *uniform* disk. Instead a school of thought was developed in which the belief in a uniform disk became sacrosanct. Hence the 'Laplacian models' assume a two-step process: (1) The evolution of an interstellar cloud into a uniform disk. (2) The evolution of this disk into the present structure, which differs drastically from uniformity.

Formation of a Uniform Disk

Centuries of research have clearly demonstrated that it is impossible to find physically reasonable processes by which an interstellar cloud develops into a more or less *uniform* disk. This seems now to be generally recognized by all who work out theories in a scientifically stringent way. The result is that the Laplacian* approach is nowadays exclusively advocated by people who sub-

*Of course the criticism of the Laplacian school is not criticism of Laplace. Being a first class scientist he would no doubt have retracted his theory had he lived when it became obvious that his interpretation of Herschel's observations were not correct.

stitute an empirically based analysis by purely hypothetical calculations.

The Laplacian school also violates the scientific tradition that objections to a theory must be taken up to discussion. The difficulties about the angular momentum distribution have never been solved, only swept under the rug. The obvious fact that the planetary system as well as the satellite systems has a pronounced *banded* structure and hence is very far from a uniform disk (Alfvén, 1943, 1954) has never been mentioned. The breakthrough of magnetohydrodynamics around 1950 had only a transient effect on the Laplacian tradition (Hoyle's theory, which because of a neglect of the already well-known kink instability led to obviously erroneous results) and it has now retreated to the pre-hydromagnetic era (Cameron, 1976; Safranov, 1972; Goldreich and Ward, 1973). The motivation given for the neglect of plasma effects is that the ionization due to cosmic rays and radioactivity in a dark cloud is not enough to given an appreciable conductivity. This is irrelevant because a main cause of ionization in cosmic physics is the transfer of kinetic energy to electromagnetic energy. For example, the ionization in interplanetary space, solar corona and chromosphere derives ultimately from the kinetic energy in photospheric convection, which supplies the energy for the heating of the corona, etc.

It is difficult to see how similar phenomena can be avoided at the contraction of an interstellar cloud which even representatives of the Laplacian approach now admit to be ionized. The contraction must be opposed by electromagnetic effects, as pointed out especially by Spitzer (1968). This leads necessarily to a conversion of kinetic (ultimately gravitational) energy into electromagnetic energy, so that the contraction results in an *increase* in magnetic energy. Magnetic fields are necessarily associated with electric currents (except in the special case of curl-free fields which is not of interest here).

In principle, magnetic energy may be dissipated either by accelerating a large number of electrons only to very low velocities, so that they produce heat but no ionization, or by accelerating a smaller number of electrons to so high energies that they also ionize. It is doubtful whether the first process is very important in cosmic physics. One of the reasons is that currents normally are contracted to thin filaments, either by pinch effect or by other mechanisms (example: auroral rays).

As an example how a cosmic plasma gets ionized we can take the ionospheric-magnetospheric plasma in the auroral zone. The conductivity in the auroral zone is much higher than in the surrounding, but this excess is not produced by radioactivity, cosmic rays or solar light, the only ionizers the Laplacian school takes account of. Instead it is produced by precipitation of energetic particles which have been accelerated by processes related to the auroral current system, which derives its energy ultimately from the relative motion between the solar wind and the magnetosphere. For example the electric currents may produce electrostatic double sheaths which accelerate particles to some keV. Moreover, electric currents under cosmic conditions seem seldom to be uniform. In most, if not all, cases they flow in thin filaments (due to ion contraction or pinch effect). The result is that when magnetic energy is dissipated a large fraction normally goes into ionization.

Another example is the solar corona. It is obvious that the energy source of its very high temperature and its ionization is photospheric convection. How this energy is transferred to the corona is perhaps controversial; it may be through sound waves, hydromagnetic waves or—most likely—through electric currents. But independent of the transfer mechanism the solar corona gives another example of conversion of kinetic energy (in photospheric convection) into ionization.

There seems not to exist any competent study of the ionization at the contraction of cosmic clouds. Very large quantities of gravitational energy are released, and a large—possibly the largest— part of it must be converted into magnetic energy because magnetic fields oppose the contraction. From what we know about the behaviour of cosmic plasma it is difficult to avoid the conclusion that a large fraction of the gravitational energy goes into ionization.

Instead of analysing this problem in a serious way adherents to the Laplacian approach simply postulate an 'expulsion' of magnetic field with essentially no other motivation than that this is necessary in order to allow them to play around with pre-hydrodynamic models.

After the breakthrough of quantum mechanics people who claim that they can explain atomic structures by classical mechanics are generally considered to carry the burden of proof that their approach deserves to be taken seriously. Twenty-five years after the breakthrough of magnetohydrodynamics, those who claim that the dynamics of cosmic clouds can be treated by pre-hydromagnetic formalism have in a similar way the burden of proof.

Transition from Uniform Disk to Present State

The second step—the transition from the hypothetical Laplacian disk to the present state in the solar system also meets serious difficulties which can be overcome only by postulating a depressingly large number of often unreasonable *ad hoc* assumptions. In order to account for the almost void region between Mars and Jupiter a number of 'instabilities' are postulated, bodies which have been formed once are 'exploded', and the debris is thrown out by processes, which in some cases are violating well-known laws, in other cases are purely *ad hoc*.

A process of the latter type is the hypothesis of a 'solar gale' which should sweep out all the left-overs of the postulated processes, including a too high solar angular momentum. This hypothesis is sometimes motivated by a reference to observed outward motion of matter in T-Tauri stars. However, what is observed is a Doppler effect of neutral hydrogen indicating that the moving medium is different from the fully ionized solar wind—at velocities of 50–100 km/sec–far below solar wind velocities (Herbig, 1976). As the moving medium may be close to the star, the velocity may be below the escape velocity and there is no proof that the matter is ejected to infinity.

Another observational fact which is claimed to support the Laplacian theories is that some stars are surrounded by gas and dust which probably form a disk. However, all cosmogonic theories envisage a disk of matter in some stage of evolution. What distinguishes the Laplacian theories from others is that the disk is supposed to be *uniform*. There is no observational support for this.

The Band-structure of the Solar System

Band-structure Diagram

As pointed out already by Laplace (but forgotten by his epigones) the satellite systems of Jupiter, Saturn, and Uranus are so similar to the planetary system that they should be treated in a similar way. Such a procedure leads to the diagram of Figure 1 which shows the gravitational energy of all planets and satellites in relation to the central bodies. It is obvious that the mass distribution in both the planetary and the satellite systems is characterized by a *band-structure* (Alfvén, 1943). *All planets* and all *prograde satellites* are located in one of three or four horizontal bands. The total mass outside these bands (asteroids and the Martian satellites) is less than 10^{-6} of the total mass.

The band-structure implies that there are void (or almost void) regions in the systems. One such region is the asteroidal region where the smeared out density is about 10^{-5} of the density in the adjacent regions.

Other void regions are found in the satellite systems. In the Saturnian system there is a regular sequence of satellites from Rhea inwards, continued by the ring system, indicating a reasonably uniform mass injection into the region from the surface of the planet out to Rhea. In striking contrast to this there is a void region inside Miranda in the Uranian system, and a similar void region between Io and the surface of Jupiter (or the orbit of Amalthea, if this very small body is taken into consideration). The void gap between Rhea and Titan and the sharp outer border of the Galilean satellite group—and also the Uranian group of satellites—are other examples of striking features accounted for by the band-structure. All these features are embarrassing to the Laplacian school and hence they are never discussed. Even the remarkable fact that Saturn and Saturn alone has a ring is swept under the rug because it is a confirmation of the band-structure but not easily explained with the Laplacian approach.

Theory of the Band-structure

Although already at an early stage a theoretical explanation of the band structure was suggested, fifteen years of experimental and theoretical work has been required in order to understand it. As soon as the 'thermonuclear' technique became available a series of plasma experiments were started in Stockholm with the purpose of clarifying the band-structure. These led to the discovery of the 'critical velocity', a phenomenon which later independently

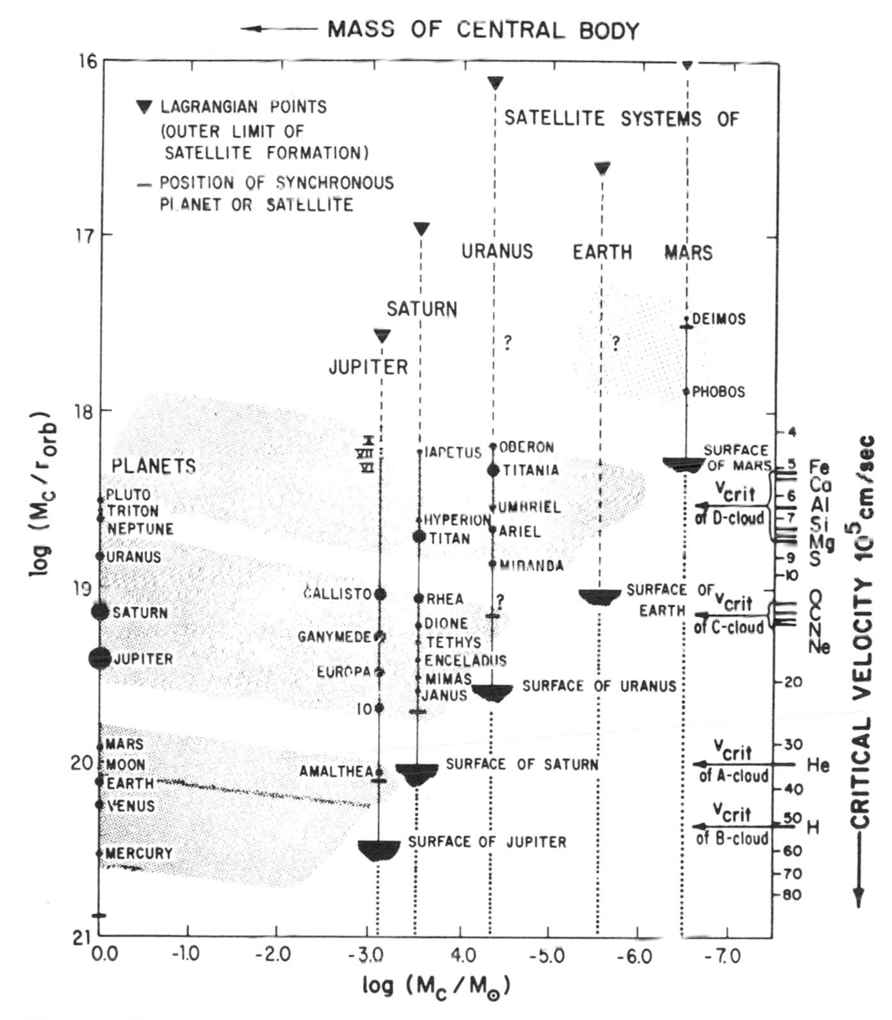

Figure 1. The gravitational energy of all planets and satellites in relation to their central bodies (Reproduced with permission from H. Alfvén and G. Arrhenius, *Structure and Evolutionary History of the Solar System*, copyrighted by D. Reidel Publishing Co., Holland, 1975)

has been discovered at several other laboratories (see Danielsson, 1973; Axnäs, 1976; Danielsson and Brenning, 1975). The theoretical explanation of the critical velocity has been very difficult and half a dozen different theories have been proposed. Sherman (1972, 1973) has given a survey of the theories, and his own theory further developed by Raadu (1975) seems to give what probably is the final explanation. Hence we now understand reasonably well the behaviour of pure atomic gases. However, the critical velocity of gas mixtures and molecular gases are still not completely understood.

We know by now enough about the relation between plasma experiments in the laboratory and in the cosmos to be able to conclude that the critical velocity should be of importance also in cosmic physics. To some extent this is confirmed by observations of certain lunar phenomena (Manka *et al.*, 1972).

The Complete Matrix of Planet–Satellite Groups

The application of the critical velocity phenomenon to the cosmogonic problem leads to the conclusion that the solar system should consist of groups of bodies produced by four clouds of different chemical composition. This is summarized by the matrix of Figure 2. Whenever a cloud falling in towards a central body, which is magnetized, reaches the critical velocity at a sufficiently large distance from the surface of the central body, a group of secondary bodies is formed. The structure of a group of bodies is completely defined by two parameters:

1. The mass of the central body
2. The spin of the central body

Furthermore the central body must have had a magnetic field above a certain limit at the time of the cosmogonic processes.

Groups of secondary bodies are found at *all* places where we theoretically

Central body	τ 10^5 sec	T_{ion}/τ for secondary bodies in cloud B	A	C	D
Sun	21.3	8.5 ☿–⊕	28 ☾–♂	520 ♃–♆	5000 P
Jupiter	0.354	0.50	1.6 Amalthea	29 Galilean satellites	286
Saturn	0.368		0.45	8.4 Inner satellites	80 Outer satellites
Uranus	0.389			1.3	12 Uranian satellites
Neptune	0.504			1.0	9.5
Earth (prior to capture of the Moon)	0.14(?)				2.2(?)

According to the theory, bodies are produced only in the groups above the line in the table.

Figure 2. Values of τ_{ion}/T where τ_{ion} is the Kepler period of a body at the ionization distance and T is the period of axial rotation of the central body (Reproduced with permission from H. Alfvén and G. Arrhenius, *Structure and Evolutionary History of the Solar System*, copyrighted by D. Reidel Publishing Co., Holland, 1975) expect them to be found and have the expected structure (except for two of the uttermost groups which are 'irregular'). 'Double planets' (planets with abnormally massive satellites) are found in the regions where two bands overlap. This accounts for the Earth–Moon and the Neptune–Triton systems. However, the details of the capture process are far from clear.

At its present state of development the theory gives a framework and defines boundary conditions for a number of models of the cosmogonic processes which are still not worked out in detail. In several cases this cannot be done until space research has supplied us with more empirical data.

As far as one can judge from the present state of development of the theory one may hope to explain the present structure of the solar system as the direct result of an evolution from an interstellar cloud without introducing the unnecessary and drastically misleading intermediate step of formation of a uniform disc.

In contrast to the Laplacian theories there is no obvious need for *ad hoc* assumptions. The planets and satellites are formed in approximately the present state, and hypothesis about 'protoplanets' several times more massive than the present planets are not necessary or rather counter indicated (by the isochronism of spins). No planet need to be 'exploded', no ejection of superfluous bodies need to be postulated, nor is any 'solar gale' necessary in order to clean our region of space from the vestiges of the Laplacian disk.

References

This paper is a summary of a monograph Alfvén, H., and Arrhenius, G. (1975), *Structure and Evolutionary History of the Solar System*. D. Reidel Publ. Co.

Alfvén, H. (1942–1945). 'On the cosmogony of the solar system.' *Stockholms Obs. Ann.*, **14**, No. 2; **14**, No. 5; **14**, No. 9.

Alfvén, H. (1954). *On the Origin of the Solar System*. Clarendon Press, Oxford.

Alfvén, H., and Fälthammar, C.-G. (1963). *Cosmical Electrodynamics*. Clarendon Press, Oxford.

Alfvén, H., and Arrhenius, G. (1976). *NASA Special Publication*, 345.

Axnäs, I. (1976). *Velocity Limitations in Coaxial Plasma Gun Experiments with Gas Mixtures*. Royal Institute of Technology, Stockholm, report TRITA-EPP-76-02.

Cameron, A. G. W. (1973). 'Accumulation processes in the primitive solar nebulae.' *Icarus*, **18**, 407.

Danielsson, L. (1973). 'Review of the critical velocity of a gas–plasma interaction—experimental observations.' *Astrophys. and Space Science*, **24**, 459, 1973.

Danielsson, L., and Brenning, N. (1975). *Experiment on the Interaction between a Plasma and a Neutral Gas. II. Phys. Fluids*, **18**, 661.

Goldreich, P., and Ward, W. R. (1973). 'The formation of planetesimals.' *Astrophys. J.*, **183**, p. 1051–1061.

Herbig, G. H. (1976). 'Star formation.' Lecture given at Nato Advanced Study Institute, *The Origin of the Solar System*, Newcastle.

Manka, R. H. *et al.* (1972). 'Evidence for acceleration of lunar ions.' In C. Watkins (ed.), *Lunar Science Vol. III*, Lunar Science Contribution No. 88, The Lunar Science Inst. Houston, Texas, p. 504.

Safranov, V. S. (1972). *Nice Symposium on the Origin of the Solar System*, C.N.R.S. 89.
Sherman, J. C. (1972). 'The critical velocity of gas–plasma interaction and its possible hetegonic relevance.' Nobel Symp. 21. *From Plasma to Planet*. Ed. A. Elvius. Wiley–Interscience.
Sherman, J. C. (1973). 'Review of the critical velocity of gas-plasma interaction, II Theory.' *Astrophys. Space Sci.*, **24**, 487–510.
Raadu, M. A. (1975). 'Critical ionization velocity and electrostatic instabilities.' *TRITA-EPP*-75–28 *Electron and Plasma Physics*. Royal Institute of Technology, Stockholm.
Spitzer, L. (1968). 'Nebulae and interstellar matter.' In Middlehurst, B. and Aller, L. H. (ed.), *Nebulae and Interstellar Matter*, Wiley, New York.

Conclusions about the behaviour of the preplanetary medium should not be based on speculations but on a transfer of knowledge from laboratory experiments and space observations. Important references are found in:

Fälthammar, C. -G. (1974). 'Laboratory experiments of magnetospheric interest.' *Space Science Rev.*, **15**, 803.

THE PRIMITIVE SOLAR ACCRETION DISK AND THE FORMATION OF THE PLANETS

A. G. W. CAMERON
Center for Astrophysics, Harvard College Observatory and Smithsonian Astrophysical Observatory, Cambridge, Massachusetts, U.S.A.

Star Formation

Typical high resolution studies of a spiral galaxy show that the massive blue stars, presumably those most recently formed in the galaxy, are concentrated in the spiral arms and appear close to the inner edges. Thus, the observations clearly indicate that there is something associated with spiral arms which induces star formation. Since our galaxy is older than the solar system by a factor of two or three, it is likely that the general morphology of our galaxy was much the same at the time of solar system formation as it is today. Thus, it is a logical first assumption that the solar system should have been born along with other stars in a spiral arm.

According to the density wave theory of spiral arms, such an arm represents a travelling gravitational trough to which stars are attracted and in which they spend a fairly large amount of time before their orbits take them out of the arm (see for example Lin and Shu, 1964). The resulting clustering of many stars which are attracted to the arm provides the gravitational potential which forms the trough. The disturbance travels through the stellar distribution with a pattern speed which, in the vicinity of the solar system in our galaxy, is expected to travel about half as fast as a star in a circular orbit about the centre of the galaxy. The interstellar gas is also attracted towards these travelling gravitational troughs, and it is expected to undergo a shock deceleration as it moves into the trough. This causes an increase of the pressure and density of the interstellar medium within the trough by a factor of a few. The actual factor is somewhat uncertain, because if the gas contains a magnetic field with a significant transverse component, the magnetic field will prove resistant to strong compression.

The interstellar medium is normally expected to exist in two phases; either hydrogen is neutral and the corresponding phase is of low temperature and fairly high density, or hydrogen is ionized and the medium is of higher temperature and lower density. These phases are often called the clouds and the intercloud medium, respectively. The temperatures and densities of the gas phases are inversely correlated in such a way that the gas pressures approach approximate homogeneity throughout the medium. This approach to an equal pressure everywhere can only be an approximate one, because the formation of massive ultra-violet emitting stars and the explosion of supernovae create local sources of high pressure which require hydrodynamic readjustments of the medium in the form of fluid flows.

Interstellar clouds exist with a wide variation of parameters. Their masses typically lie in the range of a few hundred to a few thousand solar masses. Such clouds are quite stable against collapse to form stars; they would undergo a large degree of expansion were it not for the fact that they are in approximate pressure equilibrium with the neighbouring gas.

Jura (1976) has recently shown that the mass threshold for collapse to form stars is a strong function of the pressure in the interstellar medium. Under normal conditions, this threshold would be a mass slightly in excess of 10^4 solar masses. However, if the pressure in the interstellar medium should rise by a factor of ten, then the threshold for a collapse to form stars would be reduced to around 100 solar masses.

By this criterion, the normal interstellar clouds would be stable against star formation, but when the interstellar gas moves into a spiral arm the increase in the general pressure of the medium by a factor of a few may trigger a few of the more massive clouds to undergo collapse and star formation.

Recent developments in meteorite research have pointed to the possibility that star formation can take place in a chain reaction fashion. Lee, Papanastassiou, and Wasserburg (1976) have found evidence for the presence of a short-lived extinct radioactivity, ^{26}Al, in the early solar system. This substance has a half life of $7{\cdot}2 \times 10^5$ years. Its presence in solar system materials indicates that only a few million years can have elapsed between the time that the ^{26}Al was formed and the time that it was incorporated into solid phases within solar system material. I do not consider it to be possible that the ^{26}Al was formed by energetic particle bombardment within the primitive solar nebula, in part because of the prohibitive energy requirements, and in part because the bombardment would lead to other isotopic anomalies which have not been detected. ^{26}Al should be formed with relatively low abundance during the process of explosive nucleosynthesis taking place in a supernova explosion. This has led Professor J. W. Truran and me to suggest that the close association of a supernova with the formation of the solar system was not an accident, but rather represented a causal association (Cameron and Truran, 1976).

In this picture, there will exist supernova explosions in spiral arms of the galaxy which represent the destruction of massive stars formed either from instabilities that are triggered into collapse in massive cloud complexes by the compression of the flow going into the spiral arms, or by the chain reaction of cloud collapses that is presently being described. When a massive star is formed, the intense ultraviolet radiation which is emitted from it causes heating and ionization of the surrounding gas, thereby tending to increase the local pressure. This initial increase of pressure tends to compress the interstellar clouds lying near the massive star which has been formed. When the massive star undergoes a supernova explosion, a further sharp increase of the pressure in the interstellar medium is initiated, and when the heated supernova ejecta flows over the surfaces of the compressed interstellar clouds, the pressure becomes sufficiently high that it is likely that several of the nearby clouds will be caused to collapse and to undergo star formation. In turn, some of these new stars are likely to be massive, and they will undergo supernova explosions and spread further the chain reaction of star formation.

When the fast-moving supernova ejecta impinge upon the forward face of an interstellar cloud, a Rayleigh–Taylor instability will occur, in which tongues of the supernova ejecta will manage to penetrate to the heart of the interstellar cloud. Fresh radioactivities which have been created in the supernova explosion can thus be introduced into the cloud during the earliest stages of the dynamic collapse, and in this way it is perhaps not too surprising that short-lived radioactivities were present in early solar system material.

Professor Truran and I have suggested that the nucleosynthesis products formed in such a single supernova, which triggers the formation of the solar system, may be present in sufficient abundance to cause isotopic anomalies within solar system material, without at the same time representing processes which are the principal contributors to nucleosynthesis on the galactic scale. Without going into detail on these, let me simply indicate that the candidate isotopic anomalies are those in oxygen, neon, krypton, and xenon, in addition to ^{26}Al. These effects may also include the local production of ^{244}Pu, of another extinct radioactivity, ^{202}Pb, and of the longer-lived radioactivity ^{40}K. We do not yet know enough about the detailed structure of supernovae to make a very precise prediction on these matters, but we believe that the study of extinct radioactivities and isotopic anomalies in early solar system material is likely to produce a great deal of further information concerning processes that went on in the primitive solar nebula, and we believe that the supernova trigger hypothesis may prove to be useful in providing a unifying concept which assists in the interpretations.

If these views are correct in a general way, they suggest that the formation of the solar system was triggered by the explosion of a supernova near a compressed interstellar cloud; the cloud was further compressed by the supernova ejecta until it went over the threshold for gravitational collapse. During the collapse it is expected that the cloud would fragment into much smaller pieces.

Interstellar Cloud Fragment Collapse

An interstellar cloud mass of perhaps 1000 solar masses is sufficient to crush the magnetic field which threads through it, starting from an initially somewhat enhanced density which would be characteristic of a cloud near a massive star. However, fragmentation could not progress very far before the magnetic field energy in a fragment would exceed the gravitational potential energy. Therefore, fragments of the order of a solar mass can only be formed if the magnetic field is largely able to escape from the interior of the collapsing interstellar cloud and its fragments. The process of escape of the lines of magnetic force is called ambipolar diffusion; it has been studied to some extent over a period of several years. The most extensive recent studies are those conducted by Nakano and his colleagues. A particularly basic paper in these studies was by Nakano and Tademaru (1972); in which the authors studied the escape of the magnetic field from a collapsing cloud as a whole. In this paper they assumed that the cosmic rays incident upon the cloud would travel straight in toward the centre, and hence they estimated the relative degree of ionization in the interior of the cloud after it had become compressed enough to exclude starlight from the interior, but where it was still subject to ionization by cosmic rays and by such radioactivities as ^{40}K. This was clearly an exceedingly conservative set of assumptions, and the authors found that the magnetic field would escape from the gas by the time that the density in the gas had risen in free fall collapse to about 10^9 particles per cubic centimetre. In reality the cosmic ray particles would spiral around the magnetic lines of force, would be reflected from the magnetic mirrors, and in general would produce considerably less ionization near the centre of the cloud. On the basis of the work by Nakano and Tademaru alone, I would estimate that the magnetic field is more likely to escape from the gas at a density of about 10^6 particles per cubic centimetre. More recently Scalo (1976) has shown that recombination due to molecular dissociation, which had previously been neglected, can significantly reduce the ion densities and hence decrease the ambipolar diffusion time scale. Furthermore, recently Nakano (1976) has shown that fragmentation can take place in the presence of the magnetic field, with ambipolar diffusion allowing the magnetic lines of force to slip out of the growing fragments. These various studies lead to the expectation that fragmentation of the interstellar cloud can take place without a great deal of magnetic hindrance.

From this it should not be inferred that the magnetic field in the forming solar system should slip back to values typical of the interstellar medium. Even though the magnetic field is slipping through the collapsing gas, nevertheless the dynamic motion of the gas can cause a significant compression of the magnetic field, and I estimate that magnetic field strength of the order of a significant fraction of a gauss was probably present in the primitive solar nebula for at least part of the time of its formation. This is the kind of seed field that may be expected to be present near the centre of the solar nebula at the time matter was accumulated there in sufficient mass to produce thermal ionization and to start magnetic amplification by dynamo action.

In the process of cloud fragmentation, it is expected that fragments will form around density irregularities, so that the centres of the cloud fragments are likely to have a somewhat greater density than the mean density in a fragment as a whole.

This means that in the course of free-fall collapse, the centres of the fragments will fall in more rapidly, so that the amount of gas that was associated with the earliest stage of formation of the primitive solar nebula would involve only a relatively small part of the total cloud fragment. From the motions and dimensions involved, I estimate it to be likely that the infall time associated with the material in the cloud fragment will lie in the range 10^4 to 10^5 years. This time interval is consistent with the spread of time in the star formation process found by Larson (1969). Most importantly, this time is very significantly longer than the two characteristic dissipation times for the primitive solar nebula found by Cameron and Pine (1973) in their study of models of a full-scale primitive solar nebula. These dissipation processes were radiation of the internal energy in the nebula and transport of angular momentum throughout the nebula as a result of meridional circulation currents. Thus, since the dissipation times associated with the primitive solar nebula appeared to be considerably shorter than the formation time, modelling of the primitive solar nebula can only proceed by taking into account the internal dissipation and angular momentum redistribution as the accretion of matter onto the solar nebula occurs.

In a previous study (Cameron, 1973) I have examined these sources of angular momentum likely to be present in the matter in the primitive solar nebula. Such considerations must start by noticing that the interstellar cloud was brought to the threshold of collapse in a dynamically violent manner. Such rapid and uneven compression of the gas will also stir it, and hence lead to the presence of extensive turbulence throughout the collapsing interstellar cloud. As the collapse of the cloud progresses, the temperature in the gas falls to a low value of the order of 5 or 10 K. On the other hand, the compression amplifies the turbulent velocities, so that the turbulent velocities should increase until they approach the speed of sound, at which point extensive dissipation will occur, leaving turbulent velocities in the vicinity of the sound speed. To the extent that a full spectrum of turbulence is produced, the entire collapsing cloud will contain internal motions in the form of a varying shear field. When cloud fragmentation takes place, the fragments are thus likely to have a net angular momentum derived from the shear velocity field, and I estimate that this turbulent contribution to the angular momentum of the fragment is likely to exceed any contribution resulting from the initial spin of the cloud prior to collapse. As shown in the paper by Cameron and Pine, this amount of angular momentum contained in a collapsing cloud fragment of two solar masses is likely to produce a primitive solar nebula with a radial extent of the order of 100 astronomical units. In subsequent discussion I will call this a 'nominal' or 'reasonable' amount of angular momentum to be associated with the primitive solar nebula. By that I shall not mean that all of this amount of angular momentum is contained in the primitive solar nebula at any one time, but rather that it is contained in the collapsing interstellar-gas cloud fragment.

There are some dynamical complications associated with the beginning of the solar nebula formation. The first clue about this came when Larson (1972) carried out some crude two-dimensional hydrodynamic calculations of a collapsing rotating gaseous sphere. He found that, instead of collapsing to form a flattened disk with a density maximum at the centre, the gas instead formed a ring. The gas near the spin axis was pulled out to become a part of the ring. These results were confirmed by the two-dimensional hydro-dynamic calculations carried out with a n uch larger number of mesh points by Black and Bodenheimer (1976). Both in the calculations due to Larson and in many of those due to Black and Bodenheimer, the angular momentum was considerably greater than the nominal value mentioned above, and the authors found that a great deal of the gas in the collapsing sphere participated in the ring formation, and that the ring was formed at a very large distance from the spin axis. When Black and Bodenheimer used values of the angular momentum of the collapsing sphere closer to nominal, they found that only a relatively small part of the mass of the sphere participated in the ring formation and the ring was formed much closer to the spin axis.

I believe that these instabilities are the first indication of some general instabilities that I will discuss later in connection with the primitive solar nebula. It is perhaps questionable that the collapse of the interstellar fragment will involve quite as much axial symmetry as is assumed in the calculations, so that it may perhaps be a bar rather than a ring which is formed near the centre of the collapsing gas. In either case, it is likely that two objects with masses comparable to a few times the mass of Jupiter will be formed near the centre of what will become the primitive solar nebula, and these two objects will be in orbit around each other. If we assume that a gaseous disk is formed around the binary pair of objects, and that as a result of dissipation in the disk, there is a mass flow inward leading to mass accretion onto these two objects, then to a first approximation we can suppose that the objects become more massive while retaining approximately the same value of this specific angular momentum. As their mass increases, the distance between the two objects will be decreased in order that angular momentum can be conserved, and it is likely that they will merge together in the course of time. This becomes a justification for treating the primitive solar nebula as though it had a single mass point at the centre. In collapsing cloud fragments with larger amounts of angular momentum than nominal, the two components which are formed in hydrostatic equilibrium may be considerably further apart, so that in the course of dissipation of any mutual flattened gaseous disk surrounding them, they do not succeed in merging together and are left as a close binary pair.

The presence of turbulence in the collapsing interstellar-cloud fragment will not have a significant impact upon the temperature of the gas during most of the collapse, which will stay near 5 or 10 K, but it may have interesting effects on the interstellar grains which are contained within the gas. In previous studies (Cameron, 1973, 1975), I have examined the effect of turbulence in the collapsing gas on the motions of the interstellar grains. The turbulence stirs up the grains and causes them to collide. An interstellar grain is probably a rather fluffy, fragile object, so that two grains are likely to stick together if their collisions are not too destructive, and the studies indicated that repeated collisions among the grains and grain clusters may lead to a significant amount of mass of the interstellar grains extending into a size range up to perhaps a few millimetres. If there is a significant fragmentation probability as well as a significant amalgamation probability, what is likely to result is a spectrum of grain sizes extending all the way from individual grains to the high end of the mass range. This is likely to be the situation concerning condensed solids in the infalling gases which accrete onto the primitive solar nebula.

The Primitive Solar Accretion Disk

For some years my reasearch has been aimed toward understanding the evolution of the solar nebula. The first stage in these studies was to construct a numerical model of the primitive solar nebula, and this was achieved with the models constructed by Cameron and Pine (1973). Within these models it was possible to evaluate the dissipation time scales for the primitive solar nebula, and, as discussed above, these turned out to be less than the characteristic formation time for the primitive solar nebula. Thus, the principal goal of the research shifted to the calculation of sequences of models of the primitive solar nebula in which dissipation occurred during the formation. This goal was achieved during the winter 1975–76.

Contributing in a very material way to the achievement of this goal was the development of the theory of the viscous accretion disk. The basic elements of such a theory were published many years ago by Lust (1952), but this important pioneering paper did not become well known. The theory was twice again independently worked out, by Lynden-Bell and Pringle, but these authors discovered each others' work and recently published a joint paper outlining the theory (Lynden-Bell and Pringle, 1974).

The theory of Lynden-Bell and Pringle applies to the case where there is some central mass near the spin axis which is surrounded by a flattened rotating disk of gas. In general, the disk will not rotate with uniform angular velocity, and therefore there will be shear between adjacent layers rotating around the central spin axis. The theory of Lynden-Bell and Pringle assumes that there is some kind of friction between these adjacent layers; the form of the friction is not specified by them, so that their theory is quite general. As a result of the friction, transport processes are effective within the disk and dissipation takes place. Angular momentum and energy are transported away from the central spin axis as a result of the friction acting on adjacent layers within the gas. This results in a readjustment of mass within the disk; in general, mass is transported inwards near the spin axis and it is transported outwards at large distances from the spin axis.

I have applied the theory of Lynden-Bell and Pringle in the construction of a series of models of the primitive solar nebula. The details of this will be published elsewhere, and here I give a qualitative description of the assumptions and procedures and the principal results.

First consider for a moment what would happen if the primitive solar nebula were to accrete without friction and hence dissipation. This would effectively be the case if the only source of friction were molecular viscosity. Under these circumstances the mass would collect into a disk, and there would be no appreciable mass concentration towards the central spin axis. As shown by Cameron and Pine (1973), the nebula would cool quite quickly and become very thin. Under these circumstances, the gas in the disk would become unstable against many different types of perturbation, both with and without axial symmetry. The details of this are likely to be rather chaotic, but it may be seen that this does not lead to any kind of conventional picture of the behaviour of the primitive solar nebula. We shall return to this situation later, for, as it will be seen, one of the principal conclusions of these calculations is that growth of instabilities of this type cannot be escaped even if there is efficient friction between adjacent layers in the gas.

In the calculations it was assumed that the principal source of friction was turbulent viscosity. This is reasonable because the primitive solar nebula would be a system with a very large Reynolds number, and it must be expected that the gas will be vigorously stirred. I can identify three mechanisms which will contribute to the stirring.

The first of these is meridional circulation. This has already been discussed by Cameron and Pine. In a perfectly spherical star in hydrostatic equilibrium, the surfaces of constant gravitational potential are also surfaces of constant density, temperature, and pressure. If the star rotates, it is no longer possible for these quantities to remain constant over the same surfaces, and the small force imbalances which are created set into motion meridional circulation currents which have extremely small velocities in the case of a star like the sun. However, if the star rotates very rapidly, the meridional circulation currents should travel much faster. These currents can also be expected to travel very rapidly in the case of a rotationally flattened configuration such as the primitive solar nebula, in which there is still a very significant thickness, about 20 or 30 per cent of the radial distance. However, in the limit in which the thickness of the disk would shrink to zero, the meridional circulation currents would vanish. Therefore, there is a maximum in the current velocity for a thick rotationally flattened configuration, and the primitive solar nebula should lie somewhere close to this maximum. A crude dimensional analysis indicated that the meridional circulation current in the primitive solar nebula should have a velocity comparable to sound speed. This should indeed provide a very vigorous stirring of the nebula.

A second source of stirring is expected to be associated with the infall of material which forms the primitive solar nebula while the latter is undergoing dissipation. As discussed previously, the infalling gas is expected to be turbulent, and therefore at any given spot on the nebular surface the accreting matter will exhibit a time-varying angular momentum. Furthermore, there is no reason to expect the mean angular momentum of the infalling material to match that at the point of impact, since the dissipation of the nebula results in the redistribution of the internal angular momentum. The process of sorting out the angular momentum of the merged fluid should thus produce additional significant stirring.

As already noted above, it turns out to be an expectation that the primitive solar nebula will become unstable against gravitational condensations on a rather large scale. The motion of these through the primitive solar nebula produces gravitational fields varying with time, and these will also contribute to the stirring of the gas.

The turbulent viscosity was calculated as though the largest eddies in the flow of the fluid were molecules traversing the gas. The characteristic distance associated with the fluid motion was taken to be the half-thickness of the disk. In preliminary studies, the velocity of these largest fluid eddies was taken to be equal to the sound speed, but in the results that are reported here this was reduced to one-third of the sound speed. In making this reduction I was following the advice of S. A. Colgate, who remarked that the hard-driven fluid motions associated with major explosions in the atmosphere generally did not exceed one-third of sound speed. I believe that this representation of the turbulent viscosity is a reasonable expectation for the primitive solar nebula, but it should be noticed that it represents the largest amount of friction which can be physically argued to be present within the gas.

The magnitude of the viscous couple between adjacent fluid layers is proportional to the semi-thickness of the disk over which the couple is exerted. The local dissipation of energy also depends on this semi-thickness (a majority of the energy which is dissipated at any local position is generated further in towards the spin axis and transported outwards to the point of dissipation). The surface temperature, which governs the rate at which energy is radiated into space, is proportional to the one-fourth power of the dissipation. In turn the semi-thickness of the disk and its internal temperature structure is dependent upon the value of the surface temperature. It may be seen that these quantities are interdependent, and it was necessary to determine the proper values of these quantities by an iterative process.

The vertical structure of the disk, perpendicular to the plane of flattening, was taken to be isothermal. This constant temperature approximation is good to about five or ten per cent over much of the disk. The argument for the approximation is a fairly simple one. One can compare two typical time scales. One of these is the time required for the physical transport of energy from the mid-plane of the disk to the radiating surface at one-third of the sound speed.

The other time scale is for the random-walk transport of energy by photons diffusing from the mid-plane of the disk to the radiating surface. The latter time scale is much shorter than the former time scale. Thus, it follows that the turbulent motion of the gas is not isentropic motion, and the expansion of the gas to match the lower density in the region of the radiating surface is not accompanied by adiabatic cooling, but rather the gases are kept hot by photon diffusion.

It must be expected that the upper atmosphere of the fluid, beyond the radiating photosphere, has a structure largely governed by the shock deceleration of the infalling material from the collapsing interstellar cloud fragment. If it were not for this, the structure would have exhibited a very interesting property. The turbulent motion of the gas will at all times be generating large mechanical disturbances which propagate as waves through the fluid. As such waves advance into the upper atmosphere, the conservation of energy within the fluid disturbance causes the amplitude of the motion to increase, until the internal motion with the wave approaches sound speed. At this point an efficient dissipation of the wave energy into local heat takes place, and shock dissipation will occur if the motions exceed the speed of sound. This heating would create a hot coronal layer in the upper atmosphere of the disk, similar in many respects to the hot coronal upper atmosphere of the sun. Just as the presence of a hot corona in the sun causes a hydrodynamic flow of mass away from the sun, so it must also be expected here that hot coronal layers high in the atmosphere of the disk would cause mass flow away from the disk. I have estimated that the efficiency for channelling turbulent energy into heating the upper atmosphere of the disk should be much greater than the efficiency with which similar processes occur in the sun. The conclusion, therefore, is that if it were not for the accretion flows of gas onto the surfaces of the disk, there would be a rapid mass loss from the disk due to the heating of its upper coronal layer.

It is unlikely that the accretion of the gas onto the surfaces of the disk can suppress the hot corona and the disk mass loss indefinitely. When most of the mass in the collapsing interstellar cloud fragment has accreted onto the disk the mass inflow rate must slacken. At the same time, the disk has become fully developed and the tendency toward forming a hot corona is increasing. At some point, perhaps due to a fluctuation in the infalling material, a hot corona should be able to form over some local patch of the disk. I have estimated that when this can occur, the outward expansive pressure of the gas in the corona should exceed the downward ram pressure of the descending gas, so that the hot corona should spread and drive back any residual gas which is trying to fall onto the surface of the disk. Thus it must be expected that the history of development of the primitive solar nebula will see a transition from a phase of mass accretion from the outside to a phase of mass loss generated from internal turbulent motions.

In the calculations which I carried out, this transition was arbitrarily triggered at a time when the thermodynamic conditions within the disk were interesting in terms of the formation of the planets in the solar system. For most of the cases which were investigated, the transition took place after approximately two solar masses of material had accreted into the disk, and much of this had further flowed to the centre of the disk to collect in the form of a compact object, whose properties were not explicitly treated in these calculations.

Evolution of the Primitive Solar Nebula

A preliminary investigation, which I shall not discuss in detail here, explored the effect of variation of various parameters on the properties of the primitive solar nebula. The important parameters in this investigation were the total angular momentum of the accreting material, the mass infall rate, and the mass of the central object about which the disk was formed. These preliminary calculations confirmed that a mass infall rate of about one solar mass per 30,000 years produced the most interesting conditions for the thermodynamic environment of planet formation; they showed that for nominal values of the total angular momentum, the mass and radius of the disk were very large, and this led in a natural way to a new theory of comet formation, which I shall describe below. On the other hand, if the total angular momentum of the infalling gas was only one-tenth of nominal, then the radius of the disk was still large, but the mass of gas in the disk was relatively small, and one could not get a new theory for comet formation. Accordingly, I shall discuss here only the evolutionary calculations which correspond to the indicated rate of mass accretion and to the nominal value for the total angular momentum of the infalling gas. Other evolutionary histories were run, in which the parameters that were varied included the total angular momentum of the accreting material, the mass infall rate, and the efficiency of the final stage of mass loss from the coronal layer in the atmosphere above the disk. The general results of these histories were qualitatively similar.

The mass of the disk as a function of time is shown in Figure 1 for the standard case with the standard parameters. The mass of the disk increases as long as accretion is taking place onto it. After nearly 50,000 years, the switch was made to the mass loss regime, and then the mass decreased very rapidly as can be seen in Figure 1. Although the evolution of the disk was followed until most of the mass had gone, the latter portion of this evolution cannot be considered to be determined by the procedures outlined here, since when the mass becomes very small the centre of the disk becomes optically thin, and then some of the assumptions I have described become questionable. The important thing to notice is that the total time required for both the growth and the dissipation of the disk is quite short, less than 200,000 years.

The variation with time of the radius of the disk is shown in Figure 2. As long as mass accretion is occurring, the radius of the disk increases steadily with time, since angular momentum is transported always toward the outer

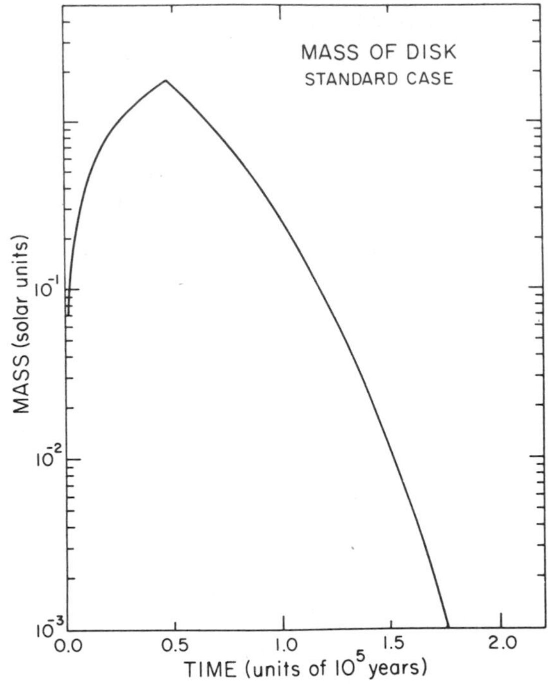

Figure 1. Variation of the mass of the primitive solar nebula with time in the evolutionary study using standard parameters. The mass of the central object is not included in the disk mass

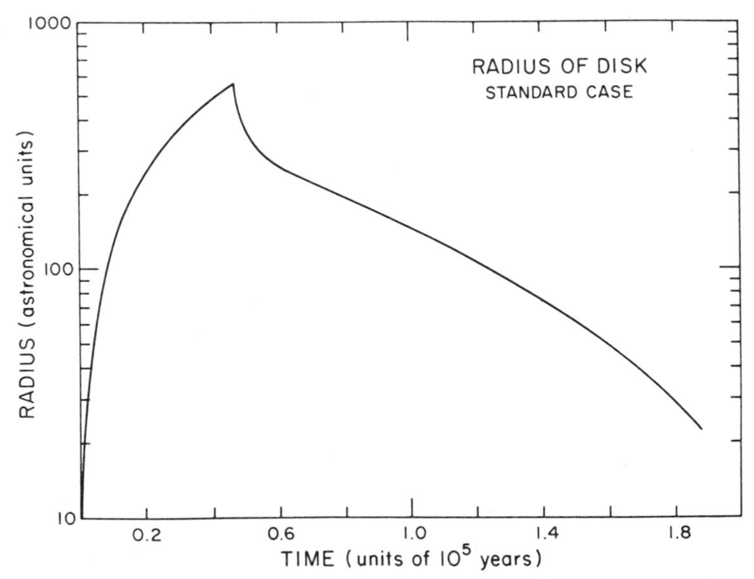

Figure 2. Variation of the radius of the primitive solar nebula with time in the evolutionary study using standard parameters

edge of the disk, and mass is also transported toward the outer edge in these solutions to the equations of Lynden-Bell and Pringle. However, when mass loss begins, the rate of mass loss from the outer edge of the disk is extremely rapid, and the radius of the disk decreases very rapidly after the onset of mass loss. Thereafter, the radius slowly retreats, and this has been followed until the radius of the disk is significantly less than the orbital radius of Neptune. The important thing to note from this figure is that the radius probably reached a value of several hundred astronomical units at the maximum extent.

Most of the mass lost from the disk flows off into space. However, typically about 20 per cent of the mass of the disk is lost into the central object. That this should be so is not surprising. At the time that the disk has reached its maximum extent, a pattern has been set up in which the mass flow is inwards throughout the region of planet formation in the primitive solar nebula. The transition to mass loss from the disk hardly affects this pattern of mass flow in the inner regions of the disk. During the mass loss regime, the flow remains inward throughout most of the region of planet formation until fairly late in the dissipation stage, at which point the radius has greatly decreased, and mass flows outwards in the region of the outer planets to make up for some of the mass lost further out. Thus, this period of the evolution is one of continued mass accretion onto the central object, even while most of the mass is being lost to outer space.

From these two figures one can see the basis for a new theory of comet formation. Notice in Figure 1 that the mass in the disk reaches very large values. Notice in Figure 2 that the radius of the disk also reaches very large values; in fact more than 90 per cent of the mass in the disk lies beyond the region of formation of Neptune at the time of maximum extent of the disk. Most of this mass is lost into space. Consider the effect of this upon a body formed in a circular orbit at several hundred astronomical units. The thermodynamic conditions in the disk at such a large distance imply a very low temperature and small density, so that such a body must be an accumulation of interstellar grains which have been preserved at low temperatures. Such a body is likely to be formed by the Goldreich-Ward gravitational instability mechanism (Goldreich and Ward, 1973), when clumps of interstellar grains descend toward mid-plane of the nebula as turbulence dies away with the onset of the mass loss. I estimate, by applying this theory to the conditions in the models under such circumstances, that bodies formed at these large distances are likely to have masses of the order of 10^{16} grams. These are obviously comet candidates.

For simplicity, imagine two stars mutually bound together in a circular orbit. If the mass of one of these stars rapidly decreases, the loss taking place in a time short compared to an orbital period, then the remaining mass will no longer be bound together into a binary system, provided the total mass loss from the system is more than 50 per cent of the initial mass present. We refer to this as 'impulsive' mass loss. In the case in question, we deal not with two point masses, but rather with a point mass and a distribution of mass around it. The situation remains qualitatively the same, but only about one-third of the total mass of the system need be lost in order to unbind a mass point lying near the outer edge of the distribution of mass, providing that the loss of mass from the system is from the distributed mass source.

A comet formed at several hundred astronomical units will have an orbital period of tens of thousands of years. The loss of mass shown in Figure 1 is indeed impulsive for such a body. It is therefore to be expected that the comets formed closest to the outer edge of the disk will be put onto hyperbolic orbits which will escape from the forming solar system. Those formed somewhat further in, so that some of the mass distribution lies outside them, do not receive as much benefit from the mass loss, and these comets will go into very long elliptical orbits, extending out towards 100,000 astronomical units. For comets formed at still smaller distances within the disk, the semi-major axis of the resulting elliptical orbit will progressively decrease. Cometary bodies formed at sufficiently small distances will have periods comparable to the mass loss, so that the mass loss cannot be considered impulsive for them, and their orbits are transformed only by spiralling outwards; the orbits remain of small eccentricity.

In this way we may expect that a great deal of condensed matter in the form of cometary bodies will be put into highly elliptical orbits extending towards 100,000 astronomical units, the region populated by comets in the 'Oort reservoir' around the solar system. Initially, these orbits are all coplanar, but Oort has shown that the orbital elements will become isotropic over the age of the solar system due to perturbations by passing stars.

In the initial stages of accretion onto the primitive solar nebula, relatively little mass is present, and a body having the specific angular momentum possessed today by the planet Mercury would lie at several astronomical units from the spin axis. As time goes by, mass would flow inwards toward the spin axis past such a body, and would accumulate on the central object. This increases gravitational attraction toward the centre, so that a body forming with the specific angular momentum of Mercury would gradually spiral inward towards its present radial distance from the centre. In fact, since the sun forming out of the central body is likely to have substantially more than one solar mass initially, the proto-Mercury would spiral inward to a distance smaller than its present distance from the sun. As the sun lost mass in its initial T-Tauri stage, the Mercurian planet would gradually spiral outwards again towards its present position. A similar type of behaviour can be expected for the other planets.

This helps to explain the behaviour shown in Figure 3, which shows the temperature in the regions of planet formation as a function of time. It should be noted that in each of the regions of planetary formation, the temperature is initially very low. The temperature gradually rises until it has the value shown at the vertical line in Figure 3, where the transition takes place from mass accretion onto the disk to mass loss from the disk. From this figure it should be noted that the thermodynamic conditions in the regions of formation of the

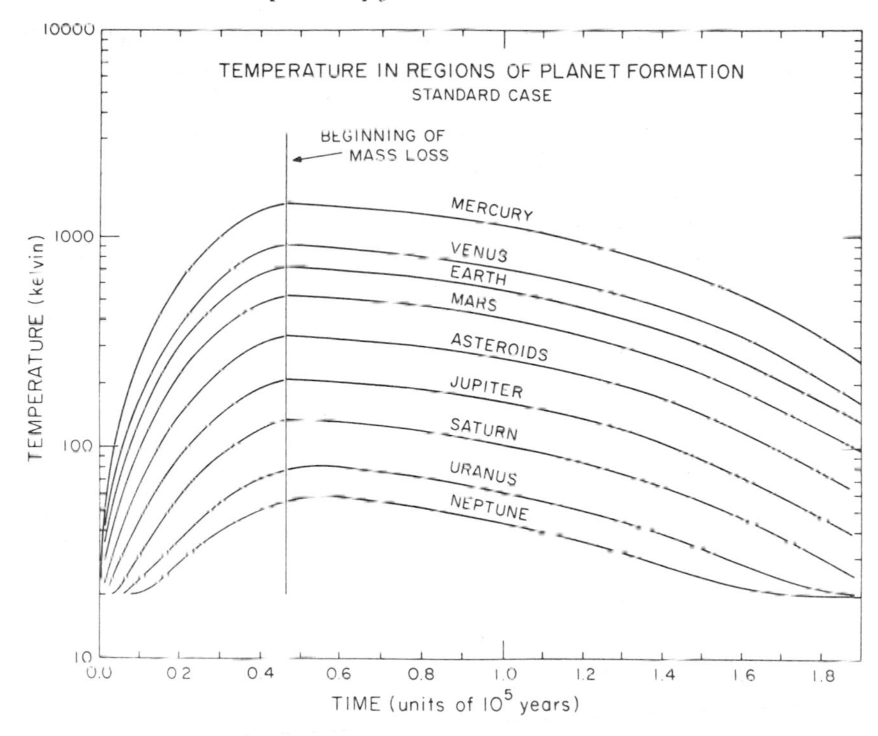

Figure 3. Variation of the temperature of the gas in the unperturbed primitive solar nebula with time in the various regions of planetary formation

inner planets are consistent with the presence of rocky materials unaccompanied by condensed ices whereas in the regions of formation of the outer planets the thermodynamic conditions favour the presence of condensed ices. At no time, anywhere in the solar nebula, anywhere outwards from the orbit of formation of Mercury, is the temperature in the unperturbed solar nebula ever high enough to evaporate completely the solid materials contained in interstellar grains. For some time a number of people have argued that the entire solar nebula started out at high temperature and cooled while solids underwent a sequence of condensation processes. In fact, there is no available energy source for any such high temperatures to have been initially present. The temperatures that are usually characteristic of bulk planetary compositions in the primitive solar nebula are the highest temperatures ever reached in the unperturbed nebula.

With these evolutionary histories of the primitive solar nebula in hand, it was possible to examine the conditions in the disk for possible instabilities. The first instability examined was the local 'patch' instability against the formation of gravitational condensations of the gas directly within the primitive solar nebula. It was found that no local patch instabilities formed anywhere during these evolutionary calculations, although conditions at times approached rather close to the point at which such instabilities would occur.

An examination was then made for axisymmetric instabilities, the formation of 'rings' within the nebula. The criterion used for this examination was the criterion developed by Yabushita (1966, 1969) for the instability of Saturn's rings. In order to apply this instability, one must treat a portion of the disk which can be considered thin compared to the radial interval examined. This forces one to consider radial intervals approaching a factor of two in the ratio of the outer to the inner radii, comparable to the ratio of successive planetary orbits, which averages a factor of 1·72. Taking this ratio as a criterion, it can be concluded that global instability against ring formation will occur if the solar nebula acquires a mass in such a radial interval greater than about 1 per cent of the mass interior to that radial interval. This critical ratio was exceeded in each one of the regions of planetary formation during the course of evolution of the primitive solar nebula, and the maximum factor by which this threshold was exceeded was typically about a factor of 20 at some point in the evolution of the region of planetary formation. Hence it is not possible to escape the conclusion that each of the regions of planetary formation should have, in turn, become unstable against the formation of rings within the primitive solar nebula. Since this conclusion is made for a sequence of nebula models in which the dissipation is highly efficient, and hence the mass flows go about as fast as possible, it follows that any less-efficient evolution of the solar nebula would lead to an even greater instability against the formation of rings.

The presence of these global instabilities is the most important conclusion to be drawn from this entire set of calculations. It indicates that the evolutionary histories studied here are at best crude approximations to the true history of the primitive solar nebula, since the effects of these instabilities have not been explicitly included in the calculations. However, it seems necessary to have gone through this exercise, in order to demonstrate that the instabilities must be present, because further progress is dependent upon understanding the properties of the gravitational condensations that are produced by these instabilities.

As a ring of matter is formed and mass flows into it, the shrinkage of the ring toward itself can only go on so long before there are bound to be perturbations which will grow, forming a set of independent blobs of gas distributed around a circle where the ring formation took place. This is an unstable situation, and it is likely that most of these blobs of gas will collect together by collision around the ring. However, it is possible that one or two of the blobs might be perturbed into some kind of resonant orbit which prevents a collision from taking place, much like the situation of Neptune and Pluto today. The minimum mass of a ring is about one per cent of the mass of the sun, or about ten Jupiter masses. The ring mass may have been significantly larger than this. If there are typically ten blobs formed out of a given ring, then it appears that masses of the blobs as low as about the mass of Jupiter must be considered.

I call these giant gaseous protoplanets, and I believe that it is necessary to investigate the properties of such protoplanets in the mass range from about 1 to about 30 times the mass of Jupiter.

Tentative Properties of Giant Protoplanets

In order that the effects of giant gaseous protoplanets upon the evolution of the primitive solar nebula can be properly understood, it is first necessary to study the evolution of such objects. This requires a study to be carried out which will borrow the techniques used to study the evolution of stars, although it will be somewhat more complicated than an ordinary stellar evolution study, owing to some complications which will be discussed below. I have not yet carried out such a study, but I have taken a very first crude look at some of the probable properties of giant gaseous protoplanets, preparatory to a more extensive study, and I will give a brief discussion here concerning these properties.

When a ring forms in the primitive solar nebula, and then further subdivides into blobs, to a first approximation we may regard the matter forming the blob to have undergone an adiabatic compression from the temperature and density conditions in the unperturbed solar nebula. If the collapse upon itself of the blob is accompanied by the formation of shocks, then the entropy of the matter in the interior of the blob is increased relative to that which would be obtained by a simple adiabatic compression. When two or more blobs collide to form larger gaseous protoplanets, then shocks are almost certainly involved in the collision, and there will be a further increase in the entropy of the gas. The higher the mean entropy of the material in a protoplanet, the hotter the interior and hence the larger the structure is likely to be.

I took a quick, conservative, look at the structure of such protoplanets by constructing isentropic models, utilizing a typical entropy of the material in the solar nebula in the region of formation of the inner planets at a time when the instability against ring formation would be near its maximum. The structure of such a model was entirely determined by the choice of the central density. These models were checked for their stability against collapse.

When the central temperature rose somewhat above 2000 K a significant dissociation of hydrogen molecules occurred within the protoplanet, and the models were unstable against collapse. Technically, this occurred because the ratio of the specific heats over a sufficient part of the interior became less than four-thirds. Once such a collapse is initiated, the density in the interior must increase by a very large factor, before the pressure of the material rises high enough to provide hydrostatic support to the overlying layers. We are concerned here with the structures of giant gaseous protoplanets in which the interior temperatures are low enough so that collapse has not occurred.

The crude protoplanet models constructed in this way were very interesting. In the range of protoplanet masses up to a few per cent of the solar mass, the protoplanet radii were all substantially in excess of one astronomical unit. Furthermore, the smaller the mass, the larger the radius. This means that if a tidal disruption process were to strip away the outer layer of a protoplanet, the adiabatic expansion of the underlying layers would produce a radius larger than that initially present. This is a property of considerable importance to which I shall return.

A second series of isentropic models was then calculated, this time with a higher internal adiabat. The general features of these models were qualitatively similar to those previously calculated, except that, as expected, a protoplanet of given mass now had a larger radius and smaller values of the central temperature and density. This is merely a demonstration that a giant gaseous protoplanet should evolve by decreasing its interior entropy and increasing the central pressure and density. In the course of time, this will lead the protoplanet toward the collapse instability and the central density will then shrink by several orders of magnitude.

However, the actual evolution of a giant protoplanet will not take place by a transition of the interior from one adiabat to another. This would only be possible if the entire protoplanet were in a state of thermally-driven convection, which in general will not be true. In particular, some interesting things happen near the centre.

Before the central temperature can rise high enough to produce a significant dissociation of hydrogen molecules, all condensed phases in the interior must evaporate. The most important of these condensed phases, from our present point of view, is metallic iron. At pressures below about 10 bars, metallic iron is the principal contributor to the interior opacity, and if the iron were to evaporate, the opacity of the material would decrease by several orders of magnitude. This abrupt decrease in the opacity would take place during the course of a temperature rise of only a few degrees. If such a sudden drop in opacity were to occur, then the radiation trapped in the interior regions would be free to stream quite readily through the matter near the centre of the protoplanet, thereby cooling the interior region back to the point at which condensation of metallic iron could occur. In practice, this means that the evolution of the protoplanet will lead to a structural readjustment near the centre such that a great deal of the central region of the protoplanet must have combinations of temperature and density lying close to the iron saturation vapour pressure curve. Only when the central pressure has risen appreciably above 10 bars will the structure no longer be governed by the presence of metallic iron, for at high pressures the pressure-broadening of the lines of triatomic molecules such as water vapour produce opacities in the material which are comparable to those of metallic iron. This argument indicates that the central pressures in a protoplanet are several orders of magnitude higher when molecular hydrogen dissociation takes place and the structure goes into collapse than would have been the case in the absence of the iron opacity effect.

There is a very important practical consequence of this. Because of the high pressure in the interior of the protoplanet, condensed phases exist at temperatures in excess of their melting point, and hence will form liquid droplets. The condensed phases which should initially be present inside a giant gaseous protoplanet are interstellar grains from which the icy materials will have evaporated, and which are possibly clustered together into objects up to a few millimetres in size. These condensed phases are not large enough to have an appreciable terminal velocity of fall through the gas of the protoplanet, and as long as the materials are solid there is no special reason to expect that the materials will stick together when they should happen to touch one another. However, near the centre, the liquid droplets can be expected to coalesce upon collision, just as raindrops do. As larger droplets are formed by coalescence their terminal velocity through the gas will increase, and the raining of such droplets should produce a body of substantial size at the centre of the protoplanet in a relatively short period of time. One can expect to get about one earth mass of liquid material collecting at the centre for every Jupiter mass of gas near the centre of the protoplanet from which the coalescence and rain-out has occurred.

One other possible property of a giant protoplanet should be mentioned. In constructing models of possibly more realistic envelopes for giant protoplanets, I have noticed that if a convection zone extends up towards the protoplanetary photosphere, then the low density of the gas requires that the characteristic convective velocity should approach sound speed. Hence, if a protoplanet ever develops convection near the photosphere, there must be a very efficient generation of mechanical waves, which will create a hot protoplanetary corona. This would lead to the hydrodynamic flow of mass away from the protoplanet, and it is conceivable that such mass loss could play a significant evolutionary role if the structure of the protoplanet evolves into a configuration with this property.

Formation of the Planets

During the solar nebula evolutionary study described above, the instability against ring formation in the region of formation of the inner terrestrial planets occurred at a very early stage, when only a small part of the total mass had fallen onto the primitive solar nebula. I define the region of formation of a planet to be that region in the primitive solar nebula in which the gas has the same specific angular momentum as the planet does today in the solar system. Because of the relatively small amount of mass present in the solar nebula at the time the 'Mercury' ring would become unstable, the radius of the ring is well over one astronomical unit. Hence, it is possible for a large giant gaseous protoplanet to be formed from the 'Mercury' ring, and for this protoplanet to fit into the available space in the solar nebula at that time. However, it is clear that the protoplanet does not have a great deal of space to spare. With the continuing dissipation of the gas in the nebula, gas will flow toward the centre of the disk, and will accrete onto the central objects at the spin axis and also perhaps onto the 'Mercury' protoplanet. In any case, the increase of mass near the centre leads to an increased gravitational attraction, and the 'Mercury' protoplanet would spiral toward the centre. The point of force balance between the central portion of the primitive solar nebula and the 'Mercury' protoplanet can be called the 'inner Lagrangian point' by analogy to double star systems. As the 'Mercury' protoplanet spirals toward the centre, it can be expected that the 'inner Lagrangian point' will approach the surface of this protoplanet, and at some point in time it should penetrate beneath the surface. This means that the outer layer of the 'Mercury' protoplanet will be stripped away, with the gas returning to that portion of the primitive solar nebula which is circulating about the central spin axis. We have already noted that such a stripping-away of the outer layer by this tidal effect will cause an adiabatic expansion of the underlying layers, which should produce an object with an even larger radius than before. But the stripping away of some mass from the protoplanet has moved the inner Lagrangian point even farther below the original surface, so that the stripping away of matter from the outer layers of the protoplanet is thereby accelerated. In general, it can be expected that the gaseous protoplanet will expand at its internal sound speed to flow through the inner Lagrangian point and be lost into the primitive solar nebula. This rapid tidal disruption should continue until essentially all of the gas has been stripped away from any condensed object which might have been formed in the core of the protoplanet. Furthermore, the dispersal of this fairly large protoplanetary mass should make the gravitational potential in the primitive solar nebula much more axisymmetric than before, so that the disintegration of one giant gaseous protoplanet in the inner solar system should enable another one to be formed shortly thereafter.

I am assuming here that the disintegration of the protoplanet occurred sufficiently soon after its formation that there was not sufficient time for the protoplanet to evolve into its collapse stage, but that nevertheless there was sufficient time for the precipitation of a significant amount of condensed mass at the centre of the protoplanet.

Therefore I advance the hypothesis that a substantial portion of the condensed materials present in the four inner terrestrial planets were formed in this way at the centres of giant gaseous protoplanets. There may have been additional protoplanets formed in the region of the inner solar system, leading to the formation of additional planetary cores which have disappeared by collision with one of the terrestrial planets.

Further out in the solar nebula, much more time would be available for the evolution of the protoplanet. I therefore advance the hypothesis that the collapse of a giant gaseous protoplanet has played a significant role in the formation of each of the outer planets. It is evident that, if this is to be the case, a considerable amount of mass must have been lost from these outer protoplanets, either before or after the collapse. In particular, since Uranus and Neptune contain very little hydrogen and helium, the process of mass loss must have been particularly efficient. A considerable amount of investigation of evolutionary model sequences will be required to test this hypothesis.

One aspect of these investigations may be particularly interesting. At the time of collapse of a protoplanet, it is not unlikely that a significant amount of rotation will exist in the protoplanet. The conservation of angular momentum during the collapse should spread out much of the mass to form a disk surrounding some sort of central body in hydrostatic equilibrium. This may have played a significant role in the formation of the regular satellite systems of Jupiter and Saturn. A similar sort of disk seems to be required in the case of the regular satellites of Uranus, but in this case some sort of major collision seems to have been involved at some point in the evolutionary history, in order to tilt the spin axis by the observed large amount. Whether such massive bodies as Pluto and Triton were formed from small protoplanets must remain a matter of speculation at the present time.

At the time of the switch of the primitive solar nebula from mass accretion to mass loss, we may thus expect that the inner solar nebula would contain several hot molten rocky bodies of planetary mass, and the outer solar nebula would have contained several collapsed gaseous protoplanets. In the immediate vicinity of these planetary bodies the temperature is likely to be very high, because of the rapid rate of formation of these bodies and the trapping of heat of formation in their interiors. Elsewhere in the primitive solar nebula, the temperature was probably in the general range indicated in Figure 3. This means that the finely-divided solids present in the gas will have a much higher content of the more volatile elements than will the planetary bodies which were formed in association with protoplanets.

Because of the vigorous turbulence present in the primitive solar nebula, the finely-divided solids will be continuously stirred up by the motions of the gas, and will be unable to settle toward the mid-plane. Only when the dissipation of the primitive solar nebula has gone a long way, so that the optical depth near the mid-plane has become small, is it likely that turbulence will die away and the solids can settle toward mid-plane. In the region of planetary formation, the general flow of the gas was inward to form the sun, and the finely-divided solids will have largely been swept out from the region of planetary formation and into the sun.

When turbulence dies away and the solid materials can settle toward the mid-plane of the nebula, then it is possible for the Goldreich–Ward instability mechanism to operate (Goldreich and Ward, 1973).

This instability mechanism is expected to gather together the finely-divided solids into bodies generally in the asteroidal mass range. The subsequent collision among the asteroidal bodies should lead both to amalgamation into a few bodies of larger size and into shattering with the production of a very large number of bodies of smaller size. This should not only have produced the asteroids and their fragments which are observed today, but it should also have produced a large number of these bodies in the vicinity of the inner and outer planets. Gravitational perturbations will have subsequently caused these small bodies either to merge with the inner terrestrial planets by collision or to have been deflected into orbits which bring them under the control of Jupiter, with which they may collide or be ejected from the solar system. A substantial fraction of the mass of the earth, perhaps ten or twenty per cent, was probably acquired from these smaller bodies, which would bring in the more volatile elements to the earth and the other inner terrestrial planets. The smaller bodies formed in the region of the outer planets were probably also largely eliminated by collision or by ejection from the solar system, with some few exceptions which are still visible today as the Trojan asteroids which are associated with Jupiter, and by a number of small captured satellites. Large amounts of mass in the asteroidal or cometary size range probably exist at significant distances beyond the orbit of Neptune. We have already seen that the cometary bodies formed farther out should have been deflected into very elliptical orbits which take them out to populate the Oort reservoir of comets.

Origin of the Moon

The moon is a sufficiently large body that it has been recognized for a long time as occupying an anomalous position with the solar system. Its anomalous chemical composition, largely devoid of metallic iron and also of the more volatile elements, presents a twin puzzle on which all of the conventional theories of lunar formation founder.

My approach to this problem, in collaboration with W. R. Ward, starts with a simple observation. About one-third of the total angular momentum of the earth–moon system (excluding the motion about the sun) lies in the rotation of the earth. The remainder is in the orbital motion of the moon. However, if we go backwards in time so that the moon would be much closer to the earth, perhaps near the Roche limit just less than three earth radii from the centre of the earth, then nearly all of the angular momentum of the earth–moon system would lie in the spin of the earth. At such a time, the earth would be spinning much more rapidly than the other planets are observed to do.

It is instructive to ask what size of body must strike the earth a glancing blow in order to set it spinning with this high rate. It turns out that it requires a body with about the mass of Mars to do this. It may seem absurd to expect that something with the mass of Mars could have struck and amalgamated with the early earth, but in view of my remarks above about the formation of additional protoplanets in the inner solar system, some of which have disappeared, I do not regard the idea as improbable. The only thing that would have been completely a matter of chance would be the impact point of the Mars-like object on the proto-earth. If the collision had been a more central one, then the earth would not have been set to spinning so rapidly.

Consider what happens to the fragments formed in such a collision. Such fragments would tend to be ejected in elliptical orbits which, at some point in their trajectories, intersect the proto-earth. This is because such trajectories originate below the surface of the proto-earth. Hence, if only the laws of celestial mechanics were to govern the situation, virtually no debris would be left in orbit above the proto-earth.

However, the collision would take place at a relative velocity of approximately eleven kilometres per second. At these velocities, rocky materials are vaporized upon shock unloading. Hence, the material which ejected from the site of the most intense part of the collision between the two bodies will be gaseous and will obey the laws of hydrodynamics as well of those of celestial mechanics. Pressure gradients will modify the motion of the gas compared to the purely celestial mechanical motions, and hence some of the gas, which recondenses into small bodies upon expansion, will be placed into orbit about the proto-earth. Ward and I have estimated that perhaps about two lunar masses of material may be put into orbit initially by this mechanism.

Collisions among the small bodies will rapidly form them into a flattened thin disk. The Goldreich–Ward instability mechanism is particularly effective under these circumstances, at least on that part of the disk which rapidly spreads beyond the Roche limit, and it appears likely that an object of lunar size should be formed in this way in the course of just a few months. Because of the short period of formation, the proto-moon will be formed very hot and hence the more volatile elements should be rapidly vaporized and evaporated away from the lunar body. Not very much metallic iron should get into the lunar body, because the bodies taking part in the original collision have probably had time to undergo chemical differentiation, in which the iron is gathered into their cores. Not very much material from the core of the Mars-like colliding body is expected to get into the cloud of debris that ultimately forms into the moon.

The process of the accumulation of the remaining material from the Mars-like body onto the proto-earth may be sufficiently asymmetric, that it is not obvious that the spin momentum vector of the resulting proto-earth must be perfectly aligned with the orbital angular momentum vector of the moon.

The Formation of the Sun

I mentioned earlier that at the centre of the primitive solar nebula there is probably a formation of a pair of bodies rather than a single body near the central spin axis, according to the calculations of Larson and of Black and Bodenheimer. With the inward dissipation of mass which had gone to these bodies, they should gradually spiral together and merge into a single object. Once merger has commenced, angular momentum can be fairly efficiently extracted from the single object by turbulent viscosity, since this object is in continuous fluid contact with the primitive solar nebula that is feeding mass toward it.

It must be remarked that some interesting problems arise in connection with the low entropy with which the central object is formed. If it is reasonable to start with the entropy of the gas in the primitive solar nebula which is fed inwards to form the central object, then a consideration of the conditions reached upon adiabatic compression of the gas proves to be very interesting (Perri and Cameron, 1973).

After this gas is compressed adiabatically to a density of the order of $100 \, \text{gm cm}^{-3}$, approximately the conditions in the centre of the sun today, the temperature has not risen beyond about 10^6 K. Since there will have been very little deuterium present in the gas, the thermonuclear destruction of the deuterium will not make a significant modification of this result. Thus it is evident that the shrinkage of the material to form a solar core at densities comparable to that present in the sun today does not lead to the ignition of hydrogen thermonuclear reactions. The entropy is too low.

There appear to be two possibilities, which have not yet been investigated. One possibility is that the central region of the forming sun increases in density beyond $10^4 \, \text{gm cm}^{-3}$, until hydrogen burning can begin by pycnonuclear ignition. Central hydrogen burning under these circumstances could gradually raise the internal entropy until thermonuclear hydrogen burning could take place under conditions of normal density. Alternatively, the entropy of the material added to the protosun at larger distances from the centre might be sufficiently higher than that with which the central regions formed, that upon shrinkage of the central regions, ignition could occur off-centre. This type of burning with a hydrogen shell source might form an expanded configuration, until the thermal conduction of heat toward the centre had heated the central regions enough to participate in the central hydrogen burning.

The existence of this large uncertainty about the way in which nuclear reactions turn on in the sun is an indication that the pre-main sequence evolution of the sun is not presently understood. It also indicates that this early evolution of the sun is intimately tied to the properties of the primitive solar nebula.

The above considerations would not prevent the formation of a reasonably compact object at the central spin axis of the primitive solar nebula, which should be capable of radiating energy at a very significant rate. Hence it is likely that there will be a central source with a rather high luminosity regardless of whether or not the nuclear reactions have turned on in the interior. A considerable structural readjustment and spectral change in the emitted radiation may accompany the onset of the central nuclear reactions.

It was mentioned earlier that the infall of gas to form the primitive solar nebula may result in a significant compression of the magnetic lines of force from the interstellar medium, perhaps producing magnetic fields of the order of 0·1 gauss in the central regions of the primitive solar nebula. As the sun forms and has a significant degree of internal thermal ionization, this magnetic field can act as a seed which is greatly amplified by dynamo action. It is quite possible that the early solar magnetic fields were greatly enhanced over the present magnetic fields in this way. Since the early sun is expected to have an outer convection zone extending upwards nearly to the photosphere, it is possible that the generation of hydromagnetic waves resulted in a much greater efficiency in heating the solar corona in the early solar system than is true today. This was probably responsible for the generation of the fairly high rates of mass loss forming the early T-Tauri phase of solar evolution. It seems likely that there would be a smooth transition from the disk coronal mass loss described earlier to the final T-Tauri mass loss from the central solar object. The disk mass loss would cease as turbulence died out in the depleted solar nebula. However, by that time so little mass should be left in the solar nebula that the T-Tauri phase of solar mass loss can probably sweep the residual gas out of the solar system in a relatively short period of time.

This research has been supported in part by a grant from the National Aeronautics and Space Administration. I have been indebted to many colleagues for interesting discussions which have helped to shape this work, and in particular I wish to thank F. H. Seguin, W. R. Ward, F. L. Whipple, and M. Lecar.

References

Black, D. C., and Bodenheimer, P. (1976). 'Evolution of rotating interstellar clouds. II. The collapse of protostars of 1, 2, and 5 $M_\odot$.' *Astrophys. J.*, **206**, 138–149.

Cameron, A. G. W. (1973). 'Accumulation processes in the primitive solar nebula.' *Icarus*, **18**, 407–450.

Cameron, A. G. W. (1975). 'Clumping of interstellar grains during formation of the primitive solar nebula.' *Icarus*, **24**, 128–133.

Cameron, A. G. W., and Pine, M. R. (1973). 'Numerical models of the primitive solar nebula.' *Icarus*, **18**, 377–406.

Cameron, A. G. W., and Truran, J. W. (1976). 'The supernova trigger for the formation of the solar system.' *Icarus*, in press.

Goldreich, P., and Ward, W. R. (1973). 'The formation of planetesimals.' *Astrophys. J.*, **183**, 1051–1061.

Jura, M. (1976). 'Cloud collapse and star formation.' *Astron. J.*, **81**, 178–181.

Larson, R. B. (1969). 'Numerical calculations of the dynamics of a collapsing protostar.' *Mon. Not. Roy. Astron. Soc.*, **145**, 271–295.

Larson, R. B. (1972). 'The collapse of a rotating cloud.' *Mon. Not. Roy. Astron. Soc.*, **156**, 437–458.

Lee, T., Papanastassiou, D. A., and Wasserburg, G. J. (1976). 'Demonstration of ^{26}Mg excess in Allende and evidence for ^{26}Al.' *Geophys. Res. Lett.*, **3**, 109–112.

Lin, C. C., and Shu, F. H. (1964). 'On the spiral structure of disk galaxies.' *Astrophys. J.*, **140**, 646–655.

Lust, R. (1952). 'Die Entwicklung einer um einen Zentralkoerper rotierenden Gasmasse, I. Loesungen der Hydrodynamischen Gleichungen mit turbulenter Reibung.' *Zeit. f. Naturforschung*, **7a**, 87–98.

Lynden-Bell, D., and Pringle, J. E. (1964). 'The evolution of viscous disks and the origin of the nebular variables.' *Mon. Not. Roy. Astron. Soc.*, **168**, 603–637.

Nakano, T. (1976). 'Fragmentation of magnetic interstellar clouds by ambipolar diffusion. II. Fragments of 1, 10, and 100 $M_\odot$ in various conditions.' Preprint.

Nakano, T., and Tademaru, E. (1972). 'Decoupling of magnetic fields in dense clouds with angular momentum.' *Astrophys. J.*, **173**, 87–101.

Perri, F., and Cameron, A. G. W. (1973). 'Hydrogen flash in stars.' *Nature*, **242**, 395–396.

Scalo, J. M. (1976). 'Heating of dense interstellar clouds by magnetic ion slip: support for early fragmentation.' Preprint.

Yabushita, S. (1969). 'Stability analysis of Saturn's rings with differential rotation.' *Mon. Not. Roy. Astron. Soc.*, **133**, 247–263.

Yabushita, S. (1969). 'Stability analysis of Saturn's rings with differential rotation—II.' *Mon Not. Roy. Astron. Soc.*, **142**, 201–212.

THE FORMATION OF THE SOLAR SYSTEM: A PROTOPLANET THEORY

WILLIAM H. MCCREA
University of Sussex, U.K.

Abstract

The origin of the solar system is studied in the setting of the formation of a new stellar cluster from turbulent interstellar material traversing a compression-lane in the Galaxy. Attention is focused upon a condensation destined to become the Sun and upon phenomena in the surrounding region extending about half-way to neighbouring condensations. It is essential to the model that turbulent elements continually move in and out of the region; this is represented by imagining it as enclosed by a boundary perfectly reflecting to such elements. It is inferred that most of the matter in the region finds its way into the Sun, but leaves behind most of the angular momentum in material which proceeds to form a flattened distribution wherein protoplanetary condensations arise. If a protoplanet then contracts faster than its component materials become segregated, it may be treated as yielding differentially rotating matter near its equatorial plane, around a uniformly rotating central body that ultimately undergoes fissional break-up. The main outcome, it is suggested, is a major planet, 'droplets' from the break-up are main satellites, while material near the equatorial plane produces minor satellites. If the heavy component of a protoplanet's material becomes segregated faster than the body contracts as a whole, after imparting most of its original angular momentum to the light component, the heavy component evolves analogously to the whole protoplanet in the first case. Here the main results are identified as terrestrial planets; the Moon might have been another 'droplet', or it might have condensed from some of the discarded material. The values of only three parameters have to be got by fitting; most general properties of the solar system are then predicted in order-of-magnitude agreement with observation.

Introduction

The purpose is to construct a theoretical model for the formation and basic evolution of the Sun and the solar system (SS). The plan is to make the assumptions explicit, to make the derivation of their consequences as strict as possible in principle, even if much simplified mathematically, to give the model as much predictive power as possible, and to test the predictions as fully as can be.

The astronomical setting is in the current picture of the structure and working of the Galaxy. This is inferred to have existed with about its present observed structure for about 15×10^9y; the formation of the stars of the present disk-population (Population I) has proceeded throughout most of that time and still continues; the raw material has always been interstellar matter (ISM) generally like what is now observed—the only significant difference being some progressive change in the chemical composition; in its circulation round the Galaxy, when such ISM approaches a spiral arm it enters a shock region where most of it undergoes temporary compression, but occasionally a cloud of the ISM is compressed to the point of condensation into a cluster of a few hundred new stars; some such survive as recognized 'galactic' clusters, but most disperse to constitute the rest of the disk-stars; the Sun is one of these, formed in this way about 5×10^9y ago so that it has existed for only about one-third the past lifetime of the Galaxy. So far as it goes, all this may apparently now be regarded as empirical knowledge not depending upon particular theories of galactic dynamics and star formation.

Solar System

We recall as briefly as possible those general features of the SS that appear to be salient for our purpose.

The Sun is a normal main-sequence star of its class.

The Planets are in orbits near to one plane, so that it has meaning to say they are described in the same sense, this sense then defining direct (prograde) revolution and rotation in the system. The orbits are generally of small eccentricity. They are spaced out from the Sun at generally increasing intervals to distance about $0{\cdot}5 \times 10^{15}$ cm. The total mass of the planets is about 1/750 solar mass, but they carry nearly 200 times as much angular momentum as the Sun.

There are six *principal* planets: the *terrestrial* planets Venus, Earth, the *major* planets Jupiter, Saturn, the *outer* planets Uranus, Neptune. There are three *lesser* planets Mercury, Mars, Pluto.

The principal planets all contain comparable—probably within a factor about 3—amounts of the chemical elements heavier than the 'light' elements hydrogen and helium. By adding appropriate amounts of these light elements there would therefore result six bodies of comparable mass having the same chemical composition, which is also about the composition of the Sun. Or, of course, starting with these six similar bodies, two would have about the mass and composition of the major planets; removing about 90 per cent of the light elements from another two would leave bodies of the mass and composition of the outer planets; removing about 99 per cent of the light elements from the remaining two would leave bodies of about the mass and composition of the principal terrestrial planets. There appears to be ample evidence that these statements are roughly valid, which is all that we wish to claim. Certainly, there are significant departures from exactness, some of which will be mentioned later.

The three lesser planets have masses about ten times smaller than Venus or Earth; they are evidently made principally of heavy elements, but differ somewhat amongst themselves in chemical composition.

The spin axes of the principal planets have various inclinations to their orbital planes, and these result in four having prograde rotation and two retrograde. The rotation-periods of all the planets have median about 15 hours, and the known periods that have not obviously been affected predominantly by tidal action of other bodies are in the range about 10 to 25 hours.

Regarding satellite systems, the six principal planets provide six systems, that of Venus alone being null. There are seven main satellites: Moon (Earth), Io, Europa, Ganymede, Callisto (Jupiter), Titan (Saturn), Triton (Neptune). It is to be noted that they are associated with all three types of planet. These are all comparable with the Moon, and they are the most standard members of the SS. Their orbits are between about $\frac{1}{3}$ and 2 million km from the planets with which they are associated; some orbits are oblique to the planet's equatorial plane; some have significant eccentricity. The total orbital angular momentum of the four main satellites of Jupiter is only about 1/100 the spin momentum of Jupiter, while the Moon's orbital momentum about Earth is about 5 times the spin momentum of Earth. There are 12 regular minor satellites and one ring-system: the innermost satellite (Amalthea) of Jupiter, the six inner satellites of Saturn and its rings, and the five satellites of Uranus. It is to be noted that these are associated with the three principal planets of lowest density and fastest rotation. Their orbits are all between 0·1 and 0·6 million kilometres from their planets, they are all close to the equatorial planes of the planets and all have very small eccentricity. These satellites evidently constitute a single species although their masses range through a factor about 500; their total mass is less than about 2 per cent of that of the main satellites.

There are also 14 other known miscellaneous small natural satellites. Mainly in its inner region, the SS contains asteroids, meteoroids, interplanetary dust and plasma. Reaching out to distances comparable to those of nearby stars there are also the comets. All of these comprise only a tiny fraction of the matter in the system.

This particular sketch of familiar features is an attempt to delineate the essential characteristics of the problem in hand. There are *standardized features* to do with principal planets and their main satellites; there are quite numerous *regularities*, but they are significantly *not* strict regularities, e.g. planetary orbits are fairly near to one plane, but they are significantly non-coplanar. There are pronounced *irregularities*, e.g. the spin-axes of the planets point in all sorts of directions, planetary densities show highly irregular variation with distances from the Sun, and so forth. The numerous small bodies in the system may in themselves be negligible members of the cosmos, but they serve certainly as a general indication that the SS is complex, and probably as essential *clues* to its provenance.

Method

It is impossible to discover the origin of the solar system by observing it now, and working steadily backwards in time in order to infer the whole of its past history. This method cannot be applied to a strongly dissipative system, e.g. if we make every possible observation of the landing of a parachutist, we should be unable to infer anything of interest about the circumstances of his jump. This is why we have to resort to the usual method of constructing a theoretical model. The one we shall discuss, in common with most others, leads to the inference that planetary systems are continually being produced. This implies that the model may in principle be tested by direct observation. Were this not the case, the scientific status of the procedure would be in doubt. However, in practice there is no prospect of testing a model by observing the actual processes of formation and evolution of any such system. We can test only whether it correctly predicts the present properties of our own system. Having only the one planetary system by which to make the tests is an unfortunate restriction. However, it is somewhat redeemed by having five or six satellite systems by which to test the model in regard to the formation of satellites. That is why we pay a good deal of attention to satellite systems. It is not because a satellite system is formed by the same processes as a planetary system; it is simply because we should have much enhanced confidence in a model that appears to account for the existence and properties of one planetary system if it is found to account also for the existence and properties of half-a-dozen satellite systems as well. If it accounts also for asteroids, etc., so much the better.

Star Formation

Consider a cloud of interstellar matter that has been compressed in a shock region of the Galaxy and is about to condense into stars. We make the following hypotheses:

(a) The material has about 'solar' composition, i.e. by mass about
70 per cent hydrogen 28 per cent helium 2 per cent heavier elements.

(b) The material being 'cold' interstellar matter of mean density σ and mean kinetic temperature T, it retains this temperature throughout the processes to be considered (unless otherwise stated). The hydrogen is mostly molecular. About 1 per cent of the mass is in the form of grains made of metals, silicates etc., and possibly coated with ice. The mean molecular weight is $\mu \approx 2$.

(c) The material is in a state of *supersonic turbulence* that is treated by supposing it to be composed of 'floccules' each of mass m, density ρ_0, radius s, moving with mean speed V and having mean free path L between encounters with other floccules.

It is to be emphasized that we are discussing a transient state of the matter, brought about by compression and loss of energy by radiation—the assembly of floccules is a representation of a passing phase. A floccule is an idealization of any part of the diffuse material that happens to be moving coherently at this phase, the motion of the material as a whole being chaotic. The 'floccule' description of supersonic turbulence is intended to have about the status of the 'mixing-length' description of subsonic turbulence; it seems appropriate because the characteristic feature of supersonic motion in general is that the compressibility of the material becomes significant, so that we regard turbulent supersonic motion as characterized by non-uniformities of density. Also, since the phase occurs on the way to the production of gravitationally bound condensations, it is part of the hypothesis that a floccule is not itself a gravitationally bound body of the material. We shall see that the values of the parameters that arise do bear out the hypothesis in this regard.

The foregoing were the conditions I first described in I (McCrea, 1960, to be referred to as I). Discussion at the 'Institute' of which the present volume is the outcome appeared to show that the picture is now fairly widely accepted. The new feature since 1960 is the recognition of the rôle of compression regions in the Galaxy. As regards turbulence, if some degree of turbulence is present in the cloud before compression, so far as the turbulent motion is concerned the compression is largely 'adiabatic' and so the mean floccule speed might be expected to increase with density as about $\sigma^{1/3}$. This gives plenty of scope for attaining high speeds. However, it is difficult to push speeds much beyond the speed of sound. So it is natural to take V to be somewhere about sound-speed in the material at temperature T.

By definition we are here concerned with a cloud at a stage when most of its material is soon to be resolved into stars. Therefore there are locations scattered through the cloud that are about to be occupied by these stars. Throughout the cloud, or at anyrate throughout some region of interest to us, there is some significant mean distance apart of those locations. Call this $2S_0$, and let O be one of the locations. Imagine a sphere $\mathscr{S}$ of radius S_0 and centre O, that is perfectly reflecting for floccules and transparent to radiation. What would go on inside $\mathscr{S}$ is practically the same as if it were not there, for a reflection of a floccule at $\mathscr{S}$ can represent the passage of one floccule out of the volume and the passage of another into it, while by hypothesis $\mathscr{S}$ has no effect upon the transmission of radiation. Let $M = Nm$ be the mass of cloud material inside $\mathscr{S}$. We ask what will happen to this material isolated within this imagined sphere.

The only purpose of $\mathscr{S}$ is to make it simpler to describe what happens. If we do proceed in this way, $\mathscr{S}$ has to have radius about equal to S_0 with similar treatment of other locations like O, because otherwise we should be ignoring some of the material, or else reckoning some of it twice over.

As a consequence of the random motions of the floccules, the material in $\mathscr{S}$ has a resultant angular momentum $\mathbf{H}$; in the postulated conditions $\mathbf{H}$ will be conserved in the subsequent behaviour of this material. I showed in I that, to a sufficiently good approximation, the expected value of $|\mathbf{H}|$ is

$$H = \tfrac{1}{2} m V S_0 N^{1/2} = \tfrac{1}{2} M V S_0 N^{-1/2} \tag{1}$$

with, of course, a considerable dispersion. Even were the values of the parameters fixed, there would be a spread of values of H because of the 'random walk' nature of the result, but in any application there would be also a spread the values of m, V. The direction of the vector $\mathbf{H}$ is random. It is characteristic of the type of model under discussion that it predicts angular momenta with such dispersions in magnitude and direction. On the other hand, in each application the order of magnitude is in little doubt. For, having regard to the last expression in (1), values of M, V, S_0 are determined almost by the definition of the system, while the quantity N that is least well-determined enters only by its square root.

Brosche (1970) has considered the expectation value of H^2; the square root of his value would give $(0{\cdot}4)^{1/2} \approx 0{\cdot}63$ in place of the coefficient 0·5 in (1). As he says, however, the two definitions are not quite the same. Here I continue to use (1) and results previously derived from it.

According to the specification of the system, most of the mass M is going to form a stellar condensation centred at O. Consider how it happens. Any condensation of the given material is perforce an aggregation of floccules. So the early stage of the destined condensation is simply the chance coming together at O of some floccules. At this stage there may be other such incipient condensations within $\mathscr{S}$ but by prescription the only one that grows considerably is that at O. When it attains the Jeans critical mass M_J for the material at temperature T and density ρ_0, by definition of M_J it begins to contract upon itself under its self-gravitation—we neglect any change of density before this stage is reached. As it proceeds to contract, it tends to offer a geometrically smaller target for further infall of floccules, but as others do fall in it offers an increasingly massive gravitational target. Floccules either fall in individually, or else they collide somewhere within $\mathscr{S}$ and some or all of their material then falls in, until nearly all the material in $\mathscr{S}$ has arrived in the central condensation. This is finally a star of mass M_s and radius R_s, say, I have shown in I that the expected angular momentum is approximately

$$H_s = \frac{2}{3}\left(\frac{1}{3} G m M_s^2 R_s\right)^{1/2} \quad (G = \text{gravitation constant}) \tag{2}$$

Noting that $M_s \approx M$ and writing $2GM/S_0 = V_0^2$, so that V_0 is the escape speed from distance S_0, (1), (2) give

$$\frac{H_s}{H} \approx \left(\frac{8}{27} \frac{V_0^2}{V^2} \frac{R_s}{S_0}\right)^{1/2} \tag{3}$$

When the numerical values considered below are inserted, (3) shows that H_s/H is quite small. Thus, even when nearly all the mass M has fallen into the central condensation (the 'star'), almost the whole of the angular momentum is still carried by what are left of the floccules still moving about throughout the whole interior of $\mathscr{S}$. Let their number be N_1 and their mean speed V_1. Evaluating their angular momentum in the same way as before, since as just stated this must again give about the value H, we have

$$m V_1 S_0 N_1^{1/2} \approx 2H \approx m V S_0 N^{1/2} \tag{4}$$

giving

$$N_1/N \approx (V/V_1)^2 \tag{5}$$

Each of these N_1 floccules is moving under the gravitational field of $M_s \approx M$ at the centre of $\mathscr{S}$, which implies that its mean speed V_1 is roughly of order V_0 though somewhat greater, say $V_1 \approx 2V_0$, and as the figures will show, (5) then implies $N_1/N \approx 1/100$. At this stage, therefore, we have to envisage N_1 floccules (something about 1 per cent of the original number) moving about at random inside $\mathscr{S}$, these random motions happening to produce resultant angular momentum about equal to H.

There is no doubt now as to what tends to be the next stage. Just as, for example, the material of Saturn's rings has settled into orbital motion in a plane, the material of the N_1 floccules will tend through inelastic interaction to leave the mass of, say, N_2 floccules in orbital motion about the star, all moving in the same sense in a plane through O. If this happens it is a process of passing from random motion to systematic motion with about the same resultant angular momentum H. So we are left with mass mN_2 circulating with mean speed V_2 at mean distance S_2, say. With suitable definitions of the means we could take V_2 to be the circular orbital speed at radius S_2, i.e. $V_2^2 = GM/S_2$ and so we have

$$H \approx m(GMS_2)^{1/2} N_2 \tag{6}$$

If we use (4) with the estimate $V_1 \approx 2V_0$, along with (6) we have

$$S_2^{1/2} N_2 \approx S_0^{1/2} (2N_1)^{1/2} \tag{7}$$

There are two possibilities that are evidently about the extremes that could be achieved:

(a) $S_2 \approx S_0$ i.e. after the flattening, the material is about as far as it can be from O, then from (7)

$$N_2 \approx (2N_1)^{1/2} \tag{8}$$

The general character of this result of going from a random to a systematic state without much change in other parameters—the removal of the square root— is what would be expected. With a value of N_1 of the order to be estimated later, (8) implies $N_2 \ll N_1$, and therefore most of the mass mN_1 would be removed from the region as a whole. This could happen only by its falling into the star as a result of the succession of inelastic encounters. This would add relatively little to the mass of the star while removing relatively little of the angular momentum from the still orbiting material. It is to be noted that inelastic encounters result in dissipation of mechanical energy and this results in the 'orbiting' material becoming indeed gravitationally bound to the star, as has been implied in the discussion.

(b) $N_2 \approx N_1$ i.e. after flattening, there is still about the same quantity of material moving round the star, but its motion has become systematic instead of random. In this case (7) would imply

$$S_2 \approx S_0(2/N_1)^{1/2} \tag{9}$$

This implies $S_2 \ll S_0$ and this again is what would be expected, i.e. in systematic motion each floccule would have much less angular momentum and so would have to be on average much nearer to O. And this result would again be achieved by the dissipation of mechanical energy.

Protoplanets

According to the model, the original condensation that later grew into the central star of our discussion started as a chance agglomeration of a few floccules (n, say) that happened to reach mass M_J sufficient to pull itself together by its self-gravitation. This particular agglomeration then had the luck to continue to grow thereafter. In the turbulent material considered the tendency to form such initial agglomerations must always be present.

In particular, the assembly of N_1 floccules throughout the region within $\mathcal{S}$, that becomes ultimately a flattened system composed of the material of N_2 floccules, achieves this development as a result of inelastic encounters between floccules. So here we certainly still have the possibility of forming agglomerations. Any that fail to reach mass M_J must disperse again. On the other hand, once any such agglomeration has reached mass about M_J it will hold together, but it will also proceed to contract under its own gravitation and this will hinder further growth. Also, unlike the situation of the early phase of the stellar condensation, there now remains very little material to supply any further growth.

It is expected, therefore, that the mass mN_2 becomes resolved into bodies p

in number where

$$p \approx mN_2/M_J \tag{10}$$

that are all of *about* mass M_J, and when formed are composed of the original raw material at temperature about T and density about ρ_0 and are all in orbit around O in about the same plane. These are the *protoplanets* of the present theory.

A *protoplanet* of this sort is thus a body of mass about M_J where

$$M_J = \left(\frac{5\mathcal{R}T}{\mu G}\right)^{3/2} \left(\frac{4}{3}\pi\rho_0\right)^{-1/2} \quad (\mathcal{R} = \text{gas constant}) \tag{11}$$

formed by the coming together at random of about n floccules of mass m, so that

$$n = M_J/m \tag{12}$$

to form a body of initial radius about R_0 where

$$\tfrac{4}{3}\pi R_0^3 \rho_0 = nm \tag{13}$$

The escape speed v_0 from the surface of such a body is given by

$$v_0^2 = 2Gnm/R_0 = 2G[\tfrac{4}{3}\pi\rho_0(nm)^2]^{1/3} \tag{14}$$

using (4). It will be found that the value of v_0 is not much different from the value adopted for V. So V is roughly the speed of arrival of floccules to compose the protoplanet. As in (1) the expected value of the spin-momentum H_0 of the protoplanet is given by

$$H_0 \approx \tfrac{1}{2}mVR_0n^{1/2} \tag{15}$$

Returning to the 'flattening', in the case of Saturn's rings this is very precise because it involves an enormous number of small bodies. In the present case, even if the material is still largely in the form of floccules, rather than protoplanets, until the flattening is well advanced, we have relatively few bodies of vastly larger extent. So the flattening could not proceed further than to within a thickness of the general order of a floccule diameter $2s$. It is a gratifying feature that the protoplanets are thus not predicted to move in exactly the same plane.

Actually Aust and Woolfson (1971) have given reasons for doubting whether the flattening would occur before the protoplanets were formed. In that case, some of the foregoing estimates might have to be somewhat modified. As against this, however, it is sometimes conjectured that flattening of a planetary system can proceed even after it is composed of normal planets.

Referring to page 82, cases (a), (b) with properties (8), (9) and using (10), our flattened system initially comprises p protoplanets, where

$$m(2N_1)^{1/2} \lesssim M_J p \lesssim mN_1 \tag{16}$$

and it has radius S where

$$S_0 > S > S_0(2/N_1)^{1/2} \tag{17}$$

Times

The times required for the various stages in any model impose crucial constraints upon it. It is a grave criticism of some current models that they afford so little possibility of estimating rates, and so do nothing to show that they can operate in plausible times.

In I, I gave estimates of the times for the processes so far discussed, and McCrea and Williams (1965) gave estimates for some of those that are concerned in the processes mentioned below. While improved estimates would be desirable, those referred to appear to be adequate to show that the model does not encounter difficulties on this score. So I shall not consider this aspect further in the present treatment.

Angular Momentum Properties

If a cloud of ISM that is in process of forming stars is looked upon in the foregoing manner, then the angular momentum of the material associated with each particular condensation, such as that at position O, about its mass-centre at the epoch of interest is fully accounted for in the treatment described. For our immediate purpose we are not called upon to keep account of the total angular momentum of the whole cloud about its mass-centre; but clearly the extension of the present concepts to that problem would show how most of that momentum goes into the translational motion of the resulting stars rather than into the spins of the individual stars. Nowhere do we have the notion of some one portion of the original cloud contracting upon itself to form a single body with spin angular momentum equal to the original angular momentum of that material. The phenomenon of turbulence given by the floccule picture excludes that notion. A particular floccule goes into a particular condensation for the very reason that it has not got too much angular momentum about this condensation—if it has too much, it simply goes into a different condensation. Even when we find it convenient to employ the idea of turbulent matter enclosed within an imagined surface like $\mathscr{S}$, this is only a device to ensure that we keep account of all the material in the cloud—a floccule imagined as being reflected at $\mathscr{S}$ actually depicts one going out of the volume and another going in, the effect being statistically the same. Moreover, even within $\mathscr{S}$ itself, we have the picture of the condensation at the centre O being formed by floccules falling into it, not by a coherent contraction of the material as a whole within $\mathscr{S}$. This behaviour achieves a filtering out of the angular momentum within $\mathscr{S}$ so that only a small fraction gets into the central condensation; most of the angular momentum is held back in the fraction of material that does not fall into the condensation.

This last result is basic, and it is indubitable. If a lot of floccules are moving about at random throughout a volume bounded by $\mathcal{S}$, they have an expected resultant angular momentum $\mathbf{H}$. If a condensation at O accumulates in the manner described, it has an expected angular momentum $\mathbf{H}_s$. With the parameters of the system to be noted below, it is found that $H_s \ll H$. The difference $\mathbf{H}-\mathbf{H}_s \gtrsim \mathbf{H}$ can only be carried by material within $\mathcal{S}$ that does not enter the condensation. This is the material that ultimately resolves itself into protoplanets. Nearly all the angular momentum $\mathbf{H}-\mathbf{H}_s$ goes into their orbital motion round the star at O. In the process, the protoplanets acquire their own spin-momenta like H_0 in (15); these will be found to be of much interest in themselves but their share of the angular momentum of the system as a whole is negligible.

Predictions: Stellar Rotation: T-Tauri Phenomenon

Even before numerical estimates are discussed, it is possible to derive some general qualitative predictions of the model having considerable significance. We still speak in terms of a system associated with a particular star like that at O; 'the protoplanets' means those in the one system, and so forth.

(a) According to the discussion in the section on *Protoplanets*, all the protoplanets when formed have approximately the same mass.

(b) The discussion shows also that when formed they would have comparable spin-momenta; but these being the results of "random-walk" compositions of fairly small numbers of random contributions the spread in values is expected to be greater than in the case of, say, the masses.

(c) From the discussions in the sections on *Star Formation and Protoplanets*, the protoplanets are expected initially to move in approximately circular orbits in approximately one plane, all in the same sense and so defining a 'direct' sense, for the system. This is, of course, the sense of the resultant angular momentum $\mathbf{H}$. As explained, the *approximate* character of the predictions is an essential feature.

(d) As derived, the spin-momenta of the star and the protoplanets are *not* correlated with one-another or with the resultant angular momentum of the system.

(e) It was seen that as the system becomes flattened, some residual material must fall into the star. This might give the star some better chance of being prograde rather than retrograde in its spin. Also if the star loses some of its original spin through electromagnetic transfer resulting from a stellar wind, this would mean that the present spin differs somewhat from the original spin in magnitude and direction. Again, if a protoplanet is assembled mainly after the flattening of the system is fairly advanced, this might result in some bias in the spin. In the model these are secondary effects; they are mentioned simply to show that, while (a) describes the general expectations, there are factors that could introduce some modification. Thus in an actual case, were all the spin-axes to favour no particular direction, this would present no problem for the model; were all the spin-axes strictly aligned, this would seriously conflict with the model; were there, however, simply some degree of bias in favour of direct motion, this need not conflict.

(f) In any particular new stellar cluster formed in accordance with the model, we should expect stars formed by the process described in the section on *Star Formation*, all to have roughly similar masses. However, the *formation* process has nothing to do with the properties that enable a condensation in due course to perform successfully as a star. In fact, we do not know what ultimately determines the mass of a condensation—this is a fundamental problem. However, it is not directly our concern at the moment; granting that the process yields comparable bodies in any one case, by definition we are dealing with the case where those bodies happen to be or become stars generally like the Sun.

(g) Consider one such case in which a cluster of a few hundred such stars are formed by the envisaged process, and consider a stage when these may be regarded as homologous main-sequence stars. Although as expected in (f) these are comparable bodies, the mass M_s must vary through some significant range of values. Then a rough empirical rule for such stars is (McCrea, 1950)

$$R_s \sim M_s^{3/4} \tag{18}$$

Any one star is formed from floccules arriving from various regions of the cloud; so we may assume that the floccule mass has the same mean value m for all the stars. Then (2) requires

$$H_s \sim M_s R_s^{1/2} \tag{19}$$

Let Ω_s be the uniform angular velocity that would produce angular momentum H_s and let $w_s = R_s \Omega_s$ be the associated equatorial speed of rotation. Then

$$H_s \sim M_s R_s w_s \sim M_s R_s^2 \Omega_s \tag{20}$$

Using (18) – (20) we have

$$w_s \sim R_s^{-1/2} \sim M_s^{-3/8} \tag{21}$$

$$\Omega_s \sim R_s^{-3/2} \sim M_s^{-9/8} \tag{22}$$

The first prediction from (21) is that the calculated equatorial speed decreases rather weakly with increasing stellar mass. Unfortunately, the rotational speed of a main-sequence star of solar mass or less is known for no star other than the Sun. So this prediction cannot be tested at present for cases of most interest.

The second prediction from (22) is that, supposing the star to grow through a sequence of homologous states as its mass is built up to M_s, the mean angular speed inside radius R increases inwards as about $R^{-3/2}$ (so long as R is not too small, since R has to be large enough for the mass inside R to behave as in a star). *Thus a star constructed in this manner would tend to be rotating faster inside than as a whole.*

The situation is complicated, however, because as the star is built up by random additions of floccules, the spin axis will change. Unless there is sufficient viscosity to cause internal rotation to adopt some orderly arrangement, the equatorial speed calculated above may not serve to predict the observed speed, and the observed speed may not give a good indication of the mean angular speed.

(h) Our picture of the formation of a star at O is that floccules fall towards the vicinity of O; those that hit the growing condensation around O become incorporated in it, the rest pass by; according to our use of the surface $\mathscr{S}$, in due course most of these fall back again towards the centre; in the long run most are in fact incorporated. [In an actual cloud without the imagined surface $\mathscr{S}$ some floccules would pass out of the region and others would come in, but an observer near O would notice no difference between $\mathscr{S}$ being there or not.] A rough calculation for a case of interest shows that at any epoch there is on average one floccule within about 10 stellar radii of O, and any such has a chance of only about 1 in 10 of being incorporated before getting further away again. That is, nearly every floccule that gets even within 10 stellar radii of O gets away again. We shall see that the normal radius of a floccule is in fact about 10 stellar radii; so if a floccule is within about that distance of O there is a good chance that any remote observer who sees the star will see it through the floccule.

If any floccule that is not incorporated in the condensation passes it by without suffering any change, then the appearance of the system to the remote observer is a star with about the same amounts of matter going in and coming out. On the other hand, if the passage of a floccule fairly close to the star affects its state in some way, there will not be symmetry between the evidence of inflow and outflow.

T-Tauri stars are usually regarded as stars seen in process of forming; at the same time it is often claimed that there is observational evidence for strong stellar winds blowing away from them. But the concept of anything being formed and yet losing matter seems to be absurd. The present model avoids this absurdity even though, as it stands, it may not fit all the observations. Actually the empirical description of the T-Tauri star R U Lupi recently given by Gahm and others (1975) is rather like the model.

The feature that emerges here of a floccule being possibly incorporated into a star much smaller in geometrical size than itself actually presents no difficulty, for an infalling floccule becomes gravitationally focused by the star. In I, I showed in the case of the Sun that the 'impact parameter' for striking the Sun is not only many times the geometrical radius of the Sun, but it is also about 50 times the radius of a floccule. So a floccule approaching the Sun along a direction passing within that distance is focused on to the surface of the Sun. A floccule making a near miss would undergo comparable focusing; if for any reason such focusing is irreversible, this would imply a dissymmetry between in-going and out-going floccules. This might mean that the model really could account for the T-Tauri phenomenon—as nothing else seems to do.

Numerical Values: Solar System

In the rest of the work the model is applied to the case of the Sun and the solar system.

There are two sorts of numerical values of parameters that have been defined in the course of the discussion:

Values Arising from the Definition of the System

There is nothing arbitrary about these quantities, but there is uncertainty about some of them either because there is actually a range of values, as in the case of T, or because the quantity cannot be directly measured, as in the case of V. It might therefore seem desirable to tabulate results for several values of each parameter, but this soon becomes unwieldy. Simplicity and clarity seem to be best achieved by selecting one most plausible or typical value for each basic parameter and presenting the corresponding results; in general it is easy to see from the formulae how any result would be affected by taking a different value for any particular parameter.

The values introduced in this section are not much different from those I used in I, most of the changes being by way of rounding off a few key quantities in order to simplify their replacement by other values if required.

Quantities used are defined in preceding sections, mainly in '*Star Formation*'. Adopted values for those in the present category are shown in Table 1. By definition M_s is to be a solar mass and we have seen $M \gtrsim M_s$, hence the value taken; we are dealing with cold interstellar material that is inferred to be largely molecular and so we take T to be towards the lower end of the temperature range usually quoted [Allen (1973) quotes 40 K to 120 K]; sound speed in molecular hydrogen at 50 K is about $\frac{1}{2}$ km^{-1}, and so $V = 1$ km^{-1} is a typical supersonic speed; the value of μ is that for molecular hydrogen, a somewhat higher mean value might be more appropriate, but by keeping to $\mu = 2$ we allow for some hydrogen being possibly still in atomic form. Whether or not I have taken the most appropriate values, it is to be repeated that the values are determined from the outset and they are not at our disposal.

Table 1. Adopted parameters

Mass of system	$M = 2 \times 10^{33}$ g
Temperature of raw material	$T = 50$ K
Mean speed of supersonic turbulence	$V = 1$ km s^{-1}
Mean molecular weight of raw material	$\mu = 2$

Chemical composition. As in the *Star Formation* section (a), (b).

Table 2. Postulated parameters

Floccule—number for system	$N = 10^5$
Radius for system (sphere $\mathscr{S}$)	$S_0 = 10^{15}$ cm
Density of floccule-material	$\rho_0 = 10^{-9}$ g cm^{-3}

Values of Parameters of the Model

In order to specify the model quantitatively we have certain parameters at our disposal. It is in the nature of cosmogony that no particular prediction is exact; consequently parameters cannot be evaluated by comparing a few crucial predictions with observation. Instead, we have to try to find a set of values that produce reasonably good agreement with as many observational properties as possible, while making no prediction in clear conflict with observation. The plainest way in which to present such a procedure is simply to postulate a set of values, obviously got by a certain amount of trial and error, and to work out the consequences and compare them with all available empirical evidence. The results do appear to be encouraging—otherwise there would be little worth in describing them. In the light of them it should be possible to go back and find an improved set of values, but we shall not attempt this.

It will be found that we can do all that is needed with only three disposable parameters; those selected and their postulated values are given in Table 2. Their merits will be discussed later.

A few basic parameters derived from those in Tables 1, 2 are given in Table 3.

Predictions: Quantitative

We proceed to consider the values resulting from the parameters in the section on *Numerical Values* for various quantities discussed in the sections *Star Formation, Protoplanets, Times, Angular Momentum Properties* and *Predictions* above.

In the initial state the sphere $\mathscr{S}$ at any instant contains N floccules, Conse-

Table 3. Derived parameters

Floccule mass (mean)	$m = MN^{-1} = 2 \times 10^{28}$ g
Floccule radius (corresponding to mean mass)	$s = m^{1/3}(\frac{4}{3}\pi\rho_0)^{-1/3} \approx 1{\cdot}7 \times 10^{12}$ cm
Jeans mass for floccule material	$M_J = \left(\frac{5\mathscr{R}T}{\mu G}\right)^{3/2}\left(\frac{4}{3}\pi\rho_0\right)^{-1/2}$ $\approx 9{\cdot}5 \times 10^{29}$ g $\approx 48m$

quently a cylinder of radius $2s$ and length S_0 contains on average the centre of

$$3N(s/S_0)^2 = 0{\cdot}85$$

floccule, or very roughly a single floccule. This shows S_0 is about the mean free path of a floccule. A well known formula for the mean free path: (Jeans, 1925, p. 37) gives the value

$$L = \frac{1}{3\sqrt{2}}\left(\frac{S_0}{s}\right)^2 \frac{S_0}{N} = 0{\cdot}83\, S_0$$

Now the only length associated with the material as a whole characterized by the parameters mentioned in the last section in this mean free path. So the quantity S_0 introduced in the section *Star Formation* can be only a multiple of L of order unity. Thus S_0 in Table 2 is a reasonable estimate of the radius of the sphere $\mathscr{S}$ with its postulated astronomical significance.

Substituting in (1) we get for the angular momentum of the whole system

$$H \approx 3{\cdot}2 \times 10^{50}\,\text{cgs units} \tag{23}$$

For the angular momentum of the Sun we predict from (2)

$$H_s \approx 7{\cdot}4 \times 10^{48}\,\text{cgs units} \tag{24}$$

The definition of V_0 following (2) leads to

$$V_0 \approx 5{\cdot}2\,\text{km s}^{-1} \tag{25}$$

whence (3) gives

$$H_s/H \approx 1/43 \tag{26}$$

in agreement with the ratio of H_s, H as evaluated.

At the stage when there are N_1 floccules moving at random inside the surface $\mathscr{S}$ we expect the mean speed to be of the order of the circular orbital speed at some distance. A convenient estimate in agreement with $V_1 \approx 2V_0$ is

$$V_1 \approx 10\,\text{km s}^{-1} \tag{27}$$

which is the speed in a circular orbit at distance about $1{\cdot}3 \times 10^{14}$cm $\approx S_0/7{\cdot}5$ from mass 2×10^{33}g. Then (5) yields

$$N_1/N \approx 1/100 \text{ giving } \quad N_1 \approx 1000, \quad N_1^{1/2} \approx 32 \tag{28}$$

From the statement concerning (12) and the value of M_J in Table 3, the initial mass of a protoplanet is conveniently taken to be

$$M_0 \approx 10^{30}\text{g} \approx 50m \tag{29}$$

so that

$$n \approx 50 \tag{30}$$

From (13), (14), we then have

$$R_0 \approx 6{\cdot}2 \times 10^{12} \text{cm} \tag{31}$$

$$v_0 \approx 1{\cdot}5 \,\text{km s}^{-1} \tag{32}$$

and using (15), with $V = 1 \,\text{km s}^{-1}$ as before, gives

$$H_0 \approx 4{\cdot}4 \times 10^{46} \text{ cgs units} \tag{33}$$

Using (16) but with M_0 in place of M_J we find from (28)

$$0{\cdot}89 < p < 20 \tag{34}$$

and using (17)

$$10^{15} > S > 4{\cdot}5 \times 10^{13} \,\text{cm} \tag{35}$$

or

$$67 > S > 3 \text{ astronomical units} \tag{36}$$

From these last two results we predict that at the stage when protoplanets first appear upon the scene there are between 1 and 20 of them and the radius of the system is between 10^{15} and 3×10^{13} cm.

Comparison with Observation: Provisional

Table 4 shows the comparison of the predictions of the last section with the empirical properties of the solar system. The predicted values are for protoplanets, whose evolution leading to normal planets has still to be considered, and so the comparison is provisional at this stage.

Table 4. Observed and predicted values

	Empirical	Predicted
Sun		
Mass (g)	$1{\cdot}99 \times 10^{33}$	2×10^{33}
Angular momentum (g cm^2 s^{-1})	$1{\cdot}6 \times 10^{48}$	$7{\cdot}4 \times 10^{48}$
Planetary system		
Total mass (g)	$2{\cdot}7 \times 10^{30}$	$0{\cdot}9$ to 20×10^{30}
[a]Radius (AU)	39	67 to 3
Total angular momentum (g cm^2 s^{-1})	$3{\cdot}1 \times 10^{50}$	$3{\cdot}2 \times 10^{50}$
Principal planets (number)	6	1 to 20
[b]*Principal planet*		
Mass (g)	$1{\cdot}9 \times 10^{30}$	10^{30}
Angular momentum (g cm^2 s^{-1})	$0{\cdot}42 \times 10^{46}$	$4{\cdot}4 \times 10^{46}$

[a]Empirical value is mean distance of Pluto
[b]Empirical values are for Jupiter

These results are to be considered along with the qualitative predictions in the section on *Stellar Rotation.* They can be claimed to be highly satisfactory, so far as they go. Where the predictions require or allow a value significantly different from an observed value the difference is in the right direction; for example, the expected angular momentum of a protoplanet is more than the angular momentum of the existing Jupiter, and we shall see that the ratio is itself predicted.

While the values in Table 2 were chosen so as to produce some of the agreement, it is important again to remember that there are only 3 disposable parameters, while Table 4 includes 8 independently observable quantities.

Protoplanet Evolution: First Case

A protoplanet in the model is initially a mass M_0 of the specified raw material pictured as a sphere of radius R_0, having uniform density ρ_0 and angular momentum H_0. We consider how it will evolve, supposing for the time being it is left to itself in space, apart from being in orbit round the Sun.

Since by definition $M_0 \gtrsim M_J$, the protoplanet tends to contract as a whole under its self-gravitation, and to lose energy by radiation. The material is such that, at the same time, a portion—to be called 'dust'—tends to settle under gravity through the remainder—to be called 'gas'. The dust consists of grains—metals, graphite, silicates, ices—together with any gaseous material they pick up physically or chemically; for present purposes, the gas is the rest of the original gas together with any grains that remain in suspension instead of sharing in the sedimentation. Such sedimentation has been extensively investigated (McCrea and Williams, 1965; Williams and Crampin, 1971; and many subsequent papers).

As extreme cases, we consider those in which, during the evolutionary phases to be discussed in what follows: (I) sedimentation is so slow compared with overall contraction that sedimentation may be disregarded; (II) overall contraction is so slow compared with sedimentation that overall contraction may be disregarded. In subsequent phases, the processes disregarded here may, of course, be significant. In this section we study case (I).

Case (I) Undifferentiated material. All the material is treated as having the same composition throughout the body.

Phase A Contraction. $M = M_0$, $R_0 > R > R_E$, $\Omega_0 < \Omega_E$. Starting in a diffuse state with mass M_0, radius R_0, the body contracts under its own gravitation. For simplicity we treat it as remaining spherical, radius R, of uniform density ρ, and of uniform angular velocity Ω, so that the angular momentum H is

$$H = 0{\cdot}4\, M\, R^2 \Omega \qquad (37)$$

During this phase mass M and angular momentum H are conserved so that

$$M = M_0 \qquad H = H_0 \qquad (38)$$

Therefore, the angular velocity Ω increases from its initial values Ω_0 according to

$$\Omega/\Omega_0 = (R_0/R)^2 \tag{39}$$

[Note: From here all the calculations concern planets and satellites, and it ought not to be confusing to use a few symbols with new meanings, e.g. M, H now apply to the protoplanet under consideration, not to the whole original system.]

At some stage $R = R_E$, $\Omega = \Omega_E$ let a particle on the surface at distance ϖ from the spin-axis be such that centrifugal force is just balanced by the component of gravity towards the axis; then

$$\varpi\Omega_E^2 = GM_0\varpi/R_E^3 \tag{40}$$

Thus the condition is independent of ϖ. If the body is rotating in this way, and if it then contracts a little further, material at the equator is left behind, because it is in free circulation under gravity. Material elsewhere in the surface is not immediately left behind but falls a little towards the equatorial plane under the action of the component of gravity in that direction. Therefore, this state 'E' is regarded in the present treatment as the start of rotational break-up.

From (37), (38) and the conservation of angular momentum, we have

$$H_0 = 0{\cdot}4(GM_0^3R_E)^{1/2} \tag{41}$$

It is to be remarked that the last stable Jacobi ellipsoid for a uniform fluid [Lyttleton, 1960; equation (16)] is described by the same equation, except that the numerical factor is 0·390 instead of 0·4 and R_E is defined as the radius of the sphere of equal volume—as, of course, it is here as well. Thus if $R = R_E$ be taken as the condition for rotational instability, then it can be said to give a valid result in the case where it can be checked by an exact theory.

It is easy to verify that phase A is one of decreasing total energy, gravitational plus kinetic. As already stated, the surplus is assumed to be mostly radiated away.

Phase B. Equatorial shedding. $M_0 > M > M_P$, $R_E > R > R_P$. The protoplanet itself will still be assumed always to adjust itself so that it can be treated as a uniform sphere, mass M, radius R, uniform density ρ, and uniform rotation Ω. The only possible behaviour compatible with this, so long as the density is increasing, is at each instant for the rotation to be critical as in (41) so that

$$\Omega^2 = GM/R^3 \tag{42}$$

which is

$$\Omega^2 = \tfrac{4}{3}\pi G\rho \tag{43}$$

showing that the critical speed depends only on the density. Therefore, as contraction proceeds, matter is shed into the equatorial plane and deposited in the appropriate circular orbit.

Consider a change from M, R, Ω, H to $M + \mathrm{d}\,M$, $R + \mathrm{d}\,R$, etc. Differentiating (37), (42) we have

$$\frac{\mathrm{d}H}{H} = \frac{\mathrm{d}M}{M} + \frac{2\mathrm{d}R}{R} + \frac{\mathrm{d}\Omega}{\Omega} \qquad \frac{2\mathrm{d}\Omega}{\Omega} = \frac{\mathrm{d}M}{M} - \frac{3\mathrm{d}R}{R} \tag{44}$$

Since $\mathrm{d}H$ is the angular momentum of mass $\mathrm{d}M$ deposited in the circular orbit of radius R under the attraction of mass M, and which thus has angular velocity given by (43), we have

$$\mathrm{d}H = R^2\Omega\mathrm{d}M \tag{45}$$

Again using (37), (45) gives

$$\frac{\mathrm{d}H}{H} = \frac{5\mathrm{d}M}{2M} \tag{46}$$

Solving (44), (46) for $\mathrm{d}M/M$ etc., and integrating, we find

$$\frac{M}{M_\mathrm{E}} = \left(\frac{R}{R_\mathrm{E}}\right)^{1/2} \quad \frac{\Omega}{\Omega_\mathrm{E}} = \left(\frac{R_\mathrm{E}}{R}\right)^{5/4} \quad \frac{H}{H_\mathrm{E}} = \left(\frac{M}{M_\mathrm{E}}\right)^{5/2} = \left(\frac{R}{R_\mathrm{E}}\right)^{5/4} \quad \frac{\rho}{\rho_\mathrm{E}} = \left(\frac{R_\mathrm{E}}{R}\right)^{5/2} \tag{47}$$

In particular, these yield

$$\frac{H/M}{H_\mathrm{E}/M_\mathrm{E}} = \left(\frac{M}{M_\mathrm{E}}\right)^{3/2} = \left(\frac{\rho_\mathrm{E}}{\rho}\right)^{3/10} \tag{48}$$

We see that the angular velocity of the protoplanet continues to increase with decreasing radius, though less rapidly than in phase A. Also, in phase A since M, H were conserved angular momentum per unit mass was conserved, whereas in phase B in accordance with (48) the angular momentum per unit mass of the protoplanet itself decreases. Thus, during phase B, the protoplanet loses angular momentum much more rapidly than it loses mass.

As the material becomes re-distributed, the gravitational field at a given radius in the equatorial plane will change a little. So the matter deposited in that plane may not remain circulating at precisely the radius at which it is first deposited. But there is no need to allow for this in the present simplified treatment. On this understanding, we see from (47) that the mass deposited between R and $R + \mathrm{d}R$ is

$$\mathrm{d}M = \frac{1}{2}\frac{M_\mathrm{E}\mathrm{d}R}{(R_\mathrm{E}R)^{1/2}} \quad (R < R_\mathrm{E}) \tag{49}$$

Thus the mass deposited in a ring of given width $\mathrm{d}R$ increases slowly with decreasing R.

As one is aware, according to current ideas on the subject, after about radius R_E has been reached, further contraction of the material of a body such as we are considering may produce various configurations which have been studied extensively. However, much of the work has been developed for the study of highly rotating stars, or possibly of the whole of a hypothetical solar nebula, in which viscosity is treated as small throughout. In the present case of a protoplanet, on the other hand, whatever part survives as a planet must be rotating uniformly, or nearly so. This is here allowed for by treating the rotation of the material in the spherical body as being always uniform; at the same time, the differentially rotating material in, or near, the equatorial plane may be viewed as a simplified representation of the flattened differentially rotating configuration of some more elaborate treatment. Furthermore, the concern of the present work is with the manner in which a system evolves as its material contracts under self-gravitation; it is not with the possible steady state that the material might adopt were the evolution halted at some stage.

On such grounds it appears that the present simple treatment may be closer to the physical realities of the problem than an adaptation of any existing mathematically more ambitious model could be. It bears some resemblance to 'equatorial break-up' long ago discussed by Jeans (1928), but the comparison should not be pressed.

State P *Incompressibility* $\rho = \rho_P$. Given any body of material having mass of the order under consideration, there is an approximate maximum mean density. This is about the density of an actual planet made of the same material; it will be called *planetary density* ρ_P for the material. Phase B is taken to terminate in state P when $\rho = \rho_P$ and we write $M = M_P$, $R = R_P$, and so on.

From (47), remembering $M_E = M_0$,

$$\frac{M_P}{M_0} = \left(\frac{\rho_E}{\rho_P}\right)^{1/5} = \left(\frac{M_0}{\frac{4}{3}\pi R_E^3 \rho_P}\right)^{1/5} \tag{50}$$

State P has the following important simple properties:

(a) We shall find that the density ρ_E at which rotation first becomes critical is quite small compared with ρ_P, but nevertheless when density ρ_P is attained there can still be a considerable proportion of the original mass remaining in the protoplanet. This is because in (50) the ratio (ρ_E/ρ_P) enters only to power 1/5; physically, it is because in reaching state P the body has shed angular momentum faster than it has shed mass.

(b) The discarded material of amount $M_0 - M_P$ is spread out in a relatively thin layer about the equatorial plane out to radius about R_E. We say 'about' because of the small readjustment of the layer already mentioned.

(c) The protoplanet is now an almost incompressible body of density ρ_P. It is still in the critical rotational state expressed by

$$H_P = 0{\cdot}4(G M_P^3 R_P)^{1/2} \tag{51}$$

This corresponds to (41); from the remark following (41), we infer that state P is effectively that of the last stable Jacobi ellipsoid—which could not be said of earlier states in phase B because it is not until state P is reached that the material becomes effectively incompressible.

(d) As seen from (43), the angular velocity in phase B depends only on the density; since we are assuming standard density ρ_P for state P, the angular velocity is the same for all protoplanets composed of the same material when in state P. The rotation period Π_P is

$$\Pi_P = \frac{2\pi}{\Omega_P} = \left(\frac{3\pi}{G\rho_P}\right)^{1/2} \tag{52}$$

if we continue to treat the body as spherical. The ellipsoid having the same angular momentum has somewhat longer period because the relevant moment of inertia is greater.

Phase C. Fissional break-up. $M = M_P$. The ensuing phase is precisely the fissional break-up of an effectively incompressible rotating body considered by Lyttleton (1960) and depicted in Figure 4 of that paper. We are glad to employ his inferences so far as they go – as he says, details of the behaviour have not been fully investigated. They apply to the rotating protoplanet itself at this stage, ignoring any effect of material discarded in the preceding phase. Briefly, they are:

The body breaks into two main portions, each with density ρ_P.
The mass-ratio of these portions must be at least about 8:1, but must be not greatly more than this.
The two portions cannot remain gravitationally bound to each other, they must separate at more than mutual escape-speed. (Lyttleton conjectures that the eccentricity of the relative hyperbolic orbit may be about $e = 3$).
If such fission occurs at about the Earth's distance from the Sun or less both portions will go into independent orbits about the Sun.
If fission occurs at about Jupiter's distance, or more, the larger portion remains in orbit about the Sun, while the smaller portion escapes from the solar system.

The picture of the fission process envisaged by Lyttleton is the break-down of the stability of the rotating ellipsoid by the development of a neck between what are about to become the two unequal new bodies. At the instant of actual fission, the two bodies have the same angular velocity Ω_F, which is also the instantaneous angular velocity of one about the other. The angular momentum of the resulting system consists then of the spin-momenta of the two bodies, plus the moments of their orbital momenta about the mass-centre, the sum being equal to the angular momentum H_P just before fission. Estimates show that most of the angular momentum H_P goes in fact into the relative motion of the new bodies, not into their spins. After fission, the rotation of these bodies is therefore not near the critical value, so it is a good approximation to treat them as spherical after they have been formed—not merely a convenient simplification. Also, since they are effectively incompressible, there is no possibility of further contraction leading to any further rotational disturbance. Thus the dynamical evolution of the protoplanet has come to an end—apart from any effects that may at any time be produced by other bodies.

The example that Lyttleton offers as typical suggests that about one-third of the angular momentum H_P remains in the spins implying

$$\Omega_F \approx \tfrac{1}{3}\Omega_P \tag{53}$$

However, the fraction depends sensitively upon the distance of separation of the parts at the instant of fission, and Ω_F could well be less than in (53).

The material composing the neck at the instant of fission would immediately afterwards collect into what Lyttleton calls 'droplets'. He has made extensive calculations on the three-body problem of their ensuing motion; he finds some may:

(a) fall back on to one or other of the two main bodies, (b) escape, (c) go into orbit around the larger body; none will go into orbit about the smaller body.

Numerical Values: Major Planets

In the model, the main bodies finally produced in the case just described have effectively the composition of an original protoplanet, i.e. 'solar' composition. We shall call them the *major planets* of the model. We proceed to give numerical results in two illustrative examples, which are designed for comparison with the actual major planets.

These examples are those having $M_0 = 2 \times 10^{30}$, 3×10^{30}g. We let observation tell us values for planetary density $\rho_P = \frac{2}{3}, \frac{4}{3}$g cm^{-3}. I first calculate R_0, H_E, R_E from (13), (15), (41). However, instead of proceeding with the calculated values of R_E, I adopt somewhat different values R_E 'postulated', for a reason that will shortly appear. Using these values, I recalculate H_E from (41) and obtain M_P/M_0 from (50) and thence H_P/H_0 from (47), and $3\Pi_P$ from (52). The results are shown in Table 5.

Also shown are the stated empirical properties of the two major planets of the SS.

We shall suggest in the section *Minor Satellites* below that the parameter R_E of the model is about the radius of the regular minor satellite system of the resulting planet. So we can treat this radius as a new (alternative) postulated parameter. This is equivalent to adopting the present model only from the beginning of phase B, and then the previous postulates are not needed.

Inferences. A major planet of the model is the larger portion after the fissional break-up of M_P, but its mass is most of M_P which may be taken, therefore, as about the predicted planetary mass. Table 5 shows that this mass is quite like the actual mass in the cases of Saturn and Jupiter. All that this implies, on the model, is that this mass could have come from the protoplanetary mass M_0 as given, and the values of M_0 are acceptable on the model.

The value of the critical period Π_P depends only upon the postulated density ρ_P which has been chosen to be about the empirical value. Any theory would

Table 5. Protoplanet without differentiation: comparison with major planets.

Model: examples		
Protoplanet mass	$M_0 = M_E = 2 \times 10^{30}$	3×10^{30} g
Floccule number	$n = 100$	150
Initial density	$\rho_0 = 10^{-9}$	10^{-9} g cm^{-3}
Mean floccule speed	$V = 10^5$	10^5 cm s^{-1}
Initial radius (13)	$R_0 = 7{\cdot}8 \times 10^{12}$	$8{\cdot}9 \times 10^{12}$ cm
Intial angular momentum (15)	$H_0 = H_E = 7{\cdot}8 \times 10^{46}$	11×10^{46} g cm^{2}s^{-1}
Critical radius (41)	$R_E = 7{\cdot}2 \times 10^{10}$	$4{\cdot}2 \times 10^{10}$ cm
Planetary density	$\rho_P = \frac{2}{3}$	$\frac{4}{3}$ g cm^{-3}
Critical radius postulated	$R_E = 5 \times 10^{10}$	2×10^{10} cm
Angular momentum (41)	$H_E = 6{\cdot}5 \times 10^{46}$	$7{\cdot}6 \times 10^{46}$ g cm^2 s^{-1}
Mass-fraction attaining density ρ_P (50)	$M_P/M_E = 0{\cdot}36$	0·58
Mass at fission	$M_P = 0{\cdot}71 \times 10^{30}$	$1{\cdot}75 \times 10^{30}$ g
Angular momentum—fraction at density ρ_P (47)	$H_P/H_E = 0{\cdot}076$	0·26
Angular momentum of resulting planet	$\frac{1}{3}H_P = 1{\cdot}6 \times 10^{45}$	$6{\cdot}6 \times 10^{45}$ g cm^2 s^{-1}
Rotation period of resulting planet (52)	$3\prod_P = 12$	8·6 hours
Empirical	**Saturn**	**Jupiter**
Mass ($\sim 0{\cdot}9\ M_P$)	$0{\cdot}57 \times 10^{30}$	$1{\cdot}9 \times 10^{30}$ g
Mean density (ρ_P)	0·63	1·25 g cm^{-3}
Rotation period ($3\prod_P$)	10·2	9·8 hours
Angular momentum ($\frac{1}{3}H_P$)	$0{\cdot}77 \times 10^{45}$	$4{\cdot}2 \times 10^{45}$ g cm^2 s^{-1}
Minor satellite system: outer radius (R_E)	(S5) $5{\cdot}27 \times 10^{10}$	(J 5) $1{\cdot}81 \times 10^{10}$ cm

Quantities in brackets are theoretical predictions being tested.

give about Π_P for the critical period. But no other theory has offered any simple reason for the actual period being a few times this critical value.

It is important to note that the values of R_E, H_E got in two different ways are actually in remarkably good agreement. This is the best test of the model at this stage. However, the full force of this becomes more apparent when we come to discuss satellite systems.

Briefly stated at the present stage in the argument, it is seen that the model predicts the existence of major planets of about the observed mass and spin.

Protoplanet Evolution : Second Case

Case (II) Differentiated material. This is the extreme case in which, during the time-interval concerned, the gas is assumed to retain uniform density ρ_1, throughout the volume of the sphere of radius R_0.

Phase a. Dust contraction. $\mathcal{M} = \mathcal{M}_0$, $R_0 > r > r_e$. At the outset, as before the total mass M_0 has uniform density ρ_0; let the gas, dust have uniform density ρ_1, δ_0 so that

$$\rho_0 = \rho_1 + \delta_0, \quad \text{where, say} \quad \alpha = \delta_0/\rho_1 \tag{54}$$

Let the whole have uniform angular velocity Ω_0.

According to the specification of this case, the dust proceeds to settle towards the centre. In keeping with the simplified treatment in Case (1), we represent the process by treating the dust at any subsequent instant as having uniform density δ within a sphere of radius r. Throughout phase a the mass of dust $\mathcal{M}$ inside radius r is constant and equal to $\mathcal{M}_0$, say. As contraction of the dust-component proceeds, its rotation tends to speed up. However, the dust and the gas in which it is immersed at any instant will rotate together—there is relative radial motion under gravity, but no relative transverse motion. Therefore we treat all the material inside radius r, of density $\rho_1 + \delta$, as having uniform angular velocity $\omega(r)$.

As the dust contracts by a small amount dr, the gas it leaves behind in the shell of this thickness is left rotating with angular velocity $\omega(r)$. In this way the dust imparts some of its original angular momentum to the gas through which it settles, the angular momentum of the material as a whole being conserved. This conservation is expressed by

$$\mathrm{d}\left[(\mathcal{M}_0 + \tfrac{4}{3}\pi r^3 \rho_1)\tfrac{2}{5} r^2 \omega\right] = 4\pi r^2 \,\mathrm{d}r\, \rho_1 \tfrac{2}{3} r^2 \omega \tag{55}$$

Using α from (54) and

$$\mathcal{M}_0 = \tfrac{4}{3}\pi \delta_0 R_0^3, \qquad \omega(R_0) = \Omega_0$$

(55) integrates to yield

$$\frac{\omega}{\Omega_0} = \left(\frac{r^3 + \alpha R_0^3}{1 + \alpha}\right)^{2/3} \frac{1}{r^2} \tag{56}$$

It can be verified that this gives the correct total angular momentum H_0 for all r.

Phase a is taken as terminating when the rotation of the body of radius r becomes critical. Let then $r = r_e$, $\delta = \delta_e$, $\omega = \omega_e$, $h = h_e$, etc., where h is the angular momentum of the dust. As in (43) the critical condition is

$$\omega_e^2 = \tfrac{4}{3}\pi G(\rho_1 + \delta_e) \tag{57}$$

By conservation of mass

$$\mathcal{M}_e = \mathcal{M}_0 \text{ giving } \quad \delta_e = \alpha \rho_1 (R_0/r_e)^3 \tag{58}$$

Then from (56)–(58) we find as the equation for r_e

$$[1 + \alpha(R_0/r_e)^3]^{1/3}\Omega_0^2 = \tfrac{4}{3}\pi G \rho_1 (1 + \alpha)^{4/3} \tag{59}$$

Using (39), (42) with $R = R_{\mathrm{E}}$, we have

$$\Omega_0^2 = \tfrac{4}{3}\pi G \rho_0 (R_{\mathrm{E}}/R_0) \quad \rho_0 = \rho_1(1+\alpha) \tag{60}$$

After a little algebra, (59), (60) lead to

$$r_e^3[1 + \alpha - (R_{\mathrm{E}}/R_0)^3] = \alpha R_{\mathrm{E}}^3 \tag{61}$$

In the examples in Table 5 above, $(R_{\mathrm{E}}/R_0)^3 < 10^{-6}$, and this should be typical. Also we are going to take $\alpha \gtrsim 0{\cdot}01$. So from (61) we shall now adopt the approximation (good to about one-third of one per cent)

$$r_e \gtrsim \alpha^{1/3} R_{\mathrm{E}} \tag{62}$$

by (58) we then have

$$\delta_e \gtrsim \rho_1 \left(\frac{R_0}{R_{\mathrm{E}}}\right)^3 \gg \rho_1 \tag{63}$$

Thus, by the stage at which the critical condition is reached, the critical body consists effectively of dust only.
Using (62), (56) gives

$$\omega_e \gtrsim \left(\frac{R_0}{R_{\mathrm{E}}}\right)^2 \Omega_0 \tag{64}$$

As regards the angular momentum of the dust, we find

$$\frac{h_e}{h_0} = \frac{r_e^2 \omega_e}{R_0^2 \Omega_0} \gtrsim \left(\frac{r_e}{R_0}\frac{R_0}{R_{\mathrm{E}}}\right)^2 \gtrsim \alpha^{2/3} \tag{65}$$

i.e. by the time the critical state is reached, the dust retains only a few per cent of its original angular momentum. Most of this momentum has been deposited in the surrounding gas, but this makes very little difference (~ one per cent) to the total angular momentum of the gas. It is interesting to notice how the approximations (62), (65) do not depend upon the value of R_0. This is because, when the dust contracts through the gas and loses momentum to it, most of the loss occurs for r near to r_e, and so the initial value of r has very little effect.

Along with the foregoing approximations, we may take

$$\mathcal{M}_0 \gtrsim \alpha M_0 \tag{66}$$

and thence

$$\frac{\delta_e}{\rho_{\mathrm{E}}} = \frac{\mathcal{M}_0/r_e^3}{M_0/R_{\mathrm{E}}^3} \gtrsim \frac{\mathcal{M}_0}{M_0}\frac{1}{\alpha} \gtrsim 1 \tag{67}$$

From all these results, to the approximation given, state 'e' corresponds to state 'E' with

$$\mathcal{M}_e = \mathcal{M}_0 \gtrsim \alpha M_{\mathrm{E}} \quad \delta_e \gtrsim \rho_{\mathrm{E}} \quad r_e \gtrsim \alpha^{1/3} R_{\mathrm{E}} \quad \omega_e \gtrsim \Omega_{\mathrm{E}}$$

corresponding to

$$M_e = M_0 \qquad \rho_E \qquad R_E \qquad \Omega_E \tag{68}$$

The results that the critical body is effectively pure dust, and that its density and spin are effectively the same as in the corresponding undifferentiated body, are rather accidental consequences of the fact that the initial state is highly diffuse and the fact that the dust is only a small fraction of the initial mass. The simplifications were not *á priori* obvious.

Phase b. Dust only: equatorial shedding. $\mathcal{M}_0 > \mathcal{M} > \mathcal{M}_p$, $r_e > r > r_p$. This phase is exactly analogous to case (I), phase B. It starts with state 'e'; the dust continues to contract under its self-gravitation—more easily, indeed, than the undifferentiated material of case (I). The contraction is through critical states, with equatorial shedding, until state 'p', say analogous to case (I), state 'P', is reached.

State 'p'. Incompressibility. $\delta = \delta_p$ Whereas ρ_P was planetary density for the original material composed largely of light elements, δ_p is planetary density for dust material, with $\delta_p \approx 5\rho_P$, as we shall see. The mass $\mathcal{M}_p$ of the resulting body, analogously to (50), is given by

$$\frac{\mathcal{M}_p}{\mathcal{M}_e} = \left(\frac{\delta_e}{\delta_p}\right)^{1/5} = \left(\frac{\delta_e}{\delta_p}\frac{\rho_P}{\rho_E}\right)^{1/5}\frac{M_P}{M_E} \approx \left(\frac{\rho_P}{\delta_p}\right)^{1/5}\frac{M_P}{M_E} \tag{69}$$

using (67). Since the ratio $(\rho_P/\delta_p)^{1/5} \approx \frac{3}{4}$, the ratio $\mathcal{M}_p/\mathcal{M}_e$, which is the fraction of the original dust remaining in the body, is not very different from the ratio M_P/M_E in case (I). Thus the mass of this body is a little less than α times the mass M_P of the corresponding body in case (I). The rest of the dust material $\mathcal{M}_0 - \mathcal{M}_p$ is spread out in a layer near the equatorial plane to radius about r_e; it is roughly α times the mass similarly discarded in case (I); also it is less than α times the amount of material left behind in phase a which, so far as the present case is concerned is still a diffuse distribution of density ρ_1 enveloping the rest.

From (52) and its analogue we have

$$\pi_p/\Pi_P = (\rho_P/\delta_p)^{1/2} \tag{70}$$

whence the critical period is roughly half that in case (I).

Phase c. Fissional break-up. $\mathcal{M} = \mathcal{M}_p$, The break-up of the mass $\mathcal{M}_p$ is exactly like that of M_P in case (I), phase C.

Numerical Values: Terrestrial Planets

The main bodies finally produced in case (II) have the composition of the dust-component of the original proto-planet. We shall call them the *terrestrial planets* of the model. In Table 6, I give numerical results for one of the same two protoplanets as in Table 5 together with some properties of the actual terrestrial planets. It is seen that proto-planets that would produce bodies closely like major planets, if their material is undifferentiated, will produce ones like

Table 6. Protoplanet with differentiation : comparison with terrestrial planets

Model : example having same initial state as first example in Table 5 specified by		
Protoplanet mass	$M_0 = 2 \times 10^{30}$ g	
Angular momentum	$H_0 = 6{\cdot}5 \times 10^{46}$ g cm^2 s^{-1}	
Dust fraction	$\alpha = 0{\cdot}01$	
Planetary density : segregated material	$\delta_p = 5{\cdot}25$ g cm^{-3}	
Dust mass (αM_0)	$\mathscr{M}_e = 2 \times 10^{28}$ g	
Critical radius (62)	$r_e = 1{\cdot}1 \times 10^{10}$ cm	
Mass-fraction attaining density δ_p (69)	$\mathscr{M}_p/\mathscr{M}_e = 0{\cdot}24$	
Mass at fission	$\mathscr{M}_p = 4{\cdot}7 \times 10^{27}$ g	
Angular momentum at radius r_e (as (41))	$h_e = 3{\cdot}0 \times 10^{43}$ g cm^2 s^{-1}	
Angular momentum fraction at density δ_p (as (47))	$h_p/h_e = 0{\cdot}027$	
Angular momentum at density δ_p	$h_p = 8{\cdot}2 \times 10^{41}$ g cm^2 s^{-1}	
Rotation period of resulting planet (as (52))	$3\pi_p = 4{\cdot}3$ hours	
Empirical	**Venus plus Mercury**	**Earth plus Mars**
Mass combined ($\mathscr{M}_p$)	$5{\cdot}20 \times 10^{27}$	$6{\cdot}62 \times 10^{27}$ g
Equivalent radius (r_p)	$6{\cdot}18 \times 10^8$	$6{\cdot}68 \times 10^8$ cm
Critical angular momentum (as (51)) (h_p)	$9{\cdot}6 \times 10^{41}$	$14{\cdot}4 \times 10^{41}$ g cm^2 s^{-1}
Spin momentum	—	$0{\cdot}606 \times 10^{41}$ g cm^2 s^{-1}
Rotation period ($3\pi_p$)	—	Earth 23·93 Mars 24·62 hours

Quantities in brackets are theoretical prediction being tested

terrestrial planets, if their material is differentiated. This greatly strengthens the plausibility of the model in either case separately.

Inferences. We shall now more specifically identify as the two main bodies in each of two such processes:

(i)	Venus	Mercury
(ii)	Earth	Mars

Lyttleton appears to have been the first to suggest this possibility. Table 7 gives some relevant properties; the masses and volumes are from the latest published measurements, and the iron–nickel fractions are those quoted by Kaula (1968).

In the first place we see that the mass-ratios of about 15 and 9 are greater than 8 as required by Lyttleton's condition quoted on page 96. In the second place, the hypothesis requires the two bodies before fission to have been similar. If Mercury and Venus, Earth and Mars, were put together without change of volume we should get bodies of masses 52 and 66 × 10^{26}g, mean densities both about 5·3 and metal content both about 30 per cent. So the requirement is rather strikingly satisfied, and no other combination of the bodies provides a similar result. Finally, were

Table 7. Terrestrial planets

	Volume 10^{25} cm^3	Mass 10^{26} g	Mass-ratio approx.	Density g cm^{-3}	Iron-nickel Fraction	Mass 10^{26} g
Mercury	6·08	3·30	15:1	5·43	0·65	2·15
Venus	92·85	48·69		5·24	0·265	12·90
Combined	98·93	51·99		5·26	0·29	15·05
Earth	108·32	59·74	9:1	5·52	0·315	18·82
Mars	16·30	6·42		3·94	0·19	1·22
Combined	124·62	66·16		5·31	0·30	20·04

Attention is called particularly to the values underlined

Earth and Mars produced by the fissional break-up of a single body, then directly after break-up the two bodies ought to have the same angular velocity. In fact Earth and Mars have rotation periods of 23·93, 24·62 hours, and their equators are inclined at about 23·5 and 24 degrees to their orbital planes. So the condition is quite well satisfied even now, which is surprising for several reasons. In the case of Venus and Mercury, the present rotations of both bodies have evidently been much affected by the tidal action of other bodies in the SS; so they provide no test of the hypothesis.

Outer Planets

In applying the theory of the section *Proto-planet Evolution: Second Case* to the terrestrial planets in the previous section we suppose the gas left behind in phase a to disperse without playing much further part in the evolution of the planets concerned. There may be other applications in which some of the gas remains and ultimately forms part of a resulting planet. We call such a body an *outer planet* in the model.

In fissional break-up, Lyttleton estimates the recoil speed of the smaller portion to be of the order of the escape speed from the surface of the larger portion. Since the smaller portion actually breaks away at a greater distance than this from the larger portion, it necessarily escapes from the latter, and may escape from the SS, always supposing there to be no other material in the immediate vicinity. However, if the fissional break-up occurs inside a considerable envelope of more diffuse material, the gravitational attraction of that material retards the escape of the smaller portion. Applying these considerations to a rough model of the actual outer planet Neptune shows that the retardation may indeed be important. It is, therefore, suggested that Pluto is the smaller portion from the break-up of a central dense part of the proto-Neptune, which, because of the retardation, did not escape from the SS. In that case, Pluto would be comparable with Mercury and Mars; its mass is usually conjectured to be about one-tenth the mass of the Earth, which agrees with the inference.

Main Satellites

Predictions. According to the model as applied here, all the present planets are main portions from fission processes, the six (principal) Venus, Earth, Jupiter, Saturn, Uranus, Neptune being 'larger' portions, the three (lesser) Mercury, Mars, Pluto being 'smaller' portions. There would have been smaller portions of proto-Jupiter, proto-Saturn, proto-Uranus that escaped. Following Lyttleton's account of the fission process, 'droplets' could have gone into orbit round any of the larger portions, but none of the smaller. We call these *main satellites* in the model which thus predicts their existence.

A main satellite of this sort is a mass of the material at about 'planetary' density that draws itself together by its self-gravitation. Now the Jeans mass $\mathscr{M}_J$ for material with density between 1 and 5 g cm^{-3}, temperature between 50 K and 200 K, which should cover the relevant possibilities, is in the range

$$1{\cdot}3 \times 10^{25} < \mathscr{M}_J < 2{\cdot}5 \times 10^{26}\text{g} \qquad (71)$$

provided we still assume mean molecular weight $\mu \approx 2$. I have considered a similar problem in relation to the origin of comets (McCrea, 1975): if there is a considerable admixture of gas, unless this is itself held together by gravitation, it must disperse and in so doing it will in general carry the dust with it (my suggestion about comets being that they arise in special circumstances in which this general conclusion can be eluded). At any rate in our case (I) of undifferentiated material, plenty of gas is certainly present. Satellites formed in this way must exceed the relevant critical mass $\mathscr{M}_J$. Having regard to (71) a main satellite is expected to have mass of order 10^{26}g.

Such a body going into orbit around a planet is subject to the effect of tidal friction. In general, the long-term quantitative effects are unpredictable from present knowledge of planetary physics. In particular, however, were some extraneous action to slow the planet's rotation to a period longer than the satellite's orbital period, we can say tidal friction would almost certainly drag the satellite into the planet.

Main satellites of solar system. Table 8 shows properties of the seven actual main satellites of the solar system [Triton's mass being very uncertain, any general statement is subject to reservation so far as that may be involved.] The planets and satellites are classified in Figure 1. Even as such satellites exist at the present time, they are the most standardized bodies in the SS. The parameter in which they do now differ most is mean density. Several pointers indicate that a density $\sim 3{\cdot}5$ g cm^{-3} implies lunar composition, while a lower density implies an admixture of 'ice'—chiefly water-ice—with possibly some of the corresponding liquid phase. The simplest model having a core of density 3·5 and a mantle of density 1 g cm^{-3} gives the values shown in Table 8. Consolmagno

Table 8. Main satellites

Planet	Satellite			'Core'	
	Radius (km)	Mass (10^{25} g)	Mean density (g cm^{-3})	Radius (km)	Mass (10^{25} g)
Earth					
E 1 Moon	1738	7·35	3·34		
Jupiter					
J 1 Io	1820	8·92	3·53		
J 2 Europa	1533	4·87	3·23		
J 3 Ganymede	2608	14·90	2·00	1925	10·45
J 4 Callisto	2445	10·64	1·74	1628	6·32
Saturn					
S 6 Titan	2500	14·01	2·14	1926	10·45
Neptume					
N 1 Triton	~1900			$\lesssim$ 1900	

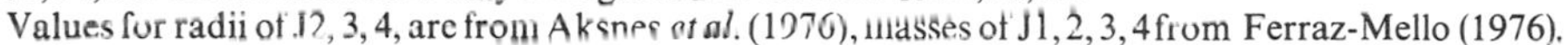

'Core' of density 3·5 surrounded by mantle of density 1 reproduces mass and external radius of J3, J4, S6. The satellite itself may be regarded as the core of E1, J1, J2.
Values for radii of J2, 3, 4, are from Aksnes *et al.* (1976), masses of J1, 2, 3, 4 from Ferraz-Mello (1976).

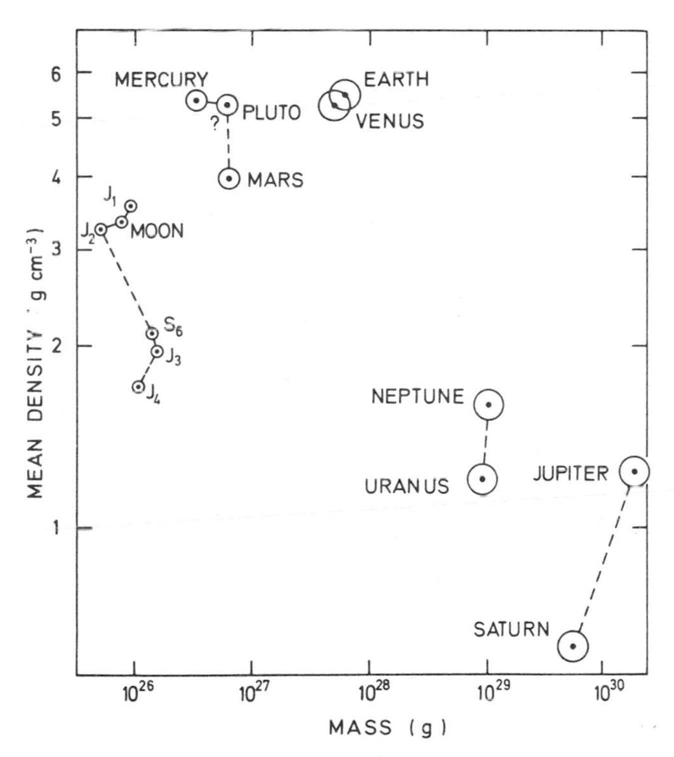

Figure 1. Planets and satellites: classification

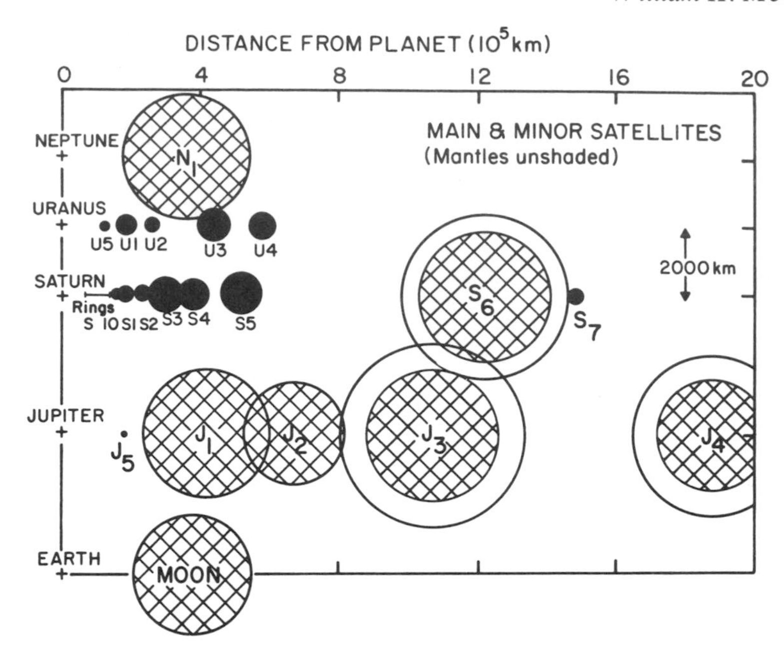

Figure 2. Satellites: main and minor

and Lewis (1976) have studied more elaborate models. The simple model is adequate to show that the core is a yet more standardized object in the SS.

The theoretical predictions agree well with the empirical properties (Figure 2). Main satellites are associated only with principal planets, but the slowly rotating Venus now possesses no satellite, as also predicted. The masses also agree in order of magnitude. A lesser planet has no main satellite, according to the model, simply because a possible orbit for a 'droplet' about such a body would intersect its surface. The 'droplet' itself would splash on to the surface, but droplets of the droplet could continue in orbit—thus offering a possible origin for Mars' tiny satellites Phobos and Deimos.

Lyttleton (1960) showed also how there could be some droplets that would escape from both portions of a dividing protoplanet. So the model implies that the early SS contained a few moonlike objects not attached to any planet. Inevitably, one speculates as to whether the *asteroids* were produced by the collisional break-up of one or more of these; it would yield material about right in total mass, physical state, and chemical composition.

Minor Satellites

Predictions. At the end of phase B in case (I) of the model, mass $M_0—M_P$, amounting to roughly half the original material of the protoplanet, remains circulating near the equatorial plane out to about the distance R_E (large compared with the final radius of the planet). Therefore, around the evolving planet, we have qualitatively the sort of structure that other theories envisage around the evolving Sun as a so-called *Solar nebula* (SN). Let us call it a *protoplanetary nebula* (PN)—regretting that 'planetary nebula', is not available.

So far as I can see, there is no evidence for the past existence of a solar nebula of the sort usually postulated, and I think it would anyhow not have produced the known SS. On the other hand, I shall suggest that there is evidence for the past existence of protoplanetary nebulae, and that they produced the regular minor satellite systems in something like the way in which other theories suppose a solar nebula to have produced a planetary system. This allows us to invoke certain results of those theories, since they deal with generally similar initial conditions. In particular, I refer to the work of Goldreich and Ward (1973)—which has features in common with that of Lyttleton (1972)— and its interpretation by Hayashi *et al.* (1976).

The authors infer that the dust in the material will settle towards the equatorial plane and that, in the resulting layer of relatively dense material, gravitational instability will lead to the formation of *planetisimals* of mass $\lesssim 10^{19}$g. They infer that these will then grow as a result of mutual collisions, the majority apparently to the order of about 10^{21}g but a few may be to 10^{25} or 10^{26}g. [Subsequent development in the SN case is supposed to produce planets, but we are dealing with much less material.] In our PN model we, therefore, very tentatively infer that there will likewise occur a certain number of bodies of dust-material in the mass-range about 10^{21} to 10^{25}g. Being formed as indicated from material symmetrically situated near the equatorial plane in circular motion under gravity, these bodies will be in the plane in direct circular motion under the gravitational attraction of the central (planetary) mass. They will occur in the distance range out to R_E, their spacing depending upon the complicated dynamics of the accumulation process. None can occur within the appropriate Roche limiting distance from the planet, although bodies held together by means other than gravitational may, of course, do so. Hence we predict that some of the PN material may remain in a divided state in circulation near the planet. The processes involved in all this are likely to be inefficient, so that the bodies and rings are expected to add up to only a small fraction of the available raw material. We call the bodies *minor satellites* of the model, and the divided material *rings*.

The processes have to accommodate the fissional break-up of the central body. Presumably fission would occur before minor satellites were formed from the PN, the greater part of which would remain around the larger of the two main pieces resulting from the break-up.

Turning to case (II), by its definition the material left behind in phase a is light gas, and so it cannot subsequently produce bodies having satellite character. On the other hand, the material of amount $\mathcal{M}_e - \mathcal{M}_p$ deposited near the equatorial plane in phase b is dust material only. Whereas in the PN of case (I) the dust had to settle through the gas to form a relatively thin layer, taken literally the present case provides an already differentiated layer of earth-like material and of earth-like mass spread out to a distance r_e ($\sim$ 10 to 20 Earth radii) and so of thickness $\sim 1/100$ Earth radius. More realistically, there would presumably be a fair amount of gas still present to fluidize this material so that the critical mass, like $\mathscr{M}_J$ above, would again be of moon-like size. Possibly, therefore, a moon-like satellite would condense out of this material; such an origin would be somewhat in line with certain suggestions made by others.

Minor satellites of solar system. Table 9 summarizes properties of the 12 minor regular satellites of the SS; the values are mostly from Morrison and Cruikshank (1974). At present only Saturn 1–3 yield useful estimates of the mean density, all being in the narrow interval 1·1 to 1·3 g cm^{-3}. So far as they go, they encourage the plausible conjecture that, while main satellites are composed predominantly of 'core' material—largely silicates— minor satellites are predominantly 'mantle' material—ices.

The model leads us to identify its minor satellites with these actual ones. The predictions are well fulfilled. Such satellites occur only for planets that would have had a PN. In each case, satellites are in circular orbits in the equatorial plane. The masses are of the general order expected, but also as expected they are not standardized like those of main satellites. The systems do extend to appropriate values of R_E; the work shows how the properties of the satellite systems and those of the planets themselves are all compatible with the R_E values employed and the interpretations we give to them. The material for Saturn's rings is accounted for, and it is well known that the rings and the innermost satellite orbit are consistent with what was said about the Roche distance.

Table 9. Minor regular satellites

Planet	Satellite	Mean distance from planet	Radius km	Mass 10^{23} g	Mean density g cm^{-3}
Jupiter	J 5 Amalthea	181	[210 to 35]		—
Saturn	Rings	73 to 137			
	S 10 Janus	159	$\sim$ 110		—
	S 1 Mimas	186	$\sim$ 200	$0·37 \pm 0·01$	1·1
	S 2 Enceladus	238	$\sim$ 250	$0·85 \pm 0·03$	1·3
	S 3 Tethys	295	$\sim$ 500	$6·26 \pm 0·11$	1·2
	S 4 Dione	377	575 ± 100	$11·6 \pm 0·3$	$1·45 \pm 0·8$
	S 5 Rhea	527	800 ± 125	18·2	—
Uranus	U 5 Miranda	130	[650 to 110]	[0·9]	
	U 1 Ariel	192	[1700 to 300]	[13·1]	
	U 2 Umbriel	267	[1100 to 200]	[5·2]	
	U 3 Titania	438	[2000 to 360]	[44·7]	
	U 4 Oberon	586	[1900 to 110]	[25·3]	

From tables of Allen (1973) Duncombe *et al.* (1973) Morrison *et al.* (1974).

The model accounts not only for the minor satellites where they do occur, but also for terrestrial planets not having them. There remains the possibility, however, that the Moon was formed in a disk of dust (and some gas) about the Earth in the manner described, and may be not after all as a droplet in the style of the other main satellites in the model. This would resolve certain known difficulties about the spins of the Earth and Mars and about the angular momentum of the Moon.

The apparent success of the model in accounting for the minor satellites may be quoted as evidence against the analogous solar nebula model for the formation of planets. It has often been argued that such a model would impose too much regularity upon the system—all orbits circular and coplanar, all spin-axes parallel, and so on. If the present work is valid, it does indeed require this regularity in entire agreement with observation in the cases (of minor satellites) to which it applies. Therefore, it cannot apply to the planetary system with all its obvious departures from strict regularity.

Other satellites. Besides the 21 satellites covered by the discussion in the last two sections, there are 12 known miscellaneous small natural satellites. Doubtless they will help to elucidate later dynamical evolution of the SS, but they appear to be not directly relevant to the present discussion.

Comment

The theory is highly tentative. There are various unsolved problems that have not been mentioned in this account. Most of them, as for example the disposal of various residues of material, occur in some form or other in all theories. If one may be permitted a general claim for the present theory it is that, whatever its crudities of formal development, it has predictive power flowing naturally to all aspects of the solar system.

References

Aksnes, K., and Franklin, F. A. (1976). 'Mutual phenomena of Galilean satellites in 1973. III.' *Astron. J.*, **81**, 464–481.

Allen, C. W. (1973). *Astrophysical Quantities*, 3rd. ed. (London : Athlone Press).

Aust, C., and Woolfson, M.M. (1971). 'Floccule theory of planetary formation.' *Mon. Not. R. astron. S.*, **153**, 21P–25P.

Brosche, P. (1970). 'Model for the early evolution of galaxies.' *Astron. Astrophys.*, **6**, 240–253.

Consolmagno, G. J., and Lewis, J. S. (1976). 'Structural and thermal models of icy Galilean Satellites.' In *Jupiter* ed. T. Gehrels (University of Arizona Press) 1035–1051.

Duncombe, R. L. Klepczynski, W.J., and Seidelmann, P. K. (1973). *Fund. Cosmic Phys.*, **1**, 119–165.

Ferraz-Mello, S. (1976). 'Masses of the Galilean Satellites of Jupiter.' *Science*, **192**, 1127–1128.

Gahm, G. F., Nordh, H. L., and Olofsson S. G. (1975). 'The T-Tauri star R. U. Lupi and its circumstellar surrounding.' *Icarus*, **24**, 372–378.

Goldreich, P., and Ward, W. R. (1973). 'Formation of planetesimals.' *Astrophys. J.*, **183**, 1051–1061.

Hartmann, W. K., Davis, D. R., Chapman, C. R., Soter, S., and Greenberg, R. (1975). 'Mars : satellite origin and angular momentum.' *Icarus*, **25**, 588–594.

Hayashi C., Adachi, I., and Nakazawa, K. (1976). *Formation of the Planets* (private communication).

Jeans J. H. (1925). *Dynamical Theory of Gases*, 4th ed. (Cambridge U.P.)

Jeans J. H. (1928). *Astronomy and Cosmogony* (Cambridge U.P.)

Kaula, W. M. (1968). *Introduction to Planetary Physics : the Terrestrial Planets* (New York : John Wiley) Table 9.5

Lyttleton, R. A. (1960). 'Dynamical calculations relating to the origin of the solar system.' *Month. Not. R. astr. S.*, **121**, 551–569.

Lyttleton, R. A. (1972). 'Formation of planets from a solar nebula.' *Mon. Not. R. astr. S.*, **158**, 463–483.

McCrea, W. H. (1950). *Physics of the Sun and Stars* (London : Hutchinson).

McCrea, W. H. (1960). 'Origin of the solar system.' *Proc. R. Soc.*, **A 256**, 245–266.

McCrea, W. H. (1975). 'Solar system as space probe.' *Observatory*, **95**, 239–255.

McCrea, W. H. and Williams, I. P. (1965). 'Segregation of materials in cosmogony.' *Proc. R. Soc.*, **A 287**, 143–164.

Morrison, D. and Cruikshank, D. P. (1974). 'Physical properties of the natural satellites.' *Space Sci. Rev.*, **15**, 641–739.

Williams, I. P. and Crampin, D. J. (1966). 'Segregation of material with reference to the formation of terrestrial planets.' *Mon. Not. R. astr. S.*, **152**, 261–275.

TOWARDS A MODERN LAPLACIAN THEORY FOR THE FORMATION OF THE SOLAR SYSTEM

A. J. R. PRENTICE
Department of Mathematics, Monash University, Clayton, Victoria, Australia

Abstract

An outline of a theory for the formation of the solar system is presented which supports the validity of the original Laplacian hypothesis. Following a brief survey of the difficulties of star formation we suggest that the protosun was created from a cold dark cloud of solid grains which safely contracted to about the orbit of Neptune. As there can exist no true equilibrium for a uniform cloud of this size, we propose that the inner 3 per cent of the mass collapsed freely to stellar size, liberating sufficient energy to stabilize the rest of the infalling mass. Next a large supersonic turbulent stress arising from overshooting convective motions causes the stabilized envelope to rotate nearly uniformly and discretely abandon at its equator a concentric system of orbiting gaseous rings. The masses and chemical compositions of these rings are in accord with the observed distribution of planetary mass and their orbital radii form a rough geometric progression similar to the Titius–Bode law of planetary distances. We examine the condensation sequence in each ring and show how when gas drag is included the condensing particles first settle onto the circular Keplerian axis of the ring to form a concentrated stream of planetesimals which then aggregate into a single protoplanetary cluster. After 3×10^5 yr the protosun enters its overluminous phase and we propose that the gaseous rings are dispersed, thereby terminating the aggregation process. Capture of the residual gases of each ring by the dense planetary cores admits a Laplacian origin for the regular satellite systems of the major planets.

Introduction

The origin of the solar system is a subject which has fascinated cosmogonists for many centuries. Although there has always been a small band who think it came into being purely as a matter of chance, the majority of workers in this field feel that there should exist a well-defined physical explanation for the creation, especially since it may not be uncommon to find other planetary systems like ours throughout the rest of the galaxy. In fact the tide of observational evidence has and continues to move heavily in favour of a highly-ordered beginning. The remarkable regularities which exist in the solar system, such as the near circularity and coplanarity of the planetary orbits, cannot be attributed to chance and the fact that the planets all revolve around the sun in the same sense as the sun's own rotation points to an ordered and cogenetic origin. Recent studies of nuclear isotopic abundances in material on the solar surface and in rocks and meteorites suggest that the whole system came into being some $4{\cdot}6 \times 10^9$ yr ago (Reeves, 1975). In addition, as Alfvén (1954) has stressed, we cannot overlook the fact that many of the regularities which are found in the planetary system are also to be seen in the regular satellite systems of the major planets. The mean distances r_n of these satellites from their primaries for example, when numbered from the centre, form a rough geometric progression

$$r_n \simeq r_0 \Gamma^n, \quad n = 0, 1, 2, \ldots, \tag{1}$$

which is precisely similar to the Titius–Bode law of planetary distances, for which the constant Γ is about $1{\cdot}73$ (ter Haar, 1948). This suggests that the same cosmogonic process must have been responsible for the origin of both types of system.

The Laplacian Hypothesis

It is not the purpose of this lecture to review the numerous theories for the origin of the solar system which have so far been advanced. These theories have been adequately described elsewhere (ter Haar and Cameron, 1963; ter Haar, 1967; Williams and Cremin, 1968; Nieto, 1972). Instead we wish to draw attention to the simplest and perhaps most attractive of these hypotheses, which seems to meet the basic observational requirements and yet which appears to have been largely ignored. This is the celebrated nebula hypothesis of Laplace, first advanced in 1796 and which is pictorially illustrated in Figure 1. Laplace was very impressed by the remarkable orderliness of the planetary motions and proposed that in the beginning the sun was a huge spinning gaseous globe having a size larger than the orbit of Neptune. As the great globe cooled off it contracted inwards and because of the conservation of angular momentum began to spin faster and faster. Eventually the stage was reached when the centrifugal force at the equator overcame the gravitational force and a ring of matter was left behind, suspended in a perfectly circular Keplerian orbit. By shedding mass and angular momentum at the equator the sun was able to safely contract a little further until the centrifugal force again matched the gravitational force and a new ring was formed. This whole process repeated itself until the sun reached its present size. Later by some unspecified process the planets were supposed to have condensed from the concentric system of

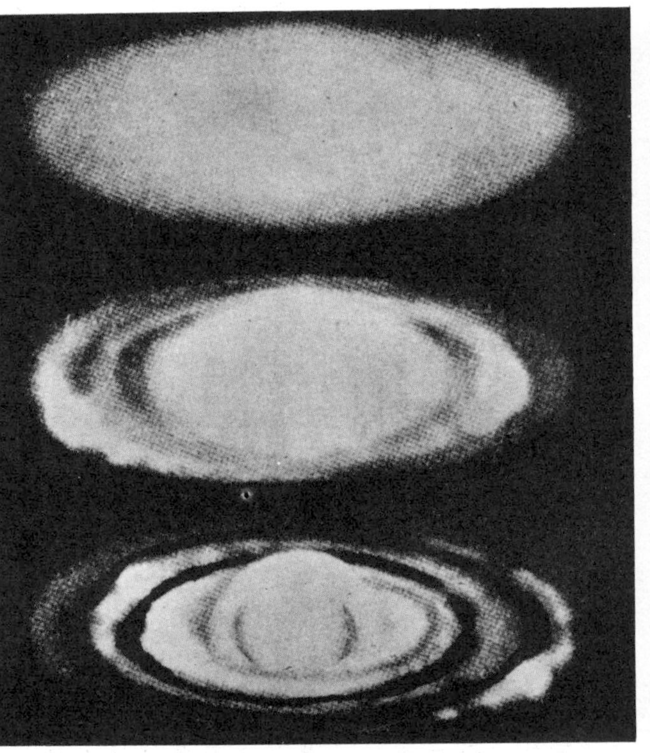

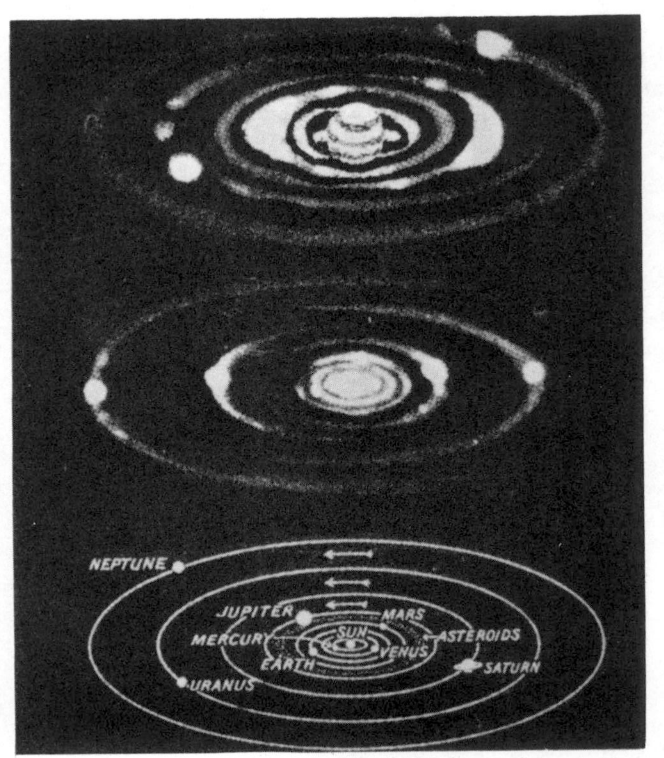

Figure 1. A visual impression of the Laplacian nebula hypothesis. The young contracting and rotating protosun sheds a concentric system of gaseous rings from which the planets later condensed (Drawings by Scriven Bolton, F.R.A.S.; Figure 158 of F. L. Whipple, *Earth, Moon, and Planets*, Harvard University Press, Cambridge, Massachusetts, 1963)

gaseous rings, thus accounting for the common sense of rotation as well as the near circularity of the planetary orbits.

Early Difficulties

Apart from the problem of explaining how the planets condensed from the gaseous rings, the main difficulty which led to the abandonment of the Laplacian hypothesis by the end of the last century was the recognition of the extraordinary character of the distribution of mass and angular momentum in the solar system. As Babinet (1861) pointed out, not only is the sun spinning far too slowly, with a period of some 25 days instead of the few hours expected for a Laplacian theory, but the planetary system is far too light to have taken up the angular momentum of the contracting protosun. Indeed if the sun were to be re-expanded to the orbit of Neptune and the angular momentum of the planets stored back inside, the centrifugal force at the equator would barely rise to one per cent of the required amount.

In 1960 Hoyle pointed out that the objection of the type raised by Babinet overlooked the fact that the present mass of the planetary system may represent only a very small fraction of the total mass of material which would have been present at the time of the formation. This is because most of the planets consist of material which is quite rare by normal cosmic standards. If then the planets were created from solar material which consists predominantly of H and He, it is necessary to take into account this large fraction of accompanying light gas. Table 1 gives the masses of the broad chemical constituents in $1000\,M_\oplus$ ($M_\oplus$ = Earth mass) of solar material, derived from the compilation of Engvold and Hauge (1974). In order to construct planets like Uranus and Neptune which are thought to consist mostly of ices and have masses of order $15M_\oplus$ we see we shall require about $10^3\,M_\oplus$ of solar material for each planet. Similarly for the terrestrial planets, which are mostly rock-like, we shall again require a mass perhaps as large as $10^3\,M_\oplus$ at each orbit, especially if, as Hoyle and Wickramasinge (1968) have mentioned, a large fraction of the rock component in each ring condensed out as a fine smoke which was later swept away. If, therefore, we suggest that the planets each condensed from roughly the same amount of solar material, namely $1000\,M_\oplus$, and the sun shed as many gaseous rings between the orbit of Mercury and its present size as between Neptune

Table 1. Broad chemical constituents of $1000\ M_\oplus$ of solar material

Type	Components	Units of Mass ($M_\oplus$)
gaseous	H, He	982
ice-like	CNO (as hydrides), Ne, Ar	13
rocky	Mg, Al, Si, Fe (as oxides) Ni etc.	5

and Mercury, we can estimate the mass of the primitive solar nebula. We obtain simply

$$m_{\text{neb}} \simeq 0{\cdot}05\, M_{\odot}. \tag{2}$$

This value agrees closely with estimates derived independently by Urey (1951), Kuiper (1951), Whipple (1971) and other cosmochemists.

The Real Problem

The real angular momentum difficulty can now be restated as follows. How is it possible for a protosun which first becomes rotationally unstable at the orbit of Neptune to safely contract all the way through to solar size, giving up essentially all of its angular momentum at the expense of losing only 5 per cent of its mass? Jeans (1929, Chapter 16) observed that such an event would be possible only if the mass distribution inside the protosun was greatly concentrated towards the centre and the whole structure rotated like a solid body, with a uniform angular velocity. The equations which govern the contraction of such a globe have been written down by Schatzman (1949). Letting R_e denote the equatorial radius and $M(R_e)$ the mass, the angular momentum of the globe is $L = MR_e^2 f\omega_e$ where f is called the moment-of-inertia coefficient and ω_e the angular velocity. Setting $\omega_e = \sqrt{GM/R_e^3}$, the Kepler value at the equator, and where G is the gravitational constant, Schatzman showed that M and L must scale according to the formulae

$$M(R_e) = M_0[R_e/R_0]^{f/(2-3f)}, \quad L(R_e) = L_0[R_e/R_0]^{1/(2-3f)}, \tag{3}$$

where R_0, M_0, L_0 denote the initial values. Hence starting at the orbit of Neptune and setting $M(R_0) - M(R_{\odot}) = m_{\text{neb}}$ we see we shall require

$$f \simeq \frac{2m_{\text{neb}}}{M_{\odot}} \Big/ \ln(R_0/R_{\odot}) \sim 0{\cdot}01. \tag{4}$$

Such a small value of f implies that the protosun must indeed have been very centrally condensed.

The first problem which faces the modern Laplacian theorist therefore is to construct a model of the protosun which has a very low f and can maintain a uniform angular velocity during contraction. If, as is usually supposed, the interior is fully convective and turbulent stress is ignored then the problem is that f is very large, typically 0·135 (Auer and Woolf, 1965). Secondly in order for the protosun to rid the angular momentum from its interior in the manner implied by equation (3) it is necessary for the contraction to occur quite slowly at a rate much less than the free fall rate. Otherwise the interior of the cloud would not have sufficient time to respond to the changes taking place at the surface. Unfortunately both Hayashi (1961) and Cameron (1962) have shown that there can be no quasi-static equilibrium for the protosun in the region of the planetary system since there is insufficient gravitational energy available to maintain a true thermodynamic equilibrium. This result remains true even if the convection is strongly supersonic (Schatzman, 1971). Thirdly, there is the problem of explaining why the protosun shed a discrete system of rings rather than a continuous disk-like nebula, as most authors commonly suppose. And lastly, of course, there is the problem of explaining how the planets aggregated from each gaseous ring.

Towards a Solution

It is the purpose of this lecture to consider each of the above difficulties in turn and present an outline of a modern Laplacian theory for the origin of the solar system. The analysis commences with the properties of the interstellar medium where we consider the general problem of star formation. After suggesting that stars form through the segregation of solid grains we follow in detail the collapse of a cloud fragment of solar mass. We show how the cloud can be brought into quasi-hydrostatic equilibrium first at the orbit of Neptune if a small dense star-like core forms at the centre. We then consider the influence of a supersonic turbulent stress on the rest of the cloud. We show how this stress leads to the formation and detachment of a discrete system of gaseous rings in the same manner as suggested by Laplace. In the final part of the lecture we consider the chemical condensation sequence in each ring and show how gas drag leads to the aggregation of the planetesimals into single planetary objects.

Very Early Stages of the Sun's Formation

It is widely believed that stars are somehow formed out of the diffuse material which makes up the interstellar medium in our galaxy (Mestel, 1965a, b). Let us therefore take a close look at the physical conditions inside a typical interstellar cloud.

The Interstellar Medium

A typical gas cloud in the solar neighbourhood of the galaxy has a mean density $n_{\mathrm{H}} \simeq 10$–$20\,\mathrm{H}$ atom/cm^3, a mean temperature $T = 125\,\mathrm{K}$, an intrinsic angular velocity $\omega_{\mathrm{i}} \simeq 3 \times 10^{-16}\,\mathrm{s}^{-1}$, and a magnetic field $B \simeq 3 \times 10^{-6}$ gauss (Spitzer, 1968). In order for a roughly uniform cloud to be self-gravitationally bound it is necessary that its gravitational energy exceed the sum of the thermal, magnetic and rotational energies, or equivalently

$$\frac{3GM}{5\ R} > \frac{\nu \mathscr{R} T}{2\ \mu} + \frac{B^2}{2\mu_{\mathrm{m}}\rho} + \frac{1}{5}\omega^2 R^2, \tag{5}$$

where μ refers to the mean molecular weight, $\mathscr{R}$ the gas constant, μ_{m} the magnetic permeability, R the mean cloud size, and ν the number of degrees of atomic freedom. Inserting the typical values above we find that the cloud mass M must exceed $10^4 M_\odot$ if the magnetic energy is ignored, and about 10^5–$10^6 M_\odot$ if this term is included. The rotational energy is negligible. Clearly, therefore, it is erroneous to speak of isolated cloud portions of solar mass existing in the interstellar medium, or for stars to form through the gravitational collapse of such portions.

Another serious difficulty which confronts the theory of star formation is the angular momentum problem. For the interstellar cloud we have just considered the spin angular momentum per solar mass is $L_i = \frac{2}{5} MR^2 \omega_i \simeq 2 \times 10^{54}$ g cm^2 s^{-1}. This value is more than two orders of magnitude larger than the angular momentum of a protosun which first becomes rotationally unstable at the periphery of the planetary system, taking $f = 0{\cdot}01$, and more than four orders of magnitude larger than that of a sun of normal density which is rotating so rapidly that the centrifugal force at the equator balances the gravitational force there. The present angular momentum of the sun is only 2×10^{48} g cm^2 s^{-1}.

We see therefore that a diffuse cloud portion of solar mass must give up essentially all of its angular momentum before it can contract either to stellar density or to the size of the planetary system. To overcome this problem McCrea (1960) and Dr. Black (this volume) have appealed to the turbulent character of the motions in the interstellar gas to suggest that stars form only in 'pockets' where the local angular momentum happens to be zero. If this were the case, however, and we supposed that the magnetic pressure problem could also be overcome in the same manner, then star formation would be an unacceptably inefficient process, since less than 1 part in 10^8 or so of the cloud mass would be available for collapse at any one time.

The Grain Cloud Hypothesis

A suggestion of Reddish and Wickramasinghe (1969) which seems to directly overcome the above difficulties is that stars are instead formed through the segregation of clouds of solid grains. Such grains may form in the cool dark regions of large interstellar clouds of mass $10^5 M_\odot$ or more which are sufficiently massive to contract to the densities $n_H \gtrsim 10^4$ H atom/cm^3 required for the obscuration of starlight. If the temperature drops to $\lesssim 5$ K the H, including He as impurity atoms, may condense out to form solid grains of radius about 10^{-4} cm. Grains rich in CNO ices condense out much sooner (Duley, 1974). In any event, by allowing the gas to condense into the solid phase the thermal pressure difficulty is immediately overcome. The magnetic pressure problem is also overcome since the presence of a few free ions or electrons on the grain surface makes no difference to the motion of a grain containing 10^{12} or so neutral atoms.

The angular momentum difficulty can also be resolved within the grain cloud hypothesis if one takes into account the frictional resistance between the residual uncondensing gases and the moving grains (Prentice and ter Haar, 1971). As the grains migrate towards their common centre of mass, dynamical friction constrains their azimuthal velocity to match that of the uniformly rotating gas cloud at each stage of their descent. The azimuthal stress exerted on the gas is transmitted out of the cloud along the magnetic field lines which are coupled to the gas via the latter's ionic component. The net effect is that the grain cloud contracts with a constant angular velocity, equal to that of the gaseous component, until it becomes so compressed that it sweeps up the residual neutral atoms interior to it and goes into free fall. This transition to free fall, which occurs when the grain cloud radius has shrunk to a fraction s_1 about 1/10th of its initial size, has been confirmed in detail by Krautschneider (1976, 1977) with numerical model calculations. The angular momentum of the grain cloud of solar mass left over at this stage is

$$L_f = s_1^2 L_i \sim 2 \times 10^{52} \, \mathrm{g\,cm^2\,s^{-1}}. \tag{6}$$

Thus two orders of magnitude of angular momentum are lost through the grain braking process, leaving over an amount just comparable with that required for a fully rotating protosun of radius $10^4 R_\odot$ and moment-of-inertia factor $f = 0{\cdot}01$. Since the ionic component of the gas cloud is not swept up until the grain cloud radius is about $10^4 R_\odot$ too, the residual magnetic energy is negligible compared with the gravitational energy there and so can be ignored. That is, although magnetic fields play a very important role during the grain braking period, they probably have very little to do with the later stages of the protosun's evolution.

Formation of a Central Chemical Inhomogeneity

There is one additional feature of the grain segregation hypothesis which is worth mentioning here. The equation of motion of a single grain of density ρ_s and radius a situated at distance r from the centre of mass of a gaseous globe of density ρ and temperature T is

$$\ddot{r} = -\frac{4\pi G \rho r}{3} - \frac{3\rho}{\rho_s a}\left(\frac{\mathcal{R}T}{2\pi\mu}\right)^{1/2} \dot{r}. \tag{7}$$

The solution of this equation is

$$r(t) = r(0)\exp(-t/t_{\mathrm{mig}}), \tag{8}$$

where the characteristic migration time t_{mig} is given by

$$t_{\mathrm{mig}} = \frac{9}{4\pi}\left(\frac{\mathcal{R}T}{2\pi\mu}\right)^{1/2} / aG\rho_s \simeq 2 \times 10^7 \, \mathrm{yr}, \tag{9}$$

taking the values $\rho = 1{\cdot}67 \times 10^{-20} \, \mathrm{g\,cm^{-3}}$, $T = 5\,\mathrm{K}$, $a = 10^{-4} \, \mathrm{cm}$, $\rho_s = 1$, $\mu = 2$ and noting that $t_{\mathrm{mig}} \gg t_{\mathrm{ff}} = \sqrt{3/4\pi G\rho} \simeq 5 \times 10^5$ yr, where t_{ff} is the cloud free-fall time.

The important feature to note about equations (8) and (9) is that dense grains sink to the cloud centre faster than light grains owing to the smaller drag per unit mass which they experience. Now since grains rich in CNO ices are at least 10 times denser than those consisting solely of solid H_2 ($\rho_{H^2} \sim 0{\cdot}1\,g\,cm^{-3}$), we see that for any given initial distribution of grain compositions a relative segregation of grain material according to chemical composition will occur as time proceeds, with the CNO-rich grains soon dominating the centre of the cloud. The formation of such a central chemical inhomogeneity has drastic consequences both for the main sequence nuclear evolution of the sun and the present solar neutrino flux, as well as for the geological evolution of the Earth and the development of life (Prentice, 1976).

Energetic Stabilization of the Protosolar Envelope

When the central density of the collapsing grain cloud and imprisoned gases rises to about $10^{-13}\,g\,cm^{-3}$, corresponding to a radius of $2 \times 10^4\,R_\odot$ for a uniform cloud of solar mass, the material becomes opaque to the free escape of heat (Gaustad, 1963). Since the gravitational energy $GM^2/R \sim 2 \times 10^{44}$ erg released through the compression of the gas is many times larger than the latent heat of vaporization of the solid H_2 grains, $Q_{vap} \sim 7 \times 10^{42}$ erg, the grains are readily evaporated. During the subsequent adiabatic compression the pressure $p \propto \rho^\gamma$ increases as $R^{-3\gamma}$, where γ is the ratio of specific heats, while the pressure required for hydrostatic equilibrium scales only as R^{-4}. Thus the free collapse is arrested and hydrostatic equilibrium is attained as long as $\gamma > 4/3$.

Now the value of γ depends on the state of dissociation and ionization of the gas, which in turn depends on the density, temperature and composition. If there is insufficient gravitational energy available to supply the dissociational and ionizational energies required for maintaining a complete thermodynamic equilibrium, then γ falls below 4/3 and the cloud proceeds to collapse uncontrollably. The calculations performed by Hayashi (1961, 1966), Cameron (1962) and others show that this certainly is the case for a protosolar cloud having radius in the interval $10^2\,R_\odot$ to $10^4\,R_\odot$, even if the convective motions prove to be supersonic (Schatzman, 1967, 1971). That is, it would seem that we cannot meet the basic quasi-static requirement for a Laplacian theory.

Larson's Model

In 1969 Larson pointed out that the uniform cloud calculations of Hayashi (1961) etc. overlooked the fact that the gravitational collapse of a free falling gaseous mass probably occurs extremely non-homologously. That is, as the collapse proceeds the density profile becomes progressively more peaked towards the centre rather than preserving its shape. Larson found that the innermost region of the cloud collapsed all the way through to stellar size to form a small compact luminous core of radius 2–5 $R_\odot$ long before the rest of the cloud had fallen very far. As the core grew in mass, maintaining much the same radius, the rest of the infalling cloud was heated by the gravitational energy liberated by the infall of fresh material onto the core. Most of this heat, however, readily escaped from the cloud owing to the low density of the infalling material.

Consider now the implication of Larson's work for our gaseous protosolar mass of mean density $10^{-13}\,\mathrm{g\,cm^{-3}}$ which we suppose to be roughly uniform, except for an enhanced CNO abundance at the centre. Since such a cloud is unable to attain a complete thermodynamic equilibrium at this density we propose that its innermost regions continue to collapse freely, but non-homologously, in the manner suggested by Larson, to form a small dense luminous core of radius about $3\,R_\odot$. The heat released through the formation of this core is then imprisoned by the rest of the infalling cloud owing to the latter's high mean density. The mass of the core grows until the gravitational energy liberated at its edge is sufficient to meet the dissociational and ionizational energy requirements of the rest of the cloud, thereby arresting the collapse. The resulting model of the protosun then consists of a small dense stellar-like core, rich in CNO, surrounded by a vast, tenuous, but optically thick, turbulent protosolar envelope of mean density still about $10^{-13}\,\mathrm{g\,cm^{-3}}$.

If the above ideas are correct then the core mass m_c needed to stabilize a cloud of solar mass and equatorial radius R_e is given by

$$m_c = \left[\frac{2r_c}{G}(E(R_e) - E_\infty)\right]^{1/2} \tag{10}$$

where

$$E(R_e) = \Omega_{\mathrm{grav}} + U_{\mathrm{th}} + U_{\mathrm{turb}} + U_{\mathrm{rot}} + \mathscr{E}_{d-i} \tag{11}$$

is the total energy of the envelope. Here Ω_{grav} is the gravitational potential energy, U_{th}, U_{turb}, U_{rot} are the thermal, turbulent, and rotational kinetic energies, and $\mathscr{E}_{d-i}$ is the sum of the dissociational and ionizational energies. The total initial energy of the cloud E_∞ is usually zero.

Calculations

We have computed $E(R_e)$ and m_c for various fully rotating turbulent polytropic structures which are defined in more detail in the section *Supersonic Turbulent Stress*. The total pressure at any point inside such a structure is the sum of the gas pressure $\rho\mathscr{R}T/\mu$ and a turbulent stress $<\rho_t v_t^2>$, described in that section, whose strength is measured in terms of a turbulence parameter β (Prentice, 1973). The results of the calculations are displayed in Figure 2. In the absence of a core the condition for hydrostatic equilibrium to be attained is $E(R_e) - E_\infty \leqslant 0$. We see from the figure that for the non-turbulent cloud ($\beta = 0$) this condition cannot be met until the cloud radius has shrunk to about $70\,R_\odot$ which is much less than the dimension of the planetary system and even less than the orbit of Mercury, as Cameron (1962) first noted. By putting some

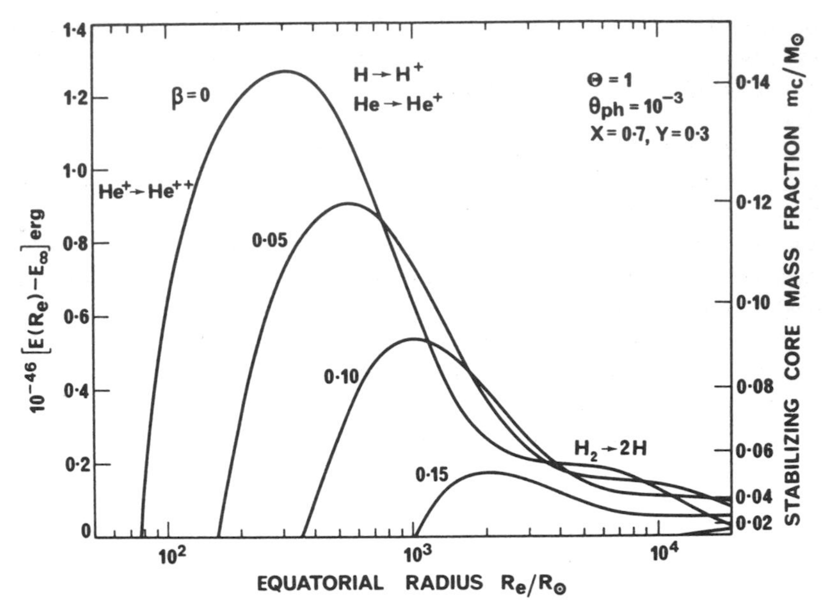

Figure 2. Total energy requirement $[E(R_e) - E_\infty]$ of the fully rotating turbulent polytropic protosun plotted against equatorial radius R_e. In order to achieve a complete hydrostatic and thermodynamic equilibrium in regions where $E(R_e) - E_\infty > 0$ it is necessary that a small compact core of mass m_c, given as the right hand ordinate, accumulate at the centre of the cloud, thus releasing some gravitational energy $Gm_c^2/2r_c$

turbulent stress into the cloud the situation is greatly improved, as Schatzman (1967) first observed, since the star is able to support itself at a lower gas kinetic temperature. At the same time, as we shall later see, the protosun becomes more centrally condensed, enhancing its apparent store of potential energy Ω_{grav}.

We note from Figure 2 that even with the assistance of turbulent stress the protosun is unable to attain a complete equilibrium in the interval $10^2 R_\odot \leqslant R_e \leqslant 10^4 R_\odot$ if no central core forms. Typically we require a value β of 0·12 to account for the Titius–Bode law (see Section 7) and for this value the protosun without a core does not become stabilized until its radius has shrunk to about $500 R_\odot$.

The Stabilizing Core Mass Fraction $m_c/M_\odot$

Consider now the vertical ordinate on the right hand side of the diagram which gives the stabilizing mass fraction $m_c/M_\odot$ corresponding to each energy $E(R_e)$. We see that when the protosun first becomes optically thick at radius $R_e \simeq 2 \times 10^4 R_\odot$, a mass $m_c \simeq (0·03 \pm 0·01)\ M_\odot$ is sufficient to fully stabilize the envelope and this value is essentially independent of β as well as the precise starting radius. As soon as the protosolar envelope has been stabilized it pro-

Table 2. Rate of contraction of the protosolar envelope (QS: quasistatic mode K–H: Kelvin–Helmholtz mode)

$R_e(R_\odot)$	1·5	10	10^2	10^3	10^4
T_e(K)	4500	4000	1000	150	20
v_e/v_{ff}	10^{-11}	10^{-8}	10^{-6}	10^{-5}	10^{-2}
τ_e(yr)	4×10^7	2×10^5	5×10^4	1×10^5	5×10^3
mode	K–H	K–H	K–H	K–H	QS

ceeds to contract quasi-statically at a rate governed by the rate of accumulation of material at the centre until the point is reached at about $R_e \sim 10^3\ R_\odot$ where $E(R_e)$ passes through a maximum. During this period of core growth the rate of collapse at the surface v_e behaves as $v_e \sim (m_c/M_\odot)^{3/2} v_{ff} \sim 0{\cdot}01\ v_{ff}$ where $v_{ff} = \sqrt{(GM_\odot/R_e)}$ is the free fall speed (Prentice, 1977). The contraction therefore is indeed quasi-static.

After reaching the point where $E(R_e)_{max}$ is a maximum, the envelope no longer requires energy from the core and in fact can contract no further without radiating away its excess energy at the surface. The contraction now proceeds much more slowly on a Kelvin–Helmholtz timescale given by $t_{K-H} = GM^2/2{\cdot}82\pi\sigma_s R_e^3 T_e^4$ where T_e is the surface temperature and $2{\cdot}82\pi R_e^2$ the surface area of the fully rotating structure. Table 2 shows v_e/v_{ff} and the local contraction time $\tau_c = R_e/\dot{R}_e$ at various stages of the contraction for surface temperatures T_e typical of our turbulent polytropic protosolar models. The mode of collapse is also indicated. We see that the rate of collapse slows very markedly during the final stages of contraction. This is because T_e is limited by photospheric conditions to level off at about 4500 K as the zero age main sequence (ZAMS) is approached (Ezer and Cameron, 1965). In all it takes some 2×10^7 yr to contract to the radius 1·5 $R_\odot$ corresponding to a fully rotating ZAMS sun. About 99 per cent of this time is spent in the last portion of the contraction $R_e \leqslant 10\ R_\odot$. In contrast, the contraction through the region of the planetary system is quite a brief affair taking only some 2×10^5 yr.

Influence of Supersonic Turbulent Stress

T-Tauri Stars

Some while ago I suggested (Prentice, 1973) that if the convective motions inside young contracting protostars were strongly supersonic one could account for many of the observed features of T-Tauri stars as well as the solar system. T-Tauri stars are thought to be young suns which may be going through or have just completed the phase of planetary formation. According to Herbig (1962, see also this volume) the emission features of these objects are quite erratic and are typically characterized by a velocity spectrum which is negatively displaced by some 100–200 km s^{-1}, indicating the presence at their surfaces of material rising upwards with these speeds. Since the sound speed of the gas $v_s = \sqrt{\gamma \mathscr{R} T/\mu}$ is of order only $\sim 10\,\text{km s}^{-1}$, the observed velocities are indeed strongly supersonic. As there is no evidence of any material returning to the star with the same speeds, a number of authors (e.g. Kuhi, 1964) have suggested that the T-Tauri stars are rapidly losing mass (about $10^{-6} M_\odot \text{yr}^{-1}$) despite the fact that the speeds are less than the escape speed and the fact that the nearly circular Keplerian orbits of the planets suggest that the sun probably did not lose much mass at all at the time of its formation.

Supersonic Turbulent Convection

An alternative picture of the T-Tauri star mass 'ejection' phenomenon is that the observed rising motions are those of long and needle-like convective elements which originate in the deep interior of the star and become visible upon moving across the photosurface. As soon as the supersonically moving eddies emerge from the positively buoyant convective interior and enter the negatively buoyant atmosphere beyond the photosurface, they proceed to decelerate to rest and eventually return to the star in the form of a stable dense slowly descending return wind which has a speed $\sim 10\,\text{km s}^{-1}$ which is indis-distinguishable from the background thermal velocity. Since all convective elements are formed in the convective interior of the star, negatively buoyant motions go completely unobserved since they do not acquire supersonic speeds until they have descended many optical depths.

A major consequence of this steady-state model of T-Tauri stars is that there should exist in their interiors a large supersonic turbulent stress $\langle \rho_t v_t^2 \rangle$ arising from the motions of the convective elements. This non-thermal stress assists the gas pressure in the support of the star against its own gravity, as McCrea (1929) first observed in his study of the support of the solar chromosphere. It also induces massive changes in the structure of the star as we shall shortly see. Further evidence for the existence of a large-scale mechanical stress in T-Tauri stars has been garnered by Thomas (1973).

Since the convective motions are driven by buoyancy, it is a simple matter to show that the turbulent stress has the form $\langle \rho_t v_t^2 \rangle = \beta \rho G M(r)/r$ where for a non-rotating star $M(r)$ is the mass interior to radius r. The total radial stress in the star may then be written as

$$p_{\text{tot}} = \rho \mathscr{R} T/\mu + \beta \rho G M(r)/r, \tag{12}$$

where the turbulence parameter $\beta = \frac{2}{3} f_e \langle |1 - \rho_e/\rho| \rangle k$ is a dimensionless number which depends on the mean density excess factor of the eddies $\langle |1 - \rho_e/\rho| \rangle$, the mean eddy mass fraction f_e, and the mean ratio $k = \lambda/r$ of eddy acceleration length λ to local radius r (Prentice, 1973). Typically we find $\beta \simeq 0{\cdot}1$ but if the convection is very strong, say when the eddy mass fraction $f_e \to 1$, then β may become quite large (Prentice, 1976). In any event the eddies can acquire supersonic speeds only by retaining their thermal identity over distances comparable with the thickness of the convective zone so that overshooting into the surrounding radiatively stable zones occurs. Thus for a fully convective star where there is only an upper radiative zone, namely the region above the photosurface, only positively buoyant elements can be accelerated to supersonic speeds whilst the negatively buoyant motions are quenched.

Our picture of the young contracting protosun thus consists of a swarm of positively buoyant needle-like elements continuously rising to the photosurface then decelerating to rest and returning to their points of origin in the form of a stable return flow. Embedded at the centre of the vast turbulent protosolar envelope of initial mean density about $10^{-13}\,\mathrm{g\,cm^{-3}}$ is the dense metal-rich core of mass $0{\cdot}03\,M_{\odot}$. Because of its compactness and much higher density $\sim 10^{-3}\,\mathrm{g\,cm^{-3}}$, we may assume that the core essentially isolates itself from the activities of the envelope once the latter has become energetically stabilized. Since the core has no azimuthal coupling which can influence the rotational evolution of the envelope we shall henceforth ignore its presence in the remainder of this lecture.

Turbulent Polytropic Structures

Inserting equation (12) into the equation of hydrostatic support

$$\mathrm{d}p_{\mathrm{tot}}/\mathrm{d}r = -\rho G M(r)/r^2, \tag{13}$$

we can compute the influence of turbulent stress on the structure of a non-rotating protosun. The results of some calculations performed by myself (Prentice, 1973) for the case of turbulent polytropic structures, defined by the equation of state

$$\rho/\rho_{\mathrm{c}} = \left(\frac{T}{\mu}\bigg/\frac{T_{\mathrm{c}}}{\mu_{\mathrm{c}}}\right)^{n}, \quad n = \frac{1}{\gamma - 1}, \tag{14}$$

where ρ_{c}, T_{c}, μ_{c} refer to the central density, temperature, and molecular weight and n is the polytropic index are shown in Figures 3, 4 and 5 below. For the case $\beta = 0$ the surface radius R_{ph} is normally defined by the zero-stress condition $T_{\mathrm{ph}} = 0$. For turbulent polytropes, however, there exist no solutions of finite radius having $T_{\mathrm{ph}} = 0$ so that it is necessary to introduce a non-zero surface parameter

$$\theta_{\mathrm{ph}} = \frac{T_{\mathrm{ph}}}{\mu_{\mathrm{ph}}}\bigg/\frac{T_{\mathrm{c}}}{\mu_{\mathrm{c}}} \simeq O(10^{-3}), \tag{15}$$

where T_{ph}, μ_{ph} refer to the temperature and molecular weight at the photosurface, to terminate the integrations.

Figure 3 shows the variation with β of the moment-of-inertia factor of the non-rotating cloud, given by

$$f(0) = 2\left[\int_0^{R_{\mathrm{ph}}} \rho r^4 \mathrm{d}r\right]\bigg/ 3R_{\mathrm{ph}}^2\left[\int_0^{R_{\mathrm{ph}}} \rho r^2 \mathrm{d}r\right]. \tag{16}$$

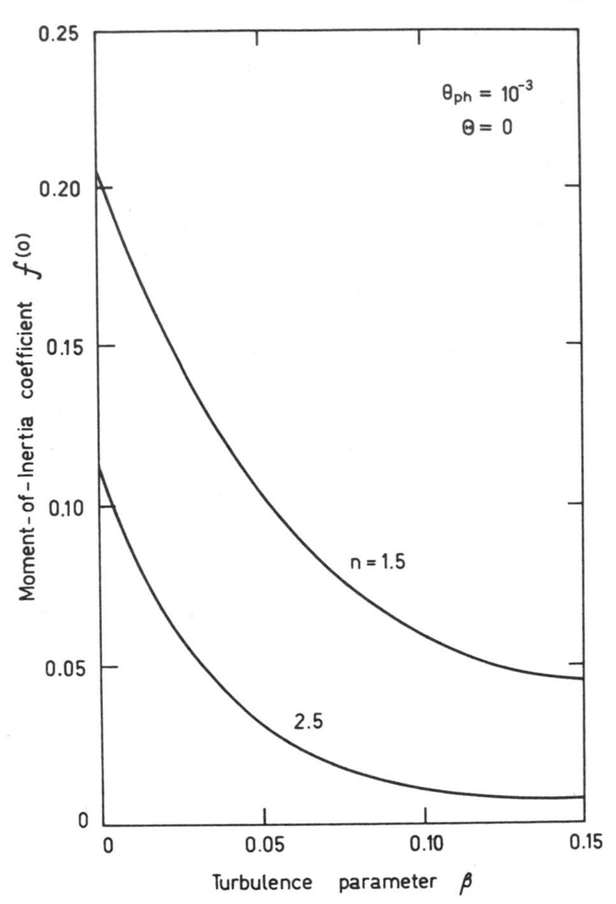

Figure 3. Moment-of-inertia coefficient $f(0)$ of non-rotating turbulent polytropic structures of index n plotted as a function of the turbulence parameter β

We observe that as β increases f steadily declines. That is, the introduction of turbulent stress causes the star to appear much more centrally condensed. This reduction in f is achieved through the non-uniform spatial distribution of turbulent stress which causes the outer less dense layers of the star to be pushed outwards proportionally very much further than the inner dense regions which are barely disturbed at all. In fact, we note from equation (12) that the turbulent stress vanishes altogether at the centre of the cloud since $M(r)/r \to 0$ as $r \to 0$. The ratio of turbulent stress to gas pressure increases progressively as one moves away from the centre rising to a maximum value

$$P_{ph} = \beta \mu_{ph} GM/\mathcal{R}\, T_{ph} R_{ph} \tag{17}$$

at the photosurface.

Figure 4 illustrates the variation of P_{ph} with β, again for the case $\theta_{ph} = 10^{-3}$. We see that P_{ph} very quickly rises to a level of about 100 for the typical case $n = 1{\cdot}5$, corresponding to a fully ionized cloud in adiabatic convective equilibrium. For the case $n = 2{\cdot}5$, which corresponds to a fully undissociated cloud of molecular hydrogen, the plateau level for P_{ph} is about 40. In both instances such a large value of P_{ph} implies the existence of non-thermal motions at the photosurface which have velocities about 10 times the adiabatic sound speed v_s, yielding values of order $100\,\mathrm{km\,s^{-1}}$ for stars of solar mass and radius, in agreement with the T-Tauri stars observations. Choosing a smaller value for θ_{ph} than 10^{-3} results in a correspondingly higher value of $P_{ph} \propto \theta_{ph}^{-1}$, so that the choice $\theta_{ph} = 10^{-3}$ is in fact the correct characteristic value for this parameter.

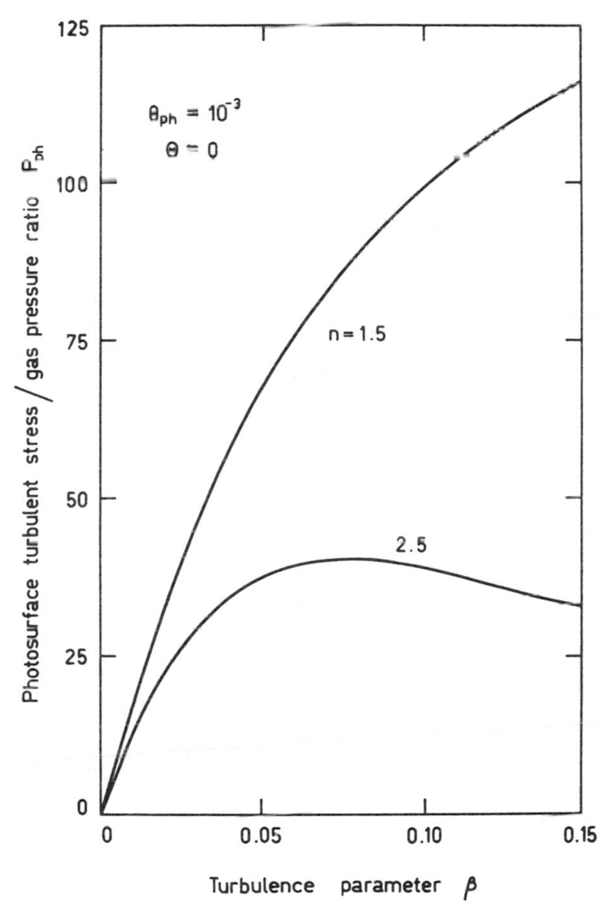

Figure 4. The ratio P_{ph} of turbulent stress $\beta\rho GM(r)/r$ to gas pressure $\rho RT/\mu$ at the photosurface of non-rotating turbulent polytropic structures, plotted against turbulence parameter β

The Overshoot Zone

The distribution of density and temperature with dimensionless radius x, expressed in Emden units, in the outermost layers of a non-rotating turbulent polytrope of index $n = 1{\cdot}5$ and turbulence parameter $\beta = 0{\cdot}01$ are shown in Figure 5. The position of the photosphere is denoted by the point x_{ph}, whilst x_s denotes the radius of the sphere beyond which all convective overshooting ceases and the material becomes non-turbulent and stationary. Since the excess heat of the convective elements can be freely liberated as soon as they cross the photosphere, we shall assume that the width of the overshoot zone $R_s - R_{ph}$ does not exceed more than a few photospheric scale heights h_{ph} given by

$$h_{ph} = \mathcal{R}\, T_{ph} R_{ph}^2/\mu GM = \beta R_{ph}/P_{ph} \simeq 10^{-3} R_{ph}, \tag{18}$$

using equation (17). The thickness of the overshoot zone is therefore assumed

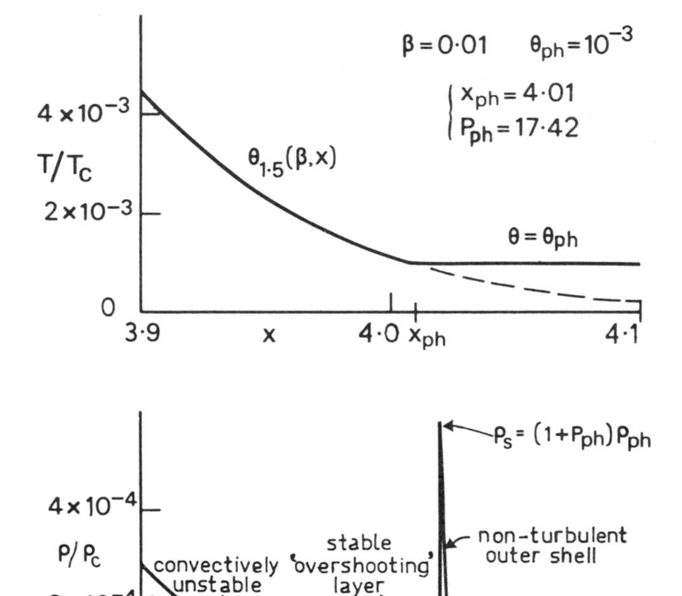

Figure 5. The temperature and density profiles in the outermost layers of a non-rotating turbulent polytropic protosun. When the temperature levels off at the photosurface, indicated by the point x_{ph}, the turbulent convective stress dies down and a very steep density inversion takes place at the top of the overshoot layer, at x_s, in order to maintain overall pressure equilibrium

to be quite negligible compared to the radius of the star. Physically this means that the deceleration of the rising supersonic convective elements occurs very strongly and non-adiabatically once they cross the photosurface, thus resulting in the liberation of large quantities of heat. This picture for a very powerful dumping of mechanical energy at the surfaces of T-Tauri stars is supported by observations (Dr. Gähm, private communication).

The Dense Outer Shell of Non-Turbulent Gas

Consider now the density distribution in the overshoot interval and beyond. The temperature distribution beyond the photosurface may be treated as being roughly isothermal as indicated in the top half of Figure 5. If we suppose for the sake of argument that the ratio of turbulent stress to gas pressure $P_t = \langle \rho_t v_t^2 \rangle \mu / \rho \mathscr{R} T$ dies down roughly linearly with distance in the overshoot zone $R_{ph} \leqslant r \leqslant R_s$, the equation of hydrostatic support can be integrated exactly yielding the solution

$$\rho(r) = \rho_{ph} \left[\frac{1 + P_{ph}}{1 + P_{ph}(R_s - r)/(R_s - R_{ph})} \right], \tag{19}$$

where ρ_{ph} denotes the density at the photosurface. Inspecting this solution we observe, as shown in the Figure, that if $P_{ph} \gg 1$ the density remains practically constant over the whole overshoot zone then rises very sharply at the end to the value

$$\rho_s = \rho_{ph}[1 + P_{ph}] \simeq 10^2 \rho_{ph} \tag{20}$$

over a very short distance $\Delta r \simeq h_{ph}/\ln P_{ph} \sim 10^{-4} R_{ph}$. That is, when the turbulent stress dies down a very steep density inversion occurs. The reason for this density inversion is to allow for the gas pressure $\rho \mathscr{R} T/\mu$ to compensate for the drop in turbulent stress so that total pressure equilibrium is maintained across the overshoot layer.

Beyond the stopping radius R_s the gas is wholly non-turbulent and integration of equation (13), setting $p_{tot} = \rho \mathscr{R} T/\mu$ and $M(r) \simeq M$, yields the atmospheric solution

$$\rho(r) =^{\pm \geqslant R_s} \rho_s \exp[-(r - R_s)/h_{ph}]. \tag{21}$$

That is, the density falls away exponentially over a scaling height h_{ph} given by equation (18).

Conclusions

In conclusion, therefore, we have seen that the presence of a supersonic turbulent stress $\beta \rho\, GM(r)/r$

(i) leads to a drastic lowering of the moment-of-inertia coefficient of the protostar,

(ii) leads to the formation of a very dense but thin shell of non-turbulent gas above the photosurface.

The mass of the non-turbulent outer shell is

$$m_{sh} = 4\pi\rho_{ph}(1 + P_{ph})R_{ph}^2 h_{ph} \simeq 4\pi\beta\rho_{ph}R_{ph}^3. \tag{22}$$

Structure of the Rotating Turbulent Protosun

Consider now the influence of rotation on the structure and evolution of the energetically stabilized protosolar envelope as it contracts to the radius R_0, corresponding to the orbit of Neptune, where the centrifugal force at the equator first balances the gravitational force there.

Turbulent Viscosity

Now as soon as the envelope has become energetically stabilized and acquires a supersonic convective equilibrium, the supersonic turbulent motions of the convective elements creates a turbulent viscosity

$$\eta_t = \tfrac{1}{3}\rho_t v_t \lambda = \tfrac{1}{3}\rho[\beta f_e GM(r)/r]^{1/2}, \tag{23}$$

where the various symbols have the same meaning as in equation (12). This viscosity tends to eliminate differential rotations in the convective interior over a timescale

$$\tau_t = \rho r^2/\eta_t = [3r^3/f_e\beta GM(r)]^{1/2} \tag{24}$$

which is of the same order as the free fall time scale $\tau_{ff} = \sqrt{\{r^3/GM(r)\}}$. Hence as long as the rate of contraction at the surface occurs much more slowly than the free fall rate, which the calculations presented earlier certainly indicate (see Table 2 above) we may conclude that the convective interior of the protosun and its upper overshoot zone rotate like a rigid body with a nearly uniform angular velocity, $\omega_e(R_e)$ say. Here the equatorial radius R_e is defined to be that of the outer edge of the overshoot layer, discussed previously, where all turbulent motions cease.

The Uniformly Rotating Interior

For a rotating centrally condensed envelope the positions of the outermost surfaces of constant density, pressure etc. can be determined from those of the non-rotating model using the atmospheric approximation as follows. The mass interior to any surface in the outer layers is nearly a constant, M say. The equation of hydrostatic support becomes

$$\frac{1}{\rho}\nabla p(s,z) = -\frac{GM}{R_e^2}\left[\frac{R_e^2\hat{\mathbf{r}}}{r^2} - \frac{\Theta\mathbf{s}}{R_e}\right], \tag{25}$$

where (s, z) are cylindrical polar co-ordinates referred to the axis of rotation, so $r = \sqrt{s^2 + z^2}$, and

$$\Theta = \omega_e^2 R_e^3 / GM \tag{26}$$

is the so-called rotation parameter. This parameter measures the ratio of centrifugal force to gravitational force at the equator. The solution of equation (25) may be formally written

$$p = p(\psi), \quad \rho = \rho(\psi), \quad T = T(\psi), \tag{27}$$

where the equipotential function $\psi = \psi(s, z)$ is given by

$$\psi(s, z) = \left(\frac{3}{2 + \Theta}\right)\left[\frac{R_e}{r} - 1 + \tfrac{1}{2}\Theta\left(\frac{s^2}{R_e^2} - 1\right)\right]. \tag{28}$$

Since $\psi(R_e, 0) = 0$, the surface of the protosolar envelope is defined by equipotential $\psi = 0$ as indicated by the heavy line in Figure 6. It also follows from equation (28) that the equatorial radius R_e exceeds the polar radius R_p by the factor

$$R_e/R_p = [1 + \Theta/2]. \tag{29}$$

That is, rotation causes the envelope to bulge outwards at the equator whilst

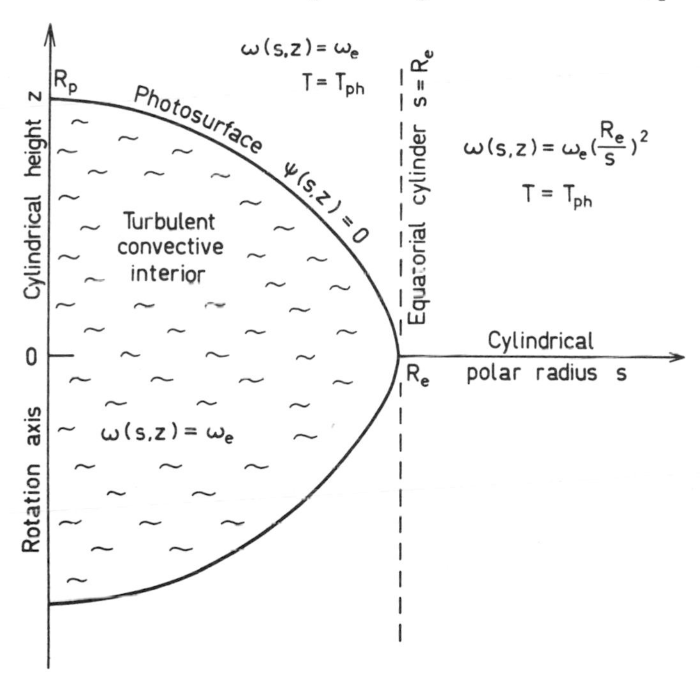

Figure 6. The cylindrical co-ordinate system (s, z) of the rotating protosun and the specifications of temperature and angular velocity $\omega(s, z)$ outside the photosurface $\psi(s, z) = 0$

the polar regions remain hardly changed. This result is valid as long as the mass distribution of the envelope is very centrally condensed so that the atmospheric approximation applies (Monaghan and Roxburgh, 1965).

Lastly, since the innermost regions of the centrally condensed envelope are barely disturbed by the rotation, the moment-of-inertia coefficient f declines with Θ according to the formula

$$f(\Theta) = f(0)\left(\frac{R_p}{R_e}\right)^2 = f(0)/(1+\Theta/2)^2. \tag{30}$$

Thus for the fully rotating ($\Theta = 1$) turbulent polytropic structures we considered in the previous section, with the typical parameters $\beta = 0{\cdot}1, \theta_{ph} = 10^{-3}$, we have

$$0{\cdot}005 < f(1) < 0{\cdot}026 \tag{31}$$

according as $2{\cdot}5 > n > 1{\cdot}5$. A typical mean value for $f(1)$ is thus 0·01, precisely in accord with that required for a modern Laplacian hypothesis.

The Outer Rotating Shell

Since the dense shell of non-turbulent gas outside the protosolar surface $\psi = 0$ is free of turbulent viscosity, we cannot assume that it will co-rotate with the convective interior. In fact since the turbulent viscosity is zero there, any fresh material extruded across the outer surface during the contraction will conserve its angular momentum once it leaves the protosun. Thus for material which crosses the surface $\psi = 0$ and moves parallel to the axis of rotation we have $\omega(s, z) = \omega_e(R_e)$, the same as the protosun's own angular velocity, and for material which crosses the equatorial cylinder $s = R_e$ and moves away from the axis of rotation we have $\omega(s, z) = \omega_e R_e^2/s^2$, as indicated in Figure 6. Hence if all of the material which makes up the non-turbulent atmosphere originates from the interior, a good representation of the angular velocity distribution is given by

$$\omega(s, z) = \begin{cases} \omega_e & \text{if } s \leqslant R_e \\ \omega_e R_e^2/s^2 & \text{if } s > R_e. \end{cases} \tag{32}$$

Next noting that the temperature outside the protosun is sensibly constant everywhere compared to the changes which occur in the density, the equation of hydrostatic support may be integrated exactly to give

$$\rho(s, z) = \rho_e(R_e) \exp\left[\alpha(1)\psi\right], \tag{33}$$

where for $s \geqslant R_e$ we have

$$\psi(s, z) = \left(\frac{3}{2+\Theta}\right)\left[\frac{R_e}{r} - 1 + \tfrac{1}{2}\Theta(1 - R_e^2/s^2)\right], \tag{34}$$

whilst for $s < R_e$, $\psi(s, z)$ has the same form as in equation (28). In equation (33),

ρ_e refers to the density of non-turbulent gas at the equator and is the same as ρ_s in equation (20). The dimensionless parameter $\alpha(1)$ is defined by the equations

$$\alpha(\Theta) = \mu GM/\mathscr{R} T_e R_e = P_{ph}/\beta[1 + \Theta/2] \sim 10^3, \tag{35}$$

where P_{ph} is the ratio of turbulent stress to gas pressure at the polar photosurface and is the same as that for the non-rotating star of the same radius $R_{ph} = R_p$ and surface temperature $T_e = T_{ph}$. During a homologous contraction T_e varies as $1/R_p$ so that $\alpha(1)$ is a constant independent of R_p.

Formation of a Dense Equatorial Belt

Meridional profiles of various equipotential surfaces of the cloud for various degrees of rotation corresponding to different stages of the contraction, described in detail shortly, are shown in Figure 7. As the degree of rotation increases during the contraction, the original nearly spherical shell of non-turbulent gas distends outwards at the equator acquiring a belt-like appearance. The volume enclosed between neighbouring equipotential surfaces increases dramatically as $\Theta \to 1$, corresponding to the curves marked $R_p/R_{po} = 1$ in the diagram, so that a lot of fresh material has to be extruded from the convective interior of the star to the equator during the contraction in order to maintain pressure equilibrium at the surface. It is this constant extrusion of gas which we feel justifies our use of the angular velocity law given by equation (32). For the fully rotating protostar, the density distribution outside the equatorial cylinder can be expressed in terms of the local co-ordinates at the equator approximately as

$$\rho(s, z) \simeq^{s \geqslant R_e} \rho_e \exp[-\tfrac{1}{2}\alpha(1)[(s - R_e)^2 + z^2]/R_e^2]. \tag{36}$$

This yields an estimate of the total mass of the equatorial belt, viz.

$$m(1) = 2\pi^2 \rho_e R_e^3/\alpha(1) = \frac{81\pi}{32} m_{sh}, \tag{37}$$

where m_{sh}, given by equation (22), is the total mass of the shell of non-turbulent gas surrounding the non-rotating star of the same polar radius. Thus the mass of the equatorial belt of non-turbulent gases of the fully rotating star exceeds that of the entire spherical shell of the non-rotating star by nearly an order of magnitude. The mass of non-turbulent material interior to the equatorial cylinder is comparable with $m(1)$.

We conclude that rotation not only causes the original spherical shell to evolve into a belt-like structure at the equator, but it also greatly enhances the mass of non-turbulent gas which can be stored outside the protostellar surface.

Evolution to the First Point of Rotational Instability

We are now in a position to compute the evolutionary rotational locus followed by the protosun during its contraction to equatorial radius R_0 where

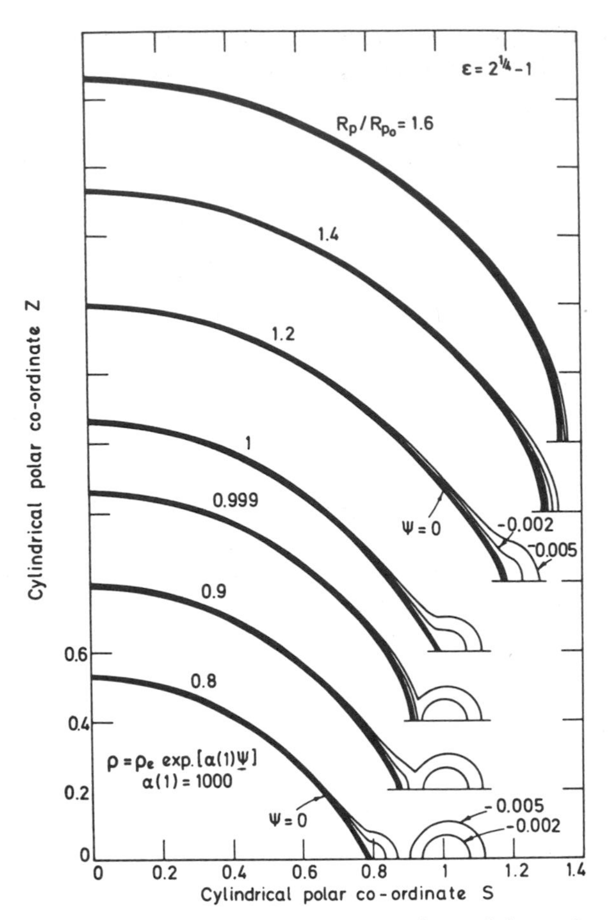

Figure 7. Meridional (polar) cross sections of the equipotential surfaces $\psi(s, z)$ of the non-turbulent outer atmosphere of the rotating protosolar envelope at various stages of the contraction, measured in units of the dimensionless polar radius R_p/R_{po}. As $R_p/R_{po} \to 1$, the rotation parameter Θ of the envelope tends to unity and the dense outer non-turbulent shell of gas evolves into a belt-like structure at the equator. This belt of gas is then discontinuously abandoned by the protosun as R_p shrinks below R_{po}

$\Theta(R_0) = 1$. The equations which determine this locus are those of the conservation of the total mass and angular momentum. Letting $M(\Theta)$ and $f(\Theta)$ denote the total mass and moment-of-inertia coefficient of the material interior to the equatorial cylinder and $m(\Theta)$ the mass exterior to this surface, we have using equation (32)

$$M(\Theta) + m(\Theta) = M_i \tag{38}$$

$$[M(\Theta)f(\Theta) + m(\Theta)]\omega_e R_e^2 = L_i, \tag{39}$$

where M_i, L_i are the total initial mass and angular momentum. Since R_e depends on Θ, through equation (29), it is convenient to choose the polar radius R_p as the independent variable governing the size of the configuration.

For the protosolar models of physical interest we have $f(1) \ll 1, m(1)/M(1) \ll 1$ and so $M(\Theta) \simeq M_i = M$ say. Now eliminating ω_e and R_e using equations (26) and (29), equation (39) may be cast in the form

$$\Theta(R_p) = \frac{R_{po}}{R_p}\left(\frac{3}{2+\Theta}\right)\left\{\frac{1+\varepsilon}{f(\Theta)/f(1) + \varepsilon m(\Theta)/m(1)}\right\}^2, \tag{40}$$

where we define

$$\varepsilon = m(1)/Mf(1) \tag{41}$$

and $R_{p_0} = \frac{2}{3} R_0$ is the polar radius corresponding to the fully rotating state $\Theta(R_0) = 1$. The mass $m(\Theta)$ can be computed from equations (33) and (34) by expanding $\psi(s, z)$ to second order in powers of the local co-ordinates at the equator, yielding the result

$$m(\Theta) \simeq m(1)\left(\frac{2+\Theta}{3}\right)^4 \exp\left[\frac{\alpha(\Theta)}{2}(1-\Theta)^2\left\{1 - \text{erf}\left[\left(\frac{\alpha(\Theta)}{2}\right)^{1/2}(1-\Theta)\right]\right\}\right], \tag{42}$$

where $\text{erf}(x)$ is the error function.

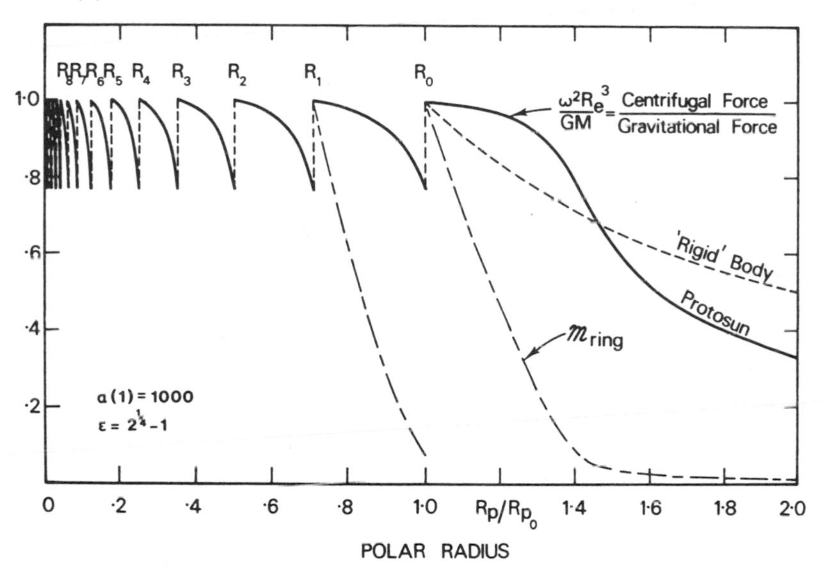

Figure 8. The equatorial rotation parameter $\Theta = \omega_e^2 R_e^3/GM$ and non-turbulent equatorial ring mass $m_{ring} = m(\Theta)/m(1)$ plotted against dimensionless polar various R_p/R_{p_0} at various stages of the contraction. (Reproduced from A. J. R. Prentice, 'Formation of planetary systems' in J. P. Wild (ed.), *In the Beginning...*, by permission of Australian Academy of Science, 1974)

Figure 8 shows the run of Θ and $m(\Theta)/m(1)$ with decreasing polar radius R_p, normalized against R_{po}, for the typical case $\alpha(1) = 1000$ and $\varepsilon = 2^{1/4} - 1$. If neither $m(\Theta)$ or $f(\Theta)$ were to change during the contraction then Θ would follow the locus given by the dotted line in the diagram, being the function R_{po}/R_p. This curve is the zeroth-order solution of equation (40) corresponding to the contraction of a protostar which alters neither its shape or mass distribution. In fact, both $m(\Theta)$ and $f(\Theta)$ do change during the contraction as the increase in the degree of rotation causes the protosun to evolve from being nearly spherical at polar radius $2R_{po}$ to lenticular in shape at radius R_{po}. The growth in Θ as $R_p \to R_{po}$ is in fact heavily curtailed in the final stages by the copious transferral of mass to the equator to build up $m(\Theta)$, as mentioned earlier. It is this outward transferral, or extrusion, of mass from the deep interior of the protostar to the equator under conditions of nearly uniform rotation maintained by the supersonic turbulent viscosity, which allows the protosun to rid its angular momentum and be rotationally braked.

Shedding of the System of Gaseous Rings

After the protosun has reached the critically rotating configuration $\Theta = 1$, the central regions of the cloud continue to contract safely inwards since they are barely disturbed by the centrifugal force. We assume that the gravitational energy released through the contraction of these central regions maintains the supersonic convective equilibrium throughout the protosolar interior, so that the internal distributions of density with pressure are preserved, apart from modest changes due to the steady dissociation of the H_2. This means that the contraction of the convective interior may be considered to occur homologously, with the density everywhere scaling as R_p^{-3} and the temperature as R_p^{-1}.

The Pressure Equilibrium Condition

Consider now Figure 9 which schematically illustrates the run of total pressure with radial distance in the equatorial plane at various stages of contraction near the radius $R_e = R_0$. If no new material were to be extruded across the equatorial cylinder $s = R_e$ then the non-turbulent gases existing outside the cylinder $s = R_0$ would freely expand into the cylindrical annular cavity $R_e < s < R_0$ vacated by the contracting protosun. The angular velocity of this inflowing material would adopt the distribution

$$\omega_{\rm ring}(s, z) = \omega_0 R_0^2/s^2, \quad \omega_0^2 = GM/R_0^3, \tag{43}$$

appropriate to the conservation of local angular momentum, the same as in equation (32). The density distribution would then be the same as in equations (33) and (34), setting $\Theta = 1$, yielding

$$\rho_{\rm ring}(s, z) \simeq \rho_0 \exp(-\tfrac{1}{2}\alpha_0 \xi^2/R_0^2), \quad \alpha_0 = \frac{\mu GM}{\mathcal{R} T_0 R_0}, \tag{44}$$

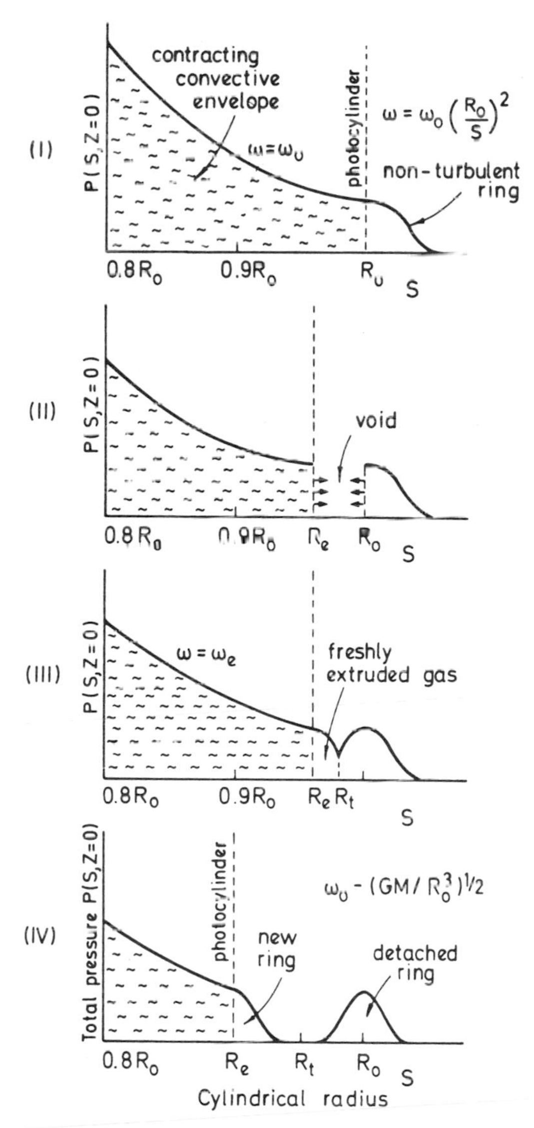

Figure 9. Schematic view of the pressure changes which take place in the equatorial plane plotted against cylindrical distance s from the centre as the protosun contracts through the first critically rotating state, given when the equatorial radius $R_e = R_0$. (Reproduced from A.J.R. Prentice, 'Formation of planetary systems' in J. P. Wild (ed.), *In the Beginning...*, by permission of Australian Academy of Science, 1974)

where $\xi = \sqrt{(s - R_0)^2 + z^2}$ is the local meridional distance from the circular Keplerian orbit $s = R_0$. That is, the material of the original equatorial belt of gas existing beyond the cylinder $s = R_0$ distributes itself uniformly about this radius to form a gaseous torus or ring. The density distribution in this ring falls off with distance ξ from the maximum ρ_0 on the circular Keplerian axis $s = R_0$, $z = 0$ in a nearly gaussian manner. The effective minor radius of the ring is thus $\xi_0 = R_0/\sqrt{\alpha_0} \simeq 0{\cdot}03\, R_0$.

The temperature T_0 on the circular orbit R_0 hardly differs from that T_e, at the protosolar equator, for R_e close to R_0 though for large separations it is necessary to use a black body law $T_0 \propto R_p^{-1/2}$. Even so, the spatial variation in T is negligible compared to those in ρ and p, so that outside the protosun the density and pressure $p = \rho \mathscr{R} T/\mu$ are really one and the same function. Now since $\rho_{\rm ring}(s, z) < \rho_0 < \rho_e(R_0)$ and $\rho_e(R_0) < \rho_e(R_e)$ if $R_e < R_0$, where $\rho_e(R_e)$ is the density at the equator, it follows that there cannot exist a pressure equilibrium between the gases of the ring and those of the protosun on the equatorial cylinder $s = R_e$. To achieve an equilibrium it is necessary for the protosun to extrude a new layer of material into the region between the equator and the gaseous ring. If we let $s = R_t$ denote the cylindrical radius of the outer boundary of this new layer, as shown in Figure 9 above, and let $\omega_n(s, z)$, $\rho_n(s, z)$ denote its angular velocity and density distributions, we have

$$\omega_n(s, z) = \omega_e(R_e) R_e^2/s^2 \tag{45}$$

$$\rho_n(s, z) = \rho_e(R_e) \exp\left[\alpha(1)\psi\right]. \tag{46}$$

where $\alpha(1)$, $\psi = \psi(s, z)$ are defined in equations (34), (35).

The location of the cylindrical surface $s = R_t$ separating the new atmosphere of non-turbulent gas from the gaseous ring can be determined from the pressure equilibrium condition $\rho_n(R_t, z) = \rho_{\rm ring}(R_t, z)$ with the result

$$R_t = R_0 \left[\frac{R_e(1 - \Theta R_e/R_0)}{R_0(2 - \Theta - R_e/R_0)} \right]^{1/2}. \tag{47}$$

For Θ close to unity we have the approximate relation $R_t \simeq \sqrt{(R_e R_0)}$, which means that the separating surface $s = R_t$ lies roughly midway between equatorial cylinder $s = R_e$ and the cylinder $s = R_0$.

The Equations of Mass and Angular Momentum

Letting

$$m_t = \int_{-\infty}^{\infty} \int_{R_e}^{R_t} \rho_n(s, z)\, s\, \mathrm{d}s\, \mathrm{d}z, \tag{48}$$

denote the mass of newly extruded material, the equations which govern the rotational evolution of the protosun are, as before,

$$M(\Theta) + m_t(\Theta) = M_0 \tag{49}$$

$$[M(\Theta)f(\Theta) + m_t(\Theta)]\omega_e R_e^2 = L_0, \tag{50}$$

where M_0, L_0 are the total residual mass and angular momentum of the protosun interior to the equatorial cylinder when $R_e = R_0$. From equations (39) and (43) it follows that

$$L_0 = Mf(1)\sqrt{(GMR_0)} = L_i/[1 + m(1)/Mf(1)] \tag{51}$$

which means that in extruding its first gaseous ring during the contraction to radius R_0 the protosun reduced its angular momentum by the factor $1/[1 + m(1)/Mf(1)]$.

Proceeding now as in the section about *The Rotating Turbulent Protosun* with the assumption $m_t \ll M_0 \simeq M$ we arrive at a single equation for Θ in terms of the polar radius R_p, viz.

$$\Theta(R_p) = \frac{R_{po}}{R_p}\left(\frac{3}{2+\Theta}\right)\left\{\frac{f(\Theta)}{f(1)} + \varepsilon\frac{m_t(\Theta)}{m(1)}\right\}^{-2}, \tag{52}$$

where ε is defined by equation (41). The numerical solution of this equation is shown in Figure 8 above as the set of curves between the points marked R_1, R_0, for the case $\varepsilon = 2^{1/4} - 1$ and $\alpha(1) = 1000$.

We observe with interest that as the polar radius R_p contracts infinitesimally below the first critical point R_{po}, where $\Theta = 1$, the rotation parameter Θ switches discontinuously to a new equilibrium value $\sim 0{\cdot}8$. That is, as R_p passes through R_{po} the envelope of the protosun undergoes a massive and abrupt change, with the equatorial regions withdrawing cataclysmically to a new equilibrium radius $R_e \simeq 0{\cdot}9\,R_0$, using equation (29), as illustrated in Figure 7.

The Catastropic Ring Detachment Process

The physical reason for this catastropic behaviour lies in the critical nature of the fully rotating protosun. When a small amount of gas Δm_t is transferred to the equator of the fully rotating star, the angular momentum of the convective interior is reduced by the amount $\Delta L = L_0 \Delta m_t/Mf(1)$ whilst Θ, using equations (30) and (52), is reduced by the amount $\Delta\Theta = (6\Delta m_t/Mf(1)^{1/2}$. That is, for $\Theta = 1$ a second order change in Δm_t induces a first order change in Θ and hence R_e. But for pressure equilibrium to be maintained at the equatorial cylinder we require $\Delta m_t \propto (R_t - R_e)\rho_e(R_e)R_e^2 \propto \Delta\Theta\, m(1)$, using equations (37) and (47), which is of the order of $\Delta\Theta$ not $(\Delta\Theta)^{1/2}$. Thus as $\Delta m_t \rightarrow 0$ there can exist no continuous solution for Θ, with the property $\Theta \rightarrow 1$, which simultaneously satisfies both the equations of conservation of angular momentum and pressure equilibrium. Therefore Θ jumps to a different solution branch.

As the equatorial regions of the convective envelope withdraw dynamically on the turbulent response timescale τ_t, given by equation (24), to the new equilibrium radius R_e, the dense belt of non-turbulent gas of mass $m(1)$ existing outside the cylinder $s = R_0$ expands freely inwards to form a uniform circular gaseous ring, as we described earlier. The protostar adjusts to the new equilibrium state by extruding sufficient material m_t, shown in Figure 8, to establish pressure equilibrium at its equator, rotationally stabilizing itself in the process. It is a simple matter to show that the density ρ_t at the interface $s = R_t$ between the new atmosphere m_t and the gaseous ring is negligible compared to the central ring density ρ_0. The protosun therefore literally detaches itself from the dense equatorial belt of non-turbulent gases as soon as it passes through the critical point R_0. Nevertheless, unless the turbulence in the protosun is sufficiently strong (see below) m_t is never large enough for this discrete detachment to occur and the interior of the star becomes rotationally disrupted.

The Titius–Bode Law

After the protosun has shed its first gaseous ring it resumes its quasistatic contraction until the stage is reached when the rotation parameter again rises to unity and a second ring of gas is disposed. The equatorial radius R_1 at the second unstable point can be computed directly from equations (50) and (51), setting $\omega_e^2 = GM/R_1^3$, and is given by

$$R_0/R_1 = [1 + m(1)/Mf(1)]^2 = [1 + \varepsilon]^2, \tag{53}$$

noting that $m_t(1) \simeq m(1)$ if the collapse is homologous. The process of ring growth and detachment continues repeatedly until the radial collapse of the protosun is halted at normal stellar size.

Hence as long as the gravitational contraction of the turbulent protosun proceeds more or less homologously, meaning that $m(1)/M$ and $f(1)$ remain constant, then the ratio of the orbital radii R_n of successively disposed gaseous rings is a constant, viz.

$$R_n/R_{n+1} = \left[1 + \frac{m(1)}{Mf(1)}\right]^2 = \text{constant}, \tag{54}$$

as seen in Figure 8 above, where $R_n/R_{n+1} = 1/\sqrt{2}$. The sequence of orbital radii $R_n (n = 0, 1, 2, 3, \ldots)$ form a geometric progression precisely similar to the Titius–Bode law of planetary distances, given in equation (1). Relating $\varepsilon = m(1)/Mf(1)$ to the physical characteristics of the non-rotating turbulent polytrope of index n we obtain

$$\varepsilon = m(1)/Mf(1) = \pi\left(\frac{3}{2}\right)^7 \left[\beta + 1/\alpha(0)\right] \frac{\rho_c}{\bar{\rho}} \frac{\theta_{ph}^n}{f(0)}, \tag{55}$$

where $\rho_c/\bar{\rho}$ is the ratio of central to mean densities.

The Role of Turbulent Stress

Figure 10 shows the run of the Titius–Bode constant $\Gamma = (1 + \varepsilon)^2$ with turbulence parameter β for the typical case $\theta_{ph} = 10^{-3}$ and $n = 1{\cdot}5$, $2{\cdot}5$. We see that in order to obtain a value R_n/R_{n+1} equal to the observed mean value $\langle R_n/R_{n+1}\rangle_{\text{obs}} = 1{\cdot}73$, we shall require a value of β in the range 0·1 to 0·13. This lies nicely within the characteristic range of this parameter expected in our theory of supersonic turbulent convection. More importantly, letting $\beta \to 0$ we see that the spacing between adjacent rings steadily shrinks to essentially zero (i.e. $R_n/R_{n+1} \to 1{\cdot}001$). Thus in the absence of turbulent stress the protosun is unable to shed a discrete system of rings. In fact unless $\beta \gtrsim 0{\cdot}05$ we find that the cloud is unable to rid enough angular momentum through the process of equatorial mass shedding to remain rotationally stable. Instead, the outer convective layers of the weakly turbulent star become progressively stripped away leaving behind a vast gaseous disk with virtually no mass remaining at the centre (cf. Cameron, 1962).

For the turbulent protosun the mass and angular momentum left over after the shedding of the nth gaseous ring scale as

$$M_n = M_0\left(1 - \frac{m(1)}{M}\right)^n \simeq M_0\left(\frac{R_n}{R_0}\right)^{f(1)/2}$$

$$L_n = L_0/\left(1 + \frac{m(1)}{Mf(1)}\right)^n = L_0\left(\frac{R_n}{R_0}\right)^{1/2}$$

Taking $f(1) = 0{\cdot}01$, $R_n/R_0 = 10^{-4}$ we see that the centrally condensed protosun gives up 99 per cent of the angular momentum at the expense of losing only about 5 per cent of its original mass during the entire contraction. Supersonic turbulent stress is therefore the crucial element to understanding how the

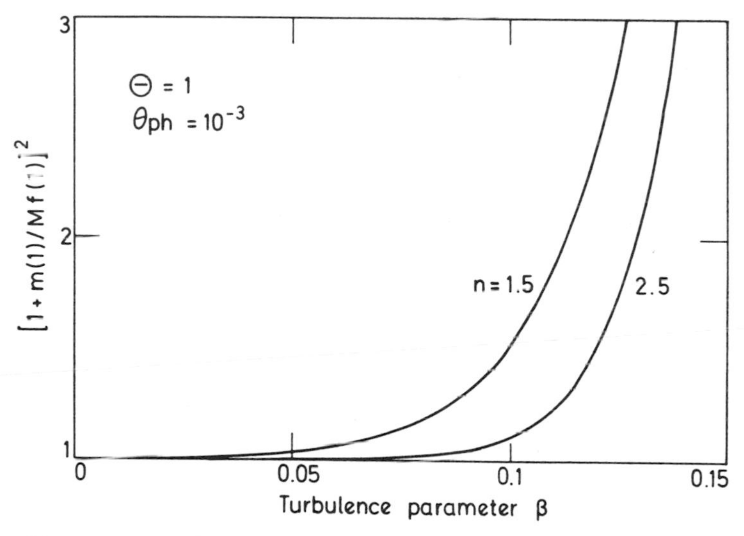

Figure 10. The Titius–Bode or ring spacing parameter $\Gamma = R_n/R_{n+1} = [1 + m(1)/Mf(1)]^2$ of fully rotating turbulent polytropic structures, plotted against turbulence parameter β

protosun both shed a discrete system of gaseous rings and contracted to its present size retaining most of its mass.

Structure of the Gaseous Rings

From equation (54) it follows that the masses of the gaseous rings are each about the same, namely

$$m_n = m(1) = 0{\cdot}003 M, \tag{56}$$

taking $f = 0{\cdot}01$, $R_n/R_{n+1} = 1{\cdot}73$. From equation (37) it follows that the central density on the circular axis of each ring is

$$\rho_0(R_n) = m(1)\alpha_n/4\pi R_n^3, \quad \alpha_n = \mu GM/\mathcal{R}\, T_n R_n, \tag{57}$$

whilst from equations (35) and (44) the effective minor radius or semi-thickness of each ring is

$$\xi_n = R_n/\sqrt{\alpha_n} \simeq 0{\cdot}03\, R_n(R_n/R_e)^{1/4}, \tag{58}$$

Here we have set $T_n \simeq T_e(R_e/R_n)^{1/2}$, using the familiar black body rule, where T_e is the temperature of the protosolar surface. The effective ring minor radius is thus initially only a few per cent of the orbital radius R_n but as the protosun contracts and gets hotter ($T_e \propto R_e^{-1}$) the gaseous rings warm up and expand as we observe in Figure 11. This diagram shows the meridional profile of the rotating protosun and its system of rings at various stages of the collapse, taking $\varepsilon = 2^{1/4} - 1$, $\alpha(1) = 10^3$. The shaded areas correspond to the regions interior to the equipotential surface $\psi = -0{\cdot}002$. For the protosun, ψ is defined by equations (28) and (34), whilst for the nth gaseous ring we have

$$\rho_{\text{ring}}(s, z) = \rho_0(R_n)\exp(\alpha_n \psi_n), \quad \omega_{\text{ring}}(s, z) = \omega_n R_n^2/s^2, \tag{59}$$

where

$$\psi_n(s, z) = R_n/r - R_n^2/2s^2 - \tfrac{1}{2}, \quad \omega_n^2 = GM/R_n^3. \tag{60}$$

We feel that Figure 11 provides a direct computational verification of the validity of the original Laplacian hypothesis.

The Chemical Condensation Sequence

After each gaseous ring has been detached by the protosun, condensation of the various elemental and chemical species takes place. Detailed models of the chemical condensation sequence in material of solar composition cooled to various temperatures and pressures have been calculated by a number of authors, notably Larimer (1967), Hoyle and Wickramasinghe (1968), Grossman (1972) and Lewis (1972 a, b).

In Table 3 we have listed the physical properties of each of the gaseous rings at the time of their detachment t_n from the protosun. This time is measured

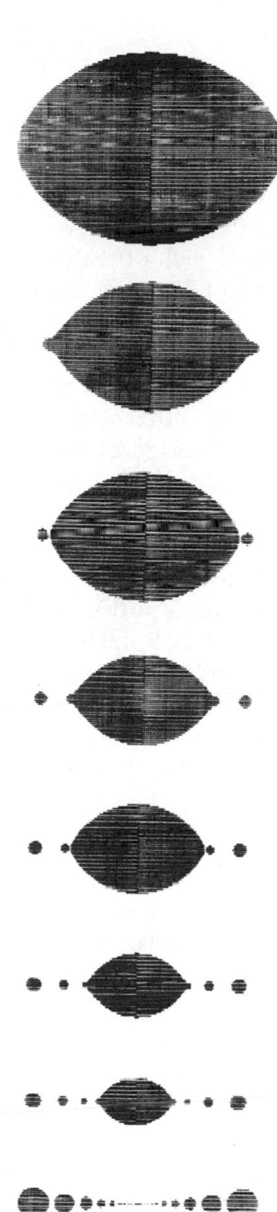

Figure 11. Meridional (polar) cross-sections of the turbulent contracting protosun and its concentric system of orbiting gaseous rings at various stages of the contraction. The shaded areas correspond to the regions interior to the equipotential surface $\psi(s, z) = -0{\cdot}002$, as defined in Figure 7 and equations (28), (34) and (60). (Reproduced from A. J. R. Prentice, 'Formation of planetary systems' in J. P. Wild (ed.), *In the Beginning...*, by permission of Australian Academy of Science, 1974)

from the protosun's contraction from radius $10^4\,R_\odot$ and is given by

$$t_n = t(R_n) = \int_{R_n}^{10^4 R_\odot} \mathrm{d}R_e / v_e(R_e) \tag{61}$$

where $v_e(R_e)$ is the inward radial velocity at the equator, discussed earlier in the section on *The Energetic Stabilization of the Protosolar Envelope.* Figures 12 and 13 show the black body temperature at each of the planetary orbits as a function of the elapsed time $t(R_e)$ and the vapour condensation temperatures for the major condensing species, derived from the data of Lewis. We assume that a gaseous ring was detached at the orbits of each of the planets, except Pluto, which is thought to be an escaped satellite of Neptune. We have also included a ring at the orbit of the fictitious planet Vulcan whose distance from the sun is supposed to be roughly half that of Mercury. According to our theory the protosun shed several rings interior to the orbit of Mercury so that it is necessary to consider why planets did not condense out at those orbits too.

The temperatures, densities, etc. given in Table 3 and the Figures are those arising from a fully rotating turbulent polytropic protosolar model specified by the parameters $\beta = 0{\cdot}12$, $\theta_{ph} = 10^{-3}$. This choice of parameters ensures that the model has a moment-of-inertia coefficient $f(1) \simeq 0{\cdot}01$ and Titius–Bode constant Γ close to the observed value 1·73. We assume a composition having hydrogen mass fraction $X = 0{\cdot}7$ and helium mass fraction $Y = 0{\cdot}3$, close to the observed values. The interior of each model of given radius R_e then consists of an inner dissociated zone of polytropic index $n = 1{\cdot}5$ surrounded by an outer zone of index $n_2 = 1{\cdot}5 + 2X/(1 + X)$ in which the H is mostly undissociated. The mass fraction of the inner dissociated zone steadily increases during the contraction as the protostar warms up. As a consequence, both $f(1)$ and Γ vary slightly as the collapse proceeds.

Table 3. Physical and chemical properties of the gaseous rings at the moment of detachment

Orbit	R_n ($R_\odot$)	t_n (yr)	T_n (K)	$\rho_0(R_n)$ (g cm^{-3})	$\log_{10} p$ (bar)	Condensing species Rocks	Condensing species Ice	m_c ($M_\oplus$)
Vulcan	40	3·0(5)[a]	2350	1·2(−5)	0·0	—	—	0
Mercury	83	2·6(5)	1260	1·0(−6)	−1·4	√	—	4
Venus	155	2·2(5)	710	1·1(−7)	−2·6	√	—	4
Earth	215	1·9(5)	530	3·5(−8)	−3·2	√	—	4
Mars	326	1·5(5)	370	0·8(−8)	−4·0	√	—	4
Asteroids	596	8(5)	215	1·1(−9)	−5·1	√	—	4
Jupiter	1120	9(3)	122	1·1(−10)	−6·3	√	√	11
Saturn	2050	3(3)	71	1·4(−11)	−7·4	√	√	12
Uranus	4120	2(3)	39	1·3(−12)	−8·7	√	√	13
Neptune	6460	1(3)	26	2·9(−13)	−9·6	√	√	13

[a]Numbers in brackets refer to power of ten., e.g. 3·0(5) means $3{\cdot}0 \times 10^5$

As the protosun approaches normal solar size (viz. $R_e = 1{\cdot}5\,R_\odot$) we find that it is not possible to maintain the condition $\theta_{ph} = 10^{-3}$ if the photosurface temperature T_e is to level off to the value of order 4500 K suggested by the calculations of Ezer and Cameron (1965). Thus at radius $R_e \simeq 20\,R_\odot$ the constraints $\theta_{ph} = 10^{-3}$, $\beta = 0{\cdot}12$ are relaxed in a manner such that the protostar moves onto the ZAMS both preserving its moment-of-inertia coefficient and with T_e approaching 4500 K. The right hand columns of Table 3 list the broad condensing chemical species at the moment of ring detachment and the total mass of those condensates, assuming a mean ring mass $m(1) = 1000\,M_\oplus$, and using Figures 12 and 13 and the heavy element abundances of Engvold and Hauge (1974).

We observe with interest that at the orbit of Vulcan the temperature is far too hot for any compound whatever to condense out. Thus for the gaseous rings interior to the orbit of Mercury we can account for the absence of any planets simply on the basis that it was too hot for any condensates to form. It is not till we move out to the orbit of Mercury that it is cool enough for refractory compounds like CaO, TiO_2, and Al_2O_3 and metallic Fe and Ni to condense out. The alkali metals Na and K can condense out as oxides at the orbit of Venus, whilst at Earth the retention of water first becomes possible

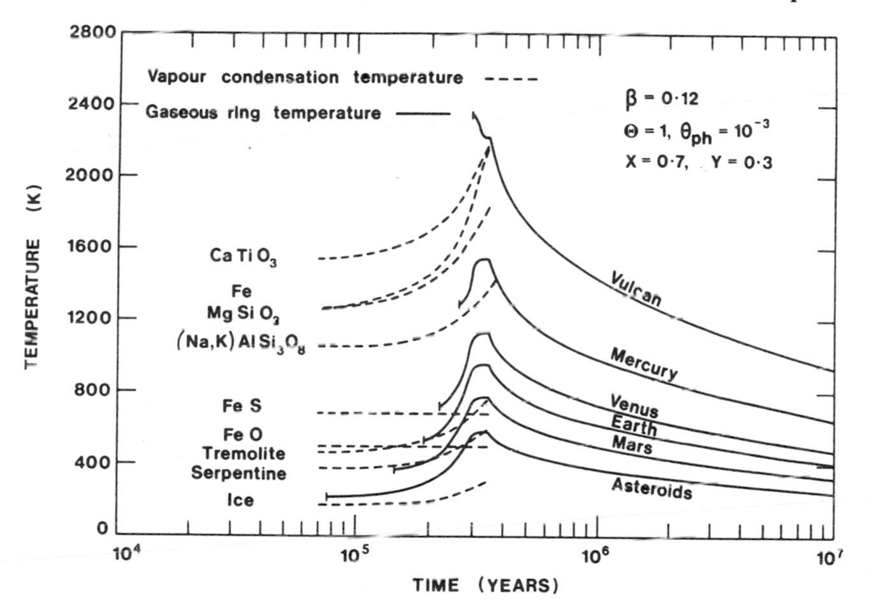

Figure 12. Temperature distributions in the inner regions of the early solar system. The full curves in the diagram give the black body temperature at the orbits of each of the terrestrial planets plotted against the time elapsed since the protosun's contraction from initial radius $10^4\,R_\odot$. The dashed lines correspond to vapour condensation temperatures of the major condensing chemical species derived from the data of Lewis (1972b), using the gas pressure shown in Table 3 characteristic of our turbulent rotating protosolar model

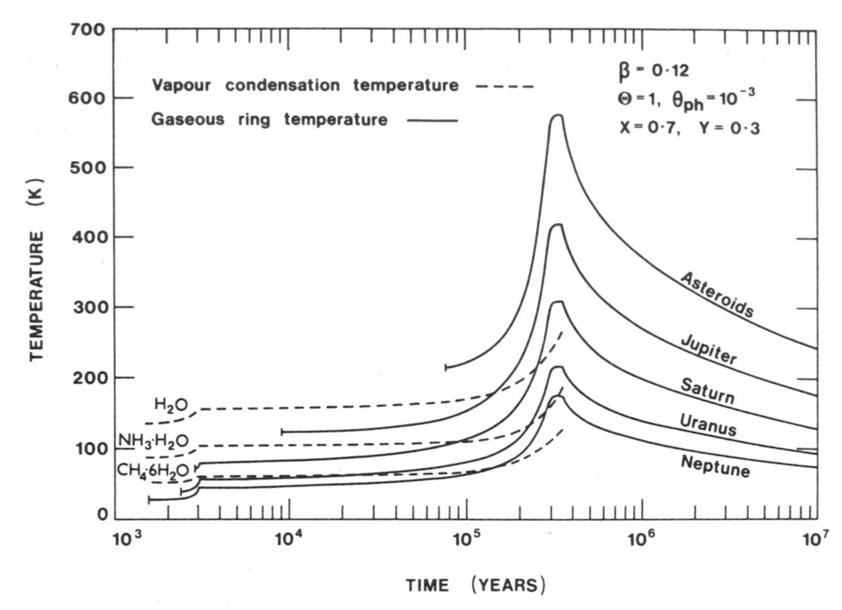

Figure 13. The temperature distributions in the outer regions of the early solar system. The notation is the same as for Figure 12, using our turbulent protosolar model and the low temperature condensation data of Lewis (1972a)

through the formation of the hydrous mineral assemblage tremolite. Sulphur condenses out here with Fe to form troilite but further out, at the orbit of Mars, and beyond, all the Fe is transformed to the oxide. Appreciable quantities of H_2O become trapped in the mineral assemblage talc at these orbits allowing us to account for the low mass density of asteroidal rock ($\rho_{ast} \simeq 2{\cdot}4\,g\,cm^{-3}$).

Stability Field of Water Ice and the Separation of Major and Minor Planets

The edge of the stability field of water ice lies between the orbits of the asteroids and Jupiter. Therein we feel lies the fundamental difference in the formation process of the inner and outer planets. Of all condensates H_2O ice is by far the most abundant, as we observe in the final column of Table 3. If all the ice and rock component in each of the outer gaseous rings were to collect together then we should find at each orbit a planetary core of mass some 10–15 $M_\oplus$, which is precisely comparable with the masses of Neptune and Uranus. This suggests that the first stage in the formation process of the outer planets was in fact the aggregation of such a planetary core. It is interesting to note that Dr. Stevenson (this volume) has reached a similar conclusion from entirely different considerations.

The fact that the masses of the terrestrial planets are appreciably less than the available amount of rock $\sim 4\,M_\oplus$ suggests that a large fraction of the rock component in the inner solar system may have remained suspended as a fine dust in the gas. The ice, however, could form large particles or flakes which would readily segregate to the axis of each ring. We shall discuss this point in more detail in the next section.

Conclusion

We conclude that the temperatures and densities provided by our protosolar model very satisfactorily account for the broad chemical features of the solar system.

Segregation of the Condensates onto the Circular Keplerian Axes $s = R_n$

Consider now the equation of motion of the condensates in each gaseous ring. As soon as a group of atoms condense out of the gas they cease to experience the gas pressure gradient which maintains the circular orbital motion of each gaseous element with the angular velocity and density distributions given by equations (59) and (60). Instead, for particles which lie above or below the equatorial plane $z = 0$ there is an unbalanced normal component of the

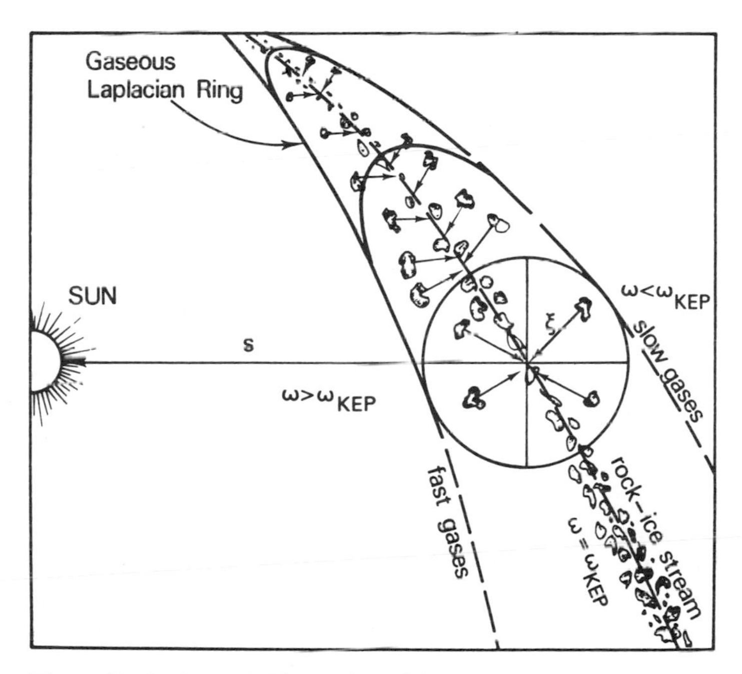

Figure 14. A schematic illustration of the gravitational settling of the condensed particles onto the circular Keplerian orbital axis of the gaseous ring to form a concentrated stream of planetesimals. (Reproduced from A. J. R. Prentice, 'Formation of planetary systems' in J. P. Wild (ed.), *In the Beginning...*, by permission of Australian Academy of Science, 1974)

sun's gravitational force, namely $-GMz/r^3$ per unit mass, which causes the particle to settle down onto the plane. And for particles of the nth gaseous ring which are born at cylindrical polar radius s which differs from the central value R_n, on which the angular velocity equals the circular Keplerian value $\omega_n = \sqrt{(GM/R_n^3)}$, there is a difference between the centrifugal force and component of gravitational force parallel to the plane, viz. $\omega^2 s - GMs/r^3 \simeq GM(R_n - s)/r^3$, which causes the particle to move towards the cylinder $s = R_n$. Thus all of the particles in each ring migrate towards their respective circular Keplerian axis $s = R_n$, $z = 0$, as indicated in Figure 14.

Influence of Gas Drag

To compute the timescale for segregation onto each Keplerian axis we have to take into account the resistive drag experienced by the particles in their passage through the gas. This aspect of the segregation process has been studied by Goldreich and Ward (1973). The type of gas drag experienced by a particle of radius 'a' depends on the gas mean free path length

$$\lambda_g \simeq 3 \times 10^{-10}/\rho_g = \begin{cases} 0{\cdot}01 \text{ cm—Earth orbit} \\ 30 \quad \text{cm—Jovian orbit,} \end{cases} \tag{62}$$

where ρ_g is the gas density, given in Table 3. For particle sizes smaller than λ_g, which is certainly the case for the early stage of segregation, the gas drag is due to independent atomic collisions and the equation of motion for a particle at position $\mathbf{r} = (s, \phi, z)$ is

$$\ddot{\mathbf{r}} = -(GM/r^2)\hat{\mathbf{r}} + \gamma_1(\mathbf{v}_g - \dot{\mathbf{r}}) \tag{63}$$

$$\gamma_1 = 3\left(\frac{\mathcal{R}T}{2\pi\mu}\right)^{1/2} \frac{\rho_g}{\rho_s a}, \quad \mathbf{v}_g = \left(\frac{h_n}{s}\right)\hat{\boldsymbol{\phi}}, \tag{64}$$

where γ_1 is called the dynamical friction coefficient, ρ_s is the particle mass density, $\mathbf{v}_g$ is the local gas velocity at the position $\mathbf{v}$, and $h_n = \sqrt{GMR_n}$ is the uniform angular momentum per unit mass of the nth gaseous ring. In writing equation (63) we have noted the gravitational force acting on the particle is primarily due to the protosun, owing to the lightness condition $m(1) \ll M$.

The azimuthal (or ϕ) component of equation (63) may be expressed in terms of the particle angular momentum per unit mass $h_\phi = v_\phi s$ as $\mathrm{d}h_\phi/\mathrm{d}t = \gamma_1(h_n - h_\phi)$. Thus since initially $h_\phi(0) = h_n$ it follows that $h_\phi(t) = h_n$ for all time. That is, the particle moves through the gas retaining its angular momentum and with angular velocity $\dot{\phi}$ matching that of the gas at each position.

Settling of the Particles onto the Circular Keplerian Orbital Axis of Each Gaseous Ring

Next letting $\boldsymbol{\xi} = (s - R_n, z)$ denote the meridional position of the particle relative to the circular Keplerian axis of the gaseous ring, as shown in Figure 14, the equation of motion in the meridional plane becomes

$$\ddot{\xi} = -\omega_n^2 \xi + \gamma_1 \dot{\xi} \tag{65}$$

Provided $\omega_n/\gamma_1 \ll 1$, the solution of this equation is simply

$$\xi(t) = \xi(0) \exp(-\omega_n^2 t/\gamma_1), \tag{66}$$

where $\xi(0)$ denotes the initial position. The particle thus sinks directly down onto the circular axis $s = R_n$ without any change in direction, as we have illustrated. The time scale for precipitation onto the axis is

$$t_{seg} = \frac{\gamma_1}{\omega_n^2} \simeq \frac{1{\cdot}1 \times 10^4}{a\rho_s}\left(\frac{R_\odot}{R_n}\right)^{1/2} \text{yr}, \tag{67}$$

where we have used equations (56) and (57), noting from Table 3 that $T_n \simeq 1{\cdot}3 \times 10^5 (R_\odot/R_n)$ K at the times of ring detachment.

Segregation Times and the Masses of the Planets

In Table 4 we have estimated t_{mig} at the typical orbits of Earth and Jupiter for particles of various sizes and appropriate compositions which are consistent with $a \lesssim \lambda_g$ and $\omega_n/\gamma_1 \ll 1$. We observe that in the case of the terrestrial orbits, where one might expect most of the rock condensate to first appear in very fine particulate form (Hoyle and Wickramasinghe, 1968), only a small fraction of the material can segregate to the circular axis of each gaseous ring in a time of order 3×10^5 yr. In the case of the outer planets, however, where one typically might expect to find ice particles (snowflakes) to have sizes in the range 0·01—1 cm, all of the available condensates, including the dust which acts as nucleation centres for the ice, will have segregated to the ring axis by 3×10^5 yr. If, therefore, as we suggest in the next section all planetary formation ceases at about the time

$$t_{max} \simeq 3 \times 10^5 \text{ yr} \tag{68}$$

when the protosun enters its most luminous phase, as shown in Figure 15, then we can understand how the major planets managed to acquire their full share of condensates while the terrestrial planets acquired only a small fraction. That is, we propose that the masses of each of the planets or planetary cores, as in the case of the major planets, are a direct function of how much of the particulate condensate material can segregate to the respective circular Keplerian orbital axis of each gaseous ring by the time $t_{max} \simeq 3 \times 10^5$ yr.

Table 4. Segregation times of the condensate material

Orbit	Condensate	ρ_s (g cm^{-3})	a (cm)	ω_n/γ_1	t_{seg} (yr)
Earth	rock	3	$10^{-4} - 10^{-2}$	$10^{-8} - 10^{-6}$	$3 \times 10^5 - 3 \times 10^7$
Jupiter	ice	1	$10^{-2} - 10^{-0}$	$10^{-5} - 10^{-3}$	$3 \times 10^3 - 3 \times 10^5$

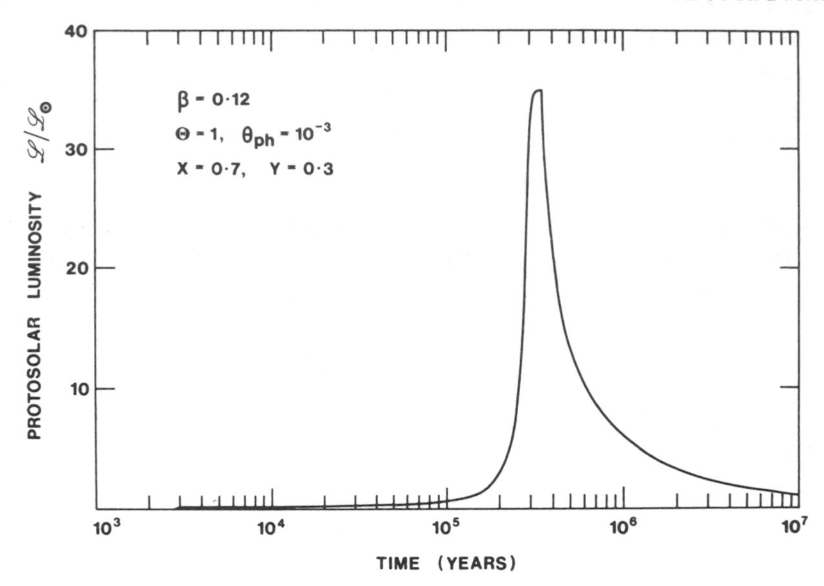

Figure 15. The protosolar luminosity as a function of elapsed time

Aggregation of the Planetesimals

We have just seen how the condensing particles in each gaseous ring precipitate down onto their respective circular Keplerian axes $s = R_n$ to form a concentrated orbiting stream of condensates, or planetesimals. Consider now the possible aggregation of the stream of planetesimals along the mean circular orbit to form single planetary objects.

As a typical planetesimal moves around the sun it interacts with the other planetesimals in its ring both through their mutual gravitational attraction and through direct collisions. The effect of collisions, in the main, is to scatter the particles off the mean orbit thus frustrating the aggregation process (Maxwell, 1859; Brahic, 1975). If we take into account the presence of the gas, however, the situation is drastically altered owing to frictional drag. As we shall see, this drag redirects or focuses the wandering particles back onto the circular Keplerian orbital axis of each gaseous ring, at the same time disposing of their excess kinetic energy.

Equation of Motion of a Single Planetesimal

Consider then in detail the equation of motion of a planetesimal which has been disturbed from the mean orbit $s = R_n$. Neglecting the mutual gravitational attraction force of neighbouring particles, for the moment, we have

$$\ddot{\mathbf{r}} = -(GM/r^2)\hat{\mathbf{r}} + \gamma_2 |\mathbf{v}_g - \dot{\mathbf{r}}|(\mathbf{v}_g - \dot{\mathbf{r}}), \tag{69}$$

where $\mathbf{v}_g = (h_n/s)\hat{\boldsymbol{\phi}}$, as before, denotes the local gas velocity, taking cylindrical polar co-ordinates (s, ϕ, z) with axis parallel to the sun's rotational axis, and where

$$\gamma_2 = \tfrac{3}{8}\frac{C_D}{a}\frac{\rho_g}{\rho_s} \tag{70}$$

is the inverse gas drag length. Here we suppose the planetesimal, of mass density ρ_s, to have a mean radius 'a' much larger than λ_g, defined by equation (62), so that the drag arises through the formation of a turbulent wake at large Reynolds number and C_D, the drag coefficient, is of order 0·3 (Goldreich and Ward, 1973).

Letting $\boldsymbol{\xi} = (s - R_n, z)$ denote the meridional co-ordinates of the planetesimal relative to the circular axis of the gaseous ring we consider, by way of illustration, the case where the particle is deflected from this axis by a collision which involves no change in its orbital angular momentum per unit mass, $h = h_n = \sqrt{(GMR_n)}$. In that case the angular velocity of the particle matches that of the gas at each point and the equation for the meridional component of motion becomes

$$\ddot{\boldsymbol{\xi}} = -\omega_n^2 \boldsymbol{\xi} - \gamma_2 |\dot{\boldsymbol{\xi}}| \dot{\boldsymbol{\xi}}, \quad \omega_n^2 = GM/R_n^3. \tag{71}$$

Integrating this equation it follows that the planetesimal performs approximately simple damped harmonic motion about the mean orbital position $\xi = 0$, with solution $\xi(t)$ eventually approaching the form

$$\xi(t)^{\omega_n t > 1} \simeq \frac{3\pi}{4\gamma_2} \frac{\cos(\omega_n t + \phi_0)}{\omega_n t}, \tag{72}$$

where ϕ_0 is a constant of order unity. We note that this asymptotic solution does not depend on the size of the initial disturbance $\xi(0)$. That is, the particle soon forgets the collision as its relaxes back onto the mean circular Keplerian orbit $s = R_n$.

Jeans Instability and Fragmentation of the Uniform Planetesimal Stream

After the various initial and collisional disturbances have died down the process of aggregation can occur. Since the total planetesimal mass m_p in any gaseous ring is so small compared to the gaseous ring mass $m(1) \simeq 10^3 M_\oplus$, as well as to $M_\odot \simeq 3{\cdot}3 \times 10^5 M_\oplus$, the planetesimals are able to gravitationally respond with one another only through their net component of force which is parallel to the circular orbit, and for which the components of the force due to the sun and the gas is zero by axial symmetry. Such a non-zero component of force arises through variations $\Delta\rho_{p\phi}$ in the azimuthal line distribution $\rho_{p\phi}$ of planetesimal mass. The system of planetesimals is in fact unstable against the formation of such line density inhomogeneities since particles are preferentially attracted towards regions of higher density from those which are sparsely populated, thus enhancing the size of the variations $\Delta\rho_{p\phi}$. That is, we have a Jeans type instability.

Hence we hypothesize that as long as the planetesimals remain bound to the circular orbital axis of the ring, fragmentation or bunching in the initially nearly uniform stream of planetesimal mass will occur leading to the formation of isolated groups of planetesimals. The distribution of these groups, or clusters, is in turn gravitationally unstable towards fragmentation leading to the formation of larger groups or superclusters. Next, provided one group of clusters outgrows the others so that a runaway takes place, the end result might be the emergence of a single large protoplanetary cloud which can proceed to 'swallow up' residual clusters and stray planetesimals.

A necessary condition for the aggregation process we have just outlined to occur is for the Jeans instability criterion to be satisfied. Letting $\dot{\eta} = v_\phi - \sqrt{(GM/R_n)}$ denote the azimuthal velocity difference between the planetesimal and the mean orbital Kepler speed, this criterion reads simply

$$\tfrac{1}{2}\langle \dot{\xi}^2 + \dot{\eta}^2 \rangle \lesssim \langle |\Delta\rho_{p\phi}|/\rho_{p\phi} \rangle Gm_p/R_n, \tag{73}$$

where the right hand side denotes the azimuthal binding energy per unit mass of the orbiting stream of planetesimals. Noting that gas drag dissipates the azimuthal velocity disturbances $\dot{\eta}$ as well as the meridional ones at the same rate, it follows from equations (72) and (73) that no fragmentation of the initial stream of planetesimals can occur until a dissipation time

$$t_{\text{diss}} \simeq [\langle \rho_{p\phi}/|\Delta\rho_{p\phi}| \rangle_0 R_n^3/Gm_p]^{1/2}/\gamma_2 R_n \tag{74}$$

has elapsed, where $\langle |\Delta\rho_{p\phi}|/\rho_{p\phi} \rangle_0$ is the initial inhomogeneity factor. Thereafter the system can respond to its own self gravitation.

The Aggregation Time

Next letting η denote the distance between a given planetesimal and the azimuthal centre of mass of the rest of the system, measured around the mean orbit $s = R_n$, the equation of motion for the migration of the particle towards this centre becomes

$$\ddot{\eta} \simeq -\langle |\Delta\rho_{p\phi}|/\rho_{p\phi} \rangle_0 Gm_p/\eta^2 + \gamma_2 \dot{\eta}^2 \tag{75}$$

Setting $\ddot{\eta} = 0$ equation (75) may be integrated exactly to yield the characteristic migration time

$$t_{\text{mig}} \simeq \tfrac{1}{2}\eta^2(0)[\langle \rho_{p\phi}/|\Delta\rho_{p\phi}| \rangle_0 \gamma_2/Gm_p]^{1/2}, \tag{76}$$

where $\eta(0)$ is the initial separation distance. Taking $\eta(0) \sim R_n$, we finally obtain an estimate of the aggregation time, viz.

$$t_{\text{agg}} = t_{\text{diss}} + t_{\text{mig}} = \left[\frac{R_n^3}{Gm_p}\left\langle\frac{\rho_{p\phi}}{|\Delta\rho_{p\phi}|}\right\rangle_0\right]^{1/2} O\left(\frac{1}{\gamma_2 R_n} + \sqrt{(\gamma_2 R_n)}\right). \tag{77}$$

We see that t_{agg} depends on the total planetesimal mass in the ring as well as the dimensionless friction parameter $\gamma_2 R_n$, which depends on the mean particle

Table 5. Aggregation times of the planetesimals

Planet	m_p $(M_\oplus)$	ρ_s $(\mathrm{g\,cm^{-3}})$	a_1 cm	$t_{agg}(a_1)$ (yr)
Mercury	0·055	3	2(5)	2(3)
Venus	0·815	3	5(4)	1(3)
Earth	1·000	3	2(4)	2(3)
Mars	0·107	3	7(3)	1(4)
Asteroids	$2{\cdot}8 \times 10^{-4}$	3	2(3)	5(5)
Jupiter	10	1	1(3)	7(3)
Saturn	10	1	2(2)	2(4)
Uranus	10	1	~ 2(2)	5(4)
Neptune	10	1	~ 1(3)	9(4)

size '*a*' as well as the gas density ρ_g. The aggregation proceeds most rapidly for particles of size

$$a_1 = 3C_D R_n \rho_g / 8\rho_s \tag{78}$$

for which $\gamma_2 R_n \simeq 1$ and t_{agg} is a minimum. In Table 5 we have calculated a_1 and $t_{agg}(a_1)$ at each of the planetary orbits, setting m_p equal to the observed mass of planetary condensate. Thus in the case of the major planets we have $m_p \simeq 10\,M_\oplus$, as discussed earlier. We have also set the initial inhomogeneity factor $\langle |\Delta\rho_{p\phi}|/\rho_{p\phi}\rangle_0 = 0{\cdot}01$.

Results

We see from the table that aggregation proceeds most rapidly in the inner regions of the solar system where particles with size in the range 0·1–1 km can theoretically form a single protoplanetary cluster in about 10^3 yr or so. As one moves out through the nebula the aggregation time lengthens to the order of 10^4–10^5 yr, due to the increasing mean separation between the planetesimals, whilst a_1 drops to the 1–10m range owing to the lower gas density ρ_g which weakens the frictional drag. In the case of the Uranus and Neptunean orbits, a_1 turns out to be smaller than the gas mean free path length λ_g so that equation (78) is inapplicable and we have set $a_1 = \lambda_g$.

Life Time of the Gaseous Rings and the Cessation of Aggregation

In any event, excepting the asteroids, t_{agg} is much shorter than the expected lifetime of the gaseous rings, viz. $t_{max} \sim 3 \times 10^5$ yr, determined when the sun passes through its most active T-Tauri phase as we see in Figure 15 above. At that time the outermost rings of the gas, which are the ones most weakly bound to the solar system, are thermally evaporated away leaving behind just the right proportion of H and He at each orbit to account for masses of the four major planets (Hoyle, 1962; Prentice, 1977). The more tightly bound and dense inner rings of the nebula, however, are probably disrupted by turbulent convective motions generated in the gas by the intense heat bath of the overluminous contracting sun, which tends to destroy the uniform angular momentum distribution of the gas. In either instance all aggregation would cease since the planetesimals remain bound to their respective circular Keplerian orbital axes $s = R_n$ only as long as the gaseous rings remain intact with angular velocity distribution given by equation (59).

Origin of the Asteroidal Belt

We therefore suggest that all segregation and aggregation of the planetary material ceases at time t_{max}. Since the aggregation time for the asteroids is of the same order as t_{max} it therefore appears that they just failed to aggregate in time before their gaseous ring disintegrated. As a consequence the larger asteroids were left strewn around their mean orbit whilst the smaller ones were dragged away with the escaping gas (Prentice, 1977).

Importance of Collisional Drag

An important question which remains is to explain how those planetesimals whose sizes greatly exceed a_1 (i.e. $\gamma_2 R \ll 1$) succeed in aggregating with their protoplanetary clusters in time. Such planetesimals, whose radii might range up to 300 km judging from the size distribution of the asteroids, are hardly influenced by gas drag at all. The answer to this question we feel lies in collisional drag provided by the smaller particles of radius $a < a_1$ (i.e. $\gamma_2 R > 1$). These latter planetesimals are strongly bound to the gas and fairly quickly form a concentrated stream along the mean circular orbit $s = R_n$, as we saw in Figure 14 above. They therefore provide an additional source of drag for the larger ones ($a > a_1$) by essentially increasing the effective ρ_g, and hence γ_2, as in equation (70). Of course as soon as the gaseous rings disappear so will the collisional drag since the smaller particles, including any of the unsegregated dust, is swept away with the gas.

In conclusion I should like to mention that a detailed numerical study of the aggregation process in the gaseous rings is being presently undertaken by Mr. K. Hourigan. Hopefully this study will substantiate some of the ideas that have been outlined here.

The Final Stages of Planetary Formation

Formation of the Planetary Cores

If the planetesimals in each gaseous ring succeed in grouping together into a single protoplanetary cluster, the next stage in the formation process is the gravitational accretion of the cluster into a compact object or core of normal planetary size. During this phase the accreting mass is heated by the gravitational energy liberated both through its overall contraction and through the sweeping up and capture of fresh material as it moves around its orbit. Ringwood (1970) has suggested that Earth's surface temperature T_s may have risen to more than 2000 °C during the final stages of its accretion.

Consider then the contraction of a uniform protoplanetary cluster of mass m_c and radius R_s over a time scale t_{con} which is much longer than the thermal adjustment time $\tau_{therm} = m_c \mathscr{R} T / 4\pi\mu\sigma_s T_s^4 R_s^2 \sim 2 \times 10^3$ yr. In that case, the equation for energy balance at its surface reads

$$-\frac{3}{5}\frac{Gm_c^2}{R_s^2}\dot{R}_s = 4\pi\sigma_s R_s^2 T_s^4 . \tag{79}$$

Supposing that T_s increases roughly homologously during the contraction as $T_s \propto R_s^{-1}$, this equation may be integrated exactly to give the time taken to contract from initial density ρ_i to final density ρ_f, viz.

$$t_{con} = \frac{Gm_c}{5\sigma_s}\frac{\rho_f}{T_f^4}\left[\left(\frac{\rho_f}{\rho_i}\right)^{1/3} - 1\right]. \tag{80}$$

We have calculated t_{con} for each of the planets in Table 6 taking $\rho_i = 10^{-3}$ g cm^{-3}, $T_f = 2500$ K. In the case of the major planets we have again set $m_c \simeq 10M_\oplus$ and $\rho_f \simeq 2$ g cm^{-3} as suggested by the densities of Uranus and Neptune.

We observe from the Table that the terrestrial planets were, on average, able to contract to their present size in a much shorter time than the cores of the major planets, primarily because of their much smaller mass. The cores of the outer planets take about 3×10^5 yr to reach normal size, or even longer if $T_f < 2500$ K. Nonetheless if we incorporate the fact that the inner gaseous rings were detached much later than the outer ones and add to t_{con} the ring detachment time t_n given in Table 3 above, we find that all of the planets probably accreted to normal size at around the same time, viz. 3×10^5 yr, which incidentally is about the same as the time t_{max} when the protosun passes through its most overluminous phase.

Table 6. Contraction times and atmospheric masses of the planetary cores

Planet	m_c ($M_\oplus$)	ρ_f (g cm^{-3})	t_{con} (yr)	m_{atm}/m_c $T_f = 2500$ K	$T_f = T_{neb}$
Mercury	0·055	5·44	6(3)	4(−5)	2(−6)
Venus	0·815	5·27	8(4)	2(−5)	4(−3)
Earth	1·000	5·52	1(5)	2(−5)	2(−1)
Mars	0·107	3·95	7(3)	2(−8)	2(−7)
Jupiter	10	~2	3(5)	1(−1)	1(50)
Saturn	10	~2	3(5)	1(−2)	1(70)
Uranus	10	~2	3(5)	1(−3)	1(100)
Neptune	10	~2	3(5)	3(−4)	1(130)

Gravitational Capture of the Light Gases

After the planetary cores have accreted to normal planetary size and cooled down ($t_{cool} \sim 2 \times 10^3$ yr) it becomes possible for them to capture the light gases (H and He) of their parent gaseous rings, if the latter have not already dispersed. The intense gravitational field of the dense core draws in the gases at its edge like a gravitational sink (Perri and Cameron, 1974; Prentice, 1974). In the simplest approximation, when both the gravitational field of the gas and its intrinsic spin angular momentum is ignored, the solution of the isothermal hydrostatic support equation for a core of mass m_c, density ρ_c and radius r_c immersed in a uniform gas of undisturbed density ρ_∞ is

$$\rho(r) = \rho_\infty \exp(\mu G m_c / \mathscr{R} T r). \tag{81}$$

The mass of atmosphere accreted onto the core is then seen to be

$$m_{atm} = 4\pi \int_{r_c}^{\infty} (\rho - \rho_\infty) r^2 \, dr \simeq 3 m_c \left(\frac{\rho_\infty}{\rho_c} \right) \frac{e^{\alpha_c}}{\alpha_c}, \tag{82}$$

where the dimensionless parameter α_c is given by

$$\alpha_c = \mu G m_c / \mathscr{R} T r_c = 1{\cdot}0 \times 10^4 \left(\frac{m_c}{M_\oplus} \right)^{2/3} \frac{\rho_c}{T}, \tag{83}$$

expressing m_c in Earth masses, ρ_c in g cm^{-3}, and T in degrees K. Equation (81) is valid as long as $m_{atm} \ll m_c$.

In Table 6 we have calculated m_{atm}/m_c both during the cluster contraction phase t_{con}, when the temperature T of the gas is assumed to equal that at the core's edge T_s, and at time $t_{max} = 3 \times 10^5$ yr when the core has assumed to have cooled off so $T = T_{neb}$ as in Figures 12 and 13 above. Quite remarkably we observe that the terrestrial planets at no stage are able to accrete more than a minor atmosphere of gas. This is partly because the gas is always far too hot, owing to its proximity to the protosun, but primarily because the minor planets are far too light to attract much gas.

In contrast to the terrestrial planets, the cores of the major planets are grossly unstable towards the capture of the gas, except during the core contraction phase where they are far too hot. As soon as the major planetary cores have stopped contracting and the gas temperature drops to the background black body value determined by the protosun, all of the gas left over in each ring can be swallowed up. Since the masses of the gaseous rings are initially assumed to be each about the same, namely $1000\,M_\oplus$, the problem now arises to explain why the masses of the major planets differ so markedly. The answer to this problem, as Hoyle (1960) first noted, lies in the evaporation of gases from the solar system. In a detailed calculation (Prentice, 1977) I find that when the protosun passes through its most luminous phase at the time t_{max}, nearly all of the gas at the orbits of Neptune and Uranus is evaporated away, whilst at Saturn's orbit 70 per cent is lost and for Jupiter only some 20 per cent disappears.

These fractions occur in precisely the same proportion as the observed masses of these planets. We conclude that the cores of the major planets had not sufficiently cooled down to capture their gaseous rings until after the time t_{max} had passed.

The Planetary Spins

Finally we should briefly discuss the spins of the planets and the formation of satellite systems within the framework of this theory. Because of the frictional drag exerted by the gas, the net spin angular momentum of the protoplanetary clusters and planetary cores is expected to be zero, or at least very small allowing for some turbulence in the planetesimal motions. In addition if, as equations (59) and (60) indicate, the angular momentum per unit mass h of the gas in each ring is everywhere constant and equal to the Kepler value $h_n = \sqrt{GMR_n}$ on the circular orbital axis $s = R_n$, then none of the gas accreting onto the core will acquire any spin angular momentum either. A detailed study of the ring detachment process, however, shows that the mean angular momentum of the gas exceeds the orbital value h_n by a small fraction $p \simeq 0{\cdot}02 - 0{\cdot}03$ (Prentice, 1977). This means that the accreting gas inherits a net prograde spin angular momentum and the cores themselves are favoured to have a prograde spin relative to their orbital motion.

Consider now the rotational evolution of the accreting gas. If all the mass $m(1)$ in each gaseous ring were to coalesce into a single uniform globe of the same mean density, the ratio of centrifugal force to gravitational force at the equator becomes

$$\Theta = 6{\cdot}25 p^2 [M_\odot/m(1)](\alpha_n/3\pi)^{1/3} \sim 0(1), \tag{84}$$

using equations (56), (57), & (58). That is, the gas is rotationally unstable even in the diffuse state. Hence in order for the gas to accrete onto the planetary cores some means need to be found for the disposing of its excess spin angular momentum.

Formation of the Regular Satellite Systems

We therefore propose that the contraction of the primitive hydrogenic atmospheres of the major planets occurred in the same manner as the protosun's own contraction, namely through the development of a large supersonic turbulent convective stress and the shedding of a discrete system of gaseous rings. The masses m_n and orbital radii r_n of these rings relative to their parent planet would satisfy a geometric Titius–Bode relationship the same as in equation (54) yielding a constant mean ring mass

$$\langle m_n \rangle = m_{atm} f [\sqrt{\langle r_n/r_{n+1} \rangle} - 1], \tag{85}$$

where m_{atm} denotes the total mass of the contracting atmosphere.

Next taking the composition of the gas to be the same as that of the protosun, we can use equation (85) and Table 1 above to predict the masses of the regular satellite systems:

Jupiter. We consider the four Galilean satellites where $\langle r_n/r_{n+1} \rangle_{obs} = 1{\cdot}65$. Taking $m_{atm} = M_{Jup} - 10\,M_{\oplus}$ and $f = 0{\cdot}02$, the condensate masses in each ring are

$$m_{rock} \simeq 5 \times 10^{25}\text{g}, \quad m_{ice} \simeq 13 \times 10^{25}\text{g} \tag{86}$$

Quite remarkably, the masses of Io and Europa, whose mean densities $\rho_c \simeq 3{\cdot}5\,\text{g cm}^{-3}$ suggest a rocky composition, are 9×10^{25} g and 5×10^{25} g respectively, whilst the ice-like satellites Ganymede and Callisto have masses 15×10^{25}g and 11×10^{25}g.

Saturn. Here the regular inner group stretching from Mimas to Rhea yields a mean ratio $\langle r_n/r_{n+1} \rangle = 1{\cdot}3$ so we expect a mean satellite mass $m_{ice} \simeq 2 \times 10^{25}$g, which is at least an order of magnitude larger than any of the observed values. In view of the very sharp decrease in the observed mass distribution of these satellites, moving inwards from Rhea to Mimas, it is quite possible that the inner protosaturnian gaseous rings may have evaporated away before the planet became cool enough for the ice particles to condense out. This is an exciting possibility requiring further study.

Uranus and Neptune. We mentioned earlier that when the protosun passes through its most luminous phase, at time $t_{max} \simeq 3 \times 10^5$yr, only a few earth masses of H and He are left over, unevaporated, at the orbit of Uranus, whilst nothing at all remains for Neptune. Thus for Neptune we expect no regular satellite system at all if the Laplacian hypothesis is correct whilst for Uranus, taking $m_{atm} = M_{Uran} - 10\,M_{\oplus}$ as above, we predict a mean satellite mass $m_{ice} \simeq 2 \times 10^{24}$g. This coincides precisely with the observed value!

Conclusion

We have shown how if supersonic turbulent stress is taken into account, a gravitationally contracting rotating gaseous mass M disposes of its excess angular momentum through the shedding of a discrete system of Laplacian rings, each of about the same mass. The orbital radii R_n of these rings form a geometric progression

$$R_n/R_{n+1} = [1 + m(1)/Mf(1)]^2$$

which is closely equivalent to the Titius–Bode law for planetary and satellite distances. If the turbulence is sufficiently strong, the masses $m(1)$ of the rings and moment-of-inertia coefficient $f(1)$ of the fully rotating cloud are sufficiently small to account for the lightness of the planetary system as well as the regular satellite systems of the major planets. Typically, for turbulence parameter $\beta = 0{\cdot}1$ we find $f(1) = 0{\cdot}01$ and $m(1) = 1000\, M_{\oplus}$, taking $M = M_{\odot}$.

Next we have considered the chemical condensation sequence in the solar 'nebula' and shown how the rock component ($\sim 4 M_{\oplus}$) in each ring condenses out at the orbits of Mercury and beyond, whilst the ice component ($\sim 8 M_{\oplus}$) precipitates at Jupiter's orbit and beyond. The modern Laplacian theory is therefore able to account for the broad chemistry of the solar system as well as its basic physical features. Next we saw how as a result of the distribution of velocity in the gaseous ring, all condensing particles settle towards the circular Keplerian orbit $s = R_n$ of each ring to form a concentrated orbiting stream of planetesimals. Nonetheless, if the rock component condenses out as a fine smoke then a large fraction of it remains suspended in the gas. The planetesimals which form on the central orbit of each ring are maintained in that position by gaseous drag.

We have suggested that gravitational instabilities in the distribution $\rho_{p\phi}$ of planetesimal mass along each circular ring axis will lead to the formation of isolated groups of planetesimals and eventually to the appearance of a single protoplanetary cluster. This process, which remains to be formally verified, can occur only as long as the gaseous rings remain intact and so act as a dissipative sink for the excess kinetic energy of the aggregating planetesimals. The aggregation time $t_{agg} = O[\langle|\Delta\rho_{p\phi}|/\rho_{p\phi}\rangle_0 R_n^3/Gm_c]^{1/2}$ turns out to be $< 10^5$ yr for all of the planets except the asteroids, where a much longer time $\sim 5 \times 10^5$ yr is expected, owing to their much smaller mass m_c. At about time $t_{max} \simeq 3 \times 10^5$ yr the contracting protosun enters its most luminous phase, with the luminosity rising to more than 30 $L_{\odot}$. We have suggested that the gaseous rings are dispersed at this point, thereby terminating the aggregation process. If this is so it follows that the asteroids just failed to aggregate in time. Large-scale thermal evaporation of the outer gaseous rings of the solar nebula also occurs at this time.

The final stage of the formation process is the contraction of the protoplanetary clusters into compact cores of normal planetary size, followed by the gravitational capture of the residual gases in each ring. Owing to the heat liberated through the release of contractional energy, we find that the cores of the major planets, each about $10\, M_{\oplus}$, are far too hot to accrete much gas for the first 3×10^5 yr or so, whilst the terrestrial planets are not massive enough to capture much gas anyway. We have shown that provided the protosun passes through its most luminous phase before the major planetary cores have cooled down, then the residual mass of unevaporated gas left over at each orbit occurs in precisely the right proportion to account for the masses of those planets and their regular satellites.

It would therefore appear that a modern version of the original Laplacian hypothesis, taking into account the effects of supersonic turbulent convection and dissipative planetesimal drag, is capable of providing a very satisfactory account of the formation of the solar system.

Acknowledgement

The author is much indebted to Dr. D. ter Haar for introducing him to this area of astronomy and for providing generous hospitality and support during a number of visits to the Department of Theoretical Physics, University of Oxford. Part of this lecture was written in Oxford with the support of a S.R.C. Senior Visiting Scientist position.

References

Alfvén, H. (1954). *On the Origin of the Solar System.* Oxford Univ. Press, London and New York.

Auer, L. H., and Woolf, N. J. (1965). 'Mass loss and the formation of white-dwarf stars.' *Astrophys. J.*, **142**, 182–188.

Babinet, M. (1861). 'Note sur un point de la cosmogonie de Laplace.' *Comptes Rendus*, **52**, 481–484.

Brahic, Andre (1977). 'A numerical study of gravitating systems of colliding particles: Applications to the dynamics of Saturn's rings and to the formation of the solar system.' *Icarus*, **25**, 452–457.

Cameron, A. G. W. (1962). 'The formation of the sun and planets.' *Icarus*, **1**, 13–69.

Duley, W. W. (1974). 'Composition of grain mantles in interstellar clouds.' *Astrophys. Space Sci.*, **26**, 199–205.

Engvold, O., and Hauge, Ø. (1974). 'Elemental abundances, isotope ratios and molecular compounds in the solar atmosphere.' Institute of theoretical astrophysics, Blindern-Oslo. Report No. 39.

Ezer, D., and Cameron, A. G. W. (1965). 'A study of solar evolution.' *Can. J. Phys.*, **43**, 1497–1517.

Gaustad, John E. (1963). 'The opacity of diffuse cosmic matter and the early stages of star formation.' *Astrophys. J.*, **138**, 1050–1073.

Goldreich, Peter, and Ward, William R. (1973). 'The formation of Planetesimals.' *Astrophys. J.*, **183**, 1051–1061.

Grossman, Lawrence (1972). 'Condensation in the primitive solar nebula.' *Geochim. Cosmochim. Acta*, **36**, 597–619.

Hayashi, C. (1961). 'Stellar evolution in early phases of gravitational contraction.' *Publ. Astron. Soc. Japan*, **13**, 450–452.

Hayashi, C. (1966). 'Evolution of protostars.' *Ann. Rev. Astron. Astrophys.*, **4**, 171–192.

Herbig, G. H. (1962). 'The properties and problems of T-Tauri stars and related objects.' *Advances in Astron. Astrophys.* (Ed. Z. Kopal), **1**, 47–103.

Hoyle, F. (1960). 'On the origin of the solar nebula.' *Quart. J. Roy. Astron. Soc.*, **1**, 28–55.

Hoyle, F. and Wickramasinghe, N. C. (1968). 'Condensation of the planets.' *Nature*, **217**, 415–418.

Jeans, J. (1929). *Astronomy and Cosmogony.* Cambridge Univ. Press, Cambridge.

Krautschneider, M. J. (1976). 'Interstellar grain cloud dynamics'. *Unpublished M.Sc. Thesis*, Monash Univ. Australia.

Krautschneider, M. J. (1977). 'On star formation via the gravitational contraction of grain clouds.' *Astron. Astrophys*, **57**, 291–302.

Kuiper, G. P. (1951). In *Astrophysics* (Ed. J. A. Hynek). McGraw-Hill, New York, Chap. 8.

Laplace, P. S. de (1796). *Exposition du Système du Monde*, Paris.

Larimer, John W. (1967). 'Chemical fractionations in meteorites—I. Condensations of the elements.' *Geochim. Cosmochim. Acta*, **31**, 1215–1238.

Larson, R. B. (1969). 'Numerical calculations of the dynamics of a collapsing proto-star.' *Mon. Not. Roy. Astron. Soc.*, **145**, 271–295.

Lewis, John S. (1972a). 'Low temperature condensation from the solar nebula.' *Icarus*, **16**, 241–252.
Lewis, John S. (1972b). 'Metal/silicate fractionation in the solar system.' *Earth Plan. Sci. Letts.*, **15**, 286–290.
McCrea, W. H. (1929). 'The mechanics of the chromosphere.' *Mon. Not. Roy. Astron. Soc.*, **29**, 718–730.
McCrea, W. H. (1960). 'The origin of the solar system.' *Proc. Roy. Soc.*, **A256**, 245–266.
Maxwell, J. C. (1859). 'On the stability of the motion of Saturn's rings.' in *Scientific Papers of J. C. Maxwell*. C.U.P. Cambridge, Chap. XIX.
Mestel, L. (1965a). 'Problems of star formation—I.' *Quart. J. Roy. Astron. Soc.*, **6**, 161–198.
Mestel, L. (1965b). 'Problems of star formation—II.' *Quart. J. Roy Astron. Soc.*, **6**, 265–298.
Monaghan, J. J. and Roxburgh, I. W. (1965). 'The structure of rapidly rotating polytropes.' *Mon. Not. Roy. Astron. Soc.*, **131**, 13–21.
Nieto, M. M. (1972). *The Titius–Bode Law of Planetary Distances: Its History and Theory*. Pergamon Press, Oxford.
Perri, Fausto and Cameron, A. G. W. (1974). 'Hydrodynamic instability of the solar nebula in the presence of a planetary core.' *Icarus*, **22**, 416–425.
Prentice, A. J. R. and ter Haar, D. (1971). 'On the angular momentum problem in star formation.' *Mon. Not. Roy. Astron. Soc.*, **151**, 177–184.
Prentice, A. J. R. (1973). 'On turbulent stress and the structure of young convective stars.' *Astron. Astrophys.*, **27**, 237–248.
Prentice, A. J. R. (1974). 'The formation of planetary systems.' in *In the Beginning ...* (Ed. J. P. Wild). Australian Academy of Science, Canberra.
Prentice, A. J. R. (1976). 'Supersonic turbulent convection, inhomogeneities of chemical composition, and the solar neutrino problem.' *Astron. Astrophys.* **50**, 59–70.
Prentice, A. J. R. (1977). 'A modern Laplacian theory for the formation of the solar system.' *Icarus*—to be submitted.
Reddish, V. C., and Wickramasinghe, N. C. (1969). 'Star formation in clouds of solid hydrogen grains.' *Mon. Not. Roy. Astron. Soc.*, **143**, 189–208.
Reeves, H. (1975). 'L'origine du système solaire.' *Recherche* **6**, 808–817.
Ringwood, A. E. (1970). 'Origin of the moon: the precipitation hypothesis.' *Earth Plan. Sci. Letts.*, **8**, 131–140.
Schatzman, E. (1949). 'On certain paths of stellar evolution. I Preliminary remarks.' *Bull. Acad. Roy. Belgique*, **35**, 1141–1152.
Schatzman, E. (1967). 'Cosmogony of the solar system and origin of the deuterium.' *Annales D'Astrophysique*, **30**, 963–973.
Schatzman, E. (1971). in *Highlights of Astronomy* (Ed. C. de Jager). D. Reidel, Dordrecht-Holland, Vol. 2, p. 197.
Spitzer, L. (1968). *Diffuse Matter in Space*. Interscience publishers, New York, p. 170.
ter Haar, D. (1948). 'Studies on the origin of the solar system.' *Proc. Roy. Danish Acad. Sci.*, **25**, No. 3.
ter Haar, D., and Cameron, A. G. W. (1963). 'Historical review of theories of the origin of the solar system.' In *Origin of the Solar System*. (Eds. R. Jastrow and A. G. W. Cameron). Academic Press, New York.
ter Haar, D. (1967). 'On the origin of the solar system.' *Ann. Rev. Astron. Astrophys.*, **5**, 267–278.
Thomas, R. N. (1973). 'A scheme of stellar atmospheric regions.' *Astron. Astrophys.*, **29**, 297–307.
Urey, H. C. (1951). 'The origin and development of the earth and other terrestrial planets.' *Geochim. Cosmochim. Acta*, **1**, 209–277.
Whipple, F. L. (1971). *Earth, Moon and Planets*. Harvard Univ. Press, Cambridge, Mass.
Williams, I. P., and Cremin, A. W. (1968). 'A survey of theories relating to the origin of the solar system.' *Quart. J. Roy. Astron. Soc.*, **9**, 40–62.

STAR FORMATION AND INTERACTIONS BETWEEN STARS

M. M. WOOLFSON
Department of Physics, University of York, Heslington, York, U.K.

Abstract

A model for the formation of stars in a galactic cluster is considered. Equations are developed for the collapse of a gas cloud within which there is turbulence and for the generation of turbulence by the energy released in the collapse of the cloud. At a certain stage of the process, the collision of the turbulent elements can give rise to high-density regions within which stars can form. Subsequent to their initial formation some stars accrete further cloud material so giving rise to more massive, including O and B, stars. The model agrees with the observation of young stellar clusters in terms of the mass–age correlation of stars and also with the increase in rate of production with time. It is shown that stars formed by the collision of turbulent streams should have no intrinsic rotation although those which acquire mass by accretion should rotate rapidly. The results agree qualitatively with observation in suggesting that the angular momentum per unit mass of a star increases with mass and with reasonable assumptions, there is also good quantitative agreement.

The probability of a stellar interaction in which a condensed star and a diffuse less-massive protostar are involved in such a way that the protostar is tidally disrupted is calculated. It is concluded that the possibility of several such encounters during the early stages of the evolution of a cluster cannot be discounted.

Introduction

The capture theory of the origin of the solar system (Woolfson, 1964) involves an interaction between two stars and this quite naturally provokes the question of how common such interactions are likely to be. The now-defunct Jeans–Jeffreys tidal theory (Jeans, 1916; Jeffreys, 1916, 1918) was attacked on the grounds that interactions between stars should be extremely rare. However, while the tidal theory was discredited for a number of perfectly valid reasons the probability of stellar encounters should not have been one of them.

According to current ideas many, if not most, stars originate in associations of one type or another. The most common association observed is a galactic cluster containing several hundred to a thousand stars with a total mass up to about 1000 $M_{\odot}$. Most modern theorists agree that the processes of fragmentation and turbulence have important roles to play in cluster star formation and, indeed, there is some visual evidence for supersonic turbulent motion in clouds within which stars are forming. The model which will be presented here for the formation of a galactic cluster has been much influenced by the work of Grzedzielski (1966) although there are important differences in some of the assumptions.

The Basic Equation of Collapse

We start by applying the virial theorem to a collapsing spherically symmetric cloud of gas. This is taken in the form

$$\frac{1}{2}\frac{\mathrm{d}^2 I}{\mathrm{d}t^2} = 2\tau + 3(\gamma - 1)U + \Omega \tag{1}$$

where $I = 4\pi \int_0^R \rho(r) r^4 \,\mathrm{d}r$, the moment of inertia about the origin, $\rho(r)$ is the density at distance r from the origin, R is the radius of the cloud, τ the kinetic energy associated with mass motion of its material, U the internal energy, γ the ratio of specific heats of the gas and Ω the self-gravitational potential energy of the cloud. Other forms of energy, such as those associated with magnetic or electric fields may be included in the virial theorem but we shall assume that they play a negligible role in the dynamics of collapse of the cloud.

The case we are going to consider is the somewhat artificial one of the homologous collapse of a uniform cloud. While such a model has some obviously-unreal features it should give a behaviour pattern of similar quality to that expected in a more realistic mode of collapse.

For a spherical cloud of mass M and uniform density

$$I = \tfrac{3}{5} M R^2 \tag{2}$$

giving

$$\frac{1}{2}\frac{\mathrm{d}^2 I}{\mathrm{d}t^2} = \frac{3}{5} M \left(\frac{\mathrm{d}R}{\mathrm{d}t}\right)^2 + \frac{3}{5} M R \frac{\mathrm{d}^2 R}{\mathrm{d}t^2}. \tag{3}$$

If the material of the cloud has linear-wave flow i.e. each element moves towards the centre with a speed proportional to r, its distance from the centre, then the associated kinetic energy is

$$\tau_{\mathrm{L}} = \frac{3}{10} M \left(\frac{\mathrm{d}R}{\mathrm{d}t}\right)^2. \tag{4}$$

Any generalized mass motion of the material may be resolved into components of linear-wave flow plus random motion. If the random motion has a mean-square speed w^2 then the kinetic energy of random motion is

$$\tau_R = \tfrac{1}{2} M w^2 \tag{5}$$

and

$$\tau = \tau_R + \tau_L. \tag{6}$$

For a monatomic gas

$$3(\gamma - 1) U = (3MkT)/m_g \tag{7}$$

where k is Boltzmann's constant and m_g is the mass of a gas molecule.

Finally the self-gravitational energy of the uniform cloud is

$$\Omega = -\tfrac{3}{5} GM^2/R. \tag{8}$$

Assembling all these terms we find the differential equation governing the collapse of the cloud in the presence of turbulence

$$\frac{d^2R}{dt^2} = \frac{5w^2}{3R} + \frac{5kT}{m_g R} - \frac{GM}{R^2}. \tag{9}$$

The first two terms on the right-hand side are due respectively to turbulent pressure and thermal pressure and act to expand the cloud while the final term, due to gravitation, tends to cause its collapse.

Turbulence Effects

In the collision of two turbulent streams of gas there will be compression and heating of the material involved. This process has been analysed mathematically by Aust (1977) and may be followed by reference to Figure 1. The two

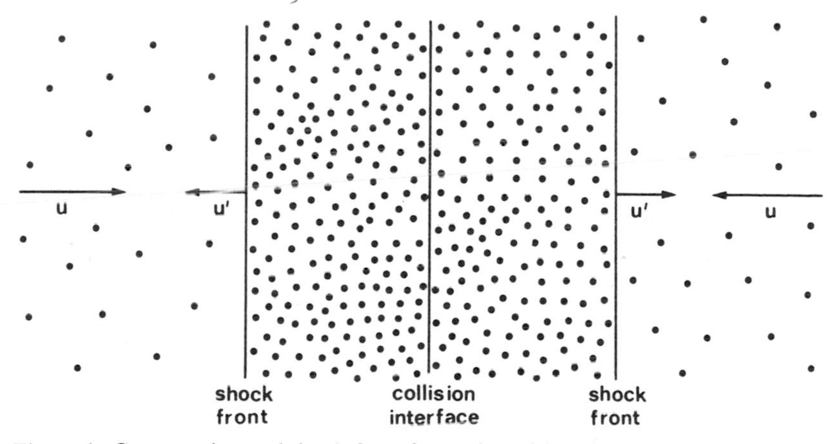

Figure 1. Compression and shock-front formation with colliding turbulent streams

colliding elements, of equal density and infinite extent, meet at the collision interface. The gas streams move in at speed u and compressed gas is formed in the shocked region. As the collision progresses the shock front moves steadily outwards, with speed u', from the collision interface. The compression ratio is

$$\frac{\rho_2}{\rho_1} = \frac{\sqrt{\left\{\left(\frac{\gamma+1}{4}\right)^2 \frac{u^2}{c^2} + \gamma\right\}} + \left(\frac{\gamma+1}{4}\right)\frac{u}{c}}{\sqrt{\left\{\left(\frac{\gamma+1}{4}\right)^2 \frac{u^2}{c^2} + \gamma\right\}} - \left(\frac{3-\gamma}{4}\right)\frac{u}{c}} \tag{10}$$

where c is the speed of sound in the gas. The form of this relationship for various values of γ and u/c is shown in Figure 2.

We now consider the heating effects that take place. In an optically-thin dusty cloud the characteristic timescale for heating and cooling is much less than that for dynamical effects in the cloud. When the turbulent speeds are high there will certainly be some increase of temperature of the compressed gas but it will not correspond to that for an adiabatic compression. This can be

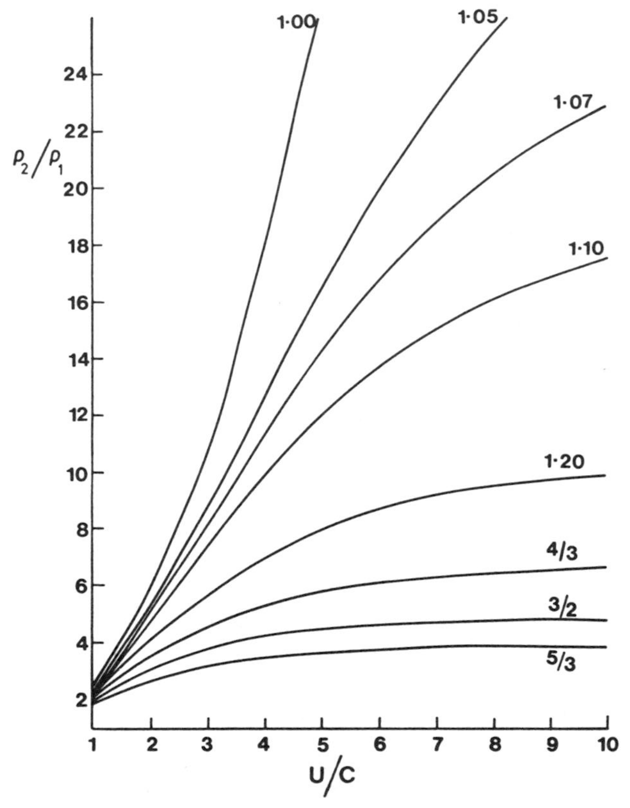

Figure 2. Compression ratios for various γ as a function of the Mach number of the colliding streams

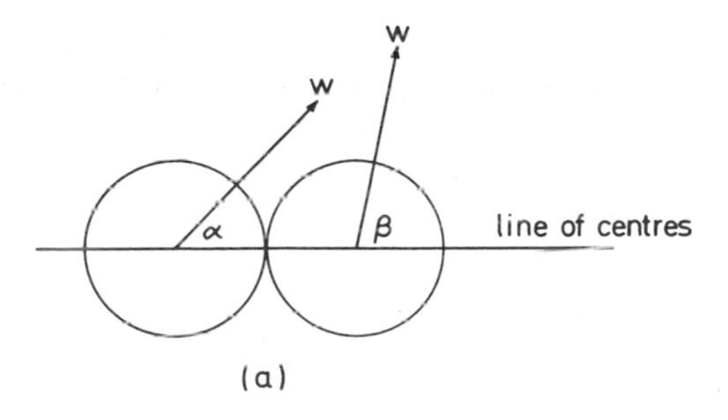

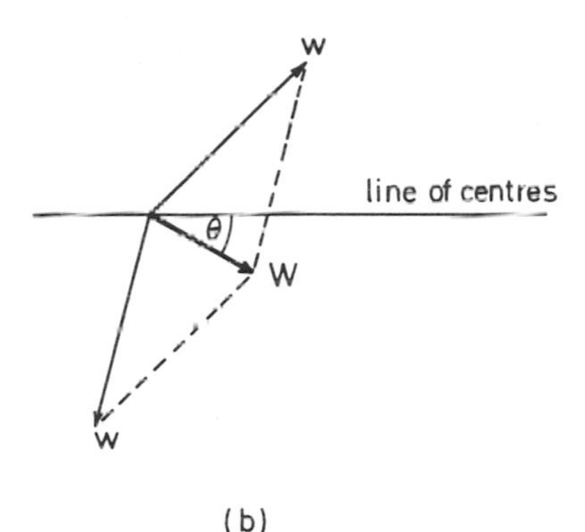

Figure 3. (a) The motions of two turbulent elements. (b) The magnitude and direction of the relative velocity of the colliding elements

simulated in equation (10) by taking an effective ratio of specific heats, γ', somewhere between the actual value and unity (which corresponds to the isothermal case). An important consequence of the radiation of heat is that it represents a loss of energy of the system as a whole.

As a simple model of turbulence we imagine that the cloud is divided into regions of uniform size moving in random directions but each with a speed w. Two such regions are illustrated in Figure 3. If $\beta > \alpha$ then some gas will be compressed and the relative speed of the regions is

$$W = \{2[1 - \cos(\beta - \alpha)]\}^{1/2} w. \tag{11}$$

W makes an angle θ with the line of centres of the regions where

$$\theta = \frac{\pi}{2} - \frac{1}{2}(\alpha + \beta). \tag{12}$$

The affected material will (i) be compressed fairly quickly with an effective γ', (ii) cool to the prevailing external equilibrium temperature with negligible re-expansion and (iii) re-expand isothermally. Because the relative motion of the regions is not along the line of centres, only a proportion δ of the total material is compressed. A reasonable approximation to δ is cos θ. The net loss of radiated energy in the three stages described above is found to be

$$\delta E = \frac{2\delta M_c kT}{m_g}\left[\frac{1}{\gamma' - 1}\left\{\left(\frac{\rho_2}{\rho_1}\right)^{\gamma' - 1} - 1\right\} - \ln\left(\frac{\rho_2}{\rho_1}\right)\right] \tag{13}$$

where M_c is the mass of one turbulent region.

In calculating W and θ in equations (11) and (12) it was assumed that the directions of motion of the two regions and the line joining the region centres were all coplanar. A proper three-dimensional treatment requires three angles to be defined and these could be α and β, the angles between the directions of motion and the line of centres, plus one twist angle ϕ.Then the average loss of energy per interacting pair of regions is

$$\overline{\Delta E} = \int_{\alpha=0}^{\pi}\int_{\beta=\alpha}^{\pi}\int_{\phi=0}^{2\pi} \delta E(\alpha, \beta, \phi)\, P(\alpha, \beta, \phi)\, \mathrm{d}\alpha \mathrm{d}\beta \mathrm{d}\phi \tag{14}$$

where $P(\alpha, \beta, \phi)$ is the probability density for the three angular coordinates.

We may now find an expression for the total rate of loss of energy due to turbulence. Let there be n turbulent regions in the cloud distributed in, say, a close-packed hexagonal array. Then each region has twelve nearest neighbours and, ignoring boundary effects, this implies $6n$ neighbouring interacting pairs.

In the usual treatment of the fragmentation of a cloud it is assumed that a theory by Jeans is applicable. The cloud will break up into regions of radius r_c where

$$r_c = \left(\frac{3kT}{2\pi A m_g \rho G}\right)^{1/2} \tag{15}$$

where A is a geometrical factor (3/5 for a homogeneous sphere) and G the gravitational constant. The corresponding mass is

$$M_c = 4\pi r_c^3 \rho/3 \tag{16}$$

In the present theory the fragmentation mass and radius are interpreted as characteristics of turbulent regions. The number of regions

$$n = \frac{M}{M_c} \tag{17}$$

and since δE contains a factor M_c there is no direct dependence of ΔE on M_c. To obtain a *rate* of energy loss due to turbulence we further assume that the characteristic time for one interaction to take place is r_c/w, i.e. the time for a turbulent stream to travel a distance equal to the radius of a region. We can think of r_c/w as a 'coherence time'; the cloud viewed after an interval much less than r_c/w will seem recognizably similar to its original state—if viewed after a long interval compared to r_c/w then there will be little similarity.

Putting these results together the rate of loss of turbulent energy becomes

$$\frac{d\tau_R}{dt} = -6\frac{M}{M_c}\frac{w}{r_0}\overline{\Delta E} \tag{18}$$

We must now consider how the turbulence is sustained.

The Generation of Turbulence

The gravitational energy released by the collapse of the cloud is converted in the first instance into the kinetic energy of its matter. Initially most of the energy will be fed into linear-wave flow but eventually some of it goes into the creation of turbulence. The factors which support turbulence are (1) high speed of motion of the fluid and (2) a high rate of input of energy into the fluid. We might also expect, and numerical tests support the idea, that the pre-existence of turbulence predisposes the generation of new turbulence.

The rate of creation of turbulence may be expressed as

$$\frac{d\tau_n}{dt} = -f\frac{d\Omega}{dt} \tag{19}$$

where f is a fraction, initially zero, that probably approaches unity when very turbulent conditions prevail. It has been decided to use a parameter- controlled

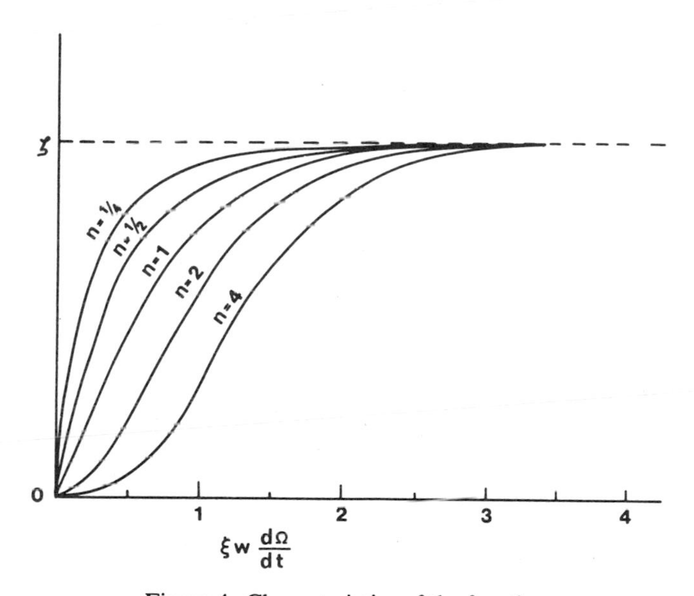

Figure 4. Characteristics of the function

$$f = \zeta\left\{\tanh\left(\xi w\frac{d\Omega}{dt}\right)\right\}^n$$

mathematical model for f of the form

$$f = \zeta\left\{\tanh\left(\xi w\frac{\mathrm{d}\Omega}{\mathrm{d}t}\right)\right\}^n \tag{20}$$

The parameter ζ controls the maximum value of f and may be unity if all the gravitational energy is eventually fed into turbulence. The constant ξ controls the rate at which f increases with increasing $w(\mathrm{d}\Omega/\mathrm{d}t)$ and the index n mainly controls the shape of the relationship curve. This last point is shown in Figure 4.

Combining (18), (19) and (20) we find the turbulence equation

$$\frac{\mathrm{d}}{\mathrm{d}t}\left(\frac{1}{2}Mw^2\right) = -\frac{3}{5}f\frac{GM^2}{R^2}\frac{\mathrm{d}R}{\mathrm{d}t} - 6\frac{M}{M_c}\frac{w}{r_c}\overline{\Delta E}$$

or, with $w^2 = \varepsilon$

$$\frac{\mathrm{d}\varepsilon}{\mathrm{d}t} = -\frac{6fGM}{5R^2}\frac{\mathrm{d}R}{\mathrm{d}t} - \frac{12\varepsilon^{1/2}\overline{\Delta E}}{M_c r_c} \tag{21}$$

This equation, together with (9) in the form

$$\frac{\mathrm{d}^2R}{\mathrm{d}t^2} = \frac{5\varepsilon}{3R} + \frac{5k}{m_c R} - \frac{GM}{R^2} \tag{22}$$

provide a pair of coupled differential equations which can be numerically solved for R and ε ($= w^2$) as functions of t, given initial boundary conditions.

Conditions for Star Formation

The collision of turbulent elements of the collapsing cloud together with rapid cooling can produce a region of high density and low temperature. Such a region will form a star if it collapses and maintains its identity until it reaches a comparatively high density. It is instructive to consider this problem in terms of the free-fall collapse time which, for a mass of density ρ, is

$$t_f(\rho) = \left(\frac{3\pi}{32G\rho}\right)^{1/2} \tag{23}$$

In fact a gaseous body will never collapse in a time t_f; the collapse is always modified by the generation of heat especially when the opacity becomes high. However for an initially diffuse body with a mass substantially greater than the initial mass, M_c, given in equation (16), the initial collapse will be close to free-fall, certainly down to the stage where the body is compact enough to be considered a protostar.

The condition for a star to be formed from a region of density ρ_2 surrounded by a medium of density ρ_1 is suggested as

$$t_f(\rho_2) < \psi t_f(\rho_1) \tag{24}$$

Table 1. Critical masses, critical radii, free-fall times, and coherence times at various densities for atomic hydrogen at 100 K

ρ_1(g cm^{-3})	$M_c(M_\odot)$	r_c(cm)	$t_f(\rho_1)$ (s)	r_c/c (s)
10^{-16}	6·15	$3{\cdot}17 \times 10^{16}$	$2{\cdot}1 \times 10^{11}$	$3{\cdot}3 \times 10^{11}$
10^{-15}	2·10	$1{\cdot}00 \times 10^{16}$	$6{\cdot}6 \times 10^{10}$	$1{\cdot}1 \times 10^{11}$
10^{-14}	0·61	$3{\cdot}17 \times 10^{15}$	$2{\cdot}1 \times 10^{10}$	$3{\cdot}3 \times 10^{10}$
10^{-13}	0·21	$1{\cdot}00 \times 10^{15}$	$6{\cdot}6 \times 10^{9}$	$1{\cdot}1 \times 10^{10}$
10^{-12}	0·06	$3{\cdot}17 \times 10^{14}$	$2{\cdot}1 \times 10^{9}$	$3{\cdot}3 \times 10^{9}$
10^{-11}	0·02	$1{\cdot}00 \times 10^{14}$	$6{\cdot}6 \times 10^{0}$	$1{\cdot}1 \times 10^{9}$
10^{-10}	0·006	$3{\cdot}17 \times 10^{13}$	$2{\cdot}1 \times 10^{8}$	$3{\cdot}3 \times 10^{8}$

where ψ must be postulated but may be somewhere in the range 0·1 to 0·5. From (23) this implies

$$\frac{\rho_2}{\rho_1} > \frac{1}{\psi^2} \tag{25}$$

or, from (10),

$$\frac{\sqrt{\left\{\left(\frac{\gamma'+1}{4}\right)^2\frac{u^2}{c^2}+\gamma'\right\}}+\left(\frac{\gamma'+1}{4}\right)\frac{u}{c}}{\sqrt{\left\{\left(\frac{\gamma'+1}{4}\right)^2\frac{u^2}{c^2}+\gamma'\right\}}-\left(\frac{3-\gamma'}{4}\right)\frac{u}{c}} > \frac{1}{\psi^2} \tag{26}$$

where γ' is the effective ratio of specific heats during gas compression with radiation of heat taken into account. Equating the two sides of (26) gives a lower limit of u for star formation, u_{lim}.

It is instructive to examine Table 1 to gain further insight into the conditions for star formation. With temperature arbitrarily fixed at 100 K one finds the mass of a region, M_c, and its radius, r_c. The value of $t_f(\rho_2)$ should be appreciably shorter than the coherence time r_c/w otherwise the compressed region may all be stirred up into the cloud before a star actually forms. The value of $t_f(\rho_2)$ will be less than the listed $t_f(\rho_1)$ by a factor ψ while for supersonic collisions r_c/w will be less than the r_c/c listed. This suggests that ψ should be made dependent on w/c in some way.

Rate of Star Formation

The mathematical apparatus assembled so far enables the collapse of a cloud to be studied together with the progress of star formation. The general steps involved in this process are:

(i) Set up the initial configuration of the cloud and select the variable parameters.

(ii) Solve equations (21) and (22) for R, $\dot{R}$, and w. This involves the evaluation of integral (14) at each stage of integrating the differential equations.

(iii) As soon as $w \geqslant u_{\text{lim}}$ the possibility of star formation exists. As α, β, and ϕ are varied in the evaluation of (14) so W, the relative velocity of the two regions, is found. When $W/2 > u_{\text{lim}}$ then star formation takes place, the stellar mass, M_*, being approximately $2\delta M_c$.

(iv) The number of pairs of regions interacting in a coherence time w/r_c is $6M/M_c$ so that the rate of formation of stars in the mass range $2M_c\delta_1$ to $2M_c\delta_2$ is

$$\frac{dN}{dt} = 6\frac{M}{M_c}\frac{w}{r_c}\underset{\substack{\frac{1}{2}W > u_{\text{lim}} \\ \text{and} \\ \delta_1 < \delta < \delta_2}}{\int_{\alpha=0}^{\pi}\int_{\beta=\alpha}^{\pi}\int_{\phi=0}^{2\pi}} P(\alpha, \beta, \phi)\, d\alpha\, d\beta\, d\phi \tag{27}$$

As cloud material forms into stars the average density of the cloud is corrected for the loss of material, but the value of M appearing in equations (21) and (22) is not changed.

The pattern to be expected from this calculation is that initially, when there is little turbulence, no stars will form. Eventually stars will just be able to form under the most favourable conditions—when clouds collide head on. Later, stars of smaller mass will form although at all stages, due to variations of δ, a range of stellar masses will result.

Observations of young stellar clusters seem to indicate that the first stars to be produced are those of about one solar mass. Later, stars of lesser mass and also of greater mass are formed and the rate of star formation increases with time (Williams and Cremin, 1969).

The present model certainly agrees with part of this pattern, especially concerning the rate of formation, but as developed so far it indicates that the stellar masses should steadily decline as the cluster evolves.

The Formation of Massive Stars

After a star has been formed it will be moving in a direction which depends on the motions of the colliding regions which produced it. There will be a strong tendency for new stars to fall towards the centre of the cloud but there will also be a considerable random component of the motion. It is to be expected that the star will move through different regions of turbulent motion and that the density of cloud material and the turbulent velocity will, in general, increase with time. These last conclusions need to be qualified since we know that at some stage star formation ceases. This is not only because material gets used up in forming stars but also because gaseous material is eventually blown out of the

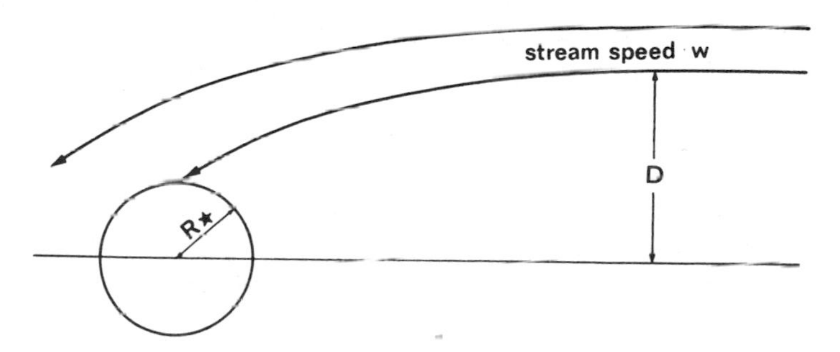

Figure 5. Motion of streams of material relative to a star in a turbulent region

system especially due to radiation from very hot and massive O and B stars.

To simplify our calculations let us assume that ρ and w are constant and that our star of mass M_* and radius R_* is moving in the cloud at a speed much less than w. In Figure 5 is shown the star and two streams of cloud material, one of which intersects the body of the star and one which does not. It can be shown that the critical interaction parameter, D_L, for which collision with the star takes place is given by

$$D_L^2 = R_*^2 + \frac{2GM_*R_*}{w^2} \tag{28}$$

Hence the rate at which new material joins the star is

$$\frac{dM_*}{dt} = \pi D_L^2 \rho w = \pi R_*^2 \rho w + \frac{2\pi G M_* R_* \rho}{w} \tag{29}$$

For most cases of interest the first term on the right hand side of (29) may be ignored and also since R_* varies slowly with M_* it may be considered constant.

Integrating (29) under these conditions we find

$$M_* = (M_*)_0 \exp\left(\frac{2\pi G R_* \rho}{w} t\right) \tag{30}$$

Calculations so far suggest that star formation is associated with turbulent speeds of about $10^6\,\mathrm{cm\,s^{-1}}$ or greater and that star formation takes place over a period of 10^{14}s or so. However the limiting time for stellar accretion will be how long it spends within the cluster and, even moving slowly, this is likely to be in the range 10^{10}–10^{14}s.

We now try various sets of values.

(a) $w = 10^6\,\mathrm{cm\,s^{-1}}, \rho = 10^{-14}\,\mathrm{g\,cm^{-3}}, R_* = 10^{11}\,\mathrm{cm}, t = 10^{10}\,\mathrm{s}$ giving

$$\frac{M_*}{(M_*)_0} = 1{\cdot}000000$$

(b) $w = 10^6\,\mathrm{cm\,s^{-1}}$, $\rho = 10^{-12}\,\mathrm{g\,cm^{-3}}$, $R_* = 10^{12}\,\mathrm{cm}$, $t = 10^{14}\,\mathrm{s}$ giving

$$\frac{M_*}{(M_*)_0} = 1{\cdot}59 \times 10^{18}\,!$$

The conclusion from these two extreme examples is that, depending on the density of material through which the star moves and the duration of its passage, it may either be virtually unchanged in mass or it may increase manyfold. The rather excessive factor found in (b) is quite unrealistic and comes from having a density which could not be maintained over such a long period.

It seems probable that the mass-increase factor for most stars will be close to unity but that occasional favourable factors, combining together, may give stars up to some tens of solar masses as are actually observed for O and B stars. Since these more massive stars are formed by accretion over fairly long periods it would be expected that they should be the youngest stars in a cluster, which agrees with the observations of Williams and Cremin.

Stellar Rotation

The rotation of stars seems to be intimately related to their spectral class and therefore to their mass. In Figure 6 there is shown the approximate relationship between mass and equatorial speed, the quantity measured by Doppler-shift methods. The slow rotation of the sun and similar stars has been considered as of central importance by a number of theorists on the origin of the solar system, notably by Hoyle (1960) and by McCrea (1960). The main brunt of their argu-

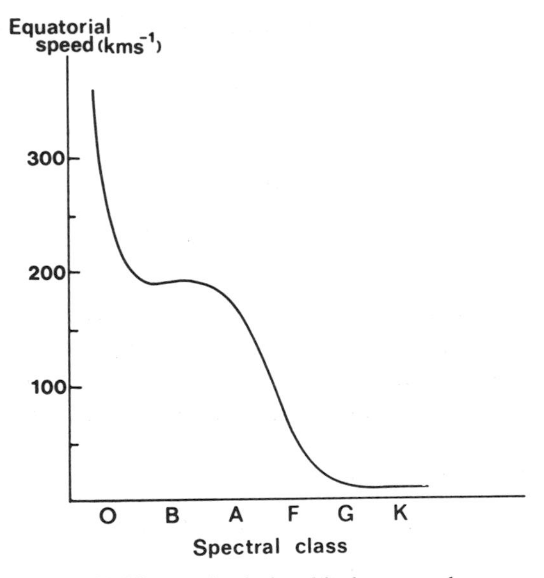

Figure 6. Observed relationship between the mass of a star and its equatorial velocity

ment was that the contraction of low-density galactic material possessing the intrinsic rotation of the galaxy would, by conservation of angular momentum, lead to stars with equatorial velocities exceeding the speed of light. They therefore argued that some part of the mechanism of star formation negates this conclusion and that this mechanism is probably also associated with the production of a planetary family.

In this work we are ascribing the formation of a star to the collision of two turbulent elements of a cloud. In Figure 7(a) two streams of gas, one representing such elements, are shown. These streams are shown in motion relative to the centre of the cloud; in Figure 7(b) their motions are shown relative to their

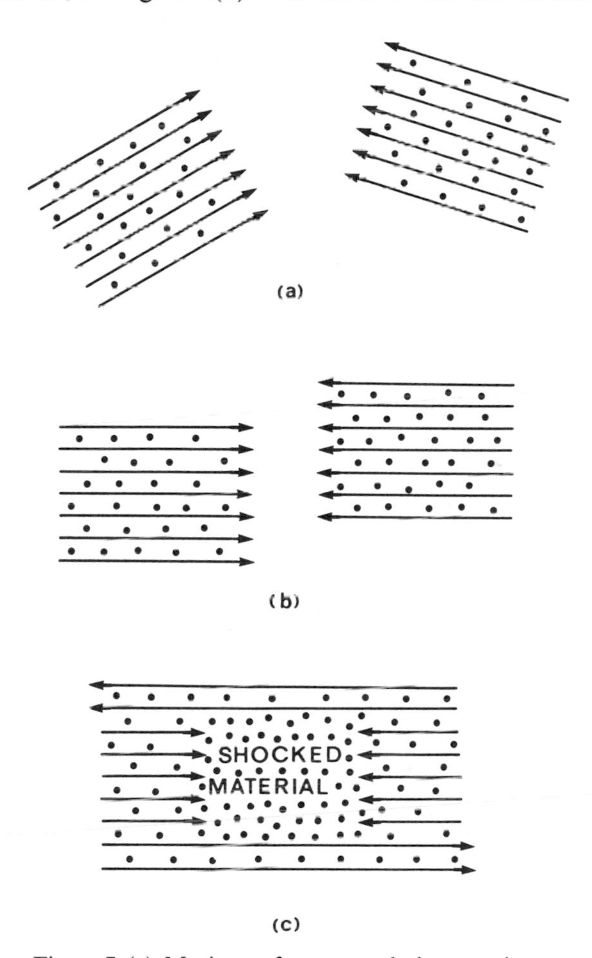

Figure 7. (a) Motion of two turbulent regions relative to centre of mass of the cloud. (b) Motion of two turbulent regions relative to their combined centre of mass. (c) The formation of a compressed region without associated angular momentum

own combined centre of mass. This picture is important. With respect to the centre of mass they must be in head-on collision although they may be offset so that not all their substance interacts. The effect of such a collision is shown in Figure 7(c). The shocked region, in which gas is compressed, has involved gas streams in head-on collision *with no resultant angular momentum.* Any resultant angular momentum due to the offset of the streams is conserved in steams of material which play no part in gas compression. There is so produced a star with little if any rotation. The slow rotation of stars now appears, in agreement with the ideas of Hoyle and of McCrea, as a consequence of the mode of formation of stars, particularly those of solar mass or less.

Why then do more massive stars rotate more quickly? It is because such stars acquire angular momentum as they gain mass and we can approximately calculate how fast they should rotate. Figure 8 shows the growing star in the path of a turbulent element of the cloud. There will always be some velocity gradient across a stream and we may take this as of order w/r_c so that passage through a turbulent element may completely reverse the direction of flow.

The angular momentum acquired by the star per unit gain of mass will be

$$\frac{\mathrm{d}H}{\mathrm{d}M_*} \simeq \phi D \delta w \tag{31}$$

where ϕ is a geometrical factor which is found to be about $\frac{1}{4}$.
Replacing δw by Dw/r_c and calling on equation (28) without the R_*^2 term

$$\frac{\mathrm{d}H}{\mathrm{d}M_*} = \frac{GM_*R_*}{2wr_c} \tag{32}$$

or, after integration with R_* taken as constant,

$$H = \frac{GR_*}{4wr_c}\left\{M_*^2 - (M_*)_0^2\right\} \tag{33}$$

This result assumes that all the mass is acquired from a single turbulent

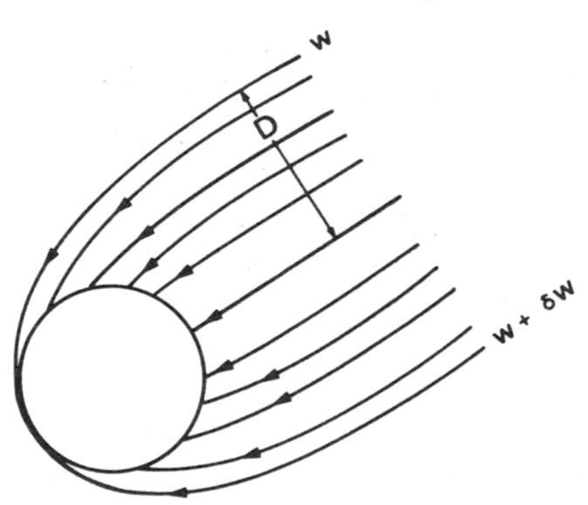

Figure 8. The accretion by a star of material with a velocity gradient

stream. If N streams are involved in random orientation then the result expected is

$$H = \frac{GR*}{4wr_c\sqrt{N}}\left\{M^2 - (M*)_0^2\right\} \tag{34}$$

Some numbers suggested by preliminary calculations are: $w = 10^6$ cm s^{-1}, $r_c = 5 \times 10^{14}$ cm. With $R* = 10^{12}$ cm, $M* = 10M_\odot$, $(M*)_0 = 1M_\odot$ and $N = 10$ then $H = 4{\cdot}2 \times 10^{51}$g cm^2 s^{-1}. Because of the uncertainties in the figures used not too much significance should be placed on the exact value found. However it does represent some 250 times more angular momentum per unit mass than has solar material and translated into an equatorial velocity, with the usually accepted radius for such a star ($\sim 5 \times 10^{11}$ cm), it gives 25 km s^{-1}, a quite reasonable figure.

One other feature of the acquisition of angular momentum in the way suggested here is that the final direction of the angular momentum vector should be completely random, which is in concordance with observation. Theories which link the rotation of stars with the general rotation of the galaxy would give a tendency for parallelism of stellar-rotation axes with each other and with the galaxy itself.

Stellar Interactions

The capture theory of the origin of planetary systems considers possible interactions between condensed stars and diffuse protostars of some lesser mass. For the solar system the mass of the protostar might be about 0·25 $M_\odot$ with an interaction distance of up to 5×10^{14} cm, leading to tidal disruption of the protostar. This model is described in more detail elsewhere; for now we are going to consider how likely such events are.

According to the present theory, in the final stages of the evolution of a cluster the stars are produced in a fairly compact region, of volume about 10^{49} cm^3. Observations of young clusters suggest that most of the least massive stars are born in a period of about 10^{13}s. The collapse time for a small star (see Table 1 above) will be of order 5×10^9s and most of this time will be spent in a highly-diffuse state. If, say, 500 fractional-solar-mass protostars are produced in all then we come to the conclusion that at any instant within the period of 10^{13}s there is a probability of 0·25 of finding a diffuse star.

Most of the stars produced earlier, when the radius of the cloud was up to 10^{17}cm would fall in towards the centre and pass through the region in which new stars were forming. In fact they would move in quasi-Keplerian orbits taking them repeatedly into the central regions of the system and then out again into the regions in which they were first formed. At each passage through the central region, of radius about 10^{16}cm, they would sweep out a volume of about 10^{46} cm^3 such that any diffuse star within this volume would be within 5×10^{14}cm and so be disrupted. With a total volume of 10^{49} cm^3 within which stars are forming, and a likelihood of 0·25 of finding a diffuse star in the central region at any time, this gives the probability per passage of an interaction of the required type of about 1 in 4000. However in a period of 10^{13}s there would be some tens of passages through the central region. We thus conclude that the probability of a massive star interacting with a protostar in the prescribed way is about 0·01.

Once again there are many uncertainties in this analysis so that the numerical result cannot be taken too literally—it might be an overestimate but, equally, it could be an underestimate. However, it is concluded that, given the number of stars in a cluster, the possibility of encounters of the type envisaged cannot be discounted on the basis of observation or of theory.

Final Comments

So far only a limited number of computer calculations have been performed on this model. While those which have been done, and are sometimes referred to, are possibly sound for giving orders of magnitude of various quantities there can be little doubt that estimates of quantities will change as the project develops. Some changes in the theoretical treatment are also possible, especially improvements in estimating the dissipative losses of the system. However there are no indications that the qualitative behaviour of the model should differ from that described in the present work.

References

Aust, C. (1977). In preparation.

Grzedzielski, S. (1966). 'On the possibility of fragmentation of a pregalaxy with large internal velocity differences.' *Mon. Not. R. astr. Soc.*, **134**, 109–34.

Hoyle, F. (1960). 'On the origin of the solar nebula.' *Q.J.R. astr. Soc.*, **1**, 28–55.

Jeans, J. H. (1916). 'The part played by rotation in cosmic evolution.' *Mon. Not. R. astr. Soc.*, **77**, 186–99.

Jeffreys, H. (1916). 'On certain possible distributions of meteoric bodies in the solar system.' *Mon. Not. R. astr. Soc.*, **77**, 84–112.

Jeffreys, H. (1918). 'On the early history of the solar system.' *Mon. Not. R. astr. Soc.*, **78**, 424–42.

McCrea, W. H. (1960). 'The origin of the solar system.' *Proc. R. Soc. A.*, **256**, 245–65.

Williams, I. P., and Cremin, A. W. (1969). 'Young stellar clusters.' *Mon. Not. R. astr. Soc.*, **144**, 359–73.

Woolfson, M. M. (1964). 'A capture theory of the origin of the solar system.' *Proc. R. Soc. A*, **282**, 485–507.

THE CAPTURE THEORY AND THE ORIGIN OF THE SOLAR SYSTEM

M. M. WOOLFSON
Department of Physics, University of York, Heslington, York, U.K.

Abstract

The reasons for the downfall of the Jeans–Jeffreys tidal theory are reviewed. The capture theory is described whereby a diffuse protostar loses a filament of material in a near passage past a more massive condensed star and the lost material is captured by the latter body. The ideas are tested by numerical analysis with point-mass models and the plausibility of the capture process is established. The application of theory by Jeans to a filament from a stellar interaction designed to give an outcome like the solar system, suggests that there ought to have been six planets, all fairly massive. A study of the condensation of protoplanets indicates that, while they all lose considerable quantities of material, stable cores should form which will not disrupt under solar tidal forces. However it seems unlikely that planets could form in the terrestrial region. The collapsing planetary cores will lack an axis of symmetry and it is shown that as they collapse a filament of matter should be left behind. Condensations in this filament can give rise to satellite families and approximate calculations give results consistent with the orbital characteristics of Jupiter's satellites.

The process of forming protoplanets from the stellar filament releases a great deal of material around the sun. This leads to round-off of initially highly eccentric orbits on a time scale in the range $10^4 - 10^8$ years, the time depending on the mass of the planet and the assumed density of the material.

Tidal Theories

The first tidal theory of great note, proposed by Jeans (1916) and Jeffreys (1916, 1918), is illustrated in Figure 1. A massive star has drawn out a tidal filament of solar material. The material of the filament is attracted by the retreating star and so acquires orbital angular momentum around the sun. Protoplanetary condensations form within the cigar-shaped filament, with more massive planets in the centre of the system.

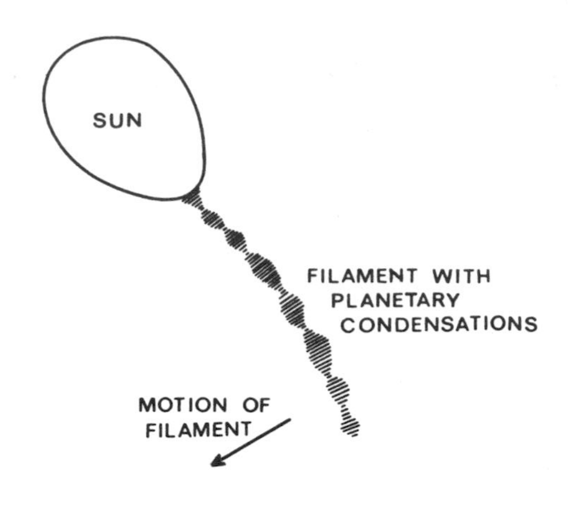

Figure 1. The Jeans–Jeffreys theory

The Jeans–Jeffreys theory was found to be untenable for a number of reasons.

(i) The proposed mechanism could not give planets sufficiently far from the sun (Russell, 1935).
(ii) Solar material is too hot to condense into planets (Spitzer, 1939).
(iii) The abundance of lithium on Earth suggests that it originated at a comparatively low temperature (less than 10^6 K).

However, Jeans subjected many aspects of his tidal theory to detailed analysis, in particular the formation of a tidal filament and its break-up into protoplanetary condensations and these analyses are quite valid.

The downfall of the tidal theory was all the more spectacular because the theory, which was simple and intuitively appealing, had enjoyed such widespread support in the scientific community. The associated trauma fairly naturally induced a deep suspicion of all tidal theories, no matter how different from the original one, and the objective evaluation of the capture theory has been somewhat delayed for this reason.

The Capture-theory; First Ideas

Without the benefit of an analysis of stellar clusters, but merely armed with the idea that stellar interactions may be common in a cluster, the author began,

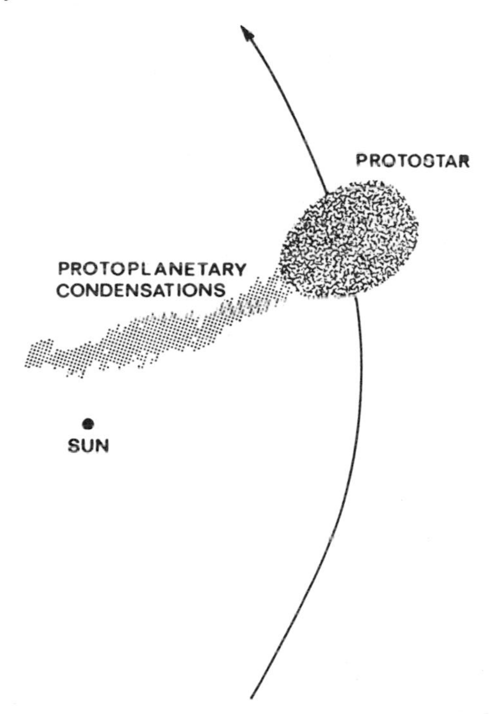

Figure 2. The basic capture theory. The filament of material is left in orbit around the sun

in 1963, an investigation of the capture-theory idea, illustrated in Figure 2. This shows the passage of a diffuse protostar past the sun. The protostar emits a tidal filament which has broken up into protoplanetary condensations—much as predicted by Jeans' analysis. However the important difference between this theory and that of Jeans and Jeffreys is that here the protoplanetary condensations coming from the diffuse star are captured by the sun. Since the protostar is cool and moving around the sun at a great distance, none of the previously-described difficulties of the original tidal theory apply to this one. Nevertheless it still remains to be shown that the capture mechanism is valid.

It was decided to test the capture idea by numerical analysis with a model star (Woolfson, 1964). The actual model used is shown in Figure 3a and it had the following characteristics:

(i) It was a point-mass model with interactions between points to simulate the various types of forces operating.
(ii) The model was two dimensional. This was not very realistic but it was all that could be managed at the time.
(iii) Since tidal forces do not greatly affect the central regions of a star the points were taken in an annular region with one central mass point.

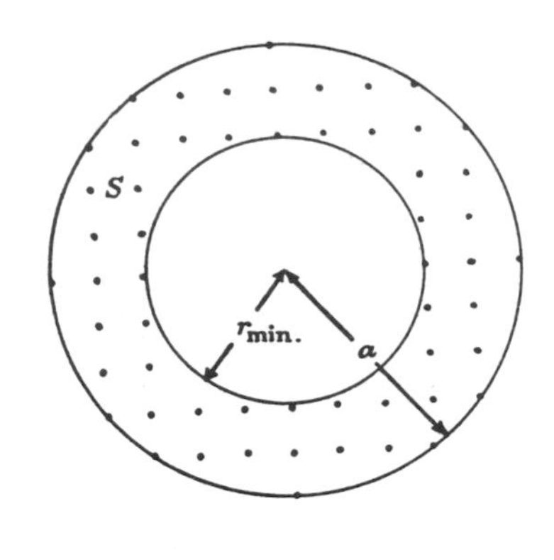

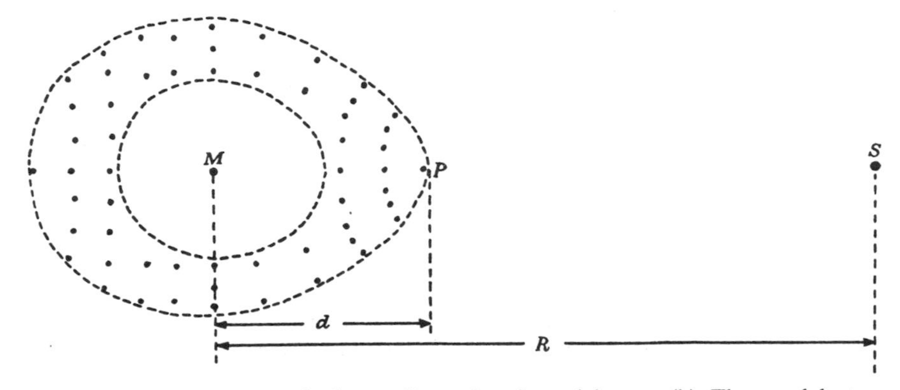

Figure 3. (a) The unperturbed two-dimensional model star. (b) The model star subjected to tidal forces (Reproduced from M. M. Woolfson, *Proc. R. Soc. A*, **282**, 485–507 (1964) by permission of The Royal Society)

(iv) The gravitational force on each point due to the protostar was assumed to act towards the centre. There was no point-to-point gravitation.

(v) The protostar was made stable by introducing a pressure-gradient simulating force. This was of the form ξ/r^m where r is the distance of the point from the centre and $m > 2$. This force was cut off if the point left the body of the star.

(vi) A test of the model in the static gravitational field of a nearby star gave the appearance shown in Figure 3b, a result similar to that given by Jeans's analysis.

(vii) Due to a tendency for points to cluster in a rather unnatural way a pressure-simulating force was introduced in the form of a repulsion between points if they approached too close, thus:

$$\left.\begin{aligned} f_p &= -\varepsilon/y \qquad & y \leqslant y_0 \\ f_p &= 0 \qquad & y > y_0 \end{aligned}\right\} \qquad (1)$$

where y is the distance between two mass points.

This model is described, not because it is the one currently used, but because it illustrates the way in which a model can be assembled to give a particular set of characteristics. Gravity, pressure gradients, pressure, energy dissipation, viscous drag—all these can be, and have been simulated in this work.

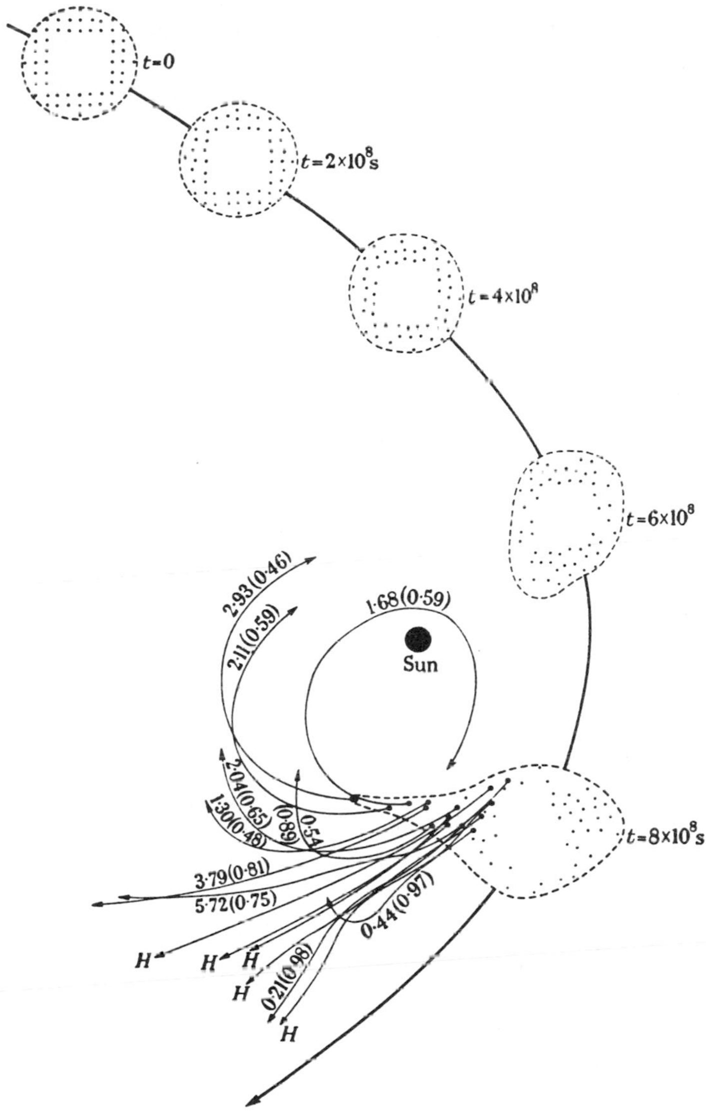

Figure 4. Disruption of a model star due to tidal effects. Captured individual points are labelled with perihelion distance (in units of 10^{14} cm) and eccentricity. Paths marked H are hyperbolic and denote mass points which are not captured (Reproduced from M. M. Woolfson, *Proc. R. Soc. A*, **282**, 485–507 (1964) by permission of The Royal Society)

The results found with this original model are reproduced in Figure 4. The mass of the protostar was 3×10^{32} g ($\sim \frac{1}{7} M_{\odot}$) and it had a mean density 7×10^{-12} g cm^{-3}. The sun–protostar orbit was parabolic with a nearest approach of $6{\cdot}67 \times 10^{14}$ cm. At the time this work was done it was thought that the rounding of planetary orbits would tend to give a final orbital radius equal to the original perihelion distance so it is the perihelion distances which are marked on the figure for individual points captured by the sun. The range of perihelion distances found was 1·4–38·4AU—a very satisfactory agreement with the range of planetary orbital radii.

A Later Model

With the experience gained from the first model, and by the use of more sophisticated techniques of numerical analysis, it has been possible to repeat the capture-theory investigation with a superior, three-dimensional model protostar (Dormand and Woolfson, 1971). In this model 75 per cent of the total mass was placed at a central point. The remainder of the mass was placed on two concentric spherical surfaces. To obtain a fairly uniform distribution of points on each surface they were placed on the normals to faces and at the vertices of regular polyhedra. For example, the hexoctahedron (3, 2, 1), shown in Figure 5(a), was used for the outer shell which formed the protostar surface. There are 48 normals to faces and 26 vertices giving 74 points in all. The inner shell, the icositetrahedron (3, 1, 1) (Figure 5(b)), similarly gave 50 points. The total number of points is 124 but because of a plane of symmetry only 70 points needed independent treatment.

In this analysis it was required to take account of variations in the gravitational field of the protostar as it changed its shape. To have done this by considering point-to-point gravitational interactions would have required the computation of 2415 forces at each stage of integrating the differential equations of motion of the points. Instead, we used the following device. An analytically-expressible geometric surface was found which had similar characteristics to the

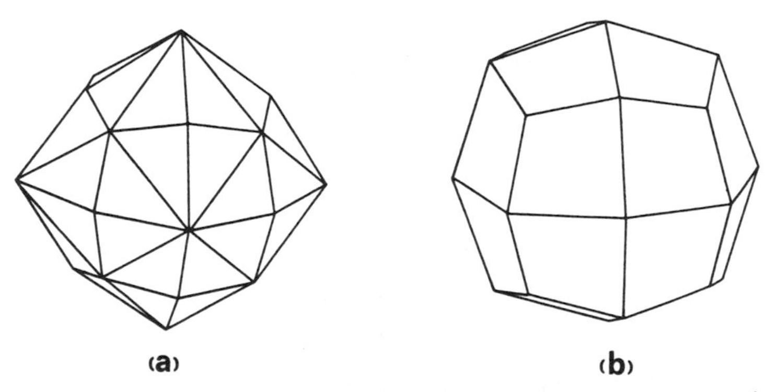

Figure 5. (a) The hexoctahedron (3, 2, 1). (b) The icositetrahedron (3, 1, 1)

expected protostar equipotential surface. The shape chosen is one based on a limocon

$$r = \frac{a(k\cos\theta - 1)}{k - 1}, |\theta| \leqslant \arccos\left(\frac{1}{k}\right) \tag{2}$$

Rotated about the axis of symmetry a surface is generated to which we have given the name limocoid. Some typical limocons are shown in Figure 6.

The gravitational field at Q (Figure 6) due to a homogeneous limocoid is a function of (R, ϕ, k) and this was computed by Gauss quadrature for a large number of positions, Q, for which $0 < R < a, 0 < \phi < \pi$ and $1 < k < 6$. At each position the centre of action, as a displacement from the centre of mass, and the ratio of the field to that of a uniform sphere at the centre of action was computed. A polynomial surface of degree (4, 9, 4) in (R, ϕ, k) was fitted

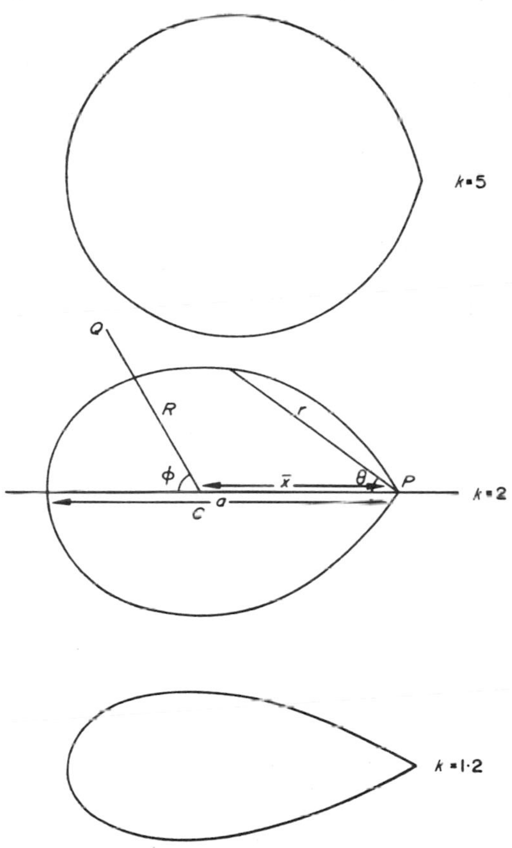

Figure 6. Limocons for various values of k (Reproduced from J. R. Dormand and M. M. Woolfson, *Mon. Not. R. astr. Soc.*, **151**, 303–331 (1971) by permission of The Royal Astronomical Society)

to the data and the coefficients were used to determine the gravitational field of the protostar at any point. At each integration step a least-squares fit of each shell of mass points was made to a limocoid surface.

The forces considered during the four-step Runge–Kutta integration of the differential equations were

(1) Gravitation due to the sun.
(2) A repulsive term to represent a pressure gradient. This was combined with the gravitational field of the protostar to give

$$\mathbf{F} = -\frac{GM_*}{R^3}\left(1 - \frac{v_c}{V}\right)R \tag{3}$$

where v_c is the actual speed of infall of the point, V its initial speed of infall, M_* the effective mass of the star and R the distance to the centre of action of the star. This form of modified gravitation tends to maintain the rate of collapse of the protostar constant.
(3) The limocoid fitting does not include escaped points. Gravitational forces are calculated individually for points at distances from the protostar centre greater than $1{\cdot}4a$.
(4) After each integration step 'velocity-sharing' was carried out. This consists of replacing the velocity of each point by a weighted mean of its own velocity and that of its neighbours. Thus the new velocity of the ith point is given by

$$\mathbf{v}_i' = \sum_{j=1}^{n} w_{ij}\mathbf{v}_j \Big/ \sum_{j=1}^{n} w_{ij} \tag{4}$$

where

$$w_{ij} = 1 \text{ if } r_{ij} \leqslant r_m$$

$$w_{ij} = (r_m/r_{ij})^4 \text{ if } r_m < r_{ij} \leqslant r_c$$

and

$$w_{ij} = 0 \text{ if } r_{ij} > r_c.$$

The parameters r_m and r_c are adjustable. The effect of velocity sharing is

(a) to prevent pooling of points,
(b) to smooth out the fluctuations one tends to get with point-mass models, and
(c) to represent drag between neighbouring regions of protostar material.

The results of the computation for various interaction situations follow the expected pattern. A filament is drawn out of the protostar, and captured individual mass points take up a wide range of orbits about the sun. The results of one quite typical calculation are illustrated in Figure 7.

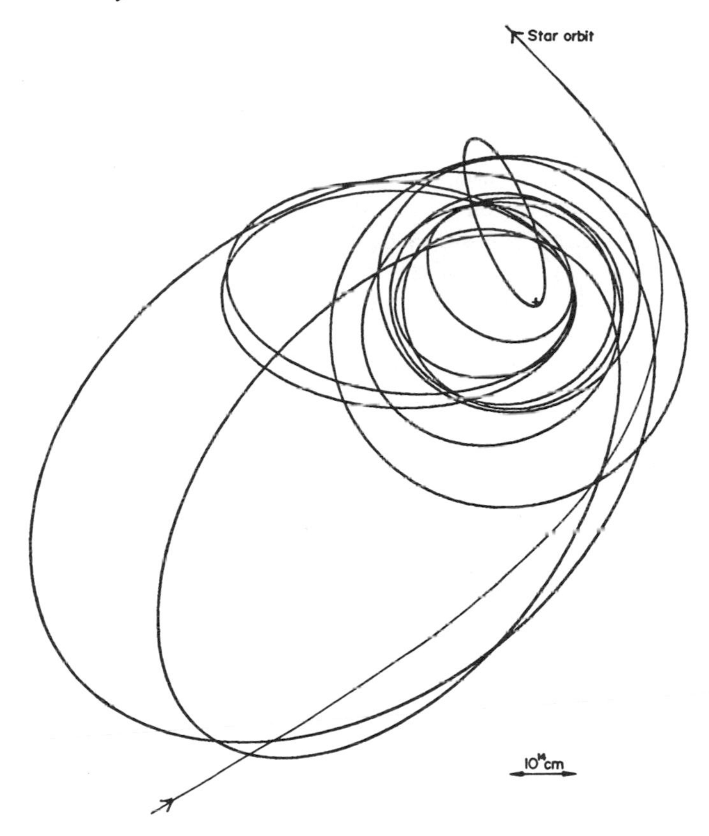

Figure 7. The range of orbits of points captured from the three-dimensional model star (Reproduced from J. R. Dormand and M. M. Woolfson, *Mon. Not. R. astr. Soc.*, **151**, 303–331 (1971) by permission of The Royal Astronomical Society)

The actual capture mechanism now seems to be well established on the basis of a large number of model investigations. The capture process itself is not very sensitive to the parameters of the model although to produce an outcome with particular characteristics—those of the solar system for example—restricts the possible range of parameters.

Within the last few years the idea of capture has been applied to a number of astronomical problems. Öpik (1972) and Mitler (1974) have evoked a capture mechanism for the formation of the Moon. Their idea is that a large body swept by the Earth, was broken by tidal forces and that some part of it was retained by the Earth.

At the other extreme, Yabushita (1971) has appealed to the capture idea for the formation of bridges between galaxies and the capture of material from one galaxy by another.

The Formation of Planets

It is a necessary condition for a satisfactory theory of the origin of the solar system to produce material around the sun with total mass and angular momentum consistent with what is required for the planets. It is a necessary but not a sufficient condition—it must also be shown that on a reasonably short time-scale such material will actually form planets.

The point-mass models indicate that material would escape from the star in the form of a filament. The model is too crude to show how the filament would break up into a series of blobs but this is not too worrying as here we have a well-proved theory to guide us—that of Jeans.

Our model is good enough to suggest the general characteristics of the filament. In Figure 8 there is shown the formation of a filament from a protostar of mass 0·25 $M_\odot$ and radius $2{\cdot}5 \times 10^{14}$ cm (Dormand and Woolfson, 1971). It will be seen that about one sixth of the mass associated with the mass points is contained in a filament of length about $2{\cdot}5 \times 10^{14}$ cm and thickness about 5×10^{13}cm. This suggests a mass of filament about 2×10^{31}g with density about $4{\cdot}0 \times 10^{-11}$g cm^{-3}. According to Jeans a condensation will form in a length of filament

$$l = \tfrac{1}{2} c \sqrt{\frac{\pi}{\gamma G \rho}} \tag{5}$$

where c is the velocity of sound in the gaseous material and γ is the ratio of specific heats. For atomic hydrogen at 100 K the value of c is about 10^5cm s^{-1} and with $\gamma = 5/3$ this gives $l = 4 \times 10^{13}$cm. This suggests about six condensations of average mass 3×10^{30}g with the masses tending to be greater for the central condensations.

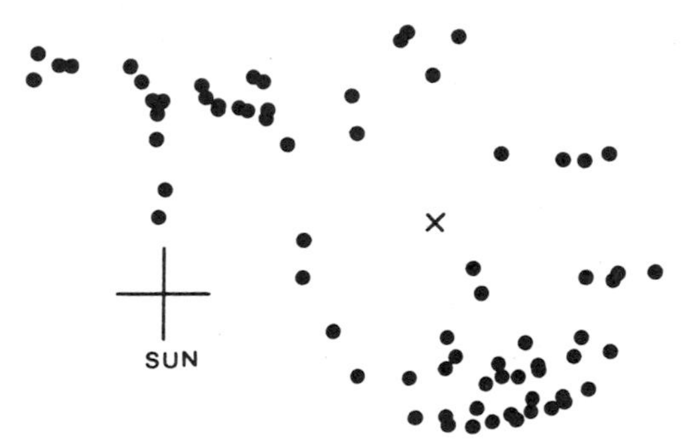

Figure 8. Intermediate stage in the disruption of the three-dimensional model star showing filament formation (Reproduced from J. R. Dormand and M. M. Woolfson, *Mon. Not. R. astr. Soc.*, **151**, 303–331 (1971) by permission of The Royal Astronomical Society)

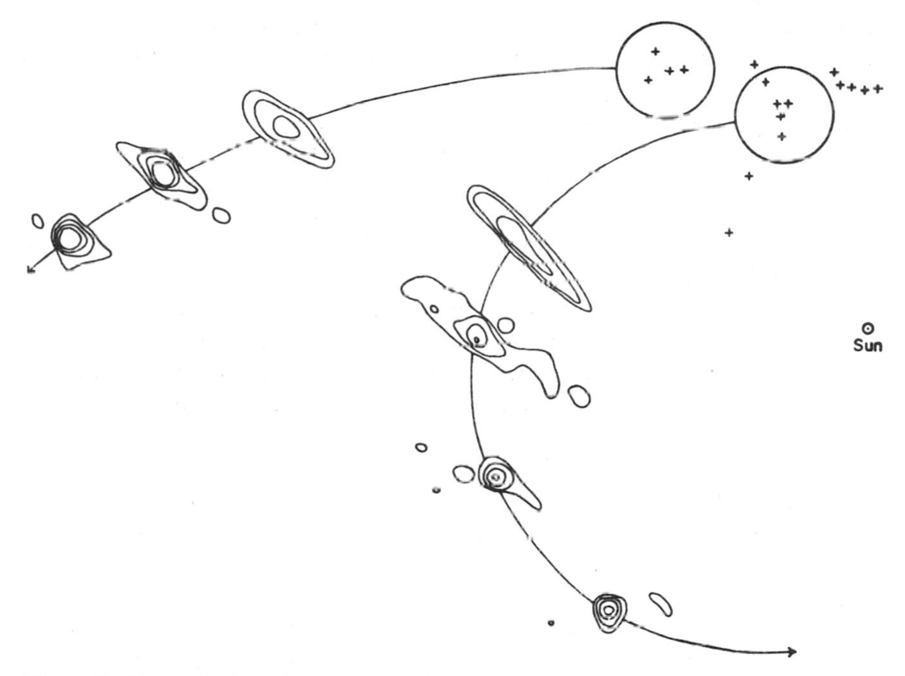

Figure 9. The evolution of two captured protoplanets. The relation is shown between the filament mass points (crosses) and the initial model planet configuration. Density map contours are drawn at 10, 20, 40, and 80 units (Reproduced from J. R. Dormand and M. M. Woolfson, *Mon. Not. R. astr. Soc.*, **151**, 303–331 (1971) by permission of The Royal Astronomical Society)

The way in which the collapse of a protoplanet was examined was again to use a point mass model. In Figure 9 there are shown in the filament two initial spherical blobs. The matter within them was rearranged over discrete mass points distributed on a cubical lattice contained within the spherical volume. The gravitational forces acting on each mass point were

(i) due to sun;
(ii) due to the protostar considered as a point mass;
(iii) due to all other mass points.

The form of interaction between mass points was made of the form

$$F_{ij} = -\frac{Gm_i m_j}{r_{ij}^2}\left[1 - \frac{\xi}{r_{ij}}\left(\frac{s_i^2 + s_j^2}{2}\right)^{1/2}\right], \tag{6}$$

where s_i is the distance of mass point i from the centre and ξ is an adjustable constant. The effect of this law is that condensation may take place without limit but also without the clumping together of mass points. In addition condensation is most favoured in the central region, a characteristic which would be expected for an actual protoplanet.

A final non-gravitational force that was used to simulate the effect of viscous forces is one which gives an acceleration $\chi \dot{s}_i / s_i^2$ to each mass point where χ is a constant.

The equations of motion of the points were integrated by a standard four-step Runge–Kutta process. If mass points approach one another then, to preserve accuracy, the time step must be reduced and the whole process may grind to a halt. To prevent this, at intervals, the mass was redistributed over a uniform cubical grid with different masses allocated to the grid points to preserve the density variations which came about.

The evolution of two protoplanets is shown in Figure 9 where the distribution of mass is displayed as projected density contours. For both protoplanets the central region collapses while outer material is shed but the central density is such that the planetary core will not be tidally disrupted during its first perihelion passage.

One result which came out of the original work by Dormand and Woolfson is that in various trials it did not seem possible to form planetary condensations such that the perihelion was less than somewhere about the orbit of Mars. At the time this was recorded as a possible difficulty of the theory being presented.

The model for planetary condensation produced here cannot be regarded as entirely satisfactory as it contains no feature corresponding to the build up of temperature in an opaque body. Some recent work by Donnison and Williams (1974) and Bodenheimer (1974) on the collapse of a proto-Jupiter has been noted, but neither of these exercises has considered the case of a protoplanet which initially has several times a Jeans' critical mass. A more detailed analysis of protoplanetary collapse is at present being undertaken by the author and N. Schofield.

The Formation of Satellites

If we examine the collapsing protoplanets in Figure 9 we notice that, due to tidal effects, they take up a distinctly non-spherical shape. In the final stages of collapse, when the density has become reasonably high, tidal effects will have become insignificant and one can study the collapse as though the body was in isolation. An idealised picture of a high density protoplanet just free of tidal effects is shown in Figure 10. We assume a density of 10^{-7}g cm^3 for a body of mass $2{\cdot}0 \times 10^{30}$g. The projection of the tidal bulge is $0{\cdot}4\,R$, where R is the radius of the protoplanet, since analysis and calculation show that this is about the maximum distortion attained without actual disruption of the body. Calculation of the free fall of various boundary points of the planet with the simplifying assumption that all its mass is concentrated at the centre, leads to the result shown of Figure 10. The tip of the bulge is progressively left further behind as the collapse continues and a filament is formed. Jeans' analysis will apply to this filament, as it does to a tidal filament, and we may expect a series of condensations to form.

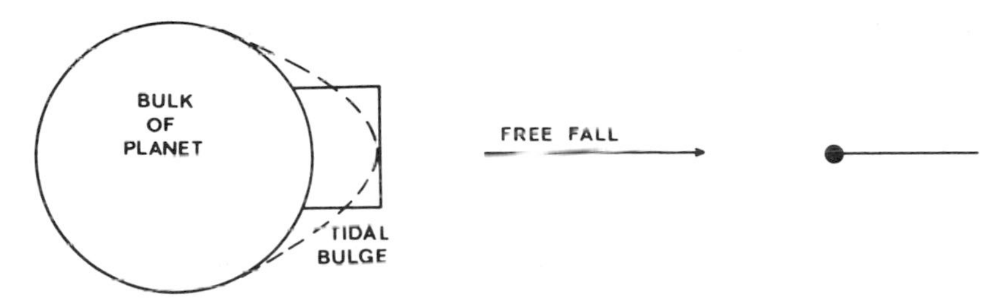

Figure 10. The collapse of a model planet with tidal bulge under the influence of gravitational forces, showing filament formation

From results on orbital round-off given later it appears that the initial perihelion distance of Jupiter might be about 3 AU with an orbital eccentricity of about 0·8. The tidal effect on the protoplanet would be expected to be a maximum at perihelion and the angular momentum per unit mass imparted should be greatest in the bulge region. If we assume that the tidal tip takes on a rotation angular velocity equal to the perihelion rate, then this is about $5 \times 10^{-8}\ s^{-1}$. The angular momentum per unit mass of tidal tip material may be now calculated as $2{\cdot}8 \times 10^{17}\ cm^2\ s^{-1}$ compared with $1{\cdot}6 \times 10^{17}\ cm^2\ s^{-1}$ for Callisto, the outermost Galilean satellite.

In view of all the assumptions the calculation is no more than suggestive and a more detailed model of this event is under active investigation. However, it can be seen that the idea that satellites are the result of rotational disruption of a collapsing protoplanet is a feasible one. Those planets with the highest rotational angular velocities are also those with the most extensive families of large satellites and these satellites are strictly in the equatorial plane, as this mode of formation would suggest.

The downfall of Laplace's nebula theory, which attributed the formation of planets to rotational instability of a protosun, was due to a comparison of the distribution of mass and angular momentum among the members of the solar system. This was, and still is, considered a crucial argument. Yet it is argued by some theorists today that the relationship of satellites to planets should be ascribed to the same mechanism that gave rise to planets in relation

Table 1.

$$r_m = \frac{\text{total mass of secondary bodies}}{\text{mass of primary body}}$$

$$r_A = \frac{\text{total orbital angular momentum of secondary bodies}}{\text{rotational angular momentum of primary body}}$$

Primary	r_m	r_A
Sun	0·0013	99
Jupiter	0·0002	0·007

to the sun. That assumption should be considered in relation to the contents of Table 1.

The Round-off of Planets

All the mass contained in the planetary filament would not end up in condensed planets; indeed we have predicted the filament as having some six times the total mass of the planets. The model of planetary formation indicates that large amounts of outer protoplanetary material are lost while the central core is increasing in density. Hanbury and Williams (1975) have described a process whereby a planet such as Uranus could have lost most of the hydrogen even from a dense core. Thus, given some initial inefficiency in converting a filament into protoplanetary condensation, plus subsequent losses from the protoplanets by various mechanisms we may conclude that a great deal of unattached material was released in the solar system to form a resisting medium.

For a planet moving in a resisting medium there are the two types of resisting mechanism. The first of these, described by Dodd and McCrea (1952), is due to the motion, and hence energy, imparted to the medium by the gravitational field of the planet. This energy comes from the planet's own motion and gives a resisting law of the form

$$\mathscr{R} = -2\pi\rho\frac{G^2M_p}{W^3}\ln\left(1+\frac{S^2W^4}{G^2M_p^2}\right)\mathbf{W}, \tag{7}$$

where $\mathscr{R}$ is the acceleration of the planet of mass M_p
$S = r(M_p/2M_\odot)^{1/3}$, the radius of the planet's sphere of influence
r is the distance of the planet from the sun,
$\mathbf{W}$ is the velocity of the planet relative to the medium and ρ is the density of the medium.

Another form of resistance is due to the accretion of the medium by the planet. This comes from 'momentum sharing' between the planet and accreted material and the resistance law which results is

$$\mathscr{R}_a = -\frac{\pi\rho D^2 W}{M_p}\mathbf{W} \tag{8}$$

where

$$D^2 = R_p^2 + \frac{2GM_pR_p}{W^2} \tag{9}$$

and R_p is the planetary radius.

Combining (8) and (9) there is found a composite resistance law

$$\mathscr{R} = -\frac{\pi\rho}{M_p}\left[D^2W + \frac{2(GM_p)^2}{W^3}\ln\left\{\frac{1+S^2W^4/G^2M_p^2}{1+D^2W^4/G^2M_p^2}\right\}\right]\mathbf{W} \tag{10}$$

The modification of the logarithmic term is due to the fact that matter which contributes to accretion resistance does not also contribute to momentum-transfer resistance.

These resistance laws were applied to various planetary orbits as indicated by the capture theory (Dormand and Woolfson, 1974). In order to test for the relative efficiency of the two mechanisms the equations of motion of planets were integrated both for resistance without accretion and then for resistance with accretion. For the tests the medium was taken of uniform density and the results are shown in Figure 11. The interesting feature of these results is the comparatively small difference in rounding time in the two cases.

Other calculations were made to test for the effect of various density distributions. The general run of results is shown in Table 2. In each case the resisting medium was assumed to be in the form of a disk of thickness 10^{14} cm with no variation of density across the thickness. In the plane of the orbit, density varied with distance from the sun as shown by the formulae in Table 2, one being a Gaussian and the other an exponential distribution. The main conclusions from this work, in which accretion was neglected, are:

(i) The rounding times are inversely proportional to the average density of medium through which the planet passes but only marginally dependent on the distribution of density—within reason.

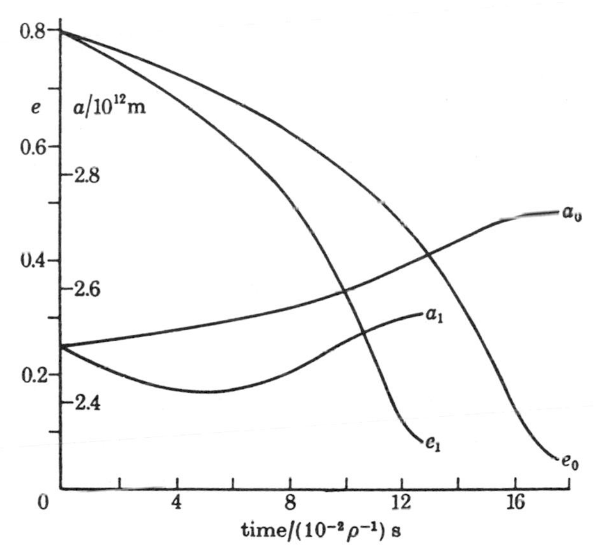

Figure 11. The action of a resisting medium with accretion (a_1, e_1) and without accretion (a_0, e_0). Density is constant and planetary mass is 10^{27} kg for no accretion but increases to 4×10^{27} kg with accretion (Reproduced from J. R. Dormand and M. M. Woolfson, *Proc. R. Soc. A*, **340**, 349–365 (1974) by permission of The Royal Society)

Table 2

Rounding times for various planetary masses and distributions of resisting medium

(a, e) initial	$= (2{\cdot}5 \times 10^{14}$ cm, $0{\cdot}8)$, Mass of medium $= 10^{31}$ g.
Situation A	$\rho = \rho_0 \exp\{-[(r - 10^{14}\text{ cm})/5 \times 10^{13}\text{ cm}]^2\}$
Situation B	$\rho = \rho_0 \exp\{-[(r - 10^{14}\text{ cm})/10^{14}\text{ cm}]^2\}$
Situation C	$\rho = \rho_0 \exp(-\alpha r)$, $\alpha = 2 \times 10^{-14}\text{ cm}^{-1}$
Situation D	$\rho = \rho_0 \exp(-\alpha r)$, $\alpha = 10^{-14}\text{ cm}^{-1}$

	Rounding times ($e < 0{\cdot}1$) in units 10^6 years		Planetary mass (g)		
Situation	10^{30}	10^{29}	10^{28}	10^{27}	10^{26}
A	0·056	0·39	3·0	24	200
B	0·062	0·42	3·2	26	220
C	0·076	0·48	3·6	29	250
D	0·069	0·43	3·2	26	220

(ii) Rounding times are approximately inversely proportional to the mass of the planet.
(iii) The eccentricity of the orbit varies slowly at first and then more rapidly as the round-off develops.
(iv) The characteristic of the changing orbit which varies comparatively little is the semi-latus rectum.

More recently the effect of the resisting medium on the inclination and other orbital characteristics has been examined (Dormand and Woolfson, 1977). For this work the gaseous part of the resisting medium was considered as well spread out and in a spherically symmetric distribution around the sun of the form

$$\rho = \rho_0 \exp\{-r^2/\alpha^2\} \qquad (11)$$

with $\alpha = 10^{14}$cm. The solid part of the medium, some 2 per cent of the whole mass, is imagined as spread into a disk of radius 8×10^{14}cm. The important feature of this model medium is that not only is there the normal resistance force but that there is also an extra non-central gravitational force acting on the planet when it is out of the plane of the disk. These two types of force will be present in any realistic situation and the results obtained do not depend sensitively on the details of how they are modelled. In Figure 12 the characteristics of a planetary orbit are defined. The line of nodes AD lies in the plane of the dust disk as indeed must the sun. The orientation of the orbit with respect to the line of nodes is defined by α which is measured in the plane of the orbit. The absolute orientation of the orbit is defined by ϕ, the angle measured in the plane of the disk between the line of nodes and a line in the disk fixed relative to the stars. Finally ψ is the angle between the plane of the orbit and the plane of the disk.

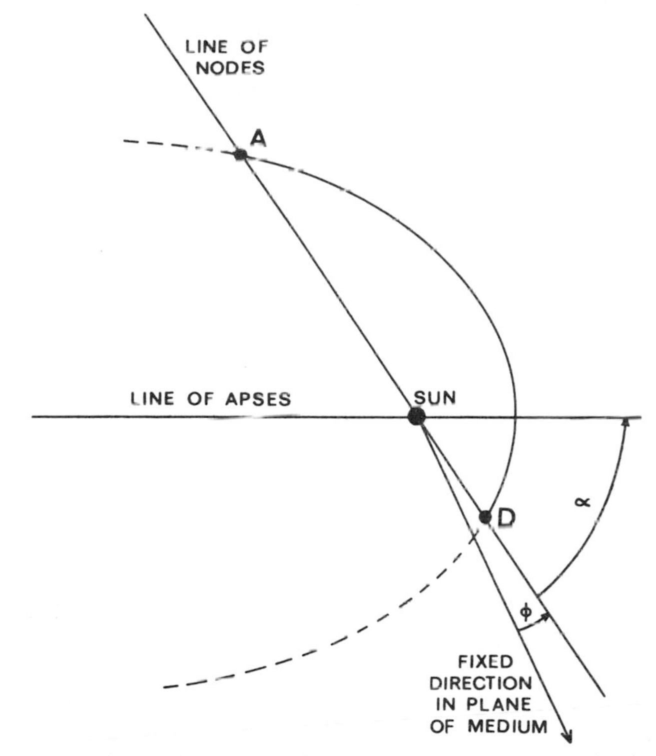

Figure 12. The description of a planetary orbit inclined to the dust disk

In Figure 13 are shown some results for a planet of mass 10^{29}g with a radius of 2×10^9cm in a medium with the distribution given by (11) and with total mass 2×10^{30}g. The results can be roughly scaled for other masses of planet and medium as previously explained. It will be seen that round-off from $e =$ 0·95 takes about 3×10^6 years. The line of nodes rotates about 4π during this period while the angle α changes much more rapidly. The inclination, ψ, has an interesting behaviour pattern with a general decline following the fall pattern of eccentricity but with a short-period fluctuation superimposed.

For the round-off of planets of mass 10^{28} g and upwards the requirement is the existence of a resisting medium for a period of 10^7–10^8 years. Various estimates by Hoyle (1960) and Dormand and Woolfson (1974) indicate that 10^8 years is a lower limit for the lifetime of a resisting medium and therefore, under the conditions postulated by the capture theory, that round-off of planetary orbits should occur.

Orbital Radii

As previously stated the point-mass model is too crude to indicate the fine details of the solar system. Nevertheless, by examining the behaviour of the

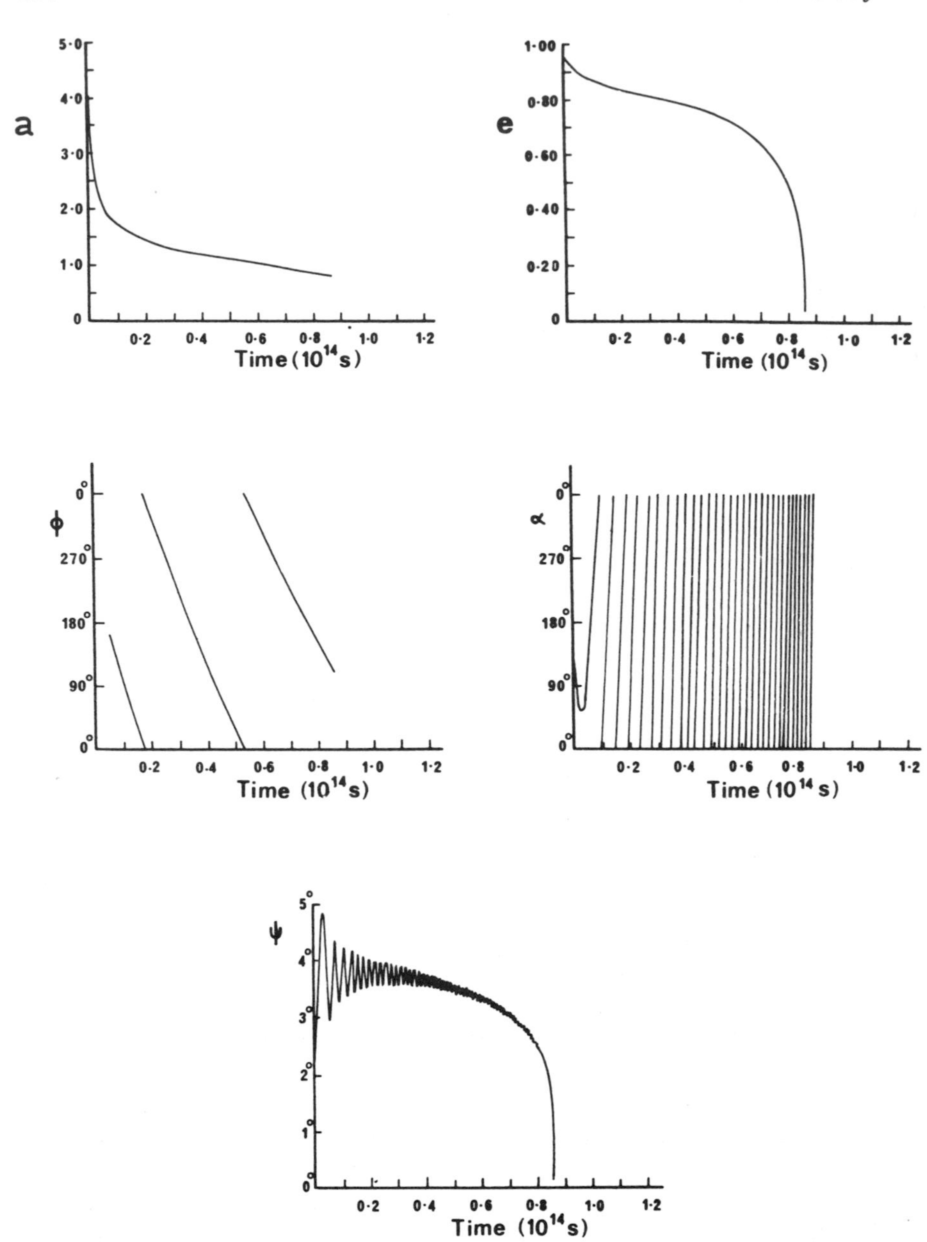

Figure 13. The effect of a resisting medium and a dust disk on the orbital characteristics of a planet

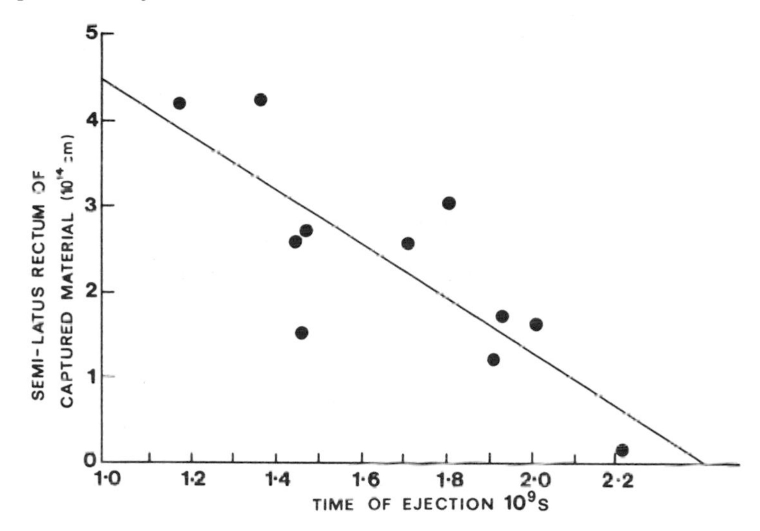

Figure 14. The relationship between semi-latus rectum of captured points and their time of ejection from the three-dimensional model protostar

mass points in the filament and accepting that some number of neighbouring mass points would represent a protoplanet it is possible to make a reasonable assessment of the orbital radii predicted by the model.

We may define a 'time of ejection' of a mass point as that time at which its distance from the centre of the protostar is 1·4 of the average radius. In Figure 14 there is recorded the semi-latus rectum ($\sim$ round-off radius) of escaping points against the time of ejection. This shows that the planets are formed with the outermost first. In addition there is a very satisfactory range of values of semi-latus rectum comparing well with range of orbital radii of the planets ($r_{\text{Neptune}} = 4{\cdot}6 \times 10^{14}$cm).

Conclusion

The capture theory is seen to give a satisfactory account of the formation of planets and of satellites and also to explain well the existence of near-circular orbits with the range of values observed in the solar system.

References

Bodenheimer, P. (1974). 'Calculations of the early evolution of Jupiter.' *Icarus*, **23**, 319–25.

Dodd, K. N., and McCrea, W. H. (1952). 'On the effect of interstellar matter on the motion of a star.' *Mon. Not. R. astr. Soc.*, **112**, 205–214.

Donnison, J. R., and Williams, I. P. (1974). 'The evolution of Jupiter from a protoplanet.' *Astrophysics and Space Science*, **29**, 387–396.

Dormand, J. R., and Woolfson, M. M. (1971). 'The capture theory and planetary condensation.' *Mon. Not. R. astr. Soc.*, **151**, 303–331.

Dormand, J. R., and Woolfson, M. M. (1974). 'The evolution of planetary orbits.' *Proc. R. Soc. A.*, **340**, 349–365.
Dormand, J. R., and Woolfson, M. M. (1977). 'Interactions in the early solar system'. *Mon. Not. R. astr. Soc.*, **180**, 243–279.
Hanbury, M. J., and Williams, I. P. (1975). 'The formation of the outer planets.' *Astrophysics and Space Science*, **38**, 29–37.
Hoyle, F. (1960). 'On the origin of the solar nebula.' *Q.J.R. astr. Soc.*, **1**, 28–55.
Jeans, J. H. (1916). 'The part played by rotation in cosmic evolution.' *Mon. Not. R. astr. Soc.*, **77**, 186–199.
Jeffreys, H. (1916). 'On certain possible distributions of meteoric bodies in the solar system.' *Mon. Not. R. astr. Soc.*, **77**, 84–112.
Jeffreys, H. (1918). 'On the early history of the solar system.' *Mon. Not. R. astr. Soc.*, **78,** 424–442.
Mitler, H. E. (1975). 'Formation of an iron-poor Moon by partial capture or: Yet another exotic theory of lunar origin.' *Icarus*, **24**, 256–268.
Öpik, E. J. (1972). 'Comments on lunar origin.' *Irish. Astron. J.*, **10**,190–238.
Russell, H. N. (1935). *The Solar System and its Origin.* Macmillan, New York.
Spitzer, L. (1939). 'The dissipation of planetary filaments.' *Astrophys. J.*, **90**, 675–688.
Woolfson, M. M. (1964). 'A capture theory of the origin of the solar system.' *Proc. R. Soc. A*, **282**, 485–507.
Yabushita, S. (1971). 'The possibility of capture in the restricted problem of three bodies and the formation of bridges between galaxies.' *Mon. Not. R. astr. Soc.*, **153**, 97–109.

THE EVOLUTION OF THE SOLAR SYSTEM

M. M. WOOLFSON
Department of Physics, University of York, Heslington, York, U.K.

Abstract

According to the capture theory of the origin of the solar system protoplanets would have formed moving in elliptical orbits of high eccentricity. Before the orbits had rounded off there would be the possibility of significant planetary interactions and the characteristic times are found for interactions leading either to direct collisions or to the escape of one or other of the planets. The consequences of a collision between two planets are considered. It is found that the larger planet could be expelled from the solar system and that the smaller, breaking into two major almost-equal fragments, could have given the terrestrial planets Venus and Earth. It is also shown that one satellite of the larger planet could be captured by the Earth fragment to give the Earth–Moon system while another satellite, released into an independent heliocentric orbit, could give the planet Mars. The features of asteroids, meteorites and comets are discussed in relation to their origin as debris from the collision.

Introduction

Planets produced by the capture process would, initially, have fairly eccentric orbits. The round-off caused by the resisting medium takes a time which depends on the mass of the planet and on the density and distribution of the resisting medium. The initial planetary orbits will tend to be coplanar, the plane being defined by the star–sun orbit, although we shall assume that owing to various factors, some intrinsic rotation of the protostar for example, they might be inclined to the mean plane by up to one-tenth of a radian, say. Nevertheless, in projection, the initial orbits will intersect and there is some possibility of appreciable interactions between the planets while round-off is in progress. Of course the planets interact gravitationally all the time; what we shall be concerned with here are interactions which give substantial modifications of the orbits of the involved planets or which give a direct collision of two planets.

Planetary Interactions in the Early Solar System

There is a fairly simple way of looking at the interaction of two planets which are in independent heliocentric orbits but approach each other closely. The characteristic of the orbits may be used to find the approach velocities of the two bodies in a sun-centred system. During the close approach of the bodies perturbation by the sun may be ignored and the interaction of the bodies is considered in a centre-of-mass coordinate system for the two bodies.

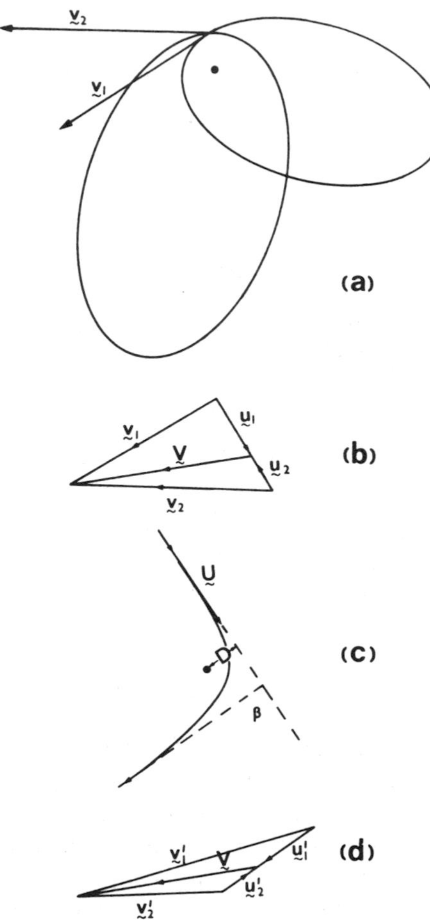

Figure 1. (a) Two intersecting orbits with velocities relative to the sun. (b) Composition of $\mathbf{v}_1$ and $\mathbf{v}_2$ to give velocities $\mathbf{u}_1$ and $\mathbf{u}_2$ relative to the centre of mass of the planets. (c) Relative motion of planets during the interaction. The relative velocity is rotated by an angle β.(d) The new sun-centred velocities, $\mathbf{v}'_1$ and $\mathbf{v}'_2$ after the interaction

Finally, when they are sufficiently removed from one another, one may transfer back to the sun-centred system and determine the new planetary orbits.

We shall examine this process and also get an insight as to what may be the result of a planetary interaction by looking at the example illustrated in Figure 1. In Figure 1(a) there are shown two intersecting orbits, for now assumed coplanar. The velocities of approach in the sun-centred system are $\mathbf{v}_1$ and $\mathbf{v}_2$. In Figure 1(b) we transfer our attention to the region of interaction. The centre of mass of the two bodies is at velocity $\mathbf{V}$ with respect to the sun while the velocities of the planets in a mutual centre-of-mass system are $\mathbf{u}_1$ and $\mathbf{u}_2$ where

$$\mathbf{v}_1 = \mathbf{V} + \mathbf{u}_1 \tag{1a}$$

and

$$\mathbf{v}_2 = \mathbf{V} + \mathbf{u}_2 \tag{1b}$$

In Figure 1 (c) the motion of one of the planets is shown relative to the other. The approach velocity $\mathbf{U}$ $(= \mathbf{u}_1 + \mathbf{u}_2)$ is along a line passing within a distance D of the other planet. We shall refer to D as the *interaction parameter*. It is easily shown that the relative motion is rotated by an angle β where

$$\tan\frac{\beta}{2} = \frac{G(M_1 + M_2)}{U^2 D} \tag{2}$$

Now we transfer back to the sun-centred system after the interaction is complete in Figure 1(d). The centre-of-mass velocity, $\mathbf{V}$, is unchanged. However as compared with the situation in Figure 1(b) the vectors $\mathbf{u}_1$ and $\mathbf{u}_2$ have been rotated through an angle β to become $\mathbf{u}_1'$ and $\mathbf{u}_2'$. The final velocities of the planets relative to the sun are

$$\mathbf{v}_1' = \mathbf{V} + \mathbf{u}_1' \tag{3a}$$

and

$$\mathbf{v}_2' = \mathbf{V} + \mathbf{u}_2' \tag{3b}$$

What will be clear from this example is that the vectors $\mathbf{v}_1'$ and $\mathbf{v}_2'$ may differ markedly from $\mathbf{v}_1$ and $\mathbf{v}_2$. In particular the magnitudes of these vectors can change greatly to the point, in fact, where one of the planets can be thrown completely out of the solar system.

We may find what is the condition for the escape of planet 1. In Figure 2 we see the velocities $\mathbf{V}$ and $\mathbf{u}_1$ the sum of which is $\mathbf{v}_1$, the initial velocity of planet 1 with respect to the sun. The effect of the planetary interaction is to swing $\mathbf{u}_1$ about the point A to give a new vector $\mathbf{u}_1'$. The possible termini of $\mathbf{u}_1'$ describe the arc XY. We draw a circle, centre O, with radius v_e, the escape speed for the region where the planets interact. It will be seen that the limits for vector $\mathbf{u}_1'$ are $\overrightarrow{AX}$ and $\overrightarrow{AY}$ corresponding to angles β_1 and β_2. From equa-

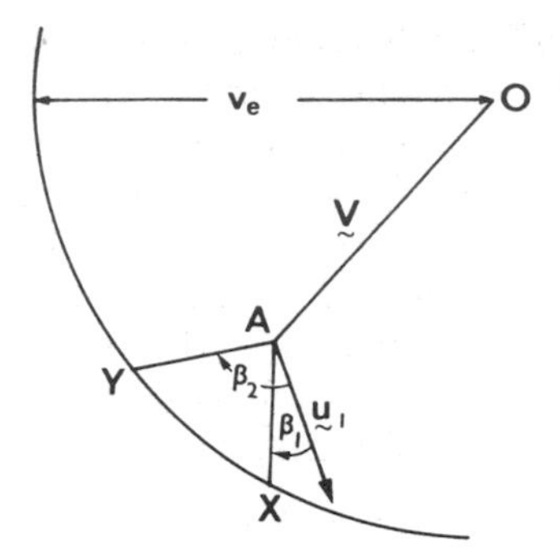

Figure 2. Construction for determining β range giving escape of a planet

tion (2) there are found two limiting interaction parameters

$$D_1 = \frac{G(M_1 + M_2)}{U^2} \cot \frac{\beta_1}{2} \tag{4a}$$

and

$$D_2 = \frac{G(M_1 + M_2)}{U^2} \cot \frac{\beta_2}{2} \tag{b}$$

Planet 1 will escape for any interaction parameter, D, such that

$$D_1 > D > D_2. \tag{5}$$

Another possibility of some interest is that of collision of the two bodies. For an interaction parameter D and an approach speed, U, the nearest approach of the two bodies, R, is given by

$$D^2 = R^2 + \frac{2G(M_1 + M_2)R}{U^2} \tag{6}$$

If the radii of the planets are a_1 and a_2 then there will be a collision if

$$R < a_1 + a_2 \tag{7}$$

This gives as the condition for a collision

$$D \leqslant D_L \tag{8}$$

where

$$D_L = (a_1 + a_2)^2 + \frac{2G(M_1 + M_2)(a_1 + a_2)}{U^2} \tag{9}$$

The outcome of an interaction with coplanar orbits is seen to depend on U and D. The former quantity depends on the geometry of the orbits and is easily calculable. The quantity D depends on how the planets approach each other at the point of orbital intersection. In Figure 3(a) we see a sequence of positions of the two planets on their orbits without disturbance by mutual interactions. Figure 3(b) shows the path of planet 2 relative to planet 1 and the interaction parameter D is marked. If the starting point of planet 2 is advanced by small increments as shown in Figure 3(a) then the relative paths of the two planets are shown in Figure 3(c). Thus with the starting point for planet 1 fixed at A there is a range of starting points for planet 2 which will satisfy condition (5), and so lead to the loss of planet 1 from the system, and another range of starting points leading to condition (8), which results in a collision. Since, in general, the planets have different and non-commensurate orbital periods then over the course of time a large variety of interaction situations will arise and it is possible to calculate from probability theory a timescale for any particular type of outcome.

The inclusion of a third dimension obviously complicates the model. The orbits, being inclined to one other, may not have an intersection at all and, indeed, the nearest approach of two planets may never be close enough to give any appreciable perturbation of either one of them. However, here we

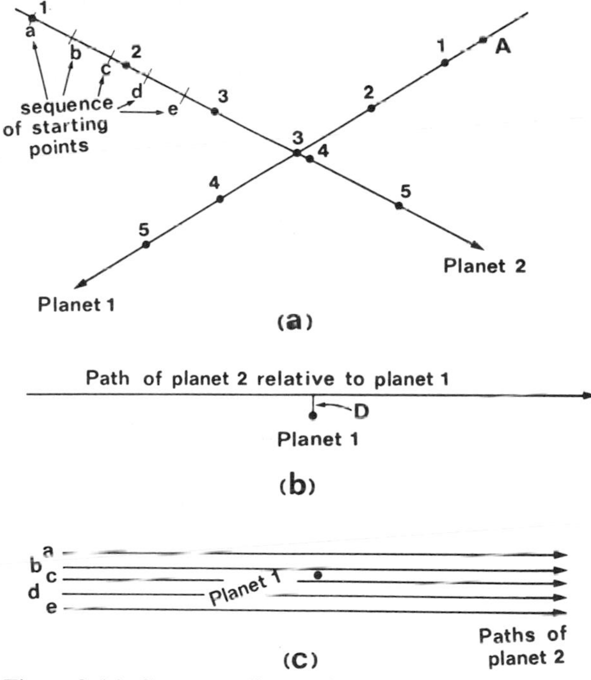

Figure 3. (a) Corresponding points for two planets on intersecting paths. (b) The path of planet 2 relative to planet 1 for the points shown in (a). (c) The effect on the relative path of starting 2 at the sequence of points shown in (a)

must recall the effect of a resisting medium which imposes a non-central force on the planets. The line of nodes of the orbits will rotate relative to a fixed direction and the line of apses will rotate fairly rapidly relative to the line of nodes. As time passes the spatial relationship of the planets will change continuously and we may assume that over a sufficient period of time, with eccentricities and inclinations not changing greatly, all possible relative configurations will occur.

It is not possible here to reproduce all the mathematics which enables the characteristic times to be calculated for the escape of planets or for a collision to take place. A full account of the analysis is given by Dormand and Woolfson (1977). However, there are some results given here and these enable conclusions to be drawn concerning the likelihood of particular types of event taking place.

In Table 1 there are shown the characteristics of six initial planets, suggested by present observation of the solar system and by the capture-theory model. The masses of the planets A and B are speculative and are approximately $30\,M_{\oplus}$ and $5\,M_{\oplus}$. The densities of all the planets are somewhat lower than those observed but it is assumed that the final collapse of planets to their present densities may take some time. The sequence of eccentricities is suggested by the capture-theory model and the inclinations are chosen fairly randomly in a range up to 3°. The rounding times are, again, suggestive rather than precise, and they could easily be an order of magnitude greater than those given.

In Table 2 there are shown the characteristic times, τ, for various outcomes for interactions between pairs of planets. The values of τ are such that the probability of the event taking place in time t is

$$p = 1 - e^{-t/\tau} \tag{10}$$

the characteristic time τ_{ME} is found from

$$\frac{1}{\tau_{ME}} = \frac{1}{(\tau_{esc})_1} + \frac{1}{(\tau_{esc})_2} + \frac{1}{\tau_{col}}. \tag{11}$$

Table 1. Suggested initial properties and orbits for a system of planets based on the capture-theory model and on observation

Planet	Mass (kg)	Density ($\times 10^3$ kgm^{-3})	Semi major axis ($\times 10^9$ km)	Eccentricity	Semi-latus rectum (AU)	Inclination	Rounding time (years)
Neptune	$1{\cdot}09 \times 10^{26}$	1·20	9·34	0·720	30·20	3°	2×10^6
Uranus	$9{\cdot}25 \times 10^{25}$	1·20	5·34	0·690	18·80	$2\frac{1}{2}$°	2×10^6
Saturn	$5{\cdot}97 \times 10^{26}$	0·50	2·79	0·680	10·07	$1\frac{1}{2}$°	3×10^5
Jupiter	$2{\cdot}00 \times 10^{27}$	1·00	2·19	0·800	5·29	2°	10^5
A	$2{\cdot}00 \times 10^{26}$	1·20	1·83	0·874	2·90	1°	2×10^6
B	$3{\cdot}15 \times 10^{25}$	0·80	1·36	0·908	1·60	1°	6×10^6

Table 2. Characteristic times for interactions between planets with properties and orbits as given in Table 1
$(\tau_{esc})_1$ is c.t. for the escape of planet 1
τ_{col} is c.t. for a planetary collision
τ_{ME} is c.t. for some major event

Planet 1	Planet 2	$(\tau_{esc})_1$	$(\tau_{esc})_2$	τ_{col}	τ_{ME}
B	A	$2{\cdot}41 \times 10^6$	$1{\cdot}79 \times 10^9$	$3{\cdot}33 \times 10^7$	$2{\cdot}24 \times 10^6$
B	Jupiter	$1{\cdot}08 \times 10^5$	∞	$1{\cdot}31 \times 10^7$	$1{\cdot}07 \times 10^5$
B	Saturn	$2{\cdot}53 \times 10^6$	∞	$4{\cdot}12 \times 10^7$	$2{\cdot}38 \times 10^6$
A	Jupiter	$9{\cdot}04 \times 10^4$	∞	$5{\cdot}91 \times 10^6$	$8{\cdot}90 \times 10^4$
A	Saturn	$2{\cdot}56 \times 10^6$	$3{\cdot}39 \times 10^8$	$2{\cdot}81 \times 10^7$	$2{\cdot}33 \times 10^6$
A	Uranus	∞	$1{\cdot}83 \times 10^8$	$4{\cdot}65 \times 10^8$	$1{\cdot}31 \times 10^8$
Jupiter	Saturn	$4{\cdot}53 \times 10^8$	$2{\cdot}15 \times 10^5$	$1{\cdot}39 \times 10^8$	$2{\cdot}11 \times 10^5$
Jupiter	Uranus	∞	$3{\cdot}41 \times 10^5$	$1{\cdot}11 \times 10^8$	$3{\cdot}41 \times 10^6$
Jupiter	Neptune	∞	$9{\cdot}47 \times 10^5$	$3{\cdot}27 \times 10^8$	$9{\cdot}44 \times 10^5$
Saturn	Uranus	∞	$4{\cdot}40 \times 10^6$	$3{\cdot}14 \times 10^8$	$4{\cdot}32 \times 10^6$
Saturn	Neptune	∞	$1{\cdot}78 \times 10^7$	$1{\cdot}15 \times 10^9$	$1{\cdot}75 \times 10^7$
Uranus	Neptune	$3{\cdot}42 \times 10^9$	$2{\cdot}19 \times 10^9$	$5{\cdot}24 \times 10^9$	$1{\cdot}06 \times 10^9$

All times given in years

A comparison of rounding times and values of τ is very informative and it may be readily seen that planetary interactions during the rounding process are not unlikely. It is also clear that once the orbits of Jupiter and Saturn have rounded off, which happens quickly, they cannot interact with each other or with the outer planets. For this reason the most likely interactions are those which take place in the inner regions of the solar system.

It is impossible to ascertain what should be expected in the way of planetary interactions. Once an interaction takes place the orbits are modified and the probabilities of future events are changed. Another difficulty concerns the interpretation of the very short characteristic times. The calculations which give them assume that an average is taken over all possible spatial relationships of the two orbits and this means averaging over a fairly long period of time. For events with large τ this is a valid assumption and for a timescale of interest all configurations will be met, but the assumption will break down for the very short characteristic times. In such a case the characteristic time will be mainly dependent on the actual initial spatial relationship of the orbits. If the initial orbits do not give any near approach then there will be no interaction for a long time. Conversely, if the initial orbits do intersect, or nearly so, an interaction will occur in a very short time.

A Planetary Collision

On the basis of the analysis of planetary interactions there is no way of saying what must have happened but only what might have happened. However, it is possible to look at the solar system against the background of the analysis to see whether the postulation of one major event enables a number of apparently unrelated phenomena in the solar system to be simultaneously explained. If this can be done then such a postulate should, at least, be considered seriously.

The postulate we shall examine is a collision between the planets A and B. Here we are moving into unexplored territory for, while collisions have been considered on all scales up to the impacts which caused lunar mare, no previous analysis has been made for the collision of planetary-size bodies.

For a small-scale, or more precisely a low-speed, impact it is known that only about one half of the original energy reappears as kinetic energy of ejecta material. However, the collision of two planets such as A and B would be characterised by kinetic energies per unit mass more than two orders of magnitude greater than for a body falling on to the lunar surface. Whereas for a projectile falling on the moon there is considerable *melting* of lunar and pro-

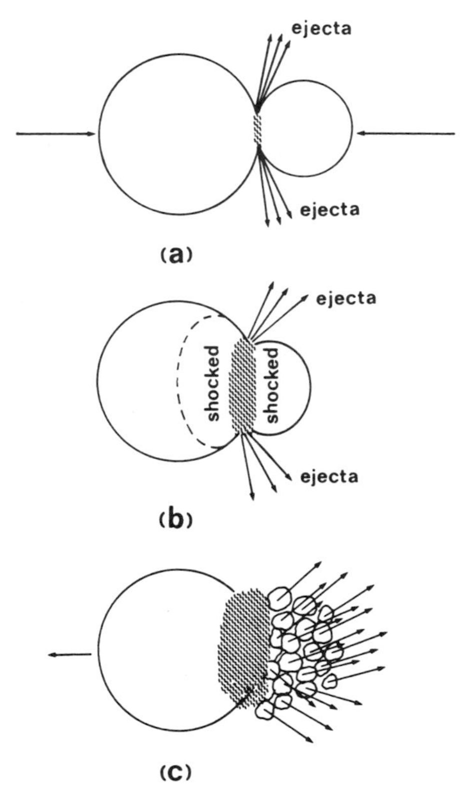

Figure 4. (a) Ejecta produced soon after first contact of planets. (b) Some time after first impact with region of vaporized material being formed and material of both planets in a shocked condition. (c) The expansion of vaporized material and fragmentation of the smaller planet

jectile material there is very little vaporization. For our planets this is not so. The kinetic energy at the moment of impact is

$$E = \frac{GM_1 M_2}{a_1 + a_2} \tag{12}$$

where a_1 and a_2 are the planetary radii.
The kinetic energy per unit mass is therefore

$$H = \frac{GM_1 M_2}{(M_1 + M_2)(a_1 + a_2)} \tag{13}$$

and from the characteristics given in Table 1 this is found to be $3{\cdot}3 \times 10^7\ \mathrm{J\,kg^{-1}}$, an amount more than enough to vaporize *all* the planetary material. Actually not all the kinetic energy at impact is transformed into heat but, equally, not all of the planetary mass shares in the heat energy that is released. The form of a head-on planetary collision is shown in Figure 4. At the moment of impact, ejecta are thrown out from the collision region as shown (Figure 4(a)). A shock wave travels through both bodies while vaporized material is trapped and compressed by the infalling planets. The situation at the time when the shock front has reached the extremity of the smaller body is shown in Figure 4(b). Finally, as illustrated in Figure 4(c), the bodies are pushed apart again by the expanding vaporized material. The smaller body, which is subjected to the largest accelerating forces, is fragmented during this stage.

This picture of events is supported by experimental evidence with small-scale collisions and by the numerical analysis of several kinds of model (Dormand and Woolfson, 1977). It is difficult to avoid the conclusion that most of the original kinetic energy, perhaps 85 per cent or more, reappears as kinetic energy after the collision.

One type of model is illustrated in Figure 5(a). The speeds of the centres of mass of the planets at any time are denoted by αV_1 and αV_2 where V_1 and V_2 are the speeds at the moment of contact. The vaporized region has a thickness δ when the distance apart of the centres of mass is x. It may be shown that the time development of α and x are linked by coupled differential equations

$$\frac{\mathrm{d}x}{\mathrm{d}t} = -\alpha(V_1 + V_2) \tag{14a}$$

and

$$\frac{\mathrm{d}\alpha}{\mathrm{d}t} = \frac{GM_1 M_2 (V_1 + V_2)}{2Ex^2} - \frac{2}{3M_1 V_1 \delta}\left\{(1 - \alpha^2)E + \frac{GM_1 M_2}{x} - \frac{GM_1 M_2}{a_1 + a_2}\right\} \tag{14b}$$

where

$$E = \tfrac{1}{2}M_1 V_1^2 + \tfrac{1}{2}M_2 V_2^2. \tag{14c}$$

The doubtful quantity in these equations is δ which is a function of t. It

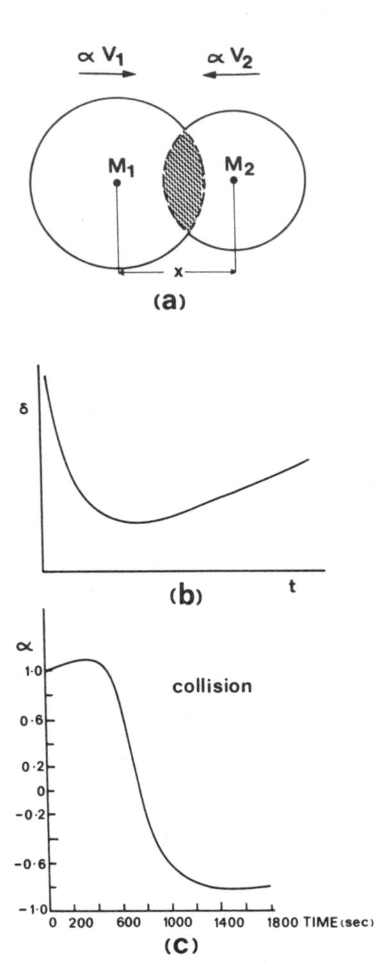

Figure 5. (a) The collision of two planets. (b) Variation of the effective width of the vaporized region with time. (c) Relative velocity of centres of mass of two planets as a function of time

describes the growth of the vaporized interface by heat energy travelling outward while the bodies moving inwards are tending to compress it. The smaller δ is, the larger is the force tending to push the bodies apart. When the bodies have only just made contact, any vaporized material will tend to escape sideways together with the ejecta, as shown in Figure 4(a). This may be simulated in our model by a large value of δ, i.e. little force being exerted. Then the vapour is trapped and the effective δ decreases. Eventually the value of δ must increase again as the vapour expands, pushing apart the planets. An analytical representation of this form of variation which has been used is

$$\delta = \frac{a}{t} + bt + c \tag{15}$$

which is illustrated in Figure 5(b).

The problem of properly representing δ is under active consideration. Various values of constants in expression (15) have been used with b tending to be close to the velocity of sound in the planetary material. No justification for any particular set of constants can be given at present but what does emerge is a surprising insensitivity of behaviour pattern with choice of parameters. Figure 5(c) shows a typical outcome with

$$\delta = \left(\frac{5{\cdot}0 \times 10^6}{t} + 2{\cdot}7 \times 10^5 t + 1{\cdot}3 \times 10^6\right) \text{ cm.}$$

The coefficient of restitution of this collision is 0·79 indicating that about 56 per cent of the initial kinetic energy reappears as kinetic energy at the end of the process.

A model for the collision of the particular planets A and B in their orbits has been described by Dormand and Woolfson (1974). A schematic view of the planets at the moment of impact is shown in Figure 6. The parameters used were:

planetary masses	$M_1 = 2{\cdot}0 \times 10^{26}$kg; $M_2 = 3{\cdot}15 \times 10^{25}$kg
speeds at collision	$U_1 = 32{\cdot}1$ km s^{-1}; $U_2 = 34{\cdot}2$ km s^{-1}
directions of motion	$\theta_1 = 100°$; $\theta_2 = 50°$
line of centres	$\alpha = 32°$
planet densities	$\rho = 0{\cdot}8 \times 10^3$ kg m^{-3}
coefficient of restitution	$\varepsilon = 0{\cdot}75$

The approach orbits of the two planets had characteristics

semi-major axis	$a_1 = 1{\cdot}83 \times 10^{12}$ m;	$a_2 = 1{\cdot}36 \times 10^{12}$ m
eccentricity	$e_1 = 0{\cdot}874$;	$e_2 = 0{\cdot}908$
semi-latus rectum	$p_1 = 2{\cdot}87$ AU;	$p_2 = 1{\cdot}60$ AU.

After collision the characteristics were

semi-major axis	a_1' (hyperbolic orbit); $a_2' = 1{\cdot}36 \times 10^{11}$ m
eccentricity	$e_1' = 1{\cdot}045$; $e_2' = 0{\cdot}750$
semi-latus rectum	$p_1' = 3{\cdot}22$ AU; $p_2' = 0{\cdot}40$ AU.

The result shows that it is possible for the more massive of the planets to have been expelled from the solar system, the energy for this being provided by the other planet which falls in towards the sun. Further implications of this are discussed later.

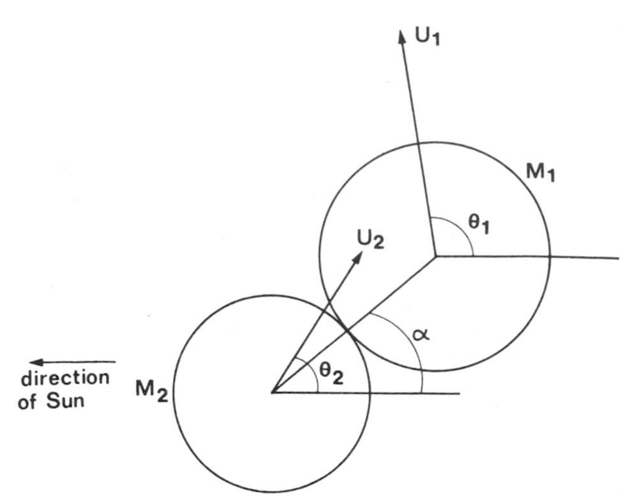

Figure 6. The parameters describing the collision of two planets (Reproduced from J. R. Dormand and M. M. Woolfson, *Proc. R. Soc. A*, **340**,349–365 (1974) by permission of The Royal Society)

It will be noted that this collision corresponds to an oblique impact of the planets. This gives an addition of shear forces to the explosive forces which contribute to the break-up of the smaller planet (B). Dormand and Woolfson (1977) have explored the consequences of this extra force on the break-up mode and conclude that a behaviour pattern similar to that shown in Figure 4(c) will occur. The debris goes off in two separate streams, with the average velocity vector of each stream in the plane defined by the relative velocity of the two planets at collision and their line of centres. Taking into account the mutual gravitational attractions of the material contained in each stream (a consideration not encountered in small-scale ballistics) it seems that the most likely net outcome of the collision for planet B is that it is broken up into two major fragments. An idea which has been examined by Dormand and Woolfson (1977) is that these two fragments may form a proto-Earth and proto-Venus. In Figure 7 is shown the basics of the model used. It is similar to that shown in Figure 6 except that planet 2 is shown schematically as broken into two equal parts. To the post-collision velocity of the smaller planet as a whole there are added two equal and opposite (momentum conserving) velocities to the two fragments at right angles to the line of centres. The collision parameters are a little different from those previously used:

$M_1 = 2{\cdot}00 \times 10^{26}\,\mathrm{kg}$; $M = 3{\cdot}15 \times 10^{25}\,\mathrm{kg}$

$U_1 = 32{\cdot}2\,\mathrm{km\,s^{-1}}$; $U_2 = 35{\cdot}8\,\mathrm{km\,s^{-1}}$

$\theta_1 = 100{\cdot}1^\circ$; $\theta_2 = 49{\cdot}7^\circ$

$\alpha = 60^\circ$ $\varepsilon = 0{\cdot}75$.

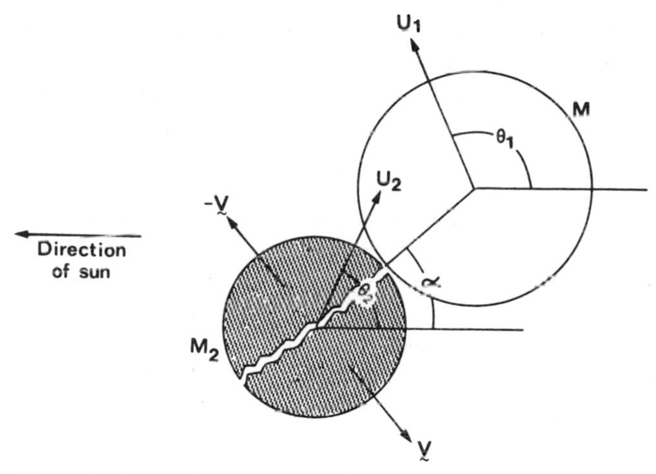

Figure 7. The collision of two planets with the smaller planet broken into two

The value of v was chosen so that 31 per cent of the initial kinetic energy was taken up by the moving apart of the fragments.

The pre-collision orbital parameters were

$a_1 = 2{\cdot}63 \times 10^9$ km; $e_1 = 0{\cdot}91$; $p_1 = 2{\cdot}93$ AU

$a_2 = 1{\cdot}51 \times 10^9$ km; $e_2 = 0{\cdot}92$; $p_2 = 1{\cdot}58$ AU

and the post-collision parameters were:

$a_1' =$ (hyperbolic orbit); $e_1' = 1{\cdot}002$

fragment 1

$a_2 = 1{\cdot}06 \times 10^8$ km; $e_2' = 0{\cdot}79$; $p_2' = 0{\cdot}70$ AU

fragment 2

$a_2'' = 1{\cdot}59 \times 10^8$ km; $e_2'' = 0{\cdot}64$; $p_2'' = 1{\cdot}06$ AU.

Of course, the parameters were chosen to obtain a preselected result—two fragments which would round off to orbits similar to those of the Earth and Venus. Nevertheless it is an important conclusion that *it is possible to expel the larger planet from the solar system and to create two small planets which, after orbital round-off, would correspond closely to the Earth and Venus.*

The calculation given here is only illustrative and we do not pretend that it portrays the *precise* details of an event in the history of the solar system. What it does show, however, is that such an event is both possible and plausible.

Asteroids, Meteorites, and Comets

Following the line proposed previously of explaining as many features of the solar system as possible in terms of a single event, it is now proposed that asteroids, meteorites and comets are various types of debris from the planetary collision. The idea that asteroids might be part of a broken-up or exploded planet is not a new one but previous suggestions have been bedevilled by two difficulties. The first is to find a source of energy which could cause an isolated planet to explode and the second, to dispose of the greater part of the mass so as to leave only a fraction of an Earth mass in the form of asteroids (Napier and Dodd, 1973). These problems do not occur with the mechanism suggested here.

Taking into account inter-asteroid collisions, the effect of perturbations, especially by Jupiter, and possible effects of a resisting medium, there seems to be no inconsistency between modern asteroid orbits and the collision hypothesis. Some larger asteroids could conceivably be ex-satellites of one or other of the colliding planets but the irregular shapes of the smaller bodies indicates a more probable origin as collision fragments.

Many meteorites may similarly be regarded as collision fragments, possibly produced by break-up of asteroids. Iron meteorites, stony-irons and achondrites have properties which might be expected for material contained in a large and fairly well segregated planet. Chondrites do not fit this pattern. They show distinct signs of having condensed from the vapour phase and carbonaceous chondrites contain appreciable quantities of carbon, water, and various volatile elements.

It seems that the most appealing aspect of the solar-nebula theory is that it postulates the existence of hot vaporized material; a great deal of work has been done on condensation sequences from a hot vapour which explain the structures and composition of the various types of chondritic meteorites quite well. It will be noted that a planetary collision also gives rise to a great deal of vaporized material and that matter coming from the outer regions of the planet would have contained much volatile substance, including carbon compounds. The theories of condensation sequences may equally well be applied to this vaporized material and will be equally satisfactory.

One of the puzzles posed by meteorites is that of the Widmanstatten patterns seen in iron meteorites. These are due to the solid state diffusion of nickel between two iron-nickel alloys, taenite and kamacite and they imply the maintenance for a long time of a temperature regime in which the material is solid but which is high enough to maintain a reasonably rapid diffusion rate. The iron must initially be molten, then it must solidify and cool at a rate somewhere between 1 K per 10^6 years to some ten to one hundred times faster, depending on the assumptions made. The energy available for melting iron on a cosmogenic scale can be either gravitational or radioactive. The former of these requires fairly rapid accretion of at least lunar-size bodies. This will melt iron but then the bodies are so large that the cooling rates would be too low so that they must now be broken up into asteroid-size pieces. This scenario presents many difficulties, especially that of rapidly forming lunar-size bodies from nebular material. The alternative, radioactive heating, does not offer any extant candidate for the role of heating agent. However, there is a great deal of current interest in the idea of heating by ^{26}Al which has a very short half-life.

Recently, in an inclusion of the Allende meteorite, there has been found a ^{26}Mg anomaly suggested as evidence for the onetime presence of ^{26}Al but this conclusion is not certain. If ^{26}Al was the heating agent then there must have been a lot of it around. One notes with interest its remarkable coyness in revealing itself. Another problem here is that heating by ^{26}Al could only be effective for a short time (a few million years) after synthesis of the element and this places a very tight time constraint on the origin of the solar system.

The postulate of the origin of iron meteorites as the debris from a planetary collision does not suffer from any of the objections raised to other ideas. Planets are formed directly with an abundance of gravitational energy to melt material and the break-up into asteroid-size bodies later provides the conditions for cooling at the required rate. This type of idea is supported by other evidence that chondritic material has at some stage been differentiated in a planetary-size body (Hutchison and Graham, 1975).

Planetary surface material remote from the collision would not have vaporized although, through spallation, much of it would have broken away from the parent body. Such material from the larger planet would have gone into orbits of high eccentricity, some even more hyperbolic than that of the planet but other parts of it would have gone into orbits of eccentricity close to but less than unity. This could provide the material for long-period comets.

The dissipation of comets at each perihelion passage gives them a finite lifetime—of the order of one thousand orbital periods. Observations of long period comets suggest that their orbits have semi-major axes in the range 15,000–25,000 AU corresponding to periods of $2–4 \times 10^6$y. Thus it is reasonable to conjecture that long-period comets of greatest original masses and periods could have survived for the lifetime of the solar system (c. $4{\cdot}6 \times 10^9$y).

Planetary perturbation could occasionally modify the orbit of a long-period comet to give one of short period and limited lifetime. Some comets have been observed to have hyperbolic orbits of eccentricity close to unity and these could be produced from long-period comets by perturbation from either planets or some of the nearer stars.

The most widely-held theory of cometary origin, that of Oort (1951), postulates that there is a reservoir of comets in roughly circular orbits at large distances from the sun (20,000–100,000 AU). According to Oort's ideas, comets arose as minor condensations within the solar system at the same time as the planets were formed. Some of these were captured by planets but others were thrown out by interactions with the planets and now move in extensive orbits with perihelia in the outer regions of the solar system. They spend most of their time at extreme distances from the sun and can occasionally be deflected by perturbations of nearby stars into orbits taking them close to the sun.

While it has been suggested here that the material may have been thrown into these large orbits as a direct result of the collision, equally a collision may have just provided the raw material on which Oort's mechanism operated.

The Moon, Mars, and Mercury

There are three main types of idea which have been put forward from time to time to explain the origin of the Moon.

(a) The Moon was produced by rotational instability of a rapidly-rotating proto-Earth. This idea has been found to be dynamically untenable.
(b) The Moon was in an independent heliocentric orbit and was captured in a close encounter with the Earth. This idea also presents considerable dynamical difficulties although these may not be present with a capture-theory-like proposal of Öpik (1972).
(c) The Moon was assembled from small particles in orbit around the Earth. There is some evidence from the thermal characteristics of the Moon that it would need to be assembled quite quickly in order to store the gravitational energy of accretion. It has not been convincingly shown that accretion could occur rapidly enough.

We examine here an alternative hypothesis, that the Moon was an ex-satellite of planet A (for which a family of large satellites would be expected) which was captured by the proto-Earth at the time of the planetary-collision event. This

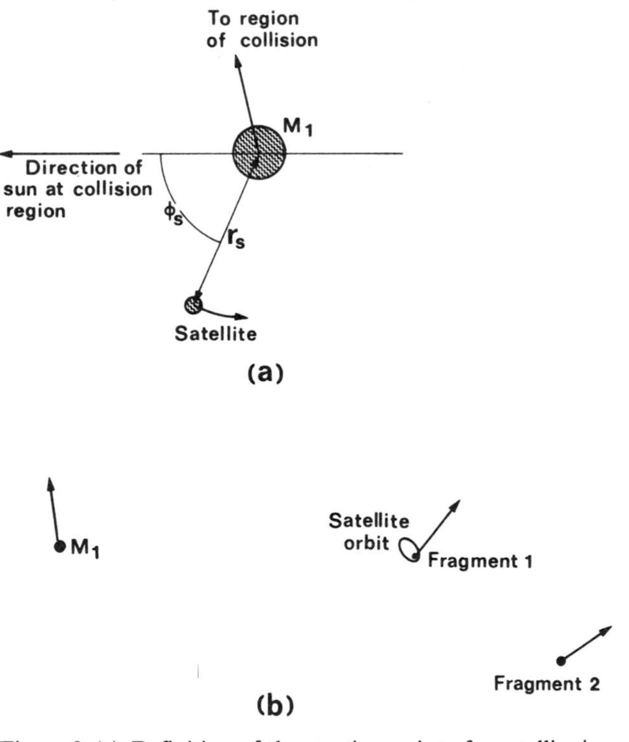

Figure 8. (a) Definition of the starting point of a satellite in a circular orbit about planet A. (b) The final outcome of the interaction with the Moon captured by the proto-Earth

possibility is easily examined by numerical analysis. The satellite is started in a circular orbit at a point, as shown in Figure 8(a), defined by the orbital radius, r_s, and phase-angle, ϕ_s. This point corresponds to a time, taken as 5×10^6 s prior to the collision. The planets, approach, collide, the smaller one fragments and then the various resultant masses move apart under the influence of the gravitational fields of the sun and of each other. During all this period the equations of motion for the satellite are integrated and it is found that three types of outcome may result:

(i) it may be carried off with planet A and leave the solar system,
(ii) it may be released into an independent heliocentric orbit, or
(iii) it may be captured by one or other of the fragments of planet B.

For the planetary collision previously described and with $r_s = 3 \times 10^5$ km and $\phi_s = 1{\cdot}37$ radians the satellite is captured by the Earth-like fragment into a stable, low-energy orbit with semi-major axis $5{\cdot}2 \times 10^5$km and eccentricity 0·74 (Figure 8(b)). In other trials semi major axes as low as $2{\cdot}3 \times 10^5$km and eccentricities down to 0·41 have been found.

There are no dynamical problems with this theory. The energy-exchange possibilities in a many-body system enable a capture to take place quite comfortably and numerical experiments confirm that this can happen over a wide range of initial conditions.

A feature of the Moon which evokes a great deal of interest and speculation is the asymmetry of its surface features. That side which faces the Earth bears evidence of collisions with large bodies which gave rise to the mare basins. On the other hand mare are conspicuous by their rarity on the far side of the Moon although for craters produced by smaller projectiles, the surface has been peppered much more uniformly.

It is the wont of planetary satellites to circle their primaries such that one face is presented to the parent body. This habit is related to asymmetry in the figure of the satellite and the mass distribution within it so that this type of coupling between its axial rotation and orbital period gives a state of dynamical stability. This would have been the relationship of the Moon to planet A. The first and fastest collision debris would have struck that side which faced the planet—and therefore faced the collision. The dynamical considerations which caused that side to face planet A would equally cause it to face the Earth, after capture and the influence for some period of tidal forces. The Earth and its new satellite would have travelled in an orbit which repeatedly passed through that region where the collision occurred, as would all the debris. Until round-off and perturbation of orbits took its toll both bodies would suffer repeated intense and uniform bombardment, although never again with projectiles as fast as those of the original deluge.

Mars is a body about which, so far, no mention has been made although its place in the solar system seems to have been usurped by planet B. Mars is a very small planet, with one-ninth of the Earth's mass, and with a density, $3{\cdot}90\ \mathrm{g\ cm^{-3}}$ more akin to that of the Moon ($3{\cdot}34\ \mathrm{g\ cm^{-3}}$) than that of the Earth ($5{\cdot}51\ \mathrm{g\ cm^{-3}}$) It seems to fit into the general scheme of the planetary collision hypothesis as another of the satellites of planet A. This planet would have been subjected to greater tidal forces than Jupiter at first perihelion passage after its formation and that it should have a satellite family with members even more massive than the Galilean satellites is quite plausible. Thus the low mass and low density of Mars are quite well accounted for. So indeed might be its suface features. It too, like the Moon, has two hemispheres of remarkably different appearance. No attempt has yet been made to interpret the Martian surface as the outcome of a short-term asymmetric bombardment by large projectiles followed by longer lasting symmetric bombardment by smaller debris. Such an interpretation needs to take into account the turbulent thermal history of Mars, as evinced by its volcanic structures, but an approach to a study of Mars from this point of view may be fruitful.

The final body to be considered is Mercury with the lowest mass and the highest intrinsic density (allowing for compression effects) of any planet. Again we seek to interpret this body as a byproduct of the planetary collision and it seems sensible to consider it as a high-density fragment of planet B. The density $\sim 5{\cdot}5\ \mathrm{g\ cm^{-3}}$, and the mass are thus related only incidentally and this apparently anomalous relationship presents no problem with this interpretation.

Conclusions

In the nature of the problem we can never be certain what is a true theory of the origin of the solar system—we can only recognize false theories because they fail certain crucial tests. A theory may survive for some time almost by virtue of its inability to explain anything in detail. Such theories are flexible enough to be defended by hand-waving arguments against any attack. Yet, thought about rationally and without prejudice, no theory should be considered more than a curiosity until it gives a convincing explanation of the more obvious and large-scale features of the solar system, in particular the existence of the planets and of the satellites, even if one takes as given the existence of the sun. The weaving of intricate and cunning mathematical patterns and descriptive theories, however entertaining, are no substitute for hard results.

Beyond the stage of explaining the gross features of the system one should then look at the details and the apparent anomalies for these contain a mine of information about evolutionary features of the system and perhaps even about the mechanism of its origin. For example a gross feature of the system is that it is planar. This may tell us something about its origin or perhaps only something about its evolution—we cannot be sure. But wait—let us look again! The system is *not* planar; there are important departures from planarity, not least the rotation of the sun itself. What does this tell us? One thing it certainly tells us and that is that any theory that initially gives a highly planar system, with all angular momentum vectors of rotation and orbit exactly parallel, requires a subsidiary mechanism to give what we actually observe in the solar system today.

Another characteristic that one might expect for a respectable and highly-developed theory is that it does not produce *ad hoc* explanations for every feature of the solar system. As presented, a notable feature of the capture theory is that the postulation of one event, a planetary collision, that analysis shows to be not at all unlikely, gives ready explanations for many features—in fact for virtually all the major features of the inner part of the system.

Other aspects of the solar system have been satisfactorily investigated in terms of the capture theory but not reported here for reasons of space. Other aspects are about to be investigated. All this work will be reported in due course, given the willingness of journals to publish it, including, when it happens, a report of the failure of the crucial test!

References

Dormand, J. R., and Woolfson, M. M. (1974). 'The evolution of planetary orbits.' *Proc. R. Soc. A*, **340**, 349–365.

Dormand, J. R., and Woolfson, M. M. (1977). 'Interactions in the early solar system.' *Mon. Not. R. astr. Soc.*, **180**, 243–279.

Hutchison, R., and Graham, A. L. (1975). 'Significance of calcium-rich differentiates in chondritic meteorites.' *Nature*, **255**, 471.

Napier, W. McD., and Dodd, R. J. (1973). 'The missing planet.' *Nature*, **242**, 250–251.

Oort, J. H. (1951). 'Origin and development of comets.' *The Observatory*, **71**, 129–144.

Öpik, E. J. (1972). 'Comments on lunar origin.' *Irish Astron. J.*, **10**, 190–238.

SOME ASPECTS OF EARLY STELLAR EVOLUTION THAT MAY BE RELEVANT TO THE ORIGIN OF THE SOLAR SYSTEM

G. H. HERBIG
Lick Observatory, Board of Studies in Astronomy and Astrophysics, University of Calfornia, Santa Cruz, U.S.A.

The Properties of Young Solar-type Stars, and Their Relevance to the Origin of the Solar System

The presence of 'star formation' on a program devoted to the origin of the solar system is evidence of the widely-held belief that our planetary system was an immediate by-product of the formation of our sun, and that we may be able to learn something of what happened at that time by looking at stars in process of formation today. That may be quite true, and yet our own case may not be at all typical. It is imaginable that something special happened when the sun was formed that sets it apart from many or most other G2 dwarfs. One can conceive a number of disquieting possible complications: for example, perhaps the sun was originally formed as one member of a double star, or in a star cluster since dissolved, or in an event triggered by a nearby supernova. Who can say? The only practical thing for one to do is to ask if the sun differs from other stars in any of those physical characteristics that seem to be relevant to the possession of planets.

I intend to return to this specific question after considering some of the properties of young stars. Not very many of these observational characteristics clearly bear upon the planetary problem, but there appear to be a few. I think that as confusing or unwelcome as these may be, it is unscientific to ignore them. And one is not free to pick and choose: if the 'T-Tauri solar wind' is found useful in scouring out the solar nebula, then one must also accept and contend with the other properties of T-Tauri stars.

At about 1 solar mass, there are available for our inspection stars of the following age categories:

Very young (For very young stars, the phase of $t = 0$ is not clearly defined except by appeal to some theory. In clusters, the important issue of whether all stars formed at essentially the same time is frequently debated, but remains unresolved): 10^6 to 10^7 years; T-Tauri stars, still associated with parent nebulosity; ages are estimated by cross-motion arguments or by locating them on theoretical evolutional tracks or by membership in clusters that are themselves dated from the theory of upper main sequence evolution.

Young: to 10^8 years, no longer associated with nebulosity; dated by membership in young clusters (such as the Pleiades).

Medium to Old: 10^8 to $> 10^9$ years; dated by cluster membership (Hyades) and in the field (roughly) by space motion.

First, several negative generalizations may be made directly from observation:

(1) There is no evidence for a preferential orientation of angular momentum vectors. In particular, the residual AM in stars of binary systems does not derive from differential galactic rotation. This is demonstrated by these facts: the orientation of the orbit planes of visual binaries is at random; the spin velocities of stars seen near the galactic poles are not obviously less than those seen in other directions; the invariable plane of the solar system is inclined at about 60° to the galactic plane. Thus, however it is done, the AM problem of a contracting cloud is solved completely, and what we see in the stars must have some other origin.

(2) The motions of very young, low-mass stars with respect to each other and with respect to the nebular material are now known for a few nearby aggregates. These r.m.s. velocities are small: in the neighbourhood of 1–2 km/sec, an amount characteristic of internal velocities in cool H I clouds. Thus even if such young stars are found today in H II regions such as the Orion Nebula, they cannot have been formed in that hot, turbulent, ionized gas but rather at an earlier time when the cloud was cool and neutral, or less probably, in some cool un-ionized region of the present hot cloud.

(3) In the formation of stars of this kind, there is no obvious evidence that a hierarchical fragmentation of a much larger mass ($\sim 1000 \odot$) has to take place. In the larger aggregates such as the Taurus–Auriga clouds, there are no dense clusterings of T-Tauri stars (Herbig and Peimbert, 1966). This conclusion may be drawn from optical observation, but also from the recently discovered infrared sources in dark clouds. Many or most of the 2μ sources are probably embedded stars, and on the average may be younger than the T-Tauri stars seen near the cloud surfaces, but again there is no obvious tendency to form tight clusterings. Furthermore there exist a few small dark nebulae in which a total of only one or two T-Tauri stars have been found. It is difficult to avoid the conclusion that in this mass range, multiple fragmentation plays no significant role. Presumably densities and/or temperatures within such clouds are able to attain values which cross the thresholds contained in the Jeans criterion for collapse at 1–2 solar masses.

(4) There is no doubt but that as stars contract from interstellar clouds

down to stellar size, they must pass through dimensions of $\sim 10^4$ AU. There is no evidence, however, that the Bok globules of that size which we see in some H II regions are to be identified with such early contracting configurations: these globules seem to be no more than transient agglomerations of dust in process of dissipation in the neighbourhood of hot, high-luminosity stars (Herbig, 1974).

These are all negative points: they deprecate phenomena once considered seriously, but now seem not to be of central concern. Now consider some *unanticipated* properties of young stars that are revealed by direct observation:

(1) The characteristic signature of T-Tauri stars is a set of emission lines superposed upon the conventional absorption line spectrum of a solar-type (or somewhat cooler) star, which resemble the emission spectrum of the solar chromosphere. In the sun, this region and particularly the temperature rise above it are believed to be excited by mechanical energy generated in a turbulent region below the stellar surface. It has been speculated that perhaps a similar mechanism is active in the T-Tauri stars. It is observed that solar-type stars in clusters of known age show a lower level of the same phenomenon, but of progressively lower intensity with greater age, while in the sun the chromospheric phenomenon is very weak. This decline with age encourages the suspicion that within the T-Tauri stars, emission line strength is also an index of age, the weak-emission-line stars being the older. But it can also be contended that the effect is not a tight one, but more statistical in nature, so that some stars might always show stronger or weaker emission lines than the norm, for some other reason.

In the sun, the corona is supported and excited by the same energy source that is responsible for the chromosphere. On this account, one might expect that the coronal lines might be detectable in T-Tauri stars, but despite an extensive search, they have not been found.

(2) Superposed upon the stronger T-Tauri emission lines are a set of absorption lines, usually displaced to shorter wavelengths, which are most obviously understood as due to rising cooler material. Displacements of 50 to 150 km/sec are typical, but in a few stars the absorbing material is near zero velocity with respect to the underlying star. These velocities are not constant for a given star, but change with time. This is the so-called 'T-Tauri solar wind' of which so much has been made. Elementary modelling of the line profiles (Kuhi, 1964) leads to mass losses, in the most powerful cases, of several $10^{-8}_{\odot}$/yr. But I must emphasize first that the material rising is most prominent in lines of neutral H, and hence is not a highly-ionized wind like the sun's. And second, although there is no spectroscopic evidence of returning material, there is also no proof that this gas really leaves the star, and hence that this is a true mass 'loss'. And if it did, even at the maximum dm/dt of $10^{-8}_{\odot}$ yr^{-1} for about $\sim 10^7$ yr, such a star would lose only a small fraction of a solar mass in its T-Tauri phase.

I must mention however that many T-Tauri stars are surrounded by a tenous envelope of gas showing a forbidden line spectrum, from which one infers that $n \sim 10^4/\text{cm}^3$. Since $n \sim 10^{10}/\text{cm}^3$ at the base of the region of rising material, this tenuous envelope must lie much higher; indeed in a few stars it can be seen directly. It has conventionally been assumed that this material is supplied from below, and this gives some support to the idea that material does actually leave the star.

There has been some concern among theoreticians about this evidence for rising material in T-Tauri stars, where one might like to see some sign of infalling interstellar material. In the last year, R. Ulrich has argued that these profiles have been interpreted incorrectly, and that they really demonstrate that material is falling in. In Ulrich's view, the emission line structure is not a broad emission divided by displaced absorption but two overlapping emission lines, one from accreted material flowing equatorward over the surface of the star, the other produced by material streaming approximately radially inward over the surface of an accretion disk. This interpretation cannot be ruled out (given the usual rather rough spectral data) when the (presumed) absorption reversal is weak. However in a number of T-Tauri stars the reversal is very strong and clearly descends well below the continuum level, which cannot be explained by Ulrich's model. Furthermore there are a few T-Tauri stars having extremely strong, high velocity absorption components which lie at the extreme edge of their emission lines. This is certainly mass ejection in the P Cygni sense. Thus the maximum claim that could be made is that for T-Tauri stars with weak reversals, either ejection and accretion could account for the observations, while for the more extreme examples, accretion cannot be responsible. If accretion is to be considered seriously for the first group, then two different phenomena would have to be in operation in different stars. This seems to me to be a complication to be resisted unless the evidence for it is more convincing.

(3) About ten years ago it was discovered by Mendoza (1966) that the T-Tauri stars are anomalously bright in the infrared, with respect to the amount of energy expected from a normal star of the type corresponding to the absorption spectrum. The interpretation of that excess has become slightly more controversial since the original discovery, because it was originally believed that the excess was produced entirely by hot circumstellar dust, which of course is reminiscent of a 'solar nebula'. Recently it has been proposed by Rydgren, Strom, and Strom (1976) that the excess energy at shorter infrared wavelengths is actually due to emission by the hot gases of the stellar chromosphere, and that only the component beyond 10 μ is produced by dust at $T \sim 200$ K. Until this controversy is settled, one cannot be sure whether or not hotter dust (at $T \sim 1000$ K) also contributes. It is surprisingly difficult to settle such questions because so many contributing uncertainties are involved: not only the frequency function of circumstellar particle sizes enters, but the particle compositions also enter into both emissivity and extinction corrections.

The presence of solid matter near these stars is demonstrated also by appreciable plane polarization of their light. This dust is often distributed asymmetrically around the star, as shown by the polarization as well as by the double-fan reflection nebulosity seen directly at several stars. These fans must be nearby dust illuminated directly by the star in the directions of the poles of a flattened circumstellar disk. The dimensions of these fans are large—of the order of 10^4 a.u. or more—so that one does not regard them as *ejected* dust. However, until radial velocities are measured in these fans, one cannot be certain of this point.

The infrared evidence for the presence of circumstellar dust is strong for the T-Tauri stars, weak in a few objects whose T-Tauri characteristics are subdued, but lacking in solar-type stars that have reached the main sequence. This looks very much like direct evidence of the dissipation of the solar nebula, although as yet the data are meagre. If so, a $1/e$ time of the order of 10^7 to 10^8 years is suggested.

(4) One of the most striking characteristics of the T-Tauri stars is their high surface abundance of lithium (Table 1). Li can be destroyed by (p, α) reactions at temperatures near 3×10^6 K, and in stars having surface convection zones that deep, one would expect the Li abundance to be very low. Solar-type stars may be very deeply convective in their early stages but then the central temperature is $\lesssim 10^6$ K. It is a delicate matter to predict how much Li will survive as evolution proceeds, and the convection zone withdraws toward the surface, but the central temperature rises. This subject has a long history. What is clear is that the Li surface abundance does decay on a long time scale, of $1/e \gtrsim 1{\cdot}0 \times 10^9$ yr, so that the sun now has about 1/200 of its original Li. It appears

Table 1. Lithium abundances in material having various histories

	$\log \frac{N(\text{Li})}{N(\text{H})}$	Age [yr]	Note
T-Tauri stars	$-9{\cdot}0 \pm 0{\cdot}3$	$\sim 10^6$	a
NGC 2264	$-9{\cdot}0$	3×10^6	b
Pleiades	$-9{\cdot}2$	7×10^7	c
Hyades	$-9{\cdot}5$	9×10^8	c
Field stars	$-9{\cdot}6$ to $\leqslant -11{\cdot}0$	Various	d
Sun	$-10{\cdot}9$	$(4{\cdot}6 \times 10^9)$	e
Carbonaceous chondrites:			
Type I	$-8{\cdot}7$		f
Type II	$-8{\cdot}8$	$4{\cdot}6 \times 10^9$	f
Type III	$-8{\cdot}8$	—	f
Interstellar	$-9{\cdot}6$	—	g
	$-9{\cdot}1$	—	h

Notes:
[a]The data are summarized by Zappala (1972); the spread indicated is that of individual stars
[b]Zappala (1972)
[c]The data are Zappala's, but for those stars having approximately the temperature of the sun
[d]Herbig (1965)
[e]Stellmacher and Wiehr (1971)
[f]Nichiporuk and Moore (1974)
[g]Traub and Carleton (1973)
[h]Field (1974): The data of reference *g* rediscussed

incidentally that this is a most useful way of dating solar type stars, because the surface Li concentration decays along with the chromospheric activity, while the mass ejection and the infrared excess weaken more rapidly.

Up until recently it was not sure where this Li came from; there were early speculations on it being produced by fast-proton spallation in the early pre-solar interstellar cloud, and then being swept up at the last moment in the young sun. The exceedingly weak lines on Li I in the interstellar gas were first detected in 1973, and although there are technical difficulties in converting these data to an interstellar Li/H ratio, it appears that the interstellar Li abundance is very much like that in T-Tauri stars. This was forecast by Meneguzzi, Audouze, and Reeves (1971) who predicted that Li should be produced there by spallation of heavier nuclei by cosmic rays.

We therefore regard the approximate equality of the chondritic Li abundance with that of the T-Tauri stars as a consequence of the chondrites being formed from the same material as the young sun, although the sun has long since burned away most of its original complement of surface Li.

One is always seeking some proof (as distinct from presumption) that the planets and the sun were formed at the same time. It seems that the Li abundances in the chondrites show us that they cannot have come out of the sun after it was more than perhaps 10^9 years old as the result, say, of an encounter with a passing star, because by then the Li concentration in the solar convection zone would be down by a factor of 3 or more. But of course this is of no use if the aged sun made planets by somehow picking up interstellar material, in which Li is presumed to be maintained everywhere at the chondritic level. (H. Reeves pointed out during the meeting that the presence of deuterium in stony meteorites set a much more stringent condition upon the history of that material: certainly it could never have been at a temperature higher than about 7×10^5 K.)

(5) The last property of T-Tauri stars that I intend to discuss is the phenomenon responsible for the Doppler broadening of their absorption lines. Ordinary main-sequence (i.e. H-burning) stars of these spectral types have sharp lines, indicative of a low level of atmospheric mass motion and a low speed of axial rotation. It has been known for a long time that the T-Tauri stars have broad lines which, if interpreted as axial rotation, correspond to (projected) equatorial velocities up to about 65 km/sec; in comparison the sun rotates at 2 km/sec. The only slender encouragement to the notion that this is rotation and not some kind of atmospheric activity is that if these stars were rotating, the rotation periods would be of the order of several days or a week. It has been found that a few T-Tauri stars (unfortunately, not the ones for which line breadths are known) are cyclically variable in light with just such periods, as if their surfaces were spotted and the spots were being carried around regularly by axial rotation.

If one assumes that the line widths do in fact indicate rotation, then evolutionary theory gives the change in radius as the stars complete their contraction.

If they rotate as solid bodies, then the surface velocity will increase in inverse proportion to the radius, and one can compare these predicted rotational velocities with the rotations observed for normal main sequence stars. In all cases, there is acceptable agreement: the T-Tauri stars for which the line-width measurements are possible will then spin at appropriate rates. But the crucial observation has not yet been made: these particular T-Tauris have large enough masses to bring them to the main sequence above the point where there is an abrupt decline in rotational velocities, near 1 solar mass. It is of great importance to repeat this observation for a few low-mass T-Tauris: they should have very sharp lines if this simple picture is correct; if they do not, something is wrong.

At this point, now that the familial characteristics of the T-Tauri stars have been described, let us return to that initial question: what can we see in these

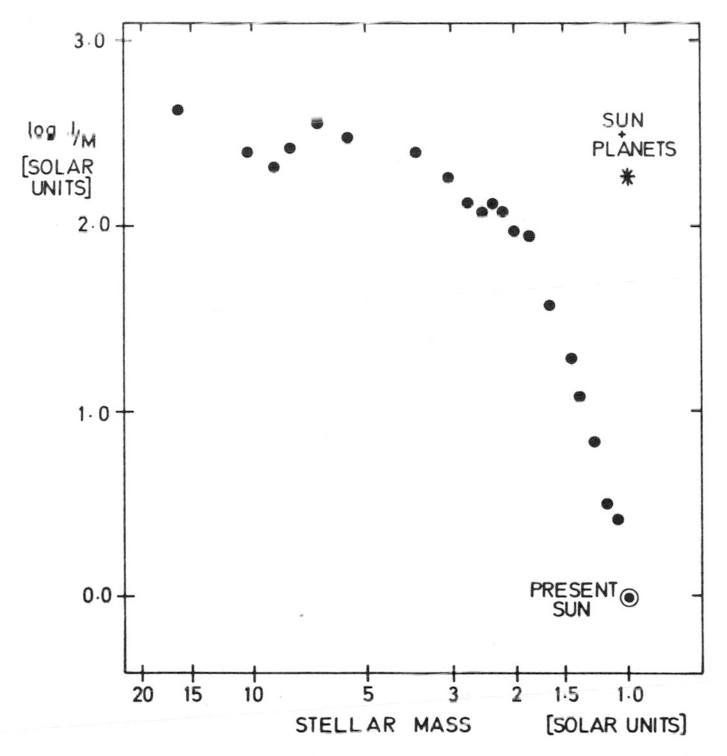

Figure 1. The logarithm of specific angular momentum J/m (in units of the solar value) *vs* stellar mass (in solar units) for main sequence stars. No information is available for stars of less than the solar mass. The assumption made throughout is, of course, that the stars rotate as solid bodies. Omitted is a factor which varies only slowly with mass across this diagram, the stellar radius of gyration. The discontinuity in J/m near 2 solar masses is apparent. The velocities of axial rotation used in this diagram are taken from a discussion by Bernacca and Perinotto (1974).

young stars that is directly relevant to the formation or the existence of planets, as distinct from phenomena that planetary theorists may find useful? It appears to me that the most encouragement, to which we have all held tenaciously for a long time, is provided by that famous diagram constructed from observations of stellar rotational velocity along the main sequence, of specific angular momentum *vs.* mass, in which a major decline between 1 and $2_{\odot}$ is striking. It has often been pointed out that if the angular momentum of the planets were returned to the sun, the sun would be raised in this diagram to about the level of its more massive neighbours. This correspondence has been urged as evidence of the ubiquity of planetary systems around stars less massive than $1–2_{\odot}$.

There are at least two objections to this cheerful idea. First, if the 'reconstituted' mass of the planets is about 20 times their present mass, why should all the missing solar angular momentum reside in a 5 per cent residue of the total mass? Second, there is some evidence from young star clusters that when young solar-type stars first complete their contraction, they may be rotating 6–7 times as rapidly as the present sun. An interesting paper by Dicke (1972) argues that over a long time the surface angular momentum is carried away by the solar wind, leaving a slowing rotating shell superposed on a rapidly rotating core having a radius of about 55 per cent of the sun. It should be to the angular momentum augmented by these corrections that we add the planetary contribution, and this would clearly place the sun well above the normal J/M relationship. It seems to me that these considerations confuse the argument of the last paragraph to a degree that we are left with no proplanetary leg to stand on. We are free to speculate as we like on the relevance of the astronomical information to the formation of the solar system, but it seems to me that we proceed largely as an act of faith.

The Fuors: A New Phenomenon of Early Stellar Evolution

The term 'fuors' was coined by Ambartsumian (1971). The earlier work on *FU* Orionis itself is described in Herbig (1966). Much of the material in this section was presented as the Henry Norris Russell lecture of the American Astronomical Society, in December 1975. A detailed paper on the subject is in the press.

It is not meant to imply that the phenomenon represented by *FU* Orionis is newly recognised: astronomers have been puzzling over the behaviour of that object since its discovery in 1939. What is new is the deepening belief that this star, and two others of the same type discovered more recently, represent a violently eruptive phase of early stellar evolution that may occur in many young stars. In particular if the sun behaved in this way in its early youth, that fact would have major implications for the thermal history of meteorites and the dissipation of the solar nebula.

For some years prior to the eruption, *FU* Orionis was a slightly variable sixteenth-magnitude star lying in a small dark cloud at the edge of a large H II region excited by λ Orionis. With no apparent warning, in late 1936 the star brightened by 6 magnitudes (a factor of about 250) over an interval of less than a year. Since that time, it has faded only slightly (about 0·2 mag, or 20 per cent), although there have been some minor fluctuations. At first believed to be a nova, the spectrum is very peculiar and quite unique. Among other unusual features it reveals a high surface abundance of lithium, comparable to those of the T-Tauri stars although at maximum light the spectrum is completely unlike those objects. Nothing is known of its pre-outburst spectrum. The association of *FU* Orionis with the dust cloud is shown by the appearance at maximum light of a curious reflection nebulosity close about the star.

In 1970, a faint star in a dark nebula in the H II region NGC 7000 (the North America Nebula) was struck by the same phenomenon. This object, V1057 Cygni, like *FU* Orionis had been an irregular variable of no particular distinction for some 70 years until it suddenly rose in brightness, again through a range of about 6 magnitudes, the rise taking place in about 400 days. At maximum light, it showed essentially the same spectrum as *FU* Orionis, and a similar arc-shaped reflection nebula also appeared. But in the case of V1057 Cygni one important piece of new information was available: completely by chance, a spectrogram of the star had been taken 12 years before the outburst: it showed that then V1057 Cygni was a conventional T-Tauri star. No further proof was required that these outbursts are a phenomenon of early stellar evolution.

In the case of *FU* Orionis, spectroscopic observations were not begun until about 2 years following the initial brightening, but work on V1057 Cygni began just 4 months after maximum light and has continued intensively to the present time. As a result there is now a great deal of detailed information on this event.

In the past year, a third example has been discovered: a faint star near the reflection nebula NGC 6914 in which the same phenomenon is currently unfolding, but much more slowly. This star has been brightening at least since 1949. There is some problem with the photometry: it is not certain whether the star reached peak brightness about 1960, or is still rising slowly. Despite the difference in time scales, this star has a spectrum very similar to those of *FU* Orionis and V1057 Cygni, and illuminates a tiny arc-shaped nebula as well. It is interesting that *FU* Orionis, following a sudden drop in brightness of 0·6 mag. about 2 years after the initial rise, has declined only slightly in 36 years, while V1057 Cygni has faded 1·5 mag. in 5 years. At that rate, the latter star will have returned to the pre-outburst brightness in about 15 more years. It is not clear what will happen to the star near NGC 6914.

The small reflection nebulae near these 3 objects were clearly not ejected at the time of the recent eruptions. Both direct photographs and CO spectroscopy show that the arcs are not in detectable motion. Probably they represent sections of circumferential dust rings, illuminated as the result of the star's brightening, centred not on the present position of the star but on its location at some time in the past. They could represent a piling-up of dust that was driven back by high-velocity gas ejected from the star at an earlier eruption. As will be seen, there is reason to suspect that these outbursts may occur at intervals of roughly 10^4 years. The cross-velocity of V1057 Cygni implied by its displacement from a central position is only 5 km/sec, an entirely reasonable value.

The spectrograms of V1057 Cygni show that the star ejected at least one shell of gas, at a velocity of 500 km/sec when first observed; ejection continues to the present time but now at a velocity of only about 300 km/sec. The initial flareup can be explained as the combination of an increase in surface brightness and a rather moderate increase in radius, the latter by a factor of 3 to 5. Since the outburst, the surface temperature of the star has fallen slowly.

As in the T-Tauri stars, V1057 Cygni shows a powerful excess in the infrared peaking near 18 μ, which has been slowly fading since 1970. This is presumed to be due to pre-existing dust very near the star—perhaps the 'solar nebula' of V1057 Cygni—which was abruptly reheated at the time of the eruption. *FU* Orionis, 36 years after its outburst, has only a weak infrared excess while the third star has an excess of intermediate amount. The similarity of the three stars thus extends even to this phenomenon.

It is hardly possible to draw very firm conclusions as to how often these events take place from a sample of 3, but some interesting estimates are possible; it appears likely that this phenomenon is endemic among T-Tauri stars; although direct proof that these three were originally T-Tauri stars is available only for V1057 Cygni, the idea is supported by the pre-outburst irregular variability of *FU* Orionis which is typical for T-Tauri stars, and the location of all 3 stars in dust clouds containing other T-Tauri stars. It is interesting that all three had approximately the same pre-outburst luminosity (2 to 4 times that of the sun). Assume therefore that these eruptions are equally probable in all T-Tauri stars of about that luminostiy or brighter, not that these three objects were special in some unknown way. Now, there are about 630 such T-Tauri stars within 1 kiloparsec of the sun, which is approximately the distance to which the data are available and within which we might expect with reasonable certainty to detect such an outburst. If in addition the type of photographic sky survey that would be expected to detect such activity has been in operation about 60 years, then the mean interval between such events in a given T-Tauri star is $60 \times 630/3$ or about 10^4 years. If some examples have been missed, the mean interval is shorter; if less luminous T-Tauri stars are also susceptible, it is longer. Although this precise estimate cannot be taken too literally, it seems that the frequency of these events is surprisingly high.

It has been suggested that the fuor phenomenon represents the permanent removal of a T-Tauri star to some other state. This idea seemed more plausible before the rapid decline of V1057 Cygni became apparent, and the very slow fading of *FU* Orionis was noted. There are other arguments as well which indicate that these eruptions represent only a temporary event.

It is quite likely that such eruptions may occur more than once in a given star. If the events are only superficial, and probably repetitive, then perhaps in a given star the level of this activity tapers off as the object ages. There are in fact several T-Tauri stars which exhibit sporadic or repeated flareups that are very reminiscent of the fuor phenomenon on another scale.

There have been many suggestions as to the physical mechanism reponsible for the fuor outbursts. Those based on the notion that a foreground screen of dust was suddenly dissipated are contradicted by the spectroscopic evidence of violent mass ejection and physical change in the star. Those which assume a permanent internal adjustment in the star have difficulty with the decline of V1057 Cygni, and the fact that T-Tauri stars still exist in clusters 10^6 years old, although the mean time between outbursts is apparently only 10^4 years. Another suggestion has been that a large object (like Jupiter) fell to the surface of the star; the kinetic energy released would indeed be enough to power *FU* Orionis for a decade but the initial flareup would be very sudden, not of the durations observed. Another proposal involves the abrupt release of nuclear energy: if a large quantity of D could be produced by surface activity and then stored in a shallow convection zone, it is imaginable that this could be ignited suddenly and cause the star to brighten. This mechanism does have the advantage of permitting repeated eruptions until the electromagnetic activity in the stellar surface subsided with age. But Reeves pointed out at this conference that the surface production of D, like that of Li (Ryter, Reeves, Gradsztajn, and Audouze, 1970) is very inefficient and would require an enormous amount of energy. Ambartsumian (1971) suggested that the outburst represented the thermalization, by an ejected shell, of the kinetic energy of a corona of high-speed particles which envelopes the star at all times, but is normally undetectable. The observations of V1057 Cygni show that such a shell was indeed ejected, but it is not obvious that the additional hypothesis of a permanent hot corona is also required to explain the observation.

The fuor episode, that is the sporadic brightening of the star and its ejection of a high-velocity shell of gas, may be of interest in the context of the early history of the solar system. If the sun erupted in this way, perhaps more than once, one can imagine the local consequences. The inner region of the circumstellar nebula that contained the embryo planetary system would be purged of its gas and finer solids. The radiation field alone of V1057 Cygni would raise the black-sphere temperature to above 1000 K within a radius of about 4 AU. This would be adequate to re-melt small solids in the inner solar system; it might account for chondrules, which appear to have frozen directly from a liquid phase. It would be interesting to inquire whether the occurrence of such events, perhaps repeatedly, might account for other properties of the meteorites.

The Origin of Interstellar Dust

Some of the issues raised here are discussed in much greater detail by Aannestad and Purcell (1973) and in Field and Cameron (1975).

Like the primeval oceans from which terrestrial life is believed to have sprung, the dust concentrations in the interstellar medium work mark the cool, dense clouds where stars are born. In the last half-decade it has been recognized that the interstellar molecules are heavily concentrated in these same cool clouds, one reason surely being that only there are they sheltered from rapid dissociation by the galactic radiation field. The dust thus helps to lower the temperature of the gas, with the result that quite moderate masses can cross the Jeans threshold leading to gravitational collapse. As the result of radio studies, it is now recognized that these 'molecular clouds' are cooler and much denser than the conditions assumed for interstellar clouds in classical studies of star formation. Furthermore it not only protects the molecules but also probably provides catalytic surfaces for the formation of the simplest species (Hollenbach and Salpeter, 1971; Watons and Salpeter, 1972a, b). The question naturally arises: where did the dust come from, and how did it so conveniently come to be gathered up in these dense clouds? I propose to review the various suggestions as to origin that have been made, which in one case provides an answer to the second question as well.

1. In 1935, B. Lindblad suggested that the dust was a condensation product of the interstellar gas. The idea that, given condensation nuclei, dust would build up from the condensible gas led to a picture of particles of frozen ices (principally H_2O, CH_4, NH_3) that would accumulate indefinitely. A halt to growth could come about by exhaustion of the condensibles, which seemed to conflict with the observed lack of depletion of elements such as O in some clouds, or by the dissipation of the cloud. It was not certain that cloud collisions, the most obvious means of such dissipation, would always occur so conveniently as to interrupt particle growth at just the dimensions required by observation. This classical picture is by Oort and Van de Hulst (1946); a modern verison was described by Coradini at this conference. Such hypotheses face the additional difficulties that they do not explain the large quantities of very small particles required by modern observations of the far ultraviolet, they do not account for interstellar non-volatile solids such as silicates, or for the selective depletion pattern of the interstellar gas, and are embarassed by the weakness of the $3{\cdot}07\,\mu$ H_2O ice absorption in the interstellar dust at large.

2. The possibility that dust could form in the atmospheres of cool stars and then be ejected into the interstellar medium was first proposed by Hoyle and Wickramasinghe in 1962 as a source of interstellar graphite (the idea and its background are described by Wickramasinghe 1967). The subsequent discovery of a broad feature near $10\,\mu$ in some high-luminosity M-type stars and its ascription to a silicate reinforced the idea that particle could indeed form in stellar atmospheres. A thermal continuum found in the infrared of carbon-rich C stars has been ascribed to some other solid particle, possibly graphite. These observations have been reviewed by Woolf (1973), while the nucleation, growth, and outflow processes have most recently been studied

by Salpeter (1974a, b) and by Kwok (1975). The difficulty lies in a lack of direct demonstration that this dust is ejected into the interstellar medium in significant amounts. It is not disputed that in the *M* stars gas ejection does take place to some degree, as shown by the structure off the OH maser lines in some *M* supergiants, by high-resolution interferometry of a few examples, and by displaced circumstellar lines in the optical spectra of many *M* stars. There are clearly some difficulties that have to be explained: for example, (a) the major concentrations of interstellar dust bear no relationship to the space distribution of *M* and C-type stars. (b) Furthermore, this type of star is rarely associated with reflection nebulae (local dust illuminated by the stellar radiation), which is curious if they are to be dust sources. It should be mentioned, however, that of the hundreds of C stars known, there is one (IRC + 10216) which is closely enveloped in a dust and molecular cloud apparently of its own making. However there have as yet been no calculations to determine whether the surface brightness of such nebulae would be above the detection threshold. (c) The interstellar C^{12}/C^{13} ratio lies between 40 and the terrestrial value of 90, yet the evolved *M* and C stars that are presumed to be ejecting large quantities of dust into the interstellar medium have C^{12}/C^{13} values of about 5–10. This however may be explicable in terms of subsequent large-scale mixing (Audouze, Lequeux, and Vigroux, 1975). (d) If C stars eject solid carbon particles, then presumably SiC_2, C_2, and C_3 molecules are driven out of the stars as well; but none of these have been detected in the interstellar gas, although again, the interstellar radiation field may have imposed another abundance pattern upon the ejected material.

3. Planetary nebulae are expanding shells of hot ($T_{\text{kinetic}} \sim 10^4$ K) gas ejected from stars in a late stage of stellar evolution. Infrared observations in the 10 μ region have shown a continuum in a number of these nebulae that is believed to be produced by warm dust mixed with the hot gas. Some of the broad structure in this continuum may be due to emissivity variations in the solids, and on this basis it has been suggested that the most conspicuous maximum (near 11·2 μ) is produced by $MgCO_3$ (Gillett *et al.*, 1973; Bregman and Rank, 1975).

This particulate matter could have originally been present in the atmosphere of the cool *M* star that is believed to have been the evolutionary progenitor of the planetary nebula. In this case, it seems certain that the dust is carried out into the interstellar medium, but again the distribution of planetary nebulae and interstellar dust do not match at all. This concern aside, the amount of dust produced could be of galactic significance, as the following estimate shows.

If the total mass of the interstellar medium at the present time is about $5 \times 10^{9}{}_{\odot}$, then that of its dust component must be about $5 \times 10^{7}{}_{\odot}$. It has been estimated (Osterbrock, 1973) that planetary nebulae eject a total of about $5_{\odot}\ \text{yr}^{-1}$, of which about 1 per cent is condensible as refractories under the assumption of conventional composition. If the residence time of such material in space before it is taken up in star formation is about 10^9 yr, then there should be about $5 \times 10^{7}{}_{\odot}$ of interstellar dust having this origin, which is about the amount observed. The correspondence between these two values should not be taken too seriously, however, in view of the major uncertainties concealed in the estimates; it shows only that the planetaries are a promising possibility. One should mention that a rough estimate of the mass loss in one type of M-type star leads to a dust production rate by those objects of the same order (Osterbrock, 1973; Audouze *et al.*, 1976).

4. It is amusing that supernovae, which over the years have been called upon by theoreticians to serve as agents for a variety of astrophysical phenomena, have now been put forward as the sources of interstellar dust as well. Hoyle and Wickramasinghe (1970) originally considered particle formation in the cooling shell, where the temperature drops through $T \sim 2000\,\mathrm{K}$ and a particle density of about $10^{11}/\mathrm{cm}^3$ within a year after the outburst. A difficulty with this suggestion is that supernovae are estimated (Osterbrock, 1973) to eject only about 0·07 $_{\odot}$ /yr, which would fall short by a factor of at least 5 of accounting for the observed amount of dust. Salpeter (1974c) have avoided this difficulty by considering the formation of particles in the compressed (10^4 atoms/cm^3 at 100 K or less) interstellar gas that is swept through the blast wave of the explosion, but 10^5 or 10^6 years after and perhaps 100 light years distant from the original outburst. In addition, Clayton and Hoyle (1976) have considered particle formation in the cooling shell of a conventional nova, pointing to the huge infrared excess observed in Nova Serpentis 1970 as possible evidence for the process in action. The authors of such proposals can often show that, under plausible but generally optimistic assumptions, a surprisingly large amount of particulate matter *could* be produced and ejected. But direct evidence in support is sparse: again, neither novae or supernovae (at least of Type 1) are distributed as are the interstellar dust clouds; nor to the limited extent that information is available, do novae or supernovae shells exhibit evidence of internal extinction.

5. The final hypothesis to be discussed here is that dust is produced in 'solar nebulae' like that which enveloped the early solar system, and is soon thereafter ejected back into the interstellar medium. In order to account for all the interstellar dust, one requires that in the vicinity of each new star, following its formation, an additional 20 per cent of its mass be processed so as to extract all the condensibles as solids (Herbig, 1970, 1971). Of course, it is unlikely that the process is completely efficient. Furthermore, only the inner region of the nebula can have been raised to a sufficiently high temperature to enable condensation from a hot gas to take place. For such reasons, the total mass required for the average solar nebula must be substantially higher. The 'reconstituted mass' of our own planets gives only a lower limit on the amount of material that once occupied our inner solar nebula, about 0·03 $_{\odot}$; there is no indication of how much more dust was produced and lost. An attractive feature of this proposal is that repeated cycling of interstellar gas through such nebula explains the depletions of Ca, Ti, Al in the residual gas, since these elements would be drawn off preferentially in high-temperature condensates. The process has been

examined in some detail by Field (1974, 1975). Silk (1976) has more recently considered the formation of solids in the atmosphere of a more massive ($5_{\odot}$) protostar, and their ejection by radiation pressure.

Although the degree of depletion appears to vary from place to place in the interstellar medium, the general similarity of the pattern demonstrates that somewhere, a major fraction of the interstellar gas must have been cycled through such an equilibrium condensation process. The solar nebula is only one of several such sites but does have certain advantages over others. It is probably unique that it produces dust in just those regions around young stars where dust is found to occur. (The "chicken or the egg" issue will not be discussed here). Also, it depends not upon hypothetical phenomena to form dust: dust having such a depletion pattern was clearly produced in copious quantities in the early solar system, and a great deal of material was certainly ejected. If one is willing to assume that the early history of the sun was typical in these respects, then a certain fraction of the interstellar dust must necessarily have been formed in this way. The only argument can be whether it is a major fraction or not.

A specific objection to this mechanism has been raised by Cameron (1975), who believes that dust particles grown from the solar nebula would quickly grow to such sizes that they could not be ejected from the system. This of course involves not only assumptions based upon some model of the nebula, but also upon knowledge of the nature of solar activity at that early time. Neither of these, it appears to me, are as yet very well established.

The original proposal also considered the possibility that complex molecules of the types found in carbonaceous chondrites would be formed in the solar nebula and expelled along with the dust (Herbig, 1970, 1971; Anders, 1973). This would be an effective means of enriching the molecular clouds in which stars form. It must be possible, however, for the simplest interstellar molecules to be produced by other means. This is clearly shown for example by the presence of OH, which has a very short lifetime in the interstellar radiation field, in low-density dust clouds (Davies and Matthews, 1972). These thin clouds are certainly too transparent to be considered as sites of star formation at the present time.

In summary, there are now a number of mechanisms that seem capable of generating interstellar dust, perhaps in the quantities required. The fact that some have not been firmly ruled out long ago shows how disappointingly weak are the astronomical constraints, and how elastic the predictions of most theories.

References

Aannestad, P. A., and Purcell, E. (1973). 'Interstellar grains.' *Ann. Rev. Astr. Astrophys.*, **11**, 309–362.

Ambartsumian, V. (1971). 'Fours.' *Astrofiz.*, **7**, 557–572.

Anders, E. (1973). 'Interstellar Molecules: Formation in Solar Nebulae.' *Molecules in the Galactic Environment*, M. A. Gordon and L. E. Snyder, eds. (John Wiley and Sons), pp. 430–442.

Audouze, J., Lequeux, J., and Vigroux, L. (1975). 'Isotopes of carbon, nitrogen and oxygen as probes of nucleosynthesis, stellar mass losses and galactic evolution.' *Astr. Ap.*, **43**, 71–83.

Bregman, J. D., and Rank, D. M. (1975). 'Identification of the 890 cm^{-1} carbonate signature in NGC 7027.' *Ap. J. Letters*, **195**, L125–126.

Cameron, A. G. W. (1975). Field and Cameron, 1975, pp. 1–31.

Clayton, D. D., and Hoyle, F. (1976). 'Grains of anomalous composition from novae.' *Ap. J.*, **203**, 490–496.

Davies, R. D., and Matthews, H. E. (1972). 'Physical conditions in interstellar hydroxyl and formaldehyde clouds.' *Mon. Not. R.A.S.*, **156**, 253–262.

Dicke, R. H. (1972). 'Rotation and lithium abundance in solar-type stars.' *Ap. J.*, **171**, 331–362.

Field, G. B. (1974). 'Interstellar abundances: gas and dust.' *Ap. J.*, **187**, 453–459.

Field, G. B. (1975). 'The composition of interstellar dust.' In Field and Cameron, 1975, pp. 89–105.

Field, G. B., and Cameron, A. G. W., eds. (1975). *The Dusty Universe* (Smithsonian Ap. Obs.)

Gillett, F. C., Forrest, W. J., and Merrill, K. M. (1973). '8–13 micron spectra of NGC 7027, BD + 30° 3639, and NGC 6572.' *Ap. J.*, **183**, 87–93.

Herbig, G. H. (1965). 'Lithium abundances in F5-G8 Dwarfs.' *Ap. J.*, **141**, 588–609.

Herbig, G. H. (1966). 'On the Interpretation of *FU* Orionis.' *Vistas in Astr.*, **8**, 109–125.

Herbig, G. H. (1970). *Optical Spectroscopy of Interstellar Molecules.* Talk given at Boulder, Colo. meeting of American Astronomical Society (June 1970); Introductory Remarks to Évolution Stellaire avant la Sequence Principale: *Mem. Soc. Roy. Sci, Liège*, 5th ser., **19**, 13–26.

Herbig, G. H. (1971). 'Interstellaren Staub als Nebenprodukt der Sternentstehung.' *Sterne und Weltraum*, **10**, 4–8.

Herbig, G. H. (1974). 'On the nature of the small dark globules in the Rosette Nebula.' *Publ. Astr. Soc. Pacific*, **86**, 604–608.

Herbig, G. H., and Peimbert, M. (1966). 'The distribution of emission-line stars in the Taurus Dark Nebulae.' *Trans. Int. Astr. Union*, **12B**, 412–415.

Hollenbach, D., and Salpeter, E. E. (1971). 'Surface recombination of hydrogen molecules, *Ap. J.*, **163**, 155–164.

Hoyle, F., and Wickramasinghe, N. C. (1962). 'On graphite particles as interstellar grains.' *Mon. Not. R.A.S.*, **124**, 417–433.

Hoyle, F., and Wickramasinghe, N. C. (1970). 'Dust in supernova explosions.' *Nature*, **226**, 62–63.

Kuhi, L. V. (1964). 'Mass loss from T-Tauri Stars.' *Ap. J.*, **140**, 1409–1433.

Kwok, Sun (1975). 'Radiation pressure on grains as a mechanism for mass loss in Red Giants.' *Ap. J.*, **198**, 583–591.

Mendoza V. E. E. (1966). 'Infrared photometry of T-Tauri Stars and related objects.' *Ap. J.*, **143**, 1010–1014.

Meneguzzi, M., Audouze, J., and Reeves, H. (1971). 'The production of the elements Li, Be, B by galactic cosmic rays in space and its relation with stellar observations.' *Astr. Ap.*, **15**, 337–359.

Nichiporuk, W., and Moore, C. B. (1974). 'Lithium, sodium, and potassium abundances in carbonaceous chondrites.' *Geoch. et Cosmoch. Acta*, **38**, 1691–1701.

Oort, J. H., and Van de Hulst, H. C. (1946). 'Gas and smoke in interstellar space.' *Bull. Astr. Inst. Netherlands*, **10**, 187–204.

Osterbrock, D. E. (1973). 'The origin and evolution of planetary nebulae.' *Les Nébuleuses Planétaires*: *Mem. Soc. Roy. Sci. Liège*, 6 ser., **5**, 391–402.

Rydgren, A. E., Strom, S. E., and Strom, K. M. (1976). 'The nature of the objects of Joy: A study of the T-Tauri phenomenon.' *Ap. J. Suppl.*, **30**, 307–336.

Ryter, C., Reeves, H., Gradsztajn, E., and Audouze, J. (1970). 'The energetics of L-Nuclei formation in stellar atmospheres and its relevance to X-ray astronomy.' *Astr. Ap.*, **8**, 389–397.

Salpeter, E. E. (1974a). 'Nucleation and growth of dust grains.' *Ap. J.*, **193**, 579–584.

Salpeter, E. E. (1974b). 'Formation and flow of dust grains in cool stellar atmospheres.' *Ap. J.*, **193**, 585–592.

Salpeter, E. E. (1974c). 'Dying stars and reborn dust.' *Rev. Mod. Phys.*, **46**, 433–436.

Silk, J. (1976). 'The protostellar origin of interstellar grains.' *Far Infrared Astronomy*, ed. M. Rowan-Robinson (Pergamon Press), pp. 309–319.

Stellmacher, G. and Wiehr, E. (1971). 'The influence of the sunspot model on the Li-Abundance.' *Solar Phys.*, **21**, 96–100.

Traub, W. A., and Carleton, N. P. (1973). 'Detection of interstellar lithium.' *Ap. J. Letters*, **184**, L11–14.

Ulrich, R. K. (1976). 'An infall model for the T-Tauri phenomenon.' *Ap. J.*, **210**, 377–391.

Watson, W. D., and Salpeter, E. E. (1972a). 'Molecule formation on interstellar grains.' *Ap. J.*, **174**, 321–340.

Watson, W. D., and Salpeter, E. E. (1972b). 'On the abundances of interstellar molecules.' *Ap. J.*, **175**, 659–671.

Wickramasinghe, N. C. (1967). *Interstellar Grains*. (Chapman and Hall, Ltd.)

Woolf, N. (1973). 'Circumstellar infrared emission. I: The circumstellar origin of interstellar dust.' *Interstellar Dust and Related Topics*, J. M. Greenberg and H. C. Van de Hulst, eds., (D. Reidel Publ. Co.), pp. 485–504.

Zappala, R. R. (1972). 'Lithium abundances of stars in open clusters.' *Ap. J.*, **172**, 57–74.

COLLAPSE DYNAMICS AND COLLAPSE MODELS

R. B. LARSON
Yale University Observatory, U.S.A.

Introduction

Although we now possess a wealth of information about the properties of molecular clouds and the circumstances in which stars form, we still have little direct information about the crucial question of how the material in molecular clouds actually becomes condensed into stars. In this report we shall discuss briefly the current status of theoretical attempts to understand this problem, based on calculations of the dynamics of collapsing clouds and protostars.

Many collapse calculations have treated the collapse of individual protostars idealized as isolated objects, but a complete understanding of star formation requires also a consideration of how gravitational collapse proceeds on scales larger than that of individual protostars. The properties and evolution of protostars depend on the conditions under which they are formed, and their observable characteristics are particularly influenced by the properties of the material surrounding them. The common occurrence of young stars in close groups and multiple systems indicates that forming stars cannot in general be considered as isolated systems. Also, it is evident that any understanding of such long-standing problems as the initial stellar mass spectrum and the efficiency of star formation requires an understanding of the collapse problem in a larger context than that of individual protostars. Thus, while most collapse calculations refer to such special cases, it will be important to keep in mind their possible implications for the more general problems of how clouds collapse and fragment into stars.

The Collapse Problem

In the evolution of a collapsing interstellar cloud or cloud fragment we may, at least conceptually, distinguish two phases: at first, the cloud is not gravitationally bound but is compressed by external forces; later, the self gravity of the cloud becomes dominant over other forces (i.e., its free-fall time becomes shorter than all other time scales) and the cloud collapses gravitationally. Here we shall not consider the stages leading to the formation of a gravitationally bound cloud; instead, we are interested in how the cloud develops subsequently and how the results depend on the conditions existing when gravity takes over and collapse begins. It is not clear that such a separation into two main phases of evolution is always possible, especially on the scale of individual protostars, but this is a question on which multi-dimensional collapse calculations such as those mentioned below can potentially shed some light.

The collapse of an interstellar cloud and the formation of stars in it involve many different processes, and many questions arise when we attempt to understand these various parts of the collapse problem. During the early stages of the collapse, the most important questions concern the process of fragmentation: How do condensations form in a collapsing cloud, and what determines their properties? Do the condensations, once formed, continue to fragment indefinitely or do they generally form only a single star or close multiple system? Do large scale gas flows and shock compression in collapsing clouds play an important role in producing the conditions required for fragmentation into protostars? Considering the later stages, we wish to understand the evolution of the individual collapsing fragments or protostars: How does the material in a protostar ultimately become condensed to stellar density? What is the role of accretion phenomena, such as accretion shocks and accretion disks? What determines whether a single star or a binary is formed? How much mass eventually goes into the star(s), and how much is dispersed? How do newly formed or forming stars react back on the collapsing cloud and influence its further evolution? Finally, and perhaps most importantly, how do all of these processes work to determine the masses and the mass spectrum of the stars finally formed?

Many calculations using a variety of different assumptions and methods have been made to study the collapse of gas clouds, and while definitive answers to the above questions are not yet in hand, the results of these calculations provide valuable insights into the way in which collapse proceeds. The case of spherical collapse without rotation or magnetic fields has been relatively well explored, and the effects of initial conditions, boundary conditions, and thermal and radiative properties of the material have been investigated in a number of studies. Also, the spherical case is the only one in which it has been possible to follow the collapse all the way to stellar densities and the formation of hydrostatic pre-main sequence stars. Results for collapse with axial symmetry are more limited, and calculations of this type have so far concentrated on studying the effect of rotation and the question of whether rings are formed. Finally, some preliminary results are available for the general three-dimensional collapse problem, obtained with both a grid method and a finite-particle method; these results provide new information on the fragmentation problem, and support the general validity of certain results of the spherical and axisymmetric calculations. In the following sections we shall review the results of each of these types of calculation in turn.

Spherical Collapse

The basic features of spherical collapse have previously been discussed at some length by Larson (1973, 1974, 1975), and more recent studies by Ferraioli and Virgopia (1975), Kondo (1975), Appenzeller and Tscharnuter (1974, 1975), Westbrook and Tarter (1975), and Yorke and Krügel (1977) confirm the basic qualitative results discussed earlier, although with some quantitative differences which we shall consider later. In all cases the collapse is non-homologous and the density distribution approaches the approximate form $\rho \propto r^{-2}$ if the collapse is nearly isothermal. Eventually the inner regions become opaque and a hot dense stellar core is formed, surrounded at first by an extended infalling envelope that still contains most of the mass, and the material in the envelope later falls into the core through an accretion shock at its surface.

The non-homologous collapse found in the spherical case occurs also in the two-and three-dimensional collapse calculations (see below), and has many important implications for the later evolution of collapsing clouds and protostars. Many properties of protostars, for example, depend strongly on the existence of a remnant infalling envelope, and on how much its collapse is retarded relative to that of the core. There are two reasons for a non-uniform collapse: (1) any initial density inhomogenities in a collapsing cloud tend to be amplified because the densest regions collapse fastest, and (2) the collapse of the outer parts of the cloud is retarded relative to the collapse of the core by pressure gradients; even if not present initially, such pressure gradients are soon established by the inward propagation of a rarefaction wave from the boundary of the cloud. The importance of both of these effects depends on how closely the initial conditions satisfy the Jeans criterion. Density irregularities on scales smaller than the Jeans length will tend to be smoothed out by acoustic waves before the cloud can collapse very far, but larger scale inhomogeneities can survive and grow if the cloud size exceeds the Jeans length. On the other hand, the propagation of rarefaction waves is most important if the cloud size is comparable to the Jeans length, i.e. the sound travel time is comparable to the free fall time. In either case, the collapse is non-uniform, the dominant scale of the density inhomogeneities being comparable to the Jeans length. As will be discussed, however, quantitative results can be rather sensitive to the initial conditions, so it is important for quantitative purposes to understand in some detail just how collapsing clouds and protostars are formed and what their initial conditions are.

The predicted non-homologous collapse of spherical clouds and protostars has important implications for the dynamics of collapse in general. If some parts of a collapsing cloud become much more condensed than the rest, the dynamics of the remaining diffuse material is increasingly dominated by the effects of these dense condensations. In particular, local self-gravitational forces in the residual gas become unimportant in comparison with the tidal forces produced by the mass concentrations, and therefore further fragmentation is inhibited and the collapse of the remaining material takes on the character of an accretion process, with the dense condensations acting as the centres of accretion. This qualitative picture is supported by the results of 3-dimensional collapse calculations (below). Since the time scale for the collapse and internal evolution of the condensations is relatively short, stars can begin to form in them and react back on the remaining gas in a variety of ways; for example, radiation pressure, ionization, or stellar winds may blow away the remaining gas and prevent it from condensing into stars. Thus the masses finally attained by the forming stars and the efficiency of star formation may depend strongly on the extent to which the collapse is non-homologous, leaving behind much diffuse gas after stars have begun to form.

Axisymmetric Collapse

Non-rotating

A number of calculations of the isothermal collapse of non-rotating prolate and oblate configurations have been made by Larson (1972a). The results showed the same kind of non-homologous collapse with increasing central concentration as had previously been found in the spherical case. It was also found that if the initial conditions approximately satisfy the Jeans criterion, pressure gradients remain important and tend to prevent the growth of large deviations from spherical symmetry during an isothermal collapse. These results suggest that if the Jeans criterion is nearly satisfied, fragmentation will not occur and a single central object will form, just as in the spherical case. The only configuration which showed any tendency toward fragmentation was a cold cylinder of gas which was much longer than the Jeans length and hence was already unstable to fragmentation initially; in this case, two centres of condensation formed. While not conclusive, these results suggest that fragmentation is not likely to occur during non-rotating isothermal collapse unless the initial configuration is already unstable to fragmentation; otherwise, the collapse proceeds roughly spherically and only a single central star is formed, very much as in the spherical case.

Rotating

There has been greater interest in studying the collapse of rotating axisymmetric clouds, and such calculations have been made by Larson (1972a), Tscharnuter (1975), Fricke *et al.* (1976), Black and Bodenheimer (1976), and Nakazawa *et al.* (1976). All calculations find once again a strongly non-homologous collapse with the development of a small region of high density near the centre, even when centrifugal force and gravity are initially nearly in balance. The high density near the centre is attained largely as a result of infall of matter along the axis of rotation, where infall is not impeded by centrifugal force. However, the various calculations disagree in the detailed form of the high density region near the centre: the results of Larson, Black and Bodenheimer, and most of those of Nakazawa *et al.* show the formation of a ring structure near the centre, while Nakazawa *et al.* find that no ring is formed in a case with higher initial temperature, and Tscharunuter and Fricke *et al.*, using a different numerical method, find no tendency toward ring formation in any circumstances.

It is clear that numerical errors must be present in some (or all) of these calculations, but the results also suggest that the formation of a ring may be sensitive to the initial conditions and to the detailed distribution of angular momentum in a collapsing cloud. The grid methods of Larson, Black and Bodeinheimer, and Nakazawa *et al.* contain an inherent numerical viscosity which may lead to inaccuracies in conservation of angular momentum; on the other hand, it is not clear that the Legendre expansion method of Tscharnuter and Fricke *et al.* is adequate to represent the formation of rings (Nakazawa *et al.* 1976). In any case, all investigations have assumed inviscid flow with strict conservation of angular momentum for each fluid element, but in reality it is likely that torques due to magnetic forces, turbulent viscosity, or deviations from perfect axial symmetry would alter the results importantly. Until these various numerical and physical problems are better understood, it seems reasonable to conclude that there probably exist circumstances in which rings are formed and other circumstances, perhaps not very different, in which rings are not formed.

If a ring forms, it seems inescapable that the end result will be the fragmentation of the ring into (at least) a binary system of condesations orbiting around each other. This result is suggested by stability considerations, and has been found in preliminary 3-dimensional calculations by Black and Wilson (1977) of the evolution of such rings. In view of the very general susceptibility of rapidly rotating systems to non-axisymmetric instabilities, it even seems possible or likely that a rapidly rotating cloud will fragment directly into a binary system even before a well-defined ring is formed. This result is in fact a frequent outcome of the crude 3-dimensional collapse calculations described below. If it can be assumed that each of the resulting condensations forms a single star, this would provide an attractive explanation of the observed preponderance of binary stars (Abt and Levy, 1976).

However, it remains to be understood how the two orbiting condensations resulting from such a fragmentation process would evolve subsequently. Do they in fact form single stars, or does further fragmentation take place? As members of a binary system, they are now subject to external forces of a non-axisymmetric nature, and the idealization of axisymmetric collapse is no longer justifiable. Tidal torques will tend to transfer angular momentum from spin to orbital form within the system, thereby possibly enabling the condensations to collapse into single stars without further fragmentation. Also, if the condensations begin with small mass and accrete most of their mass from a common surrounding envelope, the accretion process may tend to select mass elements with small angular momentum relative to the two accreting centres, so that they end up with little spin angular momentum and form single stars. It will be important to try to answer these questions with 3-dimensional collapse calculations.

In summary, the axisymmetric collapse calculations suggest that in the absence of rotation, fragmentation does not generally occur unless the initial configuration is already unstable to fragmentation. In the presence of rotation, fragmentation into a binary or small multiple system seems likely to be the most common result. In neither case, however, do the calculations suggest the occurrence of an extensive hierarchial fragmentation process of the sort that has often been assumed in discussions of star formation. Instead, the results suggest that successive fragmentation occurs only to a very limited extent, generally not proceeding beyond the formation of binary systems, and that the properties of the fragments are largely determined by the initial conditions existing at the time when gravitational collapse begins.

Magnetic

Although no calculations of collapse with magnetic fields have yet been made, Mouschovias (1976a, b) has calculated sequences of equilibrium models of self-gravitating clouds partially supported by magnetic fields, which provide some indication of how collapse might proceed with magnetic fields. These models are moderately flattened and strongly centrally condensed, the degree of central condensation increasing as self-gravity becomes more important until, at a critical point, equilibrium is no longer possible and collapse must occur. This is qualitatively the same behaviour as is found without magnetic fields, and this suggests that the collapse of a magnetic cloud will take place in qualitatively the same non-homologous way as the collapse of a non-magnetic cloud, with ever increasing central concentration. Mouschovias has also emphasized the probable role of magnetic fields in retarding the collapse of the outer layers of the cloud even after the inner part has decoupled from the magnetic field; this would enhance the separation into a dense core and diffuse envelope that is already implied by the results for non-magnetic clouds. Magnetic fields may thus play an important role in reducing the efficiency of star formation.

Three-Dimensional Collapse

Grid Methods

We mention first some calculations which, although not strictly 3-dimensional, probably display many features of a realistic 3-dimensional collapse. These calculations have been made by Theys (1973) and Quirk (1973) using the 'beam scheme' to calculate the dynamics of self-gravitating thin disks of gas. Theys, in studying possible models for ring galaxies, calculated the collapse of rapidly rotating gas disks and found that rings form in some cases but not in others, depending on the assumed initial density distribution. In all cases, such rings were found to be very unstable, and rapidly fragmented into a number of dense lumps. Quirk calculated the collapse of initially nearly uniform rapidly rotating disks with different temperatures, and found rapid fragmentation into systems of orbiting condensations or knots, the sizes of the fragments being approximately consistent with the initial Jeans length. The fragments tended to become more centrally condensed with time, but showed no tendency toward further fragmentation. A considerable part of the initial mass remained as diffuse gas swirling around the condensations, some being slowly accreted by them and some being dispersed. Many of these features are expected to occur in a realistic 3-dimensional collapse; however, besides being limited to a thin disk, these calculations contain an inherent numerical viscosity which, although it may qualitatively mimic real effects, is difficult to justify quantitatively.

Preliminary calculations with a 3-dimensional grid have been made by Black and Wilson (1977) to study the evolution of self-gravitating rings such as those found in several calculations of the collapse of axisymmetric rotating clouds. The results show that the ring begins to fragment into two major condensations orbiting around each other, thus supporting the conjecture that the collapse of a rotating cloud will ultimately result in a binary system.

Finite-particle (N-Body) Method

Finally, we mention some preliminary results obtained with a 'finite-particle' method (Larson 1978a, b) in which a fluid is simulated by a system of representative particles or fluid elements of finite mass whose motions are followed individually, as in N-body calculations of stellar dynamics. In addition to a modified inverse-square gravitational force, the particle interactions include a pressure force between neighbouring particles, supplemented by an 'artificial viscosity' term simulating the effect of shock waves; these terms are analogous to those appearing in standard Lagrangian hydrodynamic codes. The form adopted for the pressure term assumes that the gas remains isothermal, but the temperature is an adjustable parameter. To avoid the necessity of calculating the very rapid internal dynamics of the tightly bound condensations that form, the particles have been merged when they come very close together.

The first applications of this method have been made with systems of 100 particles which are initially scattered randomly in a sphere and given a rigid rotation; the main parameters which have been varied are the temperature and the initial angular velocity. The collapse of this system of particles or 'cloud' is found to depend on a modified Jeans criterion similar to that of Larson (1972a) for axisymmetric collapse, in which the kinetic energies of thermal motion and rotation contribute approximately additively to counterbalancing gravity. The outcome of the collapse depends strongly on how closely the initial conditions satisfy this modified Jeans criterion, and we briefly summarize here the different types of results found.

If the initial condition nearly satisfies the Jeans criterion, i.e. if gravity is almost balanced by pressure plus centrifugal force, the cloud remains very extended and roughly spherical in overall structure and develops a single central condensation, surrounded by a diffuse envelope containing most of the mass. This is qualitatively the same result as was previously found in the spherical and axisymmetric calculations, except that here no tendency for ring formation is evident; however, it is not clear that the small number of particles involved could adequately represent a ring structure. The central condensation rapidly forms a single object containing many particles, and this central object continues to accrete material from an extended and moderately flattened 'accretion envelope' around it. The innermost part of the accretion envelope is more strongly flattened and may be described as an accretion disk. Depending on the initial angular velocity, at least 20 per cent and perhaps as much as half of the initial mass is eventually accreted on the central object, the remainder being gradually dispersed.

In these results, an important role is apparently played by the 'turbulent viscosity' produced by the interactions between the individual particles; this viscosity acts to transfer angular momentum outward in the accretion envelope, allowing the inner part to fall onto the central object. In this respect the model resembles the floccule theory of star formation proposed by McCrea (1960). While a number of mechanisms probably operate to redistribute angular momentum in a collapsing cloud, including magnetic and gravitational torques as well as viscosity due to random motions, these effects are not yet quantitatively well understood, so it is not clear to what extent the finite-particle viscosity is quantitatively realistic; thus the results must be considered as only illustrative rather than definitive. Nevertheless, they do indicate the likely importance of viscous forces in collapsing clouds, and they support the suggestion from spherical and axisymmetric calculations that fragmentation is not likely to occur if the initial conditions closely satisfy the Jeans criterion.

As the initial ratio of pressure plus centrifugal force to gravity is reduced, a point is soon reached where the cloud fragments into a binary system. There are two effects which favour fragmentation: (1) a reduction in the initial Jeans length, and (2) an overall contraction of the cloud which raises its mean density, further reducing the Jeans length in comparison with the size of the cloud. The exact mode of fragmentation into a binary seems to depend on the initial angular velocity: in a case in which the initial centrifugal force was small enough to allow a significant overall contraction, the cloud collapsed to a relatively compact bar-like configuration which then separated into two lumps, whereas in a case with higher angular momentum the cloud developed a more diffuse and elongated bar-like structure with initially only one major condensation, and the remaining material then formed a 'spiral arm' structure in which a second condensation later appeared. In either case, each of the two condensations collapses without further fragmentation to form a single condensed object, and the two objects thus formed continue to accrete matter from the remainder of the cloud. If they are well separated, each is surrounded by its own accretion envelope, but if they are close together their accretion zones overlap and they share a common accretion envelope. In most cases, there appears to be a tendency for the accretion process to approximately equalize the masses of the two accreting objects, and together they may eventually accrete as much as half of the initial mass of the cloud.

As the temperature of the cloud is further reduced, fragmentation into a larger and larger number of distinct condensations is observed. The number of condensations formed is approximately inversely proportional to the initial Jeans mass, but the amount of fragmentation taking place and the fraction of the cloud mass going into dense condensations are enhanced by an initial overall contraction of the cloud which becomes more important as the initial angular velocity is decreased. With faster initial rotation, on the other hand, a larger fraction of the mass remains in diffuse form and is eventually dispersed. In systems which form many condensed objects or 'stars', there is a strong tendency toward the formation of subgroups consisting of binary or small multiple systems, often with a hierarchial structure reminiscent of that observed for multiple stars. Binary systems generally appear to form as described above, and a hierarchial triple system forms if the temperature and Jeans mass are small enough that one of the condensations in a binary system that is beginning to form fragments further into a closer binary pair.

Another phenomenon observed in these results is that secondary condensations may sometimes form in the accretion envelopes or disks around more massive accreting objects; this can occur because of the relatively high density and high velocity of the accreted gas around a massive object, which can result in strong shock compression. This process may lead to the formation of small companion stars or large planets like Jupiter around some stars. However, the formation of secondary condensations is also inhibited by the tidal force field of the accreting object, and this process does not seem to account for a majority of the 'stars' formed in these calculations; the bulk of the fragmentation appears to occur as a one-shot process during the initial collapse of the cloud.

In summary, these crude 3-dimensional collapse calculations suggest that fragmentation is largely determined by the Jeans mass at the time when collapse begins. The condensations in a collapsing cloud are formed approximately satisfying the Jeans criterion, and depending in detail on how closely they satisfy it, they may form either single stars, binaries, or small multiple systems; however, extensive hierarchical fragmentation is not found. Provided that viscosity or other mechanisms for redistributing angular momentum are present, individual collapsing condensations or protostars evolve much as in the spherical case, forming a dense accreting core surrounded by an extended accretion envelope, which has a flattened disk-like structure in the innermost region near the accreting core. The principal difference from the spherical case is that the inflow of matter into the core may not occur in free fall but at a rate determined by the viscosity of an accretion envelope or disk. The formation of an accretion disk offers an attractive possibility for understanding the origin of planetary systems, and accretion disk models for the formation of the solar system have been studied by Cameron (herein, pp. 49–74).

We note that important implications for star formation theory follow if extensive hierarchical fragmentation does not occur in collapsing clouds, as has generally been assumed. If fragmentation is determined by the conditions existing when collapse begins, then it is important to understand the dynamical and thermal evolution of clouds prior to the onset of collapse; in particular, it is necessary to understand the processes of cloud compression and cooling which lead to the high densities and low temperatures which are observed in molecular clouds and which are necessary for the collapse of protostars in the normal stellar mass range.

Evolution of Protostellar Cores

The only existing calculations of the later stages of protostellar evolution have all assumed spherical symmetry, so we now return to consider in more detail the spherical collapse models discussed in the section above. As we have seen, these idealized spherical models are probably qualitatively correct, but may need quantitative revision to take into account various effects such as rotation which they neglect.

According to the spherical models, protostars during their later stages of evolution consist of two very distinct regions which can for many purposes be treated separately. The accreting core is essentially a hydrostatic star, and its properties and evolution depend on those of the infalling envelope primarily through the rate of infall of matter onto its surface; the detailed structure of the envelope is not important except insofar as it determines the inflow rate. On the other hand, the envelope is affected by the core only through the core's mass and luminosity, and perhaps during later stages by other effects such as stellar winds. In this and the following section we consider in turn the properties of protostellar cores and envelopes.

Low Mass Protostars

It is convenient to distinguish low mass protostars with masses less than a few solar masses from more massive protostars, since the evolution of the core is qualitatively different in the two cases. In the low mass case, radiative energy transfer remains unimportant in the core during the entire accretion process, and until an outer convection zone appears during the final stages of core accretion, each mass element in the core retains the entropy which it had immediately after passing through the accretion shock. Thus the entropy distribution and the structure of the core are determined by the thermal properties of the surface layers just inside the accretion shock, which depend in turn on the properties of the infalling matter just outside the shock.

If the protostellar envelope is extremely dense and compact and falls rapidly onto the core, the infalling material may be so opaque that no significant radiative energy losses can occur and the surface layers of the core end up with a very high specific entropy; as a result, the core of a 1 $M_\odot$ protostar attains a large radius of $\sim$ 100 $R_\odot$ and a correspondingly high luminosity of $\sim$ 500 $L_\odot$ at the end of the accretion phase. An example of this type of evolution is provided by the models of Narita *et al.* (1970).

If, on the other hand, the early stages of the collapse are very non-homologous and leave a much more diffuse and extended envelope surrounding the core, as the calculations described above lead us to expect, the envelope eventually becomes sufficiently optically thin at infrared wavelengths to allow radiation to escape from the surface of the core during the accretion process. In this case the kinetic energy of the infalling matter is largely radiated away at the surface of the core and the surface temperature is determined by the kinetic energy inflow rate, which depends in turn on the rate of mass accretion; a higher accretion rate implies a higher surface temperature and hence a higher entropy and a larger radius for the core. If the collapse is strongly non-homologous and the time scale for envelope accretion is as long as the initial free fall time, as in the models of Larson (1972b) and the 3-dimensional calculations described above, the entropy inside the accretion shock is relatively low and the core ends up with a radius of only a few $R_\odot$ and a luminosity of a few $L_\odot$ when the accretion ceases to be important. These results have been numerically confirmed within a factor of two by Appenzeller and Tscharnuter (1975), who obtained a core radius about twice as large as Larson, possibly partly because of the different opacities assumed.

The final properties of the stellar core are related in a simple way to the time scale for infall of the envelope. During the final stages of the accretion process, the internal contraction luminosity of the core becomes comparable to the luminosity produced by accretion. Since the kinetic energy of the infalling gas is comparable to the thermal kinetic energy of the material in the core, this implies that the Kelvin–Helmholtz contraction time of the core must be comparable to the time required to accrete an amount of mass equal to the core mass. Even if the core first forms with a contraction time shorter than the accretion time, it would contract rapidly while still accreting matter until the two time scales become comparable. Thus if the accretion time is long, the final core must have a long Kelvin–Helmholtz contraction time, i.e. it must have a small radius and luminosity; conversely, an object of large radius and luminosity can be formed only by a corresponding rapid accretion process.

The final properties of the core thus depend on the collapse time of the envelope, which depends on both the initial density and the degree to which the collapse is non-nomologous and leaves an extended envelope around the core. A higher initial density implies a shorter collapse time and a larger final radius; for example, Larson (1969a) found that an increase by a factor of 10^3 in the initial density increases the final radius by about a factor of 3. Much larger final radii were obtained in the calculations of Westbrook and Tarter (1975), in which pressure retardation of the outer layers of the collapsing cloud is apparently less important than in the models of Larson or Appenzeller and Tscharnuter, with the result that the bulk of the envelope falls onto the core relatively soon after the core has formed. The final core radius for a 1 $M_\odot$ model is about 90 $R_\odot$ in these calculations. While a difference in this direction from the results of Larson and of Appenzeller and Tscharnuter is expected because the lower initial temperature assumed by Westbrook and Tarter (3 K instead of 10 K) makes pressure retardation of the envelope collapse less important, several features of the Westbrook and Tarter calculations are crude and the quantitative results are difficult to interpret. In any case, the 3-dimensional collapse results described above suggest that the assumptions of Larson and of Appenzeller and Tscharnuter are more realistic. In addition, the lengthening of the accretion time scale by the effects of rotation and the formation of an accretion disc would tend to produce a smaller final core radius.

Massive Protostars

If the mass of a protostar exceeds a few solar masses, radiative transfer begins to become important in the core before the infall of the envelope is completed, and the core approaches radiative equilibrium as a conventional radiative pre-main sequence object. The occurrence of radiative energy losses from the core on a time scale shorter than the accretion time means that the Kelvin–Helmholtz contraction time becomes shorter than the accretion time and the intrinsic core luminosity begins to exceed the infall luminosity. If the core mass exceeds approximately 3–5 $M_\odot$, the core can contract all the way to the main sequence and begin nuclear burning as a main sequence star while continuing to accrete matter and grow in mass. Once the internal luminosity of the core becomes dominant, its structure and evolution are no longer determined by the properties of the accretion shock and surface layers, and are altered only by the addition of mass.

The approach to radiative equilibrium is accompanied by a rapid jump in core luminosity, which may be very important for the evolution of the envelope and the observed properties of the protostar. In the 5 $M_\odot$ and 10 $M_\odot$ models of Larson (1972b) the effect is relatively modest, but in the 60 $M_\odot$ model of

Appenzeller and Tscharnuter (1974) the outer layers of the core are heated so strongly and rapidly that hydrostatic equilibrium is destroyed and the outermost layers of the core and the entire infalling envelope are blown off when the core attains a mass of about 18 $M_\odot$. It is clear from this result that only a fraction of the material in a massive protostar actually becomes incorporated in a star, but more calculations of different cases will be necessary to provide a better understanding of this phenomenon and its possible observational implications. It seems possible, for example, that the predicted rapid flareup in luminosity may be related to the '*FU* Orionis phenomenon', but no quantitative agreement between models and observations can yet be claimed.

If an accreting protostellar core attains a mass in the upper main sequence mass range, probably having already become a main sequence star, its very high luminosity almost certainly begins to affect the dynamics of the infalling envelope. There are a number of ways in which the radiation from a massive protestellar core can cause the envelope to be blown off, and these probably set a limit of less than 100 $M_\odot$ to the mass with which a star can form (Larson and Starrfield, 1971). Larson and Starrfield estimated that the most important limiting effect would be the ionization of the envelope when the core becomes an O star in the mass range 30–60 $M_\odot$. However, in a more detailed analysis of the effect of radiation pressure, Kahn (1974) concluded that radiation pressure would be more important and would limit the mass that can be attained by accretion to about 40 $M_\odot$, depending on the properties of the dust in protostellar envelopes. The models of Westbrook and Tarter (1975) for massive protostars show that an increasing fraction of the mass is blown off by radiation pressure as the total mass is increased; for example, a 50 $M_\odot$ model loses 35 $M_\odot$ in this way, leaving a core of only 15 $M_\odot$. In a detailed numerical study of the dynamics of protostellar envelopes, Yorke and Krügel (1977) found that the most important effect limiting the core mass is radiation pressure acting on the outer part of the envelope, and in an example with a total mass of 150 $M_\odot$, a star of only 35 $M_\odot$ was formed and the remainder of the material was dispersed by radiation pressure.

From all of these studies it is clear that it is difficult to form massive stars because a variety of effects, including radiation pressure and perhaps also stellar winds, will tend to blow off the protostellar envelope before all of it has been accreted on the core. These effects could plausibly account for the steep decline in the stellar mass function at large masses and for the apparent absence of stars with masses exceeding 100 $M_\odot$; indeed, there may even be some difficulty in understanding how the most massive observed stars can form. A very dense protostellar envelope is required to overcome the effects of radiation pressure or stellar winds and allow accretion to continue; thus it may be that the most massive stars form only in regions of unusually high density. Nonspherical accretion may also play a role; if a protostellar envelope has large density inhomogeneities or develops a flattened disk-like structure, the effects of radiation pressure, etc., may more easily be overcome.

Protostellar Envelopes and Observed Properties

During most of its evolution, the protostellar core is completely obscured by the optically thick infalling envelope, and the observed properties of the system depend strongly on the structure of the envelope. The radial density variation in a spherical accretion flow has the form $\rho \propto r^{-3/2}$, and the envelope density distributions in spherical collapse models approach this form during the later stages of the collapse. Shu (1977) has shown that the development of an accretion flow with this type of density distribution can be approximately described by a similarity solution (which is of a qualitatively different nature from the similarity solution proposed for the earlier isothermal stages of collapse by Larson (1969a) and Penston (1969)). Until the protostellar envelope is almost completely accreted or dispersed, the luminosity generated in the accretion shock or in the core is absorbed by dust grains in the envelope and converted to infrared radiation, so that the emitted spectrum of the protostar depends on the transfer of infrared radiation through the extended optically thick envelope.

Approximate calculations of the spectrum of radiation emitted from protostellar envelopes with power-law density distributions have been made by Larson (1969b) and Rowan-Robinson (1975), and a more detailed numerical solution of the transfer problem has been made by Bertout (1976). In these studies the dust opacity was simply assumed to vary with a power of the wavelength, so only the gross features of the spectrum can be represented. The predicted spectrum resembles a blackbody spectrum, except that there is more emission at long wavelengths. As the protostar evolves and the envelope density decreases, the 'photosphere' or surface of optical depth unity moves inward to higher temperatures, and the wavelength of peak emission decreases. The corresponding evolutionary tracks in an infrared HR diagram (Larson 1972b) fall in the same region as a number of infrared sources which are believed to be young objects or protostars.

Qualitative but not quantitative agreement is found between the predicted spectra and observed infrared spectra, which are generally broader than a blackbody spectrum and also show absorption features near 3 and 10 microns due to water and silicates. A more detailed treatment of the dust opacity is evidently required to reproduce such features, and Finn and Simon (1977) find that the 10 micron feature can be closely reproduced by a model with silicate grains in a protostellar envelope having a power-law density distribution. Another important factor is the geometrical structure of the envelope and the distribution of dust in it. For example, Cohen (1973) has obtained better fits to the observed infrared spectra by using more elaborate double shell models. Such a 'double cocoon' structure is in fact predicted by the detailed envelope models of Yorke and Krügel (1977), which have an outer shell of ice grains at a temperature of < 200 K and an inner shell of refractory grains at a temperature of ~ 1000 K. Deviations from spherical symmetry of protostellar envelopes, such as a clumpy or flattened structure, are probably also important in producing regions with a range of temperatures and hence in making the spectrum broader than a blackbody.

When the protostellar envelope is no longer completely optically thick, a double-peaked spectrum is predicted showing contributions from both the envelope and the core, which by now is essentially a conventional hydrostatic pre-main sequence star. The predicted spectra qualitatively resemble those of a number of T-Tauri stars, such as T-Tau, R-Mon, and R-CrA, which show in varying degrees separate infrared emission peaks probably attributable to circumstellar dust. If accretion is still important, radiation from the hot layers inside the accretion shock may make an important contribution to the spectrum at short wavelengths (Walker, 1972; Ulrich, 1976), where enhanced emission is often observed.

At present it is controversial whether any features in the optical spectra of T-Tauri stars can be explained by the effects of infall from a remnant protostellar envelope. The conventional interpretation of T-Tauri emission line profiles is in terms of outflow of matter, and it has been argued that this outflow is important for the final dispersal of protostellar envelopes (Strom *et al.* 1975). Only a few stars, the so-called YY Ori stars (Walker, 1972), have been found in which the line profiles sometimes seem to indicate infall rather than outflow. However, Ulrich (1976) has challenged the conventional interpretations, and has shown by a detailed study of the kinematics and radiative transfer in an infall model that some spectral features usually attributed to outflow can also be explained as radiation from the accretion shocks in an infall model. Lynden-Bell and Pringle (1974) have also proposed an accretion-disk model to explain some of the properties of T-Tauri stars. As we have seen, the formation of accretion disks seems to be an almost inescapable result of the collapse process, and provides an attractive framework for understanding the formation of planetary systems. However, none of the available collapse calculations have incorporated the effects of stellar winds, largely for want of any quantitative understanding of this phenomenon, and it is entirely possible that such winds could eventually dominate the dynamics and reverse the infall in protostellar envelopes, perhaps even before the core becomes visible, in which case infall would never be observed.

Many interesting phenomena related to star formation appear to be associated with the final dispersal of protostellar envelopes. For example, Strom *et al.* (1975) have suggested that winds from obscured T-Tauri stars first clear out holes in the surrounding cloud material, allowing these stars to illuminate or excite patches of nebulosity which are then observed as Herbig–Haro objects. Herbig (1970), Field (1974), and Silk (1976) have suggested that the dissipation of protostellar envelopes or 'solar nebulae' may provide an important source of dust grains for the interstellar medium. The possible separation of gas and dust in protostellar envelopes has been studied by Edmunds and Wickramasinghe (1974), and the formation and dynamics of dust shells or 'cocoons' have been studied by Davidson (1970), Burke and Silk (1976), Yorke and Krügel

(1977), and Cochran and Ostriker (1977). These dust shells are driven outward by radiation pressure, and this phenomenon may play a dominant role in the final evolution of protostellar envelopes and in determining the upper mass limit for star formation. It is perhaps also in these cocoons that the conditions required for OH and H_2O maser emission are produced (de Jong 1973).

Summary

Because of the complexity of the dynamics of collapsing clouds, the existing crude and/or idealized calculations cannot be expected to provide detailed or definitive predictions of their evolution, but should be regarded as providing illustrative examples of how collapse might proceed in various idealized cases. Nevertheless, a number of qualitative features are predicted with considerable generality in many of these calculations. In all cases, if a cloud collapses at all it does so in a non-uniform fashion, developing a centrally condensed or lumpy structure. During the earliest stages of 3-dimensional collapse the cloud often shows a bar- or spiral-shaped structure, in which condensations later form. An individual collapsing region or cloud fragment may form a single dense core, but the available results suggest that a more common outcome is the formation of a binary system of two orbiting centres of condensation. Once a dense core has formed, it grows in mass by accretion from the surrounding envelope, finally becoming a conventional hydrostatic star when all of the surrounding material has been accreted or dispersed.

All calculations indicate the importance of accretion processes of some form for star formation. In cases where an accreting core is relatively isolated, it seems almost inescapable that a flattened, rotating accretion envelope or accretion disc will form around it. It is in this type of situation that many problems related to star formation must be treated; for example, the 'angular momentum problem' becomes the problem of understanding how angular momentum is transported in accretion envelopes or disks by viscous or other torques. Various studies of the dynamics of disks in other contexts may thus turn out to be relevant to star formation as well. Also, if the solar system may be viewed as a remnant of the solar accretion disk, it may ultimately yield valuable information about the way in which the sun formed.

The calculations indicate that the size of the fragments that form in a collapsing cloud is given approximately by the Jeans criterion, as expected classically. Because of the very non-homologous collapse of the fragments, little tendency for further fragmentation beyond the formation of binary or triple systems is observed. However, fragmentation can be enhanced by a significant overall collapse of the cloud; because collapse velocities are generally supersonic, this means that shock compression can be important for fragmentation into small masses. Also, because of the sensitivity of collapse results to the assumed temperature, it is clear that any cooling occuring during the collapse will be important for fragmentation; this effect is not present in most of the calculations, which have assumed isothermal collapse. The importance of understanding the dynamical and thermal evolution of clouds prior to the onset of collapse is also evident.

On the basis of rather general time-scale arguments, it may be anticipated that newly formed stars of low mass first become visible as pre-main sequence stars on the lower part of their Hayashi tracks, while massive stars do not become visible until they are already on or near the main sequence. During the very earliest stages of stellar evolution, the observable properties of stars are predicted to be dominated by the remnant protostellar envelopes around them, which absorb or scatter much of the stellar luminosity and reradiate it at infrared wavelengths. These predictions seem to be consistent in a general way with the observed properties of newly formed stars, but because of the many uncertainties and complexities of detail, more quantitative predictions and comparisons with observations can probably not yet be considered significant.

Acknowledgements

This article was first published in *Proceedings IAU Symposium No. 75* in 1977 by D. Reidel Publishing Company and is reprinted by permission of the publisher and the editor, Dr. Teije de Jong.

References

Abt, H. A., and Levy, S. G. (1976), *Astrophys. J. Suppl.*, **30**, 273.
Appenzeller, I., and Tscharnuter, W. (1974), *Astron. Astrophys.*, **30**, 423.
Appenzeller, I., and Tscharnuter, W. (1975), *Astron. Astrophys.*, **40**, 397.
Bertout, C. (1976), *Astron. Astrophys.*, **51**, 101.
Black, D. C., and Bodenheimer, P. (1976), *Astrophys. J.*, **206**, 138.
Black, D. C., and Wilson, J. R. (1977), in preparation.
Burke, J. R., and Silk, J. (1976), *Astrophys. J.*, **210**, 341.
Cochran, W. D., and Ostriker, J. P. (1977), *Astrophys. J.*, **211**, 392.
Cohen, M. (1973), *Mon. Not. Roy. astron. Soc.*, **164**, 395.
Davidson, K. (1970), *Astrophys. Space Sci.*, **6**, 422.
de Jong, T. (1973), *Astron. Astrophys.*, **26**, 297.
Edmunds, M. G., and Wickramasinghe, N. C. (1974), *Astrophys. Space Sci.* **30**, L9.
Ferraioli, F., and Virgopia, N. (1975), *Mem. Soc. Astron. Italiana* **46**, 313.
Field, G. B. (1974), *Astrophys. J.*, **187**, 453.
Finn, G. D., and Simon, T. (1977), *Astrophys. J.*, **212**, 472.
Fricke, K. J., Möllenkoff, C., and Tscharnuter, W. (1976), *Astron. Astrophys.* **47**, 407.
Herbig, G. H. (1970), in 'Evolution Stellaire Avant la Séquence Principale,' 16th Liège Symposium, *Mem. Soc. Roy. Sci. Liège*, *Ser. 5*, **19**, 13.
Kahn, F. D. (1974), *Astron. Astrophys.*, **37**, 149.
Kondo, M. (1975), *Publ. Astron. Soc. Japan*, **27**, 215.
Larson, R. B. (1969a), *Mon. Not. Roy. astron. Soc.*, **145**, 271.
Larson, R. B. (1969b), *Mon. Not. Roy. astron. Soc.*, **145**, 297.
Larson, R. B. (1972a), *Mon. Not. Roy. astron. Soc.*, **156**, 437.

Larson, R. B. (1972b), *Mon. Not. Roy. astron. Soc.*, **157**, 121.
Larson, R. B. (1973), *Ann. Rev. Astron. Astrophys.*, **11**, 219.
Larson, R. B. (1974), *Fund. Cos. Phys.*, **1**, 1.
Larson, R. B. (1975), in 'Problèmes d'Hydrodynamique Stellaire,' 19th Liège Symposium, *Mem. Soc. Roy. Sci. Liège*, *Ser. 6*, **8**, 451.
Larson, R. B. (1978a), *J. Computational Phys.*, in press.
Larson, R. B. (1978b), *Mon. Not. Roy. astron. Soc.*, in press.
Larson, R. B., and Starrfield, S. (1971), *Astron. Astrophys.* **13**, 190.
Lynden-Bell, D., and Pringle, J. E. (1974), *Mon. Not. Roy. astron. Soc.*, **168**, 603.
McCrea, W. H. (1960), *Proc. Roy. Soc. London*, **A256**, 245.
Mouschovias, T. C. (1976a), *Astrophys. J.*, **206**, 753.
Mouschovias, T. C. (1976b), *Astrophys. J.*, **207**, 141.
Nakazawa, K., Hayashi, C., and Takahara, M. (1976), *Prog. Theor. Phys.*, **56**, 515.
Narita, S., Nakano, T., and Hayashi, C. (1970), *Prog. Theor. Phys.*, **43**, 942.
Penston, M. V. (1969), *Mon. Not. Roy. astron. Soc.*, **144**, 425.
Quirk, W. J. (1973), *Bull. Amer. Astron. Soc.*, **5**, 9.
Rowan-Robinson, M. (1975), *Mon. Not. Roy. astron. Soc.*, **172**, 109.
Shu, F. H. (1977), *Astrophys. J.*, **214**, 488.
Silk, J. (1976), in *Far Infrared Astronomy*, ed. M. Rowan-Robinson, p. 309. Pergamon Press.
Strom, S. E., Strom, K. M., and Grasdalen, G. L. (1975), *Ann. Rev. Astron. Astrophys.*, **13**, 187.
Theys, J. C. (1973), *Ph.D. Thesis*, Columbia University. See also Theys, J. C., and Spiegel, E. A. (1977), *Astrophys. J.*, **212**, 616.
Tscharnuter, W. (1975), *Astron. Astrophys.*, **39**, 207.
Ulrich, R. K. (1976), *Astrophys. J.*, **210**, 377.
Walker, M. F. (1972), *Astrophys. J.*, **175**, 89.
Westbrook, C. K., and Tarter, C. B. (1975), *Astrophys. J.*, **200**, 48.
Yorke, H. W., and Krügel, E. (1977), *Astron. Astrophys.*, **54**, 183.

THE COLLAPSE OF INTERSTELLAR MATTER IN STARS

H. REEVES
Centre d'Etudes Nucleaires de Saclay, Saclay, France

It has been known for many years that a typical cloud of interstellar matter is certainly too magnetized and rotates too fast to ever become a star. This question has been discussed several times during this school. Some of the schemes presented consider the question of rotational deceleration, by appealing to a chance coincidence of exact alignment (but opposite direction) between the galactic corotational component and the 'random' component of angular momentum (Black). Others, (Larson, McCrea) appeal to a special class of slowly spinning condensed clouds for which the problem is essentially already solved. I would like to make two points:

(a) At sometime in the past the matter of any star must have been diluted to mean galactic density (a few atoms cm^{-3}) since we all believe that the galaxy was initially fully gaseous and since, furthermore, matter is constantly rejected, by supernovae remnants, to very low densities. The fact that we observe condensed clouds ($\sim 10^3$ cm^{-3}) should not prevent us from understanding how they are formed from more dilute gas. We *have* to face the problem.

(b) Any scheme of star formation which solves the problem of rotation but does not consider the problem of magnetic field is bound to be incomplete. It seems more appropriate to try to solve both problems at the same time. An interstellar cloud is *not* an isolated system since it remains linked to the surrounding matter by the lines of force of the galactic magnetic field. This link is most likely the key to the gradual loss of angular momentum during the contraction.

Conversely the rotation is very likely the source of the force which pulls the ionized component of the gas—attached to the field—through the neutral component and hence slowly decreases the magnetic flux embedded in the cloud.

How this takes place exactly is almost impossible to describe at the present time. Three-dimensional hydrodynamical collapse calculations with magnetic field are far from being tractable. Nevertheless a qualitative scheme discussed by Mestel and Spitzer (1956) and Spitzer (1968) amongst others, can be described in the following way.

Consider a mass of galactic gas, with $n_H \sim$ a few cm^{-3} $B \simeq 3 \times 10^{-6}$ gauss, $\Omega \simeq 10^{-15}$ sec^{-1} (quite typical mean galactic values). Its axis of rotation may have any orientation with the galactic magnetic field. Throughout the volume of this object (many light years in diameter and its mass is many times the solar mass) the magnetic field will not be constant but will have gradients in arbitrary directions.

Consider first a phase of gravitational contraction. The magnetic flux pressure gradient is still too weak to play any role; the angular momentum is conserved and part of the gravitational energy released is transformed into rotational energy (another part is emitted in the form of electromagnetic radiation, the molecule of CO apparently playing a major role here). The contraction is pursued until a quasi-equilibrium is reached between gravitational and centrifugal force at the equator. During this new phase the winding of magnetic lines of forces in the spinning body eventually becomes tight enough to interfere with the motion, and remind the cloud that it has not cut its 'umbilical cord' with its mother—the interstellar matter complex. Magnetic pressure gradients become stronger and stronger. Their effect, mostly on the equatorial surface of the cloud, is to slowly brake the rotation velocity—as a rubber band attached to a spinning sphere, and to some other fixed point will do. The gradual and slow evacuation of the angular momentum produces a regime in which quasi-equilibrium between gravitational and centrifugal force remains established while the structure loses more and more gravitational energy and contracts progressively.

The rate of gravitational work against the cloud is governed by the rate at which the rotational energy is lost via the magnetic link with outside. The exact way by which this loss occurs is not very well known. One important mechanism is the so called 'ambipolar diffusion' which governs, on a microscopic level, the extent to which magnetic lines of forces are really attached to a partially ionized gas. The ionized fraction of the gas, spiralling around the lines of forces, tries to 'resist' the motion of the (mostly neutral) gas, which is carried along in the rotation. This 'resistance' is accomplished, at the particle level, by multiple collisions between the neutral and the ionized fraction, leading to thermal energy release. As time goes on, the nebula becomes denser and more opaque to outside ionizing radiation; the fractional ionized mass of the gas decreases progressively and the rate of collision decreases proportionately—since it varies with the product of both fractional gas masses. The lines of force—always attached to the ionized particles—are less and less carried away by the motion. As discussed before this leads to a slow evacuation of magnetic flux, to a loss of magnetic moment for the entire structure and hence solves at the same time the 'magnetic field problem'.

There are other possible ways by which the magnetic field may lead to rotational energy dissipation. Lines of force, pulled symmetrically on each side of the equatorial plane (or any other plane perpendicular to the axis of rotation) may find themselves in such a configuration that a field in one direction is immediately adjacent to a field in an opposite direction (a hairpin configuration). Both lines may then abruptly annihilate each other, leading to a brutal loss of magnetic energy, which may well be an important source of particle acceleration (possibly similar to solar flare acceleration).

As discussed in Reeves and Cesarsky (1977, to appear) the release of fast particles by contracting (PreMain Sequence) stars may contribute significantly to the galactic cosmic rays, at least in some energy range, but, for a number of reasons, it is not certain that they are the main source of fast particles in our galaxy.

It has been known for some time that the axes of rotation of stars show no preferred alignment along the axis of galactic rotation and this has led to some discussion (Herbig, 1976; Larson, 1976; both in this volume). It will appear that this problem is best discussed if we take into account the existence of magnetic field gradients. Astronomical observations lead to a picture of the galactic magnetic field which is largely aligned along the spiral arms, but also incorporate an—almost as large—random component superimposed over the aligned one. We expect the intensity of the field embedded in a typical interstellar cloud to present spatial variations of at least a few percent. It will be clear that if the axis of rotation and the axis of the field are not *perfectly aligned* then the winding of the lines of force will lead to a *tilting* of the rotation axis. To visualize the event, consider two rubber bands; a strong one fixed on the northern hemisphere and a weaker one on the southern hemisphere, and remember that, even if the bands are almost of equal strength (within a few percent), it takes many many windings before important deceleration takes place. As a result, the required loss of angular momentum (by about two orders of magnitude) may well also result in a complete randomization of the axis of rotation of the stars, born from an interstellar cloud, even if this cloud was essentially spun by the galactic rotation.

In summary, gravitational energy is *first transformed* into rotational energy, and is *stored there* until the galactic magnetic link *dissipates it into heat or fast particle acceleration*, while extracting itself from the collapsing cloud. The magnetic link is definitely cut when the inflow of matter is reversed, that is, when the star starts to emit a stellar wind. It is generally believed, after Mestel and Schatzman, that this leads to another important reduction of the stellar angular momentum, accounting for the deceleration of late-type stars far below the rotational break-up velocity.

References

Mestel, L. and Spitzer, L. (1956). *Mon. Not. R. astr. Soc.*, **132**, 1.

Reeves, H. and Cesarsky, C. (1977). *15th International Conference on Cosmic Rays, Plovdiv, Bulgaria.*

Spitzer, L. (1968). *Diffuse Matter in Space*, p. 238, Interscience, New York.

DYNAMICAL CONSIDERATIONS OF THE ORIGIN OF THE SOLAR SYSTEM

C. L. GOUDAS,[a] G. A. KATSIARIS,[a] AND A. A. HALIOULIAS[b]
[a]*Department of Mechanics, University of Patras, Greece*
[b]*Department of Geodesy, Technical University of Athens, Greece*

Abstract

Two recent important findings seem to bear on the origin of the Solar System. These findings refer to the dynamics of systems and concern conditions and configurations that prevail (the second) and may, perhaps, have prevailed (the first) during the early stages of formation of the solar system.

The first finding can be described as *the instability of 3-star and many-star systems*, which causes disruption of such systems, usually to single and double stars.

The second finding can be described as a *general instability of 3-dimensional orbits in the restricted three body problem*. This fact, which is based on analytical and numerical work, speaks in favour of the view that the flatness of such systems can simply be the result of orbital evolution.

In view of these findings the Sun, if not born alone, as seems to be virtually certain, has undergone the mechanism of 'escape' through close encounter with at least two other members of the group of stars within which it was born. This mechanism is found by numerical integrations to be quite rapid and involves severe tidal action and partial fragmentation of the escaping body (the Sun). The escaping body and some fragments with appropriate conditions of motion after the catastrophic event of the close encounter, developed into what is today our solar system. The escape of the Sun took place more than 4·6 b. y. ago.

Introductory Remarks

Old and recent observational data and analyses of lunar rocks and meteorites permit the following brief description of the Solar System:

(a) The largest body of the system, by mass and volume, is dynamically free, i.e. is not a member of a two or three or many star system. It is a 'loner' star.

(b) The rest of the members of the Solar System move around the Sun on more or less co-planar orbits. The Solar System is practically 'flat'.

(c) Almost all the angular momentum of this system is carried by secondary bodies, such as Jupiter, Neptune, Saturn, etc. The Sun would not survive as a single star if all the momentum of the system were to be transferred to it today.

(d) The Solar System includes a star, a few planets with satellites, asteroids, comets and interplanetary matter, gaseous and meteoritic. All these bodies seem to comply with the rule of 'sharp isochronism' (Kirsten, 1976), a rule indicating that these bodies were formed within a brief period of 10^7 years, about 4·6 billion years ago.

(e) The above mentioned bodies move in a stable or very slowly changing and known configuration that we shall not describe here. This configuration depends on its initial form and the conservative or non-conservative dynamical evolution.

This brief description of the solar system, deliberately does not include other, perhaps basic, features of a physical and constitutional nature, in view of the dynamical variables upon which our attention will be focused in this paper, indicating thus the position held here that at least the features from (a) to (e), were under the control and are the result of dynamical processes that commenced some time after the formation of the Sun in a group of dynamically interacting stars.

In the past fifteen years two major contributions in celestial and stellar dynamics, with obvious connection with the origin of the Solar System, were made.

The first contribution can be described as *the instability of 3 and n-body systems*. We mention this because some tend to believe that the Sun was born alone, or was always a single member of a loose open cluster. The established fact that as a rule triple systems break into a double and an escaping single star in about 90 per cent of the cases and that the single star is usually the less massive of the three, puts forward the question of whether there was a group of stars of which the Sun was originally a member. The 'escape' of a star from a 3 or *n*-star system occurs fairly early, for example, after a few hundred revolutions in the 3-star case and is always the result of a close encounter of the three stars. Finally, the close encounter can be accompanied by strong to catastrophic tides on the departing body.

The second contribution can be described as *the extreme instability of 3-dimensional orbits in the restricted three body problem*. This fact, which is based on analytical and numerical work, speaks in favour of the view that the flatness of such systems can be the result of orbital evolution. We should add here that on the plane of the two bodies there are lots of stable regions where 3-dimensional orbits can be trapped.

The above contributions will be discussed below in two successive paragraphs while their connection with the origin of the solar systems will be discussed in the final paragraph.

The Instability of Triple and Many-Star Systems

The dynamics of triple star systems, usually called the 'general 3-body problem', when the three stars are treated as mass-points, has received many contributions by prominent names such as Whittaker's (1904), Chazy's (1918), Birkhoff's (1927), Wintner's (1941), Leimanis' (1958), Pollards's (1966), and Siegel's (1971). An important contribution of the past ten years termed by Szebehely (1973), as the 'solution' of the 3-body problem in Birkhoff's sense, is the discovery, through numerical experiments, that the class of motions termed *escape* dominates the rest (Agekyan, 1967; Szebehely, 1967). 'In other words, for arbitrary initial conditions and after a sufficiently long time the outcome of the motion is hyperbolic–elliptic' (Szebehely, 1973, p. 77).

The term 'hyperbolic–elliptic' for a solution of the 3-body problem was introduced by Chazy (1918), to indicate that for negative energy h of the system one of the three particles after some time escapes on a hyperbolic orbit, while the other two members of the trio stay on elliptic orbits around their common centre of mass. Before escape occurs the motion of the three-bodies, for initial conditions picked at random but excluding 'closed' solutions, (these will be discussed later), perform a motion called 'interplay and ejection' in which at irregular time-intervals one of the bodies, usually the one with least mass, acquires enough energy to be ejected on an elliptic motion but at some distance from the other two. This body will later join the other two for an 'interplay', which will again be followed by an 'ejection'. The ejections are preceded by an encounter of the three bodies. Successive ejections, the many numerical experiments conducted by the groups of Szebehely and Agekyan showed, lead in more than 90 per cent of the cases to hyperbolic ejection which means escape of one body and formation of a double star. The hyperbolic ejections or escapes require closer encounters of all three particles to occur before the ejections than the elliptic ejections. Such close encounters in the case of stars will in most cases be accompanied by strong and destructive tidal effects that will be discussed later.

The future of a 3-body system is prescribed by its initial state of motion. A qualitative study of the kinds of solutions to be expected based on the initial conditions, the differential equations of motion and the total energy h of the system, has been presented by Chazy (1918–1922), Birkhoff (1927) and Szebehely (1971). The basic parameter in these qualitative studies is the total energy h which is defined as

$$h = T + V,$$

where

$$T = \tfrac{1}{2} \sum_{i=1}^{3} m_i \dot{\mathbf{r}}_i^2, \quad V = -\tfrac{1}{2} \sum_{\substack{i,j=1 \\ i \neq j}}^{3} \frac{G m_i m_j}{r_{ij}},$$

with $G, m_i, \dot{\mathbf{r}}_i, r_{ij}$ defined as follows:

G is the gravitational constant,
m_i is the mass of the ith body,
$\dot{\mathbf{r}}_i$ is the position vector of the ith body,
r_{ij} is the distance between the ith and jth body.

Because stars forming within the same nebula will, as a rule, have relative velocities such that $h < 0$, the case $h > 0$ need not be discussed. However, this case ($h > 0$) is the simplest of all since the triple system will immediately be disrupted, either to three single stars (*explosion*) or to a single and a double star (*escape*).

The case of interest is when $h < 0$ and as already mentioned the numerical work of Szebehely and Agekyan suggests that in the majority of cases the system is disrupted by the escape of one body. To complete the discussion we should mention the existence of many types of stable closed (periodic) solutions of the problem which are now known (see, for example, Henon, 1976; Szebehely and Feagin, 1973; Broucke and Boggs, 1975; and many others) in the case of planar motion. Three-dimensional periodic solutions are currently under investigation and presumably many such solutions, of the stable and the unstable type, will be found.

Because new stars do not necessarily form in threes but in larger groups the N-body problem should also draw our attention. The extensive numerical work done in the field of orbital calculation of this problem for $N = 4, 5, 10, 20, 100$, and 1000 bodies (see, for example, Hayli, 1972) has shown that escapes after close encounters are frequent in such systems. A few possible ways of disintegration of a group of stars can be mentioned. For instance a group of four stars can dissolve to a double star and two singles or two double stars. Similarly, a group of five can dissolve to three single and one double star, or two doubles and one single. In all cases the close approach is the necessary event for the 'escapes' to occur.

The Instability of 3-dimensional Motions

A finding connected with property b of the Solar System originated from the study of 3-dimensional periodic orbits of the restricted 3-body problem. Both analytical and numerical investigations of the stability of such orbits have shown that for each family there is a critical inclination beyond which no stable solutions exist. The analytical treatment of Jeffreys and Moser (1966) showed, for example, that the critical inclination of a certain type of orbits is about 36°, while the numerical work of Halioulias (1974) and Halioulias *et al.* (1976) showed that there are families with critical inclinations as high as 52°. The instability of the three-dimensional solutions of the restricted problem was first pointed out by Goudas (1961).

The stability of 3-dimensional periodic orbits depends on their characteristic exponents which in the usual formulation of the problem (see, for example, Goudas, 1961), render the orbit stable when the parameters p and q satisfy the conditions

$$|p| < 2, |q| < 2$$

where

$$p + q = -[Tr\,\Delta(\mathbf{x}_0; T) - 2],$$

$$pq = \tfrac{1}{2}\{[Tr\Delta(\mathbf{x}_0; T) - 2]^2 - [Tr\Delta^2(\mathbf{x}_0; T) - 2]\} - 2,$$

$$\Delta(\mathbf{x}_0 ; T) = \left(\frac{\partial x_i}{\partial x_{0j}}\right)_{t=T},$$

$$\mathbf{x} = (x_1, x_2, x_3, x_4, x_5, x_6), \qquad \mathbf{x}_0 = (x_{01}, x_{02}, x_{03}, x_{04}, x_{05}, x_{06})$$

and T is the period of the orbit.

The inclination i of the orbit is measured by its maximum value occurring at the intersection of the orbit with the plane of motion of the two primaries (the ecliptic in the present case), and is computed from the formula

$$\tan i = \frac{x_{06}}{x_{01} + x_{05}}.$$

Of course, the concentration of matter on a plane perpendicular to the total angular momentum vector of a system was in the past mentioned, but this was a result of non-conservative events such as collisions of particles. The instability mentioned here however is purely dynamical. Its effect, nevertheless, shows only under the presence of non-conservative forces.

The conclusion that can be drawn on the basis of this finding and always under the limitations of the validity set by the adopted assumptions, is that property (b) of the solar system need not be explained by theories on its origin as a result of initial configuration or other dynamical and physical or chemical mechanism. It is now clear that solar systems that by initial conditions are not flat, will acquire this property to the extent, of course, that the restricted problem assumptions are valid for these. Furthermore, the presence today of rarified interplanetary matter in the solar system speaks for a gradual approach to the ecliptic by all bodies, mainly asteroids and comets moving on three-dimensional unstable orbits. By the way, the planets are shown on the basis of the restricted problem model to move on stable orbits (Markellos, V., *et al.*, 1975).

Discussion

If we discard the hypothesis that the Sun was formed and has spent all its time alone and dynamically free, we can reconstruct the events of its life as follows. It was formed sometime before 4·6 b.y. ago within a group of stars which apart from following the orbital course of the group in the general galactic field interacted strongly under their mutual Newtonian fields. About 4·6 b.y. ago the Sun freed itself from that group of stars after a close encounter with one, two or more of its members. The close encounter was accompanied by sufficiently strong tides on all bodies participating in the event, as to produce many nebular fragments, probably hot and of density comparable to that of a stellar atmosphere. Some of these fragments followed the Sun along its escape course and constituted the embryos of the planets or the protoplanets. These fragments at the expense of the energy and angular momentum of the bodies participating in the encounter, carried with them enough energy and angular momentum to explain the present distribution of these quantities in the solar system. Within a brief time interval, not exceeding 10 million years, the protoplanets developed into planets and satellites. This was perhaps done according to the descriptions given by Safronov (1972), or Cameron (1976), or Harris (1976). The fragments that did not follow the Sun may also have developed into planets of the single, double, or multiple star that participated in the original close encounter.

The description of the events after the close encounter given here is, of course, not complete but the reader can use the given references. The entire description of the origin of the solar system raises many questions, all difficult to answer. However it is equally difficult to ignore the fact that the rather rare phenomenon of stars unescorted by companions of comparable mass can be true for the Sun as well as the fact that the above description of events gives a simultaneous and compatible answer about the origin of properties (a) to (e). Quantitative treatment of the present model for the origin of the solar system will be reported in the future.

References

Agekyan, T. A., and Anoscova, Z. P., (1967). *Astron. Zh.*, **44**, 1261.
Birkhoff, G. (1927). *Dynamical Systems*, An. Math. Soc. Publ., Providence, R. I.
Cameron, A. G. W. This volume.
Chazy, J. (1918). *Bull. Astron.*, **35**, 321.
Chazy, J. (1922). *Ann. Sci. Ecole Norm.*, **39**, 29.
Goudas, C. L. (1961). *Bull. Soc. Math. de Grece (N. Serie)*, **2**, 1.
Harris, A. W. This volume.
Halioulias, A. A. (1974). *Doctoral Thesis*, Technical Univ. of Athens, Greece.
Halioulias, A. A., Katsiaris, G. A., and Markellos, V. V. (1976). *Astrophys. Space Sci.*, **41**, 417–422.
Jefferys, W. A., and Moser, J. (1966). *Astron. J.*, **71**, 568.
Kirsten, T. This volume.
Leimanis, E., and Minorsky, N. (1958). *Dynamics and Non-linear Mechanics*, Wiley Publ., New York.
Pollard, H., (1966). *Mathematical Introduction to Celestial Mechanics*, Prentice-Hall Publ., N. J.
Safronov, U. S. (1969). *Planetary Cloud and Formation of the Earth and the Planets*, English trans. by Israel Program for Scientific Translations, Jerusalam, 1972.
Siegel, C. L., and Moser, J. (1971). *Lectures on Celestial Mechanics*, Springer, Berlin.
Szebehely, V. (1967). *Proc. Nat. Acad. Sci. U.S.A.*, **58**, 60.
Szebehely, V. (1971). *Celest. Mech.*, **4**, 116.

Szebehely, V. (1973). *Recent Advances in Dynamical Astronomy*, p. 77, D. Reidel Publ. Co., Dordrecht Holland.
Wintner, A. (1941). *The Analytical Foundations of Celestial Mechanics*, Princeton Univ., Princeton, N. J.
Whittaker, E. T. (1904). *Analytical Dynamics*, Cambridge Univ. Press, London and New York.
Hayli, A. (1972). *Proceedings of IAU Col. No. 10, Gravitational N-Body Problem*, M. Lecar, ed., p. 73.
Hénon, M. (1976). *Celestial Mechanics*, **13**, 267–285.
Broucke, R., and Boggs, D. (1975). *Celestial Mechanics*, **11**, 13–38.
Szebehely, V., and Feagin, T. (1973). *Celestial Mechanics*, **8**, 11–23.
Markellos, V. V., Goudas, C. L., and Katsiaris, G. A. (1975). *Astrophys. Space Sci.*, **33**, 341–346.

TIME AND THE SOLAR SYSTEM

T. KIRSTEN
Max-Planck-Institut Für Kernphysik, Heidelberg, F.R. Germany

Abstract

Our present picture of the origin and evolution of the solar system is largely influenced by the achievements of cosmochronology. The various methods of radiometric age dating have enabled us to embed the most significant milestones of evolution into an absolute time scale and to unravel important genetic interrelationships. The present chapter is an attempt to review the relevant evidence with the intention of combining an outline of the general principles involved with an account of the most recent achievements. This approach implies that many topics cannot be dealt with in great detail, but reference is made to the respective literature.

The main topics are the ages of the elements, the duration and sequence of solar system formation, the ages of meteorites and their parent bodies, lunar chronology, and geochronology. Particular attention has been paid to a clear distinction of planetary ages (related to solar system formation) and rock ages (related to the individual planetary evolution).

Introduction

Time is an important parameter in elucidating the physical processes which have governed the evolution of matter from nucleo-synthesis, through solar system formation to the present. It places boundary conditions on the physical, chemical, and geological processes which are responsible for the formation and subsequent evolution of the planets, such as condensation, accretion, differentiation, metamorphism, and exogenic effects. In particular, age differences among a number of objects involved in an evolutionary sequence may be diagnostic with respect to the physical nature of this evolution. Time measurements may also decide whether or not two objects are genetically linked. The intention of this chapter is to describe the time frame for the history of the solar system as it has emerged within the last decade, largely from the investigation of isotopic anomalies in meteoritic, lunar, and terrestrial samples. Attempts will be made:

(a) to relate solar system matter to its original sources
(b) to estimate the time and duration of the formation of the solar system
(c) to determine the sequence of planetological events in time
(d) to unravel individual planetary evolutions.

The article is divided into four parts. After an outline of the principles of radiometric age dating (section on *Principles*), attention is given to the ages of the elements (section on *Ages of Elements*). Next, we discuss the age of the solar system and its members and the subsequent evolution of meteorite parent bodies, the Moon, and the Earth (section on *Planetary Ages and Rock Ages*).

The last part takes up the duration of solar system formation and the resolution of small differences between the formation times of individual objects (section on *Duration and Sequence of Solar System Formation*).

Principles

General Problems in the Measurement of Elapsed Time

In order to measure elapsed time one takes advantage of material changes that occur at measurable rates in material systems. In the following paragraph we enumerate the basic requirements for age determinations.

Independent clock. The rate of time change must be known as a function of time or it should be independent of time itself and not affected by any external influences such as state of matter, temperature, or pressure. Radioactive decay conforms to this requirement much better than any other time dependent process like for instance the accumulation of sediments or the mutation rate of fossils. The slight pressure dependence of the decay rate of K-capturing radionuclides is of no practical importance outside stellar interiors. However, if we consider geological time periods of the order of billions of years, it is not obvious whether radioactive decay constants λ have always been constant in time. Within the framework of Dirac's cosmology, one expects that decay rates involving weak interactions such as Beta decay depend on the age of the universe θ according to

$$\lambda_\beta \sim \theta^{-n}; n = \tfrac{3}{8} \pm \tfrac{1}{8} \quad (1)$$

whereas for strong interactions like α-decay, λ_α should be independent of θ (Kanasevich and Savage, 1963). This difference should lead to increasing discrepancies between α- and β-clocks with increasing age of test specimens. It has been found, however, that the degree of agreement between Rb–Sr ages (β-decay) and U–Pb ages (α-decay) does not systematically depend on the age of the samples (Kanasevich and Savage, 1963). Consequently, time independent decay constants may be assumed for all practical purposes.

Appropriate decay rate. In radioactive dating, the passage of time becomes manifest in isotopic variations of parent and daughter elements. From the experimental standpoint, useful isotopic variations result when the half life of the isotope in question is comparable to the times to be determined. In cosmochronology, the desired range is $\sim 10^5$ to 10^{10} years. Isotopes of practical importance are listed in Table 1. To some extent, an unfavourable lifetime may be compensated for by favourably high parent/daughter ratios and/or by technical refinements in mass spectrometry such that even very small ($\sim 10^{-4}$) isotopic anomalies become measurable.

Known decay rate. The present accuracy of the determination of absolute decay constants by physical counting methods is in general better than ± 1 per cent. This is not true, however, for long-lived Beta emitters such as ^{87}Rb or ^{187}Re.

Table 1. Half-lives of isotopes with cosmochronological significance

Nuclide	Isotopic abundance (%)	Decay type	Decay product	Half-life (years)	Remarks
Primordial Nuclides					
^{87}Rb	27·83	β^-	^{87}Sr	$4{\cdot}99 \times 10^{10}$	Old value
				$4{\cdot}88 \times 10^{10}$	New recommendation[b]
^{232}Th	100	α	^{208}Pb	$1{\cdot}40 \times 10^{10}$	
^{238}U	99·28	α	^{206}Pb	$4{\cdot}50 \times 10^{9}$	Old value
				$4{\cdot}47 \times 10^{9}$	New recommendation[b]
		sp. fission	$^{131-136}Xe$[a], tracks	$8{\cdot}2 \times 10^{15}$	
^{235}U	0·72	α	^{207}Pb	$7{\cdot}13 \times 10^{8}$	Old value
				$7{\cdot}04 \times 10^{8}$	New recommendation[b]
^{40}K	0·0117	(β^-) K-capture	(^{40}Ca) ^{40}Ar	$1{\cdot}25 \times 10^{9}$	Total half-life $\lambda_K/\lambda_\beta = 0{\cdot}117$
^{147}Sm	15·0	α	^{143}Nd	$1{\cdot}06 \times 10^{11}$	
^{187}Re	62·6	β^-	^{187}Os	$\sim 5 \times 10^{10}$	
Extinct Nuclides					
^{129}I	—	β^-	^{129}Xe	$1{\cdot}7 \times 10^{7}$	
^{244}Pu	—	(α), sp. fission	(^{232}Th) $^{131-136}Xe$[a] tracks	$8{\cdot}2 \times 10^{7}$	Total half-life; $\lambda_f/\lambda_\alpha = 1{\cdot}25 \times 10^{-3}$
^{26}Al	—	β^+	^{26}Mg	$7{\cdot}4 \times 10^{5}$	
^{146}Sm	—	α	^{142}Nd	$1{\cdot}03 \times 10^{8}$	
^{205}Pb	—	K-capture	^{205}Tl	2×10^{7}	
^{248}Cm	—	(α), sp. fission	(^{244}Pu) $^{131-136}Xe$[a]	$3{\cdot}7 \times 10^{5}$	
^{250}Cm	—	sp. fission	$^{131-136}Xe$	$1{\cdot}1 \times 10^{4}$	Isotopic composition of fission-Xe unknown

[a]The isotopic composition of fission-Xe from these nuclides is distinct from each other
[b]See text, Section on *Whole Rock Isochrons* and Steiger and Jäger (1977)

Long lifetimes correspond to decay energies so low that the lowest part of the beta-spectrum is seriously affected by self-absorption in the counting device. As a result, the important decay constant of ^{87}Rb is still a matter of some debate. The value actually in use was based entirely on the postulated concordance of the U–Pb and the Rb–Sr ages of a series of old rocks (Aldrich *et al.*, 1956) rather than on a physical determination of the half-life. Fortunately, this value has now been improved by an advanced absolute counting experiment (Neumann and Huster, 1976) so that one can assign independent significance to the general concordance of K–Ar, U–Pb, and Rb–Sr ages as observed for terrestrial, meteoritic, and lunar samples (see Section on *Planetary Ages and Rock Ages*).

Defined onset of time registration. The probability of radioactive decay is, of course, in no way affected by the event which we want to date. The event may, however, occur with the removal of all previously decayed nuclei from the radioactive parent. In such cases, all decay products found in the sample at present must have accumulated after the event and therefore give a measure of the time elapsed since the 'clock was reset'. Accordingly, *only those occurrences can be dated in which a fractionation between parent and daughter element occurs.*

The fractionation need not be complete, however, even though this would be most favourable. It may be sufficient to destroy the intimate parent–daughter relation in a differentiated system and to isotopically equilibrate the daughter element throughout the system. This involves local parent–daughter fractionations which then render isotopic dating possible. The point to remember is that only events which fractionate (totally or partially) the radioactive parent from its daughter can be dated. If the daughter element is not removed it must at least be isotopically equilibrated.

The nature of the fractionation determines the significance of the age. It may be a process of crystallization (geochemical fractionation: 'crystallization age'); of thermal diffusion: ('gas retention age'); of magmatic differentiation, or of physical separation of a reservoir from a larger system. The exact onset of time registration is determined by the cessation of the relevant fractionation, that is, by the isolation of the system with respect to the exchange of radioactive parent and daughter atoms for reasons other than radioactive decay.

Closed system. While the radioactive clock operates there should be no loss or gain of the parent or the daughter element from the system, whether by admixture of extraneous unequilibrated material, or by physico-chemical processes such as diffusion, chemical exchange, annealing, leaching, or contamination. Quite frequently, this condition is only partially satisfied and special measures must be taken to allow for these effects (see ^{39}Ar–^{40}Ar and ^{206}Pb–^{207}Pb methods).

Types of 'Ages'

In this section we shall introduce some formalisms and definitions relevant to the most important age determination methods in order to make the later summary of results more transparent. The basic age equation in radioactive dating is

$$t = \frac{1}{\lambda} \ln\left(\frac{{}^{i}D_{r}}{{}^{k}P} + 1\right) \tag{2}$$

where ${}^{k}P$ and ${}^{i}D_{r}$ are the atomic abundances of the radioactive parent and daughter isotopes of the elements denoted by P and D, λ is the decay constant of ^{k}P, and t is the age to be determined. The subscript r denotes 'radiogenic (in situ)', that is, produced by radioactive decay of ^{k}P during time t. Equation (2) may also be written as

$$^i\mathrm{D}_r = {}^k\mathrm{P}\,(\mathrm{e}^{\lambda t} - 1) \tag{3}$$

For samples of common age t, this equation describes a straight line in a $^i\mathrm{D}_r$ *vs.* ^{k}P-plot with slope $(\mathrm{e}^{\lambda t} - 1)$. This line is called an '*isochron*'.

The total measured quantity ^{i}D need not be entirely radiogenic if the fractionation between elements D and P was incomplete. Rather, we have

$$^i\mathrm{D} = {}^i\mathrm{D}_r + {}^i\mathrm{D}_0 \tag{4}$$

where $^i\mathrm{D}_0$ is the amount of ^{i}D present at $t = 0$ (closing time of the system). To account for this contribution, one usually makes use of a stable reference isotope ^{j}D of the same element D, and since the latter does not depend on t, we may write

$$\left(\frac{^i\mathrm{D}_r}{^j\mathrm{D}}\right) = \left(\frac{^i\mathrm{D}}{^j\mathrm{D}}\right) - \left(\frac{^i\mathrm{D}}{^j\mathrm{D}}\right)_0 \tag{5}$$

and with equation (3)

$$\left(\frac{^i\mathrm{D}}{^j\mathrm{D}}\right) - \left(\frac{^i\mathrm{D}}{^j\mathrm{D}}\right)_0 = \frac{^k\mathrm{P}}{^j\mathrm{D}}(\mathrm{e}^{\lambda t} - 1) \tag{6}$$

This equation of a straight line with slope $(\mathrm{e}^{\lambda t} - 1)$ and ordinate intercept $(^i\mathrm{D}/^j\mathrm{D})_0$ again describes an isochron. To define such an isochron strictly from experimental data, the ratio $^i\mathrm{D}/^j\mathrm{D}$ must be measured for at least two, but preferably for more, cogenetic subsamples with different chemical abundance ratios P/D. The accuracy of this '*internal isochron age*' depends partly on the occurrence of sufficient spread in this 'geochemical' ratio. If only an average sample is available, its age can be determined only if a value for $(^i\mathrm{D}/^j\mathrm{D})_0$ is assumed. Such ages are termed '*model ages*'. Internal isochron ages are most frequently determined from mineral separates of one rock ('*mineral isochrons*'). As an example, a Rb–Sr mineral isochron for the Nakhla meteorite is shown in Figure 1.

A number of whole rock samples which derived from a common source reservoir and having common ultimate origin may also define an isochron, even if they have quite distinct metamorphic or crystallization ages, provided each whole rock behaved as a closed system during crystallization. Such a line is termed a '*whole rock isochron*', and the age calculated from it is not to be confused with the '*whole rock age*'. The latter term is used interchangeably with the term model age discussed above. Whole rock *isochron* ages may in fact also have some model character if a common origin is merely a supposition.

Isochron plots for ^{238}U–^{206}Pb; ^{235}U–^{207}Pb, and ^{232}Th–^{208}Pb ages are

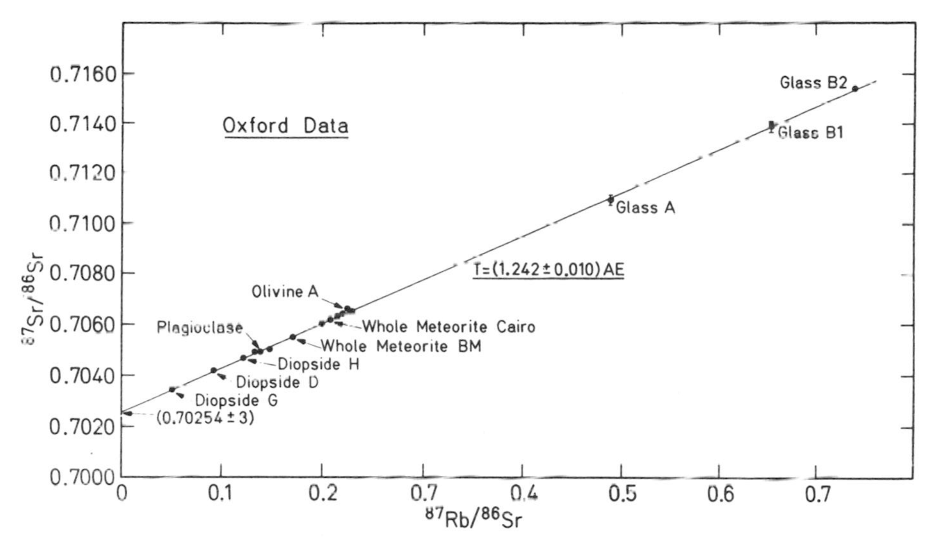

Figure 1. Example of an internal isochron showing the Rb–Sr evolution in the Nakhla meteorite as given by Gale *et al.* (1975). Data points were measured on mineral separates with differing Rb/Sr ratios. The slope of the isochron indicates that the Sr-isotopes have been equilibrated 1·24 b.y. ago. The $^{87}Sr/^{86}Sr$ ratio at that time was 0·70254 as given by the ordinate intercept. Analogous isochron plots may involve for instance $^{206}Pb/^{204}Pb$ *vs.* $^{206}Pb/^{238}U$; $^{207}Pb/^{204}Pb$ *vs.* $^{207}Pb/^{235}U$; $^{143}Nd/^{144}Nd$ *vs.* $^{147}Sm/^{144}Nd$ (Figure 14), or $^{40}Ar/^{36}Ar$ *vs.* $^{40}K/^{36}Ar$. (Reproduced with permission from Figure 1 of N. Gale *et al.*, 1975, *Earth Planet. Sci. Letters*, **26**, 195–206)

analogous to the ^{87}Rb–^{87}Sr case illustrated in Figure 1. The time independent reference isotope in these cases is ^{204}Pb.

The existence of *two* long lived uranium isotopes provides the unique opportunity to replace the D/P ratio by the present day isotopic ratio $^{238}U/^{235}U = 137{\cdot}9$. If one combines the two respective equations (6), then the ^{207}Pb–^{206}Pb isochron equation is:

$$\left(\frac{^{207}Pb}{^{204}Pb}\right)-\left(\frac{^{207}Pb}{^{204}Pb}\right)_0=\frac{1}{137{\cdot}9}\,\frac{e^{\lambda_{235}t}-1}{e^{\lambda_{238}t}-1}\left\{\left(\frac{^{206}Pb}{^{204}Pb}\right)-\left(\frac{^{206}Pb}{^{204}Pb}\right)_0\right\} \qquad (7)$$

An example of such a Pb–Pb isochron is given in Figure 2. Evidently, Pb–Pb ages can be determined from Pb-isotope measurements alone. In practice, however, Pb/U ratios are often measured since metamorphic events can cause lead losses long after the original igneous differentiation. In this case, the so-called 'concordia diagram' (Wetherill, 1956) can be usefully employed (example in Figure 3). The 'concordia' defines the position of $^{206}Pb/^{238}U$ and $^{207}Pb/^{235}U$ for samples which underwent U–Pb fractionation at various times t before present, provided that the ^{206}Pb–^{238}U and the ^{207}Pb–^{235}U ages are concordant. The curvature of the concordia reflects the increase of the $^{235}U/^{238}U$ ratio in the past due to the shorter mean life of ^{235}U as compared to ^{238}U.

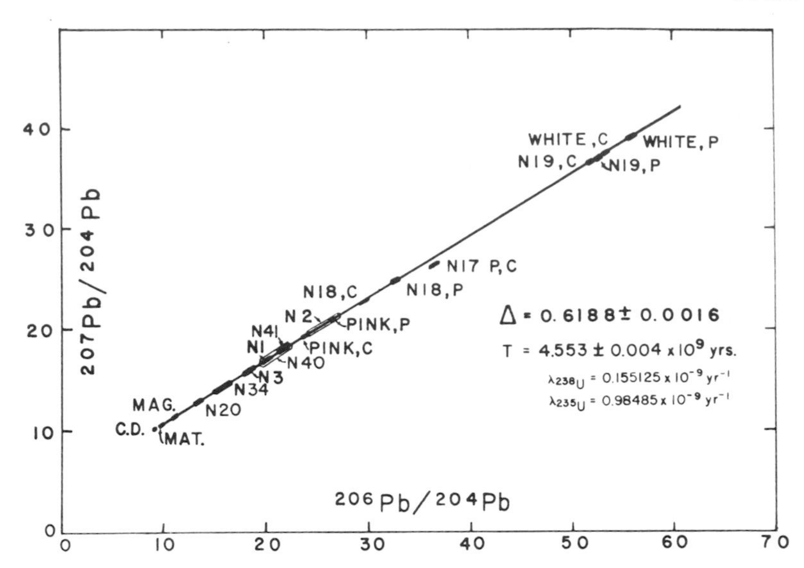

Figure 2. Example of a ^{207}Pb–^{206}Pb isochron obtained by Tatsumoto *et al.*, 1976, for various inclusions of the Allende meteorite. The slope yields an age of 4·553 b.y. The initial ^{207}Pb/^{204}Pb and ^{206}Pb/^{204}Pb ratios cannot be determined without U-analysis, but they are not required for the age calculation. Reproduced with permission from Figure 4 of N. Tatsumoto *et al.*, 1976, *Geochim. Cosmochim. Acta*, **40**, 617–634, copyrighted by Pergamon Press

If at a time t_m before present a metamorphic event has caused lead losses of various degrees in different phases of common origin (that is, common separation of the uranium and lead from the previous environment at time t_s), then ^{207}Pb and ^{206}Pb as present at t_m will be proportionally affected. This implies that samples which underwent different degrees of lead loss lie on a straight line which connects t_m with t_s (Figure 3). Consequently, a sufficient number of data points on this line will define the age t_s of the original U/Pb separation at the upper intercept with the concordia ('*upper intercept age*', corresponding to Pb-loss zero), and the time of metamorphism t_m at the lower intercept with the concordia ('*lower intercept age*', corresponding to complete lead loss at t_m). The problem of continuous lead losses will not be considered here.

In K–Ar dating, complications do not usually arise from the incomplete separation of parent and daughter elements since in most rock forming processes argon expulsion is nearly complete. On the contrary, the problem encountered is most frequently that of partial loss of radiogenic ^{40}Ar by diffusion. This problem has been tackled and partially solved by the development of the ^{39}Ar–^{40}Ar method (Merrihue and Turner, 1966).

A basic feature of this method is the extraction of the radiogenic ^{40}Ar contained in a sample in a stepwise fashion at progressively higher temperatures in order to resolve the ^{40}Ar released from various phases or lattice sites of different activation energies. In addition, one performs a simultaneous

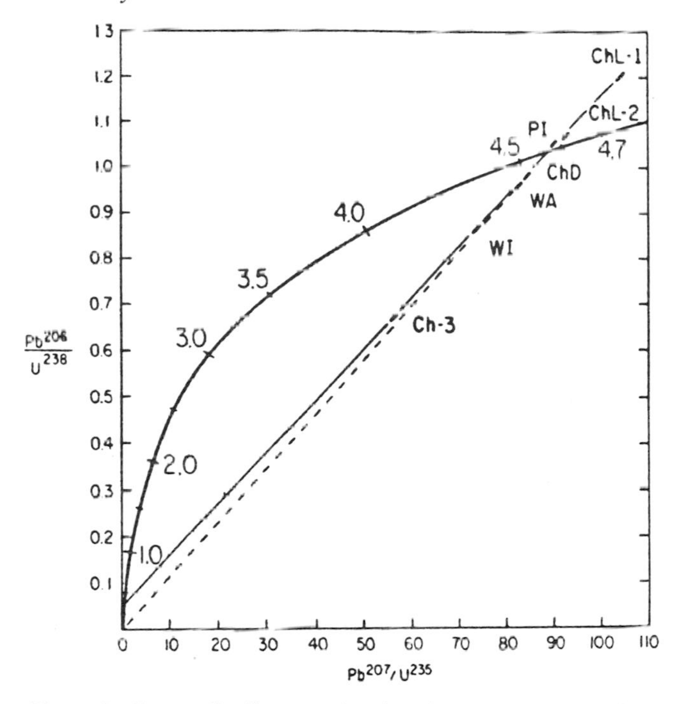

Figure 3. Concordia diagram showing the U–Pb relationships for chondrules and inclusions from the Allende chondrite (Chen and Tilton, 1976). The curved line is the 'concordia'. This theoretical curve determines the position of samples which remained closed U–Pb systems for times before present which are indicated by figures (in b.y.). The solid line is the 'discordia' which is determined by the data points for the various subsamples (thick dashes). The discordia intersects concordia at 4·57 b.y. (t_p, time of original U/Pb separation) and at 0·28 b.y. (t_m, time of partial lead loss). Recent lead loss (e.g., in a laboratory experiment) would yield the dashed line ($t_m = 0$). Reproduced with permission from Figure 2 of J. H. Chen and G. R. Tilton, 1976, *Geochim. Cosmochim. Acta*, **40**, 635–643, copyrighted by Pergamon press

determination of the ^{40}K-concentration in exactly the same sites as those which release their ^{40}Ar at a given temperature. Since the isotopic abundance of ^{40}K is well determined, this measurement can be achieved by converting a proportion of ^{39}K into ^{39}Ar (half life 269 years) by a fast neutron irradiation via the reaction $^{39}K\,(n, p)\,^{39}Ar$. Provided that the release of ^{39}Ar produced in a given type of lattice site follows the release of radiogenic ^{40}Ar from the same sites one can in principle determine the K–Ar age of each phase by a mass-spectrometric determination of $^{40}Ar/^{39}Ar$ isotopic ratios. If the less retentive phases have suffered ^{40}Ar losses during geological time, the apparent ages obtained at lower degassing temperatures will be lower than ages obtained

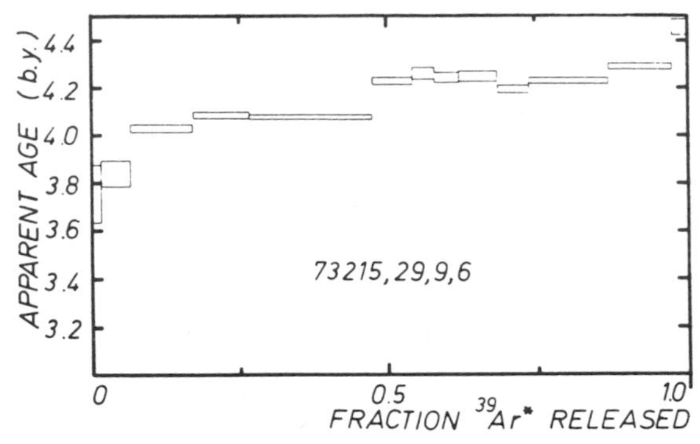

Figure 4. Example for an $^{40}Ar-^{39}Ar$ age spectrum obtained for a clast from Apollo 17 lunar breccia 73215 (Jessberger *et al.*, 1976). Each data bar represents one release fraction in the incremental heating procedure, beginning with 550 °C at the left and ending with 1415 °C at the right end. The amount of gas released in the various fractions is reflected in the widths of the bars, expressed as proportion of the total ^{39}Ar contained in the sample. The ordinate scale is already transformed into a time scale, but the actually measured quantity is the $^{40}Ar/^{39}Ar$ ratio. The relevant relation is

$$t = \frac{1}{\lambda} \ln \left[1 + \frac{(^{40}Ar/^{39}Ar)}{(^{40}Ar/^{39}Ar)_m} (e^{\lambda t_m} - 1) \right]$$

where the subscript m refers to a monitor of known age t_m irradiated together with the sample, λ is the decay constant of ^{40}K and t is the age of the rock. Analytical errors are indicated by the height of the data bars. The figure suggests a 'low temperature plateau age' at 4·06 b.y. and a 'high temperature plateau age' at 4·24 b.y. (see text)

at temperatures sufficiently high to degas those phases which did not undergo natural diffusion losses. If a sample contains at least some highly retentive phases which quantitatively retained their radiogenic ^{40}Ar then the ages obtained above a certain release temperature must approach a constant (maximal) value and give rise to a 'plateau' in a plot of age *vs.* release temperature or $^{40}Ar/^{39}Ar$-ratio *vs.* cumulative fraction of ^{39}Ar released (see Figure 4). This high-temperature *plateau age* is unaffected by natural ^{40}Ar losses and may have geochronological significance. It determines the time at which the temperature dropped below that critical level at which the rock began to retain radiogenic argon (the so-called argon 'blocking temperature').

In some cases like the example shown in Figure 4 it is possible to distinguish a 'high temperature plateau' and a 'low temperature plateau'. The two corresponding ages may then be argued to relate to the time of formation and metamorphism respectively.

Ages of the Elements

Principles

The observed abundances of certain elements place constraints on the models of nucleosynthesis. The problem is to find a set of ages for the elements, or better, to specify the time dependence of nucleosynthesis in such a way that observed elemental abundances are consistent with the time which has elapsed since the production of the elements ended. The basic approach introduced by Burbidge *et al.* (1957) is to compare the abundances of long-lived isotopes with the values expected from various models of nucleosynthesis at the time of their production.

Dating a nucleosynthesising event requires:

(a) an isotopic abundance ratio $A_{RS} = A_R/A_S$ *measured* at a known time t (e.g. $t = 0$, present). At least one isotope, A_R, must be radioactive and long-lived, the other, A_S, should preferably be stable or at least longer lived than A_R.

(b) a prediction for the production ratio $P_{RS} = P_R/P_S$ from the nuclear properties relevant to the production mechanism.

(c) known decay constants $\lambda = 1/\tau$ for the nuclides under consideration (τ is the mean life of the isotope).

The time of the event t before present is then determined by

$$t = \frac{\ln\left(\dfrac{P_{RS}}{A_{RS}}\right)}{\lambda_R - \lambda_S} \tag{8}$$

A first application of this formalism to the r-process isotope pair ^{235}U–^{238}U yields the 'single event-age' of the element uranium. $P_{235,238}$ is estimated to be $\sim 1{\cdot}5$ from the respective numbers of α-progenitors of the two isotopes under the assumption that each isobar in the mass range between 230 and 250 is about equally populated in the r-process (Seeger and Schramm, 1970; Schramm, 1974a). The result, $t_U \sim 6{\cdot}5$ b.y., immediately contradicts the assumption of heavy element production in a single event since the 'Hubble time' is ~ 18 b.y. (Sandage and Tammann, 1975) and the age of globular clusters is (14 ± 1) b.y. (Böhm–Vitense and Szokody, 1973). Such a proposition is also excluded by evidence for the presence of ^{244}Pu and ^{129}I at the time t_p (Figure 5) of the formation of planetary bodies in the solar system (see below). These isotopes are also produced in the r-process but are too short-lived ($T_{1/2} = 82$ m.y. and 17 m.y.) to survive the ~ 2 b.y. between 6·5 b.y. ago and t_p, which will be shown to be $\sim 4{\cdot}6$ b.y. (Section on *Meteorite Ages*). It must then be concluded that multiple nuclide producing events have contributed to the element reservoir from which the solar system was derived and the cosmochronological question must be rephrased into: what is the time dependence of the elemental production rate $E(t)$ or what are the *mean* ages ϑ of the elements? For the heavy elements produced in the r-process, $E(t)$ probably reflects the rate of Supernovae explosions which have contributed to the source region of the solar system. The maximum age inferred from such considerations should be very close to the age of our galaxy since the massive stars which turn into Supernovae undergo a rather fast stellar evolution. If one avoids 'plausible' model assumptions such as $E = \text{const}$ for the duration of nucleosynthesis T, or $E \sim e^{-kt}$, then the number of analytically accessible 'structures' in the function $E(t)$ depends on the number of available cosmo-chronometers (Schramm and Wasserburg, 1970). Each pair of isotopes yields an *average age* (Figure 5) and concordance is achieved by adjustment of $E(t)$.

It should be noted that the average age cannot simply be calculated from equation (8) ('single event'-equation) since one must take into account an extended period T when the production of an element is counterbalanced by its decay. For instance, in case of continuous production at a constant rate during the period T the duration of nucleosynthesis can be derived from

$$A_{\text{RS}}(t_s) = P_{\text{RS}} \frac{\lambda_s(1 - e^{-\lambda_R T})}{\lambda_R(1 - e^{-\lambda_S T})} \tag{9}$$

and the average age of stable elements follows as

$$\vartheta = \frac{T}{2} + t_s \quad \text{(Figure 5)}. \tag{10}$$

However, it has been shown by Schramm and Wasserburg (1970) that for long-lived isotopes ($\tau \gg T$) the mean age ϑ is almost independent from the production function and given by the 'single event solution'. With respect to material admixtures to the solar system, T is terminated at the time t_s at which the separation of the protosolar system reservoir from the galactic element factories took place (Figure 5). For this reason, A_{RS} in equation (9) refers to the time of the end of nucleosynthesis.

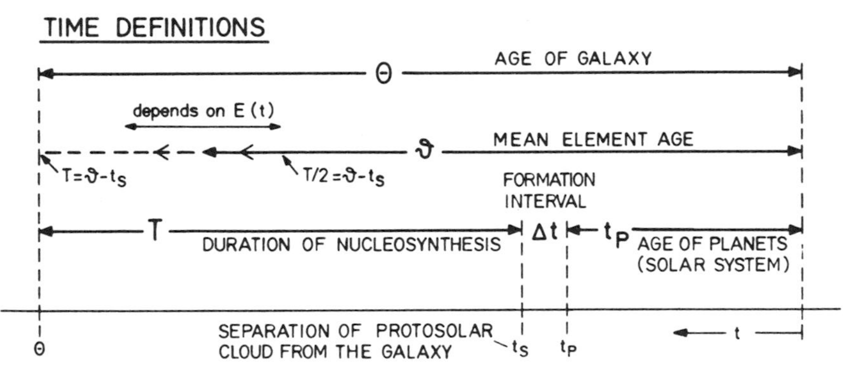

Figure 5. Time definitions used in the Section *Ages of the Elements*. If the element production rate $E(t)$ did not *de*crease backwards in time, it follows that $\vartheta - t_s < T \leqslant 2(\vartheta - t_s)$. The equal sign applies to $E(t) = \text{const}$. Present best estimates are $\vartheta \sim 11$ b.y.; $T \sim 10$ b.y.; $\theta \sim 15$ b.y.; $\Delta t \sim 0{\cdot}15$ b.y.; $t_p \sim 4{\cdot}6$ b.y. The drawing is approximately in scale, except for Δt.

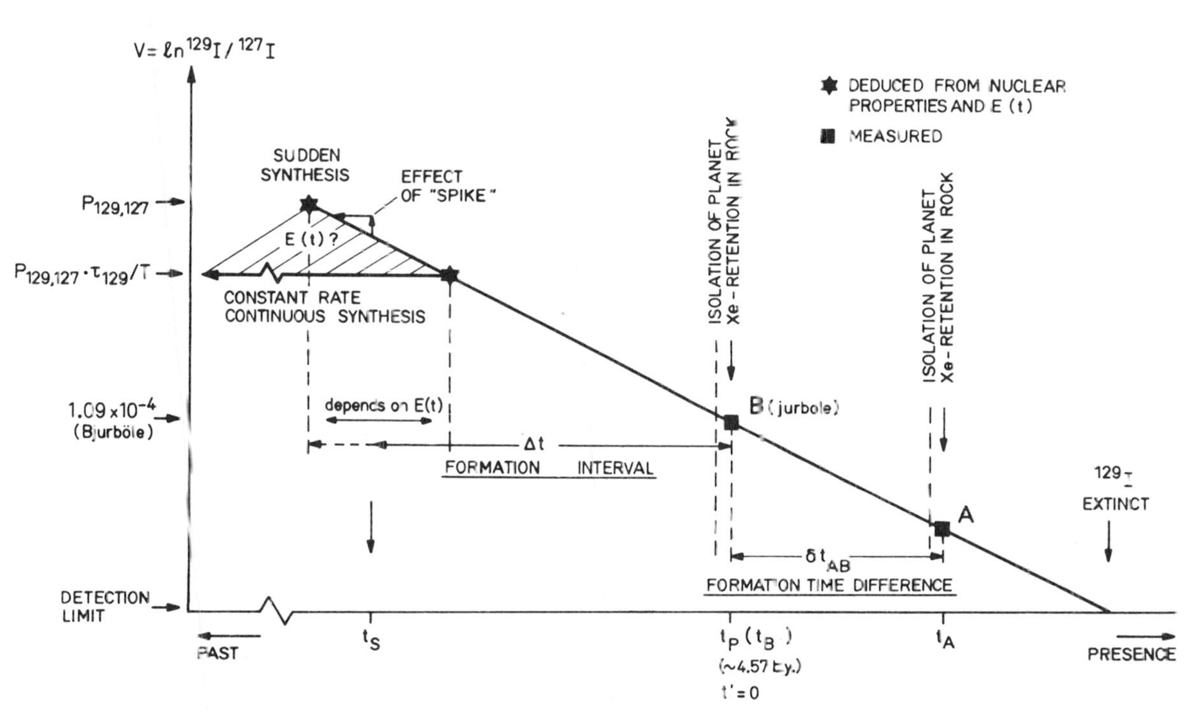

Figure 6. Illustration of 'formation interval' Δt and individual formation time differences δt as deduced from the decay of now extinct ^{129}I, normalized to stable ^{127}I ($V = {}^{129}\text{I}/{}^{127}\text{I}$). The ratios $V(t_A)$, $V(t_B)$ can be measured in meteorites A and B (section on *Measurement of Small Time Differences*). For the reference meteorite Bjurböle, $V(t_{Bj}) = 1{\cdot}09 \times 10^{-4}$ and $t_{Bj} \sim 4{\cdot}57$ b.y. $V(t_{Bj} + \Delta t) = V(t_s)$ must be deduced from *r*-process production ratios $P_{129/127}$ and the time characteristics of the element production rate $E(t)$ (see text). Note that a spike near t_s would increase Δt.

Values for $A_{RS}(t_s)$ are not directly accessible since some time Δt must have elapsed between the end of nucleosynthesis at t_s (which could possibly correspond to the isolation of the protosolar reservoir from the galaxy), and the consolidation of solid planetary material in which measurements can be made (Figure 6). If solid bodies (and supposedly the whole solar system) formed at t_p (Figure 5), then we have

$$t_s = t_p + \Delta t \tag{11}$$

and the age of the galaxy is in fact

$$\theta = T + t_s \tag{12}$$

Furthermore,

$$A_{RS}(t_s) = A_{RS}(t_p)e^{(\lambda_R - \lambda_S)\Delta t} \tag{13}$$

where $A_{RS}(t_p)$ is principally measurable in primitive objects such as meteorites. The so-called formation interval Δt may be determined from equation (13) and from a prediction for $A_{RS}(t_s)$ such as given in equation (9) for the case of continuous nucleosynthesis:

$$\Delta t = \frac{1}{\lambda_R - \lambda_S} \ln\left(\frac{P_{RS}\lambda_S(1 - e^{-\lambda_R T})}{A_{RS}(t_p)\lambda_R(1 - e^{-\lambda_S T})}\right) \tag{14}$$

Two appropriate chronometer pairs are sufficient to simultaneously solve this equation for T and Δt. For the long-lived species, $A_{RS}(t_p)$ and $A_{RS}(t_s)$ (equation (13)) can be easily calculated from the present day values $A_{RS}(0)$ and t_p ($\sim 4{\cdot}6$ b.y., see section on *Meteorite Ages*). For shorter lived, now extinct isotopes such as, for example, ^{244}Pu, $A_{RS}(t_p)$ may be inferred by measuring their decay products provided these decay products were retained in meteorites after consolidation of their parent bodies at or near t_p ('extinct radioactivities', see below). If $1/\lambda_R = \tau_R \ll T$ and $\tau_s \gg T$ then $A_{RS}(t_s)$ will be an equilibrium value and equation (14) reduces to

$$\Delta t = \tau_R \ln\left(\frac{P_{RS}\,\tau_R}{A_{RS}(t_p)\,T}\right) \tag{15}$$

Again it should be emphasized that these relations depend on the choice of $E(t)$ and this is also the case for the relation between ϑ and T.

Available Chronometers

Suitable *r*-process chronometers are ^{232}Th–^{238}U; ^{235}U–^{238}U, ^{187}Re–^{187}Os; ^{244}Pu–^{232}Th, and ^{129}I–^{127}I. Their respective half lives are listed in Table 1 above. For ^{232}Th and ^{244}Pu there exist no stable or long-lived reference isotopes and one is forced to consider ^{238}U and ^{232}Th respectively as 'quasi-isotopes' on the basis of their geochemical affinity. This may introduce uncertainties if geochemical fractionations between chronometer and reference element did in fact occur.

The required production ratios P_{RS} for the actinides are inferred from the summation of α-progenitor isobars. We have adopted the P_{RS} values of Schramm (1973), which, however, may have to be reduced due to the influence of delayed fission of neutron rich progenitors (Wene, 1975). For iodine, extrapolations are made from the neighbouring Te-isotope abundances. For the now extinct nuclides ^{129}I and ^{244}Pu it is possible to infer A_{RS} (t_p) (that is, ^{129}I/^{127}I and ^{244}Pu/^{238}U) from the measurement of their decay products in certain meteorites. It was discovered by Reynolds (1960) and Rowe and Kuroda (1965) that stony meteorites (for Pu, in particular basaltic achondrites) contain excessive amounts of ^{129}Xe and $^{131-136}$Xe respectively. These Xe-isotopes were interpreted as products from the beta decay of ^{129}I and the spontaneous fission of ^{244}Pu respectively, implying that ^{129}I and ^{244}Pu were not extinct at a time when the minerals in which the Xe-isotopes are now observed were cooled below the temperature(s) for Xe-rentention ($\sim 700°$C). In fact, it was later convincingly proved that the excess ^{129}Xe and the excess fission Xenon with the specific isotopic composition characteristic for ^{244}Pu-fission are correlated with the iodine and uranium respectively in the various meteoritic mineral phases. Neutron activation was used to transform (*in situ*) some of the ^{127}I and ^{235}U (related to ^{238}U) into Xe isotopes of distinct isotopic patterns. Subsequent stepwise heating to release the gases demonstrated the intimate relation of the extinct isotopes with their stable or long-lived companions (Jeffery and Reynolds, 1961; Podosek 1970a, 1970b; see also Figure 17 below) and it became possible to quantitatively determine A_{RS} (t_p) for the time of cooling of the meteorite. As will be shown in the section on *Planetary Ages and Rock Ages*, this time can be taken as the radiometric solidification age of the meteorites, $\sim 4{\cdot}57$ b.y.

Results

Mean Age of the Elements

Even without model assumptions for $E(t)$, certain ranges for ϑ can be calculated from the various chronometers.

The average element ages indicate a somewhat higher element production rate $E(t)$ at earlier times since there is a tendency of increasing ϑ with increasing mean life τ of the respective isotopes. However, the ranges of ϑ obtained are so large that a constant rate element production ($E = \text{const}$) is not excluded. The number of available long living chronometers is insufficient to describe $E(t)$ more accurately. This causes large uncertainties in ϑ and even larger ones in the actual value of T. The only safe limitation is that there is no secular *in*crease of $E(t)$ (Schramm, 1973, 1974a, 1974b). This implies

$$\vartheta - t_s < T < 2(\vartheta - t_s) \qquad \text{(see Figure 5)} \qquad (16)$$

From the longest living pair, ^{238}U–^{232}Th, one obtains an average age of the elements ϑ of 7^{+3}_{-1} b.y., and for the duration of nucleosynthesis, T, a value between 1·5 and 11 b.y. The age of the galaxy, θ, lies between 6 and 15·5 b.y. (Schramm, 1974a).

The degree of the $E(t)$-dependence of a chronometer could be reduced if an isotope were available for which truly $\tau \gg \vartheta$, since such an isotope would merely integrate over T and not sensitively react to the structure in $E(t)$. In this respect, the following approach may be advantageous (Clayton, 1964). ^{187}Re ($\tau \sim 7 \times 10^{10}$yr) decays to ^{187}Os. It is necessary to determine the total amount of 'cosmoradiogenic' $^{187}Os_c$, i.e. the amount decayed during $T + \Delta t$. If ϑ is the *mean* age of the elements, we have

$$^{187}Os_c = \lambda_{187}(\vartheta - t_p)^{187}Re \tag{17}$$

With $^{187}Re = {}^{187}Re_{t_p} + {}^{187}Os_c$ it follows

$$\vartheta - t_p = [\lambda_{187}(1 + {}^{187}Re_{t_p}/{}^{187}Os_c)]^{-1} \tag{18}$$

$T + \Delta t$ is limited by

$$T + \Delta t \leqslant 2(\vartheta - t_p) \tag{19}$$

(Figure 5 above) where the equal sign corresponds to $E = $ const. If one excludes chemical separation between Re and Os in iron meteorites, then the required ratio can be determined, provided the amount of non-radiogenic ^{187}Os can be subtracted. Primordial ^{187}Os and ^{186}Os are produced in the *s*-process for which there is an empirical expectation that the products of the respective abundances and thermal neutron capture cross sections should be equal. Hence, from ^{186}Os the non-radiogenic ^{187}Os may in principle be estimated and subtracted. The latest measurements of the required neutron capture cross sections (Browne and Berman, 1976) led to a mean age ϑ of $\sim$ 11 b.y., the duration of nucleosynthesis, T, between 6·5 and 13·5 b.y., and the age of the galaxy θ between 11 and 18 b.y. (Hainebach and Schramm, 1976). These numbers are higher than those inferred from the other chronometers (with shorter mean life times) but in reasonable agreement with astronomical estimates. This again indicates a decrease of $E(t)$ with time.

Extinct Radioactivities and Formation Intervals

The $(^{129}I/^{127}I)_{t_p}$-ratios inferred from I–Xe measurements in about 20 meteorites of various classes are rather uniform (see Section on *Duration and Sequence of Solar System Formation*) and still as high as $\sim 10^{-4}$. In view of a mean element age in the order of 10 b.y. it seemed surprising that ^{129}I ($T_{1/2} =$ 17 m.y.) was still present at t_p, when meteorite parent bodies consolidated. This implies a rather short formation interval $\Delta t = t_s - t_p$ between the end of nucleosynthesis and the onset of Xe retention in the meteorites (Figure 6 above). A 'last minute spike' in $E(t)$ near t_s was suggested to account for this observation. This conclusion can be avoided, however. From equation (15) it can be calculated that the data would still be compatible with a constant rate continuous element production. If we take $T = 5$ b.y., we arrive at a formation interval of $\Delta t = 105$ m.y. If $E(t)$ decreased in time, Δt would be shorter, if a last minute spike occurred, Δt would be longer (see Figure 6 above).

The ($^{244}Pu/^{232}Th$) ratios inferred from Xe produced from Pu-fission in basaltic achondrites are not as consistent as the $(^{129}I/^{127}I)_{t_p}$ ratios, probably a consequence of geochemical fractionation between Th and Pu in the various meteoritic mineral phases. The $(^{244}Pu/^{232}Th)_{t_p}$ ratios range from 0·006 to 0·015 of which the lower values are probably more reliable (Podosek, 1970a, 1972). The lower value would yield $\Delta t = 75$ m.y., the higher one would be incompatible with $E(t) =$ const (Δt would be negative). The problem is aggravated if T is 10^{10} yr as suggested from the ^{187}Re-results.

Because of its shorter mean life, the ^{129}I system is less sensitive to variations in T. For $T = 10$ b.y., Δt from iodine would be 90 m.y., instead of 105 m.y. for $T = 5$ b.y. Formally, 'concordance' of ^{244}Pu- and ^{129}I-formation intervals for $E(t) =$ const could be achieved for $T = 3{\cdot}5$ b.y., with $\Delta t = 115$ m.y. However, since the Re–Os data indicate a higher value for T and since measured $(^{244}Pu/^{232}Th)_{t_p}$-ratios do frequently exceed the value of 0·006, it is more likely that $E(t)$ was in fact not constant and a more complex time dependence of element production must be assumed. High $^{244}Pu/^{232}Th$-ratios require a late enhancement of $E(t)$ near t_s. The increase of mean element ages ϑ with the life time of the chronometers requires an early enhancement near θ. This may be approximated by

$$E = A + B + C \tag{20}$$

where A is the portion of elements produced in 'prompt synthesis' near θ, B is the portion produced at constant rate continuous synthesis during T and C is the portion produced in a 'last minute spike' near t_s.

For instance, Hohenberg (1969) has demonstrated that all available chronometers could be made compatible with $T \sim 4$ b.y. and $\Delta t \sim 175$ m.y. if $A = 85 \pm 5$, $B = 12 \pm 2$; and $C = 0 \pm 8$ per cent of the total element production and if $(^{244}Pu/^{238}U)_{t_p} = 0{\cdot}035$ and $P_{244,\,238} = 0{\cdot}75$. However, the latter two values are not very accurately known and it must be emphasized that this model calculation implies no answer as to the reality of a 'last minute spike' since the solution includes $C = 0$.

Finally, it should be emphasized that regardless of the actual form of $E(t)$, Δt cannot exceed ~ 220 m.y. since the observed ($^{129}I/^{127}I$) ratios are only 10^4 or 2^{13} times smaller than the production ratio P_{RS}. Consequently, Δt cannot exceed 13 half lives of ^{129}I, namely 220 m.y.

In conclusion, the existence of extinct radioactivities requires element production as late as ~ 200 m.y. before the formation of planets. Element production at a constant rate during T cannot be definitely excluded, but the evidence, particularly from ^{244}Pu, suggests a 'last minute spike' which contributed ~ 5 per cent of the total heavy element inventory of the solar system.

The rotation period of the galaxy is ~ 180 m.y. Reeves (1972) has suggested that the formation interval Δt actually indicates the last time prior to t_p that the protosolar gas was in a spiral arm, where the rate of Supernovae explosions is thought to be enhanced. An interpretation of Δt as the time elapsed between the collapse of the protosolar cloud and the formation of planets would be in contradiction with the much shorter time scales in recent models of solar system formation (Cameron, 1976). For a completely different interpretation of extinct radioactivities, see section on *short-lived extinct isotopes.*

Planetary Ages and Rock Ages

Samples available for establishing a time frame for planetary evolution come from the earth, the moon, and from meteorites. In general, the fractionation associated with the separation of a planet from the solar nebula will not be recorded in consolidated rocky materials, since subsequent processes can reset the radiometric clocks. For large bodies like the earth, the time required to develop a solid crust from which rocks could survive is rather long. In general, the oldest rock age defines only a lower limit for the age of the respective planet. The distribution of rock ages t_ρ from a planet reflects its individual evolution, while the planet ages t_{p_k} can be inferred in principle from the slope of whole rock isochrons.

In this section we shall summarize both rock age distributions and planet ages for meteorites, the moon, and the earth. Precise differences between the formation times of these objects will be discussed in the section *Duration and Sequence of Solar System Formation.*

Meteorite Ages

Formation Time of Meteorite Parent Bodies

The formation time of meteorite parent bodies may be derived either from whole rock isochrons of objects which are believed to originate from one and the same parent body, or from a critical evaluation of the distribution of individual mineral isochron or high temperature plateau ages which were collected for each class of meteorites. In the latter case, emphasis rests on the maximum ages of the distribution, since $t_\rho \leqslant t_{p_k}$. (Rock ages substantially lower than the maximum values are treated in the section: *Subsequent Meteorite Evolution*).

Individual meteorite ages. The determination of the ages of individual meteorites is hampered by a number of complications. Internal Rb–Sr isochrons are frequently disturbed due to partial re-equilibration of Sr isotopes in later, possibly metamorphic events. In U/Th–Pb dating, the problem of lead contamination in the process of mineral separation is most severe. ^{39}Ar–^{40}Ar dating frequently suffers from poorly understood disturbances of the age plateaus which are at least in part due to experimental artifacts. In spite of these problems, a substantial number of meteorite ages have been accumulated recently, utilizing all three major dating techniques and the more reliable data are summarized in Table 2. The table is arranged according to the chemical–petrological classification of the meteorites (Van Schmus and Wood, 1967; *see also* Scott and

Wasson, 1975) which can be read as an evolutionary sequence in which the carbonaceous chondrites represent the most primitive stage and the iron meteorites are the most differentiated objects.

U–Th–Pb data. The data for Allende (Figure 2) and Angra dos Reis represent the first self-consistent internal ^{207}Pb–^{206}Pb isochrons, the initial lead composition not being necessary for their construction. A serious problem encountered in attempts to date chondrites by the ^{207}Pb–^{206}Pb method is the lack of sufficient absolute quantities of uranium to account for the observed radiogenic lead, even when the ^{207}Pb/^{206}Pb ratios give apparently reliable ages. It is most remarkable that this discordance between U–Pb and Pb–Pb ages does not occur in Angra dos Reis (Table 2). For Allende it should be noted that not *all* its constituents lie on the *same* isochron. Many stony meteorites are of a brecciated nature and frequently contain xenolithic inclusions whose petrogenesis is unrelated to the host material. In such cases, the supposed internal isochrons are in fact whole rock isochrons for more or less unrelated materials which were 'accidentally' compacted into one meteorite in collisional processes. For methodological reasons, Table 2 also contains 'ion probe data' for Allende. In this new technique, the mineral separation is replaced by a spotwise Pb-isotope analysis *in situ* with an ion microprobe (Lovering *et al.*, 1976).

Rb–Sr data. In view of the long mean life of ^{87}Rb and the unfavourable Rb/Sr ratios in most stony meteorites the accuracy of the more recently determined Rb–Sr isochrons is very remarkable (Table 2). Data for Norton County and Krähenberg were determined some years ago and may be subject to errors larger than stated. For Norton County, it was recently found that its Sr-isotopes are not completely equilibrated (Birck *et al.*, 1975). The internal consistency of the Krähenberg isochron is very good due to the presence of extraordinarily Rb-rich phases, but some systematic calibration error cannot be excluded.

Sm–Nd. The application of the decay of ^{147}Sm ($T_{1/2} = 1.06 \times 10^{11}$ yr) to ^{143}Nd (compare Figure 14 below) in the field of isotopic dating has recently been attempted by Lugmair *et al.* (1975a) and preliminary Sm–Nd isochron ages were obtained for Angra dos Reis and Juvinas (Table 2 below).

K–Ar. Most ^{39}Ar–^{40}Ar plateau ages are from whole rock samples and the required mineral separation is replaced by a stepwise heating procedure. Only those data are listed for which the high temperature plateaus are reasonably well defined. Mineral separates were used only in the case of Mundrabilla silicates.

A few high 'old fashioned' total K–Ar and U/Th–^{4}He ages are also included in Table 2 below. Total K–Ar ages can be equivalent to ^{39}Ar–^{40}Ar plateau ages in the absence of diffusion losses. This is most likely the case if K–Ar and U/Th–^{4}He ages are the same, since ^{4}He is even more easily lost than Ar. The reason for including some high total K–Ar ages of silicate inclusions in iron meteorites is the scarcity of Ar–Ar data and the difficulty in dating the nickel–iron phase of the iron meteorites directly. The only suitable siderophile radioisotope would be ^{187}Re, but, among other difficulties, there is no way to construct an internal isochron from just one metal phase.

Table 2. Radiometric ages of individual meteorites[a]

Type[b]	Meteorite	Method	Age (b.y.)	Remarks	References
C 3	Allende	Pb–Pb internal isochron	4·565 ± 0·004	Inclusions	Chen and Tilton (1976)
		Pb–Pb internal isochron	4·553 ± 0·004		Tatsumoto *et al.* (1976)
		Pb–Pb ion probe	4·53 – 4·64	No primordial Pb	Lovering *et al.* (1976)
		Pb–Pb upper intercept	4·57 ± 0·02		Chen and Tilton (1976)
		Th–^{208}Pb internal isochron	4·565		Chen and Tilton (1976)
C 4	Karoonda	Ar–Ar plateau	4·60 ± 0·10		Podosek (1971)
L 4	Bjurböle	Ar–Ar plateau	4·60 ± 0·03		Turner (1969)
L 4	Saratov	Ar–Ar plateau	~ 4·60		Turner (1976)
L 6	Peace River	Rb–Sr internal isochron	4·56 ± 0·03	One subsample deviates	Gray *et al.* (1973)
LL 4	Soko Banja	Rb–Sr internal isochron	~ 4·50		Gopalan and Wetherill (1969)
LL 5	Krähenberg	Rb–Sr internal isochron	4·70 ± 0·02 (?)		Kempe and Müller (1969)
LL 5	Olivenza	Rb–Sr internal isochron	4·63 ± 0·16		Sanz and Wasserburg (1969)
		Ar–Ar plateau	4·58 ± 0·05		Turner and Cadogan (1973)
LL 6	St. Mesmin	Ar–Ar plateau	4·57 ± 0·05	Xenolith	Cadogan and Turner (1975)
LL 6	St. Severin	Rb–Sr internal isochron	4·55 ± 0·09		Manhes *et al.* (1975)
		Pb–Pb internal isochron	4·53 ± 0·02		Manhes *et al.* (1975)
		Ar–Ar plateau	4·50 ± 0·03		Podosek and Huneke (1973b)
L	Barwell	Ar–Ar plateau	~ 4·53		Turner (1976)
H 4	Forest Vale	Total K–Ar	4·61 ± 0·1		Kirsten *et al.* (1963)
		Total U/Th–^{4}He	4·7 ± 0·2		Kirsten *et al.* (1963)
H 4	Ochansk	Ar–Ar plateau	~ 4·57		Turner (1976)
H 5	Allegan	Ar–Ar plateau	4·60 ± 0·1		Podosek (1971)
H 6	Guarena	Ar–Ar plateau	4·50 ± 0·03		Podosek and Huneke (1973b)
		Rb–Sr internal isochron	4·56 ± 0·08		Wasserburg *et al.* (1969)
H 6	Mount Browne	Ar–Ar plateau	4·58 ± 0·05		Turner and Cadogan (1973)
H 6	Queens Mercy	Ar–Ar plateau	4·58 ± 0·05		Turner and Cadogan (1973)
H	Nadiabondi	Total K–Ar	4·56 ± 0·1		Kirsten *et al.* (1963)
		Total U/Th–^{4}He	4·55 ± 0·2		Kirsten *et al.* (1963)
H	Utzenstorf	Total K–Ar	4·5 ± 0·1		Zähringer (1968)
		Total U/Th–^{4}He	4·4 ± 0·2		Zähringer (1968)

Table 2. Radiometric ages of individual meteorites[a] (*Continued*)

Type[b]	Meteorite	Method	Age (b.y.)	Remarks	References
E 4	Adhi Kot	Total K–Ar	4·7 ± 0·1		Zähringer (1968)
		Total U/Th–^{4}He	4·7 ± 0·2		Zähringer (1968)
E 4	Indarch	Rb–Sr internal isochron	4·56 ± 0·15		Gopalan and Wetherill (1970)
Au	Norton County	Rb–Sr internal isochron	4·70 ± 0·10	Possibly not in equilibrium	Bogard *et al.* (1967a)
Di	Tatahouine	Rb–Sr internal isochron	~ 4·55		Brick *et al.* (1975)
An	Angra dos Reis	Pb–Pb internal isochron	4·548 ± 0·002		Adorables (1976)
		U–^{206}Pb internal isochron	4·61 ± 0·07		Adorables (1976)
		Th–^{208}Pb internal isochron	4·57 ± 0·10		Adorables (1976)
		Sm–Nd internal isochron	~ 4·42		Adorables (1976)
Eu	Ibitira	Rb–Sr internal isochron	4·53 ± 0·10		Birck *et al.* (1975)
	Juvinas	Rb–Sr internal isochron	4·60 ± 0·07		Birck *et al.* (1975)
		Sm–Nd internal isochron	4·56 ± 0·08		Lugmair *et al.* (1975a)
Ho	Kapoeta	Rb–Sr internal isochron	4·54 ± 0·12	Clast	Papanastassiou and Wasserburg (1976)
Iron	Mundrabilla	Ar–Ar plateau	4·62 ± 0·02	Olivine inclusions	Kirsten (1973)
		Ar–Ar plateau	4·58 ± 0·02	Plagioclase inclusions	Kirsten (1973)
Iron I A	Toluca	Total K–Ar	4·63 ± 0·1	Silicate inclusions	Bogard *et al.* (1967b)
Iron I B	Four Corners	Total K–Ar	4·58 ± 0·1	Silicate inclusions	Bogard *et al.* (1967b)
Iron II E	Colomera	Rb–Sr internal isochron	4·61 ± 0·04	Silicate inclusions	Sanz *et al.* (1970)

Notes:

[a]For meteorites with ages substantially below 4·57 b.y., see section on *Radiometric Ages* and Table 4 below.

[b]C: Carbonaceous chondrite; L: low iron chondrites (hypersthene chondrite);
H: high iron chondrites (bronzite chondrite); E: enstatite chondrite; LL: oxidized low iron chondrite;
Au: Aubrite (Ca-poor achondrite); Di: Diogenite (Ca-poor achondrite); An: Angrite (anomalous Ca-rich achondrite); Eu: Eucrite (Ca-rich achondrite); Ho: Howardite (Ca-rich achondrite); 3–6: increasing degree of metamorphism (Van Schmus and Wood, 1967); IA, IB, IIE: Iron meteorite subgroups (Scott and Wasson, 1975).

Additional Note:

Gale *et al.*(1973) have published a preliminary report on Pb–Pb isochron ages of ~ 4·8 b.y. for Appley Bridge and Parnallee. Reservations of the authors themselves have not yet been withdrawn, the data are therefore not listed.

Interpretation

The data in Table 2 above represent the three major meteorite groups; chondrites, achondrites, and irons. The most apparent feature is an upper limit of individual meteorite ages at around 4·6 b.y. In fact, practically all individual ages listed are, within the experimental errors, compatible with a mean value of $4{\cdot}57 \pm 0{\cdot}03$ b.y. The possible significance of small differences among them will be discussed in the section on *Duration and Sequence of Solar System Formation*, but the overall result clearly indicates a nearly contemporaneous formation of rather different objects at $\sim 4{\cdot}57$ b.y. ago.

It is generally believed that meteorites are fragments from a small number of different parent bodies (see Anders, 1971). If the *rock* ages of all these objects coincide, it may be concluded with confidence that the rock ages are within the experimental errors indistinguishable from the respective 'planet ages', for otherwise one would have to postulate identical cooling histories for all objects which would imply identical sizes. On the contrary, it may be concluded from the coincidence of the ages that all meteorite parent bodies cooled rather rapidly and hence stem from relatively small bodies. Chondrite parent bodies may have never been thoroughly molten, whereas melting is required to generate the differentiated achondrites and irons.

In principle, the different dating methods date different events; in practice, experimental errors do not permit their resolution on the short time-scale inferred for consolidation. Present error limits may also disguise possible small deviations of the relevant decay constants (note, however, the general agreement of the three dating methods; see in particular, the data for St. Severin in Table 2). Having concluded that the rock ages actually coincide with the parent body ages, it follows that a small number (say, ten) individual small planets were formed within ~ 60 m.y. around 4·57 b.y. ago—strong circumstantial evidence for the formation of the whole solar system at that time $t_p = (4{\cdot}57 \pm 0{\cdot}03)$ b.y.

Whole rock isochrons. In principle, a whole rock isochron should be constructed only for objects which are genetically linked. On the other hand, if a group of samples defines a reasonably good isochron, this is a retrospective indication of their generation from a common source either in a single, or a number of nearly contemporaneous processes which are not resolvable in time. The close similarity of the internal isochron ages discussed above suggests that all undisturbed meteorites may define one whole rock isochron. A somewhat more conservative expectation would be that the members of the several meteorite groups will define independent and possibly different whole rock isochrons for their respective parent bodies. Historically, Patterson (1955) had presented the first Pb–Pb whole rock isochron long before internal isochrons became available and he had inferred an age of $\sim 4{\cdot}55$ b.y. from two chondrites, one achondrite and one iron meteorite. In fact, this was the first reliable estimate for the age of the solar system. The tacit assumption of a nearly contemporaneous origin of such different objects was justified by the results themselves. Later

Table 3. Whole rock isochron ages of meteorite groups

Meteorite group	Number of Meteorites	Method	Age[a] (b.y.)	Remarks	References
L-Chondrites	6	Rb–Sr	4·54 ± 0·12	Unshocked falls	Gopalan and Wetherill (1971)
LL-Chondrites	10	Rb–Sr	4·56 ± 0·15	Falls only	Gopalan and Wetherill (1969)
H-Chondrites	12	Rb–Sr	4·69 ± 0·14	Falls only	Kaushal and Wetherill (1969)
H-Chondrites	3	^{207}Pb–^{206}Pb	4·55 ± 0·03		Tatsumoto *et al.* (1973)
C-Chondrites	5	Rb–Sr	4·46 ± 0·35		Murthy and Compston (1965)
C-Chondrites	5	Rb–Sr	~4·69	Other C-chondrites deviate	Kaushal and Wetherill (1970)
E-Chondrites	8	Rb–Sr	4·54 ± 0·13	Falls only	Gopalan and Wetheril (1970)
C + L + H-Chondrites	16	^{207}Pb–^{206}Pb	4·505 ± 0·008		Huey and Kohman (1973)
C + L + H-Chondrites	6	^{207}Pb–^{206}Pb	~4·57		Tilton (1973)
Eucrites	7	Rb–Sr	4·39 ± 0·26		Papanastassiou and Wasserburg (1969)
Eucrites	4	^{207}Pb–^{206}Pb	4·59 ± 0·02		Silver and Duke (1971)
Irons (silicate inclusions)	5	Rb–Sr	4·6 ± 0·2		Burnett and Wasserburg (1967)
Irons (metal)	14	Re–Os	4 ± 0·8		Herr *et al.* (1961)
Ch + Ach. + Irons	4	^{207}Pb–^{206}Pb	~4·5		Patterson (1955)
Ch + Ach. + Irons	33	^{207}Pb–^{206}Pb	4·50 ± 0·03	Literature survey	Kanasevich (1968)
Ch + Ach. + Irons	15	^{207}Pb–^{206}Pb	4·50 ± 0·05	Literature survey	Cumming (1969)

[a] Adapted for new U-decay constants

Pb-model ages (initial lead composition assumed) have been reported for approximately ten meteorites of various types, e.g. Silver and Duke (1971); Tatsumoto *et al.* (1973); Huey and Kohman (1973). They range from 4·50–4·57 b.y. but are not listed.

on, Rb–Sr and Pb–Pb whole rock isochrons were separately determined for essentially all major meteorite groups (Table 3). Indeed, most members of a given meteorite class did fit on a straight line. Exceptions occurred mostly for meteorite 'finds' as a consequence of weathering and contamination. For example, a whole rock isochron for L-group chondrites determined by Gopalan and Wetherill (1971) shows very good alignment of the data points for observed meteorite 'falls' (Figure 7), whereas many 'finds' deviate from this isochron (Gopalan and Wetherill, 1971).

Table 3 summarizes the present status of whole rock isochron data. It also includes a Re–Os whole rock isochron for four iron meteorites, however, with a large error due to analytical difficulties and uncertainties in the decay constant, etc. Within the experimental errors, the ages of all meteorite groups are essentially indistinguishable and compatible with an age between 4·5 and 4·6 b.y. This implies a common origin from an initial lead (or strontium) source and makes it reasonable to determine a common best fit isochron for essentially *all* meteorite groups in order to determine the time of this initial fractionation. Kanasevich (1968) considered the error distribution of the then available Pb–isotope data for a group of 15 chondrites, 3 achondrites, and 15 iron

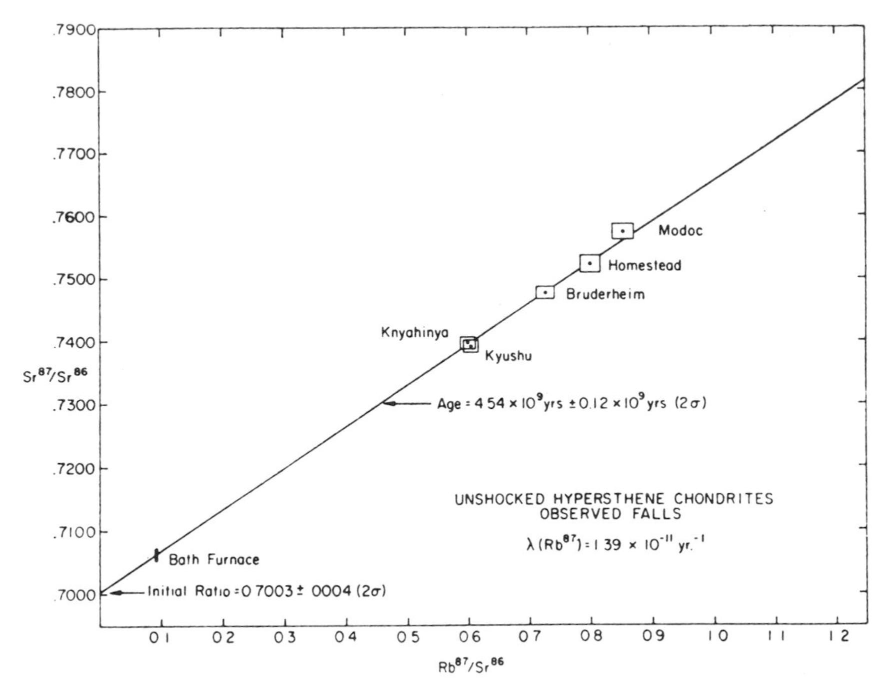

Figure 7. Rb–Sr whole rock isochron for unshocked hypersthene chondrites (Gopalan and Wetherill, 1971). The linear alignment of six different meteorites indicates contemporaneous origin at $4{\cdot}54 \pm 0{\cdot}12$ b.y. ago. (Reproduced with permission from Figure 1 of G. Gopalan and G. W. Wetherill, 1971, *J. Geophys. Res.*, **76**, 8484–8492, copyrighted by American Geophysical Union)

meteorites. He arrived at a best fit isochron of 4·56 ± 0·03 b.y. in very close agreement to the original Patterson value derived from four objects only. To compare this number with more recent data, some remarks must be made concerning the selection of decay constants. Jaffey *et al.* (1971) have redetermined the α decay constants for ^{238}U and ^{235}U with superior experimental techniques. Their results (see Table 1 above) require 0·7 and 1·4 per cent increases of the respective decay constants which were used until 1971. For Pb–Pb ages near 4·5 b.y. this implies an age reduction by ~ 0·06 b.y. The situation for the ^{87}Rb decay constant is quite similar, even though in this case the adoption of a new value has not yet been generally accepted. The value of $\lambda_{^{87}Rb} = 1{\cdot}39 \times 10^{-11}$ yr^{-1} used at present is largely based on an intercomparison of Rb–Sr and U/Th–Pb ages of a large suite of terrestrial rocks and is therefore also affected by a change of the U-decay constants. The required increase of λ by ~ 1·5 per cent coincides with a physical redetermination by Neumann and Huster (1976) (see Table 1 before). Since this change is not yet in effect, we will for the moment retain the old Rb-decay constant in order to avoid confusion. It should be remembered, however, that in intercomparisons a reduction of Rb–Sr ages by ~ 1·5 per cent must be considered a possibility.

Returning to the whole rock isochron data of Table 3, Kanasevich's best fit becomes 4·50 ± 0·03 b.y. This is in excellent agreement with Huey and Kohman's (1973) value of 4·505 ± 0·008 b.y. for 16 C, L and H chondrites. On the other hand, Tilton (1973) obtains ~ 4·57 b.y. for six chondrites (C, L, and H). From the Rb–Sr and Pb–Pb whole rock data in Table 3 it is safe to state that the original separation of the meteorite parent bodies from a common source reservoir in the solar system occurred at 4·55 ± 0·05 b.y. The agreement of this figure with the overall result from internal isochron ages, 4·57 ± 0·03 b.y. implies that this original separation occurred very close to the time when the individual meteoritic rocks became closed systems. For this reason, the age of $t_p = 4{\cdot}57 \pm 0{\cdot}03$ b.y. is commonly considered to define the time of the formation of the solar system.

Subsequent Meteorite Evolution

So far, we have restricted our discussion to *high* meteorite ages. Their frequent (but not exclusive) occurrence demonstrates that in general terms meteorites are very primitive objects which have remained isotopically closed systems since near the formation of the solar system ~ 4·57 b.y. ago. This observation severely limits the extent to which they could possibly have been subjected at a later stage to conditions which could either fractionate parent and daughter elements or equilibrate isotopes of the daughter elements. Here we shall discuss those observations which indicate that such processes have nevertheless occurred to some extent. In general, the major mechanisms leading to the opening of isotopic systems or to isotopic equilibration are thermal or mechanical strain. Such strain can be provided by initial heat and slow cooling or by internal or external (collisional) re-heating or melting by shock metamorphism, by radiation damage or similar environmental factors. The various radiometric isotope pairs respond quite differently to such conditions, and it may well be that, for example, the Rb–Sr system remains undisturbed, while the K–Ar system is drastically affected. Such effects can be used to deduce information about the history of the meteorites *after* their formation. An instructive example is the loss of radiogenic ^{4}He by solid state diffusion even at very moderate temperatures. We have already shown that a few chondrites still have U/Th–^{4}He ages of ~ 4·55 b.y. (Table 2 above), but these are rare exceptions, and most U/Th–He ages range from ~ 0·2–4 b.y. (Kirsten *et al.*, 1963; Zähringer, 1968; Taylor and Heymann, 1969). For the K–Ar system, the situation is quite similar. Many stony meteorites have total K–Ar ages far below 4 b.y. due to continuous, and/or sporadic diffusive Ar-loss and indeed it is for this reason that the ^{39}Ar–^{40}Ar method was required to infer reliable K–Ar ages from the high temperature release fractions. On the other hand, the extent of Ar loss from the less retentive minerals as deduced from the low temperature fractions may yield information on the thermal history of the meteorite. In an idealized case, a metamorphic event could completely reset the K–Ar clock of a less retentive mineral phase and have no effect on the most retentive phases. Then one could at the same time date the time of formation and the time of metamorphism, quite analogous to the 'intercept' method for metamorphic lead loss (see section above *Types of 'Ages'*). In other cases, a metamorphic event could be so intense that the whole meteorite is reset and it remains no longer a 'primitive' object. In the following we shall summarize the *radiometric* evidence for the evolution of meteorites *after* their formation.

Radiometric ages. Gas retention may be affected by processes of relatively moderate intensity, while the complete homogenization of Sr isotopes in an evolved system with coexistent phases of different Rb/Sr ratio requires, in general, a very intense metamorphism. Chemical fractionation of non-volatile elements implies the existence of igneous processes and it is for this reason that Rb–Sr data are most informative as to such events late in the history of the meteorites.

The most spectacular case is represented by the anomalous achondrite Nakhla. Internal Rb–Sr isochrons for this meteorite have been reported by two groups (see Figure 1 above) and an intense (Papanastassiou and Wasserburg, 1974) or a *complete* (Gale *et al.*, 1975) homogenization of Sr-isotopes near 1·3 b.y. ago has been inferred from the data. Gale *et al.* interpret this as evidence for an igneous differentiation of the parent material of Nakhla as late as 1·3 b.y. ago, whereas Papanastassiou and Wasserburg favour a very intense metamorphic event. The difference stems in part from the assignment of different significance to the observation that some whole rock chip data deviate from the mineral isochron, indicating an incomplete resetting. The age is further supported by well defined Ar–Ar plateau ages of ~ 1·3 b.y. both for Nakhla and Lafayette (Podosek, 1973). These are the only two members of this highly anomalous meteorite class ('Nakhlites'), which possess closer geochemical similarities to the earth than to any other meteorite group (Lafayette may in fact stem from the same fall as Nakhla). There are no shock effects observed in Nakhla and the cause for this unusual late event is not known.

Another anomaly of Nakhla stems from the fact that its whole rock model age (*ca.* 3–4 b.y., Papanastassiou and Wasserburg, 1974) does not plot on the Rb–Sr whole rock isochron of all meteorites, which it should do in spite of the re-equilibration 1·3 b.y. ago, if Nakhla evolved in a simple way from the same source material as all other meteorites, and if the homogenization 1·3 b.y. ago occurred in a closed system (see the section above *Types of Ages*). This *might* indicate another differentiation somewhere between 3 and 4 b.y. ago.

Further internal Rb–Sr isochron ages substantially below 4·57 b.y. were discovered for silicate inclusions from the iron meteorite Kodaikanal ($\sim$ 3·8 b.y.) (Burnett and Wasserburg, 1967) and for at least one clast from the Kapoeta howardite breccia ($\sim$ 3·63 b.y.; Papanastassiou *et al.*, 1974; Papanastassiou and Wasserburg, 1976). For Kodaikanal, the internal isochron age of $3{\cdot}8 \pm 0{\cdot}1$ b.y. is in accordance with a *total* K–Ar (minimum) age of 3·5 b.y. (Bogard *et al.*, 1967b). The Sr-isotope equilibration $\sim$ 3·8 b.y. ago requires either intense metamorphism and/or a chemical differentiation between Rb and Sr at that time. The latter is more likely since even the whole rock model age is $\leqslant$ 4 b.y. which implies that the Rb/Sr ratio before the 3·8 b.y. event must have been substantially lower than afterwards. This would imply that solid objects were differentiated and physically separated from pre-existing sources as late as 750 m.y. after solar system formation. The observed shock effects in Kodaikanal, the breccious nature of Kapoeta, and the well established influence of collisional processes on the ages of lunar rocks, especially in the period between 3·7 and 4·3 b.y. ago (see section *Lunar Chronology* below) make it most plausible to consider that catastrophic collisions among meteorite parent bodies were the mechanism responsible for the production of exceptionally young objects. Head-on collisions can provide sufficient energy for partial or complete re-melting of very large fragments and subsequent differentiation. In the following sections it will become evident that meteorite evolution is in fact dominated by collisional processes in the solar system which produce a continuum of effects which depend on the intensity of the respective impact processes.

Late stage alterations of less intensity can be studied with the help of radiometric clocks which involve a volatile daughter isotope such as ^{40}Ar. The time of an outgassing event can in principle be determined if a sample exhibits a high temperature ^{39}Ar–^{40}Ar 'plateau' (type I) or an intermediate temperature plateau (type II) followed by an age increase due to retentive relics which were not reset (type IIa) or only partially reset (type IIb) in the event (Figure 8). Types IIa and IIb are distinguished according to whether the maximum age of the retentive relics is close to 4·57 b.y. or substantially below that age. This implies the supposition that the formation age of all objects of type II is $\sim$ 4·57 b.y.

The occurrence of a plateau consisting of at least three subsequent release fractions is the minimum requirement to interpret an age in terms of a datable event. However, Ar-losses may also be due to multiple events or continuous

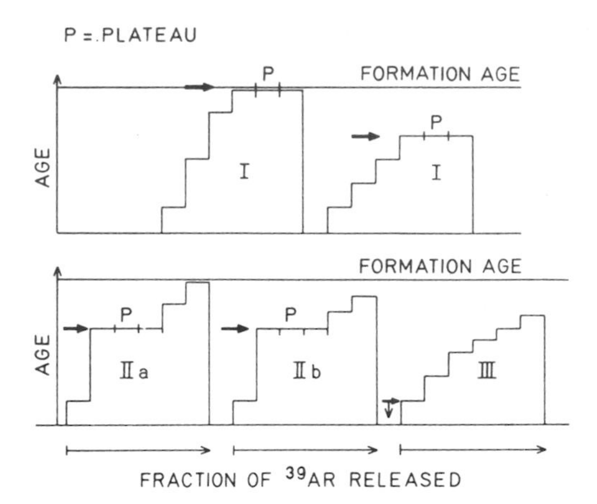

Figure 8. Schematic characterization of ^{39}Ar–^{40}Ar age spectra types. At least three consecutive release fractions are required to define an 'age plateau'. Arrows mark the time of the resetting event.
(*Type I*: High temperature plateau, complete resetting. *Type IIa*: Intermediate temperature plateau, most-retentive phases not affected in the event. *Type IIb*: Intermediate temperature plateau, most-retentive phases partially reset in the event. *Type III*: No plateau defined, maximum age without significance, minimum age of the lowest temperature fraction is an upper bound for the time of the resetting event.)

diffusion for long time periods. They can result in age spectra of type III which have no defined plateau and a maximum age significantly below 4·57 b.y. For such cases, the minimum age obtained for the *least* retentive phase could be interpreted as an upper bound on the time of the last (but possibly also the only) effective event, while the maximum age has no chronological significance.

The data available at present are summarized in Table 4. Plateau ages of type I in principle date events quite comparable in intensity to the cases of Nakhla and Kodaikanal. However, while some samples of St. Mesmin, Bununu, and Malvern belong to type I, there are other coexistent phases from these meteorites which have maximum ages near ~ 4·57 b.y. (Table 4). Considered as a whole, each meteorite belongs to group IIa. The most plausible interpretation is that violent collisions produced or injected the Malvern- and Bununu- glasses ~ 3·7 and 4·24 b.y. ago. At the same time, the K–Ar ages of the host meteorites were partially reset (see Table 4). For St. Mesmin the age difference of various clasts indicates a consolidation period of this meteorite of ~ 150 m.y. (Cadogan and Turner, 1975). Parts of Clovis No. 2 must have been nearly totally degassed within the last 40 m.y. This is a striking indication that rather violent resetting mechanisms have been effective throughout the history of the solar system.

Table 4. ^{39}Ar–^{40}Ar gas retention ages substantially below 4·57 b.y.[a]

Type[b]	Meteorite	Release Type[c]	Apparent Ages (b.y.)			Remarks	References
			Minimum	Plateau	Maximum		
L 6	Colby, Wisconsin	I		3·9			Turner (1969)
LL 6	St. Mesmin light xenolith	I		4·42		Dark xenolith: plateau age 4·57 b.y.	Cadogan and Turner (1975)
H 6	Clovis 2	I		< 0·04			Turner and Cadogan (1973)
	Clovis 2	IIb		~ 0·01	3·26		Turner (1976)
Eu	Stannern	I		~ 3·7		Scatter	Podosek and Huneke (1973a)
Na	Lafayette	I		1·33			Podosek (1973)
Na	Nakhla[d]	I		1·3		Decreasing age at highest temperatures	Podosek (1973)
Ho	Bununu glass	I		4·24		Scatter	Rajan *et al.* (1975)
	Bununu plagioclase	IIa		4·42			Rajan *et al.* (1975)
Ho	Malvern glass	I		3·7		Scatter ± 0·15 b.y.	Kirsten and Horn (1975)
	Malvern clast	IIa		3·6		Scatter	Kirsten and Horn (1975)
Eu	Pasamonte	IIa		4·1			Podosek and Huneke (1973a)
LL 6	Appley Bridge	IIa		3·9			Turner *et al.* (1974)
LL 6	Mangwendi	IIa		4·0			Turner *et al.* (1974)
L 5	Wittekrantz	IIb		~ 0·35	0·9		Turner and Cadogan (1973)
L 6	Peace River	IIb		~ 0·46	1·19		Turner (1976)

Table 4. ^{39}Ar–^{40}Ar gas retention ages substantially below 4·57 b.y.[a] (*Continued*)

Type[b]	Meteorite	Release Type[c]	Apparent Ages (b.y.)			Remarks	References
			Minimum	Plateau	Maximum		
Ho	Bholghati	IIb		~ 3·4	3·7	Scatter	Leich and Moniot (1976)
L 4	Barratta	III	0·47		4·0		Turner (1969)
L 5	Ergheo	III	0·53		2·1		Turner (1969)
L 6	Bruderheim	III	0·5		2·2		Turner (1969)
L 6	Chateau Renard	III	0·31		2·4		Turner (1969)
L 6	Rakovka	III	1·5		3·0		Turner and Cadogan (1973)
L 6	Zavid	III	0·54		3·0		Turner (1969)
L 6	Zemaitkiemis	III	0·52		2·2		Turner (1969)
L 6	Zomba	III	0·44		2·7		Turner (1969)
L	Bovedy	III	0·5		1·5		Cadogan and Turner (1975)
LL 3	Parnallee	III	1·98		4·08	Decreasing age at highest temperatures	Turner (1976)
LL 6	Ensisheim	III	1·5		3·0		Turner and Cadogan (1973)
H	Darmstadt	III	0·32		3·6		Turner (1976)
Eu	Petersburg	III	2·2		4·4		Podosek and Huneke (1973a)

[a]For Rb–Sr ages below 4·57 b.y., see section on *Radiometric Ages* (Nakhla, Kodaikanal inclusions, Kapoeta clast).
[b]see Table 2.
[c]see Figure 8 for explanation.
[d]Rb–Sr age 1·37 b.y. (Papanastassiou and Wasserburg, 1974); 1·24 b.y. (Gale *et al.*, 1975).

Collisional heating or a very close approach to the sun may be responsible in this case.

An intriguing observation is that five Ca-rich achondrites (Eucrites, Howardites) have ages between ~ 3·4 and 4·2 b.y. Perhaps, this indicates a unique thermal history of their parent body (bodies) e.g. extremely slow cooling. On the other hand, a certain parallelism with the moon's evolution during that time period is striking (see section *Lunar Chronology*). While it is most unlikely for other reasons that these objects have a direct genetic link with the moon, their ages may well indicate that the intense bombardment which affected the moon in that period was not a local occurrence but applied to the regions where the achondrites orbited as well.

Two members of another subgroup, the LL amphoterites Mangwendi and Appley Bridge also experienced a strong metamorphism ~ 4 b.y. ago.

Turning to the L-chondrites of group IIb, a rather intense event as recently as $\lesssim$ 460 m.y. ago is indicated from the plateau ages of Wittekrantz and Peace River. The same event may be responsible for the apparent age reduction of the L (hypersthene) chondrites of group III (no plateau) by somewhat less thorough resetting. Except for Rakovka and Chateau Renard, their least retentive phases have apparent ages near 500 m.y. (Table 4). A 'hypersthene chondrite parent body-event' is also indicated from the distribution of the numerous conventional total K–Ar and U/Th–He gas retention ages. Before we discuss these data, some remarks on related evidence from the Rb–Sr and Pb–Pb systematics may be appropriate. Internal isochron data for objects with reduced Ar–Ar plateau ages are scarce: Stannern has an inconclusive Rb–Sr internal isochron age of 4·1 ± 0·7 b.y. (Papanastassiou, 1970) and some phases of Peace River define an internal isochron of (4·56 ± 0·03) b.y. (Gray *et al.*, 1973). This would indicate that the Rb–Sr system did not respond to the process which caused the K–Ar age reduction. On the other hand, there are also components of Peace River which deviate from the 4·56 b.y. isochron and have model ages of ~ 4·25 b.y. (Gray *et al.*, 1973) which could be due to a limited response of the Rb–Sr system to the outgassing event.

Rb–Sr and Pb–Pb *model* ages are available for a number of meteorites with reduced Ar–Ar ages of types II and III and they are all consistent with *un*disturbed systems of primary age 4·6 ± 0·2 b.y. (Rb–Sr: Pasamonte, Ensisheim, Bruderheim (Papanastassiou and Wasserburg, 1969; Gopalan and Wetherill, 1969; 1968); Pb–Pb: Pasamonte, Bruderheim, Appley Bridge (Silver and Duke, 1971; Huey and Kohman, 1973; Gale *et al.* 1973)). However, it is a property of model ages that they are also consistent with a differentiation and re-equilibration at any later time if the system as a whole remained closed and no further conclusions can be drawn.

Gopalan and Wetherill (1971) deduced a severe metamorphism of the L-chondrites Orvinio and Farmington from their disturbed Rb–Sr patterns at about 300–500 m.y. ago, quite consistent with the Ar–Ar data for other L-chondrites.

The U–Pb intercept method has been applied to the carbonaceous chondrite Allende by Tatsumoto *et al.* (1976) and by Chen and Tilton (1976) to date metamorphism from the lower intercept age. They arrive at 280 ± 70 m.y. and 107 ± 70 m.y. respectively for the time of a late disturbance of the U–Pb system in a single event (Figure 3 above).

We shall now return to the question of conventional K–Ar and U/Th–He gas retention ages. Such data are much more readily available than Ar–Ar ages, but they have only a limited potential for a quantitative deduction of the time of metamorphic events, since sporadic and continuous diffusion losses cannot be distinguished.

Some clues can be deduced from a comparison of K–Ar and U/Th–He ages since He leaks out more readily than Ar. Consequently, concordant K–Ar and U–He ages must date the time of the last complete degassing, and insignificant losses can be assumed thereafter (Kirsten *et al.*, 1963). If concordant ages far below 4·57 b.y. occur, the time of an intense degassing event can be dated. The first extensive collection of K–Ar and U/Th–He ages revealed a cluster of such concordantly low ages between 0·4 and 1 b.y. (Kirsten *et al.*, 1963). Later it was shown by Heymann (1967) that this cluster is peculiar to hypersthene (L-group) chondrites, whose textures show indications of re-heating and shock effects in the 0·1–1 Mbar range (Taylor and Heymann, 1969). It was thus possible to deduce that at least one third of all hypersthene chondrites were involved in a collision of their parent body at $\sim 520 \pm 60$ m.y. ago (Taylor and Heymann, 1969) and this age nearly coincides with a number of observations previously discussed. Heymann has speculated that the second partner of this collision of a hypersthene chondrite parent body was the parent body of a large number of iron meteorites (octaedrites) which underwent collisional rupture ~ 600 m.y. ago. The latter figure arises from a completely different dating approach based on the interaction of galactic cosmic rays with meter-sized meteoroids which shall now be discussed.

Cosmic ray exposure ages

Principles. In interplanetary space, galactic cosmic ray protons penetrate approximately one metre into exposed solid objects. Spallation reactions of these protons with the solid matter give rise to the production of a great variety of stable and radioactive isotopes with masses below and near the target mass. The quantity of the interaction products is a measure of the time during which a meteoroid was exposed to the radiation in the form of a body whose dimensions were small compared to the mean penetration depth of the cosmic radiation (see Kirsten and Schaeffer, 1971).

Meteorites falling on earth rarely exceed one metre in length and it is therefore likely that, in general, they have acquired the bulk of their 'spallation-products' in the last period of their evolution. The exposure age t_e inferred from these spallation products determines the lifetime of the meteorite as a small object, measured backwards from the time of its fall on earth. As will be seen, meteorite exposure ages are about 1–3 orders of magnitude shorter than the meteorite formation ages of approximately 4·57 b.y. On the other hand, it is known from related meteorite investigations which will not be discussed in this article that the long-term average cosmic ray intensity has remained essentially unchanged for the last billions of years (see Schaeffer, 1975). It must then be concluded that for most of their history most meteorites were largely shielded from exposure within larger parent bodies from which they were expelled by break-up, collision or explosion in relatively recent times. The generally accepted view is that the size distribution of interplanetary objects (planets, asteroids, comets, meteoroids) is governed by mutual collisions, whereas continuous erosion by cosmic dust and radiations is relatively unimportant (Dohnanyi, 1969; see also Kirsten and Schaeffer, 1971). In this way, cosmic ray exposure ages are linked to the mean lifetime of solid objects against destruction and/or planetary capture, and this, in turn, is related to their primary orbits and hence to the question of their origin. In addition, coincident exposure ages of a number of meteorites of the same chemical or petrological type provide circumstantial evidence that these meteorites originate from one and the same parent body which must have undergone collisional break-up at a time determined by the respective exposure age.

The spallation reactions are rather complex and we shall not discuss in any detail the complications which arise from the complexity of the energy spectrum, the target chemistry, and the excitation functions (see Kirsten and Schaeffer, 1971; Herzog and Anders, 1971; Lal, 1972; Bogard and Cressy, 1973 for reviews on this topic). In principle, however, exposure ages can be determined in a rather straightforward way. In a small body, the quantity ${}^{i}D_{s}$ of a stable spallogenic product has been accumulated over the whole exposure time t_e and may be measured. One may also measure the activity ${}^{j}A_{r}$ of a radioactive isotope at the time of fall. In the case of radioactive equilibrium ($\tau \ll t_e$) this activity will be equal to the production rate. Hence

$$t_e = \frac{{}^{j}\bar{\sigma}_r\, {}^{i}D_s}{{}^{i}\bar{\sigma}_s\, {}^{j}A_r} \tag{21}$$

where ${}^{j}\bar{\sigma}_r/{}^{i}\bar{\sigma}_s$ is the effective cross section ratio for the whole energy spectrum and target composition. It is usually approximated by accelerator data at just one energy when the mass numbers i and j are not too different (see Kirsten and Schaeffer, 1971). The practical choice of D_s is dictated by the necessity of distinguishing the spallation products from the primordial concentrations of the same element by means of their distinct isotopic composition. One is therefore limited to elements with extremely low primordial concentrations such as rare gases, or, in case of iron meteorites, potassium.

Commonly used radioactive isotopes are ^{26}Al ($T_{1/2}$ = 0·74 m.y.; ^{36}Cl ($T_{1/2}$ = 0·31 m.y.); ^{3}H($T_{1/2}$ = 12·3 y); ^{39}Ar ($T_{1/2}$ = 269 y); ^{22}Na ($T_{1/2}$ = 2·6 y); ^{40}K ($T_{1/2}$ = 1·3 b.y.); ^{81}Kr ($T_{1/2}$ = 0·21 m.y.). The respective exposure age pairs are ^{21}Ne/^{26}Al; ^{36}Ar/^{36}Cl; ^{3}He/^{3}H; ^{38}Ar/^{39}Ar; ^{22}Ne/^{22}Na; ^{41}K/^{40}K, and ^{81}Kr/^{83}Kr.

With the exception of the latter two, which are determined by mass spectrometry, elaborate low level activity measurements with large samples are required. Consequently, stable rare gas data are more frequently available than activity measurements and it has been a common practice to use average production rates of stable rare gas isotopes normalized for chemical composition, if no activity measurements are available. This approach is applicable only to small objects like most stony meteorites. It has considerably improved the statistics of exposure age distributions, but individual exposure ages of this type may have errors as large as 30 per cent. We shall now summarize the major results from exposure age determinations on meteorites.

Results. The overall distribution of exposure ages is shown for chondrites in Figure 9 and for iron meteorites in Figure 10. Typical exposure ages are ~ 10 m.y. for stones and 500 m.y. for irons. Extreme values for stones are 2×10^4 yrs (Farmington; Heymann and Anders, 1967) and 80 m.y. (Norton County, Herzog *et al.*, 1976), for irons 4 m.y. (Pitts; Cobb, 1966) and 2·3 b.y. (Deep Springs; Voshage, 1967). In addition to the gross difference between the average

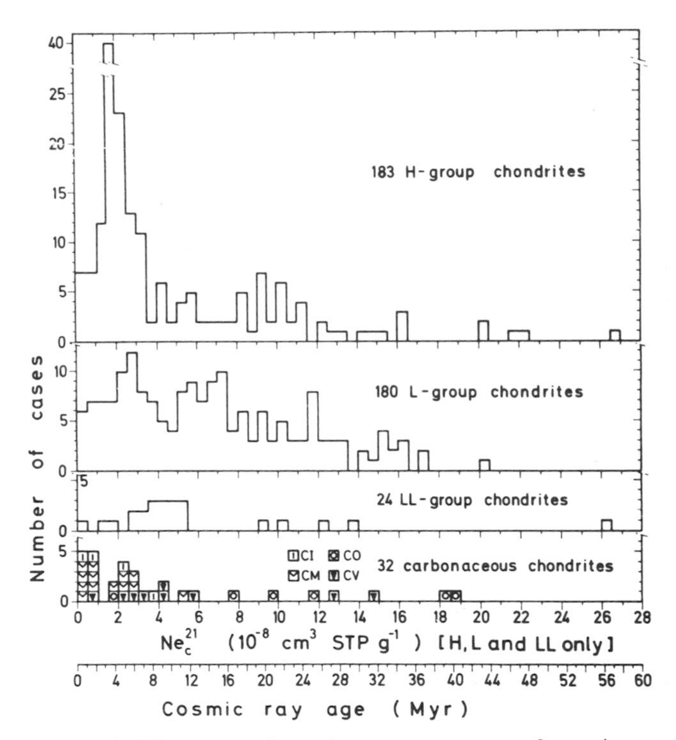

Figure 9. Histogram of cosmic ray exposure ages for various types of chondrites, based on their cosmogenic ^{21}Ne contents and on average production rates for each chemically uniform group, based on data compiled by L. Schultz (see also Schultz, 1976) and from Mazor *et al.*, 1970). (Reproduced with permission from Figure XI–2 of J. R. Wasson (1974), *Meteorites*, Springer, Berlin–Heidelberg–New York.)

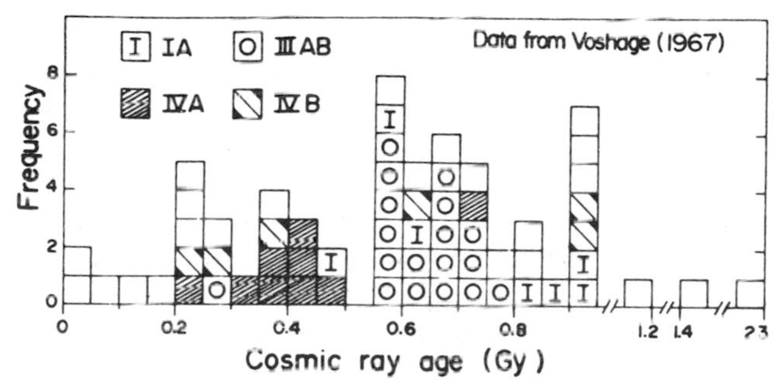

Figure 10. Histogram of cosmic ray exposure ages for 62 iron meteorites determined by the $^{40}K/^{41}K$ method (Voshage, 1967). Compilation by Scott and Wasson (1975) according to chemical classification of the irons. Note the clusters for group III and IV A and the much longer time scale as compared to stony meteorites (Figure 9); Gy = 10^9 years = 1 b.y. (Reproduced with permission from Figure 7 of E. Scott and J. Wasson, 1975, *Rev. Geophys. Space Physics*, **13**, 527–546, copyrighted by American Geophysical Union)

ages of irons and stones, certain meteorite classes show a tendency to cluster at certain values. The most prominent clusters are at 4 m.y. for H-chondrites (Figure 9) and at 650 m.y. for medium octaedrites (iron meteorites of chemical subgroup III; Figure 10). Further significant clusters appear at ~ 8 m.y. for LL chondrites, at ~ 40 m.y. for aubrites, and at 400 m.y. for octaedrites of chemical group IV A. Other, statistically less significant clusters are indicated for L-chondrites near 5 m.y., for eucrites near 5 and 11 m.y., and for diogenites near 15 m.y. The majority of chondrites has exposure ages < 10 m.y. but the majority of achondrites has exposure ages > 10 m.y. Exposure ages of carbonaceous chondrites are with very few exceptions restricted to less than 15 m.y., other cut-offs seem to occur at 200 m.y. for stony-irons (Begemann *et al.*, 1976) and at 300 m.y. for hexaedrites (chemical group IV A) (Scott and Wasson, 1975).

These observations indicate that the last step in the evolution of meteorites is governed by collisional processes. In some cases, the break-up of particular meteorites has been directly observed in that two distinct exposure ages have been measured for different specimens from the same meteorite fall. Examples are the irons Sikhote Alin, Canyon Diablo, Arispe, Odessa (Vilcsek and Wänke, 1963; Kolesnikov *et al.*, 1972; Heymann *et al.*, 1966) and the stones New Concord (Zähringer, 1966), Mighei, Ivuna and Serra de Mage (Fuse and Anders, 1969).

All these exposure ages are much shorter than formation ages; they date late events which did not significantly affect the radiometric clocks related to formation ages, except that moderate rare gas diffusion losses may occur in these collisions. A speculation about a connection between the 'hypersthene event' inferred from K–Ar dating and the 650 m.y. medium octaedrite exposure age cluster has already been mentioned. Otherwise, early events are not accessible with this dating technique, with the possible exception of a xenolith from the Weston meteorite which may have acquired a part of its spallogenic isotopes very early on the surface of a parent body before it became incorporated in the meteorite (Schultz *et al.*, 1972).

Inferences. The collisional break-up of individual parent bodies can be recognized from exposure age clusters for meteorites of common petrological type, whereas the broad distributions are due to statistical fragmentation of the surface of the parent objects by minor collisions and moderated by the probability distributions for earth capture. The absolute exposure ages reflect either the mean lifetime of metre-sized objects with initial orbits characteristic for their place of origin against ultimate capture by the earth, or the lifetime T_{coll} of these objects against mechanical destruction by collisions in their source region. If T_{coll} is shorter than the mean capture time T_{capt}, the exposure ages reflect the mean time between two collisions and the yield of earth captured meteorites is strongly depressed. Different mechanical strength could be the major reason for the roughly 100-fold higher exposure ages of irons as compared to stones, but differences in the place of origin of these two types of objects could just as well account for this difference.

The two principle alternatives for the origin of meteorites are that they originate either from asteroids or from comets. A lunar or martian origin can be dismissed on the basis of chemical dissimilarities and dynamical difficulties (Anders, 1971; Wetherill, 1971). Cometary orbits have typical lifetimes in the order of 10^6–10^7 yrs and could thus account quite well for the exposure age distribution of stony meteorites, but this can hardly be a source for irons (Wetherill, 1974). Essentially the same conclusion can be deduced from physico-chemical arguments (Anders, 1971). Comets are numerous but too small (1–10 km) to allow major chemical differentiations within them, thus they are potential sources only for undifferentiated objects, such as chondrites, particularly carbonaceous chondrites. For iron meteorite parent bodies, larger, asteroidal sizes of 10–500 km are required to account for their differentiated nature and for their experimentally determined cooling rates of 1–10 K/m.y. (Anders, 1971).

In addition, all iron meteorites fall into at most ten chemically well defined chemical subgroups which requires a few large parent bodies rather than numerous small ones. Typical lifetimes for asteroidal objects depend strongly on their initial orbits. Normal ring-asteroids have practically no chance for ejection into meteoritical orbits, but the rare Mars crossing asteroids have lifetimes against planetary capture in the order of 10^9–10^{10} yr. They could be the source of iron meteorites judging from their exposure age distribution (Wetherill, 1974). Anders (1971) favours these same objects also as the source for ordinary chondrites in order to restrict the number of parent bodies for stony meteorites. The short chondrite exposure times are then interpreted by the short lifetime of friable rocky material against collisional destruction. The earth crossing Apollo asteroids which are probably supplied by burnt-out comets have lifetimes of the order of 10^7–10^8 yr and are also a potential source of stony meteorites. Since the exposure ages contribute only one aspect to this subject, the reader is referred to the reviews by Wetherill (1974) and Anders (1971) for a more thorough discussion.

Lunar Chronology

The thermal evolution which is characteristic for a planet the size of the moon led to an early global differentiation of at least its outer few hundred kilometres. Contrary to the case of meteorite parent bodies, one does not therefore expect individual rock ages to give the planetary age. The differentiated nature of the moon is evident from a petrological inspection of all accessible lunar rock types. The two principal and chemically distinct lithologies are the so-called 'ANT'-rocks (anorthosite, norite, troctolite) which dominate the lunar highlands; and the Fe- and Mg-rich mare basalts which cover the circular mare basins and other lunar lowlands (for a general introduction see, for example, Taylor, 1975).

K, Rb, U, and Th, the elements which are so important for chronometry and also for the planetary energy balance, are all depleted in the anorthositic rocks, but they are highly enriched in the 'KREEP'-basalts frequently associated with ANT-rocks. Geochemical considerations suggest that KREEP-rocks (enriched in *K*, *R*are *E*arth *E*lements, *P*, also in U, Th) must have concentrated *below* the anorthositic crust in the process of crustal differentiation, and hence their present co-existence with anorthosites is a result of late mechanical mixing. Much of the highlands material was actually ejected from the large, multi-ring basins by asteroid-sized bodies which collided with the moon early in its history, and rocks from large depths have been exposed upon excavation. Subsequent smaller meteorite impacts have compacted rock fragments and clasts of various origin into the typical highland breccias. Uniform crystalline highland rocks (possibly impact melts) are very rare, and true bedrocks are not available.

The second major lunar rock type is crystalline mare basalts. These are igneous rocks derived from lava flows which poured into the basin depressions some time after the basin forming impact. The Fe- and Mg-rich source reservoirs of these magma flows are estimated to be at depths of $\lesssim$ 100 to $\sim$ 400 km. Basaltic magmas were derived from these sources by partial melting, and have ascended through $\sim$ 60 km of consolidated anorthositic crust, a process presumably facilitated by crustal fractures created by the giant collisions.

The task of lunar chronology is to provide the absolute ages for the scenario described, and abbreviated as follows:

(a) Accretion of the moon (age of the moon).
(b) Formation of an undifferentiated pristine solid 'egg shell'.
(c) Global differentiation of at least the outer 400 km into an anorthositic crust and the mafic *source region* of mare basalts.

(d) Bombardment and penetration of the growing anorthositic crust by objects of asteroidal size and formation of the major basins which to this day dominate the morphology of the face of the moon.
(e) Extrusion of basaltic magmas from depth into topographical lows (period of lunar volcanism).
(f) Subsequent processes.

We shall now review the most important results which relate to these processes, in the same order.

Age of the Moon and Early Crustal Differentiation

When the moon was formed, accretional heating must have melted the outermost layers. At the surface, however, the heat radiated away rather quickly and a thin 'egg shell' of crystallized undifferentiated material must have formed very shortly after, or even during the final stages of accretion. The most straightforward way to estimate the age of the moon would therefore be to search for residues of this undifferentiated crust and to determine their crystallization age. However, the chances that such a rock has survived the subsequent global differentiation and intense bombardment are practically nil. We are then left with three possible avenues towards an estimate of the age of the moon:

(a) Search for the oldest early differentiates (anorthositic crustal rocks or mafic mantle cumulates) to establish a meaningful lower limit for t_p.
(b) Determine whole rock isochron ages to infer the time of the last large scale fractionation between parent and daughter element.

It should be noted, however, that the dated event will be the differentiation into anorthositic and mafic cumulate but *not* the condensation of the protolunar material from the solar nebula.

(c) Consider initial strontium- and lead-isotope evolution (see section *Duration and Sequence of Solar System Formation*).

Ages of early lunar differentiates. The only *mafic* cumulate rock found so far is a dunite clast separated from Apollo 17 breccia 72417. Papanastassiou and Wasserburg (1975a) have determined an internal Rb–Sr isochron age for this clast of 4·55 ± 0·1 b.y., hence, the only available rock from the lunar 'mantle' is at the same time among the oldest of all known lunar rocks.

Unaltered *ANT* rocks are rare because of the intense bombardment of the early lunar crust. Relevant Rb–Sr internal isochron ages have been reported for troctolite 76535 (4·60 ± 0·09 b.y.; Papanastassiou and Wasserburg, 1976), plagioclases from breccia 76055 (~ 4·49 b.y., Tera *et al.*, 1974); and for noritic fragments from Apollo 17 boulder 77215 (4·43 ± 0·05 b.y., Nunes *et al.*, 1976). Evidence for 4·47 b.y. old lithic fragments in Apollo 16 breccia 65015 has been deduced from high temperature release fractions in ^{39}Ar–^{40}Ar dating (Jessberger *et al.*, 1974). From these data we can already conclude that the moon must have accreted more than 4·5 b.y. ago and that differentiation of its crust began very soon after its formation.

Model ages. The lunar rocks are derived from geochemically distinct source regions which were established during the primary lunar differentiation. In addition to the compositional differences between the three principal reservoirs of crust, KREEP, and mantle, regional geochemical differences must exist among each category, and the rocks which were later generated from these various source regions will reflect these differences. This implies that a suite of rocks with a certain provenance will be geochemically rather uniform but may be distinct from another suite which was generated from another source region, say, 100 km away. With respect to element ratios like Rb/Sr or U/Pb, this could mean that the gross differences among various rock types were essentially established as a result of the early global differentiation and not as a result of the later generation of rocks from their sources. If this is the case, it could imply that each source region behaved as a closed system for the parent/daughter ratio *since the initial differentiation*, irrespective of later rock forming processes and of the accompanying *internal* redistribution of these elements among the *minerals* of a given rock. As a consequence, two or more local source regions which were formed *contemporaneously* from a common 'grandparent-reservoir' and which are represented by respective whole rock samples will define a whole rock isochron which dates the age of the sources, that is, the time of differentiation because *this* is the process which has effectively fractionated the parent from the daughter element. Deviations from an isochron will occur when:

(a) Samples are from non-contemporaneous or non-cogenetic sources (e.g. incomplete homogenization of isotopes in the source region).
(b) Fractionations *did* occur when the rocks crystallized from the source region.
(c) The closed system requirement was violated during the lifetime of the rocks by metamorphic or similar processes (in this case, internal isochrons will also be disturbed).

Restriction (b) can be partially relaxed in the case of ^{207}Pb–^{206}Pb/U-dating since the partial fractionation during rock formation (Pb-loss) can be accounted for by the concordia intercept method (section *Types of 'Ages'*). The upper intercept age is the age of the source, the lower intercept age is the age of the rock. Only if the lead loss is complete does it become impossible to date the age of the source region.

How realistic the supposition of only minor element fractionation on the occasion of rock crystallization actually is, can only be decided by the actual data. If it does apply, one could expect either (a) different whole rock isochrons for the three major rock types if the primary cumulates established their geochemical individuality at appreciably different times (extended duration of crustal differentiation), or (b) just one common isochron for all rock types if the global differentiation occurred rather quickly. In this respect, lunar soils should be diagnostic since they may be considered to represent 'natural aliquots' of the lunar crustal rocks which were ground and mixed by micro- and macro-meteorite impacts.

Soils from all landing sites yield a common Rb–Sr whole rock isochron of ~ 4·4 ± 0·2 b.y. (Papanastassiou and Wasserburg 1972a, 1972b). There is appreciable scatter and anomalous model ages in excess of 4·6 b.y. do occur, but these may be due to late Rb-losses during the grinding process. In general, the existence of this isochron points towards a large scale initial differentiation between ~ 4·4 and 4·6 b.y. ago.

The collection of mare basalts from all landing sites does not define a whole rock isochron. Instead, the whole rock model ages scatter between 3·8 and 5·4 b.y. (compilation in Schonfeld, 1976). This must be explained by chemical fractionations during the crystallization of these rocks. For instance, the potassium-rich Apollo 11 basalts have a common whole rock model age of ~ 3·85 b.y. (Papanastassiou and Wasserburg, 1971b). This age has probably no chronological significance but the group behaviour indicates a common geochemical fractionation.

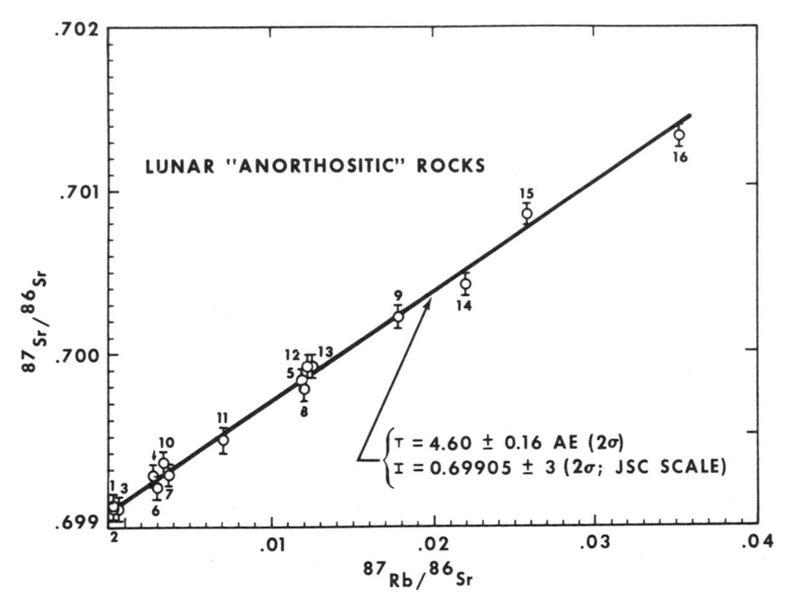

Figure 11. Rb–Sr whole rock isochron for ANT-type rocks from the Apollo 16, 17, and 15 missions after correction for KREEP-contamination from an appropriate mixing model (Schonfeld, 1976). The isochron indicates the formation of an anorthositic lunar crust 4·6 ± 0·16 b.y. ago. (Reproduced with permission from Figure 3 of E. Schonfeld, 1976; *Lunar Science VII*, 773–775.)

Granitic rock types are extremely rare on the moon but Apollo 12 breccia 12013 is a notable exception. Various lithic fragments of this highly differentiated rock gave a Rb–Sr isochron model age of 4·52 b.y. (Lunatic Asylum, 1970). Rb–Sr whole rock isochrons for ANT-rocks from various landing sites and for the KREEP suite are relatively well defined. A complication arises from the contamination of many anorthositic breccias with KREEP-type material. After an appropriate correction, Schonfeld (1976) deduced a whole rock isochron for ANT rocks of $4{\cdot}60 \pm 0{\cdot}16$ b.y. (Figure 11) and KREEP-rich fragments yield common isochrons of 4·3–4·4 b.y. (Nyquist *et al.*, 1973, 1974; Shih, 1976).

Potentially, the Sm–Nd method is capable of yielding analogous results, but at present the experimental data are limited. Lugmair *et al.* (1976) have shown that the troctolite 76535 and the mare basalt 15555 have a common 'whole rock isochron' age of $4{\cdot}53 \pm 0{\cdot}10$ b.y. Since they differ in their Sm/Nd ratio, this geochemical fractionation of rare earths must have occurred as early as $\sim 4{\cdot}53$ b.y. ago.

Important information has also come from the U–Pb systematics. In a U–Pb evolution diagram, the vast majority of terra rocks and soils as well as some mare basalts plot on a linear chord which intersects concordia at $3{\cdot}9 \pm 0{\cdot}1$ and $\sim 4{\cdot}42$ b.y. (Tera and Wasserburg, 1974; Figure 12). The significance of the lower intercept age will be discussed in the section *Lunar Rock Ages*. The upper intercept defines the time when the last major fractionation between U and Pb occurred and is best explained as the time of the initial differentiation of the lunar crust. It is particularly important to note that for some rocks (68415, 67559, and mare basalt 75055), the lead loss during their crystallization was negligible and their U–Pb ages are concordant at 4·42 b.y. (Tera and Wasserburg, 1974; Figure 12). This age could conceivably be close to the age of the moon itself if quick differentiation occurred immediately after the end of accretion, but this interpretation is in conflict with the high internal isochron ages of some cumulate rocks (section *Ages of Early Lunar Differentiates*). In addition, there is evidence for some higher upper intercept ages near 4·6 b.y. for soil 10084, breccia 10061 (Tatsumoto, 1970) and orange glass 14220 (Tera and Wasserburg, 1976). Nunes *et al.* (1975) have pointed out that the widespread occurrence of the 4·42 b.y. upper intercept age may be accidental. In their interpretation, the U–Pb systematics of most rocks are dominated by their contamination with uranium-rich KREEP material and the common upper intercept age is the natural result of the average isotopic composition of widely distributed KREEP. In this view, Pb has evolved in a multi-stage evolution since $\sim 4{\cdot}6$ b.y. and the 4·42 b.y. age is without direct chronological significance.

Initial strontium and lead. The initial $^{87}Sr/^{86}Sr$ ratios inferred from mineral isochrons of mare basalts and some ancient cumulate rocks are very low and similar to values in the solar nebula at the time of condensation. In fact, it is deduced from the minimum ratio that the protolunar material condensed shortly before the basaltic achondrites (see section *Duration and Sequence of*

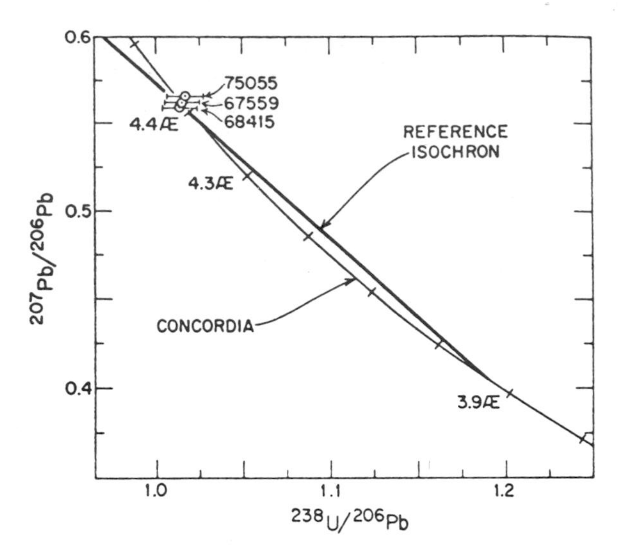

Figure 12. U–Pb evolution diagram for lunar highland rocks (Tera and Wasserburg, 1974). The diagram differs from the regular concordia plot (Figure 3) in order to better account for the *initial radiogenic* ^{207}Pb and ^{206}Pb common in lunar rocks (Tera and Wasserburg, 1972 a, b). The primordial lead (estimated from ^{204}Pb) is subtracted before the data are plotted. The time meaning of the concordia intersections is still similar to that applicable for the regular concordia plot. The line marked 'reference isochron' is experimentally defined by the vast majority of highland rocks from the Apollo 14, 15, 16, and 17 missions, but for clarity individual data points are not plotted; except for three particular rocks (including mare basalt 75055) which have concordant ^{206}Pb–^{238}U and ^{207}Pb–^{235}U-ages at 4·42 b.y. The implications of the linear alignment of numerous highland rocks and of the intersections with concordia at 4·42 and ~ 3·9 b.y. are discussed in the text. (Reproduced with permission from Figure 6 of F. Tera and G. Wasserburg, 1974, *Proc. Lunar Sci. Conf., 5th, Vol.* 2, 1571–1599, copyrighted by Pergamon Press)

Solar System Formation). A direct consequence of the low initial Sr ratios is that the source material of these rocks cannot have spent any appreciable time ($> 10^8$ yr) after 4·57 b.y. ago in a solar or chondritic Rb/Sr-environment (e.g. Gray *et al.*, 1973). Instead a very effective depletion of the volatile Rb must have occurred between 4·6 and 4·5 b.y. ago, either when the moon formed or in a subsequent differentiation. Although the moon is generally depleted in volatiles, some Rb-rich source regions must also have existed in its earliest history judging from the extremely *high* initial $^{87}Sr/^{86}Sr$ ratio in some components of rock 12013 (Lunatic Asylum, 1970). In a similar way, the existence of an early (> 4·3 b.y.) differentiate enriched in uranium can be inferred from the highly radiogenic initial lead in breccia 78155 (Nunes *et al.*, 1975).

Schonfeld (1976) has attempted to deduce a time difference between the formation of the anorthositic crust and the mafic cumulates from the slight difference in the initial $^{87}Sr/^{86}Sr$ ratios of the two respective whole rock isochrons. He arrives at 4·62 b.y. for the anorthosites if the mare basalt sources were laid down 4·38 b.y. ago. However, the initial Sr ratios of the whole rock isochrons are not sufficiently accurate to affirm the reality of this time difference.

In conclusion, the age of the moon inferred from the oldest internal isochrons is between 4·5–4·6 b.y. A global differentiation of the lunar crust occurred between ~ 4·3 and 4·6 b.y. This is inferred from ages of cumulate rocks, model ages and Sr- and Pb-isotope evolution studies.

Lunar Rock Ages

Highland rocks. The primary lunar crust was subsequently exposed to an intense bombardment with interplanetary objects including bodies of asteroidal size which have formed the large circular basins. Consequently, the age distribution for the highland breccias which were excavated and compacted in this process reflects primarily the collisional history of the moon after 4·3 b.y. ago. However, the way in which a rock age responds to these impact-related processes is not very clear and the interpretation of highland rock ages is still a matter of debate (Horn and Kirsten, 1977; Turner, 1977).

An individual rock age determines the time of the last isotopic equilibration, but the nature of this event is not clear for the lunar highland rocks. Some alternatives shall be listed:

(a) an endogenic process.
(b) heating and/or melting caused by one or a sequence of subsequent moderate impacts.
(c) deposition of rocks in the hot ejecta blanket produced in a large basin forming impact.
(d) cooling subsequent to uplift generated by a large impact from a depth in which the temperature had previously exceeded the blocking temperature of the radiometric clock.

With respect to basin stratigraphy there is the additional problem of the assignment of a particular rock to the ejecta blanket of a particular basin. Usually, such assignments are made on the basis of proximity to the nearest basin, but large scale transport processes are operative on the lunar surface and may confuse these assignments. Due to these uncertainties, it is usually not possible to date a basin from a single rock age; instead the age distribution of a given rock suite must be considered (Kirsten and Horn, 1974). If the rock ages are 'basin dominated', they should yield essentially identical ages and the highest rock age of a given suite is an upper limit for the age of the basin. If, however, the ages are (small) 'crater' dominated, the excavating event may not have influenced the isotopic system to any appreciable degree, but the age distribution reflects the influence of small but numerous local craters.

From a technical point of view, it is difficult to date anorthosites by the Rb–Sr method because of their extremely low Rb/Sr ratios (10^{-2}–10^{-3}) whereas the alkali-rich KREEP-basalts are more amenable to Rb–Sr dating. Both rock types can be dated by the ^{39}Ar–^{40}Ar method, and anorthosites are exclusively dated in this way. In addition, with the ^{39}Ar–^{40}Ar method it is possible to date very small (mm-sized) rock fragments ('coarse fines') as well as large rocks. Apart from making possible better age statistics, this is also important since the cumulative grinding of rocks in the lunar regolith may preferentially remove the oldest large rocks.

Due to the high mobility of Pb, meaningful U–Pb internal isochron rock ages are essentially non-existent. However, the lower intercept in the whole rock isochron discussed in the foregoing section is of general importance for the interpretation of highland rock ages. The whole rock line intersects concordia between 3·8 and 4 b.y. for virtually all rocks. This implies that all these materials experienced a serious U–Pb fractionation (Pb-loss) in this 200 m.y. time span and the simplest explanation would be that all rocks have been involved either in igneous processes or in large resetting impacts in this time period. It could for example, imply that essentially all major basins have been formed within only ~ 200 m.y. (so-called 'cataclysm'; Tera and Wasserburg, 1974; Tera *et al.*, 1974).

The following is a summary of the more than 150 available highland rock ages (see Turner, 1977, also for References of the original data sources) of which the majority are ^{39}Ar–^{40}Ar ages. The general agreement between ^{39}Ar–^{40}Ar and Rb–Sr ages of lunar rocks (including mare basalts) is good in all cases (Turner, 1977) except for a systematic difference of ~ 100 m.y. for the potassium-*rich* Apollo 11 basalts.

The *Apollo 14* Fra Mauro highland basalt ages cluster around 3·93 b.y. and range from 3·87 to 4·04 b.y. (Turner, 1977). A subdivision into KREEP- and non-KREEP basalts reveals that KREEP basalts (e.g., 14310) (3·87–3·91 b.y.) are clearly distinct in age from the non-KREEP highland basalts (e.g., 14053) (3·91–3·99 b.y.) (Turner, 1977) (see section *Model Ages*). The Fra Mauro formation is stratigraphically related to the Imbrium basin and it is clear that the Imbrium event must have affected the ages of the excavated rocks. However, the observed range of ages and the two age groups are inconsistent with a complete isotopic homogenization in just one event, and partial resetting of the radiometric clocks is more likely (see Horn and Kirsten, 1977). If ~ 3·93 b.y. is the age of the Imbrium excavation, then some rocks have not been totally reset on this occasion and some others have been involved in additional resetting events on a more local scale. The explanation of the age differences in terms of prolonged storage times of rocks in hot ejecta blankets is unlikely since many breccias are not thermally equilibrated.

Three Apollo 14 coarse fine fragments are exceptionally young and have ages between 3·54 and 3·63 b.y. (Kirsten *et al.*, 1972). They demonstrate the assignment problem in lunar chronology. Judged on petrological grounds, these fragments are unrelated to the Fra Mauro basalts and were thrown into the Apollo 14 site from nearby mare regions.

Apollo 15 ANT-rocks from the Appenine front have ages similar to the Apollo 14 breccias (two KREEP-basalts 3·91–3·94 b.y.) (e.g., 15386, Nyquist *et al.*, 1975; three anorthosites 3·95–3·99 b.y., Stettler *et al.*, 1973). Most ages of the Apollo 16 breccias from the Descartes region in the Central highlands fall between 3·89 and 4·04 b.y. (see Turner, 1976b), but five ages are between 4·19 and 4·30 b.y. (Schaeffer and Husain, 1973; 1974; Kirsten *et al.*, 1973) and demonstrate the survival of solidified rocks throughout the collisional history of the lunar crust. Another significant observation is the exceptionally *low* age of anorthosite 60015 (3·50 b.y., Schaeffer and Husain, 1974). As will be seen, no basin can be made responsible for this age, and it is therefore evident that local cratering is capable of resetting the isotopic systems. A glass coating on this rock is visible evidence for its production in a regional impact.

Ages for the Apollo 17 Taurus-Littrow highland rocks cluster between 3·95 and 4·05 b.y. (see Turner, 1977), but anorthositic rock ages between 4·13 and 4·28 b.y. also occur rather frequently (Schaeffer and Husain, 1974; 1975; Kirsten and Horn, 1974). They include the large rock 78155 (Turner and Cadogan, 1975) and various clasts from breccia 73215. The latter sample is exceptional in that it is not a complex regolith breccia; rather its clasts were imbedded into rapidly cooled groundmass in one single event (James *et al.*, 1975; Jessberger *et al.*, 1976), presumably the excavation of the Serenitatis basin. In this case, the usual interpretative problems are relaxed and the lowest age from within the groundmass ($\sim$ 3·9 b.y., felsite, Jessberger *et al.*, 1977) should be an upper limit for the age of Serenitatis basin. The existence of five different clasts with gas retention ages between 4·2 and 4·28 b.y. (see also Figure 4) (Jessberger *et al.*, 1976) in this breccia is a direct proof that a basin forming event did not completely reset all rocks involved in the event. Consequently, the clast age distribution is related to the stratification of the pre-Serenitatis pluton. This does not contradict the view that many Apollo 17 rocks have been influenced in the Serenitatis excavation, but mostly resetting was not complete since felsite 73215 and some other rocks (77115, 77135; 3·88–3·90 b.y.; Tatsumoto *et al.*, 1974, Stettler *et al.*, 1975) are significantly younger than most others with ages between 3·95 and 4·05 b.y.

Luna 20 highland rocks from the Apollonius mountains gave ages of $\sim$ 3·90 b.y. (Podosek *et al.*, 1973), and one older fragment of 4·3 $\pm$ 0·1 b.y. (Turner and Cadogan, 1975). These rocks may be Crisium ejecta.

In Figure 13 we have summarized the highland ages from all landing sites. From this it is quite evident that the time between 3·85 and 4·05 b.y. ago has been a very active period in shaping the lunar crust because of major basin

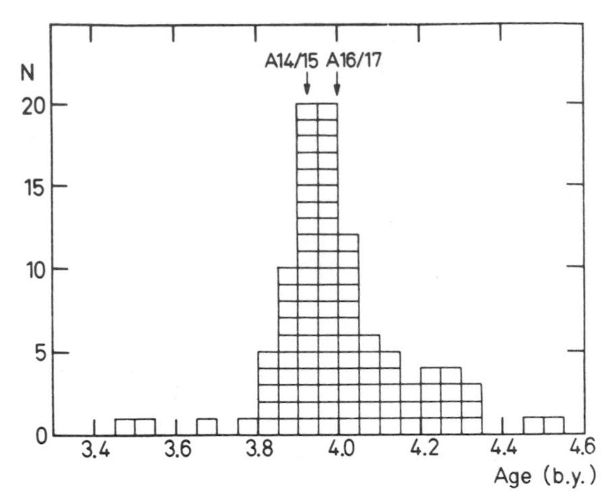

Figure 13. Histogram of lunar highland rock ages from all landing sites. Note the deficient resolution of ages for rocks from different landing sites and the occurrence of rocks with ages in excess of 4 b.y. (Reproduced with permission from Figure 2 of P. Horn and T. Kirsten, 1977; *Phil. Transactions*, **A285**, 145–150)

forming collisions at that time. This conclusion is supported by the large scale volatilization of lead inferred from the lower intercept of the whole rock U–Pb concordia diagram at $3{\cdot}9 \pm 0{\cdot}1$ b.y. From these data a period of cataclysm $\sim 3{\cdot}9$ b.y. ago has been inferred in which the impact rate of objects of asteroidal size was higher than between 4·0 and 4·3 b.y. ago (Tera and Wasserburg, 1974). However, the frequent occurrence of rock ages between 4 and 4·3 b.y. and objections to a purely basin dominated chronology (Horn and Kirsten, 1977) point more towards an impact rate which decreased continuously between $\sim 4{\cdot}5$ and 3·8 b.y. ago. In this view, the rapidly decreasing cratering rate enabled an increasing number of rocks to survive the bombardment as time proceeded. During the same period, the large basin forming impacts occurred. The lower age limit for the large impacts, 3·8 b.y., follows from the well determined age of the subfloor basalt in the Taurus–Littrow valley (Kirsten and Horn, 1974) (section *Mare Basalts*). This old mare surface postdates the youngest basin (Orientale) since no ejecta debris is found on it. The stratigraphic age sequence indicates that twelve large circular basins are younger than the oldest one, Serenitatis (Hartmann and Wood, 1971). If the age of the latter is $\sim 4{\cdot}0$ b.y. (above), then all basins must have formed between 3·8 and 4 b.y. The rock age distributions, however, are not determined but merely overprinted in these events. Because the basin excavations have limited resetting power, it is practically impossible to establish a detailed absolute basin chronology between 3·8 and 4·0 b.y. This is reflected in Figure 13 above where the age distribution of 'Imbrium dominated' Apollo 14 rocks is barely resolvable from the 'Serenitatis dominated' Apollo 17 rock age distribution. For a detailed basin stratigraphy it may be more appropriate to determine the relative basin ages by remote crater counts in photogeology (Neukum *et al.*, 1975). The relative basin ages can then be transformed into absolute ages from the crater densities in areas for which radiometric age data are available. This method is generally important for the estimation of the ages of areas not accessible to sampling. The resulting sequence is Serenitatis, Nectaris, Humorum, Crisium, Imbrium, Orientale.

Mare basalts. Partial melting of mafic cumulates below the solidified anorthositic crust generated fluid basaltic magmas which ascended from ~ 100–400 km depth through the fractured crust and filled the basins and other topographical lowlands to achieve hydrostatic equilibrium with basalt covers of ~ 1–10 km thickness. The basins were not filled at once; subsequent flows can still be recognized from visual inspection (Head, 1976). This period of lunar volcanism can be dated from the distribution of crystallization ages of the mare basalts in a rather straightforward way since the basalts are crystalline rocks which cooled and crystallized quickly after extrusion of the magma. Most crystallization ages have been determined by the ^{39}Ar–^{40}Ar method and by Rb–Sr mineral isochrons. Methodologically, it is also worthwhile to note that two U–Pb lower intercept ages (15555, 75055) (Tera and Wasserburg, 1976) and a Sm–Nd internal isochron age (75075) (Lugmair *et al.*, 1975a) (Figure 14) are consistent with the other data.

A reasonable model for the progressive growth of the solid lunar crust and

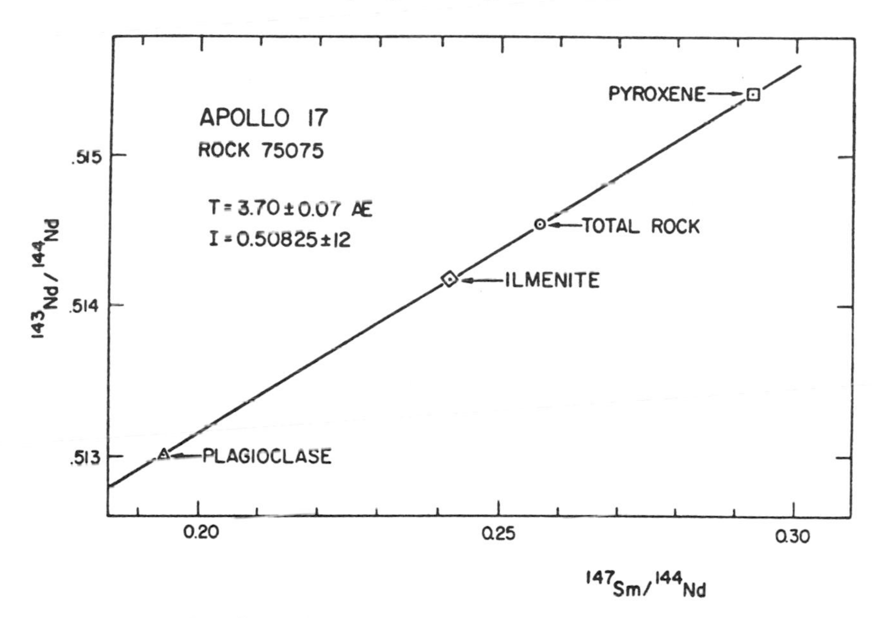

Figure 14. The first Sm–Nd internal isochron for a lunar rock (Apollo 17 basalt 75075; reproduced with permission from Figure 1 of G. Lugmair *et al.*, 1975, *Proc. Lunar Sci. Conf.* 6th, *Vol.* 2, 1419–1429, copyrighted by Pergamon Press)

upper mantle (see Taylor, 1975) divides the dated mare basalts into three groups:

(a) Pre-Imbrium mare basalts with ages $> 3{\cdot}8$ b.y. from shallow depths ($\gtrsim$ 50 km?).
(b) Early Ti-rich basalts from source depths of $\sim$ 100 km.
(c) Late basalts from source depths of $\sim$ 200–400 km.

Basalts of type (a) are rare exceptions but they are nevertheless very significant since they demonstrate that the earliest endogenic mare volcanism *predates* the end of the period of basin formation. The evidence consists of Apollo 11 basalt 10003 (Ar–Ar age 3·91 b.y.) (Stettler *et al.*, 1974); Rb–Sr age 3·84 b.y. (Papanastassiou and Wasserburg, 1975b); Apollo 14 non-KREEP-basalts 14053 and 14167 (3·95 b.y. by both methods, Turner *et al.*, 1971; Papanastassiou and Wasserburg, 1971a; Stettler *et al.*, 1973) and an Apollo 16 basaltic fragment from 66043 (Ar–Ar age: 3·79 b.y.; Schaeffer and Husain, 1973). Basalts of the second type (b) dominate the Mare Tranquillitatis (Apollo 11) and the valley floor of Mare Serenitatis (Apollo 17). At the Apollo 11 site two geochemically distinct basalt types occur which differ in their depth of origin. The 'low K'-basalt ages range from 3·63–3·82 b.y., the high K-ages from 3·51–3·68 b.y. (Turner, 1970; Papanastassiou *et al.*, 1970; Stettler *et al.*, 1974). Apollo 17 basalts cover the time period from 3·66–3·84 b.y. (see Turner, 1977).

In general, mare fillings are distinctly younger than the basins themselves which we have seen to be older than 3·8 b.y. The age ranges are larger than the individual analytical errors and indicate that the mare filling of a large basin has required extended time periods of some 10^8 years. These observations indicate that the basin forming impact itself is *not* the energy source for the generation of the magmas which must be of truly endogenic origin.

Type (c) basalts are represented by the samples from Mare Fecunditatis (Luna 16, 3·42–3·50 b.y.; Huneke *et al.*, 1972; Papanastassiou and Wasserburg, 1972a; Cadogan and Turner, 1976); Mare Imbrium (Apollo 15, 3·13–3·44 b.y., Husain, 1974; Papanastassiou and Wasserburg, 1973), and Oceanus Procellarum (Apollo 12, 3·11–3·30 b.y., see Turner, 1977). Younger basaltic flows ($\sim$ 2·5–3 b.y.) are not documented by accessible samples but their occurrence in Mare Serenitatis and Oceanus Procellarum has been inferred from photogeological observation (Boyce *et al.*, 1974).

In conclusion, the period of mare volcanism lasted at least 800 m.y. from 3·9 to 3·1 b.y., possibly somewhat longer. The individual basins were filled at distinctly different times, very late after their excavation. No evidence exists for widespread endogenic activity on the moon more recently than $\sim$ 3 b.y.

Glasses and soils. The few ages below 3 b.y. which have been found are clearly related to impact processes. Apollo 12 KREEP-glass has ^{39}Ar–^{40}Ar and U–Pb ages of $\sim$ 850 m.y. and was most likely formed when the Crater Copernicus was excavated (Eberhardt *et al.*, 1973; Tatsumoto *et al.*, 1971). Impact produced glass veins and splashes have been dated by the ^{39}Ar–^{40}Ar method and plateau ages are ~ 1·15 b.y. (glass vein in rock 77017, Kirsten and Horn, 1974), 1·3 b.y. (66043, Schaeffer and Husain, 1973); 1·1 b.y. (15465 glass coating, Husain, 1972) and 1·0 b.y. (15015 and 15286 glass coatings, Schaeffer and Husain, 1973). In addition, large scale outgassing events are indicated from the reduced total K–Ar ages of bulk soils. They are frequently lower than the ages of the rocks from which they must have originated and range from 2·6 b.y. (15531) to 4·2 b.y. (14149) (Alexander *et al.*, 1976). The soil ages cluster between 2·5–3 b.y. and between 3·7 and 4·2 b.y. This has been taken as an indication of periods of widespread thermal stress in large regions of the moon, but endogenic processes are most likely not responsible (Pepin *et al.*, 1972).

Consequences for the thermal evolution of the moon. Models for the thermal evolution of the moon (Toksöz and Solomon, 1973; Toksöz *et al.*, 1973; Toksöz and Johnston, 1974) must account for the early chemical differentiation of the crust which began ≥ 4·5 b.y. ago, and for the extrusion of mare basalts up until 1·5 b.y. afterwards. It can be shown that radioactive heat sources in an initially cold moon cannot melt the source rocks of mare basalts within 1·5 b.y. (Toksöz and Solomon, 1973; Toksöz *et al.*, 1973; Toksöz and Johnston, 1974). Moreover, impact generated heat produced during and after the final stage of accretion is insufficient to generate deep seated melts as late as 1 b.y. after formation. On the other hand, if the moon were initially completely molten, all the radioactive heat sources would quickly become enriched in the outer layers. In consequence, the moon would have cooled very rapidly and the majority of crustal rocks would have ages largely in excess of 4 b.y.; mare basalts younger than 3·5 b.y. could not exist. Toksöz and Solomon (1973) have proposed that accretion of the moon occurred quite rapidly (~ 10^3 yr). As a result, the interior remained relatively cool, while gravitational energy released in the final accretion period melted the outer few hundred kilometres. The molten shell differentiated and a solid lithosphere developed on top of it as a result of surface heat losses, giving rise to the oldest crustal rocks. The underlying magma subsequently erupted through the solidified crust to form the mare basalts. This period of mare volcanism overlapped the period of the intense bombardment of the moon which began with the final stages of accretion and ceased ~ 3·8 b.y. ago. Approximately 1·5 b.y. after the moon's formation the solid crust had grown to depths of ~ 400 km and became impenetrable for ascending magmas.

Radioactive heating of the centre of the moon was very slow and core formation probably postdated the end of mare volcanism in the outer lunar regions. A summary of the major epochs of lunar chronology is compiled in Figure 15.

Cosmic ray exposure ages. The interaction of cosmic rays with lunar surface rocks may be used to determine cosmic ray exposure ages in a manner analogous to that discussed in section *Subsequent Meteorite Evolution* for meteorites (Reedy and Arnold, 1972). In the case of lunar samples, exposure ages refer to the time which a rock has spent in the topmost lunar surface. The exposure may be a result of rock ejection in a (mostly local) cratering event or of a downslope

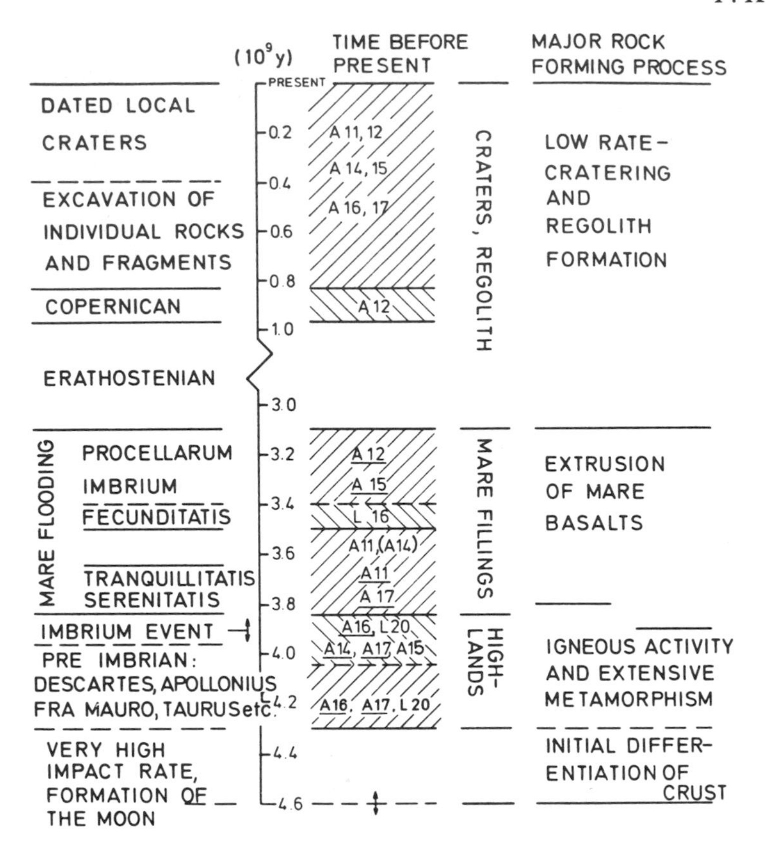

Figure 15. Summary of lunar chronology (modified with permission from Figure 1 of T. Kirsten, 1974, *Raumfahrtforschung*, **5**, 235–237)

movement in an avalanche. In this way, local craters or surface features such as ejecta rays may become datable. In general, this is a complex task since many rocks were involved in more than one local impact. In addition, the accumulated effects of transport processes in the lunar regolith as a result of micro- and macrometeorite influx may increase the effective shielding by soil covers or decrease it by erosion. Exposure ages for individual rocks are therefore model ages which may integrate over various periods of exposure. Such individual exposure ages have been reported for approximately 150 lunar rocks from all missions (for review, see Arvidson *et al.*, 1975). They are mainly ^{81}Kr–^{83}Kr or ^{38}Ar-ages (see section *Subsequent Meteorite Evolution*) and range from ~ 1 m.y. (66095) to 700 m.y. (14160) (Arvidson *et al.*, 1975). About two thirds of the ages are below 150 m.y. and one third between 150 and 700 m.y.

Another type of exposure age rests on the number of etchable radiation damage tracks which are produced by galactic heavy ions, mainly ^{56}Fe (see Walker, 1975). Since their mean range in solids is only ~ 30 g/cm^2, they register exposure only in the top ~ 10 cm of the lunar regolith, compared to galactic protons, which penetrate approximately one metre. The combined application of both methods can thus be used to detect or exclude complex exposure histories. Arvidson *et al.* (1975) have defined useful criteria which should be met if one attempts to date a crater or a lunar surface feature. Rock exposure ages should only be related to a certain feature if either multiple samples related to the same feature yield the same exposure age, or if agreement between ^{81}Kr–^{83}Kr and track exposure age indicate a simple (one stage) exposure history. Dated craters which fulfil this requirement are Apollo 14 Cone Crater (26 m.y.) Apollo 16 North Ray Crater (50 m.y.) and South Ray Crater (2 m.y.). The emplacement of two large boulders at the Apollo 17 site occurred before 22 m.y. (76315) and 28 m.y. (77135) respectively (Arvidson *et al.*, 1975). Ejecta from the crater Tycho yield a probable age for this crater of 96 m.y. (Arvidson *et al.*, 1976).

The exposure ages of lunar soils reflect the average exposure time of the lunar regolith and the degree of maturity of a soil. They range typically from 150 to 450 m.y. (Arvidson *et al.*, 1975). In a simplistic model of continuous turnover, this corresponds to mean turnover rates of 2–5 mm/m.y. Much more detailed studies have been carried out by the analysis of drill core sections (for example, Hübner *et al.*, 1975; Bogard and Hirsch, 1975), but this is not discussed in this article.

Geochronology

The Age of the Earth

Because of its larger size, the earth has evolved more slowly than the moon. Whereas the thermal energy of the moon has been decreasing for $> 2{\cdot}5$ b.y., it is presently near its maximum for the earth. Only a thin lithosphere of $\sim$ 100 km has yet been developed and the earth is tectonically very active. It is, therefore, most unlikely that one could find consolidated crustal rocks of nearly the age of the planet earth. However, the earth's age may be determined from the lead isotope evolution due to the coupled decay of ^{235}U and ^{238}U (Houtermans, 1946).

Let us suppose that a series of uranium-free lead minerals crystallized after t_i years of lead evolution in an unconsolidated source area of fixed (but unspecified) U/Pb ratio. Let us further suppose that the time of ore mineralization relative to the present, that is, the 'conventional' ore age t_ρ,can be determined from dating of rocks which are geochronologically associated with the ore mineralization. We can then apply the ^{207}Pb–^{206}Pb isochron (equation (7)) to determine the time t_i. This is possible for in ^{207}Pb–^{206}Pb dating the U/^{204}Pb ratio applicable for the time period which we want to determine need not be known (equation (7)). All that is required is that the lead evolution in the closed system source area occurred in a simple manner from a known initial Pb-composition: $(^{207}Pb/^{204}Pb)_0$; $(^{206}Pb/^{204}Pb)_0$; at t_0 up to the time t_i when the lead became quantitatively separated from the uranium such that the isotopic composition of the lead in the source area at time t_i was 'frozen' in the mineral and remained unchanged ever since because of the absence of uranium during t_ρ.

The requirement of a 'simple' or one stage evolution seems to be fulfilled for certain classes of essentially uranium-free Pb-minerals from conformable sulphide deposits (Stanton and Russell, 1959). This is concluded from the observation that their Pb-isotopic compositions depend on t_ρ in a manner consistent with the decay schemes of ^{235}U and ^{238}U at least for the experimentally accessible range of t_ρ from 0 up to 3·2 b.y. If we assume that this 'one-stage behaviour' can also be extended backwards in time then the time which was required for the ore lead to evolve from an initial lead characteristic for the whole earth in its initial stage can be calculated. For the composition of the initial lead it is probably the best guess to assume that it was the same as that of the least radiogenic lead found in sulphides from iron meteorites which became U-free very shortly after the formation of the solar system. With this meteoritic initial lead as input value, equation (7) yields t_i as the time between the earth's 'formation' and the mineralization of the conformable lead ore. The earth age is then the sum of t_i and the ore age t_ρ.

A special case would be modern lead with 'ore' age $t_\rho = 0$, and $t_i = t_{earth}$, provided that the modern lead composition (averaged from sediments and crustal rocks) subscribes on an earthwide scale to the single stage evolution requirement.

A recent survey of lead isotope data for conformable galenas and modern lead gave a single stage earth age of 4·43 b.y. (Doe and Stacey, 1974), and a similar result (4·47 ± 0·05 b.y.) was inferred by Gancarz *et al.* (1975) from the lead composition in uranium-poor feldspar from one of the most ancient known rocks, a 3·62 b.y. old Amitsoq gneiss from West Greenland. However, is 4·45 b.y. really the age of the earth? The exact meaning of this number is difficult to specify. Conceptually, $t_\rho + t_i$ measures the time taken for lead to evolve from a uniform composition of initial lead assumed to match that in meteorites 4·57 b.y. ago. In principle the dated event could refer to an earthwide differentiation and Pb-isotope homogenization which postdated the formation of the earth, but the radiogenic lead which must have been generated between 4·57 and 4·45 b.y. has not been found. In other words, terrestrial lead is not consistent with a one-stage evolution from meteoritic primordial lead for 4·57 b.y. This problem has led Stacey and Kramers (1975) to consider a two-stage lead evolution model in which a large scale differentiation occurred in the earth at some time t_d and the age of the earth is *assumed* to be contemporaneous with meteorites, 4·57 b.y. The authors have shown that the available lead isotope data is much more consistent with a two-stage evolution for 4·57 b.y. than with a single stage evolution for 4·43 b.y. The best fit of all data is obtained for $t_d \sim 3{\cdot}7$ b.y. before present. This time could refer to an early large scale mantle differentiation, but the existence of exceptional samples indicates that there was not just one earthwide differentiation. In conclusion, a 4·43 b.y. old earth only *seems* to fit' the data since initial lead at 4·43 b.y. ago must in any case be different from initial lead at 4·57 b.y. ago, a parameter which was needed in deriving the 4·43 b.y. geochron. On the other hand, the age of the earth is consistent with the age of the solar system, 4·57 b.y., even though a model assumption rather than a straightforward determination free of adaptions of a 'known' result from meteorites has led to this conclusion.

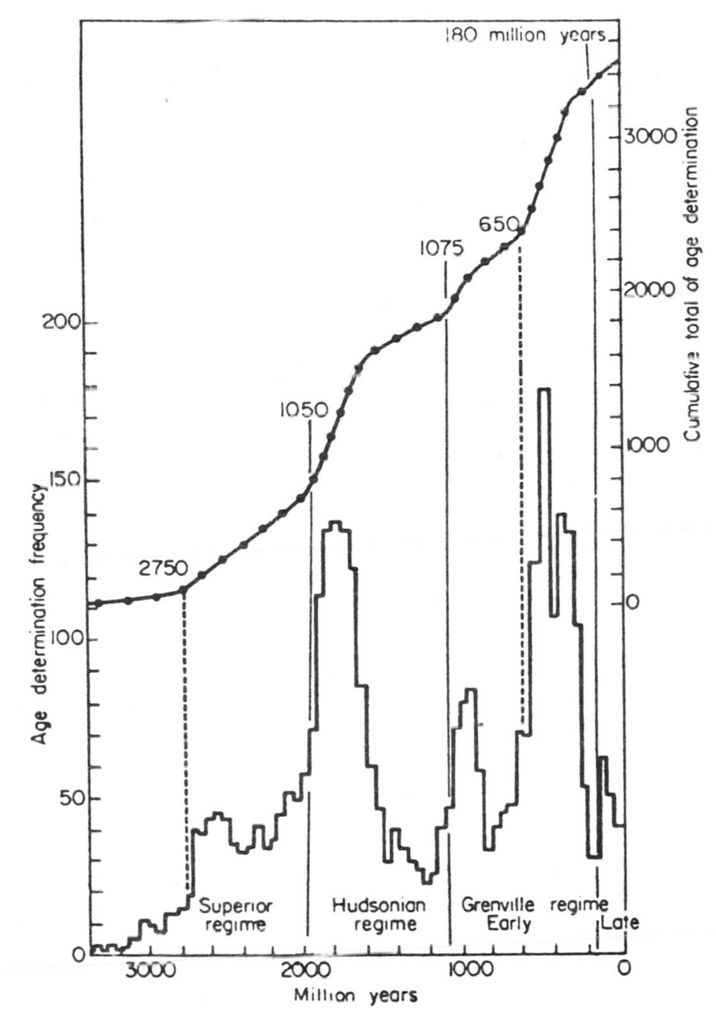

Figure 16. Frequency histogram of radiometric ages of more than 300 metamorphic and igneous terrestrial rocks from all continents (left hand scale) and cumulative number of ages $N(>t)$ (upper curve, right hand scale). Note the preponderance of certain age figures. (Reproduced with permission from Figure 1 of R. Dearnley in *The Application of Modern Physics to the Earth and Planetary Interiors* (S. K. Runcorn, ed.), Wiley–Interscience, London, pp. 103–110

Terrestrial Rock Ages

The vast majority of terrestrial crustal rocks have ages less than 2·8 b.y. (Figure 16), but rocks with ages in excess of 3 b.y. have been discovered in the Archaean shield complexes in North America, Greenland, the Balticum, and South Africa. A large suite of granites and pegmatites from the Monchegorsk pluton, Kola Peninsula, has a range of K–Ar ages between 3·2 and 3·4 b.y. (Gerling *et al.*, 1965). Ultrabasic xenoliths from the same area gave even higher apparent K–Ar ages (Gerling *et al.*, 1965; Kirsten and Müller, 1967), but a more recent investigation applying ^{39}Ar–^{40}Ar dating has revealed that they contain excess ^{40}Ar which has not decayed *in situ*, but was otherwise incorporated (Kaneoka, 1974).

U–Pb dating of zircons from Minnesota and Montana gave upper intercept ages of 3·1–3·3 b.y. (Catanzaro, 1963; Catanzaro and Kulp, 1964; Nunes and Tilton, 1971).

Whole rock Rb–Sr isochrons for gneisses and granites have firmly established an age of 3·5–3·6 b.y. for the Rhodesian basement (Hickmann, 1974, Hawkesworth *et al.*, 1975) and the lower Onverwaacht series in Zwaziland yields Rb–Sr and U–Pb ages ranging from 3·2 to 3·5 b.y. (Sinha, 1972; Jahn and Shih, 1974).* The most ancient rocks, however, have been discovered in the North Atlantic craton, particularly in the Archaean gneiss complex of western Greenland and in Labrador.

Rb–Sr whole rock isochrons for gneisses from the Saglek Bay in Labrador yield ages of 3·62 ± 0·1 b.y. (Hurst *et al.*, 1975). Comparable rocks from western Greenland are even older. Gneisses from the Godthaab area gave consistent whole rock Rb–Sr isochron ages of 3·75 ± 0·09 and 3·74 ± 0·1 b.y. (Moorbath *et al.*, 1972), a whole rock ^{207}Pb–^{206}Pb isochron age of 3·56 ± 0·1 b.y. (Black *et al.*, 1971) and an upper intercept U–Pb age for zircons of 3·59 ± 0·05 b.y. (Baadsgaard, 1973). An age closer to 3·6 b.y. is indicated from the Pb–Pb whole rock data of Baadsgaard *et al.* (1976).

So far, the *oldest* rocks of the earth are supracrustal rocks and Amitsoq type gneisses from the Isua area 150 km northeast of Godthaab. Whole rock Rb–Sr isochron ages are 3·78 ± 0·13 b.y. and 3·7 ± 0·14 b.y. (Moorbath *et al.*, 1975, 1972), and Pb–Pb whole rock isochron ages are 3·7 ± 0·07 b.y. and 3·74 ± 0·12 b.y. (Moorbath *et al.*, 1973, Moorbath *et al.*, 1975).

These data firmly establish the existence of a granitic crust on the earth 3·6–3·7 b.y. ago. At the same time the low initial $^{87}Sr/^{86}Sr$ and $^{207}Pb/^{204}Pb$ ratios in these rocks rule out the possibility that any of their igneous precursors spent more than 200 m.y. in a granitic environment. This indicates that the differentiation of the earth's mantle in this region occurred later than 3·9 b.y. ago. This seems to exclude the possibility that any crustal rocks older than 3·9 b.y. will be found in the future. Instead, we are faced with the probability that no solid rocks have survived the first 800 m.y. of the earth's history. It would then seem that

*All U-Pb ages cited in this section have been adjusted to the new decay constants; see section *Whole Rock Isochrons* and Table 1.

this is the time which was required for the earth to differentiate into core, mantle, and crust as a result of heating by long-lived radioactive isotopes in its interior.

Even after the crust was formed and the continents had developed, the earth remained (and remains) an internally active body. As a result of convective currents in the mantle (eventually influenced by a slowly growing core) the continents were subjected to considerable stress and most areas have been rejuvenated many times by later orogenic processes and volcanic activity. Hence, the ages of today's crustal rocks reflect this subsequent activity.

A statistical analysis of more than 3000 age determinations from all continents revealed the preponderance of certain ages (Figure 16). A cycle time characteristic of orogenic processes may be inferred from this earth-wide rock age distribution, with periods of very intense igneous activity followed by more quiet periods. 'Megacycles' of 800 m.y. and orogenic cycles of 200 m.y. duration seem to have shaped our continents (Sutton, 1963).

The wealth of additional information accumulated in the field of geochronology could fill a library but it will not be covered in the context of this article.

Duration and Sequence of Solar System Formation

The conventional dating techniques have revealed that the solar system formed between 4·5 and 4·6 b.y. ago, but more accurate information on the absolute duration of solar system formation and the sequence of events that occurred cannot be obtained in this way since typical errors of absolute ages are in the order of $\sim$ 50 m.y.

The immediate requirement is for sharper chronological tools to investigate that most exciting time period around 4·57 b.y. ago, which could enable us to resolve small time differences δt between the formation of various planetary objects and to estimate the duration of the formation of the solar system with significantly improved precision. In this section we shall at first give an outline of the principal avenues towards a solution of this problem. The second section is a summary of the most important results.

Measurement of Small Time Differences

Extinct Radioactivities

It was mentioned in the section *Ages of the Elements* that most meteorites contain ^{129}Xe and fission $^{131-136}$Xe as decay products of now extinct ^{129}I and ^{244}Pu and that the ratios $A_{RS} = {}^{129}\mathrm{I}/{}^{127}\mathrm{I}$ and $A_{RS} = {}^{244}\mathrm{Pu}/{}^{238}\mathrm{U}$ are experimentally accessible measures of the time at which the meteorites cooled below the blocking temperature for Xe-retention. Because of the relatively short mean life of ^{129}I and ^{244}Pu, these ratios react very sensitively to small

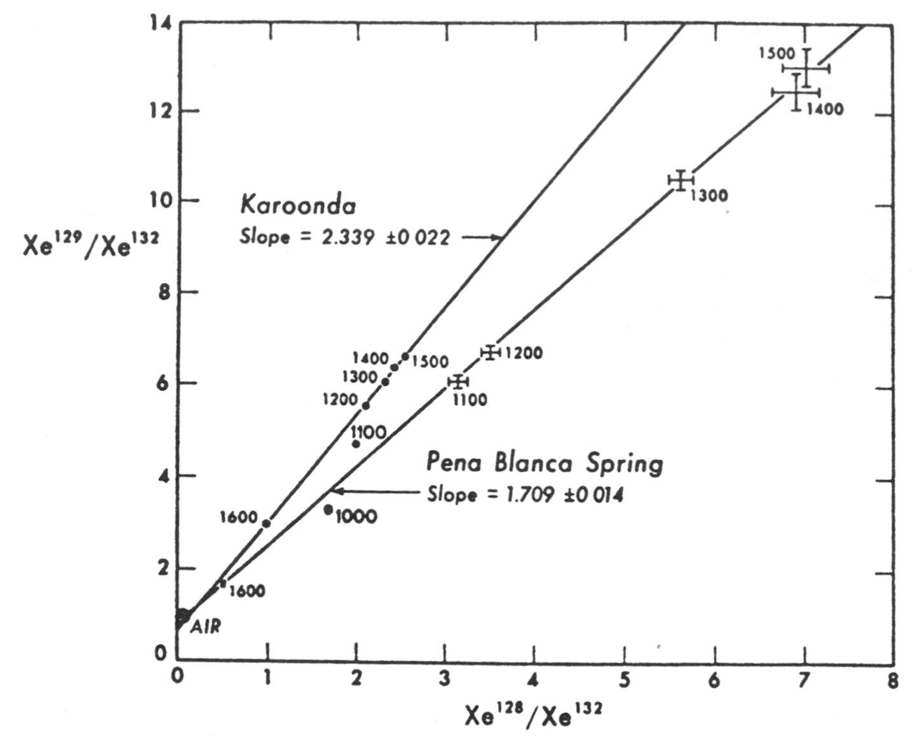

Figure 17. Correlation of ^{129}Xe from extinct ^{129}I and pile produced ^{128}Xe from stable ^{127}I both released in the high temperature (> 1100 °C) fractions of the incremental heating experiment for the Karoonda and Peña Blanca Spring meteorites (see section on *Ages of the Elements*, Podosek, 1970b). Digits indicate release temperatures in °C. ^{132}Xe-normalization serves to abstract from Xe-components unrelated to iodine isotopes. The slopes of the correlation lines yield the $^{129}I/^{127}I$ ratios in the respective meteorites at the time they cooled below their xenon retention temperature. The slope difference of the two meteorites indicates that Karoonda cooled $\delta t = 7{\cdot}5$ m.y. before Peña Blanca Spring. (Reproduced with permission from Figure 5 of F. A. Podosek, 1970, *Geochim. Cosmochim. Acta*, **34**, 341–365, copyrighted by Pergamon Press)

formation time differences δt_{AB} (Figure 17) provided they do not vary because of local inhomogeneities within the solar nebula. By analogy to equation (8) we have

$$\delta t_{AB} = \frac{\ln \dfrac{B_{RS}}{A_{RS}}}{\lambda_R - \lambda_S} \tag{22}$$

where A_{RS} and B_{RS} are the measured ratios in objects A and B. For $^{129}I/^{127}I$,

$\lambda_S = 0$, this reduces to

$$\delta t_{AB} = \tau_{129} \ln\left(\frac{B_{RS}}{A_{RS}}\right) \tag{22a}$$

Since $T_{1/2} = 17$ m.y., a time difference of 17 m.y. corresponds to a factor of 2 in the two measured $^{129}Xe/^{127}I$ ratios. In fact, Xe-isotopic ratios which differ by as little as 2 per cent are still experimentally distinguishable, corresponding to a 'formation' time difference of $\sim 0{\cdot}5$ m.y. These examples illustrate that measurements of the decay products of extinct radionuclides can indicate absolute time differences on the order of 1 m.y. around an arbitrary 'zero point' (by convention the Xenon retention age of the Bjurböle chondrite), in spite of the fact that the absolute time relation of the reference point to the presence is only approximately known (4570 ± 30 m.y.).

Initial Sr-ratios

Within the solar nebula, the isotopic composition of elements which are supported by long-lived radioactive parent isotopes is time-dependent. In principle, this provides an opportunity to resolve condensation time differences between the various planets from a determination of their initial isotopic ratios ${}^iD_0/{}^jD_0$, either in samples free of parent element, or as ordinate intercepts of isochrons (see section above *Types of 'Ages'*). The prime candidates for this approach are strontium and lead isotopes but it has been found that initial lead isotope differences are not time-correlated due to the relatively high mobility of lead and the extremely small magnitude of the effects (Tilton, 1973). On the other hand, Sr isotopes are of practical importance (for example, Papanastassiou and Wasserburg, 1976).

Suppose a sequence of planetary objects p_k condensed at times t_{p_k} from a solar nebula in which the chemical composition was homogeneous in space. Then the initial Sr-composition

$$\left(\frac{^{87}Sr}{^{86}Sr}\right)_{p_k} = \left(\frac{^{87}Sr}{^{86}Sr}\right)_{t_{p_k}}$$

depends on the condensation time t_{p_k} since ^{87}Sr in the solar nebula increases with time due to the decay of ^{87}Rb. Since $\tau_{87} \gg t_{p_k}$, we have

$$\left(\frac{^{87}Sr}{^{86}Sr}\right)_{p_k} = \left(\frac{^{87}Sr}{^{86}Sr}\right)_{po} + \lambda\delta t_{0,k}\left(\frac{^{87}Rb}{^{86}Sr}\right)_{s} \tag{23}$$

where $(^{87}Sr/^{86}Sr)_{po}$ is the initial ratio in an arbitrarily selected reference body for which $t_{po} \sim 4{\cdot}57 \pm 0{\cdot}03$ b.y.; $\delta t_{0,k}$ is the condensation time difference between the reference object and the planet k, and $(^{87}Rb/^{86}Sr)_s$ is the abundance ratio in the solar nebula. Because of $\tau \gg t_{p_k}$ the latter ratio is practically time-independent and easily calculated from the solar elemental abundance ratio Rb/Sr. The best estimate is $(^{87}Rb/^{86}Sr)_s = 1{\cdot}87$ (Lambert and Mallia, 1968; Lambert and Warner, 1968).

Uncertainties in this ratio will proportionally expand or contract the absolute δt-scale, but the relative position of different objects on this scale will not be affected. By analogy to equation (6) we have

$$\delta t_{0,k} = \tau_{87} \frac{\left(\frac{^{87}Sr}{^{86}Sr}\right)_{p_k} - \left(\frac{^{87}Sr}{^{86}Sr}\right)_{po}}{(^{87}Rb/^{86}Sr)_s} \tag{24}$$

or

$$\delta t_{0,k} = 3{\cdot}8 \times 10^4 \left\{ \left(\frac{^{87}Sr}{^{86}Sr}\right)_{p_k} - \left(\frac{^{87}Sr}{^{86}Sr}\right)_{po} \right\} \text{m.y.} \tag{24a}$$

Consequently, a time difference of 3·8 m.y. corresponds to a change in the $^{87}Sr/^{86}Sr$ ratio of 1 part in 10^4, just within the scope of extremely refined mass spectrometric techniques (Gray *et al.*, 1973). The initial ratio $(^{87}Sr/^{86}Sr)_{p_k}$ can be determined as ordinate intercept of a whole rock isochron, provided (a) there is sufficient evidence that the suite of samples was derived from the same body and (b) that either these samples were not differentiated since their condensation or that they belong to a 'cell' which remained a closed system if considered as a whole. Alternatively, $(^{87}Sr/^{86}Sr)$-ratios directly measured in essentially Rb-free phases can yield meaningful upper limits for the initial ratio. If the material underwent differentiation in an open system, t_{p_k} will refer to this differentiation which postdates condensation time. The age figure will still be meaningful in terms of early solar system chronology if there was only one major fractionation between Rb and Sr.

For individual specimens the ordinate intercept of an *internal* isochron has time significance only if there is circumstantial evidence that the sample did not undergo late differentiations in changing Rb/Sr-environments (e.g. undifferentiated chondrites).

High Precision of Classical Formation Ages

Ages near 4·5 b.y. determined by classical age determination methods, i.e. ^{39}Ar–^{40}Ar; Rb–Sr and ^{207}Pb–^{206}Pb, have typical errors of $\sim$ 50 m.y. On an absolute scale there is little hope of improving this situation because of uncertainties in the decay constants and interlaboratory bias. However, if a suite of samples is analysed by just one method in one laboratory under completely analogous conditions, (identical equipment, calibration, irradiation if applicable) it may be possible to measure significant age differences between two objects of the order of a few million years.

Early Solar System Chronology

Extinct Radioactivities

Reliable ^{129}I–^{129}Xe ages have been determined for more than 15 meteorites and the results are summarized in Figure 18. All major meteorite types are

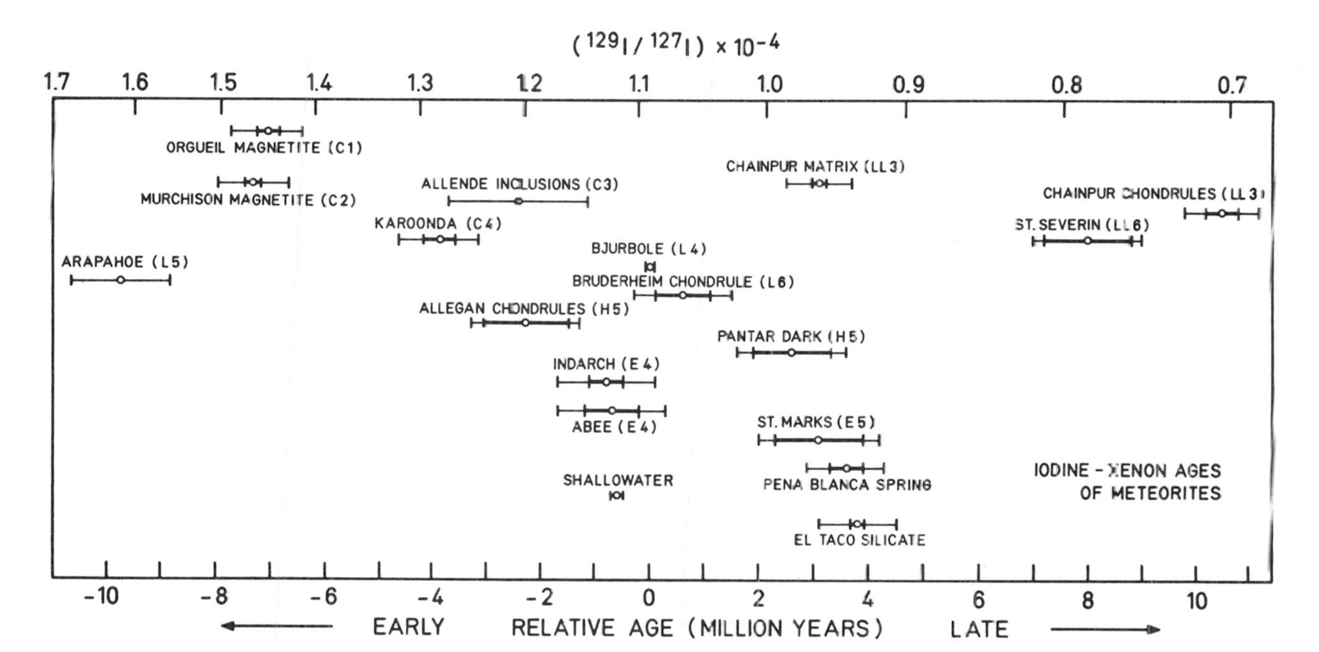

Figure 18. Summary of I–Xe ages of meteorites relative to Bjurböle, The respective $^{129}I/^{127}I$-ratios are indicated on the top scale. Basic data from Podosek 1970b, with additions for Orgueil and Murchison (Lewis and Anders, 1975); Allende (Podosek and Lewis, 1972), and Arapahoe (Drozd and Podosek, 1976). (Modified with permission from Figure 8 of F. A. Podosek, 1970, *Geochim. Cosmochim. Acta*, **34**, 341–365, copyrighted by Pergamon Press)

included, from the most primitive carbonaceous chondrite Orgueil up to the highly differentiated iron meteorite El Taco. The data indicate that the total time span for the onset of Xe-retention in all these different objects was only ~ 20m.y. It is surprising, however, that there seems to be no 'genetic' order in the age sequence (Podosek, 1970b). From the geochemical and petrographic evidence it is believed that carbonaceous chondrites like Orgueil are primitive condensates which were never heated above ~ 500 K. On the other hand, most ordinary chondrites have probably been reheated above the Xe-blocking temperature (~ 1000 K) after condensation and aggregation within their parent bodies. Depending on the size of the parent bodies and the accretion and cooling rates, they were subjected to different degrees of metamorphism which are reflected in the classification scheme (Van Schmus and Wood, 1967), where the degree of metamorphism increases with classification number from 1 to 6. The igneous textures of achondrites clearly demonstrate that the latter underwent magmatic differentiation within their parent bodies, irrespective of any earlier differentiation which may have taken place when the parent bodies condensed and accreted from the gas phase. Similar considerations apply to iron meteorites. On this basis one would expect that the I–Xe ages of the type 1 carbonaceous chondrites (e.g. Orgueil) should be the oldest since they date the condensation phase, whereas highly metamorphised chondrites (type 4–6) and differentiated meteorites should have lower I–Xe ages simply because some time must be allowed for accretion, reheating and cooling below the Xe-blocking temperature. In fact, however, the L 5 chondrite Arapahoe has an I–Xe rentention age which is 3 m.y. older than that of the C1-chondrite Orgueil (Figure 18). This could be explained by real condensation time differences of the two parent bodies and very fast accretion and metamorphism in the Arapahoe parent body. It would imply, however, that some objects were already accreted and reheated while others were still condensing, possibly in different locations of the solar nebula. In addition, the L 4 chondrite Bjurböle is 10 m.y. younger than the L 5 Arapahoe, which contradicts a metamorphism time of less than 10 m.y. if both L-chondrites are from the same parent body. Also, I–Xe ages should decrease, but not increase with increasing metamorphic grade for meteorites which are believed to originate from the same parent body. Altogether it must be stated that the influence of the post-accretional thermal history as reflected in the I–Xe ages is not understood, and it is not clear whether the different metamorphic grades are caused by reheating or by different cooling rates from the high initial temperatures supplied by the collapse of the solar nebula (Pellas and Storzer, 1976). The best approach is therefore to focus on objects which meet criteria for being early condensates, apart from the important and firm generalization that all major meteorite types have condensed, accreted and cooled contemporaneously within a time span of 20 m.y. only. It should be stressed, however, that *all* I–Xe ages are exclusively based on the highly retentive iodine-bearing phases, for which the I–Xe *in situ* correlation is preserved in the stepwise heating experiment. One may ask whether these 'magic sites' remained unaffected during the moderate thermal metamorphism experienced by ordinary chondrites and consequently still date the time of condensation. Since this is an open question, the highest confidence that I–Xe ages refer to condensation time, is achieved for the carbonaceous chondrites never heated above 500 K and for high temperature condensates like the Ca–Al-rich inclusions found in the Allende meteorite. Because of their primitive nature, these objects are often inhomogenous and unequilibrated. Intimate mixing of constituents which may have condensed at different times creates scatter in the I–Xe correlation upon thermal release and prevents the extraction of a single I–Xe age. However, all such objects for which a reliable I–Xe age has been obtained (Orgueil, Karoonda, Murchison, Allende; see Figure 18 for References) have ages between 7 and 2 m.y. before Bjurböle, hence 5 m.y. may be a good estimate for the time during which condensation took place in the solar nebula.

For the most part the subsequent consolidation phase of meteorite parent bodies has lasted approximately 20 m.y. A few objects experienced a much later metamorphism judging from their absolute ages (see section on *Radiometric Ages*). In particular, the eucrites Petersburg and Lafayette have no excess ^{129}Xe corresponding to I–Xe ages of > 20 m.y. and > 66 m.y. respectively (Podosek, 1972).

In the context of the resolution of small time differences on a m.y. scale, fission-Xe from extinct ^{244}Pu is of little value because of the relatively long half life of ^{244}Pu and because of Pu/U-fractionations. Only in late metamorphosed objects like Petersburg is it possible to infer useful Pu–Xe ages (138 ± 15 m.y. after Bjurböle; Podosek, 1972). It should be mentioned, however, that within the large range of uncertainty of the Pu–Xe data none is inconsistent with the I–Xe time scale.

Fission tracks from extinct ^{244}Pu may be used to infer cooling rates of chondrites in the temperature regime below the xenon blocking temperature on the basis of their annealing properties (Pellas and Storzer, 1976). For instance, the Pu–Xe age of a uranium rich phosphate mineral from the St. Severin meteorite is 380 m.y. higher than the Pu-fission-track age. This implies that 380 m.y. of cooling were required between the blocking temperature for Xe-retention and the track retention temperature in this mineral.

One may also ask whether Pu–Xe (and I–Xe) could be used to link the chronologies of the meteorite parent bodies, the moon, and the earth. The problem here rests in the scarcity (or absence) of samples sufficiently old to contain extinct radioactivities from *in situ* decay. In one 3·95 b.y. old lunar breccia, 14321, a Pu–Xe age (and a Pu-track age) has been measured (Marti *et al.*, 1973; Hutcheon and Price, 1972). The result is consistent with a contemporaneous origin of the moon and the meteorites but again there are large uncertainties because of the possibility of geochemical fractionation of Pu and U. With respect to the earth, excesses of ^{129}Xe and (presumably) ^{244}Pu-fission-Xe have been observed both in well gases and in the atmosphere (Butler *et al.*,

1963; Kuroda and Manuel, 1962; Boulos and Manuel, 1971; Marti *et al.*, 1970). While this fact clearly establishes that ^{129}I and ^{244}Pu were not extinct when the earth had finally accreted, involved assumptions are required to calculate I–Xe or Pu–Xe ages for the earth since the *in situ* relationship between radioactive parent and daughter is lost in the process of outgassing. However, if one considers the earth as a whole it can be estimated from the isotopic composition of atmospheric Xe that 6–10 per cent of the total ^{129}Xe is due to ^{129}I (Pepin and Phinney, 1976). Under the assumptions that the earth is nearly completely outgassed and that Xe has been quantitatively retained in the atmosphere, it follows that the earth stopped accreting 96 ± 10 m.y. after Bjurböle. A similar consideration for Pu–Xe yields 67 ± 20 m.y. (Pepin and Phinney, 1976).

Short-lived Extinct Isotopes and an Alternative Interpretation of Extinct Radioactivities

Attempts have been made to detect additional extinct radioisotopes in meteorites other than ^{129}I and ^{244}Pu. Huey and Kohman (1972) searched for ^{205}Tl from the decay of ^{205}Pb ($T_{1/2} \sim 20$ m.y.) and Lugmair *et al.* (1975b) tried to find ^{142}Nd excess from the decay of the p-process nuclide ^{146}Sm ($T_{1/2} = 106$ m.y.), but no detectable anomalies have been found in either case. On the other hand, there is one observed anomaly for which it has so far not been possible to identify an extinct parent isotope. This is a fission xenon excess of particular isotopic composition which is frequently found in carbonaceous chondrites in rather large quantities (Pepin, 1967). Apart from speculations about a superheavy progenitor, the *r*-process nuclide ^{248}Cm has been suspected as a possible explanation (Rao and Gopalan, 1973). In the meantime, Leich *et al.* (1977) have shown that the isotopic composition of ^{248}Cm fission-Xe is distinctly different from that of carbonaceous chondrite fission-Xe. Another candidate, ^{250}Cm has a relatively short half-life of $1{\cdot}1 \times 10^4$ yr, and its survival in condensed matter would contradict the I–Xe formation intervals of $\gtrsim 100$ m.y. If production of nuclides admixed to the solar reservoir ceased more than 100 m.y. before solar system formation, Cm would have been extinct long before condensation began. The only way to retain the Cm-fission hypothesis is to postulate the existence of interstellar dust grains which (a) are older than the solar system, (b) were not homogenized with the bulk of the solar material, and (c) became incorporated during accretion of the carbonaceous chondrites. This would violate the long accepted doctrine of a completely homogenized solar nebula, but in recent years it has become increasingly clear that isotopic inhomogeneities did indeed exist.

R. N. Clayton has convincingly demonstrated that carbonaceous chondrites, ordinary chondrites, and differentiated objects such as achondrites, iron meteorites, earth, and moon have been derived from sources with distinctly different isotopic composition of oxygen (R. N. Clayton and Mayeda, 1975; R. N. Clayton *et al.*, 1976). Related evidence exists for the primordial abundances of neon and xenon isotopes (Eberhardt, 1974; Manuel *et al.*, 1972). These observations require inhomogeneities in the primitive solar nebula either because of incomplete mixing of the gaseous nebula or because of various proportions of admixed interstellar grains in different regions of the nebula. D. D. Clayton (1975a, 1975b) has advocated the latter alternative and presented arguments that interstellar grains can condense on a time scale of years within the expanding envelopes of explosive stars or Supernovae in which heavy elements were synthesized (D. D. Clayton and Hoyle, 1976). This would automatically relax the time constraint on the observability of short-lived extinct nuclides imposed by the conventional interpretation of I–Xe formation intervals. In fact, D. D. Clayton suggests that essentially all the observed extinct radioactivities are due not to the *in situ* decay after condensation in the solar system, but rather to the *in situ* decay after condensation of solid grains immediately after local nucleosynthesis. This would, of course, imply that the formation intervals and the small formation time differences deduced from extinct radioactivities would need reinterpretation. If interstellar grains with isotopic compositions characteristic for the particular nucleosynthetic source occurred in different proportions in various regions of the solar nebula, it would not be ruled out that they became vaporized on a local scale. However, the observed high temperature correlation between iodine and excess ^{129}Xe and the observation of excess ^{244}Pu-fission tracks (typical track length $\sim 10\,\mu$m) would require the survival of solid interstellar grains of at least $10\,\mu$m size. So far, it has not been possible to observe such grains directly which leaves the whole subject an open problem. D. Clayton (1975a) has suggested that carbonaceous chondrite fission Xe is also of this origin, and in considering this picture, Howard *et al.* (1975) identified ^{248}Cm and ^{250}Cm as the most likely parents.

The solution to this problem would bear on the interpretation of recently discovered Mg-isotopes anomalies. Lee *et al.* (1976) have found a 1·3 per cent enrichment of ^{26}Mg in a Ca–Al-rich chondrule of the Allende meteorite, which could be due to the *in situ* decay of extinct ^{26}Al ($T_{1/2} = 0{\cdot}74$ m.y.). The subsamples exhibit a correlation between ^{26}Al-excess and the Al/Mg-ratio. Three interpretations are possible for these observations:

(a) The ^{26}Mg excess is introduced by an unhomogenized presolar component which is rich in aluminium. The material escaped subsequent recrystallization with the bulk of the meteorite. In this picture one might expect other isotopic anomalies in early condensates such as an ^{41}K excess in Ca-rich material from the decay of $1{\cdot}3 \times 10^5$ yr ^{41}Ca produced in nucleosynthesis. No such excess was found by Begemann and Stegmann (1976) in a Ca-rich inclusion of the Allende meteorite.

(b) The ^{26} Mg excess is due to *in situ* decay of ^{26}Al which was produced within the solar system by a nuclear reaction either in the gas or in condensed dust. Heymann and Dziczkaniec (1976) have shown that 10^{22} protons/cm^2 in the energy range between 1 and 10 MeV acting on the gas could achieve this goal, but it seems not possible for protons of this energy to penetrate the solar nebula. On the other hand, if dust was bombarded by low energy protons, no significant anomaly would result (Heymann and Dziczkaniec, 1976). If high energy protons incident on dust are considered, spallation reactions on other target nuclides should result in other isotopic anomalies which are not observed.

(c) The ^{26}Mg excess is due to the *in situ* decay of ^{26}Al which was synthesized shortly before the condensation of the solar nebula. This would require an admixture of a few percent of freshly synthesized elements. The measured $^{26}Al/^{27}Al$ ratio in the Allende chondrule is $0{\cdot}6 \times 10^{-4}$ (Lee *et al.*, 1976). The expected production ratio is 10^{-3}, hence the Al-formation interval cannot be longer than 4 half lives or ~ 3 m.y. The discrepancy with the I–Xe formation interval of $\gtrsim 100$ m.y. would either imply that Xe-retention began ~ 100 m.y. after the condensation of the Al-component or that ^{26}Al and ^{129}I were synthesized in independent nucleosynthetic events. If a late spike of freshly synthesized elements immediately before the onset of condensation did indeed introduce the ^{26}Mg-anomaly it could be the long-sought heat source needed to melt km-sized parent bodies and hence to account for the differentiated meteorites (achondrites, iron meteorites).

Strontium Isotopes

In the case of I–Xe dating it was necessary to distinguish between primary condensates and evolved objects. A similar situation is encountered in Sr-isotope chronology. A measured initial $(^{87}Sr/^{86}Sr)_i$ ratio may define the value of this ratio in the solar nebula at the time of condensation, or it could result from a subsequent Sr isotope equilibration, e.g. from magmatic differentiation or thermal metamorphism. In any case, the $^{87}Sr/^{86}Sr$ ratio is monotonically increasing with time, whether in the solar nebula or in condensed matter, and only the rate of increase is affected by complex Rb–Sr differentiations. For this reason, one is always hunting for the 'most primitive' strontium (that is, the lowest $(^{87}Sr/^{86}Sr)_i$ which should bear on the time of the earliest condensation.

If a possible or real differentiation occurred in an environment in which the Rb/Sr ratio was strongly depleted relative to the solar nebula, it may turn out that the increase of $(^{87}Sr/^{86}Sr)_i$ between condensation and re-equilibration is insignificant and the initial ratio of the condensation phase is practically preserved. For this reason, Rb-poor objects are the best candidates to get information about the condensation itself (Gray *et al.*, 1973). In addition, there is the practical ground that the ordinate intercept of an isochron (Rb = 0) is more accurately determined if the measured points are nearby the ordinate (quite contrary to the isochron age, for which a large spread in Rb/Sr is desirable). The ideal case would be one in which the 'ordinate intercept' is directly measured since the sample is completely free of Rb.

It is true that such initial ratios need not be very primitive if the sample became Rb-free as a result of a late differentiation from a Rb-rich source, but in

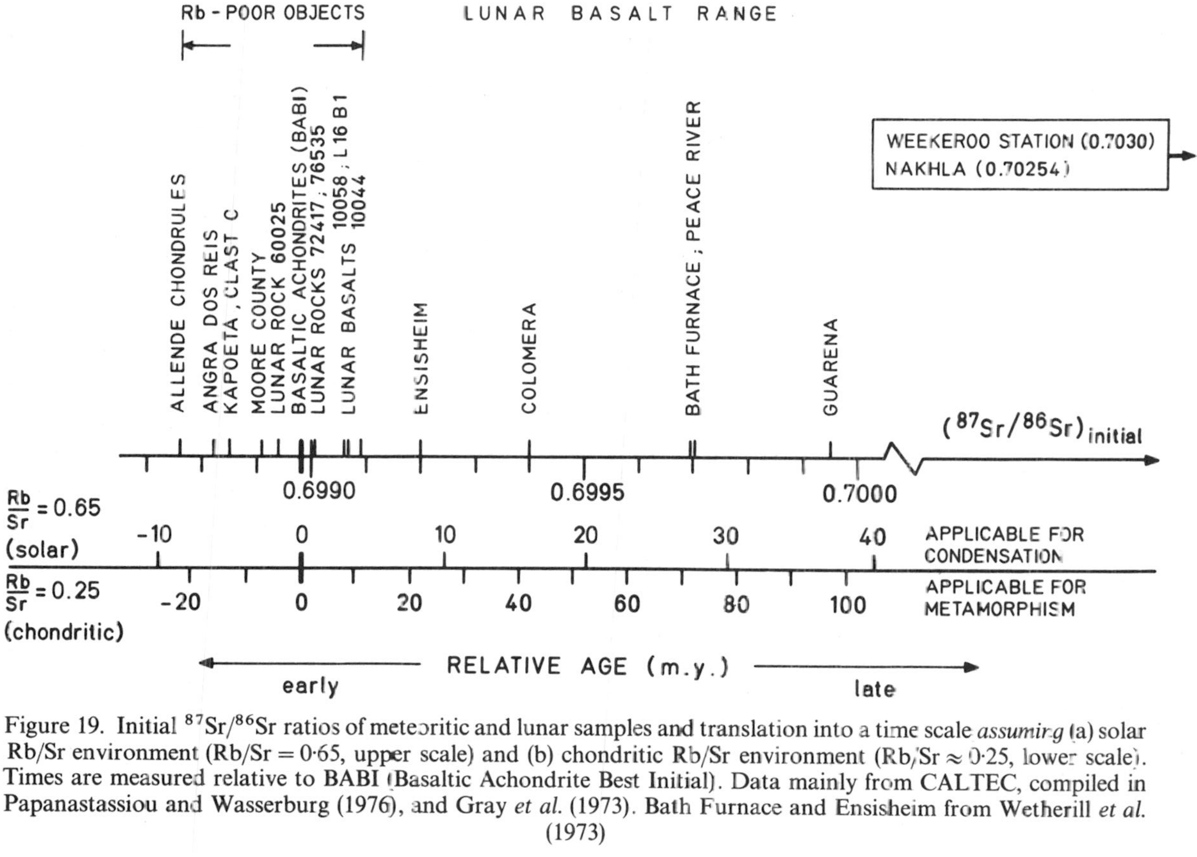

Figure 19. Initial ^{87}Sr/^{86}Sr ratios of meteoritic and lunar samples and translation into a time scale *assuming* (a) solar Rb/Sr environment (Rb/Sr = 0·65, upper scale) and (b) chondritic Rb/Sr environment (Rb/Sr ≈ 0·25, lower scale). Times are measured relative to BABI (Basaltic Achondrite Best Initial). Data mainly from CALTEC, compiled in Papanastassiou and Wasserburg (1976), and Gray *et al.* (1973). Bath Furnace and Ensisheim from Wetherill *et al.* (1973)

some cases it may indeed be very low because the Rb-depletion occurred during condensation. In addition, even if the depletion occurred in the condensed state, an informatively low value may be obtained if the evolution required only a relatively short time period. This is so since a directly measured low initial Sr ratio is always an upper limit for $^{87}Sr/^{86}Sr$ at the time of condensation of the parent material. Figure 19 is a compilation of reliably determined initial Sr-ratios for meteoritic and lunar samples. The absolute scale of this ratio is transformed into time scales according to equation (24). For the time to the left of a given object it is not clear, a priori, in what kind of Rb/Sr environment the Sr isotope evolution occurred. Two scales are given, one for the Rb/Sr ratio characteristic for the solar nebula, and one for a chondritic Rb/Sr environment. The zero point is arbitrary (like that for Bjurböle in I–Xe dating). It is the ratio defined by a whole rock isochron of seven basaltic achondrites (depleted in Rb/Sr). The acronym for this ratio is BABI (*B*asaltic *A*chondrites *B*est *I*nitial). All ratios near and below BABI apply to objects for which the Rb/Sr ratio is substantially below the solar and the chondritic value. In particular, the lunar Rb/Sr ratio is about 40 times lower than the chondritic ratio. Thus, the source region of the lunar mare basalts is characterized by rather primitive initial Sr-ratios, and the same applies to the two oldest lunar rocks, dunite 72417 and troctolite 76535 (Figure 19). The absolute lunar minimum ratio for anorthosite 60025 makes it possible to conclude that protolunar material condensed about 3 m.y. before the basaltic achondrites. Even earlier condensation must be concluded for the achondrites Moore County, Kapoeta, and Angra dos Reis. The most primitive Sr of all analysed materials was found in Rb-poor chondrules of the Allende chondrite (Gray *et al.*, 1973) which, as a whole, is not Rb-poor. This leads to the important conclusion that at least some chondritic material must have condensed very early, even before the basaltic achondrites. Consequently, the generally high initial values obtained for ordinary chondrites and silicate inclusions from iron meteorites (Figure 19) are best explained as the result of a late stage Sr-re-equilibration in high Rb (chondritic) parent bodies, some 10–100 m.y. after the onset of condensation in the solar nebula. This raises the question, however, why the total time span of ordinary chondrite evolution determined from I–Xe ages is at least five times shorter (Figure 18 above). Unfortunately, there is at present no *ordinary* chondrite for which both Sr-isotope data and I–Xe ages are determined, but cross comparisons for members of the same type are possible and it is quite probable that this discrepancy will remain a general feature. If one excludes a disturbance by global isotopic inhomogeneities within the solar nebula, it must be assumed that moderate temperatures during cooling or metamorphism are sufficient to re-equilibrate Sr isotopes, but not sufficient to initiate Xe-losses from the high temperature phases for which the I–Xe correlation is established. This observation makes it problematic to connect the Sr-isotope time scale with the I–Xe time scale, even though there is one single object for which both types of data exist, the Allende carbonaceous chondrite. On the other hand, since Allende has the lowest of all initial Sr-ratios, condensation rather than thermal metamorphism should be responsible for both ages, particularly since the high temperature I–Xe correlation was found to be *more* temperature resistant than Sr-isotopes. Altogether, the picture is quite consistent for the period of *condensation*. If this period comprises the low Rb-objects between Allende and the moon (Figure 19), it must have lasted for about 10 m.y., in reasonable agreement with the period of condensation as inferred from the I–Xe chronology.

Precise Classical Ages

If a whole rock isochron for a group of meteorites is determined with ultimate analytical precision and the analytical uncertainty assigned to an individual data point is less than its deviation from the isochron, the deviation may indicate a real age difference for that meteorite. However, as soon as one meteorite is considered individually, the respective age becomes a model age with an assumed initial ratio, and apparent age differences may in fact be caused by

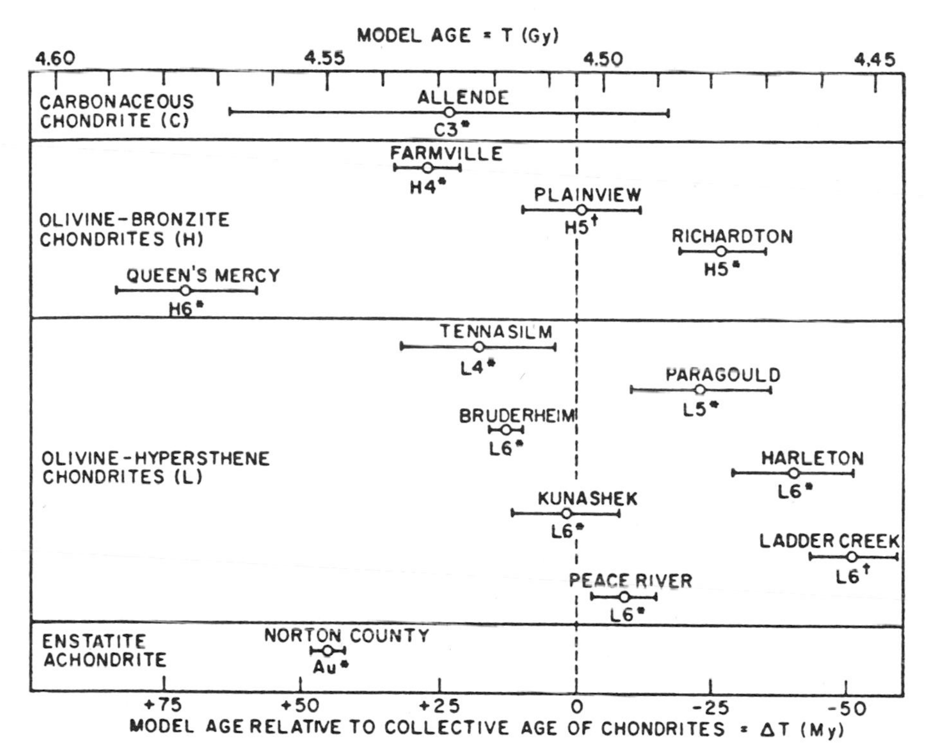

Figure 20. ^{207}Pb-^{206}Pb model ages of individual stone meteorites (upper scale, 1 Gy = 10^9 years) (Huey and Kohman, 1973). The deviation of individual Pb–Pb ages from the common whole rock isochron at 4·505 b.y. is indicated on the lower scale. (Reproduced with permission from Figure 2 of J. M. Huey and T. Kohman, 1973, *J. Geophys. Research*, **78**, 3227–3232, copyrighted by American Geophysical Union)

different initial ratios of the compared objects. This is critical in Rb–Sr dating where the total radiogenic ^{87}Sr rarely exceeds 10 per cent of the total ^{87}Sr. In U/Th dating the same problem exists for most chondrites, in particular for carbonaceous chondrites because of their very low U/Pb ratios. Huey and Kohman (1973) have calculated ^{207}Pb–^{206}Pb model ages for 12 ordinary chondrites and were able to resolve individual *apparent* age differences as small as ~ 5 m.y. (Figure 20).* The deviations of the individual Pb–Pb ages from the common whole rock isochron at $4{\cdot}505 \pm 0{\cdot}008$ b.y. cover an interval of about 50 m.y. in agreement with similar findings of Tatsumoto *et al.* (1973). The model age of the achondrite Norton County is ~ 45 m.y. older than the mean chondrite age. It is possible that the age sequence from Figure 20 reflects at least in part the response of the U/Pb system to a late stage metamorphism. The time span is comparable to that deduced from initial Sr-ratios (see section on *Strontium Isotopes*) and the degree of metamorphism increases towards younger ages. However, all this could be fortuitous and the scatter could be due to variations in the initial lead composition.

For achondrites, the situation is quite different. Tatsumoto *et al.* (1973) have

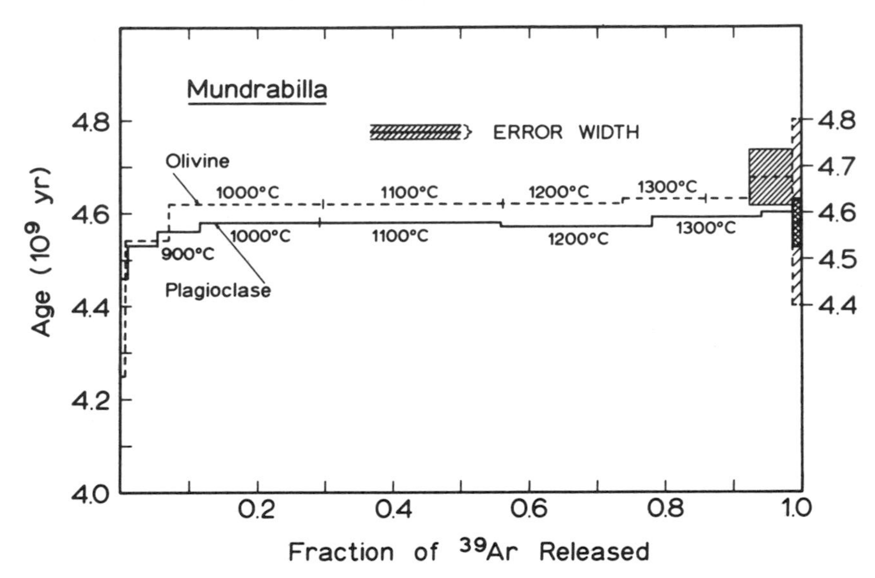

Figure 21. ^{39}Ar–^{40}Ar age spectrum (see Figure 4) for olivine and plagioclase mineral separates from the Mundrabilla iron meteorite, irradiated and measured under identical experimental conditions. The age difference between olivine ($4{\cdot}62 \pm 0{\cdot}02$) b.y. and plagioclase ($4{\cdot}58 \pm 0{\cdot}02$) b.y. is significant and is probably related to the cooling time elapsed between the Ar-retention temperatures of the two minerals. (Reproduced with permission from Figure 1 from T. Kirsten, 1973, *Meteoritics*, **8**, 400–402)

*Note added in proof: M. S. Lancet and T. Kohman have remeasured Queen's Mercy and obtained a Pb–Pb single stage model age of 4·502 b.y. The authors also feel that the 'collective age of chondrites' is too low a reference point since the highly metamorphosed ('younger') chondrites dominate the isochron (T. Kohman, private communication, and in preparation).

determined very precise whole rock model ages for Sioux County, Nuevo Laredo, and Angra dos Reis. The U/Pb-ratio in these meteorites is so high that more than 90 per cent of the lead is radiogenic and the initial lead composition becomes unimportant. For this reason, the age difference of 26 ± 8 m.y. between Angra dos Reis ($4{\cdot}555 \pm 0{\cdot}005$ b.y.) and Nuevo Laredo ($4{\cdot}529 \pm 0{\cdot}005$ b.y.) is a significant result. It compares reasonably with the Sr-isotope chronology, which gave 15 m.y. time difference between the formation of basaltic achondrites and Angra dos Reis (for chondritic Rb/Sr, Papanastassiou and Wasserburg, 1976; Figure 19 above). This concordance may indicate that the achondrites condensed in a very early stage of the solar system's evolution.

Another possibility for measuring small formation age differences lies in the determination of Ar–Ar plateau ages in samples which were irradiated together

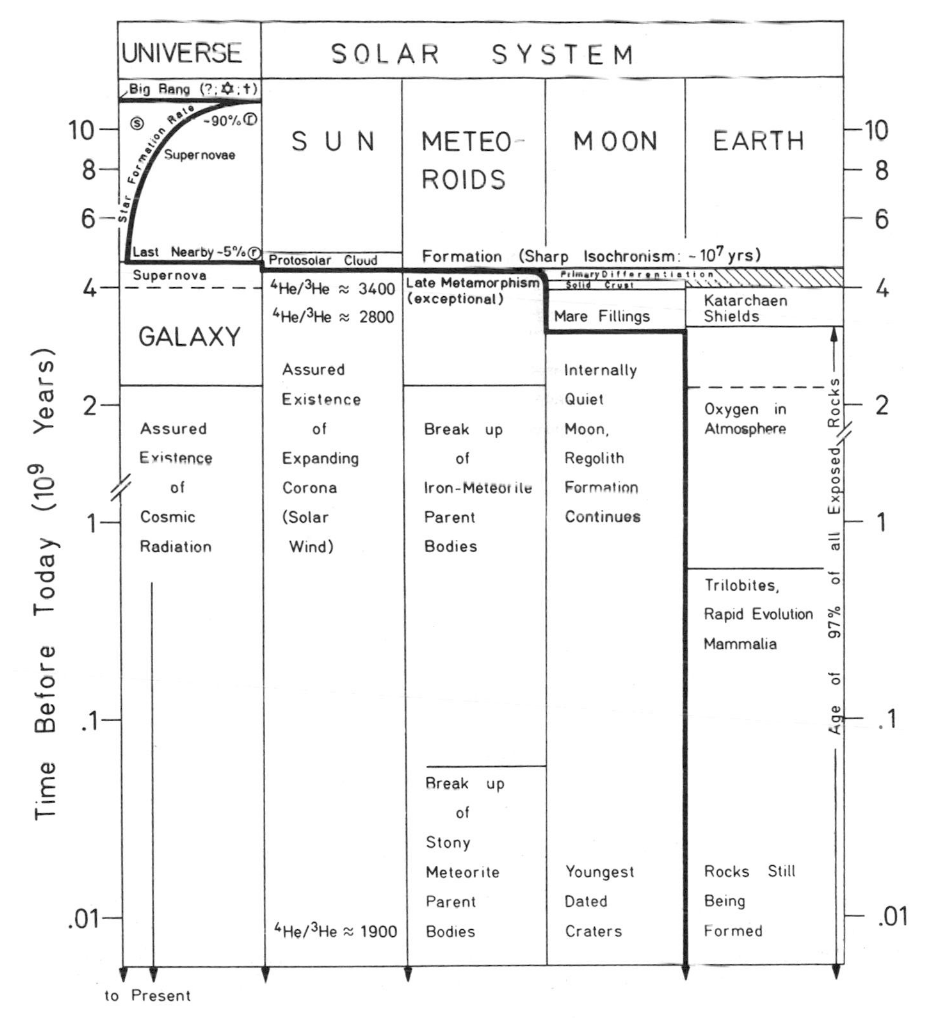

Figure 22. Schematic summary of a cosmochronological time table (logarithmic scale)

under identical conditions. Figure 21 gives an example for a significant difference of ~ 40 m.y. in the gas retention ages of olivine and plagioclase separates from the inclusions of the Mundrabilla iron meteorite. The time difference reflects the cooling time which elapsed between the Ar-blocking temperatures of olivine and plagioclase.

Concluding Remark

The foregoing chapter has been an attempt to demonstrate how radiometric dating has contributed to our present understanding of the origin and evolution of the solar system. The time scale which has emerged from such studies is summarized in Figure 22.

Acknowledgements

I wish to thank Professor Gregory Herzog for stimulating discussions and valuable comments on the manuscript, Prof. G. J. Wasserburg for helpful comments, and Heidi Urmitzer for stylistic improvements and her patience in preparing the manuscript.

Note added in proof

Before this volume went into print, a number of publications appeared which are closely related to specific topics of this chapter. Some of them which bear directly on the context are listed below together with the page number to which their content is related:

page 292: D. Bogard, and L. Husain (1977). 'A new 1·3 Aeon-young achondrite.' *Geophys. Res. Letters*, **4**, 69–71.

page 294: D. Bogard, L. Husain, and R. J. Wright (1976). '$^{40}Ar-^{39}Ar$ dating of collisional events in chondrite parent bodies.' *J. Geophys. Res.*, **81**, 5664–5678.

page 304: E. Jessberger, B. Dominik, T. Kirsten, and Th. Staudacher (1977). 'New $^{40}Ar-^{39}Ar$ ages of Apollo 16 breccias and 4·42 AE old anorthosites.' *Lunar Science*, **VIII**, 511–513.

page 314: S. Guggisberg, P. Eberhardt, J. Geiss, N. Grögler, and A. Stettler (1977). 'Youngest and oldest mare basalts: The temporal extent of mare filling.' *Lunar Science*, **VIII**, 386–388.

page 328: N. Scheinin, G. Lugmair, and K. Marti (1976). 'Sm–Nd systematics and evidence for extinct ^{146}Sm in an Allende inclusion.' *Meteoritics*, **4**, 1976; see also *Lunar Science*, **VIII**, 619–621 (1977).

page 329: D. D. Clayton (1977). 'Solar system isotopic anomalies: Supernova neighbor or presolar carriers?' Submitted to *Icarus*.

page 329: T. Lee, D. Papanastassiou, and G. Wasserburg (1977). 'Aluminum 26 in the early solar system: Fossil or fuel.' *Astrophys. J.*, **211**, L 107–110.

References

Adorables, E. (1976). 'Progress by the consorts of Angra dos Reis.' *Lunar Science*, **VII**, 443–445.

Aldrich, L., Wetherill, G., Tilton, G., and Davis, G. (1956). 'Half life of ^{87}Rb.' *Phys. Rev.*, **103**, 1045–1047.

Alexander, E., Bates, A., Coscio, M., Dragon, J., Murthy, V., Pepin, R., and Venkatasan, T. (1976). 'K/Ar dating of lunar soils II.' *Proc. Lunar Sci. Conf. 7th, Vol. 1*, 625–648.

Anders, E. (1971). 'Interrelations of meteorites, asteroids and comets.' In: *Physical Studies of Minor Planets* (T. Gehrels, ed.) NASA SP-267, Washington, D. C., 429–446.

Arvidson, R., Crozaz, G., Drozd, R., Hohenberg, C., and Morgan, C. (1975). 'Cosmic ray exposure ages of features and events at the Apollo landing sites.' *Moon*, **13**, 259–276.

Arvidson, R., Drozd, R., Guinness, E., Hohenberg, C., and Morgan, C. (1976). 'Cosmic ray exposure ages of Apollo 17 samples and the age of Tycho.' *Proc. Lunar Sci. Conf. 7th, Vol. 3*, 2817–2832.

Baadsgaard, H. (1973). U-Th-Pb dates on zircons from the early precambrian Amitsoq gneisses, Godthaab District, West Greenland. *Earth Planet. Sci. Letters*, **19**, 22–28.

Baadsgard, H., Lambert, R., and Krupicka, J. (1976). 'Mineral isotopic age relationships in the polymetamorphic Amitsoq gneisses, Godthaab District, West Greenland.' *Geochim. Cosmochim. Acta*, **40**, 513–527.

Begemann, F., and Stegmann, W. (1976). 'Implications from the absence of a ^{41}K anomaly in an Allende inclusion.' *Nature*, **259**, 549–550.

Begemann, F., Weber, H., Vilcsek, E., and Hintenberger, H. (1976). 'Rare gases and ^{36}Cl in stony iron meteorites: cosmogenic elemental production rates, exposure ages, diffusion losses and thermal histories.' *Geochim. Cosmochim. Acta*, **40**, 353–368.

Birck, J. L., Minster, J. F., and Allegre, C. (1975). '^{87}Rb–^{87}Sr chronology of achondrites.' *Meteoritics*, **10**, 364–365.

Black, L., Gale, N., Moorbath, S., Pankhurst, R., and McGregor, V. (1971). 'Isotopic dating of very early precambrian amphibolite facies gneisses from the Godthaab District, West Greenland.' *Earth Planet. Sci. Letters*, **12**, 245–259.

Bogard, D., and Cressy, P. (1973). 'Spallation production of ^{3}He, ^{21}Ne and ^{38}Ar from target elements in the Bruderheim chondrite.' *Geochim. Cosmochim. Acta*, **37**, 527–546.

Bogard, D., and Hirsch, W. (1975). 'Noble gas studies on grain size separates of Apollo 15 and 16 deep drill cores.' *Proc. Lunar Sci. Conf. 6th, Vol. 2*, 2057–2083.

Bogard, D. D., Burnett, D. S., Eberhardt, P., and Wasserburg, G. J. (1967a). '^{87}Rb–^{87}Sr isochron and ^{40}K–^{40}Ar ages of the Norton County achondrite.' *Earth Planet. Sci. Letters*, **3**, 179–189.

Bogard, D., Burnett, D., Eberhardt, P., and Wasserburg, J. (1967b). '^{40}Ar–^{40}K ages of silicate inclusions in iron meteorites. Earth Planet.' *Sci. Letters*, **3**, 275–283.

Böhm-Vitense, E., and Szokody, P. (1973). 'The interpretation of the two color and color magnitude diagrams of M 15 and M 92.' *Astrophys. J.*, **184**, 211–226.

Boulos, M., and Manuel, O. (1971). 'The xenon record of extinct radioactivities in the earth.' *Science*, **174**, 1334–1336.

Boyce, J. M., Dial, A., and Soderblom, A. (1974). 'Ages of the lunar nearside light plains and maria.' *Proc. Lunar Sci. Conf. 5th, Vol. 1*, 11–23.

Browne, J. C., and Berman B. L. (1976). 'Neutron capture cross-sections for ^{186}Os and ^{187}Os and the age of the universe.' *Nature*, **262**, 197–199.

Burbidge, E. M., Burbidge, G. R., Fowler, W., and Hoyle, F. (1957). 'Synthesis of the elements in stars.' *Rev. Mod. Phys.*, **29**, 547–650.

Burnett D. S., and Wasserburg G. J. (1967). '^{87}Rb–^{87}Sr ages of silicate inclusions in iron meteorites.' *Earth Planet. Sci. Letters*, **2**, 397–408.

Butler, W. A., Jeffrey, P., Reynolds, J. H. and Wasserburg, J. (1963). 'Isotopic variations in terrestrial xenon.' *J. Geophys. Res.*, **68**, 3283–3291.

Cadogan P. H., and Turner G. (1975). 'Luna 16 and Luna 20 revisited.' *Meteoritics*, **10**, 375–376.

Cadogan P. H., and Turner G. (1976). '^{40}Ar–^{39}Ar dating of Luna 16 and Luna 20 samples.' *Preprint*.

Cameron, A. (1976). 'The primitive solar accretion disk and the formation of the planets.' In Proceedings, *The Origin of the Solar System*, S. F. Dermott (ed.), Wiley, Chichester (preprint Cf A 666).

Catanzaro E. (1963). 'Zircon ages in Southwestern Minnesota.' *J. Geophys. Res.*, **68**, 2045–2048.

Catanzaro E., and Kulp J. (1964). 'Discordant zircons from the Little Belt (Montana), Beartooth (Montana) and Santa Catalina (Arizona) Mountains.' *Geochim. Cosmochim. Acta*, **28**, 87–124.

Chen J. H., and Tilton G. R. (1976). 'Isotopic lead investigations on the Allende carbonaceous chondrite.' *Geochim. Cosmochim. Acta*, **40**, 635–644.

Clayton, D. D. (1964). 'Cosmoradiogenic chronologies of nucleosynthesis.' *Astrophys. J.*, **139**, 637–663.

Clayton, D. D. (1975a). 'Extinct radioactivities: Trapped residuals of presolar grains.' *Astrophys. J.*, **199**, 765–769.

Clayton, D. D. (1975b). '^{22}Na, Ne–E, extinct radioactive anomalies and unsupported ^{40}Ar.' *Nature*, **257**, 36–37.

Clayton, D. D., and Hoyle, F. (1976). 'Grains of anomalous isotopic composition from Novae.' *Astrophys. J.* **203**, 490–496.

Clayton, R. N., and Mayeda, T. (1975). 'Genetic relations between the moon and meteorites.' *Proc. Lunar Sci. Conf. 6th, Vol. 2*, 1761–1769.

Clayton, R. N., Onuma, N., and Mayeda, T. (1976). 'A classification of meteorites based on oxygen isotopes.' *Earth Planet. Sci. Letters*, **30**, 10–18.

Cobb, J. C. (1966). 'Iron meteorites with low cosmic ray exposure ages.' *Science*, **151**, 1524.

Cumming, G. L. (1969). 'A recalculation of the age of the solar system.' *Canad. J. Earth Sci.*, **6**, 719–735.

Dearnley, R. (1969). 'Crustal tectonic evidence for earth expansion.' In: *The Application of Modern Physics to the Earth and Planetary Interiors*. S. K. Runcorn, ed., Wiley–Interscience, London, 103–110.

Doe, B. R., and Stacey, J. S. (1974). 'The application of lead isotopes to the problems of ore genesis and ore prospect evaluation: A review.' *Econ. Geol.*, **69**, 757–776.

Dohnanyi, J. (1969). 'Collisional model of asteroids.' *J. Geophys. Res.*, **74**, 2531–2554.

Drozd, R. J., and Podosek, F. A. (1976). 'Primordial ^{129}Xe in meteorites.' *Earth Planet. Sci. Letters*, **31**, 15–30.

Eberhardt, P. (1974). 'A neon-E rich phase in the Orgueil carbonaceous chondrite.' *Earth Planet. Sci. Letters*, **24**, 182–187.

Eberhardt, P., Geiss, J., Grögler, N., and Stettler, A. (1973). 'How old is the crater Copernicus?' *Moon*, **8**, 104–114.

Fuse, K., and Anders, E. (1969). 'Aluminum 26 in meteorites. VI-Achondrites.' *Geochim. Cosmochim. Acta*, **33**, 653–670.

Gale, N., Arden, J., and Hutchison, R. (1973). 'Uranium-lead chronology of chondritic meteorites.' *Fortschr. Mineral.*, **50**, Beiheft 3, 71–74.

Gale, N., Arden, J., and Hutchison, R. (1975). 'The chronology of the Nakhla achrondritic meteorite.' *Earth Planet. Sci. Letters*, **26**, 195–206.

Gancarz, A., Tera, F., and Wasserburg, G. (1975). '3·62 AE Amitsoq gneiss from West Greenland and a 4·45 AE age of the earth.' *Geological Soc. of America, Abstracts with Programs, Vol. 7*, 1081–1082 (Salt Lake City, 1975 Annual Meeting).

Gerling, E. K., Maslenikov, V. A., Morozova, I., Matveeva, S., and Vasiljeva, S. (1965). 'Ancient basic and ultrabasic rocks from Monche Tundra (Kola Peninsula) and new data on the absolute age of the earth's crust (in Russian).' In: *Absolute Ages of Precambrian Rocks of the USSR* (in Russian), Akademiya Nauk, Moscow, 11–34.

Gopalan, K., and Wetherill, G. W. (1968). 'Rubidium–strontium age of hypersthene (L) chondrites.' *J. Geophys. Res.*, **73**, 7133–7136.

Gopalan, K., and Wetherill, G. W. (1969). 'Rubidium–strontium age of amphoterite (LL) chondrites.' *J. Geophys. Res.*, **74**, 4349–4358.

Gopalan, K., and Wetherill, G. W. (1970). 'Rubidium–strontium studies on enstatite chondrites: Whole meteorite and mineral isochrons.' *J. Geophys. Res.*, **75**, 3457–3467.

Gopalan, G., and Wetherill, G. W. (1971). 'Rubidium–strontium studies on black hypersthene chondrites: Effects of shock and reheating.' *J. Geophys. Res.*, **76**, 8484–8492.

Gray, C., Papanastassiou, D., and Wasserburg, G. (1973). 'The identification of early condensates from the solar nebula.' *Icarus*, **20**, 213–239.

Hainebach, K. L., and Schramm, D. N. (1976). 'Galactic evolution models and the rhenium 187–osmium 187 chronometer: A greater age for the galaxy.' *Astrophys. J.*, **207**, L 79–82.

Hartmann, W., and Wood, C. (1971). 'Moon: Origin and evolution of multiring basins.' *Moon*, **3**, 3–78.

Hawkesworth, C., Moorbath, S., O'Nions, R., and Wilson, J. (1975). 'Age relationships between greenstone belts and "granites" in the Rhodesian archaean craton.' *Earth Planet. Sci. Letters*, **25**, 251–262.

Head, J. W. III (1976). 'Lunar volcanism in space and time.' *Rev. Geophys. Space Physics*, **14**, 265–300.

Herr, W., Hoffmeister, W., Hirt, B., Geiss, J., and Houtermans, F. G. (1961). 'Versuch zur Datierung von Eisenmeteoriten nach der Rhenium–Osmium-Methode.' *Z. Naturforsch.*, **16a**, 1053–1058.

Herzog, G., and Anders, E. (1971). 'Absolute scale for radiation ages of stony meteorites.' *Geochim. Cosmochim. Acta*, **35**, 605–611.

Herzog, G. F., Cressy, P. J., and Carver, E. A. (1976). 'Shielding effects in Norton County and other aubrites.' Submitted to: *J. Geophys. Res.*

Heymann, D. (1967). 'The origin of hypersthene chondrites: Ages and shock effects of black chondrites.' *Icarus*, **6**, 189–221.

Heymann, D., and Anders, E. (1967). 'Meteorites with short cosmic ray exposure ages, as determined from their ^{26}Al content.' *Geochim. Cosmochim. Acta*, **31**, 1793–1809.

Heymann, D., and Dziczkaniec, M. (1976). 'Early irradiation of matter in the solar system: Magnesium (proton, neutron) scheme.' *Science*, **191**, 79–81.

Heymann, D., Lipschutz, M., Nielsen, B., and Anders, E. (1966). 'Canyon Diablo meteorite: Metallographic and mass spectrometric study of 56 fragments.' *J. Geophys. Res.*, **71**, 619–641.

Hickman, M. H. (1974). '3500 Myr old granite in Southern Africa.' *Nature*, **251**, 295–296.

Hohenberg, C. (1969). 'Radioisotopes and the history of nucleosynthesis in the galaxy.' *Science*, **166**, 212–215.

Horn, P., and Kirsten, T. (1977). 'Lunar highland stratigraphy and radiometric dating.' *Phil. Trans. R. Soc. Lond. A*, **285**, 145–150.

Houtermans, F. (1946). 'Die Isotopenhäufigkeiten im natürlichen Blei und das Alter des Urans.' *Naturwissensch.*, **33**, 185–186.

Howard, W., Arnould, M., and Truran, J. (1975). 'On possible short-lived progenitors of fission xenon in carbonaceous chondrites.' *Astrophys. Space Sci.*, **36**, L1–L9.

Hübner, W., Kirsten, T., and Kiko, J. (1975). 'Rare gases in Apollo 17 soils with emphasis on analysis of size and mineral fractions of soil 74241.' *Proc. Lunar Sci. Conf. 6th, Vol. 2*, 2009–2026.

Huey, J., and Kohman, T. (1972). 'Search for extinct natural radioactivity of ^{205}Pb via thallium isotope anomalies in chondrites and lunar soil.' *Earth Planet. Sci. Letters*, **16**, 401–412.

Huey, J. M., and Kohman, T. (1973). '^{207}Pb–^{206}Pb isochron and age of chondrites.' *J. Geophys. Res.*, **78**, 3227–3244.

Huneke, J., Podosek, F. A., and Wasserburg, G. J. (1972). 'Gas retention and cosmic ray exposure ages of a basalt fragment from mare Fecunditatis.' *Earth Planet. Sci. Letters*, **13**, 375–383.

Hurst, R., Bridgewater, D., Collerson, K., and Wetherill, G. (1975). '3600 m.y. Rb-Sr ages from very early archaean gneisses from Saglek Bay, Labrador.' *Earth Planet. Sci. Letters*, **27**, 393–403.

Husain L. (1972). 'The ^{40}Ar–^{39}Ar and cosmic ray exposure ages of Apollo 15 crystalline rocks, breccias, and glasses.' *The Apollo 15 Lunar Samples* (J. W. Chamberlain, ed.), Lunar Science Institute, Houston, 374–377.

Husain L. (1974). '^{40}Ar–^{39}Ar chronology and cosmic ray exposure ages of the Apollo 15 samples.' *J. Geophys. Res.*, **79**, 2588–2606.

Hutcheon I., and Price P. B. (1972). 'Plutonium-244 fission tracks: Evidence in a lunar rock 3·95 billion years old.' *Science*, **176**, 909–911.

Jaffey A. H., Flynn K. F., Glendenin L., Bentley W. C., and Essling A. (1971). 'Precision measurements of half-lives and specific activities of ^{235}U and ^{238}U.' *Phys. Rev. C.*, **4**, 1889–1906.

Jahn B. -M., and Shih C. -J. (1974). 'On the age of the Onverwaacht group, Swaziland sequence, South Africa.' *Geochim. Cosmochim. Acta*, **38**, 873–885.

James O. and Consortium 73215 (1975). 'Consortium studies of matrix of light gray breccia 73215.' *Proc. Lunar Sci. Conf. 6th, Vol. 1*, 547–577.

Jeffery P. M., and Reynolds J. H. (1961). 'Origin of excess Xe^{129} in stone meteorites.' *J. Geophys. Res.*, **66**, 3582–3583.

Jessberger E. K., Huneke J., Podosek F., and Wasserburg G. (1974). 'High resolution argon analysis of neutron irradiated Apollo 16 rocks and separated minerals.' *Proc. Lunar Sci. Conf. 5th, Vol. 2*, 1419–1449.

Jessberger, E. K., Kirsten, T., and Staudacher, T. (1976). 'Argon–argon ages of consortium breccia 73215.' *Proc. Lunar Sci. Conf. 7th, Vol. 2*, 2201–2215.

Jessberger, E., Kirsten, T., and Staudacher, Th. (1977). 'One rock and many ages—further K–Ar data on Consortium breccia 73215.' *Proc. Lunar Sci. Conf. 8th, Vol. 2*, in print.

Kanasevich, E. R. (1968). 'The interpretation of lead isotopes and their geological significance.' In: *Radiometric Dating for Geologists.* (E. Hamilton, R. Farquhar, editors.) Interscience, London, pp. 147–223.

Kanasevich, E. R., and Savage, J. C. (1963). 'Dirac's cosmology and radioactive dating.' *Canad. J. Phys.*, **41**, 1911–1923.

Kaneoka, I. (1974). 'Investigation of excess argon in ultramafic rocks from the Kola peninsula by the ^{40}Ar–^{39}Ar method.' *Earth Planet. Sci. Letters*, **22**, 145–156.

Kaushal, S., and Wetherill, G. W. (1969). 'Rb^{87}–Sr^{87} age of bronzite (H-group) chondrites.' *J. Geophys. Res.*, **74**, 2717–2726.

Kaushal, S., and Wetherill, G. W. (1970). 'Rubidium 87–strontium 87 age of carbonaceous chondrites.' *J. Geophys. Res.*, **75**, 463–468.

Kempe, W., and Müller, O. (1969). 'The stony meteorite Krähenberg.' *Meteorite Research*, P. Millman (ed.) Reidel Publ. Comp., pp. 418–428.

Kirsten, T. (1973). 'Isotope studies in the Mundrabilla iron meteorite.' *Meteoritics*, **8**, 400–403.

Kirsten, T. (1974). 'Lunar chronology.' *Raumfahrtforschg.*, **5**, 235–237.

Kirsten, T., and Horn P. (1974). 'Chronology of the Taurus-Littrow region III: Ages of mare basalts and highland breccias and some remarks about the interpretation of lunar highland rock ages.' *Proc. Lunar Sci. Conf. 5th, Vol. 2*, 1451–1475.

Kirsten, T., and Horn, P. (1975). '^{39}Ar–^{40}Ar dating of basalts and rock breccias from Apollo 17 and the Malvern achondrite.' *Proc. Soviet-American Conference on Cosmochemistry of the Moon and Planets*. Nauka, Moscow 1975, pp. 386–401.

Kirsten, T., and Müller, O. (1967). 'Argon and potassium in mineral fractions of three ultramafic rocks from the Baltic Shield.' *Radioactive Dating and Methods of Low Level Counting*, IAEA, Vienna 1967, pp. 483–498.

Kirsten, T., and Schaeffer, O. A., (1971). 'High energy interactions in space.' In: *Elementary Particles; Science, Technology and Society*. L. C. Yuan (ed.) Academic Press, New York pp. 76–157.

Kirsten, T., Krankowsky, D., and Zähringer, J. (1963). 'Edelgas- und Kalium-Bestimmungen an einer größeren Zahl von Steinmeteoriten.' *Geochim. Cosmochim. Acta*, **27**, 13–42.

Kirsten, T., Deubner, J., Horn, P., Kaneoka, I., Kiko, J., Schaeffer O. A., and Thio S. K. (1972) 'The rare gas record of Apollo 14 and 15 samples.' *Proc. Lunar Sci. Conf. 3rd, Vol. 2*, pp. 1865–1889.

Kirsten, T., Horn, P., and Kiko, J. (1973). '^{39}Ar–^{40}Ar dating and rare gas analysis of Apollo 16 rocks and soils.' *Proc. Lunar Sci. Conf. 4th, Vol. 2*, 1757–1784.

Kolesnikov, E. M., Lavrukhina, A., Fesenko, A., and Levski, L. (1972). 'Radiation ages of different fragments of the Sikhote Alin meteorite fall.' *Geochim. Cosmochim. Acta*, **36**, 573–576.

Kuroda, P., and Manuel, O. (1962). 'On the chronology of the formation of the solar system. 1. Radiogenic xenon 129 in the earth's atmosphere.' *J. Geophys. Res.*, **67**, 4859–4862.

Lal, D. (1972). 'Hard rock cosmic ray archaeology.' *Space Science Rev.*, **14**, 3–102.

Lambert, D. L., and Mallia, E. A. (1968). 'The abundances of the elements in the solar photosphere. VI. Rubidium.' *Month.Not. Royal Astr. Soc.*, **140**, 13–20.

Lambert, D. L., and Warner, B. (1968). 'The abundances of the elements in the solar photosphere. V. The alkaline earths, Mg, Ca, Sr, Ba.' *Month. Not. Roy. Astr. Soc.*, **140**, 197–221.

Lee, T., Papanastassiou, D., and Wasserburg, G. (1976). 'Demonstration of ^{26}Mg excess in Allende and evidence for ^{26}Al.' *Geophys. Res. Letters*, **3**, 41–44.

Leich, D. A., and Moniot, R. (1976). 'Rare gas chronology of enstatite achondrites and the Bholghati howardite.' *Lunar Science VII*, 479–481.

Leich, D., Niemeyer, S., and Michel, M. (1977). 'Elimination of ^{248}Cm as a possible progenitor of carbonaceous chondrite fission xenon.' *Earth Planet. Sci. Letters*, **34**, 197–208.

Lewis, R., and Anders, E. (1975). 'Condensation time of the solar nebula from extinct ^{129}I in primitive meteorites.' *Proc. Nat. Acad. Sci. U.S.A.*, **72**, 268–273.

Lovering J. F., Hinthorne J. R., and Conrad R. L. (1976). 'Direct $^{207}Pb/^{206}Pb$ dating by ion microprobe of uranium–thorium rich phases in Allende calcium–aluminium-rich clasts (Carc's).' *Lunar Science VII*, 504–506.

Lugmair, G. W., Scheinin, N. B., and Marti, K. (1975a). 'Sm–Nd age and history of Apollo 17 basalt 75075: Evidence for early differentiation of the lunar exterior.' *Proc. Lunar Sci. Conf. 6th, Vol. 2*, 1419–1429.

Lugmair, G. W., Scheinin, N., and Marti, K. (1975b). 'Search for extinct ^{146}Sm. 1. The isotopic abundance of ^{142}Nd in the Juvinas meteorite.' *Earth Planet. Sci. Letters*, **27**, 79–84.

Lugmair G., Kurtz J., Marti K., and Scheinin N. (1976). 'The low-Sm/Nd region of the moon: Evolution and history of troctolite and a KREEP basalt.' *Lunar Science VII*, 509–511.

Lunatic Asylum (1970). 'Mineralogic and isotopic investigations on lunar rock 12013.' *Earth Planet. Sci. Letters*, **9**, 137–163.

Manhes G., Minster J. -F., and Allegre C. (1975). 'Lead-lead and rubidium strontium study of the St. Severin LL6 chondrite.' *Meteoritics*, **10**, 451.

Manuel O. K., Hennecke E. W., and Sabu D. (1972). 'Xenon in carbonaceous chondrites.' *Nature, Physic. Sci.*, **240**, 99–101.

Marti K., Lugmair G. W., and Urey H. C. (1970). 'Solar wind gases, cosmic ray spallation products, and the irradiation history.' *Science*, **167**, 548–550.

Marti K., Lightner B., and Lugmair G. (1973). 'On ^{244}Pu in lunar rocks from Fra Mauro and implications regarding their origin.' *Moon*, **8**, 241–250.

Mazor, E., Heymann, D., and Anders, E. (1970). 'Noble gases in carbonaceous chondrites.' *Geochim. Cosmochim. Acta*, **34**, 781–824.

Merrihue, C. M., and Turner, G. (1966). 'Potassium argon dating by activation with fast neutrons.' *J. Geophys. Res.*, **71**, 2852–2857.

Moorbath, S., O'Nions, R., Pankhurst, R., Gale, N., and McGregor, V. (1972). 'Further rubidium–strontium age determinations on the very early precambrian rocks of the Godthaab District, West Greenland.' *Nature, Physic. Sci.*, **240**, 78–83.

Moorbath, S., O'Nions, R., and Pankhurst, R. (1973). 'Early archaean age for the Isua iron formation, West Greenland.' *Nature, Physic. Sci.*, **245**, 138–139.

Moorbath, S., O'Nions, R., and Pankhurst, R. (1975). 'The evolution of early precambrian crustal rocks at Isua, West Greenland—Geochemical and isotopic evidence.' *Earth Planet. Sci. Letters*, **27**, 229–239.

Murthy, V. R., and Compston, W. (1965). 'Rb–Sr ages of chondrules and carbonaceous chondrites.' *J. Geophys. Res.*, **70**, 5297–5307.

Neukum, G., König, B., and Fechtig, H. (1975). 'Cratering in the earth–moon system: Consequences for age determination by crater counting.' *Proc. Lunar Sci. Conf. 6th, Vol. 3*, 2597–2620.

Neumann, W., and Huster, E. (1976). 'Discussion of the ^{87}Rb half life determined by absolute counting.' *Earth Planet. Sci. Letters*, **33**, 277–288.

Nunes, P., and Tilton, G. (1971). 'U–Pb ages of minerals from the Stillwater igneous complex and associated rocks, Montana.' *Geol. Soc. Am. Bull.*, **82**, 2231–2250.

Nunes, P., Tatsumoto, M., and Unruh, D. (1975). 'U–Th–Pb systematics of anorthositic gabbros 78155 and 77017—Implications for early lunar evolution.' *Proc. Lunar Sci. Conf. 6th, Vol. 2*, 1431–1444.

Nunes, P., Nakamura, N., and Tatsumoto, M. (1976). '4·4 b. y. old clast in boulder 7, Apollo 17.' *Lunar Science VII*, 631–632.

Nyquist, L. E., Hubbard, N., Gast, P., Bansal, B., Wiesmann, H., and Jahn, B. (1973). 'Rb-Sr systematics for chemically defined Apollo 15 and 16 materials.' *Proc. Lunar Sci. Conf. 4th, Vol. 2*, 1823–1846.

Nyquist, L. E., Bansal, B., Wiesmann, H., and Jahn, B. (1974). 'Taurus-Littrow chronology: Some constraints on early lunar crustal development.' *Proc. Lunar Sci. Conf. 5th, Vol. 2*, 1515–1539.

Nyquist, L. E., Bansal, B., and Wiesmann, H. (1975). 'Rb-Sr ages and initial ^{87}Sr/^{86}Sr for Apollo 17 basalts and KREEP basalt 15386.' *Proc. Lunar Sci. Conf. 6th, Vol. 2*, 1445–1465.

Papanastassiou, D. A. (1970). 'The determination of small time differences in the formation of planetary objects.' *Ph. D. Thesis*, California Institute of Technology.

Papanastassiou, D. A., and Wasserburg, G. J. (1969). 'Initial strontium isotopic abundances and the resolution of small time differences in the formation of planetary objects.' *Earth Planet. Sci. Letters*, **5**, 361–376.

Papanastassiou, D., Wasserburg, G., and Burnett, D. (1970). 'Rb-Sr ages of lunar rocks from the Sea of Tranquility.' *Earth Planet. Sci. Letters*, **8**, 1–19.

Papanastassiou, D., and Wasserburg, G. (1971a). 'Rb-Sr ages of igneous rocks from the Apollo 14 mission and the age of the Fra Mauro formation.' *Earth Planet. Sci. Letters*, **12**, 36–48.

Papanastassiou, D., and Wasserburg, G. J. (1971 b). 'Lunar chronology and evolution from Rb–Sr chronology of Apollo 11 and Apollo 12 samples.' *Earth Planet. Sci. Letters*, **11**, 37–62.

Papanastassiou, D., and Wasserburg, G. (1972a). 'Rb–Sr age of a Luna 16 basalt and the model age of lunar soils.' *Earth Planet. Sci. Letters*, **13**, 368–374.

Papanastassiou, D., and Wasserburg, G. (1972b). 'Rb-Sr systematics of Luna 20 and Apollo 16 samples.' *Earth Planet. Sci. Letters*, **17**, 52–63.

Papanastassiou, D., and Wasserburg, G. (1973). 'Sb–Sr ages and initial strontium in basalts from Apollo 15.' *Earth Planet. Sci. Letters*, **17**, 324–337.

Papanastassiou, D. A., and Wasserburg, G. J. (1974). 'Evidence for late formation and young metamorphism in the achondrite Nakhla.' *Geophys. Res. Letters*, **1**, 23–26.

Papanastassiou, D., and Wasserburg, G. (1975a). 'Rb-Sr study of a lunar dunite and evidence for early lunar differentiates.' *Proc. Lunar Sci. Conf. 6th, Vol. 2*, 1467–1489.

Papanastassiou, D., and Wasserburg, G. (1975b). 'A Rb-Sr study of Apollo 17 boulder 3: Dunite clast, microclasts, and matrix.' *Lunar Science VI*, 631–633.

Papanastassiou, D., and Wasserburg, J. (1976). 'Early lunar differentiates and lunar initial $^{87}Sr/^{86}Sr$.' *Lunar Science VII*, 665–667.

Papanastassiou, D. A., Rajan, R., and Wasserburg, G. (1974). 'Rb–Sr ages and lunar analogs in a basaltic achondrite: Implications for early solar system chronologies.' *Lunar Science V*, 583–585.

Patterson, C. (1955). 'The Pb^{207}/Pb^{206} ages of some stone meteorites.' *Geochim. Cosmochim. Acta*, **7**, 151–153.

Pellas, P., and Storzer, D. (1976). 'On the early thermal history of chondritic asteroids derived by 244-plutonium fission track thermometry.' In: *The Interrelated Origin of Comets, Asteroids and Meteorites*. (A. Delsemme, editor), Univ. of Toledo publications.

Pepin, R. O. (1967). 'Neon and xenon in carbonaceous chondrites.' In: *Origin and Distribution of the Elements*, (L. Ahrens, editor), London, Pergamon, 379–386.

Pepin, R. O., and Phinney D. (1976). 'The formation interval of the earth.' *Lunar Science VII*, 682–684.

Pepin, R., Bradley, J., Dragon, J., and Nyquist, L. (1972). 'K–Ar dating of lunar fines: Apollo 12, Apollo 14, and Luna 16.' *Proc. Lunar Sci. Conf.*, 3rd, Vol. 2, 1569–1588.

Podosek, F. A. (1970a). 'The abundance of ^{244}Pu in the early solar system.' *Earth Planet. Sci. Letters*, **8**, 183–187.

Podosek, F. A. (1970b). 'Dating of meteorites by the high temperature release of iodine correlated Xe^{129}.' *Geochim. Cosmochim. Acta*, **34**, 341–365.

Podosek, F. A. (1971). 'Neutron-activation potassium–argon dating of meteorites.' *Geochim. Cosmochim. Acta*, **35**, 157–173.

Podosek, F. A. (1972). 'Gas retention chronology of Petersburg and other meteorites.' *Geochim. Cosmochim. Acta*, **36**, 755–772.

Podosek, F. A. (1973). 'Thermal history of the nakhlites by the ^{40}Ar–^{39}Ar-method.' *Earth Planet. Sci. Letters*, **19**, 135–144.

Podosek, F. A., and Huneke, J. C. (1973a). 'Argon 40–argon 39 chronology of four calcium-rich achondrites.' *Geochim. Cosmochim. Acta*, **37**, 667–684.

Podosek, F. A., and Huneke, J. C. (1973b). 'Noble gas chronology of meteorites.' *Meteoritics*, **8**, 64.

Podosek, F. A., and Lewis, R. S. (1972). '^{129}I and ^{244}Pu abundances in white inclusions of the Allende meteorite.' *Earth Planet. Sci. Letters*, **15**, 101–109.

Podosek, F., Huneke, J., Gancarz, A., and Wasserburg G. J. (1973). 'The age and petrography of two Luna 20 fragments and inferences for widespread lunar metamorphism.' *Geochim. Cosmochim. Acta*, **37**, 887–904.

Rajan, R., Huneke, J., Smith, S., and Wasserburg, G. (1975). '^{40}Ar–^{39}Ar chronology of isolated phases from Bununu and Malvern howardites.' *Earth Planet. Sci. Letters*, **27**, 181–190.

Rao, M., and Gopalan, K. (1973). 'Curium 248 in the early solar system.' *Nature*, **245**, 304–307.

Reedy, R. C., and Arnold, J. R. (1972). 'Interaction of solar and galactic cosmic ray particles with the moon.' *J. Geophys. Res.*, **77**, 537–555.

Reeves, H. (1972). 'Spatial inhomogeneities of nucleosynthesis.' *Astron. Astrophys.*, **19**, 215–223.

Reynolds, J. H. (1960). 'Determination of the age of the elements.' *Phys. Rev. Letters*, **4**, 8–10.

Rowe, M. W., and Kuroda, P. K. (1965). 'Fissiogenic xenon from the Pasamonte meteorite.' *J. Geophys. Res.*, **70**, 709–714.

Sandage, A. R., and Tammann, G. A. (1975). 'Steps towards the Hubble constant: VI: The Hubble constant determined from redshifts and magnitudes of remote ScI galaxies: The value of q_0.' *Astrophys. J.*, **197**, 265–280.

Sanz, H. G., and Wasserburg, G. J. (1969). 'Determination of an internal ^{87}Rb–^{87}Sr isochron for the Olivenza chondrite.' *Earth Planet. Sci. Letters*, **6**, 335–345.

Sanz, H. G., Burnett, D. S., and Wasserburg, G. J. (1970). 'A precise $^{87}Rb/^{87}Sr$ age and initial $^{87}Sr/^{86}Sr$ for the Colomera iron meteorite.' *Geochim. Cosmochim. Acta*, **34**, 1227–1239.

Schaeffer, O. A. (1975). 'Constancy of galactic cosmic rays in time and space.' *Proc. 14th Int. Cosmic Ray Conf. München, Vol. 11*, 3508–3520.

Schaeffer O. A., and Husain L. (1973). 'Early lunar history: Ages of 2 to 4 mm soil fragments from the lunar highlands.' *Proc. Lunar Sci. Conf. 4th, Vol. 2*, 1847–1863.

Schaeffer O. A., and Husain L. (1974). 'Chronology of lunar basin formation.' *Proc. Lunar Sci. Conf. 5th, Vol. 2*, 1541–1555.

Schaeffer O. A., and Husain L. (1975). 'The duration of volcanism in the Taurus-Littrow region and ages of highland rocks returned by Apollo 17.' *Lunar Science VI*, 707–709.

Schonfeld E. (1976). 'Rb-Sr evolution of the lunar crust.' *Lunar Science VII*, 773–775.

Schramm D. N. (1973). 'Nucleo-cosmochronology.' *Space Science Reviews*, **15**, 51–67.

Schramm, D. (1974a). 'Nucleo-cosmochronology.' *Ann. Rev. Astron. Astrophys.*, **12**, 383–406.

Schramm, D. N. (1974b). 'The age of the elements.' *Scientific American*, **230**, No. 1, 69–77.

Schramm, D. N., and Wasserburg, G. J. (1970). 'Nucleochronologies and the mean age of the elements.' *Astrophys. J.*, **162**, 57–69.

Schultz, L. (1976). 'On the cosmic ray exposure age distribution in ordinary chondrites.' *Meteoritics*, **11**, 359–360.

Schultz, L., Signer, P., Lorin, J., and Pellas, P. (1972). 'Complex irradiation history of the Weston chondrite.' *Earth Planet. Sci. Letters*, **15**, 403–410.

Scott, E., and Wasson J. (1975). 'Classification and properties of iron meteorites.' *Rev. Geophys. Space Phys.*, **13**, 527–546.

Seeger, P., and Schramm D. (1970). 'r-process production ratios of chronologic importance.' *Astrophys. J.*, **160**, L157–L160.

Shih, C. -Y. (1976). 'On the origin of KREEP basalt.' *Lunar Science VII*, 800–802.

Silver, T., and Duke, M. (1971). 'U-Th-Pb relations in some basaltic achondrites.' *EOS, Transact. Am. Geophys. Soc.*, **52**, 269 (Abstract).

Sinha, A. K. (1972). 'U-Th-Pb systematics and the age of the Onverwaacht series, South Africa.' *Earth Planet. Sci. Letters*, **16**, 219–227.

Stacey, J., and Kramers, J. (1975). 'Approximation of terrestrial lead isotope evolution by a two-stage model.' *Earth Planet. Sci. Letters*, **26**, 207–221.

Stanton, R. L., and Russell, R. D. (1959). 'Anomalous leads and the emplacement of lead sulfide ores.' *Econ. Geol.*, **54**, 588–607.

Steiger, R., and Jäger E. (1977). 'Subcommission on Geochronology: Convention on the use of decay constants in geo- and cosmochronology.' *Earth Planet. Sci. Letters*, **36**, 359–362.

Stettler, A., Eberhardt, P., Geiss, J., Grögler, N., and Maurer, P. (1973). 'Ar^{39}—Ar^{40} ages and Ar^{37}–Ar^{38} exposure ages of lunar rocks. *Proc. Lunar Sci. Conf. 4th, Vol. 2*, 1865–1888.

Stettler, A., Eberhardt, P., Geiss, J., Grögler, N., and Maurer, P. (1974). 'On the duration of lava flow activity in Mare Tranquillitatis.' *Proc. Lunar Sci. Conf. 5th, Vol. 2*, 1557–1570.

Stettler, A., Eberhardt, P., Geiss, J., Grögler, N., and Guggisberg, S. (1975). 'Age sequence in the Apollo 17 station 7 boulder.' *Lunar Science VI*, 771–773.

Sutton, J. (1963). 'Long-term cycles in the evolution of the continents.' *Nature*, **198**, 731–735.

Tatsumoto, M. (1970). 'Age of the moon: An isotopic study of U–Th–Pb systematics of Apollo 11 lunar samples.' *Proc. Lunar Sci. Conf. 1st, Vol. 2*, 1595–1612.

Tatsumoto, M., Knight, R., and Doe, B. (1971). 'U–Th–Pb systematics of Apollo 12 lunar samples.' *Proc. Lunar Sci. Conf. 2nd, Vol. 2*, 1521–1546.

Tatsumoto, M., Knight, R., and Allegre, C. (1973). 'Time differences in the formation of meteorites as determined from the ratio of lead 207 to lead 206.' *Science*, **180**, 1279–1283.

Tatsumoto, N., Nunes, P., Knight, R., and Unruh, D. (1974). 'Rb Sr and U, Th–Pb systematics of boulders 1 and 7 Apollo 17.' *Lunar Science V*, 774–776.

Tatsumoto, M., Unruh, D., and Desborough, G. (1976. 'U–Th–Pb and Rb–Sr systematics of Allende and U–Th–Pb systematics of Orgueil.' *Geochim. Cosmochim. Acta*, **40**, 617–634.

Taylor, G. J., and Heymann, D. (1969). 'Shock, reheating and the gas retention ages of chondrites., *Earth Planet. Sci. Letters*, 151–161.

Taylor, S. R. (1975). *Lunar Science: A post-Apollo view*. Pergamon, 372 pages.

Tera, F., and Wasserburg, G. (1972a). 'U–Th–Pb systematics in three Apollo 14 basalts and the problem of initial Pb in lunar rocks.' *Earth Planet. Sci. Letters*, **14**, 281–304.

Tera, F., and Wasserburg, G. (1972b). 'U–Th–Pb systematics in lunar highland samples from the Luna 20 and Apollo 16 missions.' *Earth Planet. Sci. Letters*, **17**, 36–51.

Tera, F., and Wasserburg, G. J. (1974). 'U–Th–Pb systematics on lunar rocks and inferences about lunar evolution and the age of the moon.' *Proc. Lunar Sci. Conf. 5th, Vol. 2*, 1571–1599.

Tera, F., and Wasserburg, G. J. (1976). 'Lunar ball games and other sports.' *Lunar Science VII*, 858–860.

Tera, F., Papanastassiou, D., and Wasserburg, G. (1974). 'Isotopic evidence for a terminal lunar cataclysm.' *Earth Planet. Sci. Letters*, **22**, 1–21.

Tilton, G. R. (1973). 'Isotopic lead ages of chondritic meteorites.' *Earth Planet. Sci. Letters*, **19**, 321–329.

Toksöz, M. N., and Johnston, D. (1974). 'The evolution of the moon.' *Icarus*, **21**, 389–414.

Toksöz M. N., and Solomon S. C. (1973). 'Thermal history and evolution of the moon.' *Moon*, **7**, 251–278.

Toksöz M. N., Dainty A. M., Solomon S., and Anderson K. R. (1973). 'Velocity structure and evolution of the moon.' *Proc. Lunar Sci. Conf. 4th, Vol. 3*, 2529–2547.

Turner G. (1969). 'Thermal histories of meteorites by the ^{39}Ar–^{40}Ar-method.' *Meteorite Research*, P. Millman (ed.), Reidel publ. 407–417.

Turner G. (1970). 'Argon 40–argon 39 dating of lunar rock samples.' *Proc. Lunar Sci. Conf. 1st, Vol. 2*, 1665–1684.

Turner G. (1976). Unpublished data, courtesy Dr. Turner.

Turner G. (1977). 'Potassium-argon chronology of the moon.' *Phys. Chem. Earth*, **10**, 145–195.

Turner G., and Cadogan P. H. (1973). '^{40}Ar–^{39}Ar chronology of chondrites.' *Meteoritics*, **8**, 447–448.

Turner G., and Cadogan P. (1975). 'The history of lunar basin formation inferred from ^{40}Ar ^{39}Ar dating.' *Lunar Science VI*, 826–828.

Turner G., Huneke J., Podosek F., and Wasserburg G. (1971). '^{40}Ar–^{39}Ar ages and cosmic ray exposure ages of Apollo 14 samples.' *Earth Planet. Sci. Letters*, **12**, 19–35.

Turner G., Cadogan P., and Yonge C. (1974). 'The early chronology of the moon and meteorites.' *Lunar Science V*, 807–808.

Van Schmus W. R., Wood J. A. (1967). 'A chemical-petrologic classification for the chondritic meteorites.' *Geochim. Cosmochim. Acta*, **31**, 747–765.

Vilcsek E., Wänke H. (1963). 'Cosmic ray exposure ages and terrestrial ages of stone and iron meteorites derived from Cl^{36} and Ar^{39} measurements.' *Radioactive Dating*, IAEA, Vienna, 381–393.

Voshage H. (1967). 'Bestrahlungsalter und Herkunft der Eisenmeteorite.' *Z. Naturforschung*, **22a**, 477–506.

Walker R. M. (1975). 'Interaction of energetic nuclear particles in space with the lunar surface. *Ann. Rev. Earth Planet. Sci.*, **3**, 99–128.

Wasserburg G. J., Papanastassiou D. A., Sanz H. (1969). Initial Sr for a chondrite and the determination of a metamorphism or formation interval. *Earth Planet. Sci. Letters*, **7**, 33–43.

Wasserburg, G. J., Schramm, D. N., and Huneke, J. C. (1969). 'Nuclear chronologies for the galaxy.' *Astrophys. J.*, **157**, L 91–96.

Wasson, J. T. (1974). *Meteorites: Classification and Properties*. Springer-Publishers, Berlin-Heidelberg-New York, 316 pp.

Wene C. O. (1975). 'The effect of delayed fission in nucleo-cosmochronology.' *Astron. Astrophys.*, **44**, 233–236.

Wetherill G. W. (1956). Discordant uranium lead ages. *Am. Geophys. Union Trans.*, **37**, 320–326.

Wetherill G. W. (1971). 'Cometary versus asteroidal origin of chondritic meteorites.' In: *Physical Studies of Minor Planets*. (T. Gehrels, ed.) NASA SP-267, Washington, D.C., 447–460.

Wetherill, W. (1974). 'Solar system sources of meteorites and large meteroids.' *Ann. Rev. Earth Planet. Sci.*, **2**, 303–332.

Wetherill, G. W., Mark, R., Lee-Hu, C. (1973). 'Chondrites: Initial strontium 87/strontium 86 ratios and the early history of the solar system.' *Science*, **182**, 281–283.

Zähringer, J. (1966). 'Die Chronologie der Chondrite aufgrund von Edelgasisotopen-Analysen.' *Meteoritics*, **27**, 25–40.

Zähringer, J. (1968). 'Rare gases in stony meteorites.' *Geochim. Cosmochim. Acta*, **32**, 209–237.

METEORITES: RELICS FROM THE EARLY SOLAR SYSTEM

JOHN W. LARIMER
Department of Geology, Arizona State University, Tempe, Arizona, U.S.A.

Abstract

The origin of meteorites in small bodies, with low energy inputs, has left the meteoritic material relatively unaltered since the formation of the solar system. Chondritic meteorites retain a record of events from the cooling nebula:

(1) A high-temperature ($>$ 1400 K) fractionation of refractory elements
(2) Metal/silicate fractionation at 1000 to 500 K
(3) Partial remelting and outgassing (= chondrule formation) of the original condensate at 600 to 350 K
(4) Accretion at $T = 550$ to 350 K and $P = 10^{-6}$ to 10^{-3} atm

Carbonaceous chondrites are unique: they contain compounds of biological significance, high noble gas contents with peculiar isotopic anomalies and cm-sized aggregates whose mineralogy, elemental and isotopic composition point to a high-temperature ($>$ 1400 K) origin. Isotopic data suggests a component in these aggregates may pre-date the origin of the solar system. Enstatite chondrites, with their highly reduced mineralogy, indicate an inhomogeneous distribution of C and O in the nebula.

Differentiated meteorites provide clues to the early evolution of planetary material. Iron meteorite chemistry indicates that cores originate by a melting-gravitational segregation process. Elemental abundances in planetary silicates suggest that the composition of all bodies was controlled by the same processes which operated on chondritic material.

Introduction

The arguments for and against the origin of meteorites in small bodies were reviewed by Anders in 1964. He concluded that the evidence was decidedly in favour of small bodies, most probably the asteroids. All of the evidence accumulated since then supports this view. There is at least one important difference between asteroidal sized objects, less than 500 km in radius, and the larger bodies in the solar system. In larger bodies the combined gravitational (accretion and core formation) and radioactive (U, Th, K^{40}) energy input is sufficient to cause extensive melting and chemical rearrangements which erases the early history. By contrast, meteoritic material is relatively little altered. In fact, one of the problems in meteoritics is finding an energy source sufficient to cause the small degree of differentiation observed (cf. Tozer, 1978).

Meteorites can be broadly subdivided into three types: irons, stony-irons and stones. As the names imply, this grouping is based on the relative proportions of metal and silicate material. Though admittedly geocentric philosophically, for pedagogic purposes it is useful to draw the obvious analogy with the structure of the earth: core, core-mantle boundary, mantle. But since smaller bodies are involved, there are likely to be important differences. The core forming processes may have ceased prematurely; the mantles may be less altered, perhaps pristine in part. One type of stony meteorite, the chondrites, are quite primitive. They contain all the non-volatile elements in nearly cosmic, or solar, proportions implying little or no processing since the time of origin. Viewed in this way, chrondites are the pristine material, modified only slightly by varying degrees of thermal metamorphism. They contain a record of the physical and chemical conditions that existed in the nebula prior to, and during, accretion. The differentiated meteorites represent a tangible record of processes, such as core formation, which are inaccessible in larger bodies. The task confronting the student of meteorites is to find the clues in the mineralogy, texture and composition of meteorites from which it is possible to deduce the temperatures, pressures, time-spans and chemical environments that existed in the early solar system.

We shall begin with a brief review of meteorite terminology and mineralogical, chemical and textural data. The condensation sequence of the elements from a cooling gas of cosmic composition is covered in the second section and appropriate comparisons drawn with data from chondritic meteorites. A separate section is devoted to the extreme types of chondrites, the carbonaceous and enstatite chondrite groups. The fourth section deals with differentiated meteorites and points out some of the implications for other bodies of the solar system.

Glossary of Meteoritics

It is convenient at the outset to subdivide stony meteorites into two types, chondrites, and achondrites. *Chondrites* are stony meteorites containing *chondrules*, silicate spherules about the size of a small pea. Achondrites are stony meteorites lacking chondrules; they appear to be the products of melting and igneous differentiation processes. In chondritic meteorites, the chondrules are embedded in a *matrix*, or fine-grained groundmass. In most iron and stony-iron meteorites, the metal has segregated itself into a Ni-rich and a Ni-poor phase which are arranged in a regular pattern, named the *Widmanstätten* pattern after its discoverer.

Mineralogy of Meteorites

Only a few mineral names are necessary; where other minerals are discussed the chemical formula will be used. Fortunately, the major minerals of meteorites are few in number and easily committed to memory. Sulphides are ubiquitous but the only abundant one is *troilite*, FeS. The metal consists of two Fe–Ni alloys: *kamacite*, α-FeNi with < 8 per cent Ni; and *taenite*, γ-FeNi with > 8 per cent Ni. Kamacite crystals are cubic (hexagons) and taenite crystals resemble two 4-sided pyramids stacked base to base (octagons). The silicate mineralogy in chondrites reflects the relative abundances of non-volatile elements in cosmic matter; Fe, Mg and Si are dominant. The ions Fe^{+2} and Mg^{+2} have similar radii, which permits extensive substitution in the silicates. There are two of these: *olivine*, $(MgFe)_2\ SiO_4$, and *pyroxene*, (Mg, Fe) SiO_3. The FeO/FeO + MgO ratio varies giving rise to a variable composition in these two minerals. By meteoritical (not terrestrial) conventions the pyroxenes have three names depending on their FeO/(FeO + MgO) ratios: 0 to 0·1 = *enstatite*, 0·1 to 0·2 = *bronzite*, and 0·2 to 0·3 = *hypersthene*. Likewise the ratio determines the specific name of the olivine but here we need only two: 0 = forsterite (Fo) and 1 = fayalite (Fa). In highly oxidized meteorites a portion of the Fe is oxidized to Fe^{+3} and occurs as *magnetite* (Fe_3O_4). The silicates in these meteorites contain water in their structure but the structure, and hence the name, varies on a submicroscopic scale so we shall refer to them as *hydrated silicates*. The next most abundant elements, Al, Ca, and Na occur as *glass* (which actually is not a mineral) or *plagioclase feldspar*, a solid solution of $NaAlSi_3O_8$ and $CaAl_2Si_2O_8$. Like the pyroxenes, the plagioclase feldspars are referred to by different names depending on the relative proportions of the end members. In cosmic systems, however, only the end members are common, $NaAlSi_3O_8$ = albite and $CaAl_2Si_2O_8$ = anorthite.

Meteorite Classification

The four types of meteorites (irons, stony-irons, chondrites and achondrites) are further subdivided on the basis of texture and chemistry; in other words what they look like and what they are made of. This is reflected in Figure 1, where the four types of meteorites are subdivided according to current conventions.

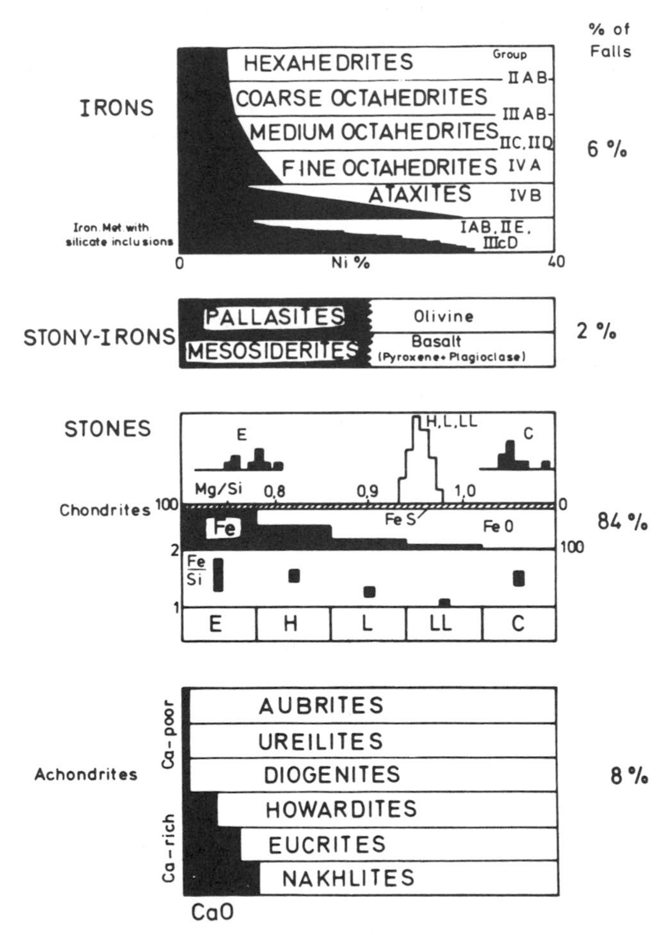

Figure 1. Meteorite classification is based on minerals present, their textural relations, and bulk chemistry

Iron Meteorites

The structural classification is best understood in terms of the phase relations in the system Fe–Ni. At high temperatures ($>900\,^{\circ}C$), octahedral taenite is the stable phase at all compositions. Upon cooling, kamacite exsolves on the crystal faces of the octahedron. If the Ni content falls below 6 per cent, all of the metal converts to cubic kamacite at lower temperatures. Iron meteorites comprised of only kamacite are named hexahedrites. If the Ni content exceeds 6 per cent, some taenite persists and the overall structure retains the octahedral pattern. The amount of kamacite exsolved depends on the Ni content. At low Ni contents, kamacite dominates and forms large crystals; at higher Ni contents the amount of kamacite, as well as the crystal size, diminish. This is reflected in the Widmanstätten pattern and provides descriptive names for these meteorites: coarse, medium, and fine octahedrites. Iron meteorites with other textures are named ataxites. A significant number of iron meteorites (~ 20 per cent) contain abundant silicate inclusions; they are considered as distinct types of iron meteorites depending on the structure and composition of the metal.

Besides the variation in Ni content, Ga and Ge contents tend to cluster into groups which originally were simply named group I (highest Ga and Ge content), group II, III and IV. Subsequently, Wasson has measured Ga, Ge, Ir, and Ni contents in nearly 500 iron meteorites (Scott and Wasson, 1975). From these data, 12 groups of 5 or more members each have been resolved. These are considered chemical groups and are named group IAB, IIAB, IIC, IID, etc. About 14 per cent of the 500 iron meteorites studied do not belong to one of the 12 groups, they are all referred to as anomalous.

Stony-iron Meteorites

The subdivision of the stony-iron meteorites into two groups is based on the mineralogy and composition of the silicate material. In pallasites, the silicate fraction consists almost entirely of olivine. In mesosiderites, it usually consists of a 50–50 mixture of pyroxene and plagioclase feldspar, though these meteorites are a rather heterogeneous group.

Chondrites

To fully illustrate the generally accepted classification of Van Schmus and Wood (1969) would require a third dimension in Figure 1 above. Fortunately, this dimension is easily comprehended. As shown in Figure 1, chondrites are divided into five groups based on various chemical criteria: the metallic to oxidized iron ratios, the Mg/Si ratio and the total Fe/Si ratio. The five groups are identified by letter: E(= enstatite), H(= high iron), L(= low iron), LL(= low iron, low metal) and C(= carbonaceous). Enstatite chondrites contain little oxidized iron. By contrast, most of the Fe in carbonaceous chondrites is oxidized. In the H, L and LL groups, sometimes lumped together and referred to as ordinary chondrites, the amount of oxidized iron increases while the total Fe/Si ratio decreases. Van Schmus and Wood used 10 petrographic (textural and mineralogical) criteria to subdivide each of the 5 chemical groups into a maximum of 6 'petrologic types' (the dimension which is not illustrated). The most easily recognized criteria is the distinctness of the chondrules against the matrix background. In types 2 and 3, the chondrules are sharply delineated against the submicron-sized grains of the matrix; in types 4 to 6, the chondrules become less distinct as the grain size of the matrix increases and, in extreme cases, there is extensive intergrowth which largely obliterates the chondrules. Petrologic Type 1 is reserved for Type I carbonaceous chondrites (= Cl), which contain few, if any, chondrules

Achondrites

These are relatively rare meteorites which, coupled with their diverse chemistry and texture, pose a pretty classification problem. In several cases, a meteorite is unique and requires a separate classification category. But the more common types are best considered by first dividing them into two groups on the basis of Ca content. The Ca-poor variety appear to be more closely related chemically to chondritic meteorites. These include: aubrites (= enstatite achondrites), ureilites and diogenites (= hypersthene achondrites). Aubrites are highly reduced; they resemble enstatite chondrite material except that metal and sulphide are missing. Ureilites consist of olivine, pyroxene and a few Ni-poor (~ 2 per cent) metal grains plus 1 per cent or more carbon. Most of the carbon is present as graphite and diamonds. The dominant mineral in diogenites is hypersthene. These meteorites resemble the silicate fraction of ordinary chondrites except that they are also deficient in olivine.

The most common types of achondrites are the Ca-rich eucrites and howardites. Eucrites resemble terrestrial and, to an even greater extent, lunar basalts. They normally contain pyroxene and plagioclase feldspar in equal amounts. Though almost all achondrites are brecciated, howardites are an extreme example. They are a mixture of different rock types compacted together. In addition, there are other, rarer, types of achondrites whose significance is yet to be resolved.

Chondrites and the Chemistry of The Solar Nebula

Chondrites appear to be the most primitive sample of planetary material available. But how primitive? If the petrologic types of Van Schmus and Wood (1967) reflect varying degrees of thermal metamorphism, as they appear to, then most chondrites have been altered to some extent (Wood, 1962; Dodd, 1969). The problem is to distinguish between those features which were established prior to, or during, accretion in the nebula and those that developed later, when the meteorites resided in their parent bodies.

In addition to the chemical differences between chondrites illustrated in Figure 1 above, it now appears that nearly all known elements are fractionated to some extent. But these fractionations are not random; instead there are definite patterns. All the elements can be placed into one of four groups: refractory, siderophile and two groups of volatile elements referred to as 'normal' and 'strongly depleted'. The elements in the refractory group all have high boiling points, or form compounds with high boiling points. Siderophile (= metal loving) elements are more noble than Fe and tend to concentrate in the metal. 'Normal' elements have moderate boiling points, they tend to be depleted in chondrites to the same extent (Anders, 1964). 'Strongly depleted' elements are the most volatile, they behave like normal elements in C-group chondrites but can be extremely depleted (X1000) in E, H, L and LL-group chondrites (Anders, 1964).

This grouping implies four processes which must in some way be related to the volatility of the elements. All elements which belong to one of the first three groups (refractory, siderophile and normal) move in unison within each chemical group of chondrites. If, for example, one refractory element is depleted by a certain factor all others are too, by the same factor. These patterns extend across petrologic types indicating that the fractionations preceded metamorphism. They must be primary features, established prior to accretion of the meteorites. The abundance of strongly depleted elements does decrease with increasing petrologic type. We shall return to the question of whether this is a causal relationship later.

If the fractionation of the elements occurred prior to accretion and is related to their volatility, a likely setting is the cooling solar nebula. Let us therefore begin by considering the condensation sequence of the elements from a cooling gas of solar composition.

Condensation of the Elements

The simplest case to consider is the condensation of a monatomic gaseous element to a pure metal, e.g. Fe-gas to Fe-solid. Condensation occurs when the partial pressure, p, in the gas equals the vapour pressure, p_0. From the Clausius–Clapeyron equation, the vapour pressure varies with temperature according to $\log p_0 = A/T + B$ where A and B equal the heat and entropy of vaporization divided by $2{\cdot}303R$ (= the gas constant, 1·98 cal/mol). The partial pressure of the gaseous species is simply the total pressure, P_T, times the mole fraction, which in a cosmic gas is the cosmic abundance of the element divided by the cosmic abundance of H_2. The condensation equation is written:

$$\log \frac{C(E)}{C(H_2)} + \log P_T = -\frac{A}{T} + B \qquad (1)$$

To begin the calculation, some total pressure must be assumed; we shall use $P_T = 10^{-4}$ atm for reasons that will become apparent.

Of course, this simple case applies to only a few elements, like iron. In most cases, elements exist as compounds in the gas and condense to sulphides, silicates or oxides. Moreover, most elements condense as impurities in the major phases (metal, sulphide, or silicate). The principles remain the same, but the algebra becomes complicated and a full discussion is beyond the scope of this paper. A more complete discussion, including references to the appropriate thermodynamic data, has recently been published (Grossman and Larimer, 1974). The solar system abundances were obtained from Cameron (1968, 1973a).

The results of the calculations are summarized in Figure 2. Two cases are illustrated. In one the system is presumed to cool rapidly; diffusion of the minor elements into the major species breaks down leading to condensation of essentially pure phases with thin surface coatings of impurities. In the second case, cooling is presumed to be slow enough that complete equilibrium in the solid state obtains. The vapour pressure of any species in solution is always less than that over the pure species; hence, in this case, minor species condense at higher temperatures. It seems that in reality some intermediate case would apply.

It is useful at this point to relate some of the key features of chondritic meteorites to the diagram. Carbonaceous chondrites contain Fe_3O_4, hydrated silicates and their full complement of volatile elements. This suggests formation at low

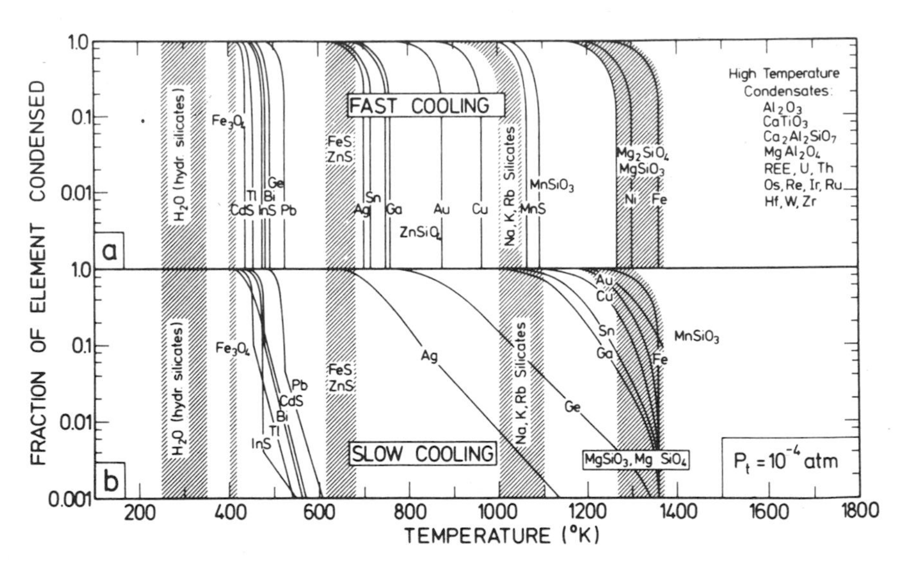

Figure 2. Condensation sequence of a gas with solar composition. In slow cooling sequence, diffusional equilibrium is assumed, with formation of solid solutions. In fast cooling sequence, diffusional equilibrium breaks down leading to condensation of pure elements and compounds. Shaded areas indicate condensation or formation of major constituents. (Reproduced with modifications from J. W. Larimer, 1967, *Geochim. Cosmochim. Acta*, **31**, 1215–1238, copyrighted by Pergamon Press)

temperatures, below 400 K, as was first noted by Urey (1952a, b, 1954). On the other hand, H, L and LL-group chondrites contain FeS, little or no Fe_3O_4 of preterrestrial origin and have variable Pb, Bi, In, and Tl contents. This points to a slightly higher formation temperature, 400 to 650 K (Larimer, 1967; Larimer and Anders, 1967). The simplest interpretation is that the various types of chondritic material became isolated from the gas, presumably by accretion, over a range of temperatures.

It is worth noting that diagrams such as Figure 2 are nothing more than phase diagrams drawn for a specific pressure and composition. As such they can be used to predict the composition and mineralogy of the condensed material at any temperature. To the extent that such predictions are observed, it can plausibly be inferred that a rather specific P–T–X condition existed in the solar nebula. Let us see what conditions are indicated, starting at high temperatures (> 1500 K) and working down toward the accretion temperatures.

Condensation and Fractionation of Refractories

The first hint of a high temperature fractionation process is in Figure 1, where it was shown that the Mg/Si ratios in chondritic material vary systematically; C/H, L, LL/E is 1·05/0·95/0·77 (von Michaelis, *et al.*, 1969). If this fractionation is due to the gain or loss of a refractory rich material then other refractory elements should fall into the same pattern. Fractionation factors for 14 elements are shown in Table 1. In all cases the pattern is the same; C > H, L, LL > E.

Table 1. Refractory element fractionation in chondrites; abundances relative to carbonaceous chondrites[a]

Element	Ordinary Chondrites	Enstatite Chondrites	Allende Inclusions
Al	0·71	0·55	21
Cu	0·67	0·53	17
Hf	0·50	0·43	23
La	0·80	0·43	26
Sc	0·72	0·63	24
Sr	0·62	0·48	15
Ti	0·74	0·55	21
U	0·68	0·50	13
Y	0·75	0·44	21
Yb	0·81	0·48	22
Zr	0·58	0·47	23
Cr	0·85	0·77	0·16
Mg	0·90	0·74	0·68

[a]The sources of the data in the first two columns are given by Larimer and Anders (1970); values in the last column are from Wänke *et al.* (1974).

While these data were being compiled (Larimer and Anders, 1970), cm-sized inclusions comprised of $MgAl_2O_4$, $CaTiO_3$, etc. were found in some carbonaceous chondrites (Christophe, 1968; Keil *et al.*, 1969). Also at about this time (Feb. 8, 1969), a large carbonaceous chondrite (Allende), which contained numerous inclusions of similar mineralogy (Marvin *et al.*, 1970), fell in northern Mexico. Nature thus obligingly supplied direct evidence that material of the right mineralogy and composition not only formed but managed to become segregated into cm-sized objects.

Since this refractory-rich material may represent the initial condensate in the solar nebula, it has been subjected to extensive study. We shall return to this topic in the next section. It is worth noting here however that the refractory material comprises about 5 wt per cent of the total condensible planetary material (= Cl chondrites) in the inner solar system. Refractory elements thus are expected to be enriched by about a factor of 20, just as observed (Table 1 above). However, Mg, which provided the first hint of a refractory element fractionation event, and Cr are not enriched in these inclusions. This presumably indicates that some other material was fractionated as well. Probably some olivine, Mg_2SiO_4, rich in Mg and, in some cases, Cr (Larimer and Holdsworth, 1975) was involved.

The mechanism by which this fractionation occurred remains a mystery. Possibly the cm-sized objects became gravitationally segregated or were moved around by large scale convection cells. In support of lateral movement in the solar system, there have been many suggestions that the earth and moon are enriched in refractory elements (see section *Differential Meteorites*). Whatever the mechanism of removal, it may be common to all nebulae. Herbig (1978) notes that relative to Na and K, interstellar gas is depleted in refractories; presumably these elements are present in the interstellar dust.

Metal–Silicate Fractionation

The variation in Fe/Si ratio among chondrites is direct evidence of a metal–silicate fractionation (Urey and Craig, 1953). Density differences among the planets provide indirect evidence (Urey, 1952a). But this observation is open to other interpretations, such as variation in oxidation state and amounts of volatiles (Ringwood, 1966; Lewis, 1972). In chondrites, the variation in Fe/Si includes all members of a chemical group, even the least metamorphosed. This indicates that the fractionation preceded metamorphism and, hence, accretion as well. Loss of metal by partially melting a parent body is inconsistent with this evidence. On the other hand, fractionation could not be the result of accretion during partial condensation of metal and silicate because the meteorites should then be devoid of FeS and other volatiles. Such an incomplete condensation process may explain the density of Mercury (Clark *et al.*, 1972; Lewis, 1972) but cannot explain the meteorite data. Instead, a fractionation of metal from silicate grains, while both were dispersed in space, is indicated.

Additional information on the conditions at the time of fractionation can be obtained in two ways. First, a plot of chondritic Fe/Mg *vs.* Ni/Mg ratios, defines the fractionation trend (Anders, 1964; Larimer and Anders, 1970; also see Y'avnel, 1966) (Magnesium is used for normalization by Larimer and Anders because the Fe/Mg ratio depends only on the oxidation state while the Fe/Si ratio depends on both oxidation state and proportion of olivine to pyroxene.) The slope and intercept give the Fe/Ni ratio in the metal and the Fe/Mg ratio in the silicate (plus sulphide) at the time of fractionation. From a least-squares analysis of the data, for the C, H, L and LL-groups:

$$\frac{\mathrm{Fe}}{\mathrm{Mg}} = (14{\cdot}50 \pm 0{\cdot}50)\frac{\mathrm{Ni}}{\mathrm{Mg}} + (0{\cdot}14 \pm 0{\cdot}015) \quad (2)$$

and for the E-group:

$$\frac{\mathrm{Fe}}{\mathrm{Mg}} = (15{\cdot}3 \pm 1{\cdot}2)\frac{\mathrm{Ni}}{\mathrm{Mg}} + (0{\cdot}0 \pm 0{\cdot}07) \quad (3)$$

The intercepts indicate that at the time of fractionation the Fe/Mg ratio in the silicate plus sulphide fraction was about 0·14 in the case of C, H, L and LL material but close to, or equal to, zero in the case of the more highly reduced E material. Second, those siderophile elements which had condensed on the metal grains before fractionation should be depleted along with Fe and Ni. Conversely, any which had not completely condensed would not be fractionated. The fractionation factors for such elements along with temperatures of almost complete condensation are shown in Table 2.

All elements which condense before Ga, which would be 90 per cent condensed at 1050 K are depleted along with Fe and Ni. This led Larimer and Anders (1970) to conclude that the fractionation occurred shortly after the metal grains became magnetic but before Ga was completely condensed. However, Chou

Table 2. Fractionation of siderophile elements in ordinary chondrules

Element	L Group/H Group	Condensation Temperature[a], K
Ir	0·57 ± 0·09	1500
Ni	0·68 ± 0·03	1310
Fe	0·74 ± 0·02	1270
Co	0·66 ± 0·05	1265
Pd	0·55 ± 0·23	1260
Au	0·61 ± 0·18	1190
Ga	0·93 ± 0·15	1050[b]
Ge	0·78 ± 0·16	810
S	1·07 ± 0·19	560

[a]For 90 per cent condensation at $P_T = 10^{-4}$ atm.
[b]The ferromagnetic Curie point of Fe–Ni alloys is 985–900 K.

and Cohen (1973) have shown that a large fraction of the Ga in the least metamorphosed chondrites is in the silicate phases, not the metal. Thus, the conclusion is less firm. Also, Larimer and Anders further argued that appreciable FeO would form before FeS, suggesting that the fractionation occurred before formation of FeS. But just the opposite now seems to be the case (Larimer, 1973). If FeS does form before FeO then some portion must have moved with the silicate fraction, possibly accounting in part for the fractionation of S between the H and L groups (Table 2). When these new data are taken into account, it appears that metal–silicate fractionation occurred after a portion of the Fe was present as FeS, below 680 K. This makes the mechanism they suggested, magnetism, look less appealing but does not rule it out.

Volatile Elements

Urey (1952a, b; 1954) first suggested that the volatile elements might be used as cosmothermometers to estimate the accretion temperatures of meteorites and the earth. The data available at the time indicated no deficiency of volatiles, except for the noble gases, H, C, N, etc., which led him to conclude that both the earth and meteorites accreted at low temperatures, less than 300 K. Since then more precise data, largely by neutron activation analyses, have shown that nearly all volatile elements are depleted in most chondritic meteorites. These data have led to new, more detailed interpretations.

Abundance Patterns

On the basis of good, but meagre, data Anders (1964) noted that in C-group chondrites all volatiles were present in maximum amounts in Cl material but were depleted by factors of about 0·5 and 0·3 in C2 and C3 chondrites. The quantity and quality of data have vastly improved since then and the trend has been toward even more constant depletion factors (Krähenbühl *et al.*, 1973a). Anders interpreted this pattern in terms of Wood's (1963) model in which chondrites are a blend of two components: a high-temperature, volatile-poor component (= chondrules) and a low-temperature, volatile-rich component (= matrix). The trend (C1/C2/C3–1·0/0·5/0·3) is explained by postulating 100, 50, and 30 per cent matrix in the three types of meteorites. This agrees with the observed proportions (Figure 3) and is supported by much additional evidence (Larimer and Anders, 1967; Anders, 1971).

In the H, L and LL-groups a somewhat different pattern is observed. Nine elements (Ag, Cu, Ga, Ge, S, Se, Sn, and Te) are again depleted by a constant, though smaller, factor of 0·23 (Krähenbühl *et al.*, 1973a). These elements, referred to as 'normal' elements (Anders, 1964), are most simply interpreted as indicating a slightly higher proportion of chondrules to matrix in these meteorites. Other volatiles, however, are depleted by larger factors, approaching 10^3 in the case of Bi, In, and T1. These elements are referred to as 'strongly depleted' (Anders, 1964).

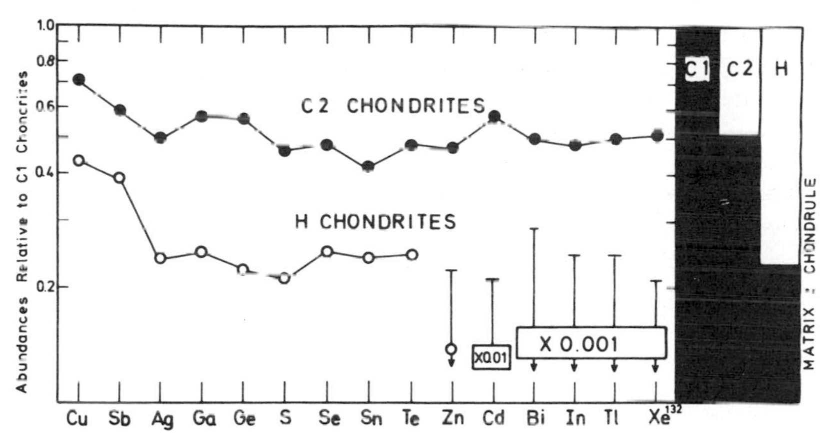

Figure 3 Many volatile elements are depleted to almost the same degree in C2 and H chondrites, relative to Cl abundances. The degree of depletions corresponds to the proportion of volatile-deficient chondrules in these meteorites. A few elements, Cu and Sb, are only partly depleted in the chondrules and, hence, the bulk meteorites also. In H chondrites, highly volatile elements are depleted by much larger factors, reflecting their deficiency even in volatile-rich matrix

Origin of the Two Components

There are many theories on the origin of chondrules. The more plausible can be divided into two broad categories: primary and secondary.

In primary theories, chondrules and matrix are both considered original condensates, differing only in their condensation paths: vapour to liquid yielding mm-sized spherules and vapour to solid yielding submicron-sized dust (Wood, 1962, 1963). The larger spherules (= chondrules) would be inefficient collectors of volatiles owing to their smaller surface area and larger diffusion depths. The difficulty with this model is that high pressures are required, 10 to 10^3 atm, if the silicates and metal are to condense as liquids (Suess, 1963). To circumvent this problem, Wood (1963) proposed that chondrules formed in transient high P–T events during the T-Tauri stage of solar evolution (see Herbig, 1976). Alternatively, Blander and Katz (1967) suggested that, because of the lower surface energy, liquid droplets might condense before solids even in the stability field of the solids.

In secondary theories, the original condensate consists of dust (= matrix), a portion of which is later reheated, melted and outgassed. Various mechanisms have been proposed; lightning discharges (Whipple 1966; Cameron, 1966); high-velocity impacts on the surface of parent bodies (Urey, 1952, King *et al.*, 1972), between dust grains (Whipple, 1972; Cameron, 1973b), and between metre-sized aggregates (Kieffer, 1975). In support of impact hypotheses, fragmented chondrules are common and at least one shows evidence of an impact event which partially melted it (Lange and Larimer, 1973).

Whatever the mechanism of chondrule formation, their spherical shape, the presence of glass and lack of volatiles clearly points to an origin as hot, molten droplets dispersed in space. Judging from Figure 2(a) above, which is the more appropriate diagram for all of the above models, the temperature of formation must be at least 1000 K ($P_T = 10^{-4}$) if the normal elements are to be volatilized. The next group of elements, alkalis and Mn, behave like normal elements in the C-group (chondrules) but are not depleted, or only partly depleted, in all other groups (Schmitt *et al.*, 1965; Larimer and Anders, 1967). The same is true of As and Sb (Case *et al.*, 1973). This suggests a higher T, or lower P_T, in the case of the C-group chondrules relative to the other groups. It is also significant that while the chondrule content in C-group chondrites varies from 0 to 70 per cent, the Al/Si, Ca/Si, Mg/Si, etc. ratios are constant. This indicates that the chondrule and matrix components must have the same non-volatile composition implying a close genetic link. It is unlikely that the two components formed in different regions of the nebula and then became mixed together.

This depletion of volatiles in chondritic material poses some interesting mass-balance problems. For example, if chondrules had been produced in the proportions in which they are observed (50 to 80 per cent) then the missing volatiles should condense on the fine-grained matrix. This would enrich the matrix in volatiles and, upon mixing, there would be no net depletion. Seemingly, the only way to avoid this paradox is to argue that chondrules were produced in relatively small amounts, less than 10 per cent, but were preferentially accreted (Larimer and Anders, 1967). Whipple (1972) points out that chondrule-sized objects might preferentially leak through the flow of gas and dust around an accreting body as it passed through the nebula. The fate of the excess dust remains a mystery. Perhaps it was swept up by the sun or blown out of the system during the T-Tauri stage (Herbig, 1978).

Accretion Conditions

Chondrules evidently ceased to maintain equilibrium with the gas below about 1000 K. The matrix, however, must have continued to equilibrate down to much lower temperatures, perhaps close to the time when accretion isolated the dust from the solar gas. The volatile content of the matrix may therefore be a useful guide to the conditions in the nebula at the time of accretion. But the fact that an element is volatile implies that it will be relatively easily remobilized during any subsequent reheating, thereby erasing its usefulness as an indicator of accretion conditions.

There have been a number of tests made to see which of the two models, accretion in a cooling nebula or metamorphism in a heating parent body, best accounts for the data (Keays *et al.*, 1971; Laul *et al.*, 1973; Case *et al.*, 1973; Kurimoto *et al.*, 1973; Larimer, 1973; Binz *et al.*, 1976). All of the available data, from these and other studies, are presented in Figure 4 together with the

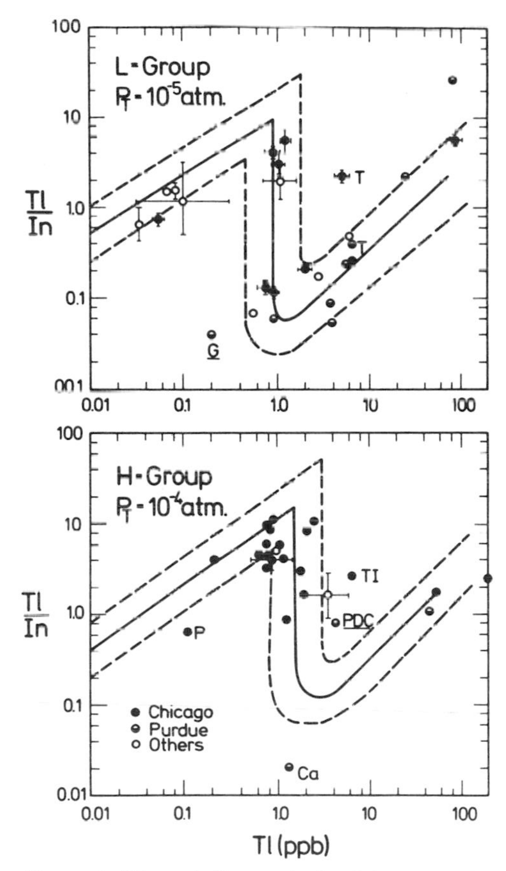

Figure 4. The relative and absolute amounts of In and Tl tend to follow the trend predicted to be established during condensation. Most samples fall within a factor of two (dashed lines) of the condensation curves. The exceptions are usually breccias (initials) or finds (underlined initials). Where two samples of such meteorites have been measured, (point T, also see Figure 5) variations by nearly a factor of 10 are observed. Data: Chicago, Keays *et al.*, 1971; Laul *et al.*, 1973; Purdue, Case *et al.*, 1973; Kurimoto *et al.*, 1973; Binz *et al.*, 1976; others, see text

trends predicted to develop in the nebular dust as it cooled from 550 to 420 K. The predicted trends have an unusual pattern owing to the fact that In, which occurs in the gas as In_2S and condenses to InS, follows a condensation path that differs from that of metallic T1 (Larimer, 1973). Both the H and L group data follow the predictions remarkably well, 43 of 51 samples fall within a factor of two to the curves. Most exceptions are brecciated meteorites or finds, where abundances vary significantly from sample to sample.

Breccias are more common among H chondrites, which is reflected in both Figure 4 and 5. One light-dark coloured H chondrite, Supuhee, contains material in its dark fraction which is very enriched in Bi and T1 and plots off the diagrams (Laul *et al.*, 1973). This material is probably present in other H chondrites, and may distort the data to some unknown extent.

All available In–Bi data are plotted in Figure 5 along with the condensation curves. Again the predictions have an unusual trend, which in this case also changes appearance at $P_T \sim 3 \times 10^{-5}$ atm. This reflects the dependence of the condensation curves on P_T; In condenses before Bi at lower P_T and after

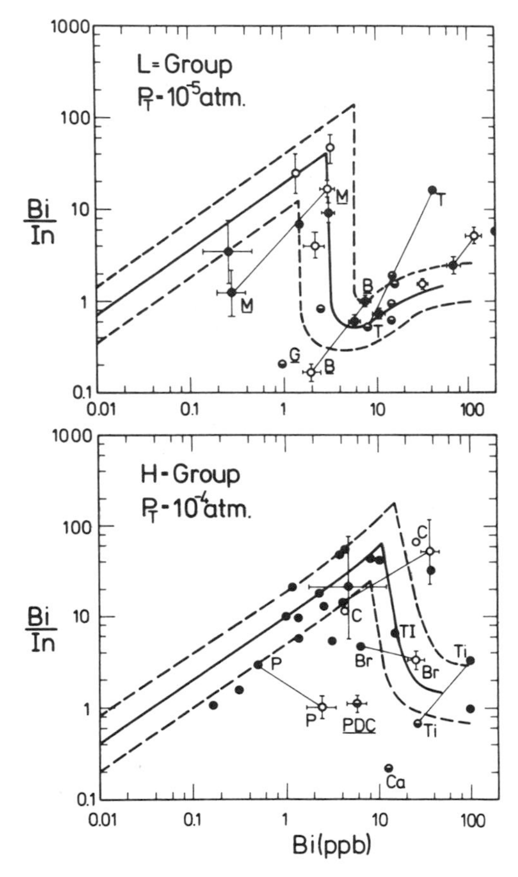

Figure 5. Bi/In ratios are frequently less than 1 in L chondrites but rarely in H chondrites. The exceptions are breccias (initials) or finds (underlined initials) whose trace element contents are suspect. This suggests a slight pressure difference between the regions where the two groups formed; pure InS condenses first at $P_T = 10^{-5}$ and pure Bi first at $P_T = 10^{-4}$ atm. For symbol definition, see caption to Figure 4

Bi at higher P_T (Larimer, 1973). If the breccias and finds are excluded, it appears that at similar Bi contents of 10 p.p.b., most L chondrites have In/Bi ratios < 1 while most H chondrites have ratios > 10. This suggests a slight pressure difference at the formation locations of these two groups.

The temperatures inferred from the Bi, In and T1 contents range from 550 to 400 K. If pressures of 10^{-5} and 10^{-4} atm are assumed, the temperatures tend to cluster around 450 to 460 K for L chondrites and 470 to 480 K for H chondrites. Obviously, data on carefully selected meteorites are required to further test these inferences.

However, there are additional thermometers and barometers which point to the same range of P–T conditions. The FeO content of the silicates points to a temperature in the range of 500 K (Larimer, 1968a, 1973). The presence of FeS and lack of Fe_3O_4 in these meteorites indicates a temperature between 405 and 550 K. Both the formation of FeS and Fe_3O_4 are pressure independent reactions while the condensation curves of the volatile elements are pressure dependent. If the trace elements condense between 550 and 405 K, then the pressure must fall between 10^{-6} and 10^{-3} atm (Anders, 1971; Larimer, 1973).

The internal consistency of the mineralogical and chemical data argues against loss or extensive migration of these volatiles during metamorphism. More likely the correlation between decreasing volatile content and increasing metamorphic grade is not a causal one. It is more simply interpreted as a natural consequence of accretion in the cooling nebula: material that accretes first is deficient in volatiles and is buried to the greatest depth where it is highly metamorphised.

The various cosmothermometers and barometers are summarized in Table 3. The last two entries in the table, C^{13}/C^{12} ratios in carbonates relative to hydrocarbons and Ar solubility in Fe_3O_4, apply only to the C chondrites. They are discussed by Lancet and Anders (1970, 1973). These independent clues, based on the mineralogy, elemental and isotopic composition, all point to the same

Table 3. Meteoritic thermometers and barometers

Thermometer/barometer	T(K)	P(atm)
FeS	$\leqslant 550$	Indep.
FeS–Pb depletion	$\leqslant 550$	$\leqslant 10^{-3}$
FeO in silicates	$500 \begin{cases} +100 \\ -\ 60 \end{cases}$	Indep.
Depleted Volatiles (Bi, Cd, In, Pb, Tl)	$550 \geqslant T \geqslant 400$	$10^{-3} \geqslant P \geqslant 10^{-6}$
No Fe_3O_4—Tl depletion	$\geqslant 400$	$\geqslant 10^{-6}$
In > Bi	460 ± 40	$< 10^{-5}$
Bi > In	460 ± 40	$> 10^{-5}$
Fe_3O_4	$\leqslant 400$	Indep.
Hydrated Silicates	375 ± 25	$10^{-3} \geqslant P \geqslant 10^{-6}$
C^{13}/C^{12} (carbonate-polymer)	360 ± 20	Indep.
Ar in Fe_3O_4	350 ± 35	10^{-6+1}

Table 4. Accretion conditions of chondrites

Group	P(atm)	T(K)
E	$\lesssim 10^{-4}$	470–500
H	$\gtrsim 10^{-4}$	470–500
L	$\sim 10^{-5}$	450–475
LL	$\lesssim 10^{-5}$	430–460
C	10^{-6}	350–400

range of $P-T$ conditions at the time the condensed material became isolated from the gas, presumably by accretion. Slight differences in $P-T$ conditions are inferred for each chemical group, perhaps reflecting increasing radial distance from the sun (Table 4); see also Larimer and Anders (1967).

Alternative Views

The possibility that volatile depletion is the result of outgassing during metamorphism remains a widely held view, despite the arguments just presented. Dodd (1969) has summarized this viewpoint. Until recently, however, there have been no quantitative predictions which could be used to test the hypothesis. Ikrumuddin *et al.* (1976) reported experiments in which samples of Allende were heated and outgassed, under vacuum, and the loss of Bi, In and T1 was determined. They argue that the experimental trends match the observed trends in E chondrites, and suggest that in this specific case outgassing in an open system occurred. But they did not include S in their experiments, which should be lost almost entirely under these conditions, yet it is not depleted in E chondrites.

Other explanations advanced to account for the chemistry of chondrites differ more fundamentally in their assumptions regarding the chemistry of the nebula. Blander and colleagues (Blander and Katz, 1973; Blander and Abdel-Gawad, 1969; Blander, 1971; Blander and Fuchs, 1976) argue that supercooling, or supersaturation, of the nebular gas played an important role during condensation. An interesting prediction is that metals, owing to their higher surface energies, will tend to supersaturate to a greater degree than silicates. For example, the activity of Fe in the vapour might increase to the point where it would condense as FeO incorporated in the silicates, or even FeS if the gas were extremely supersaturated. Unfortunately, finding an unambigious observation to support this view is not easy. We shall return to this question briefly in the next section.

Alfvén and Arrhenius (1976), on the other hand, have proposed an entirely different model. They argue that the chemical processes in the early solar system took place in a partially ionized gas that slowly accumulates around an already existing sun. In their view, any coincidences between thermodynamic calculations and observations on meteorites merely reflect the fact that the bonding properties of the elements control condensation in both cases. Also, in their model condensation occurs in a regime of rising pressure at constant $T(=500\,K)$. But if this were true an entirely different mineralogic, if not elemental, condensation sequence is predicted: Fe will condense as FeS and FeO not Fe, Ti as $FeTiO_3$ not $CaTiO_3$, Al as $CaAl_2Si_2O$ not Al_2O_3 or $MgAl_2O_4$, etc. Not only does this leave the mineralogy of the Allende inclusions unexplained, it seemingly leaves the refractory element/Si and metal/Si fractionations inexplicable.

Carbonaceous and Enstatite Chondrites—The Extremes

While all chondrites are thought to be primitive, the C and E chondrites are of special interest. They are, respectively, the most highly oxidized and reduced of all meteorites. Of the two, the C chondrites are best known and most thoroughly studied. They also fall naturally into the chemical sequence as the low-T end product of the condensation process. By contrast, the E chondrites do not fit easily into any model of chondrite genesis.

Carbonaceous Chondrites

There are many features of C chondrites which hold information in addition to those already discussed. Let us begin with the refractory-rich inclusions, briefly mentioned in the previous section.

Ca–Al-rich Inclusions

The fall of the Allende meteorite, which contains numerous inclusions rich in refractories, in early 1969 was followed later that year by Apollo 11. The first lunar samples were also rich in the same refractory elements and depleted in volatiles which led to almost immediate speculation that the moon was enriched in material resembling the inclusions (cf. Gast, 1972). This prompted extensive studies on the inclusion material because, not only may it represent the initial condensate in the nebula, it may also be important for understanding the bulk composition of the moon and planets.

Grossman (1972) computed the high-T condensation sequence in detail and showed that this sequence was reflected in the mineralogy and the textural relationships in the inclusions (Grossman and Clark, 1973). That is, the first minerals to condense were observed to be enclosed in minerals predicted to condense later, at lower temperatures. From Grossman's calculations it is also possible to predict the composition of the condensate at any temperature. There is good to excellent agreement between these predictions and the observed composition of the Ca–Al rich inclusions (Table 5). This agreement between predicted and observed compositions was also shown to extend to a number of trace elements (Grossman, 1973; Wänke *et al.*, 1974; see Table 1). The results

Table 5. Calculated *vs.* observed composition of Ca–Al-rich Allende inclusions

	Calculated condensate Composition $P_{tot} = 10^{-3}$ atm Grossman (1975) 1475 K	1450 K	Ca, Al-chondrule Clarke *et al.* (1970)	A 3 Palme *et al.* (1975)	A 17
CaO	32·31	27·23	26·76	25·3	22·1
Al_2O_3	34·81	29·22	31·61	33·1	28·8
TiO_2	1·77	1·49	0·99	1·52	1·25
MgO	9·39	16·98	10·82	11·1	12·3
SiO_2	21·71	25·09	29·79	29·5	29·1
FeO			0·37	0·98	7·2
Na_2O			0·11	0·46	0·4
	100·0	100·0	100·45	101·9	101·2

of a recent study on trace elements abundances from a larger sample of Allende inclusions is illustrated in Figure 6. The enrichment by nearly a factor 20, consistent with the fact that the initial condensates represent only 5 wt per cent of the condensible material, still holds. It even extends to some refractory

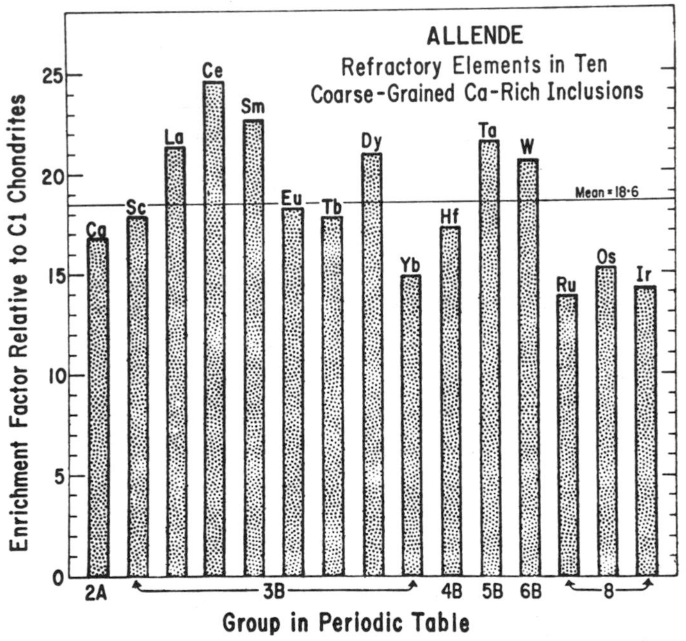

Figure 6. Refractory elements are uniformly enriched by nearly a factor of 20 in Ca–Al-rich inclusions from Allende (Reproduced with permission from L. Grossman and R. Ganapathy, 1976, *Geochim. Cosmochim. Acta*, **40**, 331–344, copyrighted by Pergamon Press)

noble metals, although their abundances are prone to vary in different samples.

The presence of refractory noble metals in these Ca–Al-rich inclusions raises the question of whether they condense as minute metal particles or as atoms dissolved in the silicate and oxide minerals. This question appears to have been answered by the discovery of small metal grains, greatly enriched in refractory metals, in several inclusions (Palme and Wlotzka, 1977; Wark and Lovering, 1976). The largest particle found is shown in Figure 7. Palme and Wlotzka compared the observed bulk composition of the particle, determined by non-destructive neutron activation analysis, to the predicted composition of metal grains in the cooling nebula (Figure 7). The Pt/Ir and Pt/Os ratios in the particle equal the cosmic ratios indicating that Pt, the most volatile of the refractory metals, had completely condensed. On the other hand, only a small fraction of the cosmic abundance of Fe and Ni is incorporated in this particle. These two elements are expected to condense in small amounts, as part of the complex alloy, above the temperature at which they would condense as pure Fe–Ni grains. If the alloy is assumed to have the properties of an ideal solution, the temperature difference between complete Pt condensation and the 90 per cent condensation of Fe–Ni is only 20 K. For example, at $P_T = 10^{-3}$atm, the permissible temperature range is from 1450 K to 1470 K. Of course, the thermodynamic properties of this peculiar alloy are not well understood so the upper temperature limit could be somewhat higher.

It does seem significant, however, that the particle contains appreciable Fe and Ni but ceased to equilibrate with the gas before the bulk of the Fe and Ni condensed. This indicates that Fe and Ni began to condense above, not below their predicted condensation temperature. Evidently, the refractory-rich metal particles served as seed nucleii to initiate the condensation of the major species. This makes supersaturation of the gas with respect to Fe vapour, as suggested by Blander and Katz (1967), a less likely possibility.

If these Ca–Al-rich inclusions represent the initial condensate in the nebula, this should be reflected in a greater age. However, the time that elapsed between the onset of condensation and accretion of the bodies in the solar system must be small compared to the $4{\cdot}5 \times 10^9$ yr that has elapsed since then (Kirsten, 1978). Determining age difference within this short time span requires methods with exceptional resolution. One method, capable of resolving events on a scale of $\pm$ 2 Myr, is the measurement of Sr^{87}/Sr^{86} ratios. This ratio will increase gradually in a homogenous gas as Rb^{87} decays to Sr^{87}. The first material to became isolated from the gas will have the smallest ratio. The measurements should preferably be made on Sr-rich, Rb-poor samples; otherwise subsequent decay of Rb will further increase the ratio. These inclusions, rich in Sr and poor in Rb are ideal. The smallest Sr^{87}/Sr^{86} ratio ever measured was on a sample from an Allende inclusion (Gray *et al.*, 1973). Other inclusions, with high Rb/Sr ratios, have younger ages, as recent as $3{\cdot}6 \times 10^9$ yr. (Gray *et al.*, 1973; Wetherill *et al.*, 1973). This may mark the time of a late addition of the Rb to the inclusion from the surrounding matrix, perhaps during the mild metamorphism experienced by Allende.

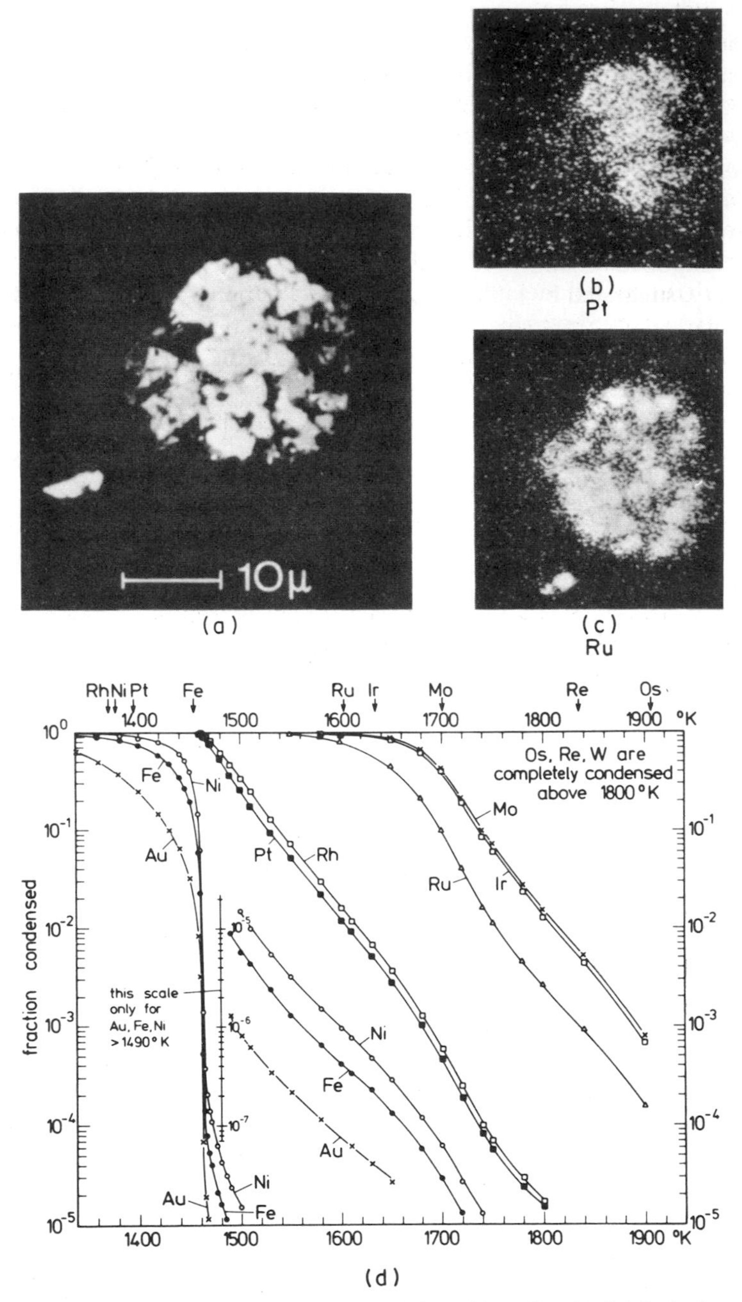

Figure 7. (a) The largest metal particle found in a Ca–Al-rich inclusion is highly enriched in all refractory metals. Electron microprobe analysis revealed that some, like Pt (b) are uniformly distributed while others, like Ru (c), appear to concentrate in discrete grains; (d) Pt appears to be fully condensed, while Fe and Ni are not, providing a means to delimit the temperature to 1500 to 1450 K

Two recent studies have yielded other, somewhat unexpected, data on the isotopic composition of the inclusions. Clayton *et al.* (1973) found the inclusions to contain unique $O^{16}/O^{17}/O^{18}$ ratios. Until this discovery, it appeared that all material from the solar system, though possessing oxygen of different isotopic composition, could have evolved from a homogeneous reservoir of uniform composition. The observed variations were explicable, and predictable, in terms of isotopic exchange during chemical reactions that occur at different temperatures. But the inclusions are enriched in O^{16} and the pattern suggests that a component of almost pure O^{16} must be present. This raises the intriguing possibility, suggested by Clayton *et al.* upon their discovery, that a component of presolar matter is present in these inclusions. Presumably almost pure O^{16}, formed by nuclear processes (α-burning) in a pre-existing star, condensed on interstellar dust grains and escaped evaporation during the formation of the solar system (cf. Cameron, 1978). A second isotopic anomaly, an enrichment of Mg^{26} has also recently been reported (Lee *et al.*, 1976). The enrichment seems to correlate with Al content, suggesting that it might be due to the decay of Al^{26} with its 7×10^5 yr half-life. The possible presence of Al^{26} in the early solar system has long been of interest because it could supply the energy needed to melt small bodies, such as asteroids, very early in their history (Urey, 1955; Fish *et al.*, 1960). Whether the Al^{26} formed within or outside the solar system is an important but, as yet, unresolved problem (Heymann and Dziczkaniec, 1976; Cameron, 1978).

Noble Gases

In addition to the clues on the high-T chemistry of the nebula, C chondrites also appear to best preserve clues on the low-T chemistry. Of all samples of solar system matter available, C chondrites are the richest in volatiles, including the noble gases. Both the isotopic and elemental abundance of the noble gases have long been regarded as important clues to the origin of meteorites as well as the earth and its atmosphere.

In meteorites, terrestrial and lunar rocks the noble gases are depleted by factors of 10^4 to 10^{12} or more relative to their inferred solar system abundance (Cameron, 1973a). Moreover, a significant fraction of the gases present are of secondary origin, the result of radioactive decay ($U, Th \rightarrow He^4, K^{40} \rightarrow Ar^{40}$) or cosmic-ray-induced reactions (He^3, Ne^{21}). This fraction of the gases is important for dating purposes (see Kirsten, 1978) but provides little information on the origin of primordial gases. Fortunately, the fraction of secondary gases can easily be subtracted using isotopic ratios. When this is done, there appear to be not one but two types of primordial gas in the meteorites. In one fraction, called 'solar' the noble gases are present in their inferred solar system proportions. The gas is found in meteorites of all types, generally associated with shock and brecciation. It was also found in lunar soils and breccias. Ingenious experiments (Hintenberger *et al.*, 1965; Eberhardt *et al.*, 1965; Pellas *et al.*,

1969) have established that this gas represents implanted and trapped solar wind particles (Wänke, 1965). The second fraction called 'planetary', is enriched in the heavier elements, similar to the earth's atmosphere (Figure 8). Evidently, it is of more ancient origin, occurring in largest amount in primitive chondrites. This fraction presumably was incorporated at the time of origin (for detailed reviews see Pepin and Signer, 1965; Heymann, 1971).

While the elemental abundances are explicable in terms of two fractions, the isotopic composition is more complex. The isotopic ratios differ between the 'solar' and 'planetary' components in meteorites while the earth's atmosphere has yet a third composition (Figure 8).

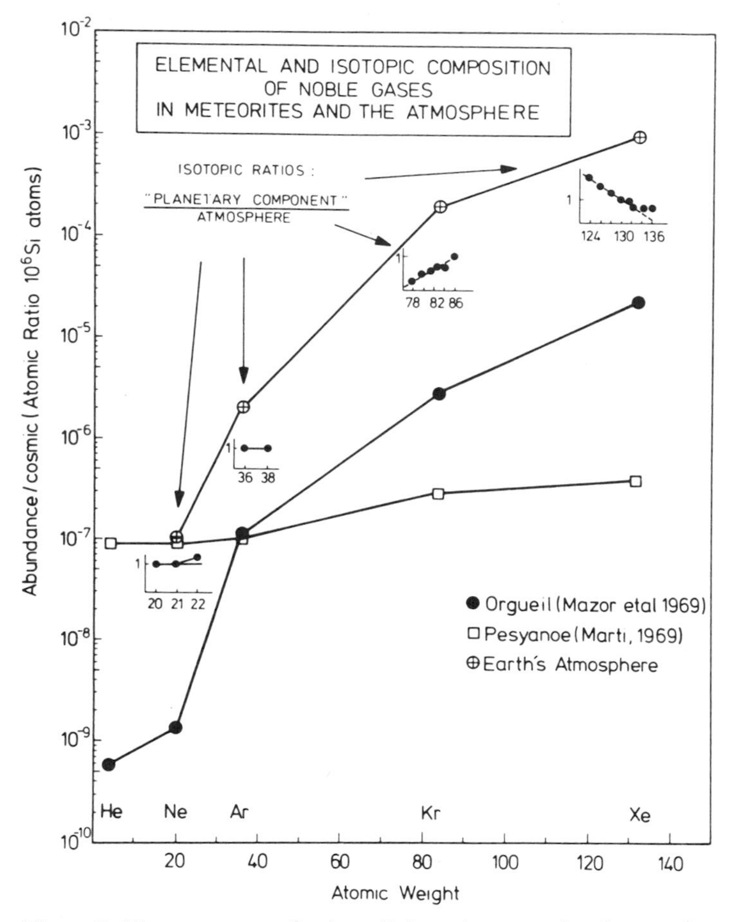

Figure 8. The two types of primordial gas in meteorites have quite different elemental abundance ratios. 'Solar gas' is little fractionated relative to cosmic abundances, it displays a nearly flat trend. 'Planetary gas' is enriched in the heavy gases, like the earth's atmosphere. But planetary gas differs from the atmosphere in isotopic composition, including a curious reversal in trend between Kr and Xe

A difference in isotopic composition between the planetary and solar components is measurable only in the case of He and Ne. The solar component is enriched in the light isotope of both elements, He^3 and Ne^{20} (Black and Pepin, 1969; Jeffery and Anders, 1970; Black, 1976). The enrichment of He^3 in the solar component can most plausibly be explained by a differential acceleration of the two isotopes in the solar wind (Geiss *et al.*, 1970). The isotopic composition of Ne has become complicated with the discovery that the planetary component may be a mixture of two or more components, enriched to varying amounts in Ne^{22} (see Black, 1976).

The differences between the meteoritic planetary component and the atmosphere are best resolved among the Kr and Xe isotopes. Curiously, the isotopes have almost exactly opposite fractionation trends (Eugster *et al.*, 1967, see Figure 8). The atmosphere is highly enriched in ^{40}Ar, derived from the decay of ^{40}K. The remaining two Ar isotopes, 36 and 38, are within experimental error identical in the meteorites and the atmosphere. As discussed, meteoritic Ne is complex; if the traditionally accepted 20/22 ratio of 8·2 in the planetary component is assumed, meteoritic Ne is slightly enriched in Ne^{22} relative to the atmosphere. Nearly all the He in the atmosphere is He^4, derived from the decay of U and Th.

Attempts to explain these patterns have led to considerable speculation. But some recent experimental data have changed the entire perspective. In a search for the mineral which contains the excess heavy Xe isotopes, 134 and 136, Lewis *et al.* (1975) made several unexpected discoveries. From previous work, a metal or sulphide mineral was suspected. An attempt was made to concentrate an Fe–Ni sulphide, reportedly insoluble in HC1, by dissolving everything else. The mineral turned out to be soluble also, but the acid-insoluble residue, amounting to only 0·5 per cent of the meteorite, contained virtually all of the planetary gas in the meteorite, including the excess heavy Xe. The residue consisted of amorphous carbon, $FeCr_2O_4$ and possibly an Fe–Cr sulphide. An attempt was then made to dissolve the Fe–Cr sulphide in various acids. Only about 4 to 8 per cent of the sample dissolved but it lost nearly all of its noble gases, except for He and Ne. Most of the He and Ne presumably resides in $FeCr_2O_4$; similar minerals have previously been shown to be enriched in the light gases (Jeffery and Anders, 1970). Essentially all of the Ar, Kr and Xe appear to reside in the minor mineral comprising 0·02 to 0·04 per cent of the meteorite. Similar experiments have now been reported by Phinney *et al.*, (1976), but failing to observe an Fe–Cr sulphide argue that the gases reside in the amorphous carbon.

The purpose of the search for a mineral containing excess Xe^{134} and Xe^{136} goes back to a suggestion made independently by Srinivasan *et al.* (1969), Anders and Heymann (1969) and Dakowski (1969). Noting theoretical predictions which indicate that some isotopes near proton number 114 and neutron number 184 might be long-lived, all suggested that the excess heavy Xe might be the fission product of a superheavy element. Anders and Heymann pointed out that the quantity of excess Xe correlated with the volatile content of meteorites, which focused suspicion on elements 112–119 whose volatilities might be similar to their light cogeners Hg, T1, Pb, Bi, etc. Anders and Larimer (1972) attempted to deduce the heat of vaporization and boiling point of the element, if it existed, and compared these inferences to those predicted from the periodicity of the elements. This narrowed the possibilities somewhat, to elements 111 to 116, with element 115, eka-Bi being the most likely. Interestingly, Bi was among the elements that went into solution along with the minor Fe–Cr sulphide phase and the noble gases in the experiments of Lewis *et al.* (Anders *et al.*, 1975).

An independent search for minerals containing noble gases in ureilites, the achondrites with graphite veins and diamonds, has turned up equally significant data. First, it was shown that the graphite veins were highly enriched in primordial noble gases (Weber *et al.*, 1971). Then, by a process of elimination, it now appears that the diamonds within these veins hold the bulk of the gases (Weber *et al.*, 1976).

Organic Matter

The carbon in C chondrites, besides being present in the amorphous form, also combines with H, N and O to form compounds of biological significance. A search for evidence of life has produced many claims and counter-claims over the past 20 years. Hayes (1967) has reviewed the problem, emphasizing the formidable analytical difficulties owing to the possibility of terrestrial contamination. The controversy led to detailed studies of carbon-bearing compounds and the discovery of adenine, quanine and cytosine, the constituent bases of DNA and RNA (Kvenvolden *et al.*, 1971; Hayatsu *et al.*, 1971, Yoshino *et al.*, 1971). But while these compounds are now known to occur in meteorites, it is generally agreed that they are of abiological origin. This, however, does not diminish their significance for it implies that abiological processes, operating in the early solar system, must have produced many of the compounds basic to life.

To understand the formation process requires an understanding of the chemistry of carbon in a solar gas. The thermodynamics have been studied extensively (Urey, 1953; Suess, 1962; Dayhoff *et al.*, 1964). At high temperatures, CO is the most stable carbon bearing compound. But it becomes less stable at lower temperatures and reacts with H_2 to form CH_4 according to the reaction

$$CO + 3H_2 \rightarrow CH_4 + H_2O \tag{4}$$

This reaction depends on P_T proceeding to the right at 750 K to 450 K at $P_T = 10^{-3}$ to 10^{-6} atm. However, Urey (1953) pointed out that this reaction may not proceed smoothly in the absence of a suitable catalyst. Indeed, if the reaction had gone to completion, all of the C in the inner solar system would have been present as gaseous CH_4 and lost along with the other volatiles.

Anders *et al.* (1973) summarized the results of a series of experiments in which mixtures of CO and H_2 were exposed to natural catalysts expected in the nebula (metal, Fe_3O_4, hydrated silicates, etc.). These catalytic reactions, known industrially as Fischer–Tropsch reactions, preferentially produce straight-chain or slightly branched hydrocarbons of the general formula C_nH_{2n+2}. This type of compound is the most abundant in C chondrites. The match between the experimentally produced compounds and those found in meteorites is impressive in its detail. In addition, Lancet and Anders (1970) have shown that during these reactions the C^{12}/C^{13} fractionation between the hydrocarbons and CO_2 is in the right direction and of the correct magnitude to account for the observed fractionation (see the Section *Chondrites and the Chemistry of the Solar Nebula*).

An important question persists: do the reactions take place in the nebula or during metamorphism in the parent bodies? In the experiments the H_2/CO ratio was a factor of 2 or more below the cosmic value. Also the $P_T \sim 1$ atm; when the pressure is dropped to 10^{-3} atm, no reaction occurs. This, however, may simply be due to the short laboratory time scale. The most likely alternative mechanism to produce hydrocarbons is the Miller–Urey reaction, which involves irradiation of nebular gases (Miller, 1953). But the free radicals produced link randomly and the process thus lacks the required selectivity to produce straight-chain hydrocarbons. Nor does it explain the C-isotopes fractionation in a straightforward manner.

Enstatite Chondrites

The nebular chemistry of carbon obviously assumes new significance with the recent developments bearing on the siting of the noble gases and the origin of hydrocarbons. It also appears that carbon chemistry may bear on the origin of E chondrites. For many years, these highly reduced meteorites were considered the extreme end product of a metamorphic-reduction process operating on material of C chondrite composition (c.f. Ringwood, 1966). But this cannot be the case. The E-group, like the other groups, contains members which are relatively unmetamorphosed. In some E-chondrites, the chondrule-matrix intergrowth is negligible, unequilibrated minerals (Keil, 1968) and unequilibrated mineral assemblages (Binns, 1969), are present, and they are as rich in volatiles as C chondrites (Anders, 1971). Clearly, these meteorites could not have been heated and reduced to the extremes required anytime after their formation. More probably, their basic chemistry was established in the nebula prior to and during accretion, just like other chondrites.

The C and E chondrites, though opposites in many ways, also share some characteristics. In both cases, the petrologic types differ chemically in ways that are inconsistent with a simple genetic relationship where the higher metamorphic grades evolve from the lower via thermal metamorphism. For example, obvious primary features such as major element ratios (Fe/Si) and chondrule/

Table 6. Enstatite *vs.* carbonaceous chondrites

Characteristic	Carbonaceous	Enstatite
Oxidation State	High FeO/MgO	FeO/MgO ~ O
Mineralogy	Hydrated silicates, Fe_3O_4	CaS, TiN, graphite
	FeS—Fe	FeS—Fe
Bulk Composition		
Mg/Si	1·05	0·7 to 0·9
Fe/Si	0·8 to 0·9	0·7 to 1·4
Volatile trace elements	High	Moderate to high

matrix ratios vary with petrologic type. The seemingly significant similarities and differences are summarized in Table 6.

If the nebula can be considered a chemical system then the uniqueness of these highly reduced chondrites must be linked to one of three variables: T, P_T or composition. Like other chondrites, these contain high-T minerals (pyroxene, olivine, metal, etc.) and high volatile contents, suggesting an origin in the same thermal regime. A high P_T origin has been proposed by both Blander (1971) and Herndon and Suess (1976), using quite different arguments. Blanders suggests that Fe vapour is more prone to supersaturation at high P_T, to the extent that FeS might condense before Fe, at 1000 K to 1200 K (compare to Figure 2). On the basis of hypothetical phase relations, Ca, Mg, and Mn are predicted to condense as sulphides dissolved in the FeS. There are a number of difficulties: the degree of supersaturation required is excessive, FeS is unstable at these temperatures and would decompose rapidly, Ca, Mg, and Mn sulphides are present as discrete phases, not exsolution products and the model does not explicitly account for nitrides, carbides or graphite. Herndon and Suess argue that at high pressures, TiN condenses before TiO_2 and CaS before CaO. But these calculations have no significance because the most stable high-T compounds ($CaTiO_3$, $CaAl_2Si_2O7$, etc.) have inexplicably been omitted. Moreover, all oxidation–reduction reactions are independent of P_T; they depend only on the H_2/H_2O ratio. A different pressure or temperature regime may account for a few, but not all, of the properties unique to E chondrites. This leaves composition as the key variable.

The overall chemistry of a solar gas is relatively insensitive to small changes in composition. A different abundance value for one element may change its condensation temperature but, in general, this has little or no effect on any other element. There are two exceptions, O and C the third and fourth most abundant elements. A change in their relative abundance by less than a factor of 2 can produce an entirely new condensation sequence.

In a study on the stability of CaS, a mineral unique to enstatite chondrites and achondrites, Larimer (1968b) pointed out that this and other minerals unique to these meteorites would become stable if the solar C/O value of 0·6 were increased to 0·9. This suggestion was followed by a study on the effect of C/O ratio on the entire condensation sequence. The preliminary results were unexpected. All the high-T oxides and silicates, including those minerals found in Allende inclusions, condense at progressively lower temperature as C/O increases (Larimer, 1975). This result, which perhaps should have been expected, has a straightforward explanation. Since CO is the most stable C- or O-bearing compound at high-T, only the element in excess is free to form other compounds. As C/O approaches 1, the amount of excess O decreases thereby lowering the condensation temperatures of all O-bearing compounds. Metals, such as Fe, are unaffected.

Of possibly even greater significance, the oxides and silicates are replaced by a new suite of high-T minerals. These include: CaS, TiN, MgS, AlN, SiC, etc. (Larimer and Bartholomay, 1976), many of which occur in E chondrites. In addition, C is stable as graphite and Fe_3C may form, both of which are observed in these meteorites. The obvious implication is that these minerals may represent the initial condensate in some region of the nebula.

Again, the chemical evidence points to a fractionation process but not to a mechanism. Gas-dust separation is perhaps the simplest and most effective process. But here both elements become appreciably condensed only at very low-T, less than 200 K, assuming a C/O ratio of 0·6. One possibility is that graphite, which becomes very refractory as C/O approaches 1 may have escaped complete evaporation during the formation of the solar system, like the O^{16}-rich component in the Allende inclusions.

Differentiated Meteorites—Implications for the Evolution of Planetary Matter

The iron, stony-iron, and achondritic meteorites are generally believed to have passed through a molten stage while residing in their parent bodies. If this is true they contain information of a different sort, not just about nebular processes but also about processes that occur during the early evolution of planetary matter.

It is convenient to divide this discussion into two parts, by first considering the metal fraction and then the silicate fraction. Besides having different chemistries, the two fractions must have separated early in the evolution of the solar system, as indicated by their great ages (cf. Kirsten, 1978), and must have had separate histories ever since.

Metal Fraction

Iron meteorites have a simple appearance that conceals a complex chemistry. Their trace element contents vary by up to a factor of 10^4, but tend to cluster or display correlations indicative of groupings. Wasson and colleagues (summarized by Scott and Wasson, 1975) have resolved 12 groups on the basis of Ga, Ge, Ir, and Ni contents in nearly 500 iron meteorites. This grouping implies

12 parent bodies. But 14 per cent of the meteorites studied do not fit into one of these groups and, while there are a few with similar chemistries, it appears the number of bodies is even larger.

The chemical trends between and within groups are complex, presumably reflecting the various processes that left their imprints. An important interpretative step was taken by Scott (1972). He first showed that the content of 15 other elements coincides with and thus supports the chemical groupings. He then points out that two fractionations must be explained: the between-group and the within-group fractionations. Within a group, pairs of elements correlate over wide concentration ranges. Such trends are best explained by a fractional crystallization process, perhaps one that operated in the cores of the parent bodies. The fractionations between groups must predate those within groups, presumably reflecting chemical differences between bodies.

Using these ideas as a guide, Kelly and Larimer (1976) took up the problem from a slightly different perspective. They traced the cosmic history of the metal phase through four stages:

(1) Condensation, nebular fractionation and accretion;
(2) Oxidation or reduction;
(3) Melting, or partial melting and segregation;
(4) Fractional crystallization during cooling.

The changes in composition at each stage are predictable and can be compared to the end product, the iron meteorites.

Chemical History of The Metal Phase

Only strongly siderophile elements (Au, Co, Cu, Fe, Ga, Ge, Ir, Mo, Ni, Os, Pd, Pt, Re, Rh, and Ru) were included in the study. This simplifies the problem, yet, because the elements do differ in volatility and melting–freezing behaviour, provides a means to resolve trends that arise at each stage in the metal's history.

The predicted compositional changes during stages 1, 2, and 4 are schematically illustrated in Figure 9. It is useful to normalize the data to a major element; Ni is the obvious choice. The elements are divided into three groups: those with low, similar, and high volatilities relative to Fe. Since volatility (or boiling point) tends to correlate with melting point, the elements should retain this grouping throughout the history.

Just below the condensation temperature of Fe, 1275 K at 10^{-5} atm, the metal is predicted to contain about 18 wt per cent Ni. It will be enriched in refractory elements (Os, Re, etc.) and depleted in volatiles (Au, Cu, etc.). Three elements, Co, Ni, and Pd, condense almost simultaneously and thus retain their cosmic proportions during condensation. With further cooling, Fe and Ni condense rapidly, achieving their cosmic proportions ($\sim$ 5·5 wt per cent Ni) within 125 K or so. The refractory element contents are simply diluted to their

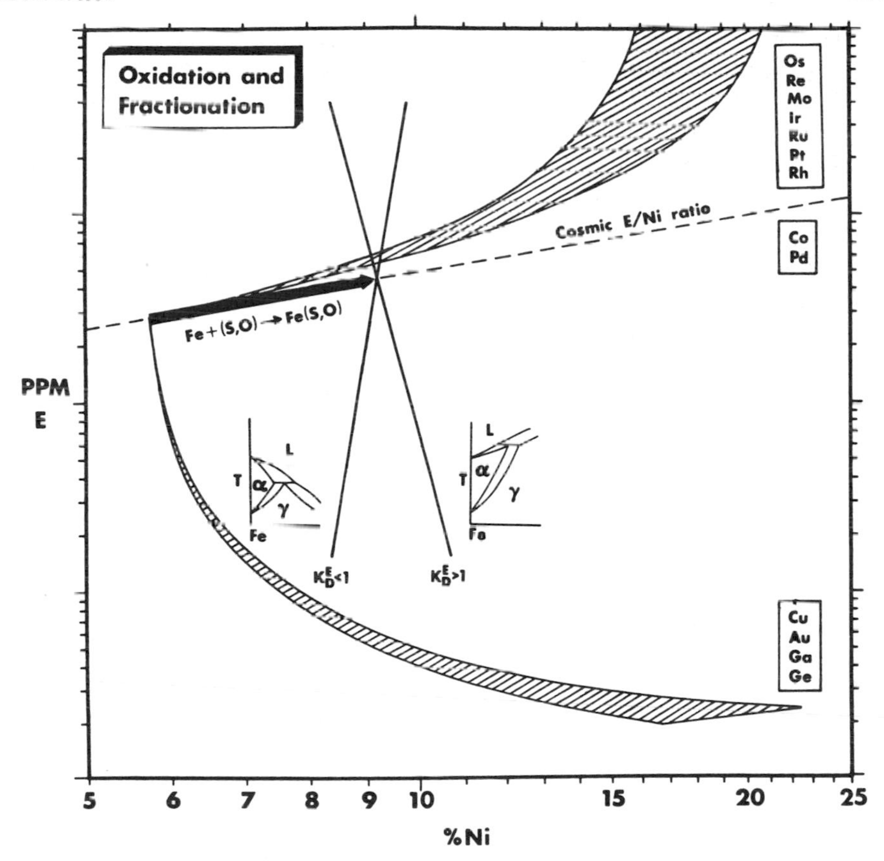

Figure 9. At the condensation temperature of Fe (1270 K, $P_T = 10^{-5}$ atm), the predicted Ni content is 19 ± 3 per cent. Refractory metals (Os, etc.) are enriched in the early condensate; their content varies along a line in the upper shaded region. Volatiles (Au, etc.) are depleted, their content varies along a line in the lower shaded region. The size of the shaded region reflects uncertainties in the thermodynamic data. Co, Pd, and Ni condense almost simultaneously, they retain their cosmic proportions during condensation.

Oxidation of Fe shifts the composition along the heavy arrow. Fractional crystallization of a melt with a composition at the tip of the arrow spreads out the trace element contents along lines of positive or negative slope depending on their phase relations with Fe

cosmic ratios over this interval (also see Figure 7) while the volatile content increases and continues to increase at even lower temperatures.

As discussed earlier, between 700 and 500 K some fraction of the Fe is converted to FeS and FeO. Since the elements considered here remain in the metal phase, they retain their cosmic proportions. The removal of Fe simply shifts the composition along the cosmic ratio line, or one parallel to it.

To keep this discussion simple, let's consider just three elements, a refractory

(Ir), a volatile (Au) and Ni. Let's further assume that they have fully condensed and are present in their cosmic proportions in a partially oxidized metal phase, similar to that found in the H or L chondrites. The body heats up, the metal sinks to form a molten core and begins to refreeze. Each element will distribute itself between solid and liquid metal according to its distribution coefficient, K_D^E, defined as

$$K_D^E = C_E^S / C_E^L$$

where C^S and C^L are the concentration of the element E in the solid and liquid phases. Though quantitative values of K_D are lacking, as Scott (1972) points out, elements with high melting point relative to Fe should concentrate in the solid ($K_D > 1$) and *vice versa*. During fractional crystallization, the concentration of an element in the solid varies according to

$$C_E^S = K_D^E C^0 (F)^{K_D^E - 1}$$

where C^0 is the bulk concentration and F is the fraction of liquid remaining ($0 \leqslant F \leqslant 1$). Combining such a relationship with a similar one for Ni yields

$$\log C_E^S = A \log C_{Ni}^S + B$$

The slope of this equation, A, is equal to $(K_D^E - 1)/(K_D^{Ni} - 1)$. The Fe–Ni phase diagram resembles the left one in Figure 9, indicating that in cosmic metal, with its low Ni content, $K_D^{Ni} < 1$. Any element with a K_D less than 1 (Au) will be correlated positively with Ni but if K_D is greater than 1 (Ir) it will be correlated negatively.

Iron Meteorites

A representative sample of the Au, Ir, and Ni data from iron meteorites are illustrated in Figure 10. For the most part, the remaining elements are consistent with these patterns (Scott, 1972; Kelly and Larimer, 1976).

One group of iron meteorites, IVB, stands out; it is enriched in refractories and depleted in volatiles. For all 15 elements considered, this group falls on or near the predicted condensation curve. The temperature indicated is surprisingly high, just below the condensation of Fe (~ 1270 K at 10^{-5} atm). No other combination of processes is capable of producing metal with this composition. It appears that in at least one case, the metal and presumably the parent body accreted at very high temperatures. Another, smaller group (5 members), IIIF, appears to have accreted at slightly lower temperatures, ~ 1180 K. Yet a third group, IVA, appears to have accreted at ~ 1220 K, though in this case the metal subsequently became fractionated as discussed below.

Several groups, IIAB and IIIAB, display fractionation trends that are consistent with a more traditional history. Here, the patterns are similar to those schematically illustrated in Figure 9; the only difference being the degree of oxidation.

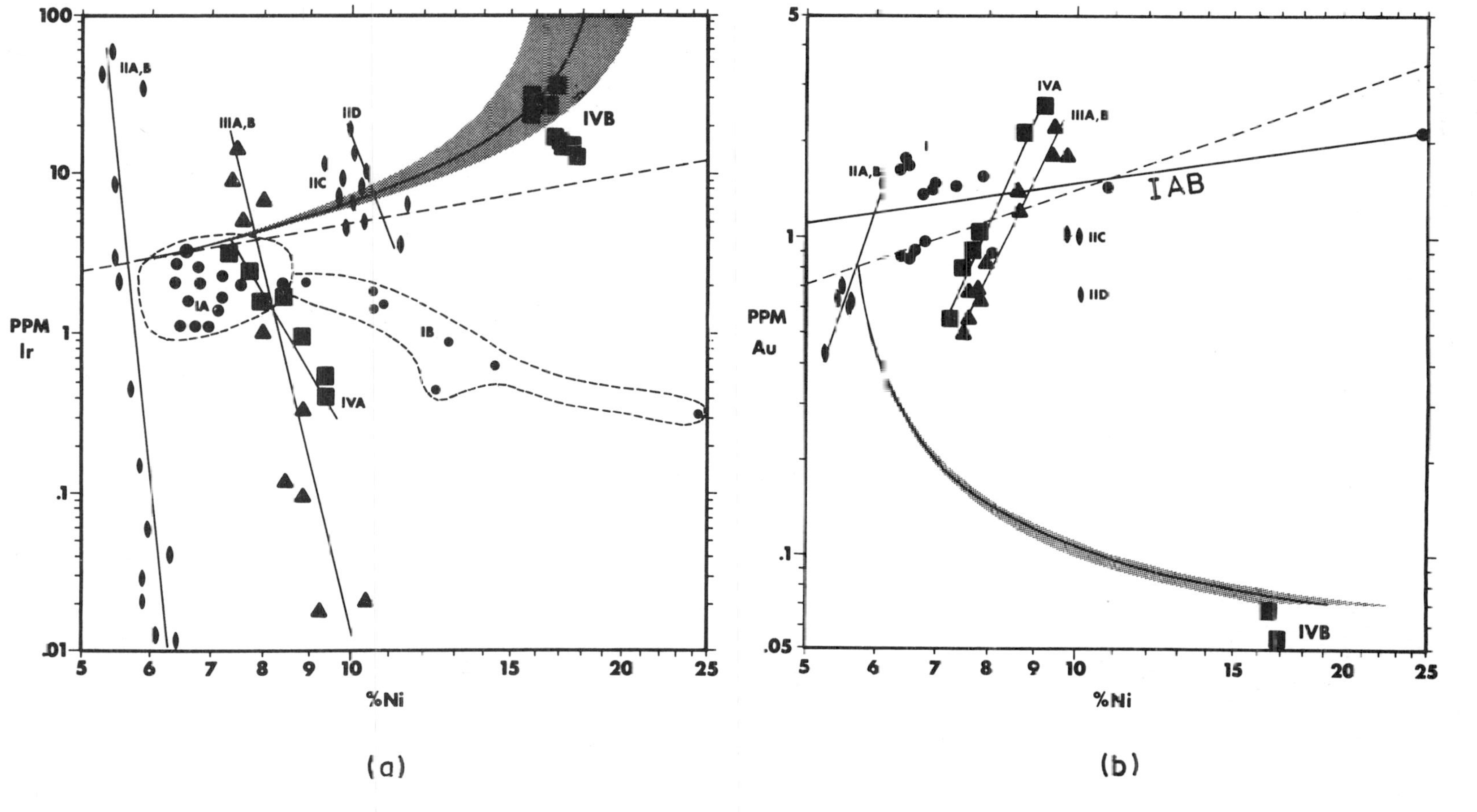

Figure 10 (a) Refractory Ir is negatively correlated with Ni in several iron meteorite groups (IIAB, IIIAB, etc.), just as predicted from Figure 9. Group IVB stands out, its composition falls on the condensation curve at high temperatures. Group IAB has a unique Ir–Ni correlation. (b) Volatile Au is positively correlated with Ni, as predicted. Group IVB again falls on the condensation curve and Group IAB displays a unique Au–Ni correlation

The Au and Ir data define unique trends in Group IAB. These meteorites are unusual in other ways. Many contain silicate inclusions, almost chondritic in composition. One possible explanation is that these meteorites are derived from a body in which the metal–silicate segregation process did not go to completion. Perhaps the body began to cool before the segregation process had run its course. These meteorites may retain a record of the partial melting process.

The composition of liquid and solid metal during partial melting can be predicted in a manner similar to the fractional crystallization stage. There are some physical differences which lead to a slightly different algebra. The problem is analogous to the terrestrial and lunar one of generating basaltic liquids by partial melting (Gast, 1968a; Shaw, 1970). Again, each element distributes itself between solid and liquid according to its distribution coefficient. Following Shaw's algebraic solution to the problem, three steps in the melting process are distinguished. These are best arranged in their probable chronological order:

(i) Melting proceeds to some finite extent before migration begins.
(ii) Once migration begins, pores and channels become available and the melt flows away continuously, as it forms.
(iii) The increments of melt generated in step ii aggregate into large masses which become homogenized.

In the first step, when fraction F has melted, the concentration of element E in the liquid, $C_{\mathrm{E}}^{\mathrm{L}}$, relative to its original concentration, $C_{\mathrm{E}}^{\mathrm{O}}$, is obtained by substituting K_{D} into a mass balance equation:

$$C_{\mathrm{E}}^{\mathrm{L}}/C_{\mathrm{E}}^{\mathrm{O}} = [K_{\mathrm{D}}^{\mathrm{E}} + F(1 - K_{\mathrm{D}}]^{-1}$$

As the melt begins to flow continuously, the composition changes continuously according to:

$$C_{\mathrm{C}}^{l}/C_{\mathrm{E}}^{\mathrm{O}} = \frac{1}{K_{\mathrm{D}}^{\mathrm{E}}}(1 - F)^{1/(K_{\mathrm{D}}-1)}$$

where C^{l} is used to distinguish this melt from that of the first step. The form of this equation is similar to the fractional crystallization equation, and combining such an equation for element E with a similar one for Ni yields a linear relationship on a log–log plot. Here, however, the slope

$$A = \left(\frac{1}{K_{\mathrm{D}}^{\mathrm{E}}} - 1\right)\Big/\left(\frac{1}{K_{\mathrm{D}}^{\mathrm{E}}} - 1\right).$$

The composition of the aggregate melt, obtained by integration varies according to

$$\bar{C}_{\mathrm{E}}^{1}/C_{\mathrm{E}}^{\mathrm{O}} = \frac{1}{F}\left[1 - (1 - F)^{1/K_{\mathrm{D}}^{\mathrm{E}}}\right]$$

where $\bar{C}^{1}$ is used to distinguish this composition from the others.

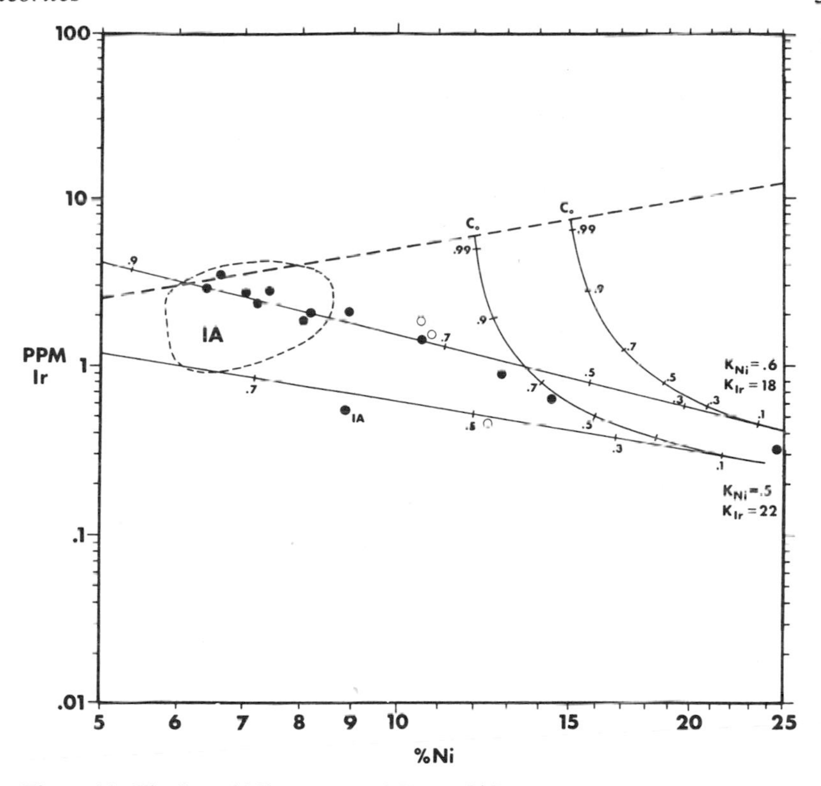

Figure 11. The Ir and Ni contents of Group IAB members, known to contain silicate inclusions (filled circles), are linearly correlated. The correlation follows the predicted fractional melting trend for partially molten metal that is continuously draining away from its residual solid. The curved lines indicate the predicted composition for aggregates of molten metal. Co is the assumed starting composition; the open circles are Group IAB members which do not contain silicate inclusions

If the metal in Group IAB refroze in the process of segregating itself from the silicates then it should have the compositional trend associated with step ii. This relationship, along with the Ir–Ni data from Group IAB are illustrated in Figure 11.

Obviously, it is of interest to look for chondrites that may have lost part of their metal. If the model is correct, the metal that remains would represent the residual solid and should have a complementary composition. The meteorite Shaw, a highly metamorphosed L chondrite which is deficient in metal and sulphide, is a likely candidate (Fredriksson and Mason, 1967). The composition of its metal together with data from average L6 chondrite metal and metal from San Cristobal, the extreme end member of Group IAB is presented in Table 7. The expected relationship holds in all cases; in fact, the melts from Shaw and San Cristobal can be shown to be quantitatively, as well as qualitatively, complimentary (Rambaldi and Larimer, 1976).

Table 7. Average composition of the metallic fractions of three L 6 chondrites (Alfianello, Leedey, and Mocs), of the L 7 chondrite, Shaw, and of the Group IB iron, San Cristobal

	Sb ppm	Ni %	Cu ppm	As ppm	Au ppm	Co %	Ga ppm	Mo ppm	Ge ppm	Ir ppm	Os ppm	Re ppm	Pt ppm	Ru ppm
Shaw Metal	0·52	14·5	595	16·8	1·70	0·69	31·7	8·7	175	9·2	9·8	0·94	18·6	10·9
Average L 6	0·92	13·8	626	18·6	1·70	0·69	12·6	6·0	121	5·2	5·9	0·64	13·7	7·9
San Cristobal (IB iron)	2·10	25·0	1000	26·5	2·20	0·61	11·0	2·2	27	0·33	0·44	0·024	<0·5	0·47

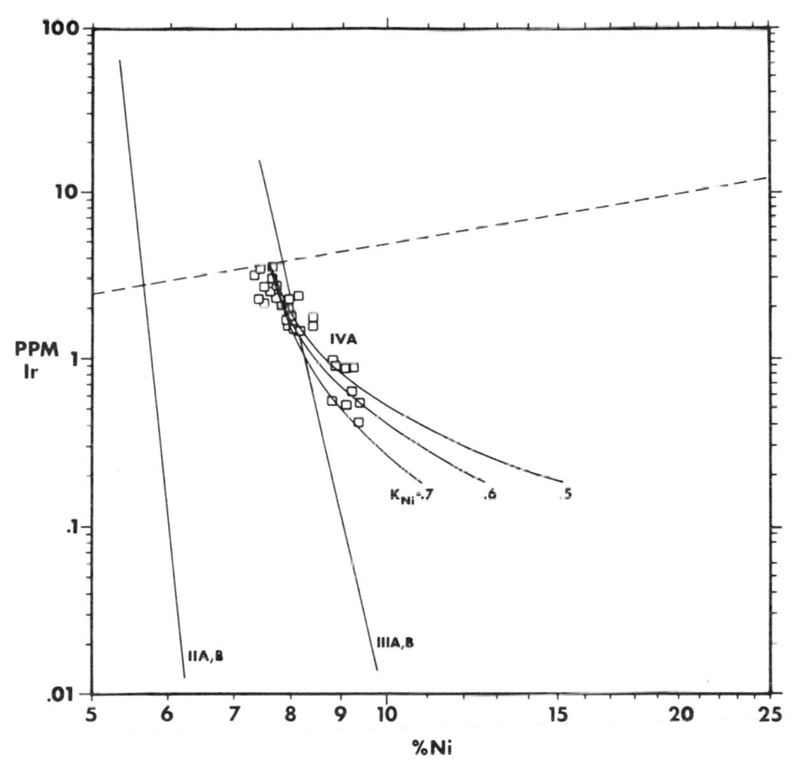

Figure 12. Ir and Ni are always negatively correlated, but along lines of different slope (see Figure 10). The correlation in Group IVA can be duplicated approximately by the trend predicted for homogenized aggregates of partial melts, where the molten metal collects into isolated masses, or raisins, distributed radially in the parent body. This model is consistent with both the chemistry and the divergent cooling rates observed in this group

Implications for Core-forming Process

Independent evidence on the formation conditions of iron meteorites comes from cooling rate estimates (Wood, 1964; Goldstein and Ogilvie, 1965; Goldstein and Short, 1972a, b). These estimates, based on the measured diffusion rates of Ni in Fe–Ni alloys, are made by measuring the Ni concentration gradient across the taenite crystals. The data indicate that most iron meteorites cooled at a rate of about 1 K to 10 K per Myr. The estimates tend to cluster for some chemical groups, implying an origin at similar depths in the parent bodies. But in other groups the rates differ suggesting an origin at different depths (Fricker *et al.*, 1970).

A problem arises in comparing the chemical trends and cooling rate estimates of Groups IIIAB and IVA. Both have similar chemical trends, except for a slight difference in slope (Figure 10 above). But the cooling rate estimates differ significantly. In Group IIIAB, the Ni-rich members tend to cluster around 1·5 K per Myr while the Ni-poor members spread out between 2 and 10 K per Myr. In Group IVA, the spread is even larger, 7 to 200 K per Myr (Scott and Wasson, 1975). This evidence strongly suggests that the members of Group IVA did not originate in a single core while many, but not all of Group IIIAB did. Yet the chemical data suggest fractionation in a single core in both cases. One solution to this problem is to find another process besides fractional crystallization, that produces nearly identical chemical trends. One possibility is that Group IVA represents objects derived from pools of metal distributed throughout the body, descriptively referred to as the raisin-bread model (Urey, 1966). These pools might represent aggregates of metal that become segregated, and homogenized, but did not sink to form a single core. The chemical trends, perhaps modified by mixing and subsequent fractional crystallization, might follow those of step III in the partial melting process. Surprisingly, such trends are nearly linear and very similar to the fractional crystallization trends (Figure 12). This model would thus be consistent with both the chemistry and cooling rates.

The obvious implication is that the core formation process can be considered in a series of steps: molten metal begins to flow and segregate itself from the surrounding silicates; it aggregates into large masses which slowly sink to form a core; and subsequent cooling leads to fractional crystallization. In the meteorites, each of these stages is represented: Group IAB, Group IVA, and Group IIIAB.

Silicate Fraction

Interpretative study of achondrites is not very advanced. One problem is their rarity; another is their complexity. Most appear to be breccias, mixtures of different rock types compacted together. An origin on the surface of parent bodies seems likely for most of these. The most abundant type, the eucrites or basaltic achondrites, are relatively little brecciated. In terms of composition, mineralogy and texture they resemble lunar basalts. Subtle differences rule out a lunar origin but the similarity remains striking. Accurate Rb–Sr dating is possible because the minerals vary considerably in their Rb–Sr ratios. All eucrites fall on the same isochron, and yield a surprisingly old age of $4{\cdot}7 \pm 0{\cdot}004 \times 10^9$ yr (Papanastassiou and Wasserburg, 1969).

Basalt formation evidently is widespread in the solar system, known to occur on the earth, moon and meteorite parent bodies, and suspected on Mercury and Mars. It is therefore of interest to compare these rocks to see how and why they differ. Presumably some of the differences are related to chemical differences among the bodies. Thus the question relates to the broader question of how and why bodies in the solar system differ chemically.

A important by-product of lunar research has been renewed interest in analysing terrestrial rocks, from which comparisons can be drawn. This has led to the discovery of several significant correlations among key elements. Until 1969, the most significant correlation was that of the K/U ratio, which remains almost constant over a 10^3 range in K content (Wasserburg *et al.*, 1964). Other similar cases, several of which are even more impressive, are shown in Figure 13.

But besides defining straight lines, of what significance are these correlations? As all students of geology rapidly learn, the processes by which igneous rocks form are quite complex. The number of variables to consider is enormous: T, P_T, P_{H_2O}, P_{O_2}, composition and mineralogy of the source rock, degree of fractional melting or crystallization, etc. It is therefore not surprising that no two terrestrial rocks are identical. Yet, in spite of differences in origin, some elements appear to be always more or less in unison. As noted by Wasserburg

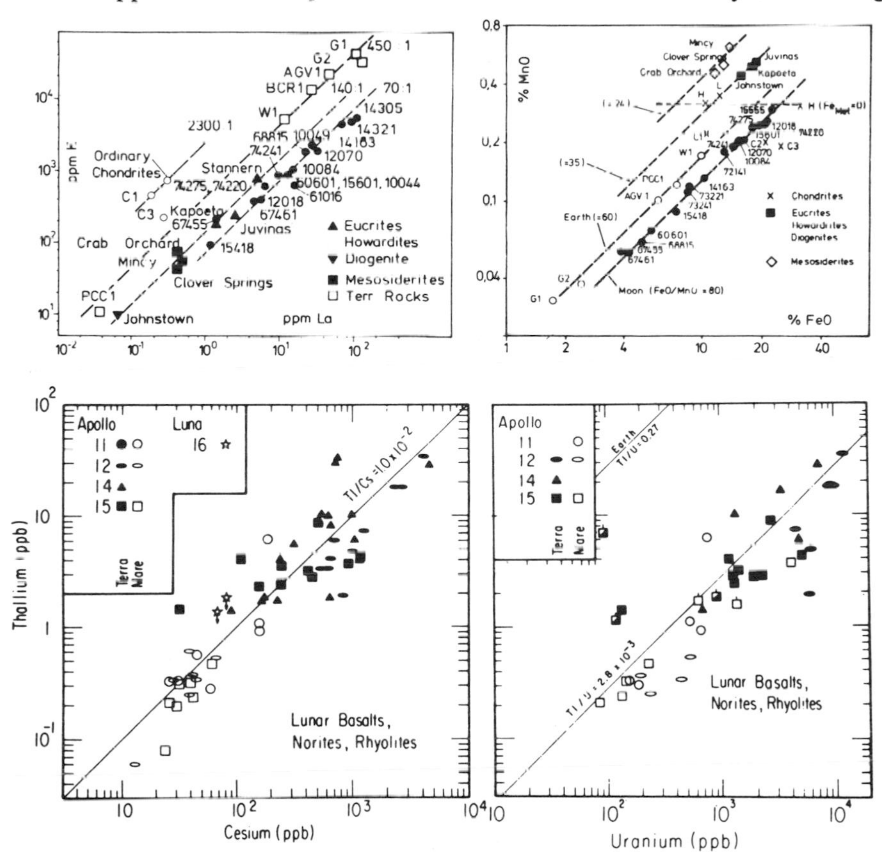

Figure 13. Various element pairs are linearly correlated in all samples from a single body, but the ratios differ between bodies. This presumably reflects differences in bulk composition: proportion of volatiles to refractories (K/La, Cs/U, Tl/U) or oxidation state (MnO/FeO) (Reproduced with permission from H. Wänke *et al.*, 1973, *Proc. Fourth Lunar Sci. Conf., Geochim. Cosmochim. Acta*, Suppl. 4, 1461–1481, and U. Kränhenbühl *et al.*, 1973, *Proc. Fourth Lunar Sci. Conf., Geochim. Cosmochim. Acta*, Suppl. 4, 1325–1348, both copyrighted by Pergamon Press)

et al. (1964) this suggests that the ratios are of fundamental significance and probably apply to the body as a whole.

If this is true then it appears that bodies differ considerably in their bulk composition. Such differences presumably date back to accretion in the nebula. Each pair of elements illustrated (Figure 13) is related to different chemical aspects of the bodies. The K/La (or K/U) and Cs/U ratios reflect the relative proportions of a moderately volatile element to a refractory element. The T1/U ratio reflects the proportions of a highly volatile to a refractory element. Manganese tends to concentrate in the silicate phase; iron is distributed among silicate, sulphide and metal. The FeO/MnO ratio is thus related to the oxidation state of the body (Laul and Schmitt, 1973). The horizontal, light-dashed line on this diagram shows how the ratio would change if all bodies had cosmic Mn/Fe ratios. The point marked XH ($Fe_{Met} = 0$) at the far right is the extreme case where all the Fe is present at FeO. To the extent that FeS and FeO are present, the ratio would shift to the left.

That the relative amounts of the elements vary from body to body, raises the question of whether the differences in ratio, for example K/La, reflect difference in refractory content, volatile content, or some combination of the two. To answer this question, the absolute abundance of at least one element in the body must be known.

For several elements, a bulk concentration can be estimated with few assumptions. For example, if all the Ar^{40} in the atmosphere is derived from K^{40} in the earth, a minimum K content of 90 ppm is required. Since Ar, being a noble gas, is likely to escape more rapidly than H_2O, nearly 1/2 of which appears to have outgassed, an upper limit of 180 ppm is established (Hurley, 1968a, b). The bulk concentration of K in the Earth is thus 135 ± 45 ppm. Similar reasoning can be used for Rb and Sr (Gast, 1960, 1968b; Hurley, 1968a, b). The absolute abundance of these 3 elements is sufficient to constrain the composition of the earth with respect to 9 other elements (Larimer, 1971: see Table 8). The comparison with chondrites is striking; in all cases except the alkalis (Na, K, Rb) which are depleted by factors of 6 to 8.

In the case of the moon, the U content can be estimated from heat flow experiments. Until recently, the data indicated a value of 59 ppb but new measurements on the thermal conductivity of the soil have reduced this estimate by nearly a factor of two (Langseth *et al.*, 1976). Using the new value, 37 ± 15 ppb, similar estimates can be made for the content of other elements in the moon (Table 8). Again, there is a similarity with chondrite abundances for all elements except alkalis and highly volatile elements, if the difference in Fe content as inferred from the moon's low density is taken into account.

It is interesting to try to relate these differences to what has been said about the fractionation process that operated on chondritic meteorite material. Several approaches have been attempted. Wänke *et al.* (1973, 1974) have developed a simple two-component model for the moon, involving a refractory-rich component and a component of slightly variable composition. Ganapathy

Table 8. Elemental Abundances in the bulk earth and moon, based on trace element ratios and fractionation patterns

Element	Ordinary chondrite	Earth	Moon	Eucrites
Na (%)	0·68	$0·07 \pm 0·06$	$0·03 \pm 0·02$	0·3
K ppm	850·0	135 ± 45	60 ± 24	360·
Rb ppm	2·8	$0·42 \pm 0·13$	$0·21 \pm 0·08$	0·24
Cs ppb	4 to 619	9 ± 4	10 ± 3	12·
Ca (%)	1·21	$1 \pm 0·8$	$4·3 \pm 1·5$	7·49
Sr ppm	11·	12 ± 4	39 ± 14	85·0
Ba ppm	4·1	$5·3 \pm 5$	$10·9 \pm 4$	35·0
Sc ppm	8·0	$7·6 \pm 3$	22 ± 8	27·0
Y ppm	2·0	$1·9 \pm 0·7$	$5·5 \pm 2$	23·0
La ppm	0·35	$0·30 \pm 0·1$	$0·86 \pm 0·3$	3·0
Th ppb	43·0	40 ± 20	130 ± 50	440·0
U ppb	12·0	12 ± 6	37 ± 15	130·0
Tl ppb	0·03 to 120	$3 \pm 1·5$	$0·1 \pm 0·03$	0·5
MnO (%)	0·33	$·058 \pm ·004$	$·023 \pm ·003$	0·50
FeO (%)	11·3	3·48	1·84	17·5

and Anders (1974) have attempted to compute the bulk composition of the earth and moon on the assumption that the chondritic fractionation processes operated throughout the solar system. They thus require seven components: refractory, unmelted and melted metal, unmelted and melted silicate, FeS and volatiles. When the estimated bulk compositions were compared to the composition of the surface basalts the fractionation patterns assumed a striking similarity. They argue that this suggests a basic sameness in the generation of basalts, the main difference being the bulk chemistry of the bodies. Using this argument, and quantifying it, Anders (1976) shows that the eucrite parent body must have a composition very similar to the moon's. He further notes that using this model, it would be possible to obtain the composition of a body from relatively few measurements (heat flow for U, chemical analyses for Fe, K, Mn and Tl, and a measurement of the density to deduce the FeO content), all of which can be done remotely.

Conclusions

Meteorites provide the best preserved record of the chemical and physical process that operated in the nebula. It now appears possible to quantify important parameters such as the P–T conditions in the nebula. Evidently, a very high-temperature regime existed, > 1400 K. The early condensate aggregated into cm-sized inclusions, which may be related to a fractionation of all refractory elements in meteorites and possibly the planets. Accretion into small bodies appears to have begun just below this temperature, when the major condensates metal and silicate appeared. But the bulk of the meteoritic material appears to have accreted at much lower temperatures, 400 to 500 K. These temperatures are still much higher than the present black-body temperature in the asteroid belt, however. The pressures inferred are quite high also, 10^{-6} to 10^{-3} atm, suggesting a massive nebula, with a low accretion efficiency.

Observational and theoretical work on protostars has now advanced to the point where some of these same parameters can be estimated (Herbig, 1978). It will be of interest to see what this comparison reveals.

Acknowledgements

Portions of this work have been supported by NASA Grant NSG 7040. I am indebted to Professor Heinrich Wänke and his group at the Max-Planck-Institut für Chemie, Mainz, Germany for their hospitality and help in preparing the manuscript.

References

Alfvén, H., and Arrhenius, G. (1978). This volume.

Anders, E. (1964). 'Origin, Age, and composition of meteorites.' *Space Sci. Rev.*, **3**, 583–714.

Anders, E. (1971). 'Meteorites and the early solar system.' *Ann. Rev. Astron. Astrophys.*, **9**, 1–34.

Anders, E. (1977). 'Chemical compositions of the Moon, Earth and Eucrite parent body.' *Phil. Trans. Roy. Soc.*, **285**, 23–40.

Anders, E., and Heymann, D. (1969). 'Elements 112 to 119: Were they present in meteorites.' *Science*, **164**, 821–823.

Anders, E., and Larimer, J. W. (1972). 'Extinct superheavy element in meteorites: Attempted characterization.' *Science*, **175**, 981–983.

Anders, E., Higuchi, H., Gros, J., Takahashi, H., and Morgan, J. W. (1975). 'Extinct superheavy element in the Allende meteorite.' *Science*, **190**, 1262–1271.

Binns, R. A. (1969). 'Olivine in enstatite chondrites.' *Am. Mineral.*, **52**, 1549–1554.

Binz, C. M., Ikramuddin, M., Rey, P., and Lipschutz, M. E. (1976). 'Trace elements in primitive meteorites—VI Abundance patterns of thirteen trace elements and interelement relationships in unequilibrated ordinary chondrites.' *Geochim Cosmochim. Acta*, **40**, 59–71.

Black, D. C., (1978). This volume.

Black, D. C., and Pepin, R. O. (1969). 'Trapped neon in meteorites, 2,' *Earth Planet. Sci. Lett.*, **6**, 395–405.

Blander, M. (1971). 'The constrained equilibrium theory: Sulphide phases in meteorites.' *Geochim. Acta*, **35**, 61–76.

Blander, M., and Abdel-Gawad, M. (1969). 'The origin of meteorites and the constrained equilibrium condensation theory.' *Geochim. Cosmochim. Acta*, **33**, 701–716.

Blander, M., and Fuchs, L. H. (1975). 'Calcium—aluminum—rich inclusions in the Allende meteorite: Evidence for a liquid origin.' *Geochim. Cosmochim Acta*, **39**, 1605–1620.

Blander, M., and Katz, J. L. (1967). 'Condensation of primordial dust.' *Geochim. Cosmochim Acta*, **31**, 1025–1024.

Cameron, A. G. W. (1966). 'The accumulation of chondritic material.' *Earth Planet. Sci. Lett.*, **1**, 93–96.

Cameron, A. G. W. (1968). 'A new table of abundances of the elements in the solar system.' In *Origin and Distribution of the Elements* (editor L. H. Ahrens) 125–143 Pergamon Press.

Cameron, A. G. W. (1973a). 'Abundances of the elements in the solar system.' *Space Sci. Rev.*, **15**, 121–146.
Cameron, A. G. W. (1973b). 'Accumulation processes in the primitive solar nebula.' *Icarus*, **18**, 407–450.
Cameron, A. G. W. (1978). This volume.
Case, D. R., Laul, J. C., Pelly, I. Z., Wechter, M. A., Schmidt-Bleck, F., and Lipschutz, M. E. (1973). 'Abundance patterns of thirteen trace elements in primitive carbonaceous and unequilibrated ordinary chondrites.' *Geochim. Cosmochim. Acta*, **37**, 19–34.
Chou, C. L., and Cohen, A. J. (1973). 'Gallium and germanium in the metal and silicates of L and LL chondrites.' *Geochim. Cosmochim. Acta*, **37**, 315–327.
Christophe, M. (1968). 'Un chondre exceptionnel dans la meteorite de Vigarano.' *Bull. Soc. Fr. Mineral. Cristallogr.*, **91**, 212–214.
Clark, S. P., Jr., Turekian, K. K., and Grossman, L. (1972). 'Model for the early history of the earth.' in *The Nature of the Solid Earth*, edited by E. C. Robertson, pp. 3–18, McGraw-Hill, New York.
Clarke, R. S., Jr., Jarosewich, E., Mason, B., Nelen, J., Gomez, M., and Hyde, J. R. (1970). 'The Allende, Mexico, meteorite shower.' *Smithson., Contrib., Earth Sci.*, **5**, 345–351.
Clayton, R. N., Grossman, L., and Mayeda, T. K. (1973). 'A component of primitive nuclear composition in carbonaceous meteorites.' *Science* **187**, 485–488.
Dakowski, M. (1969). 'The possibility of extinct superheavy elements occurring in meteorites.' *Earth Planet. Sci. Lett.*, **6**, 152–154.
Dayhoff, M. O., Lippincott, E. R., and Eck, R. V. (1964). 'Thermodynamic equilibria in prebiological atmospheres.' *Science*, **146**, 1461–1464.
Dodd R. T., Jr. (1969). 'Metamorphism of the ordinary chondrites: A review.' *Geochim. Cosmochim. Acta*, **33**, 161–203.
Eberhardt, P., Geiss, J., and Grögler, N. (1965). 'Über die Verteilung der Uredelgase im Meteoriten Khor Temiki.' *Tschermak's Mineral. Petrogr. Mitt.*, **10**, 535–551.
Eugster, O., Eberhardt, P., and Geiss, J. (1967). 'Krypton and xenon isotopic composition in three carbonaceous chondrites.' *Earth Planet. Sci. Lett.*, **3**, 249–257.
Fish, R. A., Goles, G. G., and Anders, E. (1960). 'The record in the meteorites—III on the development of meteorites in asteroidal bodies.' *Astrophys. J.*, **132**, 243–258.
Fredriksson, K., and Mason, B. (1967). 'The Shaw meteorite.' *Geochim. Cosmochim. Acta*, **31**, 1705–1709.
Fricker, P. E., Goldstein, J. I., and Summers, A. L. (1970). 'Cooling rates and thermal histories of iron and stony-iron meteorites.' *Geochim. Cosmochim. Acta*, **34**, 475–491.
Ganapathy, R., and Anders, E. (1974). 'Bulk compositions of the moon and earth, estimated from meteorites.' *Proc. Fifth Lunar Sci. Conf., Geochim. Cosmochim. Acta*, Suppl. 5, 1181–1206.
Gast, P. W. (1960). 'Limitations on the composition of the upper mantle.' *J. Geophys. Res.*, **65**, 1287–1297.
Gast, P. W. (1968a). 'Trace element fractionation and the origin of tholeitic and alkaline magma types.' *Geochim. Cosmochim. Acta*, **32**, 1057–1086.
Gast, P. W. (1968b). 'Upper mantle chemistry and evolution of the earth's crust.' In *History of the Earth's Crust* (editor R. Phinney), pp. 45–27, Princeton University Press.
Gast, P. W. (1972). 'The chemical composition and structure of the moon.' *The Moon*, **5**, 121–148.
Geiss, J., Eberhardt, P., Bühler, F., Meister, J., and Signer, P. (1970). 'Apollo 11 and 12 solar wind composition experiments: Fluxes of He and Ne isotopes.' *J. Geophys. Res.*, **75**, 5972–5979.
Goldstein, J. I., and Ogilvie, R. E. (1965). 'The growth of the Widmanstätten pattern in metallic meteorites.' *Geochim. Cosmochim. Acta*, **29**, 893–920.
Goldstein, J. I., and Short, J. M. (1967a). 'Cooling rates of 27 iron and stony-iron meteorites.' *Geochim. Cosmochim. Acta*, **31**, 1001–1023.

Goldstein, J. I., and Short, J. M. (1967b). 'The iron meteorites, their thermal history and parent bodies.' *Geochim. Cosmochim. Acta*, **31**, 1733–1770.

Gray, C. M., Papanastassiou, D. A., and Wasserburg, G. J. (1973). 'The identification of early condensates from the solar nebula.' *Icarus*, **20**, 213–239.

Grossman, L. (1972). 'Condensation in the primitive solar nebula.' *Geochim. Cosmochim. Acta*, **36**, 597–619.

Grossman, L. (1973). 'Refractory trace elements in Ca–Al-rich inclusions in the Allende meteorite.' *Geochim. Cosmochim. Acta*, **37**, 1119–1140.

Grossman, L., and Clark, S. P., Jr. (1973). 'High-temperature condensates in chondrites and the environment in which they formed.' *Geochim. Cosmochim. Acta*, **37**, 635–649.

Grossman, L., and Ganapathy, R. (1976). 'Trace elements in the Allende meteorite—I. Coarse-grained, Ca-rich inclusions.' *Geochim. Cosmochim. Acta*, **40**, 331–344.

Grossman, L., and Larimer, J. W. (1974). 'Early chemical history of the solar system.' *Rev. Geophys. Space Phys.*, **12**, 71–101.

Hayatsu, R. M., Studier, H., and Anders, E. (1971). 'Origin of organic matter in early solar system, 4, Amino acids: Confirmation of catalytic synthesis by mass spectrometry.' *Geochim. Cosmochim. Acta*, **35**, 939–951.

Hayes, J. M. (1967). 'Organic constituents of meteorites: A review.' *Geochim. Cosmochim. Acta*, **31**, 1395–1440.

Herbig, G. H. (1978). This volume.

Herndon, J. M., and Suess, H. E. (1976). 'Can enstatite meteorites form from a nebula of solar composition?' *Geochim. Cosmochim. Acta*, **40**, 395–400.

Heymann, D. (1971). 'The inert gases.' In *Handbook of Elemental Abundances in Meteorites* (editor B. Mason), 29–66, Gordon-Breach.

Heymann, D., and Dzickaniec, M. (1976). 'Early irradition of matter in the solar system: Magnesium (proton, neutron) scheme.' *Science*, **191**, 79–81.

Hintenberger, H., Vilcsek, E., and Wänke, H. (1965). 'Über die Isotopenzusammensetzung und über den sitz der leichten Uredelgase in Steinmeteoriten.' *Z. Naturforsch.*, **20a**, 939–945.

Hurley, P. M. (1968a). 'Absolute abundances and distribution of Rb, K, and Sr in the earth.' *Geochim. Cosmochim. Acta*, **32**, 273–284.

Hurley, P. M. (1968b). 'Correction to: Absolute abundances and distribution to Rb, K, and Sr in the earth.' *Geochim. Cosmochim. Acta*, **32**, 1025–1030.

Ikrumuddin, M., Binz, C. M., and Lipschutz, M. E. (1976). 'Thermal metamorphism of primitive meteorites—II Ten trace elements in Abee enstatite chondrite heated at 400–1000 °C.' *Geochim. Cosmochim. Acta*, **40**, 133–142.

Jeffery, P. M., and Anders, R. (1970). 'Primordial noble gases in separated meteoritic minerals, 1.' *Geochim. Cosmochim. Acta*, **34**, 1175–1198.

Keays, R. R., Ganapathy, R., and Anders, E. (1971). 'Chemical fractionations in meteorites, 4. Abundances of fourteen trace elements in L-chondrites; implications for cosmothermometry.' *Geochim. Cosmochim. Acta*, **35**, 337–363.

Keil, K. (1968). 'Mineralogical and chemical relationships among enstatite chondrites.' *J. Geophys. Res.*, **73**, 6945–6976.

Keil, K., Huss, G. I., and Wiik, H. B. (1969). 'The Leoville, Kansas meteorite; A polymict breccia of carbonaceous chondrites and achondrites (abstract)' in *Meteorite Research*, edited by P. M. Millman, p. 217, D. Reidel, Dordrecht, Netherlands.

Kelly, W. R., and Larimer, J. W. (1977). 'Chemical Fractionations in Meteorites—VIII. Iron meteorites and the cosmochemical history of the metal phase.' *Geochim. Cosmochim. Acta*, **41**, 93–111.

Kieffer, S. W. (1975). 'Droplet chondrules.' *Science*, **189**, 333–340.

King, E. A., Jr., Carman, M. F., and Butler, J. C. (1972). 'Chondrules in Apollo 14 samples: Implications for the origin of chondritic meteorites.' *Science*, **175**, 59–60.

Kirsten, T. (1978). This volume.

Krähenbühl, U., Morgan, J. W., Ganapathy, R., and Anders, E. (1973a). 'Abundance of 17 trace elements in carbonaceous chondrites.' *Geochim. Cosmochim. Acta*, **37**, 1353–1370.

Krähenbühl, U., Ganapathy, R., Morgan, J. W., and Anders, E. (1973b). 'Volatile elements in Apollo 16 samples: Implications for highland volcanism and accretion history of the moon.' *Proc. Fourth Lunar Sci. Conf., Geochim. Cosmochim. Acta*, Suppl. 4, 1325–1348.

Kurimoto, R. K., Pelly, I. Z., Laul, J. C., and Lipschutz, M. E. (1973). 'Interelement relationships between trace elements in primitive carbonaceous and unequilibrated ordinary chondrites.' *Geochim. Cosmochim. Acta*, **37**, 209–244.

Kvenvolden, K. A., Lawless, J. G., and Ponnamperuma, C. (1971). 'Nonprotein amino acids in the Murchison meteorite.' *Proc. Nat. Acad. Sci.*, **68**, 486–490.

Lancet, M. S., and Anders, E. (1970). 'Carbon-isotope fractionation in the Fischer-Tropsch synthesis and in meteorites.' *Science*, **170**, 980–982.

Lancet, M. S., and Anders, E. (1973). Solubilities of noble gases in magnetite: Implications for planetary gases in meteorites.' *Geochim. Cosmochim. Acta*, **37**, 1371–1388.

Lange, D. E., and Larimer, J. W. (1973). 'Chondrules: An origin by impacts between dust grains.' *Science*, **182**, 920–922.

Langseth, M. G., Keihm, S., and Peters, K. (1976). 'The revised lunar heat flow values.' In *Lunar Science VII*, 474–475, Lunar Science Institute, Houston, Texas.

Larimer, J. W. (1967). 'Chemical Fractionations in meteorites—I. Condensation of the elements.' *Geochim. Cosmochim. Acta*, **31**, 1215–1238.

Larimer, J. W. (1968a). 'Experimental studies on the system Fe–MgO–SiO–O_2 and their bearing on the petrology of chondritic meteorites.' *Geochim. Cosmochim. Acta*, **32**, 1187–1207.

Larimer, J. W. (1968b). 'An Experimental investigation of oldhamite, CaS; and the petrologic significance of oldhamite in meteorites.' *Geochim. Cosmochim. Acta*, **32**, 965–982.

Larimer, J. W. (1971). 'Composition of the earth: Chondritic or achondritic.' *Geochim. Cosmochim. Acta*, **35**, 769–786.

Larimer, J. W. (1973). 'Chemical fractionations in meteorites VII. Cosmothermometry and Cosmobarometry.' *Geochim. Cosmochim. Acta*, **37**, 1603–1623.

Larimer, J. W. (1975). 'The effect of C/O ratio on the condensation of planetary material.' *Geochim. Cosmochim. Acta*, **39**, 389–392.

Larimer, J. W., and Anders, E. (1967). 'Chemical fractionations in meteorites II. Abundance patterns and their interpretation.' *Geochim. Cosmochim. Acta*, **31**, 1239–1270.

Larimer, J. W., and Anders, E. (1970). 'Chemical fractionations in meteorites III. Major element fractionations in chondrites.' *Geochim. Cosmochim. Acta*, **34**, 367–387.

Larimer, J. W., and Bartholomay, M. (1977). 'The role of carbon and oxygen in the chemistry of cosmic gases.' *Geochim. Cosmochim. Acta*, (In press).

Larimer, J. W., and Holdsworth, E. F. (1975). 'Chromium distribution in the Allende meteorite.' Presented at *38th Ann. Meeting., Meteoritical Soc.*, Tours, France.

Laul, J. C., Ganapathy, R., Anders, E., and Morgan, J. W. (1973). 'Chemical fractionations in meteorites, 6, Accretion temperatures of H-, LL-, and E-chondrites, from abundance of volatile trace elements.' *Geochim. Cosmochim. Acta*, **37**, 329–357.

Laul, J. C., and Schmitt, R. A. (1973). 'Chemical composition of Apollo 15, 16 and 17 samples.' *Proc. Fourth Lunar Sci. Conf., Geochim. Cosmochim. Acta*, Suppl. **4**, 1349–1367.

Lee, T., Papanastassiou, D. A., and Wasserburg, G. J. (1976). 'Demonstration of ^{26}Mg excess in Allende and evidence for ^{26}Al.' *Geophys. Res. Lett.*, **3**, 41–44.

Lewis, J. S. (1972a). 'Metal/silicate fractionation in the solar system.' *Earth Planet. Sci. Lett.*, **15**, 286–290.

Lewis, R. S., Srinivasan, B., and Anders, E. (1975). 'Host phase of a strange xenon component in Allende.' *Science*, **190**, 1251–1262.

Marti, K. (1969). 'Solar type xenon: A new isotopic composition of xenon in the Pesyanoe meteorite.' *Science*, **166**, 1263–1265.

Marvin, U. B., Wood, J. A., and Dickey, J. S., Jr. (1970). 'Ca–Al-rich phases in the Allende meteorite.' *Earth Planet. Sci. Lett.*, **7**, 346–350.

Michaelis, H., Von, Ahrens, L. H., and Willis, J. P. (1969). 'The composition of stony meteorites II. The analytical data and an assessment of their quality.' *Earth Planet. Sci. Lett.*, **5**, 387–394.

Miller, S. L. (1953). 'A production of amino acids under possible primitive earth conditions.' *Science*, **117**, 528–529.

Palme, H., and Wlotska, F. (1977). 'A metal particle from a Ca–Al-rich inclusion from the meteorite Allende, and the condensation of refractory siderophile elements.' *Earth Planet. Sci. Lett.*, **33**, 45–60.

Palme, H., Spettle, B., Baddenhausen, H., and Broo, H. (1975). 'New data on the chemistry of Allende inclusions.' *Meteoritics*, **10**, 469–470.

Papanastassiou, D. A., and Wasserburg, G. J. (1969). 'Initial strontium isotopic abundances and the resolution of small time differences in the formation of planetary objects.' *Earth Plant. Sci. Lett.*, **5**, 371–376.

Pellas, P., Poupeau, G., Lorin, J. C., Reeves, H., and Audouze, J. (1969). 'Primitive low-energy particle irradiation of meteoritic crystals.' *Nature*, **223**, 272–274.

Pepin, R. O., and Signer, P. (1965). 'Primordial rare gases in meteorites.' *Science*, **149**, 253–265.

Phinney, P., Frick, U., and Reynolds, J. H. (1976). 'Rare-gas-rich separates from carbonaceous chondrites.' In *Lunar Science*, **VII**, 691–693, Lunar Science Institute, Houston, Texas.

Rambaldi, E., and Larimer, J. W. (1977). 'The Shaw Chondrite. Part 1—The case of the missing metal.' *Earth Planet. Sci. Lett.*, **33**, 61–66.

Ringwood, A. E. (1966). 'Chemical evolution of the terrestrial planets.' *Geochim. Cosmochim. Acta*, **30**, 41–104.

Schmitt, R. S., Smith, R. H., and Goles, G. G. (1965). 'Abundances of Na, Sc, Cr, Mn, Fe, Co, and Cu in 218 individual meteoritic chondrules via activation analysis, 1,' *J. Geophys. Res.*, **70**, 2419–2444.

Scott, E. R. D. (1972). 'Chemical fractionation in iron meteorites and its interpretation.' *Geochim. Cosmochim. Acta*, **36**, 1205–1236.

Scott, E. R. D., and Wasson, J. T. (1975). 'Classification and properties of iron meteorites.' *Rev. Geophys. Space Sci.*, **13**, 527–546.

Shaw, D. M. (1970). 'Trace element fractionation during anatexis.' *Geochim. Cosmochim. Acta*, **34**, 237–243.

Srinivasan, B., Alexander, E. C., Jr., Manuel, O. K. and Troutner, D. E. (1969). 'Xenon and krypton from the spontaneous fission of californium—252.' *Phys. Rev.*, **179**, 1166–1169.

Suess, H. E. (1962). 'Thermodynamic data on the formation of solid carbon and organic compounds in primitive planetary atmospheres.' *J. Geophys. Res.*, **62**, 2029–2034.

Suess, H. E. (1963). 'Comments on the origin of chondrules,' in *Origin of the Solar System*, edited by R. Jastrow and A. G. W. Cameron, pp. 154–156, Academic, New York.

Tozer, D. C. (1978). This Volume.

Urey, H. C. (1952a). *The Planets*, Yale University Press, New Haven.

Urey, H. C. (1952b). 'Chemical fractionation in the meteorites and the abundance of the elements.' *Geochim. Cosmochim. Acta*, **2**, 267–282.

Urey, H. C. (1953). 'Chemical evidence regarding the earth's origin,' in *Plenary Lectures, Thirteenth International Congress on Pure and Applied Chemistry*, pp. 188–214, International Union of Pure and Applied Chemistry, London.

Urey, H. C. (1954). 'On the dissipation of gas and volatilized elements from protoplanets.' *Astrophys, J. Suppl. Se.*, **1**, 147–173.

Urey, H. C. (1955). 'The cosmic abundances of potassium, uranium, and thorium and the heat balances of the Earth, the Moon, and Mars.' *Proc. Nat. Acad. Sci.*, **41**, 127–144.
Urey, H. C. (1966). 'Chemical evidence relative to the origin of the solar system.' *Mon. Notic. Roy. astron. Soc.*, **131**, 199–223.
Urey, H. C., and Craig, H. (1953). 'The composition of stone meteorites and the origin of the meteorites.' *Geochim. Cosmochim. Acta*, **4**, 36–82.
Van Schmus, W. R., and Wood, J. A. (1967). 'A chemical–petrologic classification for the chondritic meteorites.' *Geochim. Cosmochim. Acta*, **31**, 747–765.
Wänke, H. (1965). 'Der Sonnenwind als Quelle der Uredelgase in Steinmeteoriten.' *Z. Naturforsch.*, **20a**, 946–949.
Wänke, H., Baddenhausen, H., Palme, H., and Spettel, B. (1974). 'On the Chemistry of the Allende inclusions and their origin as high temperature condensates. *Earth Planet. Sci. Lett.*, **23**, 1–7.
Wänke, H., Baddenhausen, H., Dreibus, G., Jagoutz, E., Kruse, H., Palme, H., Spettle, B., and Teschke, F. (1973). Multielement analyses of Apollo 15, 16 and 17 samples and the bulk composition of the moon. *Proc. Fourth Lunar Sci. Conf., Geochim. Cosmochim. Acta*, Suppl. **4**, 1461–1481.
Wark, D. A., and Lovering, J. F. (1976). 'Refractory/platinum metal grains in Allende calcium–aluminum-rich clasts (CARC'S): Possible exotic presolar material.' In *Lunar Science VII*, 912–914, Lunar Science Institute, Houston, Texas.
Wasserburg, G. J., Macdonald, G. J. F., Hoyle, F., and Fowler, W. F. (1964). 'Relative contributions of uranium, thorium, and potassium to heat production in the earth.' *Science*, **143**, 465–467.
Wetherill, G. W., Mark, R. K., and Lee-Hu, C. (1973). 'Primordial strontium in an Allende inclusion (abstract).' *Eos. Trans. AGU*, **54**, 345–346.
Weber, H. W., Hintenberger, H., and Begemann, F. (1971). 'Noble gases in the Haverö ureilite.' *Earth Planet. Sci. Lett.*, **13**, 205–209.
Weber, H. W., Begemann, F., and Hintenberger, H. (1976). 'Primordial gases in graphite-diamond-kamacite inclusions from the Haverö ureilite.' *Earth Planet. Sci. Lett.*, **29**, 81–90.
Whipple, F. L. (1966). 'A suggestion as to the origin of chondrules.' *Science*, **153**, 54–56.
Whipple, F. L. (1972). 'On certain aerodynamic processes for asteroids and comets,' in *From Plasma to Planet.* edited by A. Elvius, pp. 211–232, Almqvist and Wilksell, Stockholm.
Wood, J. A. (1962). 'Metamorphism in chondrites.' *Geochim. Cosmochim. Acta*, **26**, 734–749.
Wood, J. A. (1963). 'On the origin of chondrules and chondrites.' *Icarus*, **2**, 152–180.
Wood, J. A. (1964). 'The cooling rates and parent planets of several iron meteorites.' *Icarus*, **3**, 429–459.
Y'avnel, A. A. (1966). 'On chemical fractionation in the silicate phase of meteorites.' *Geochim. Int.*, **3**, 228–237.
Yoshino, D., Hayatsu, R., and Anders, E. (1971). 'Origin of organic matter in early solar system, 3, Amino acids: Catalytic synthesis.' *Geochim. Cosmochim. Acta*, **35**, 927–938.

THE OUTER PLANETS AND THEIR SATELLITES

D. J. STEVENSON
Research School of Earth Sciences, Australian National University, Canberra, Australia

Abstract

The compositions and thermal histories of the outer planets and their satellites are discussed. Particular attention is given to how observations constrain models, and no particular cosmogonic prejudice is employed. The most important observations constraining the composition for gaseous planets are the average density, the gravitational moments, and the atmospheric composition. Thermal structure is constrained by the excess luminosity and planetary magnetic field. The hydrogen–helium planets (Jupiter and Saturn) are relatively well understood because the thermodynamics and transport properties of hydrogen–helium mixtures are amenable to first principles calculations. Jupiter appears to be almost of primordial solar composition, but the deviation is significant and may represent an order of magnitude enhancement of rock-forming elements. Saturn deviates even further from solar composition. Both planets have internal temperatures of order 10^4 K. Uranus and Neptune are much less well understood, mainly because of the paucity of relevant observations, but are predominantly 'icy'. Both planets are expected to have internal temperatures of at least one thousand (probably several thousand) K. The satellites form two groups: predominantly icy, and predominantly rocky (Moon-like). The largest icy satellites are expected to be fractionated and thermally active, with partially fluid interiors.

Introduction

The various theories for the origin of the solar system outlined in this book are remarkable for their lack of common ground. Nevertheless, these theories should be consistent with the observational facts, and straightforward implications of these facts. The purpose of this paper is to summarize the relevant observations and their implications for the solar system beyond the asteroid belt (but excluding comets). This is *not* a comprehensive survey of the type carried out by Newburn and Gulkis (1973)—and which is already in need of updating—but concentrates on the compositions and thermal histories of the outer solar system bodies, with a bias towards the more massive bodies. Secondary characteristics (e.g. ionospheres, meteorology, etc.) are not discussed since they have no apparent relevance to the cosmogony. A less justifiable neglect is the omission of any discussion on celestial mechanical constraints (such as the evolution of orbital elements). These constraints may be important for understanding the later stages of solar system evolution, but are less fundamental than the composition or thermal structure.

The elements from which the solar system formed can be loosely categorized into three classes: rock forming, ice forming, and permanent gases. The permanent gases (primarily hydrogen, helium and neon) are substances which do not condense except very near absolute zero. The ice forming elements are oxygen, carbon and nitrogen, which (in the presence of hydrogen) form water, methane and ammonia. The most abundant of the remaining elements (Si, Mg, Fe) form compounds ('rocks') whose volatility is much less than the 'gases' or 'ices' at any pressure of interest. The main difference between the major planets (Jupiter, Saturn, Uranus, Neptune) and the terrestrial planets is that the major planets have at least partly retained the volatiles (gases and ices). In the next section, it will be explained how knowledge of the densities alone is sufficient to prove the predominance of volatiles in the constitution of the major planets. Additional information on the constituents and their distribution can be obtained from the response of the planet's shape to its rotation. These constraints suggest that the total amount of rock in the outer solar system exceeds that in the inner solar system by at least an order of magnitude. This is an important constraint on the mass of the primordial solar nebula. The next section also contains a general discussion of atmospheres or surfaces as a boundary condition on planetary interiors, thermal balance and the existence of internal heat sources, and the use of planetary magnetic fields as a constraint on the dynamic and electronic state of the interior.

In the section on *The Hydrogen–Helium Planets*, the hydrogen–helium gas giants (Jupiter and Saturn) are analysed. The procedure for constructing internal models is reviewed, followed by a discussion of the relevant thermodynamics and phase diagram for hydrogen, and hydrogen–helium mixtures. Energy transport processes (conduction, convection and radiation) are assessed, and the predominance of convection established. The results for simple (i.e. predominantly homogenous) models are described and critically examined. The thermal evolutions of Jupiter and Saturn are discussed, and the internal heat sources of these planets attributed to the (as yet imperfectly understood) release of gravitational energy.

In the section following that on *Uranus and Neptune*, similar considerations are applied to Uranus and Neptune. The uncertainties are greater, in part because of inadequate observational data. Unlike Jupiter and Saturn, there is substantial ambiguity regarding the compositions of these planets. It is not yet possible to preclude substantial (gravitational) energy sources within Uranus and Neptune. In any event, internal temperatures exceeding several thousand degrees are predicted.

In the section afterwards, Pluto and the outer solar system satellites are discussed. These bodies can be loosely divided into two classes: the rocky, moon-like bodies (for example, Io, Europa); and the icy, low-density bodies (for example, Ganymede, Titan). With the remarkable exception of Titan, these bodies have very thin atmospheres. Particularly relevant to cosmogony is the existence of a density gradient within a satellite family (e.g. the Galilean satellites). Internal radio-active heating within the satellites is expected to produce melting and fractionation. Dynamo-generated magnetism and icy tactonics are intriguing possibilities.

The last section concludes with an assessment of current problems and uncertainties. Suggestions for future theoretical work and observations are made, bearing in mind the ongoing N.A.S.A. programme of unmanned space probes.

Observations and their Implications

The Gravity Field

The gravitational field of a planet rotating with a known angular velocity is the most important remote probe of a planetary interior. If the planet is in strictly hydrostatic equilibrium, then the external gravitational potential has the form

$$\phi = -\frac{GM}{r}\left[1 - \sum_{n=1}^{\infty}\left(\frac{R_e}{r}\right)^{2n} J_{2n} P_{2n}(\cos\theta)\right] \qquad (1)$$

where M is the planetary mass, R_e is the equatorial radius and θ is the angle between the rotation axis and the radial vector $\mathbf{r}$. P_{2n} are the Legendre polynomials. The J_{2n} are gravitational moments, and are approximately proportional to Ω^{2n} (where Ω is the planetary angular velocity). For $n = 1$, the proportionality constant depends primarily on the planets' moment of inertia, whereas for larger n it depends increasingly on the density distribution in the outer layers of the planet. The existence of other spherical harmonics not included in equation (1) would indicate deviations from hydrostatic equili-

Table 1. Relevant physical parameters of the major planets (Reference: Newburn and Gulkis, 1973; except where otherwise noted)

	Jupiter	Saturn	Uranus	Neptune
Mean radius (km)	69800(± 50)[a]	57800(± 100)	25200(± 300)	24600(± 200)
Mass (Earth = 1)	317·9	95·2	14·6(± 0·2)	17·2(± 0·1)
Oblateness	0·0655(± 0·0005)	0·098(± 0·001)	0·01(± 0·01)[b]	0·02(± 0·02?)
Average Density (g cm^{-3})	1·333(± 0·003)	0·705(± 0·005)	1·3(± 0·05)	1·65(± 0·03)
Rotation period (hr) (atmosphere)	$9^h 52^m$ (± 5^m variation).	$10^h 30^m$ (± 30^m variation)	11^h(± 1h?)	16h(± 1h?)
q_2 (see text and footnote[c])	0·39(± 0·005)	0·37(± 0·01)	0·30(± 0·05?)	0·27(± 0·03?)
Effective temperature (K)	125(± 5)[d]	97(± 4)	57(± 4)	45(± 3)
Excess luminosity Q_0 (see equation (8))	0·015(± 0·005)	0·04(± 0·02)	$\lesssim$ 0·04[e]	$\lesssim$ 0·025[e]
Main atmospheric constituents	H_2, He (also: NH_3, CH_4, H_2O)	H_2, He (?) (also: NH_3, CH_4)	H_2, CH_4, He?	H_2, CH_4, He?
Average intrinsic magnetic field in atmosphere (gauss)	~ 8[f]	0·5–1 ?[g]	$\lesssim$ 1 [h]	$\lesssim$ 1 [h]

[a]Calculated from measured equatorial and polar radii (Kliore *et al.*, 1975) using observations of J_2, J_4 (Null *et al.*, 1975) and third-order theory (Hubbard, 1974).
[b]Danielson *et al.* (1972). Dollfus (1970) obtained 0·03 ± 0·01.
[c]Estimated from the models described in this paper. Also see Zharkov and Trubitsyn (1972).
[d]Ingersoll *et al.* (1975).
[e]Estimated from known upper bounds for effective temperature and Bond albedo.
[f]Not known because of the substantial non-dipolar field. See Hide and Stannard (1976).
[g]Brown (1975).
[h]Kavanagh (1975).

brium. These deviations are expected to exist because of dynamic processes (e.g. convection) but are expected to be very small for a fluid planet. This is consistent with the observations for Jupiter (Null *et al.*, 1975) from the Pioneer 11 fly-by.

Table 1 lists the calculated densities of the major planets. In order to interpret these densities, consider a planet of mass M of pure composition and at zero temperature. According to the virial theorem (see, for example, Clayton, 1968)

$$E_G = -3\int P \, \mathrm{d}V \tag{2}$$

where $E_G \sim -GM^2/R$ is the total gravitational energy, and the integral extends over the volume of the planet. It follows that the central pressure in the planet is roughly of order

$$P_c \sim \frac{GM^2}{4\pi R^4} \tag{3}$$

If P_c is much less than, say, 1 Mbar ($= 10^6$ bar), then the density of matter is essentially the same as it is at zero pressure. The planet then behaves like 'ordinary matter':

$$M \simeq \tfrac{4}{3}\pi R^3 \rho_0$$
$$R \propto M^{1/3} \tag{4}$$

where ρ_0 is the zero pressure mass density, and depends only on composition.

On the other hand, if $P_c \gg 1$ Mbar, then the matter is held up against gravity by the Fermi degeneracy pressure of the electrons:

$$P \simeq 10(\rho Z/A)^{5/3} \text{ Mbar} \tag{5}$$

where ρ is in g cm^{-3}, and Z, A are the atomic number and atomic weight of the elemental composition. Note that $Z/A = 1$ for hydrogen, but $Z/A \simeq \frac{1}{2}$ for all other elements. Substitution of (5) in (3) yields

$$R \propto M^{-1/3}(Z/A)^{5/3} \tag{6}$$

which should be contrasted with equation (4). Evidently, there is some intermediate mass (and central pressure) for which R, as a function of M, goes through a maximum. Figure 1 shows some detailed calculations by Zapolsky and Salpeter (1969) which indicate that the maximum in radius occurs near the mass of Jupiter. These calculations, based on Thomas Fermi–Dirac theory for the pressure–density relationship, together with numerical solution of the equation of hydrostatic equilibrium, also indicate that hydrogen ($Z/A = 1$) has a dramatically different mass-radius relationship from any other element ($Z/A \simeq \frac{1}{2}$), as equation (6) suggests. It is this distinctive 'signature' of hydrogen, together with the fact that Jupiter and Saturn have masses near the peak in Figure 1, which enables one to incontrovertably assign a primarily hydrogenic composition to these planets. Other effects (rotation, temperature, magnetic fields)

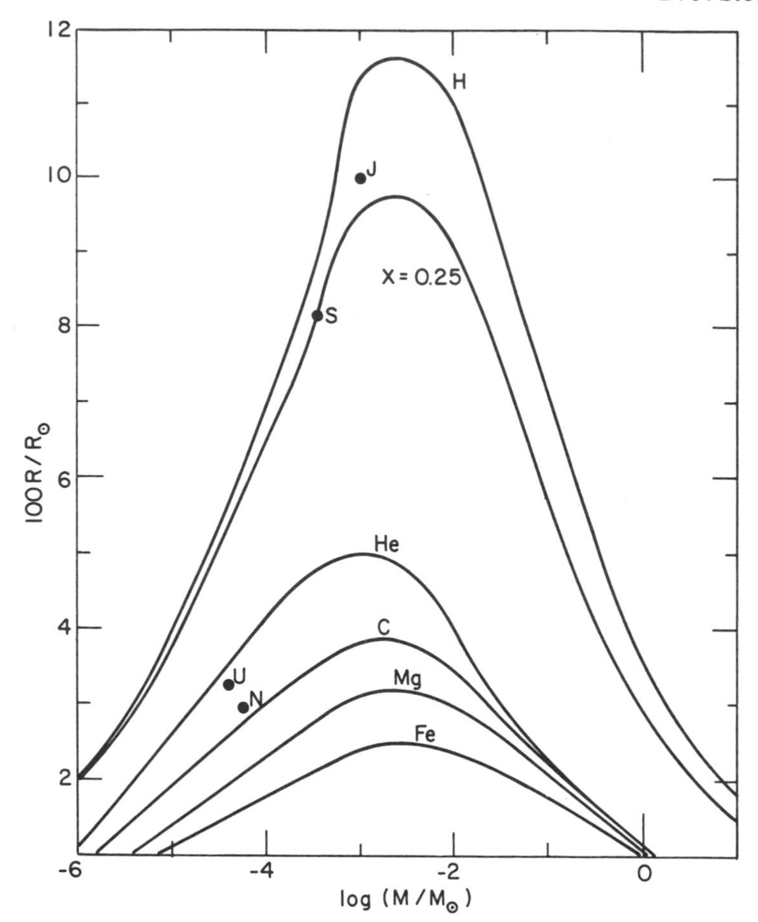

Figure 1. Mass–radius relationship for zero-temperature spheres of various chemical compositions, based on Thomas–Fermi–Dirac theory. $X = 0{\cdot}25$ refers to a 75 per cent H and 25 per cent He mixture by mass. The symbols J, S, U, N represent the giant planets. $R_{\odot}$ and $M_{\odot}$ are the solar radius and mass. (Reprinted from H. S. Zapolsky and E. Salpeter, *Astrophys. J.*, **158**, 809–813, copyrighted 1969 by The University of Chicago Press)

are all too small to change this conclusion. For example, a Jupiter constructed of helium would require a central temperature of 10^6 K. Such a body would radiate many orders of magnitude more energy than Jupiter is observed to radiate. Notice that Figure 1 indicates that Jupiter is closer to being purely hydrogen than Saturn. This is confirmed in more detailed calculations (section on *The Hydrogen–Helium Planets*).

In contrast, Uranus and Neptune are not clearly defined in composition by Figure 1. (In this respect, they resemble the terrestrial planets). It is possible, for example, to construct Uranus from almost pure, cold helium; or from warm methane. The construction of models for Uranus and Neptune must therefore be partly guided by other observations, or even by a cosmogonic prejudice. This ambiguity of composition is discussed further in the section about *Uranus and Neptune*. The only certainty is that these planets are *not* composed primarily of rock, since this would lead to average densities in excess of 4g cm^{-3}.

The rather clearly defined difference between rocks ($\bar{\rho} \gtrsim 3$ g cm^{-3}) and ices ($\bar{\rho} \sim 1$–2 g cm^{-3}) can also be used to roughly characterize the satellites. Figure 2 shows the densities and radii for these other bodies. (Pluto is excluded since the data is presently very unreliable). Io and Europa are evidently like the Earth's moon, whereas Callisto, Ganymede, and Titan are clearly 'icy'.

Table 1 also gives a tabulation of q_2, defined as

$$q_2 \equiv \frac{1}{MR_e^2} \int \rho(\mathbf{r}) r^2 \,\mathrm{d}V \tag{7}$$

where the integral is over the volume of the planet and r is a radial coordinate from the centre of the planet. If all the matter were concentrated at the centre, $q_2 = 0$; whereas a uniform density planet has $q_2 \simeq 0{\cdot}6$. The values tabulated are obtained from the observed gravitational moments J_{2n}, and are determined primarily by J_2. They are not, however, entirely independent of the type of

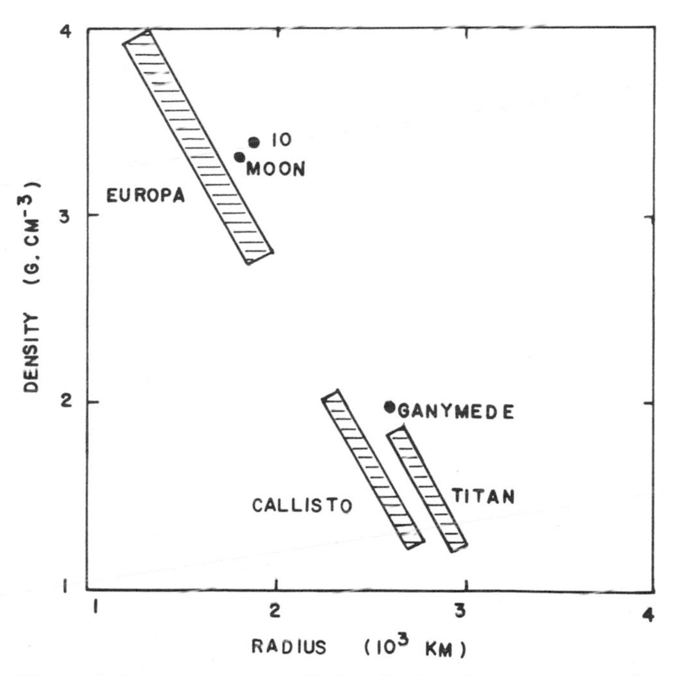

Figure 2. Densities and radii for the best known outer solar system satellites. The Earth's moon is also shown for comparison. The satellites form two groups: 'rocky' (dense and small) and 'icy' (low density and large)

compositional model proposed for the interior, and should therefore be regarded only as 'typical' values. Furthermore, the rotation rates of Uranus and Neptune are uncertain by 10 per cent or more. Nevertheless, there is a significant trend towards smaller q_2 as one proceeds from Jupiter outwards. The value of q_2 for Jupiter is almost the same as that which would be obtained for a homogeneous planet in which $P \propto \rho^2$, but detailed models (see section *The Hydrogen–Helium Planets*) suggest a slight excess concentration of matter towards the centre, suggestive of a small distinct core. Saturn appears to have a more clearly defined 'core' (or central concentration) while Uranus and Neptune have an even more pronounced central condensation.

None of the J_{2n} is known for the satellites. The next moment (J_4) is known for Jupiter and Saturn only. It provides information on the structure of the outer layers of these planets. In both Jupiter and Saturn, the value of J_4 is inconsistent with an isothermal temperature distribution below the observable atmosphere, but is consistent with an adiabatic temperature distribution (Hubbard, 1974). The significance of this is discussed in section on *The Hydrogen–Helium Planets.*

Atmospheres and Surfaces

It would be foolish to attempt to characterize terrestrial planetary interiors from the composition of their atmospheres. This is more meaningful, however, for gaseous planets because of the predominance of volatiles, and evidence (see below) that these atmospheres are apparently bottomless and convective below optically observable levels. It is presumptuous to suppose that the atmospheric composition matches the interior composition (even after allowance is made for cloud formation by condensibles), but the atmosphere is neverthless expected to be a meaningful boundary condition for interior models, in the sense that the atmosphere and interior have a well-defined (but not necessarily simple) compositional and thermal relationship.

Table 1 above lists the relevant properties of the major planetary atmospheres. First, consider Jupiter. The main constituent of the atmosphere is undoubtedly hydrogen (Newburn and Gulkis, 1973; Elliot *et al.*, 1974). Helium, expected to be the next most abundant constituent, cannot be observed spectroscopically except in the exosphere where a small amount was detected by the Pioneer UV spectrometers (Carlson and Judge, 1976). Various models of the thermal structure of the atmosphere appear to require about 20 per cent helium by mass (e.g. Orton, 1975). Nevertheless, the only definite statement that can be made at present is that the helium fraction does not exceed 50 per cent by mass. The accurate determination of helium content is the most important future observation for interior model construction, and will probably require *in situ* atmospheric sampling (planned by N.A.S.A. for the early 1980's). Many minor constituents have been observed in the atmosphere in amounts which so far do not seem inconsistent with a solar composition. Analysis of the infrared spectrum (Lacy *et al.*, 1975; Orton, 1975) indicates that the atmosphere is essentially adiabatic for $P \gtrsim 0{\cdot}5$ bars, and that $T = (175 \pm 20)$ K at $P = 1$ bar. (The latter is a commonly used boundary condition for interior models). The significance of an adiabatic temperature gradient is that it suggests convective (rather than radiative) heat transport (see the section on *The Hydrogen–Helium Planets*). The increasing brightness temperature of Jupiter at longer wavelengths indicates that the temperature increases as one goes deeper into the atmosphere (Newburn and Gulkis, 1973). This conclusion is reinforced by the recent observation (Westphal *et al.*, 1974) of 5 μm 'hotspots' corresponding to temperatures of at least 300 K.

The average input of solar radiation on Jupiter is about $1{\cdot}3 \times 10^4$ erg/cm^2 sec of which about 55 per cent is absorbed. The infrared emission of Jupiter (Ingersoll *et al.*, 1975) is about $1{\cdot}4 \times 10^4$ erg/cm^2 sec so it follows that Jupiter emits about twice as much energy as it absorbs. The implied internal energy source of 5×10^{24} erg/sec is apparently gravitational in origin, since all other proposed sources (e.g. radio-activity, accretion, thermonuclear fusion) fall short by at least two orders of magnitude (Hubbard and Smoluchowski, 1973). The implications of this very important observation are discussed in the section on *The Hydrogen–Helium Planets*.

Saturn appears to be qualitatively similar to Jupiter. The atmosphere is predominantly hydrogen, and the temperature increases as one goes deeper, consistent with an adiabat passing through $T \simeq 150$ K at $P = 1$ bar (Newburn and Gulkis, 1973; Hubbard and Smoluchowski, 1973). Saturn appears to radiate more than twice as much energy as it absorbs from the Sun (Aumann *et al.*, 1969; Nolt *et al.*, 1974; Rieke, 1975). In non-dimensional gravitational units, this excess luminosity (roughly $1{\cdot}7 \times 10^{24}$ ergs/sec) may be 'anomalously' large compared to Jupiter. To see this, consider the following measure of the excess luminosity:

$$Q_0 = \frac{L\tau}{|E_G|} \tag{8}$$

where L is the excess luminosity of the planet, τ is the age of the planet (always assumed to be $4{\cdot}6 \times 10^9$ years), and E_G is the gravitational energy. For Jupiter, $Q_0 = 0{\cdot}015 \pm 0{\cdot}005$, whereas for Saturn, $Q_0 = 0{\cdot}04 \pm 0{\cdot}02$. According to the evolutionary models discussed in the Section on *The Hydrogen–Helium Planets* the former can be explained by the gradual loss of primordial heat whereas the latter probably requires fractionation processes within the planet well after formation. It should be stressed, however, that the measurement of excess luminosity is very difficult, indirect (current methods presume independent knowledge of the planetary albedo), and susceptible to overestimation (the excess luminosity of Jupiter has 'decreased' by almost a factor of two since the first observations). The value of Q_0 for Saturn, in particular, may undergo reduction after future observations.

Much less is known about Uranus and Neptune. The very cold temperatures of their atmospheres mean that major constituents (especially water) can be frozen out to great depths. Hydrogen and methane are known to be present, and are probably the main atmospheric constituents in the observable atmosphere (Newburn and Gulkis, 1973). No excess luminosity has been detected for these planets, although it has been suggested for Neptune (Macy and Trafton, 1975). Present observations do not preclude an internal heat source comparable to the insolation. Constraints on this heat source are discussed in the section on *Uranus and Neptune*. Radio observations (Newburn and Gulkis, 1973; Briggs, 1973) suggest that the temperature increases as one goes deeper below the observable atmosphere.

Of the remaining outer solar system bodies, Titan and Saturn's rings merit special mention. Titan is remarkable because it has a dense atmosphere, major components of which are apparently hydrogen and methane (Knacke *et al.*, 1975). It is not understood why Titan is so apparently different from other comparable satellites. The rings of Saturn have been the subject of much research, recently reviewed by Pollack (1975), but are also poorly understood. The ring particles are typically of centimetre size, and water ice is a major constituent. The reason why Saturn alone has rings may be explainable by a 'condensation' theory according to which only the early Saturnian environment had the right nebular density and temperature for the nucleation and growth of ring particles from the gas phase.

Much is known about the surfaces of the solid bodies in the outer solar system (Morrison and Burns, 1976) but much caution should be exercised in using these results to infer anything about interiors. For example, the recent reported observation of methane ice on Pluto (Metz, 1976) has no implications for the overall composition, although it does impose constraints on the thermal history. Unfortunately, the internal heat flux is not known for any of these bodies. The brief discussion of interior models for the solid bodies (the section *Pluto and Satellites*) is thus constrained only by density (where available) and by plausibility arguments about the material from which these bodies condensed.

The Magnetic Field

In view of the great difficulty of performing *in situ* measurements in the outer solar system, remote sensing probes of planetary interiors are highly desirable. Two have already been mentioned (gravity field, excess luminosity); the third one presently available is the magnetic field. Many different mechanisms for planetary magnetism have been proposed (see Stevenson, 1974, for a review) but it seems likely that only the dynamo mechanism is capable of maintaining a 'substantial' internally generated field. (Exactly what is meant by 'substantial' is not yet clear: typically, a non-dynamo mechanism cannot produce global fields in excess of $\sim 10^{-2}$ gauss). The dynamo mechanism maintains the field by the action of fluid motion on the field lines in an electrically conducting fluid. The magnetic field therefore contains (poorly understood) information on the internal dynamics of a planet.

The dynamo theory has been recently reviewed by Gubbins (1974). A necessary (but not sufficient) condition for dynamo generation is that the magnetic Reynold's number R_m must exceed a critical value:

$$R_m \equiv \frac{vL}{\lambda} \gtrsim 10 \quad (9)$$

where v is a characteristic fluid velocity (relative to a rigidly rotating planet), L is a characteristic length scale (e.g. the size of the conducting fluid region) and λ is the magnetic diffusivity (proportional to the electrical resistivity). To illustrate the likely values of R_m in planets, consider first the value of v for thermal convection. According to the mixing length theory of convection (Schwarzchild, 1958)

$$v \sim \left[\left(\frac{l}{H_p}\right)\frac{F}{\rho}\right]^{1/3} \quad (10)$$

to a rough first approximation, where l is the mixing length, H_p is the pressure scale height, F is the convective heat flux and ρ is the fluid density. This formula ignores the important Coriolis effect, but is surprisingly little affected by rotation (Gierasch and Stevenson, 1977). The value of F can be as small as zero (if all the heat can be transported more efficiently by conduction or radiation) or as large as the total internal heat flux. For typical values of the latter, and for $l \sim 0{\cdot}1\ H_p$, one finds

$$v \sim 1 \text{ cm/sec} \quad (11)$$

to within an order of magnitude or two. (Since v is weakly dependent on F, it is not necessary to carefully specify the planet. To a first approximation, the Earth's core for which $F \leqslant 10$ erg/cm^2 sec and Jupiter's core for which $F \leqslant 10^4$ erg/cm^2 sec give similar convective velocities.) In subsequent sections, the likelihood of convection is explicitly considered for outer solar system bodies and it is concluded that convective thermal transport usually predominates. The analysis leading to equation (11) assumes molecular viscosity is unimportant, i.e.

$$\nu \ll vl \quad (12)$$

where ν is the kinematic viscosity. This is almost invariably satisfied.

Fluid motion might also arise from other processes. One which has received much attention is the effect of precession. Recent analysis (Rochester *et al.*, 1975) indicates that precession cannot power the Earth's dynamo; and similar conclusions may apply to other planets. Precessional generation is not considered further here.

The magnetic diffusivity is expected to be metallic-like at megabar pressures since with only one notable exception (helium) all substances become metallic at such pressures (Ross, 1972). According to equation (3), megabar pressures are exceeded in Jupiter, Saturn, Uranus and Neptune. Typical metals have $\lambda \sim 10^3$ cm^2/sec so that $R_m \sim 10^6$ if $L \sim 10^9$ cm and $v \sim 1$ cm/sec. Dynamo generation may be possible even if $\lambda \sim 10^6$ cm^2/sec, typical of good *ionic* conductors.

The sufficient conditions for dynamo generation are not known. It has been determined, however, that if fluid flow possesses a non-zero average helicity then dynamo generation is possible (Gubbins, 1974). Various authors (Parker, 1955; Steenbeck *et al.*, 1966; Gierasch and Stevenson, 1977) have pointed out that helicity may be ensured if rotation has an important dynamical effect on the flow. This requirement is satisfied if

$$R_0 \equiv \frac{v}{l\Omega} \lesssim 1 \tag{13}$$

where R_0 is known as the Rossby number, and Ω is the planetary rotation rate. One could speculate, therefore, that a sufficient condition for dynamo generation is $R_m \gtrsim 10$ *and* $R_0 \lesssim 1$. It should be noted that $R_0 \lesssim 1$ may be easily satisfied even for bodies as slowly rotating as Venus.

Observationally, only the Jovian magnetic environment has been quantitatively characterized in the outer solar system (Hide and Stannard, 1976). The magnitude, multipolarity and possible secular variation of the intrinsic Jovian field offer tantalizing clues to the internal dynamics. Radio bursts have been observed from Saturn, and have been interpreted as indicating a field that is perhaps an order of magnitude smaller than Jupiter's (Brown, 1975). Uranus and Neptune have fared poorly in the observations, and only upper bounds for the field strength are known (Kavanagh, 1975). No useful upper bounds exist for the magnetic fields of Pluto or the satellites.

The Hydrogen–Helium Planets

Model Construction

Model construction for hydrogen–helium planets is facilitated by the relative simplicity of composition (compared, for example, with the Earth) and by the confidence that can be placed in theoretical calculations for the relevant high pressure thermodynamic and transport properties. In these respects, the hydrogen–helium planets are more like stars than like terrestrial planets. Nevertheless, the models are more difficult to construct than most stellar models, because the matter is less 'ideal'.

Static models are considered first. These treat the internal heat flux as an input parameter rather than something to be derived. The heat source origin is discussed in the section on *Homogeneous Thermal Evolution*, where evolutionary models are considered. Implicit in the static model approach is the assumption that the evolutionary time scale greatly exceeds the relevant response timescale in the planetary interior, so that a thermal steady-state has been achieved. This is indeed the case for the fluid, convective planetary interiors deduced below.

The input for static models construction consists of:

1. The planetary rotation rate (assumed uniform). The coincidence of rotation rates for the magnetic field and (averaged) atmosphere for Jupiter and Saturn supports the assumption of rigid rotation. Substantial differential rotation within the metallic core cannot yet be precluded.
2. The equation for hydrostatic equilibrium, including the effect of rotation (a substantial perturbation for Jupiter and Saturn):

$$\frac{dP}{ds} = -\frac{GM(s)}{s^2}\rho(s) + \frac{2}{3}\Omega^2 s\rho(s) + O(\Omega^4) \tag{14}$$

   where s labels equipotential surfaces and is the radius of that sphere containing a volume equal to the volume within an equipotential. $M(s)$ is the mass within equipotential s, Ω is the planetary rotation rate, $\rho(s)$ the density and P the pressure.
3. The equation of state $P \equiv P\ (\rho, T, X_i)$ where $X_i = X_i(s)$ are the fractional abundances of various constituents.
4. The internal heat source and its distribution. The models are not generally sensitive to the heat source distribution, unless the energy is generated only in the outermost layers of the planet. This would be inconsistent with the gravitational energy sources expected (section on *Homogeneous Thermal Evolution*).
5. The thermal transport properties (radiation, conduction, convection) which, together with the assumed heat flux, determine the relationship between temperature and pressure, and predict the dominant thermal transport mechanism.
6. The theory of planetary figures. This is needed to relate the internal density distribution to the shape of the planet and the gravitational moments (J_2, J_4), and thus constrain possible models.

A model must fit the mass, radius, and gravitational moments and be consistent with the observable atmosphere (composition and thermal structure). It should predict the bulk composition of the planet, the state of matter within (solid or fluid, existence of phase transitions, metal or insulator), the central temperature and pressure, and be consistent with the need to generate a substantial magnetic field if required. It should make predictions which are testable (in principle) by observation; such as for higher gravitational moments, atmospheric helium abundance and normal modes for the whole planet.

In the next two sections, the input requirements for the equation of state and thermal transport properties are discussed.

Thermodynamics and the State of Matter

Models of a hydrogen–helium planet such as Jupiter require a thermodynamic description of the relevant constituents over a range of conditions extending from ideal gas behaviour (in the atmosphere) to dense, degenerate matter (in the deep interior). For simplicity, the thermodynamics of pure hydrogen are discussed first. Hydrogen–helium mixtures and minor constituents are then considered.

Consider, first, the high density (but low temperature) limit. At very high densities ($\rho \gg 1\,\mathrm{g\,cm^{-3}}$), the Pauli exclusion principle precludes the existence of molecules or localized electronic states and hydrogen is a Coulomb plasma: protons immersed in an almost uniform sea of degenerate electrons. This state, known as metallic hydrogen, is in most respects the simplest of the alkali metals, and its equation of state can be calculated accurately from first principles. The first step is to calculate the Helmholtz free energy F (or, equivalently, the partition function). At the temperatures of interest, the main contribution to F is the zero temperature energy of the uniform electron gas. At the pressures of interest, this is *not* merely the Fermi energy (the exchange and correlation energies are also important). Another important contribution to F is the electrostatic energy (Madelung energy) for the protons immersed in a uniform electron gas. Minor contributions to F arise from finite temperature corrections and the non-uniformity of the electron gas. Although the finite temperature corrections are a small fraction of F, they are crucial for evaluating certain thermodynamic derivatives (such as specific heat) and for determining whether the solid or fluid state is thermodynamically favoured at a given temperature.

Pollock and Hansen (1973) used their Monte Carlo calculations for the fluid and solid state free energies to deduce that the melting temperature of metallic hydrogen is given by

$$T_{\mathrm{M}} \simeq 1500\rho^{1/3} \quad \mathrm{K} \tag{15}$$

where ρ is in $\mathrm{g\,cm^{-3}}$. This is likely to be an upper bound since it neglects the effect of ionic screening by the electrons, which weakens the effective proton-proton interaction. This temperature is much less than the internal temperatures deduced for Jupiter and Saturn (the next section) so the following comments are limited to the more relevant fluid phase.

Two approaches to calculating the properties of the fluid state have been made. The Monte Carlo approach (Hubbard and Slattery, 1971) is more fundamental since it assumes no model for the fluid structure. A less fundamental approach, based on the perturbation theory of fluids and the hard sphere model, has been developed (Stevenson, 1975a). This offers comparable accuracy and includes some free energy contributions (such as the non-linear dielectric response of the electrons) which the Monte Carlo methods have yet to incorporate. The resulting equation of state is expected to be accurate to 1 per cent at the relevant pressures and temperatures:

$$P = 9{\cdot}95\rho^{5/3}\left[1 - \frac{0{\cdot}909}{\rho^{1/3}} + \frac{0{\cdot}164}{\rho^{2/3}} - \frac{0{\cdot}021}{\rho}\right] \tag{16}$$

where P is the pressure in Megabars at 6000 K (an arbitrary reference temperature) and ρ is the density in $\mathrm{g\,cm^{-3}}$. At other temperatures, the pressure can be evaluated using the parameter c, where

$$c = \frac{1}{Nk_{\mathrm{B}}}\left(\frac{\mathrm{d}P}{\mathrm{d}T}\right)_v \tag{17}$$

where N is the ion number density and k_{B} is Boltzmann's constant. This parameter has been tabulated (Stevenson, 1975a). It is typically larger than the ideal gas value ($c = 1$) but smaller than the ideal Debye value ($c = 3/2$).

Two other parameters of importance are

$$\gamma = \left(\frac{\mathrm{d}\ln T}{\mathrm{d}\ln \rho}\right)_s \tag{18}$$

$$C_v = \frac{k_{\mathrm{B}}c}{\gamma} \tag{19}$$

where the derivative in (18) is at constant entropy, and C_v is the heat capacity per proton. Typically, $0{\cdot}61 \lesssim \gamma \lesssim 0{\cdot}64$ and $C_v \sim 2k_{\mathrm{B}}$.

Apart from properties that depend on subtle free-energy differences (e.g. superconductivity) the thermodynamics of metallic hydrogen seem to be well understood. In contrast, the same cannot be said for molecular hydrogen. At a pressure of less than a few megabars, the molecular phase of hydrogen is thermodynamically favoured over the metallic phase. The estimated transition pressure is based primarily on theoretical arguments (Ross, 1974), since the claimed experimental verifications (Grigoryev *et al.*, 1972; Vereshchagin *et al.*, 1975) are unconvincing. The molecular phase extends from ideal gas densities to the densities of strongly interacting matter ($\rho \sim 1\,\mathrm{g\,cm^{-3}}$), where the distance between molecules is comparable to the size of a molecule. It is in this latter regime that first principles calculations are especially difficult. Some progress has been made (Friedli and Ashcroft, 1977) but for practical purposes, it is still necessary to approximate the interaction energy by a sum of pairwise-additive intermolecular potentials. The choice of potential is constrained by the experimental shock data on molecular hydrogen (Ross, 1974) but is chosen to have a form compatible with first principles calculations (McMahon *et al.*, 1974; Ree and Bender, 1974) for finite assemblages of molecules. Given an intermolecular potential, the free energy can then be calculated. This has been done for the fluid state by molecular dynamics calculations (Slattery and Hubbard, 1973) and by hard-sphere perturbation theory (Ross, 1974; Stevenson and Salpeter, 1976). The temperature and density at which the hard sphere packing fraction is approximately 0·45 gives an estimate for the melting temperature:

$$T_m \simeq 2800\,\rho^2 \quad \mathrm{K} \tag{20}$$

provided $\rho \gtrsim 0{\cdot}4\,\mathrm{g\,cm^{-3}}$. This is lower than the internal temperatures deduced for Jupiter and Saturn (section 3·3) thereby confirming that the fluid state is appropriate. It should be emphasized that because of the uncertainty in the interaction potential, the above estimate of T_m could be wrong by as much as a factor of two. The effects of these uncertainties on models have been considered (Stevenson and Salpeter, 1976) and are discussed in section *Simple Models.* The thermodynamic parameters c, γ and C_v are similarly uncertain, and also rather sensitive to density, but typical values at the highest densities ($\rho \sim 1$ g cm^{-3}) are $c \sim 1$, $\gamma \sim 0{\cdot}3$ and $C_v \sim 3\,k_B$ (per molecule). At the low density end (i.e. the atmospheres of Jupiter and Saturn) $c = 1$, $\gamma = 2/5$ and $C_v = 5/2\,k_B$, provided the rotational degrees of freedom are fully excited. No simple approximate analytical expression exists for the equation of state, but typically

$$\left(\frac{\mathrm{d}\ln P}{\mathrm{d}\ln \rho}\right)_s \simeq 1{\cdot}8 - 2{\cdot}2 \tag{21}$$

for $\rho \gtrsim 0{\cdot}05\,\mathrm{g\ cm^{-3}}$, and temperatures appropriate to Jupiter and Saturn (a few thousand degrees).

Figure 3 indicates in semiquantitative fashion the main features of the hydrogen phase diagram. Some aspects of this phase diagram are uncertain, particularly in the megabar pressure region where even the sign of the molecular-metallic phase boundary slope is not known (because the latent heat is not known). Various topological possibilities for the phase diagram have been discussed by Stevenson and Salpeter (1977a).

Planets such as Jupiter and Saturn cannot consist of hydrogen alone, and the next most abundant constituent is expected to be helium. Until recently, the thermodynamics of hydrogen–helium mixtures were estimated by appropriate interpolation between the properties of pure hydrogen and of pure helium. One such interpolation procedure is the assumption of 'Volume additivity' which states that

$$V(x,P) = x\,V(1,P) + (1-x)\,V(0,P) \tag{22}$$

where $V(x,P)$ is the volume per molecule at pressure P of a binary mixture containing a number of fraction x of one of the constituents. More recent calculations, which treat the mixture explicitly, test relationships such as (22) and also make predictions for the solubility of helium in hydrogen.

Consider, first, the metallic mixture. Although pure, crystalline helium does not become metallic until about 80 Megabars, the electronic states can be treated as being delocalized at much lower pressures, especially in the mixture (Stevenson, 1975a). Calculations using fluid perturbation theory indicate that an alloy containing 10 per cent by number of helium occupies typically 3 per cent less volume than the separated phases. This has a small but significant effect on the

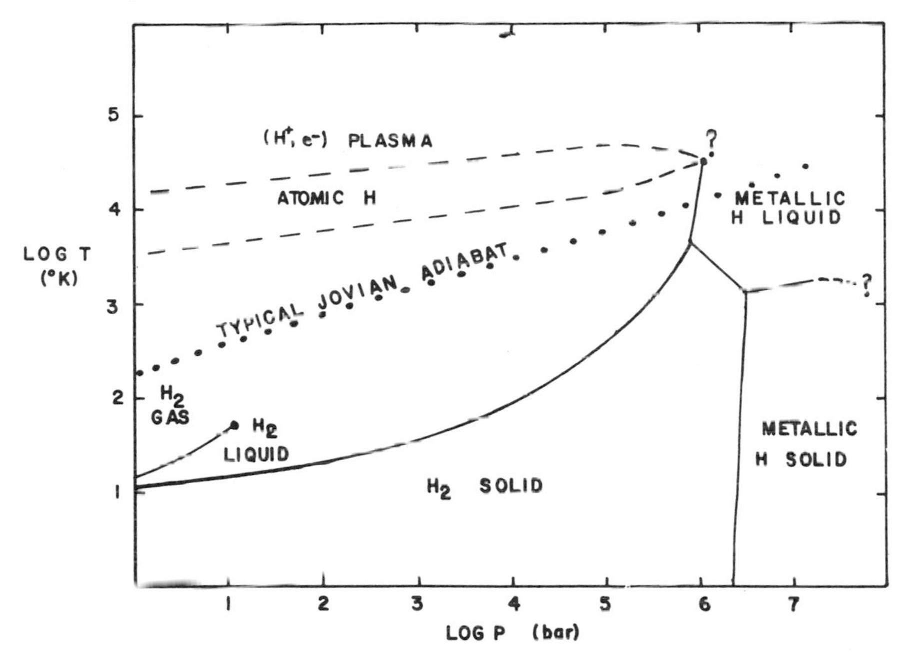

Figure 3. Hydrogen phase diagram (schematic). Superimposed is a typical Jovian adiabat passing through $T = 175$ K at $P = 1$ bar. The adiabat does not intercept any phase transition except (possibly) at very high pressures. The phase diagram is very uncertain for $P \gtrsim 10^5$ bar. The dashed lines separating fluid H_2, H and H^+, e^- plasma are poorly defined except at low densities. (Based on W. B. Hubbard and R. Smoluchowski, *Space Science Reviews*, **14**, 613 (1973), copyrighted by D. Reidel Publishing Company, Holland)

amount of helium required to construct a model planet. More dramatic is the prediction that the same alloy spontaneously separates into hydrogen-rich and helium-rich fluids when the temperature drops below about 9,000 K (Stevenson, 1975a). This has important implications for the evolution of hydrogen–helium planets (sections on *Homogeneous Thermal Evolution* and *The Helium Distribution*).

The properties of molecular hydrogen–helium mixtures are less readily predicted, even though this mixture has been experimentally investigated at low pressures (Streett, 1974). Recent calculations (Stevenson and Salpeter, 1977a) indicate that the deviation from volume additivity is very small, and that the solubility of helium is much greater in molecular hydrogen than in metallic hydrogen. This means that if a molecular and a metallic phase are coexistent, then the molecular phase must have more helium. The implications of this are briefly discussed in the section on *The Helium Distribution.* Figure 4 summarizes the theoretical estimates for the solubility of helium in hydrogen in the fluid state. The critical temperature T_c is the temperature above which

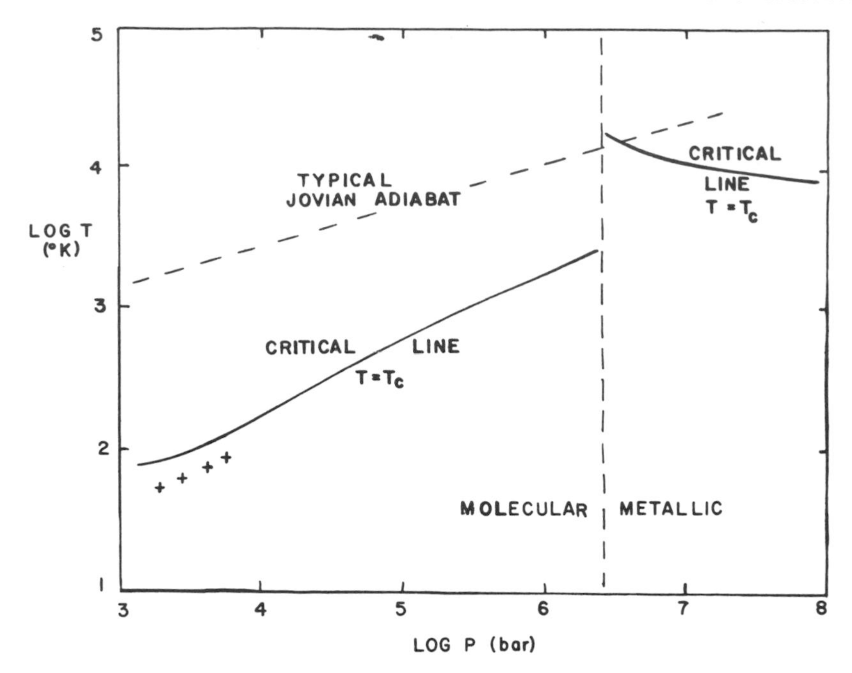

Figure 4. Theoretical estimates of the critical temperature as a function of pressure for miscibility in hydrogen–helium mixtures. The critical line is discontinuous at the molecular–metallic transition, if that transition is first order. Also shown is a typical Jovian adiabat (– – –) and Streett's (1974) experimental results (+) for T_c as a function of P.

helium is soluble in hydrogen in all proportions. Typically, a solar composition mixture would phase separate for $T \lesssim 0{\cdot}8\ T_c$.

After helium, the minor constituents are the 'ices' (H_2O, CH_4, NH_3) and the 'rocks'. Since these constitute a small fraction of the total mass, it is usually adequate to use a Thomas–Fermi–Dirac equation of state (Zapolsky and Salpeter, 1969; Podolak and Cameron, 1974) and assume volume additivity. The properties of ices and rocks are much more relevant to Uranus and Neptune, and are discussed further in the section below about these two planets.

Energy Transport and the Temperature Profile

Early models of Jupiter and Saturn assumed $T = 0\text{K}$ in the deep interior. The discovery of the excess luminosity of these planets invalidated this approach. Furthermore, models are now sufficiently accurate that the thermal expansion of the deep interior is far from negligible.

The competing mechanisms for the transport of the inferred internal heat flux are conduction (by photons or electrons or molecules) and convection. Convective transport is determined by the extent to which the temperature gradient is superadiabatic, since a subadiabatic temperature gradient (in a homogeneous fluid) is stable with respect to displacement of fluid elements. Consider the fractional superadiabaticity ε defined as

$$\varepsilon = \left[\left(\frac{\mathrm{d}T}{\mathrm{d}r}\right)_{\text{actual}} - \left(\frac{\mathrm{d}T}{\mathrm{d}r}\right)_{\text{adiabatic}}\right] \Big/ \left(\frac{\mathrm{d}T}{\mathrm{d}r}\right)_{\text{adiabatic}} \tag{23}$$

If $\varepsilon < 0$, no convection is expected. Convection is very efficient if initiated, however, and ε is much less than unity for any convective heat flux of interest. For example, $\varepsilon \sim 10^{-8}$ in Jupiter if the inhibiting effect of rotation is neglected (Hubbard and Smoluchowski, 1973) and $\varepsilon \sim 10^{-4}$ allowing for rotation (Gierasch and Stevenson, 1977). Although these estimates are based on a very crude theory of convection, they are not likely to be wrong by several orders of magnitude. Furthermore, the effect of a magnetic field is not expected to be any more inhibiting than rotation. It seems safe to assume, therefore, that if the matter has low viscosity (as is the case for liquid metallic or molecular hydrogen for which the kinematic viscosity is of order $10^{-2}\ \mathrm{cm}^2\ \mathrm{sec}^{-1}$, then the temperature gradient never substantially exceeds the adiabatic gradient in magnitude.

To demonstrate that convection is required, it is necessary to determine whether the total heat flux can be transported by conduction alone along a subadiabatic temperature gradient. Consider Jupiter (similar numbers apply to Saturn). At the deepest observable levels in the atmosphere ($P = 1$ bar, $T \simeq 175$ K) hydrogen is sufficiently opaque to ensure that convection is required to transport the heat. As one proceeds deeper into the planet, the adiabatic temperature gradient applies (and $T \propto \rho^{0\cdot45}$) since the matter continues to be too opaque. In the temperature range $400\,\mathrm{K} \lesssim T \lesssim 700\,\mathrm{K}$ hydrogen alone is insufficiently opaque, but a solar abundance of 'ices' is apparently more than adequate to prevent radiative transfer of the heat flux (Stevenson, 1976). At even deeper levels, hydrogen becomes increasingly opaque and the radiative heat flux progressively smaller (because the opacity increases more rapidly than T^4). Molecular conduction is never adequate either. Once the metallic core is reached, electronic conduction is possible, but this is too small by at least a factor of ten to transport the heat at a subadiabatic temperature gradient (Stevenson and Salpeter, 1976) except possibly in the innermost regions of the planet, where the heat flux tends to zero. The corresponding adiabat is shown superimposed on the phase diagram in Figure 3. It does not cross any gas–fluid or fluid–solid transition and is therefore self-consistent. (If a solid phase had been encountered, then the adiabatic assumption would necessarily have to be abandoned). The only phase transition encountered is the (hypothetical) one between liquid metallic hydrogen and liquid molecular hydrogen. In other words, Jupiter and Saturn have no 'surface'. Provided the relevant observations are correct (see section on *Atmospheres and Surfaces*), this conclusion is essentially inescapable.

Suppose for a moment that there were no phase transitions encountered along an adiabat at all. It would then be reasonable to equate the specific entropy in the deep atmosphere (essentially an observable) to the specific entropy in the metallic core (calculable from first principles). In this way, Hubbard (1973) estimated that the central temperature of Jupiter is $\sim 20{,}000$ K, with only about a 20 per cent uncertainty (provided the assumptions are valid). This result can be readily understood in the following way: on an adiabat $T \propto \rho^n$, with an average value of n being $\sim 0{\cdot}5$. Since the ratio of central density ($\sim 1\,\mathrm{g\,cm^{-3}}$)to the density in the uppermost convective layers is $\sim 10^4$, it follows that the corresponding temperature ratio should be $\sim 10^2$.

It is likely, however, that there is an abrupt transition at the molecular-metallic hydrogen phase change. Salpeter and Stevenson (1976) have examined this, and conclude that a sharp interface forms which inhibits convection. The temperature is continuous and entropy discontinuous across this interface, so that the specific entropies of the atmosphere and metallic core are not equal but instead differ by the latent heat of the molecular-metallic transition. Unfortunately, this latent heat is unknown in both sign and magnitude. Upper bounds can be established (Stevenson and Salpeter, 1976) indicating that the implied uncertainty in the central temperature of Jupiter could be as much as a factor of two. Similar conclusions apply to Saturn, for which an adiabatic model would predict a central temperature of around 12,000 K.

Additional complications arise if the planet is inhomogeneous, and particularly if the helium is non-uniformly distributed. This is briefly discussed in the section *The Helium Distribution.*

Simple Models

The recent flybys by Pioneers 10 and 11 have prompted an upsurge in modelling of the Jovian interior (Podolak and Cameron, 1975; Hubbard and Slattery, 1976; Stevenson and Salpeter, 1976; Zharkov and Trubitsyn, 1976). Saturn has also received attention recently (Podolak and Cameron, 1974; Zharkov *et al.*, 1975). All these models are 'simple' in the sense that they treat these planets as being, to a first approximation, homogeneous, adiabatic and of solar composition. Agreement with density, J_2 and J_4 is then obtained by enhancing the abundance of minor constituents, (e.g. by adding a 'rocky' core). The helium to hydrogen ratio is maintained at a uniform value equal to the primordial solar value. Although some unconventional theories (see Alfvén and Arrhenius, elsewhere in this book) might predict a different H/He ratio, there is at present no convincing argument for adopting other than the solar value. On the other hand, an enhancement of ices or rocks is not only plausible but may even be essential for the formation of these planets (Podolak and Cameron, 1974). The additional assumption that this solar abundance of helium remains uniformly mixed with the hydrogen after planetary formation may be wrong (see the section *The Helium Distribution*). The presently favoured value for the primordial solar abundance of helium is 24 per cent by mass (Cameron, 1973), with a value as large as 30 per cent being only marginally acceptable. There is sufficient uncertainty in this very important parameter to justify treating it as adjustable, and this is the procedure that is usually adopted. In general, one expects a trade-off between enhancing the helium content and enhancing the ice and rock content. However, if the helium is unformly distributed whereas the rock is concentrated in a core, the trade-off no longer exists and the amount of rock is primarily determined by J_2 (i.e. by the moment of inertia).

Bearing in mind the above qualifications, the following general conclusions seem to be indicated by *all* recent Jupiter models:

(a) A purely solar composition model in which the hydrogen and helium are uniformly mixed is *not possible*, regardless of the value chosen for helium abundance, and regardless of how the 'ices' and 'rocks' are distributed. (See section on *The Helium Distribution* for a brief consideration of models with a non-uniform distribution of helium).

(b) Models can be constructed which are of solar composition and uniformly mixed except for a small dense core (most plausibly consisting of rock). The mass of such a 'rocky' core is 10–20 Earth masses, which is an *order of magnitude* enhancement of the rocky component of a solar composition mix. This enhancement need not reside in a core but must be concentrated toward the centre of planet in some way (e.g. by mixing it with the metallic hydrogen core only).

(c) Models can be constructed with 24 per cent helium by mass, but about 30 per cent helium content appears to be required for the 'best' models (i.e. those which fit all known observables best).

(d) The value of J_4 is consistent with an adiabatic temperature profile in the outer envelope of the planet ($\rho \lesssim 0{\cdot}1\ \mathrm{g\ cm^{-3}}$) which passes through $150\ \mathrm{K} \lesssim T \lesssim 200\ \mathrm{K}$ at $P = 1$ bar. It is not consistent with a substantially sub-adiabatic temperature profile below the observable atmosphere, regardless of composition.

No conclusion can be reached at present about the relative abundances of ice and rock. Stevenson and Salpeter (1976) found the above conclusions even when allowance is made for the uncertainties in the thermodynamics of dense molecular hydrogen. The similarities of various recent Jupiter models are more striking than the differences. One of the more noticeable differences is that the models of Stevenson and Salpeter (1976) require less enhancement of heavy matter and/or helium. This is attributable to their allowance for volume non-additivity [i.e. deviations from equation (22)]. Podolak and Cameron (1975) require, for example, about thirty or more Earth masses of 'ice' (specifically H_2O) mixed with the hydrogen and helium. Stevenson and Salpeter require no water enhancement. Figure 5 gives a schematic representation of the interior of Jupiter for a 'typical' model and Table 2 gives details for one of the Stevenson and Salpeter (1976a) models.

The conclusions for Saturn are qualitatively similar. A solar composition model is definitely ruled out, and the value of J_2 is indicative of a dense core

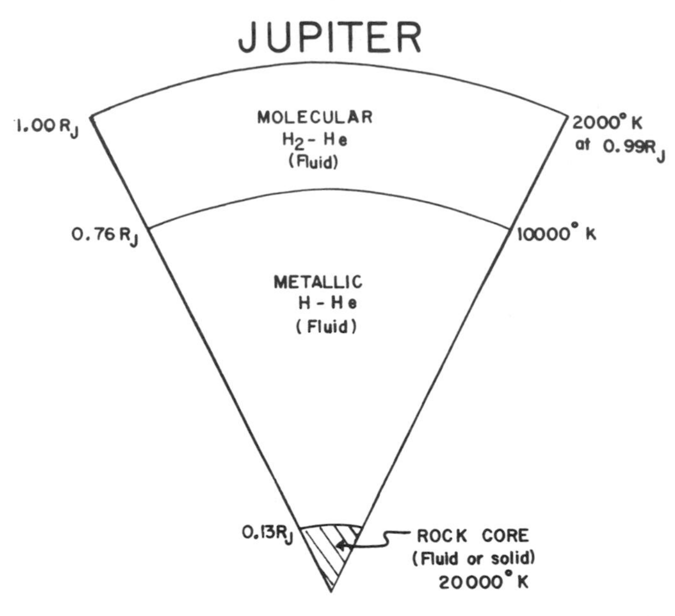

Figure 5. Schematic representation of the Jovian interior for the model detailed in Table 2.

Table 2. A 'Typical' Jupiter Model (Date from Stevenson and Salpeter, 1976)

Average Radial Distance s (10^9 cm)	Pressure (Megabars)	Density (g cm^{-3})	Temperature (K)	Comments
0·0	104·0	26·0	19760	Rock core
0·45	85·8	24·0	19760	(assumed isothermal)
0·904	44·8	18·1	19760	
0·904	44·8	4·325	19760	Metallic
1·5	35·2	3·86	18440	H–He
2·5	24·4	3·24	16450	(adiabatic)
3·5	15·3	2·60	14300	
4·5	7·40	1·89	11620	
5·307	3·09	1·33	9260	
5·307	3·09	1·09	9260	Molecular
5·9	1·29	0·80	8370	H_2–He
6·4	0·32	0·46	6250	(adiabatic)
6·8	$2{\cdot}21 \times 10^{-2}$	0·11	3290	
6·9	$2{\cdot}53 \times 10^{-3}$	$3{\cdot}37 \times 10^{-2}$	2135	
6·95	$1{\cdot}88 \times 10^{-4}$	$5{\cdot}95 \times 10^{-3}$	920	
6·98	$1{\cdot}00 \times 10^{-6}$	$1{\cdot}76 \times 10^{-4}$	160	

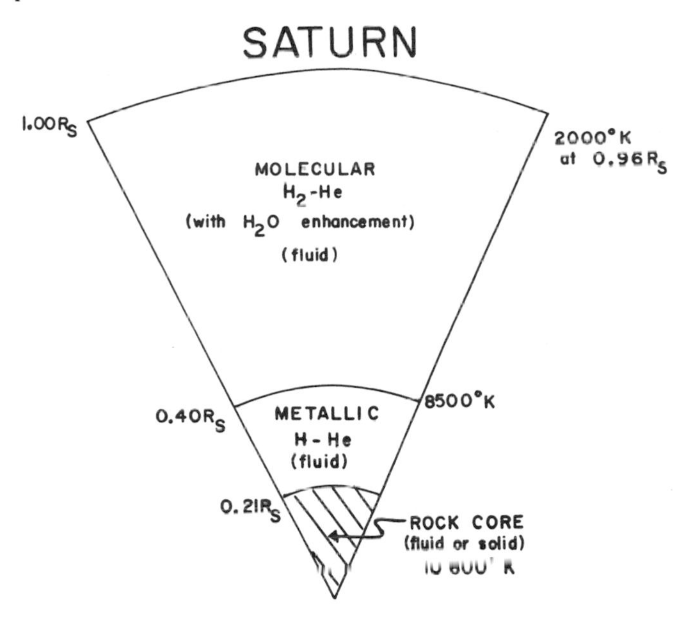

Figure 6. Schematic representation of the Saturnian interior for the model detailed in Table 3.

Table 3· A 'Typical' Saturn Model (Data from Podolak and Cameron, 1974)

Average Radial Distance s (10^9 cm)	Pressure (Megabars)	Density (g cm^{-3})	Temperature (K)	Comments
0·05	76·0	22·37	10580	Rocky core
0·50	60·0	20·29	10580	(assumed isothermal)
1·23	11·0	10·05	10580	
1·23	11·0	2·21	10580	Metallic H–He
1·54	7·77	1·92	9876	(with H_2O enhancement)
2·34	3·88	1·44	8560	(adiabatic)
2·34	3·88	1·29	8560	Molecular H_2–He
2·99	2·32	1·00	7398	(with H_2O enhancement)
3·99	0·91	0·66	5652	(adiabatic)
5·00	0·18	0·36	3615	
5·50	$1·84 \times 10^{-2}$	0·16	2254	
5·70	$3·84 \times 10^{-4}$	$1·5 \times 10^{-2}$	793	
5·78	$1·02 \times 10^{-6}$	$1·84 \times 10^{-4}$	149	

(or excess central condensation). Interestingly, the mass of this 'core' is comparable to that required in Jupiter (and is therefore a much larger fraction of the total mass). Saturn is further removed from a purely solar composition than Jupiter. As with Jupiter, the value of J_4 is indicative of a thermally expanded outer envelope. Figure 6 and Table 3 give details for a 'typical' Saturnian model based on the calculations of Podolak and Cameron (1974). Saturnian models are more sensitive than Jovian models to the present theoretical uncertainties in the high pressure properties of molecular hydrogen and 'ices' so it may be premature to attempt an evaluation of the ice/rock ratio, or to distinguish between two-layer and three-layer models.

Both Jovian and Saturnian models are essentially fluid with no 'surface' (i.e. no density discontinuity anywhere except perhaps in the deep interior). The presence of liquid, metallic, convecting cores is consistent with the observed magnetic fields. Since the metallic core of Saturn is deep within the planet, the Saturnian field observed externally would be expected to be more nearly dipolar and smaller in magnitude than is observed for Jupiter.

Homogeneous Thermal Evolution

Whereas static models assume a given heat flux, evolutionary models attempt to explain the heat flux, subject to the (assumed) initial condition that the planet is about $4{\cdot}5 \times 10^9$ years old. Only gravitationally-derived energy is a plausible source for the excess luminosities of Jupiter and Saturn, since other proposed sources (e.g., fission, fusion, accretion) are inadequate by at least two orders of magnitude. If the planet is essentially homogeneous, then the only gravitational 'source' of importance is the gradual loss of the primordial heat content that was generated during the early formation and contraction of the planet. (Jupiter and Saturn are presumably contracting even now—by less than 0·1 cm/year—but if this is a consequence of the cooling, then it is not in itself a significant luminosity source. There may be other reasons for present day contraction and these could be significant energy sources—see the section on *The Helium Distribution*). If primordial heat content is responsible then luminosity must be balanced by the rate of change of internal thermal energy in a homogeneous evolution:

$$L = 4\pi R^2 \sigma (T_{\text{eff}}^4 - T_0^4) \simeq -\frac{\mathrm{d}}{\mathrm{d}t}\left[\tfrac{4}{3}\pi R^3 C_v T_{\mathrm{i}}\right] \tag{24}$$

where L is the excess luminosity, R is the radius, σ is Stefan–Boltzmann's constant, T_{eff} is the effective temperature, T_0 is the equilibrium effective temperature in the absence of an internal heat source (non-zero because of the presence of the Sun), C_v is the average specific heat per unit volume, and T_{i} is some average internal temperature. It is assumed that the entire interior is convective so T_i is related to T_{eff} by being on the same adiabat

$$T_i \simeq T_{eff}\left(\frac{P_i}{P_{eff}}\right)^n \tag{25}$$

where P_i is a characteristic internal pressure, P_{eff} is the pressure at optical depth unity in the atmosphere (where the temperature is T_{eff}), and $n \simeq 0{\cdot}25$ is an average adiabatic index. From the virial theorem [equation (2)]

$$P_i = \frac{GM^2}{4\pi R^4} \tag{26}$$

while optical depth unity corresponds to $\rho K H_p = 1$ or equivalently:

$$P_{eff} \simeq \frac{g}{K} \tag{27}$$

where K is the effective transmission opacity for the atmosphere, and g is the acceleration due to gravity. Since T_i changes more rapidly with time than $\bar{C}_v$ or R, equation (24) becomes

$$t_{12} \simeq \frac{\bar{C}_v R}{3\sigma}\left(\frac{MK}{4\pi R^2}\right)^n \int_{T_1}^{T_2} \frac{dT}{T^4 - T_0^4} \tag{28}$$

provided K is also constant with time. The time interval t_{12} is the time taken for T_{eff} to change from T_2 to $T_1 < T_2$. According to the virial theorem, the total excess radiation over the entire evolution until now cannot exceed half the magnitude of the present gravitational energy. From equation (28), the 'age', t_0, of the planet can then be estimated by equating

$$\int_0^{t_0} L(t)\mathrm{d}t = \tfrac{1}{2}|E_G| \tag{29}$$

with $L(t_0)$ equal to the present excess luminosity. The result is

$$t_0 \simeq \frac{(0{\cdot}25)\ (\text{present heat content})}{(\text{present excess luminosity})} \tag{30}$$

for both Jupiter and Saturn. The form of this equation is obvious, but the main point of this analysis is to demonstrate that the coefficient ($\sim 0{\cdot}25$) is not very much different from unity. For the 'typical' models of the section on *Simple Models*, one finds $t_0 \simeq 4 \times 10^9$ years for Jupiter, and $t_0 \simeq 1{\cdot}5 \times 10^9$ years for Saturn; each with about a factor of two uncertainty.

The detailed evolutionary calculations for Jupiter (Graboske *et al.*, 1975) and Saturn (Pollack *et al.*, 1977) do not differ greatly from the above crude analysis. They find that K is fortuitously rather constant as T_{eff} increases, despite the fact that the main opacity source changes from H_2 to H_2O. Their approach is to start with an initially distended, adiabatic protoplanet (their proto-Jupiter has a radius thirty-five times larger than the present Jovian radius) and calculate its subsequent evolution. Hydrostatic equilibrium and an adiabatic thermal structure are assumed at all times. The early evolution is rapid and is characterized by very high luminosity (almost 1 per cent of the present solar luminosity). After only 10^6 years, the radius is within 20 per cent of the present radius. The subsequent evolution is well described by equation (28) and follows a degenerate cooling curve, given approximately by $L \propto t^{-4/3}$, where t is the age of the planet. The luminosity as a function of time is shown in Figure 7.

A very different calculation by Bodenheimer (1974) begins with a much more distended, isothermal, non-rotating gas cloud and allows for hydrodynamic effects. The subsequent evolution is complex, the main difference being that very large luminosities are never achieved (see Figure 7). The later evolution is essentially no different from Graboske *et al.* (1975). Neither of these calculations is likely to be very realistic for the formation of the planet, but it is evident that the subsequent degenerate cooling curve is insensitive to formation details. To summarize, the calculations indicate that the excess luminosity of Jupiter can be explained as primordial heat slowly leaking out, but Saturn may require additional energy generation. In the next section, gravitational layering is considered.

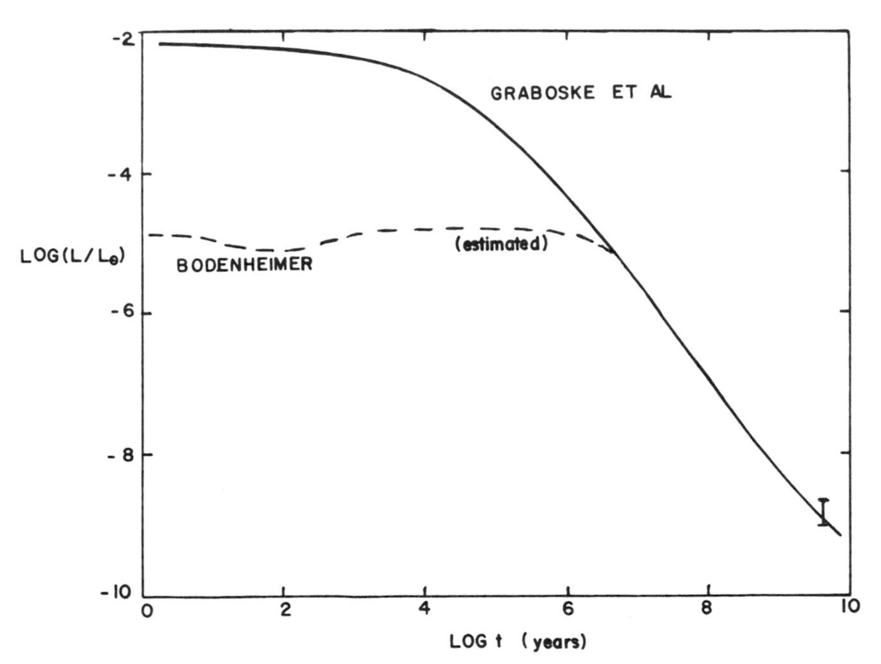

Figure 7. Luminosity of Jupiter as a function of time according to Graboske *et al.* (1975). Also shown is the hydrodynamic calculation of Bodenheimer (1974) which can be approximately extended to join the degenerate cooling curve ($\log t \gtrsim 6$). $L_\odot$ is the solar luminosity. The point marked at $\log t = 9{\cdot}66$, $\log (L/L_\odot) = -8{\cdot}90$ represents the present (observed) Jupiter

The Helium Distribution

It would be possible to construct a Jupiter model consistent with all present observations and of solar composition (except for a moderate enhancement of helium) provided one were at liberty to distribute the helium as one wished. However, this is not a valid procedure since the helium distribution is (in principle) uniquely determined by the constraints of (i) The thermodynamics of hydrogen–helium mixtures, (ii) Solute transport in an inhomogeneous convecting fluid, (iii) Thermal evolution of the planet. Although none of these constraints is fully understood, enough is known to reach some general conclusions.

Consider, first, the hypothetical case in which the molecular–metallic hydrogen transition is *not* first order. The hydrogen and helium would then be uniformly mixed until the temperature drops low enough for the mixture to become supersaturated with helium. Droplets of helium-rich fluid then nucleate from the mixture, and grow until they fall under the influence of gravity (Salpeter, 1973). The temperature at which this occurs and the composition of the droplets are determined by the hydrogen–helium phase diagram (see Figure 4 above). Some steady-state supercooling will persist, but it is not large even if homogeneous nucleation is required for droplet formation. Stevenson and Salpeter (1977b) have discussed the subsequent evolution in detail. According to the estimates of the hydrogen–helium phase diagram (Figure 4), the fractionation process may not have started yet in Jupiter, but if it has then the resulting redistribution of helium has (so far) had little effect on the density structure of the planet. The conclusion of the section on *Simple Models* that a heavy element enhancement ('rocky core') is needed is therefore probably valid. The small extent of fractionation is also consistent with the conclusions of the evolutionary studies (the section on *Homogeneous Thermal Evolution*). If much fractionation had already occurred then the total gravitational energy would be more negative and the present luminosity of Jupiter would be greater. In Saturn, however, fractionation may have been in process for a billion years or more already. In this case, the density structure of the planet is expected to be significantly affected, but since the metallic core of Saturn is much smaller than for Jupiter, the overall effect is somewhat less than it will eventually be for Jupiter. The precise extent of helium concentration towards the centre of Saturn is not known since no detailed calculations have been made yet. Significant fractionation is the most likely explanation for the unexpectedly large excess luminosity of Saturn (Graboske *et al.*, 1976). Fractionation in Saturn does not invalidate the general features of the Saturnian models summarized in the section *Simple Models*, although certain parameters (e.g. central temperature) will undoubtedly undergo significant modification in future models.

If the molecular–metallic hydrogen transition *is* first order then the Gibbs phase rule requires that the helium concentration be discontinuous at this transition. Unfortunately, relevant parameters (such as the critical temperatures for the molecular–metallic transition) are not known to better than an order of magnitude so a detailed quantitative prediction for the resulting helium distribution cannot yet be made. Various cases have been discussed (Stevenson and Salpeter, 1977b). It is clear that since helium is preferentially soluble in molecular rather than metallic hydrogen (see Figure 4), thermodynamics alone would tend to favour an *upward* redistribution of helium. This is only partially achieved in some cases and not achieved at all in other cases, because of the inherent inefficiency with which convection is able to transport helium upwards. In some of the cases discussed by Stevenson and Salpeter, a helium core can be formed because of a Rayleigh–Taylor instability: helium-rich molecular fluid becomes more dense than the underlying helium-poor metallic fluid. In any case, it is likely that the present picture of an essentially homogeneous, adiabatic Jupiter and Saturn is only a rough approximation. Much work remains to be done on the helium distribution in hydrogen–heliium planets.

Uranus and Neptune

Introduction

Uranus and Neptune receive much less space in this review because the necessary observations are either poor or non-existent. Until the observational situation improves, models of Uranus and Neptune remain poorly constrained. Furthermore, Uranus and Neptune are much further removed from a purely solar composition than Jupiter and Saturn and their composition is therefore not highly constrained by the average density alone. It is possible, for example, to construct a body with the same mass and radius as Uranus almost entirely out of cold helium, or almost entirely out of warm methane. It is nevertheless clear that Uranus and Neptune are intermediate between the terrestrial planets and the hydrogen–helium planets.

Even if the composition were known, quite large uncertainties remain in the thermodynamics, and in the equation of state particularly. The general procedure is to use an equation of state for each pure constituent which coincides with the 'low' pressure ($P \lesssim 100\,\text{kbar}$) laboratory data and the high pressure ($p \gtrsim 10\,\text{Mbar}$) Thomas–Fermi–Dirac theory, and is a plausible interpolation at intermediate pressures. This enabled Reynolds and Summers (1965) and Zharkov *et al.* (1974) to construct zero temperature equations of state for the 'ices' (H_2O, CH_4, NH_3) and rock. Finite temperature corrections are found from the Debye theory of solids. The equation of state for a mixture is estimated using the assumption of volume additivity [equation (22)]. The cumulative errors in this procedure may be as great as 10 per cent for the pressure at a given density. More subtle features of the thermodynamics (such as, for example, the mutual solubilities of CH_4, NH_3 and H_2O at very high pressures) are unknown. The electronic state of these substances at high pressures is also poorly known, but metallic conduction is expected for almost all substances at megabar pressures (see the section on *The Magnetic Field*). The metallization of water at $P \sim 1$ Mbar has recently been claimed (Vereshchagin *et al.*, 1975) and the increased *ionic* conductivity of water as the pressure is increased has been clearly established (Hamann and Linton, 1966). The possibility that a mixture of ammonia and hydrogen would form a metal at sub-megabar pressures has been proposed (Ramsey, 1951) but Stevenson (1975b) finds that megabar pressures would be required. In any event, metallic or near-metallic conductivities are expected in the deep interiors of Uranus and Neptune.

Since the excess luminosities of Uranus and Neptune are not known, static models constructed in the way described in the section on *Model Construction* are not strictly possible. It is necessary to *assume* a temperature–pressure relationship to complete the specification of a model. Two possible extremes can be considered. One is to suppose that the planet is isothermal below the observable atmosphere. With the exception of the atmosphere, which would have negligible mass, this would be equivalent to constructing a zero temperature model of the planet. The other extreme is to assume an adiabatic temperature for all of the planet except (perhaps) for a conductive core. Unlike Jupiter and Saturn, where an adiabatic structure is ensured because of the inadequacy of thermal conduction for transporting the inferred internal heat flux (see the section *Energy Transport and the Temperature Profile*), this assumption cannot be readily tested. Constraints on the thermal structure can, however, be inferred from general evolutionary considerations (see the section on *Thermal Structure*) and these indicate that the adiabatic hypothesis may be much closer to reality than the isothermal hypothesis. This has important implications for the state of matter in the deep interior, which is then predicted to be fluid rather than solid, and also implies a small (but significant) thermal expansion of the planet.

Simple Models

A 'simple' model is one in which the planet is assumed to consist of a small number of layers, each of which is homogeneous. The models thereby constructed may be inconsistent with the miscibility properties of the constituents, but in our present state of ignorance no better procedure exists. All present models of Uranus and Neptune are therefore 'simple'.

Three types of models have been considered: one-layer, two-layer and three-layer. One-layer models assume a homogeneous mixture of all constituents (gas, ice, rock). These models are found to be incompatible with the best current estimates of q_2 (Table 1 above) inferred from J_2 and the rotation rate. Two-layer models consist of an ice-rock core surrounded by a gaseous envelope. Some of the 'ice' (especially CH_4) is usually placed in the hydrogen–helium envelope rather than the core. [Observationally, methane appears to be enriched in the atmospheres of Uranus and Neptune (Owen and Cess, 1975)]. Three-layer models are similar except that the ices and gases are separated into distinct layers (see Figure 8). The two-layer and three-layer alternatives are both presently compatible with observations. None of these models make

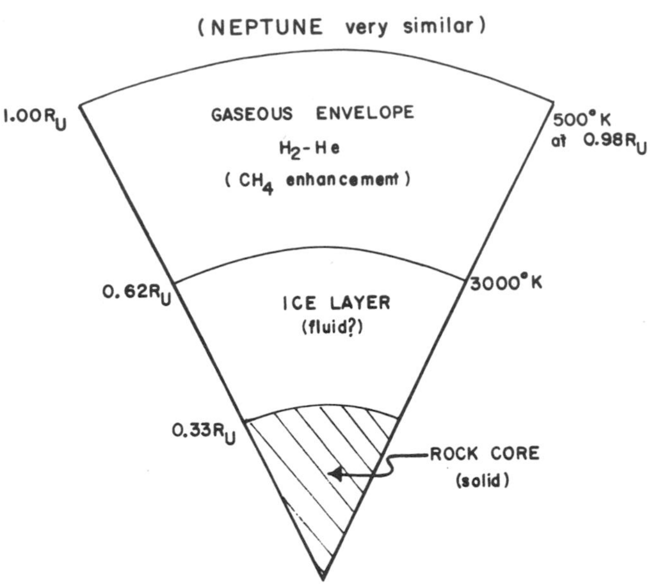

Figure 8. Schematic representation of the Uranian interior for the model outlined in Table 4. Similar three-layer models are possible for Neptune. Both Uranus and Neptune can also be adequately described by two-layer models

reliable predictions about the relative abundances of ice, rock, and gas although two points are clear: (a) A gaseous envelope is needed; (b) The rocky component cannot constitute most of the planetary mass.

A detailed description of particular models does not seem warranted, and only the general characteristics of 'typical' models are described here. Models constructed prior to 1970 are not considered because planetary radii (for example) have undergone revision since. Models have been recently constructed by Podolak and Cameron (1974), Zharkov *et al.* (1974), Podolak (1976) and Reynolds and Summers (unpublished revisions of their 1965 work).

Table 4 gives some of the characteristics of a three-layer model of Podolak and Cameron for Uranus, assuming an adiabatic gaseous envelope. In this model, the 'rock' comprises 3·8 Earth masses, the condensed ices (H_2O, NH_3) comprise 7·7 Earth masses, and the remaining mass is in the gaseous (H_2, He, CH_4) envelope. A similar model for Neptune has not been tabulated since it has very similar characteristics and differs primarily by having a greater abundance of methane (almost four Earth masses). It has a central temperature of 1800 K, lower than for Uranus because of Neptune's very low atmospheric temperature. For each planet, the temperature at one bar pressure is chosen to conform with present 'nominal' atmospheric models (Newburn and Gulkis, 1973). If the

Table 4. Main features of a three-layer Uranus Model (Data from Podolak and Cameron, 1974)

Average radial distance s (10^9 cm)	Pressure (M bar)	Density (g cm^{-3})	Temperature (K)	Comments
0·0	15·8	11·8	2850	Rock core
0·83	5·8	8·1	(ice and rock assumed isothermal)	
0·83	5·8	4·5		
1·56	0·54	2·6		Ice layer
1·56	0·54	0·82	2850	Gaseous
2·20	$6{\cdot}45 \times 10^{-2}$	0·355	1440	Envelope
2·47	$7{\cdot}58 \times 10^{-4}$	$6{\cdot}43 \times 10^{-2}$	520	(mainly H_2, He)
2·52	$1{\cdot}0 \times 10^{-6}$	$3{\cdot}3 \times 10^{-4}$	84	

adiabatic assumption is correct then the difference in average density between Uranus and Neptune is partly attributable to the difference in temperature (i.e. Uranus is more thermally expanded). Podolak and Cameron had difficulty in fitting J_2 and suggest that this might indicate errors in the current best estimates of the rotation periods. Podolak (1976) has subsequently pointed out that the best observations of J_2, rotation period and optical oblateness for Uranus are mutually incompatible. This illustrates the current difficulty in modelling these planets.

The other recently constructed models are similar to those of Podolak and Cameron, although they make different assumptions about the relative abundances of constituents. Zharkov and Trubitsyn (1972) discuss both isothermal and fully adiabatic models.

Thermal Structure

Despite the absence of excess luminosity measurements for Uranus and Neptune, it is possible to demonstrate that these planets are likely to have internal temperatures exceeding several thousand degrees. Consider, first, the highly implausible case in which all the gravitational energy of planetary formation is radiated away as the planet forms. This is unlikely because the radiative outer envelope of the protoplanet greatly limits the rate at which this large energy source can be eliminated. Even in this case, the long-lived radioactive sources in the 'rocky' component of the planet would suffice to heat the interior to in excess of a thousand degrees, assuming the 'rock' to be chondrite-like material. This heat could not readily escape because liquid 'ice' at pressures below a megabar is expected to have a thermal diffusivity characteristic of molecular fluids:

$$\kappa \lesssim 10^{-2}\ \text{cm}^2/\text{sec} \tag{31}$$

(solid ice would have an even lower diffusivity). Equating l^2/κ to $4{\cdot}5 \times 10^9$ years yields a diffusion length $l \sim 4 \times 10^7$ cm, much smaller than relevant layer dimensions. Even if conductive steady state were reached, the heat flux corresponding to a central temperature—outer temperature difference of 1000 K is less than 10^{20} erg/sec whereas the radioactivity produces 10^{21} erg/sec for a 'rocky' component of three Earth masses. (This temperature gradient would be subadiabatic so convection would not be initiated). It follows that a central temperature of one or two thousand degrees is a *lower limit*. The corresponding lower limits for the excess luminosity are of the order of 0·1–1·0 per cent of the insolation for Uranus and Neptune. (Excess luminosities as small as this are not measurable in the forseeable future).

Consider now, the more likely case in which the primordial heat arising from gravitational contraction is unable to escape. The internal temperature would then rise as the planet formed, until the adiabatic gradient was reached. Gravitational energy is much more than enough to do this. Once the adiabatic structure is reached, convection is initiated and the convective—radiative equilibrium described in the section *Homogeneous Thermal Evolution* ensues. No evolutionary calculations have been made for this, but the substitution of likely parameter values in equation (28) suggests that the present day excess luminosities of Uranus and Neptune could be *comparable* to their insolations. (In this case, they should certainly be measurable). Even if the transmission opacity of the atmosphere was once much lower than it is now, an adiabatic thermal structure once formed persists for *at least* as long as

$$t_0 \sim \frac{(\text{Planetary Heat Content})}{(\text{Conductive Heat Flux})} \tag{32}$$

where the conductive heat flux is that appropriate to an adiabatic temperature gradient. For a Uranus or Neptune which is adiabatic (except for an isothermal rocky core), $t_0 \sim 10^{11}$ years. A rocky core is expected to have metallic conductivity ($\kappa \sim 1\ \text{cm}^2\ \text{sec}^{-1}$) and be more nearly isothermal than adiabatic.

A typical central temperature for an 'adiabatic' model (with an isothermal rocky core) would be $\sim$ 5000 K. Fluid rather than solid state would prevail everywhere except in part of the rocky core. Dynamo generation of magnetic fields in the fluid 'ice' region is likely, since a good *ionic* conductivity suffices to make the magnetic Reynold's number supercritical. Typical convective velocities would be $v \sim 1$ cm/sec so that $R_m \sim 10^4$ for $\lambda \sim 10^5\ \text{cm}^2\ \text{sec}^{-1}$, and $R_0 \sim 10^{-4}$ (see the section on *The Magnetic Field*).

As with Jupiter and Saturn, primordial heat is not the only source of excess luminosity. Fractionation processes are very substantial sources of energy. Since the phase diagrams of ice or rock mixtures at megabar pressures are not known, it might seem highly speculative to invoke such processes. However, fractionation processes are inevitable for almost any conceivable thermal evolution. Uranus and Neptune are expected to be intermediate between

Jupiter and Saturn (where fractionation can take in excess of 10^{10} years) and the terrestrial planets (where fractionation may have taken as little as 10^6 or 10^7 years). Some fractionation processes may still be in progress in Uranus and Neptune.

Pluto and the Satellites

The shortness of this section is indicative of the paucity of relevant observations and does not represent a judgement of the relative importance of these bodies.

Various values for the density of Pluto have been quoted in the literature (Newburn and Gulkis, 1973). None is reliable and none is worth repeating here. In the absence of this datum, speculation on the structure of Pluto seems pointless.

Five outer solar system satellites have densities known with sufficient accuracy to merit detailed discussion (see Figure 2). They divide into two groups: Moon-like ('rocky') satellites (Io and Europa); and the 'icy' satellites (Callisto, Ganymede, and Titan). Lewis (1971) and Consolmagno and Lewis (1976) have discussed the likely structures and thermal histories for these satellites. Lewis proposes that the main cause for density differences amongst these bodies is differences in the temperatures at which they formed. At temperatures in excess of 250 K, only silicates would condense and a Moon-like body would form (e.g. Io and Europa). At lower temperatures, ices such as H_2O and hydrates of ammonia and methane would condense out. Below 80 K, methane could condense out. The temperature presumably was never low enough for H_2, He or Ne to condense, so very low density ($\rho < 1$ g cm^{-3}) satellites never formed. The densities of Ganymede, Callisto and Titan indicate that much of their mass is 'ice' (particularly H_2O) implying a formation temperature lower than 250 K. In any equilibrium condensation, the ice/rock ratio would not exceed solar, so even the 'icy' satellites would have a substantial mass fraction of silicates. If these bodies formed by homogeneous accretion, then the subsequent internal radioactive heating leads to melting and extensive fractionation. The structure of Ganymede, for example, that Lewis then envisages is a hydrous silicate core comprising roughly half the mass of the satellite, surrounded by a convecting fluid H_2O–NH_3 mix and a water ice crust. The temperature gradient in the crust would be ~ 1 K/km so that the conductive heat flux matches the radioactive heat source. The temperature thus rises rapidly as one proceeds inwards from the surface, until the melting point is reached. The temperature gradient is then adiabatic and of order 0·01 to 0·1 K/km. The central temperature of Ganymede would thus be several hundred degrees Kelvin at present. These conclusions are not greatly affected if solid state convection is allowed in the water mantle. The possibility of 'ice' tectonics is intriguing and has consequences for surface features.

The models indicate that any icy satellite whose radius exceeded 900 km would begin to fractionate. The fractionation of Ganymede, Callisto and Titan would be almost complete because of their much greater size. Io and Europa are predicted to be almost entirely silicates whereas Ganymede and Callisto are predicted to be roughly half ice and half silicates (by mass). The likelihood that Io and Europa formed at high temperatures is consistent with the high luminosity phase of the early Jupiter (Pollack and Reynolds, 1974) which evolutionary models suggest (see the section on *Homogeneous Thermal Evolution*). In this respect, the Galilean satellites may form a 'miniature solar system', but this analogy should be used with caution.

If a fluid H_2O–NH_3 mantle exists in the icy satellites, then dynamo generation may be marginally possible. Conceivable parameter values are $l \sim 10^7$ cm, $v \sim 0{\cdot}1$ cm/sec (see the section on *The Magnetic Field*), $\lambda \sim 10^5$ cm^2 sec^{-1} (for a high pressure H_2O–NH_3 fluid mixture) so that $R_m \sim 10$. Even for tidally locked satellites, $R_0 \sim 10^{-3}$ so that slow rotation need not prevent dynamo generation. The observation of a significant global field for Ganymede, Titan, or Callisto would not merely confirm the main features of Lewis's models but also impose important constraints on the internal heat flux (a quantity which may be exceedingly difficult to measure directly).

Conclusion

In some respects, the interior of Jupiter is better understood than the interior of the Earth. However, knowledge decreases rapidly as one proceeds further out in the solar system. Although a number of important theoretical calculations can be done even now, the main need at present is for improved observational data.

The important observations for compositional determination are improved values for J_2 and rotation rate (specifically for Uranus and Neptune) and determination of the atmospheric composition (especially the hydrogen to helium ratio, and the relative abundances of H_2O, NH_3 and CH_4). The important observations for thermal structure and history are the excess luminosity and magnetic field. In many instances, observations to a useful accuracy will require spacecraft flybys or atmospheric probes.

Theoretically, a better understanding of the thermodynamics and transport properties of hydrogen, helium, the 'ices', and mixtures of these is needed. It is particularly desirable to establish the main features of the phase diagrams for mixtures of these substances.

Despite the above shortcomings, the following conclusions seem justifiable:

(i) Jupiter is almost, but not quite, of solar composition. The deviation is significant, and probably represents an order of magnitude enhancement of heavy elements relative to solar composition.

(ii) If the hydrogen and helium were stripped from Jupiter, Saturn, Uranus and Neptune then the remnant would be comparable for each planet, and of order ten Earth masses.

(iii) Jupiter and Saturn are essentially adiabatic (apart from possible latent heat effects) with central temperatures of order 2×10^4 K and 1×10^4 K respectively.
(iv) In the absence of evidence to the contrary, it is much more plausible to assume that Uranus and Neptune are nearer to adiabatic than isothermal.
(v) Fractionation processes are probably in progress in the major planets even now. These are sources of excess luminosity.
(vi) Substantial magnetic fields are likely to be common. Jupiter, Saturn, Uranus, Neptune and perhaps even some of the larger icy satellites are expected to have dynamo-generated fields.
(vii) The largest icy satellites are expected to be fractionated and thermally active, with partially fluid interiors.

It can be optimistically anticipated that within ten or fifteen years, most of these assertions will be definitely confirmed or disproved.

References

Aumann, H. H., Gillespie, C. M. Jr., and Low, F. J. (1969). 'The internal powers and effective temperatures of Jupiter and Saturn.' *Astrophys. J. Lett.*, **157**, L69–72.

Bodenheimer, P. (1974). 'Calculations of the early evolution of Jupiter.' *Icarus*, **23**, 319–325.

Briggs, F. (1973). 'Observations of Jupiter and Saturn by a new method of Radio Interferometry.' *Astrophys. J.*, **182**, 999–1011.

Brown, L. W. (1975). 'Saturn Radio Emission near 1 MHz.' *Astrophys. J. Lett.*, **198**, L89–L92.

Cameron, A. G. W. (1973). 'Abundances of elements in the Solar System.' *Space Science Reviews*, **15**, 121–146.

Carlson, R. W., and Judge, D. L. (1976). 'Pioneer 10 ultraviolet photometer observations of Jupiter: The helium to hydrogen ratio,' in *Jupiter*, ed. T. Gehrels (University of Arizona Press) 418–440.

Clayton, D. D. (1968). *Principles of Stellar Evolution and Nucleosynthesis*, (publ. McGraw-Hill) p. 138.

Consolmagno, G. J., and Lewis, J. (1976). 'Structural and Thermal Models of Icy Galilean Satellites,' in *Jupiter*, ed. T. Gehrels (University of Arizona Press) 1035–1051.

Danielson, R. E., Tomasko, M. G., and Savage, B. D. (1972). 'High-resolution imagery of Uranus obtained by Stratoscope II.' *Astrophys. J.*, **178**, 887–900.

Dollfus, A. (1970). 'New Optical Measurements of the diameters of Jupiter, Saturn, Uranus and Neptune.' *Icarus*, **12**, 101–117.

Elliot, J. L., Wasserman, L. H., Veverka, J., Sagan, C., and Liller, W. (1974). 'The occultation of Beta Scorpii by Jupiter II. The hydrogen–helium abundance in the Jovian Atmosphere.' *Astrophys. J.*, **190**, 719–729.

Friedli, C., and Ashcroft, N. W. (1977). 'The band structure of molecular hydrogen at high pressure,' to appear in *Phys. Rev.*

Gierasch, P., and Stevenson, D. J. (1977). 'A simple theory for turbulent thermal convection in a rotating fluid,' unpublished.

Graboske, H. C., Pollack, J. B., Grossman, A. S., and Olness, R. J. (1975). 'The structure and evolution of Jupiter: The fluid contraction stage.' *Astrophys. J.*, **199**, 265–281.

Grigoryev, F. V., Kormer, S. B., Mikhailova, O. L., Tolochko, A-P., and Urlin, V. D. (1972). 'Experimental determination of the compressibility of hydrogen at densities 0·5–2·0 g/cm^3.' *J.E.T.P. Letters*, **16**, 201–204.
Hamann, S. D., and Linton, M. (1966). 'Electrical conductivity of water in shock compression.' *Trans. Faraday Society (GB)*, **62**, 2234–2241.
Hide, R., and Stannard, D. (1976). 'Jupiter's magnetism: observations and theory,' in *Jupiter* ed. T. Gehrels (University Arizona Press), 767–787.
Hubbard, W. B. (1974). 'Inversion of gravity data for giant planets' *Icarus*, **21**, 157–165.
Hubbard, W. B., and Slattery, W. L. (1971). 'Statistical mechanics of light elements at high pressure I. Theory and results for metallic hydrogen.' *Astrophys. J.*, **168**, 131–139.
Hubbard, W. B., and Smoluchowski, R. (1973). 'Structure of Jupiter and Saturn'. *Space Science Reviews*, **14**, 599–662.
Hubbard, W. B., and Slattery, W. L. (1976). 'Interior structure of Jupiter: theory of gravity sounding', in *Jupiter* ed. T. Gehrels (University of Arizona Press), 176–194.
Ingersoll, A. D., Münch, G., Neugebauer, A., Diner, D. J., Orton, G. S., Schupler, B., Schroeder, M., Chase, S. C., Ruiz, R. D., and Trafton, L. M. (1975). 'Pioneer 11 radiometer experiment: The global heat balance of Jupiter'. *Science*, **188**, 472–473.
Kavanagh, L. (1975). 'Synchrotron radio emission from Uranus and Neptune.' *Icarus*, **25**, 166–170.
Kliore, A., Fjeldbo, G., Seidel, B. L., Sesplaukis, T. T., Sweetnam, D. W., and Woiceshyn, P. W. (1975). 'Atmosphere of Jupiter from Pioneer 11.' *Science*, **188**, 474–476.
Knacke, R. F., Owen, T., and Joyce, R. R. (1975). 'Infrared observations of Titan.' *Icarus*, **24**, 460–464.
Lacy, J. H., Larrabee, A. I., Wollman, E. R., Geballe, T. R., Townes, C. H., Bregman, J. D., and Rank, D. M. (1975). 'Observations and analysis of the Jovian Spectrum in the ten micron band of NH_3.' *Astrophys. J.*, **198**, L145–148.
Lewis, J. (1971). 'Satellites of the outer planets.' *Icarus*, **15**, 174–183.
Macy, W. and Trafton, L. (1975). 'Neptune's atmosphere: the source of the thermal inversion.' *Icarus*, **26**, 428–436.
McMahon, A., Beck, H., and Krumhansl, J. (1974). 'Short range interaction between hydrogen molecules.' *Phys. Rev.*, **9A**, 1852–1864.
Metz, W. D. (1976). News Report, *Science*, **192**, p. 362.
Morrison, D., and Burns, J. (1976). 'The Jovian Satellites,' in *Jupiter*, ed. T. Gehrels (University Arizona Press) 991–1034.
Newburn, R. L., Jr., and Gulkis, S. (1973). 'A survey of the outer planets.' *Space Science Reviews*, **14**, 179–271.
Nolt, I. G., Radostitz, J. V., Donnelly, R. J., Murphy, R. E., and Ford, H. C. (1974). 'Thermal emission of Saturn's rings and disc at 34 microns.' *Nature*, **248**, 659–660.
Null, G. W., Anderson, J. D., and Wong, S. K. (1975). 'Gravity field of Jupiter from Pioneer 11 Tracking Data.' *Science*, **188**, 476–477.
Orton, G. (1975). 'The thermal structure of the Jovian atmosphere.' *Icarus*, **26**, 125–141.
Owen, T., and Cess, R. D. (1975). 'Methane absorption in the visible spectra of the outer planets and Titan.' *Astrophys. J. Lett.*, **197**, L37–40.
Parker, E. N. (1970). 'The origin of magnetic fields'. *Astrophys. J.*, **160**, 383–404.
Podolak, M. (1976). 'Methane rich models of uranus'. *Icarus*, **27**, 473–477.
Podolak, M., and Cameron, A. G. W. (1974). 'Models of the giant planets'. *Icarus*, **22**, 123–148.
Podolak, M., and Cameron, A. G. W. (1975). 'Further investigations of Jupiter models.' *Icarus*, **25**, 627–634.
Pollack, J. B. (1975). 'The rings of Saturn.' *Space Science Reviews*, **18**, 3–93.
Pollack, J. B., and Reynolds, R. T. (1974). 'Implications of Jupiter's early contraction history for the composition of the Galilean satellites.' *Icarus*, **21**, 248–253.
Pollack, J. B., Grossman, A. S., Moore, R., and Graboske, H. C. Jr. (1977). 'A calculation of Saturn's gravitational contraction history.' *Icarus*, **30**, 111–128.

Pollock, E. L., and Hansen, J. -P. (1973). 'Statistical mechanics of dense ionized matter II.' *Phys. Rev.*, **8A**, 3110–3122.
Ramsey, W. H. (1951). 'On the constitutions of the major planets.' *M.N.R.A.S.*, **111**, 427–447.
Ree, F. H., and Bender, C. F. (1974). 'Non-additive interaction in molecular hydrogen at high pressure.' *Phys. Rev. Lett.*, **32**, 85–88.
Reynolds, R. T., and Summers, A. L. (1965). 'Models of Uranus and Neptune.' *J. Geophys. Res.*, **70**, 199–208.
Rieke, G. (1975). 'The thermal radiation of Saturn and its rings.' *Icarus*, **26**, 37–44.
Rochester, M. G., Jacobs, J. A., Smylie, D. E. and Chong, K. F. (1975). 'Can precession power the geomagnetic dynamo?' *Geophys. J. R. astron. Soc.*, **43**, 661–678.
Ross, M. (1972). 'On the Herzfeld theory of metallization'. *J. Chem. Phys.*, **56**, 4651–4653.
Ross, M. (1974). 'A theoretical analysis of the shock compression experiments of the liquid hydrogen isotopes and a prediction of the metallic transition'. *J. Chem. Phys.*, **60**, 3634–3644.
Salpeter, E. E. (1973). 'On convection and gravitational layering in Jupiter and in stars of low mass'. *Astrophys. J. Lett.*, **181**, L83–86.
Salpeter, E. E. and Stevenson, D. J. (1976). 'Heat transport in a stratified two-phase fluid'. *Phys. Fluids*, **19**, 502–509.
Schwarzchild, M. (1958). *Structure and Evolution of Stars*, Princeton University Press, p. 47.
Slattery, W. L. and Hubbard, W. B. (1973). 'Statistical mechanics of light elements at high pressure III.' *Astrophys. J.*, **181**, 1031–1038.
Steenbeck, M., Krause, F., and Rädler, K.-H. (1966). 'A calculation of the mean electromotive force in an electrically conducting fluid in turbulent motion, under the influence of Coriolis forces.' *Z. Naturforsch.*, **21a**, 369–376.
Stevenson, D. J. (1974). 'Planetary magnetism'. *Icarus*, **22**, 403–415.
Stevenson, D. J. (1975a). 'Thermodynamics and phase separation of dense fully-ionized hydrogen–helium fluid mixtures.' *Phys. Rev.*, **B12**, 3999–4007.
Stevenson, D. J. (1975b). 'Does metallic ammonium exist?' *Nature*, **258**, p. 222.
Stevenson, D. J. (1976). *Ph.D. Thesis* (Cornell University) unpublished.
Stevenson, D. J. and Salpeter, E. E. (1976). 'Interior models of Jupiter,' in *Jupiter*, ed. T. Gehrels (University Arizona Press), 85–112.
Stevenson, D. J., and Salpeter, E. E. (1977a). 'The phase diagram and transport properties for hydrogen–helium fluid planets', to appear in *Astrophys. J. Suppl.*
Stevenson, D. J., and Salpeter, E. E. (1977b). 'The dynamics and helium distribution in hydrogen–helium fluid planets', to appear in *Astrophys. J. Suppl.*
Streett, W. B. (1974). 'Phase equilibria in molecular H_2–He mistures at high pressures'. *Astrophys. J.*, **186**, 1107–1125.
Vereshchagin, L. F., Yakovlev, E. N., and Timofeev, Yu. A. (1975). 'Possible Metallization of Hydrogen'. *Zh. Eskp. Teor. Fiz. Pis'ma (USSR)*, **21**, 190–192.
Vereshchagin, L. F., Yakovlev, E. N., and Timofeev, Yu. A. (1975). Transition of H_2O into the conducting static at static $P \sim 1$ Mbar.' *J. E. T. P. Letters*, **21**, 304–306.
Westphal, J. A., Mathews, K., and Terrile, R. J. (1974). 'Brightness of Jupiter at Five Microns.' *Astrophys. J.*, **188**, L111–113.
Zapolsky, H. S., and Salpeter, E. E. (1969). 'The mass–radius relationship for cold bodies of low mass.' *Astrophys. J.*, **158**, 809–813.
Zharkov, V. N., and Trubitsyn, V. P. (1972). 'Adiabatic models of Uranus and Neptune.' *Izv. Acad. Sci. Phys. Solid Earth (USA)*, **7**, 496–500.
Zharkov, V. N., Trubitsyn, V. P., Tsarevskiy, I. A., and Makalkin, A. B. (1974). 'The equation of state of cosmochemical materials and the structure of the major planets.' *Izv. Acad. Sci. Phys. Solid Earth (USA)*, **10**, 610–617.
Zharkov, V. N., and Trubitsyn, V. P. (1976). 'Structure, composition and gravitational field of Jupiter', in *Jupiter* ed. T. Gehrels (University of Arizona Press), 133–175.

TERRESTRIAL PLANET EVOLUTION AND THE OBSERVATIONAL CONSEQUENCES OF THEIR FORMATION

D. C. TOZER
University of Newcastle upon Tyne, England

'Whether there be, or be not, any physical traces of a state of things anterior to the commencement of our geological series of deposits is a question of no real importance.'
C. Lyell, Principles of Geology 1830

Abstract

If terrestrial planets are made of materials with rheological properties recognizably similar to those of laboratory specimens, the familiar heat conduction solutions to their thermal history problems have to be rejected on grounds of instability. In justifying the new solutions to the same problems, it is shown that for Earth they offer an attractively simple basis for understanding its present upper mantle structure and its large-scale dynamic behaviour. However, the regulation of a very large, spatially averaged viscosity—the central idea in understanding the planetary heat transfer process—forces a major reappraisal of the manner in which such large objects have become chemically differentiated. Shear heating associated with convective heat transfer is shown to provide the only acceptable explanation of how the much lower viscosities necessary for separations can be produced during the life of a planet. However, the capacity of this process to produce large local temperature perturbations rapidly decreases with diminishing planetary size and the early lunar highland differentiation is revealed as an observation particularly significant in testing theories of the initial thermal state of planets. Various arguments are advanced that lead to a rejection of the inhomogeneous accretion hypothesis and the theory of the initial thermal state that is based on 'embryo' growth. The preferred theory allows much longer and more plausible accumulation times. However, the idea of 'initial' chemical homogeneity is complicated by a demonstration that core/shell separations are potentially catastrophic in most terrestrial planets and that the immediate source of their cores were smaller cores that had separated in their respective protoplanets. These ideas have implications for the degassing and loss of primitive atmospheres, subsequent atmosphere formation and the abundance of siderophile elements in Earth's mantle. Lastly, the significance of Mercurian magnetism to the problems of accumulational heating and differentiation are discussed.

Introduction

If the papers presented at this conference show that we now talk differently about *The Origin of the Solar System*, there are also plenty of signs that even the most sensitive modern observational techniques are unable to resolve basic questions that were asked long ago about the history of planetary matter. However, before one interprets such inability as failure, one is entitled to ask: What are the objectives of this subject and are they realistic? Are questions that one may think of as basic but which are obviously so difficult to answer, not showing, by that very difficulty, their scientific unimportance? Undoubtedly, the whole enterprise was started and remains rooted in a strictly deterministic view of a system's dynamical development—a purely philosophical stance that tantalizes the imagination with visions of using the *status quo* to trace an objective history into an indefinite past. However, for those with the will to see, reality appears rather different. Evolutionary theories may give the illusion, in this day of the large computer, of being able to predict the development of a system over indefinite periods of time, but it is a brutal fact of experience in using such theories in many different but closely specified problems both inside and outside the laboratory, that they fail to make any useful contribution to understanding if the interval of prediction or retrodiction exceeds some modestly finite interval. Features that are perfectly well defined macroscopic phenomena at one moment appear to have grown from trivial or adventitious beginnings—the case of the weather is a familiar example where the interval of reliable prediction is still only a few days. In such cases, the theory is usually rescued by some remark about inadequate specification of 'initial' conditions, but whatever the truth of this matter is, it is unrealistic to use evolutionary theories to span epochs when commonplace experience suggests that the supposed states of the system would have been very prone to this sort of unstable behaviour—I am thinking particularly of fluid dynamical states involving flows at enormous Reynolds numbers. There is no scientific merit in postulating some attractively simple and symmetrical 'initial' state if its consequences for observation cannot be fairly rigorously assessed and if the susceptibility of any predictions to minor changes in the 'initial' conditions are not known.

This takes me to the central question: Is it one's aim to write history or understand what one can observe? Given that exact determinism is a mirage, some choice now has to be made. Here, I must say that the issue has not been clarified by the many observers who believe or describe their activities as 'discovering the past'. If there is one thing that is axiomatic about the past, it is its unobservability. Recognition of the observational situation as it is, rather than as one might wish or conceive it to be, is most important—the successes of modern physics testify to that. Then, not only do supposedly hard observational facts frequently vanish into thin air but feelings of frustration that so little is observable of what one may think even the simplest system to be or to have been, are alleviated. In much less exalted problems of material science than that of understanding the present configuration of planetary matter, we have learnt to live with, nay to be thankful for the fact that the observed behaviour of the system can be frequently understood in terms of its *status quo* and without reference to more than an infinitesimal part of what could be cryptically called 'its history'. Such considerations force me to adopt a different and much less ambitious view of what is involved in *The Origin of the Solar System* than most workers in the field. Rather than seeing the unravelling of an objective history as one's purpose, I see a history merely as a by-product of attempts to make sense of what has or could be observed, using a theory containing the idea of evolution. Giving reality to what is no more than an abstraction adds nothing and only makes the task of revision that much harder. I think it is the failure to see theory in this light that makes the many grand schemes of solar system development that predict nothing but what was built into them, so uninteresting. I can see no better way of defining the 'origin' of a system than as the point on one's theoretical time scale at which any close identity with the observed state has been lost and at which one judges that the 'initial' conditions compatible with our finite knowledge of the *status quo* are becoming very numerous and diverse. Judged by the differences of opinion that exist about the dynamics of planetary material immediately preceeding its existence as the familiar planetary bodies that point can be taken as the origin of the solar system.

The Preservation of the Past in Planetary Objects

The overriding constraint we place on attempts to draw a scientific history is that a system must remember the past in ways which may be quantitatively but not qualitatively different from those that have been demonstrated in controlled experiment. This experience indicates that we have basically to distinguish two kinds of memory that can be exhibited by a material body. Firstly, there is a 'global' memory that expresses the fact that as a system it will take finite times to respond to a change of environment and also to reach equilibrium among its different parts. Secondly, there is a 'local' memory of the past, preserved in the structure or properties of the material, which can be studied in samples of the system. Perhaps the distinction can be clarified by quoting the idea of a material medium which is in local thermodynamic equilibrium as one which has no local memory of the past, although it may still have a global memory if the local thermodynamic equilibrium state varies from point to point of its interior.

It is a basic result of experiment that the effective duration of any material's 'local' memory is greatly affected by quite modest changes of temperature and our problem is to reduce an enormous quantity of this memory erasure or 'annealing' data to a form that makes a plausible basis for the description of *in situ* planetary material. Although any tractable scheme is likely to be criticized as over-simplified, the task of choosing a suitable starting point is made easier by the fact that the observable consequences are not sensitively dependent on one's choice and that the significance of making any plausible assumption about it is not generally known to planetary scientists. Taking a broad view of annealing phenomena, an acceptable scheme is to attribute to *in situ* planetary material a single characteristic temperature $T_M(P)$, and use it to postulate a law of corresponding states about the rate of decay of all 'local' memory effects involving atomic movement, i.e. the rate of approach to local thermodynamic equilibrium at temperature T and pressure P. Although we shall see that the form of this law is not crucial, we may assume for the moment that the effective duration of the material's memory of its past environment, if held under these conditions, is given by a time constant $\tau = A \exp(B T_M/T)$ By suitable choice of the parameter B we can always identify $T_M(P)$ with an absolute melting or solidus temperature of *in situ* planetary material. For reasons which will later become clear, in the study of terrestrial planets we are interested in annealing behaviour for $T < T_M$ and if such an equation is fitted to laboratory observations on annealing in single phase materials in this temperature range, B is ~ 20 and A very approximately 10^{-7} sec. By extrapolation to lower temperatures, we infer that τ decreases from $> 10^8$ years to less than a few days in the temperature range $0{\cdot}4\,T_M$–$0{\cdot}9\,T_M$. Although plausible values for a T_M and the present temperatures of rocks near the surfaces of terrestrial planets suggest the formal possibility of using their 'local' memory to test hypotheses about events even scores of billions of years ago on an evolutionary time scale, it is obvious that we cannot begin to see the need for introducing hypothetical events into a history of planetary material unless we can firmly predict the observational situation in their absence. Fortunately, there is a connection between the rate of annealing of a local material memory and the speed of the heat transfer process in a body composed of that material that enables us to draw certain conclusions about the history of planet-like objects with confidence.

To understand this connection one must appreciate that the rate of annealing of past events recorded in the atomic structure of a material and its rheology have a direct relationship. For example, an ideal elastic medium is necessarily one that has a zero rate of annealing, i.e. it maintains a perfect 'memory' of the stress distribution associated with any one of its geometrical configurations. On the assumption that true thermodynamic equilibrium is only associated with hydrostatic stress, it is clear that a finite rate of annealing will always ensure that a steady rate of shear deformation $\dot{\varepsilon}$ occurs under a constant shear stress σ. We can identify the local annealing time constant τ with the ratio of an effective viscosity $\eta(=\sigma/\dot{\varepsilon})$ and an effective shear modulus $\mu(=\Delta\sigma/\Delta\varepsilon)$, where $\Delta\varepsilon$ is the 'prompt' shear strain induced for times very much smaller than τ by an increment of shear stress $\Delta\sigma$. We can now obtain a rough indication of the significance of plausibly temperature sensitive annealing rates on the thermal state of planets simply by calculating the heat source density H that is needed to steadily support, for example, twice the adiabatic temperature gradient in a homogeneous, planetary-sized mass of material when its effective viscosity is given a decreasing sequence of values, and the other constitutive functions necessary to define a theory of heat transfer for such a deformable medium are given conventional values.† Analysis shows that the heat source density increases beyond any plausible radiogenic value when the chosen values of effective viscosity decrease through the range 10^{21}–10^{19} poise—this rapid increase is due to the initiation of mass transfer as an essential part of the heat transfer process. For comparison, it can be noted that at the melting point T_M the effective viscosity of the typical laboratory material is generally in the range 10^{13}–10^{15} poise. Of course, the actual problem of a planet's heat balance is not posed like this, but as the response to the body's distributed heat sources when its external surface temperature is held by quite unrelated factors, e.g. distance from the sun, etc., at some more or less fixed value. This makes the problem somewhat more difficult to analyse (see Tozer, 1972), but subject to the reservation that these external factors do not themselves impose a viscosity $< 10^{20}$ poise on the planet, these more involved calculations confirm what is strongly suggested by the more elementary calculation—that a homogeneous, planetary-sized object made of any recognizably real material is effectively barred by its annealing properties from ever being heated by its radiogenic heat sources to any state in which its effective viscosity function, averaged throughout the interior, is less than $\sim 10^{20}$ poise. One can extend this argument, using the very non-linear nature of the response to various heat source densities (e.g. Figure 1) to infer in the case of homogeneous objects with typical meteoritic properties that if their external radius is greater than 1000 km, a convective 'core' occupies most of the interior in which the actual spatially averaged viscosity is of the same order as this effective lower bound $\sim 10^{20}$ poise. This state of very high viscosity is maintained essentially because any deviation from it is accompanied by large changes in the rate of heat transfer that tend to restore it. Indeed, we shall see that in objects larger than the Moon there may well have been a catastrophic rate of energy input at the very beginning of their existence, but so quick is the response of the heat transfer process to any disturbance from the high viscosity state ($<$ a few hundred million years—see Figure 2) that we can normally think of a planet's state as being its quasi steady response to its slowly changing radiogenic heat sources. From this explanation of a generally high viscosity state in terms of its stability one can easily understand why the lower bound on the spatially averaged viscosity remains virtually unchanged for any plausible choice of an effective viscosity function for the homogeneous material. Uncertainty about its dependence on

†The literature contains many detailed discussions of the variation with thermodynamic state of such constitutive functions as 'specific heat' C_P, 'thermal conductivity' K, 'heat source density distribution' H. While heat conduction was regarded as the theory of heat transfer such questions appeared to have great importance, but it is characteristic of the more general approach described here that the plausible variations of these functions are quite insignificant in their effect on planetary observation compared with the rapid temperature variation of viscosity.

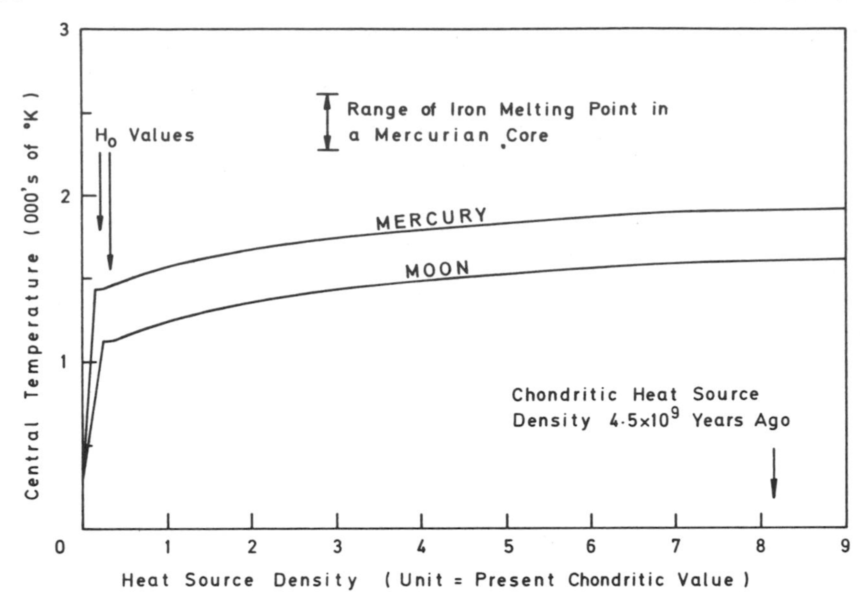

Figure 1. Dependence of the central temperatures in homogeneous models of Moon and Mercury on the rate of heat production unit volume.

The curves were based on steady state arguments, but due to the thermal time constant being $\sim 2 \times 10^8$ years (Figure 2), they would be closely followed if heat production were decaying as slowly as a chondritic radioactivity.

Note that the higher 'plateau' of the Mercury curve is mainly due to the greater effect of hydrostatic pressure in that body raising the value of T_M (see text.) The steeply rising part of the curves to the left are steady heat conduction solutions, only valid as solutions of the heat transfer problem when the effective viscosity function is everywhere $\gtrsim 10^{25}$ poise. The small H_0 values show how easily a chondritic radioactivity can support convective motions in these bodies. The melting point of iron under pressures that would exist in a Mercurian core is included to show how unlikely a liquid core would be at the present time (see text)

temperature appears in the solution of the heat transfer problem as a range of doubt about the actual temperature that accompanies this high viscosity state of the interior. This temperature uncertainty is ~ 500 °C in an object of lunar size, and rather larger in bigger objects where the effect of pressure in raising the T_M of their deep interior is well beyond that measureable in the laboratory. However, even in these cases the average temperature can confidently be assumed to be well below the solidus temperature T_M.

If one turns to consider materially inhomogeneous model objects there is, in general, no correspondingly simple way to express the action of the heat transfer process, and completely novel questions of stability have to be considered. For example, any inhomogeneity that is rapidly removed by the mixing action of the heat transfer motions is of no interest to theories of planetary evolution and as such, unacceptable as a model of the *status quo* in terrestrial planets. An interesting case of planetary inhomogeneity illustrating how a

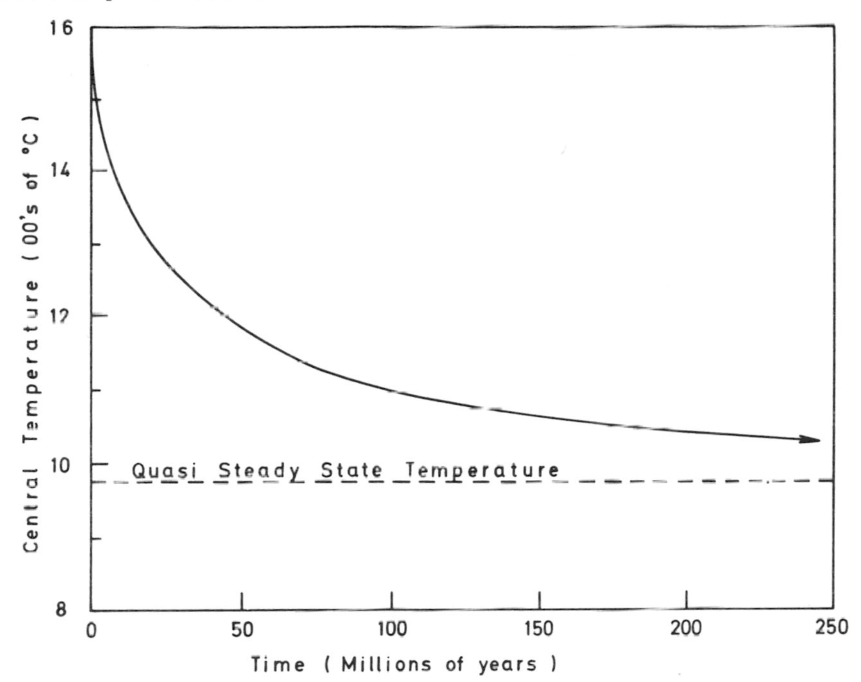

Figure 2. The change in central temperature with time for a Moon sized object initially at its melting point of 1600 °C.

This and the curves shown in Figure 1 attempt to show the effectiveness of solid state convection in regulating states of very high spatially averaged viscosity in planetary sized objects. If one had made the usual inappropriate assumption of heat conduction as the only heat transfer mechanism below the melting point, the central temperature would only have fallen by ~ 10 °C in $2{\cdot}5 \times 10^8$ years. It should be remembered that the system is highly non-linear so that the approach to steady state conditions is not entirely characterised by a single time constant. For example, the rate of approach from below the equilibrium state is fixed more by the rate of heat production than the decrease of viscosity with temperature that controls the case illustrated. However, since 'initial' temperatures are much more likely to be above than below a quasi steady state temperature distribution that can be maintained by radiogenic heating, a figure of 2×10^8 years is the relevant value to use for a thermal relaxation time

viscosity distribution very different from that described above for a homogeneous planet might be maintained by the same amount of radiogenic heating, is to suppose a core/shell configuration of materials in which the density contrast is sufficient to stabilize the situation against remixing and in which the T_M of core material is significantly less than that of the shell. Providing (T_M) core/(T_M) shell $< \sim 0{\cdot}6$ at the interface of these regions one can understand how plausible radiogenic heating may maintain the core material in a liquid condition. Models like this have an obvious relevance to the generation and maintenance of planetary magnetism and I shall return to that question later. However, I hope it will be clear from this single example that if one is dealing with any object like the terrestrial planets, in which the various materials forming the vast bulk of the object have melting points T_M several times the temperature that is maintained by external influences, the only way in which a plausible radiogenic heating rate can achieve horizontally averaged temperatures of the order of greater than the melting point is for T_M to decline significantly with increasing depth.†

At this point it is not worth pursuing specific inhomogeneous models that can represent the *status quo* of the terrestrial planets because we shall see that all in effect beg some very important questions as to how inhomogeneity could have arisen. Nevertheless, I believe it is worth digressing for a moment on the specific problems posed by the Earth's upper mantle, if only because this region has always played a key role in forming general ideas about the planetary heat transfer process. So potent is the myth that the average temperature below a depth of $\sim$ 100 km in the Earth is 'known' to be at or above the melting point of its material that the action of the heat transfer process in controlling viscosity and keeping the average temperature well below the melting point has been completely overlooked. The new understanding requires some reinterpretation of such features as the low Q, low velocity layer present under the oceanic regions depths of the order 100–300 km. Perhaps the first thing to say is that the self-regulation of viscosity at a value $\sim 10^{21}$ poise convincingly explains, i.e. without any need to guess the fraction of the Earth's radiogenic heat sources that have become concentrated in the crust, why the shell appears solid at seismic frequencies—S waves up to 10^9 seconds period could be propagated through a medium with $\eta \sim 10^{21}$ poise. More quantitatively, the predicted figure of $\sim 10^{21}$ poise for the value of any plausible effective viscosity function below a depth of some tens of kilometres is a rather critical test of the self-regulation idea. It is very satisfying, considering the colossal range of values in which it could conceivably be were such regulation not present (10^3–10^{30} poise), that the predicted value agrees so well with the estimates of viscosity assigned to the upper mantle to account for the rate of isostatic adjustment following sea level changes and ice cap removal (for a review, see Cathles, 1975). With the value of an effective viscosity directly controlled by the heat transfer process, one has all the information necessary to compute effective Rayleigh and Prandtl numbers, and hence it may be shown (Tozer, 1972a, b) that the heat transfer process accounts rather precisely for the average speed of continental drift motions and the departures from hydrostatic equilibrium that have been revealed by satellite orbit determination of the external gravity field.‡

†This situation is exactly the converse to that existing in many bodies in the outer solar system. In that sense these latter bodies are 'hot' and the terrestrial planets 'cold'. Thermal conditions in low viscosity bodies approximate those of an adiabat that passes through the state maintained at the external surface—see the chapter by Stevenson in this volume.

‡The advent of large computers has stimulated a lot of work to predict the spatial pattern of heat transfer movements at such high Rayleigh numbers (10^6 – 10^7). This approach is based on the

In reinterpreting the low velocity, low Q layer in the upper mantle below a depth of ~ 70 km, I have exploited the widely held view that primitive mantle material was a mixture of basaltic and ultrabasic material with at least a few tenths of percent of water. However, due to a differentiation process still to be discussed, I suggest that the sub-oceanic mantle is still roughly of primitive composition while the sub-continental mantle has been largely freed of its basaltic and aqueous components down to a depth of some hundreds of kilometres. In both regions, because their lateral extent is much greater than the depth of any differences in composition, the heat transfer process regulates the temperature at $\lesssim 0{\cdot}6 T_M$, which implies, contrary to static thermal models, that higher average temperatures occur at depths > 40 km in the more refractory sub-continental material. The low Q, low velocity regions that are so much better defined under the oceans, can be attributed to the thermodynamic instability of the hydrated basaltic fraction of the primitive material (an amphibolite) at pressure $\gtrsim 25$ kilobars that releases its water to form an intergranular liquid film at depths greater than about 70 km. (For phase diagram see Wyllie 1971, Green 1972). In the Earth where the pressure gradient is ~ 300 bars/km, such a film would freeze at depths ~ 400 km, but in the Moon where even the central pressure is less than 50 kb, such water of dehydration could produce a low Q 'core' region below a depth ~ 500 km (Tozer 1972a). Incidentally, it is research stimulated in large part by the lunar seismological data that has shown how exceedingly small quantities of liquid water can reduce both the speed and Q of deformation waves in rocks (Pandit and Tozer, 1970; Tittman *et al.*, 1975). Various arguments can be advanced to explain not only the generation but the persistence of the proposed arrangement of materials in the presence of heat transferring motions that one might naively suppose would have thoroughly mixed the Earth's mantle after $4{\cdot}5 \times 10^9$ years. Basically the greater creep resistance and slightly lower density of the sub-continental rocks allow one to regard the continents as 'rafts' being pushed hither and thither by motions in the more primitive sub-oceanic mantle material. In this respect I believe Wegener was nearer the correct picture than is sometimes reflected by Plate Tectonic concepts.

Whether one likes the above model or not, I have explained it in some detail in order to show that there is no necessity to believe that average temperatures in the most intensely studied part of any terrestrial planet interior are as high

myth that there is a unique solution to this problem, whereas simple experiment shows that to be untrue, and the current theoretical techniques are woefully inadequate to predict even the few statistically significant features of such high Rayleigh number laboratory scale flows. To avoid that problem, I have deliberately used only the general character of convection as a heat transport process—as measured by a Nusselt number, in my studies of self-regulation. Both experiment and theory show the independence of this quantity from the spatial details of the velocity pattern.

as, or as some would say, above T_M.† However these new solutions of the heat transfer problem emphasize a difficulty that has always been brushed aside by the builders of spherically symmetric models of planetary temperature distributions but which has been repeatedly emphasized by Jeffreys in several editions of his book *The Earth*—namely that the material ejected in volcanic eruptions is not representative of the average state of matter at any one depth in the interior. Kelvin first noted the paradox in the last century when he first assigned a shear modulus to the Earth's interior based on tidal observations, but by and large the difficulty has since been suppressed by the invention of a material that simultaneously combines the requisite rigidity and mobility! Such an unsatisfactory solution to the paradox was not regarded as a very serious omission of general planetary heat transfer theory because the loss of heat in active vulcanism is < 1 per cent of the total geothermal flow, but I now see a proper resolution of this paradox as the vital key to understanding how an otherwise extremely viscous planet can significantly differentiate. Although there now seems to be a straightforward explanation of the phenomenon, without the actual appearance of lava on Earth's external surface, nobody would have dared predict that a process which keeps the spatially averaged value of any plausible effective viscosity function for a terrestrial planet above 10^{20} poise could also generate internal regions, necessarily forming such a microscopic fraction of the total planetary volume as to be practically unobservable, in which the viscosity is more than 10^{14} times less than the well-regulated average value. Furthermore, such volumetrically trivial regions must have significant extension in a radial direction if they are to provide an environment for differentiation on a planetary scale. It has been suggested (Clark *et al.*, 1972) that the basic observed differentiation of planets has its origin in the sequence with which planetary material accumulated to form the familiar planetary bodies, but I hope to show convincingly that this solution of the problem of planetary stratification rests on an exceedingly improbable, though fashionable, picture of the planetary accumulation process. In my opinion, the chemical non-selectivity of the gravitational force and what I call a 'symmetric' view of accumulation, in which *every* body may be said to be growing at the expense of the others, compel one to examine whether the present internal structures of planets could have arisen entirely because of a process that will only work in large self-gravitating accumulations of the material. We shall see that there is some difficulty in calling some planets 'initially' homogeneous (hence the parenthesis), because one of my conclusions is that a mass of material the size of the larger terrestrial planets

† Here I must refer to an arbitrary practice prevalent among experimental petrologists of defining the solidus temperature by the appearance of a molten silicate phase. However, if one regards water just like any other rock component, I cannot see that dehydration of a phase to form a free water phase is distinguishable from melting. In that sense one could still refer to the low velocity layer of the mantle as a partially molten region in my interpretation of it in terms of amphibole dehydration, but of course the average temperature is very much lower—perhaps only 600 °C at 150 km under the oceans.

can be prone to catastrophic differentiation of the core and shell material. In other words, differentiation and accumulation may well be simultaneous processes, and this behaviour may well be the key to understanding some of the geochemical problems surrounding core formation (Murthy, 1976) that the so called 'inhomogeneous accretion theory' was meant to solve.

The Differentiation of Material in Planetary Size Objects

Before discussing how low viscosity regions ($\eta < 10^6$ poise) may be produced by the heat transfer process, it is worth examining just how critical is their existence to understanding how a planetary body initially well mixed down to a scale of centimetres or less could appreciably chemically differentiate on a billion year time scale. The texture of the most primitive meteoritic material suggests that the starting planetary material was so intimately mixed, and of course all the short range and selective forces that are normally responsible for material textures, e.g. interfacial tension, Van der Waals forces, magnetism, etc., would only dominate the non-selective gravitational interactions while the particles were very small.† Assume for the moment that planetary accumulation occurred on a $> 10^7$ year time scale. The input of energy into the accumulating material by conversion of gravitational energy to heat in interparticle collisions (see below) is then so slow that the heat transport process can maintain a very high average viscosity in the accumulating bodies. It is idle to speculate what the viscosities of all the component phases might be when the effective viscosity of the *aggregate* material is being regulated at a value $\sim 10^{20}$ poise by the global heat transfer process, but the key point to notice is that it is specifically those phases that are most prone to differentiate, i.e. those constituting a quasi-continuous network within the aggregate material, that are decisive in controlling the values of an effective viscosity function for the aggregate. Whether these network phases are the highest or lowest viscosity phases in the aggregate, it is their viscosity which will be, in a sense, selected for regulation by the global heat transport process. A necessary condition for the separation of different density phases on a planetary scale is that the relative speed of their gravitational settling exceeds that of the heat transfer motions which tend to keep the system well mixed. Since the self-regulation of viscosity keeps the mean velocity of the motions transferring radiogenic heat in the range of 10^{-7}–10^{-6} cm/sec, this

†While the particles are very small, non-gravitational forces can produce a cosmic scale chemical inhomogeneity—one has only to visualize the action of the solar wind or Poynting—Robertson effect acting on a difference in particle size of different phases in a tenuous nebula. While some such effect might help to explain compositional differences between planets, the material that subsequently accumulates to form any one planet inevitably comes under the sway of the non-selective gravitational interaction at a very early stage of the process. Homogeneity will result, unless 'new' and chemically different material is introduced at a later state of the accumulation (see below).

condition may be written:

$$gl^2 \Delta\rho > \sim 10^{14}\, \text{gm/sec}^2 \tag{1}$$

where g is the acceleration of gravity and l the length scale of density fluctuations $\Delta\rho$ from the aggregates mean density. No plausible material comes remotely near to satisfying such a condition and it is this that makes it so improbable that differentiation would ever occur as a process fairly uniformly distributed throughout a planetary interior. Perhaps one could argue that over very long periods of time spent at temperatures $\sim 0{\cdot}6T_M$ for the important phases, the starting material would undergo annealing and recrystallization that would increase the length scale of the texture. However, one hardly expects the kilometre length scale required for the satisfaction of inequality (1) to be produced in these conditions, and indeed the quality of seismic body wave transmission (Wavelength $\sim$ a few kilometres) in Earth and Moon now rule out the existence of a medium that would intensely scatter such waves.

Although such results are sufficient to rule out the popular view of differentiation as a process that only occurs when a layer of a planet is 'melted', they highlight the importance of finding the cause of active vulcanism. Although I have already said that it is only a small term in the global heat balance of the Earth, a vital clue that vulcanism is a manifestation of a truly global scale process comes from its highly non-random distribution over the Earth's surface. Moreover this distribution has a very definite relationship with the pattern of large scale relative movements of surface rocks in the last 2×10^8 years of Earth history. Since the correct average speed of these movements has now been clearly predicted as the response of the Earth to radiogenic heat sources diffusely distributed throughout it, it is natural that we should look for the necessary correlation of vulcanism and surface movement in terms of a heat source that is inseparable from the existence of these particular motions—the heat being produced by the shearing of a viscous material. Although this heat source is secondary in the sense that it must be less than the radiogenic sources responsible for the motions in the first place and does not augment the average rate of heat loss from the interior, the peculiar manner in which viscous heating is known to become concentrated when large masses of a material with an exponentially varying viscosity are subjected to shear (Gruntfest, 1963) make it immediately attractive as a possible explanation of vulcanism. If the mean rate of viscous dissipation throughout a convective flow that is induced by a uniform heat source density H is also expressed as a heat source density $\bar{H}_v$, it may be shown that for $(H - H_0) > 0$

$$\bar{H}_v \sim 0{\cdot}1 \frac{g\alpha L}{C_p}(H - H_0) \tag{2}$$

where H_0 is the value of the primary heat source density required to initiate mass transfer as an integral part of the heat transfer process in the object (see for example Figure 1), α, C_p the volume expansion coefficient and specific heat of the medium and L the length scale of the heat transferring motions induced by a positive $(H - H_0)$. For a planet the quantity H_0 may be written C/R^2 where R is its external radius and C is a parameter depending on the thermal conductivity and effective viscosity function one assigns to the object. For example, choosing material properties appropriate to a chondritic meteorite, one calculates that H_0 would equal their present radiogenic heat production when $R \sim 800$ km, i.e. spheres of this radius are the largest of this particular material that can steadily transfer their existing heat production without a concomitant mass transfer. This 'threshold' radius for steady convective movement would have been ~ 300 km $4{\cdot}5 \times 10^9$ years ago (due to enhanced radiogenic heat production), and of course if one were to suppose the material is less refractory than this material (perhaps due to water, which profoundly reduces the effective viscosity function of silicates), it would also be reduced. If one imagines a set of bodies of increasing radius but fixed material properties, when $H = H_0$ ($\propto 1/R^2$) infinitely slow convective motions are induced in a central region of the object with radius $\sim 0{\cdot}3\,R$. For larger objects the convective 'core' region grows outward to involve most of the interior when $H > \sim 2H_0$ (Tozer, 1972b) and it is then acceptable to use the approximations:

$$L \sim R\left(1 - \frac{H_0}{H}\right)^{1/2} \tag{3}$$

$$g \sim \pi G \rho L \tag{4}$$

where ρ is the mass density of the 'homogeneous' object and G the universal gravitational constant.†

It will be immediately clear from (2) and (4) on substituting a plausible value for $\alpha/C_p (\sim 2 \times 10^{-12}\,\mathrm{sec^2/cm^2})$ that $\bar{H}_v$ will only be a significant fraction of H ($> 10^{-2}$ for example) in objects that can support motions on a length scale $\gtrsim 10^3$ kilometres, which immediately warns one that the heat transfer process in planetary size objects might look very different, due to secondary heating, from any simply geometrically scaled up velocity field seen in laboratory heat transfer experiments. To understand how an $\bar{H}_v$ that is almost certainly much smaller than one's uncertainty about an H value may, in certain circumstances, introduce the factor of differentiation into planetary evolution, let us examine, as a thought experiment, the effect of making the temperature coefficient of effective viscosity progressively more negative, e.g. by making B increasingly positive in the expression $\eta = \eta_0\,\mathrm{e}\,(BT_M/T)$, but at the same time adjusting η_0 to keep the actual value of effective viscosity η fixed at some chosen value (see below) Starting with the case $\partial\eta/\partial T$ (or B) $= 0$, it may be shown that the effect of viscous dissipation is to introduce temperature perturbations

†It might appear from (3) that L would increase indefinitely if one considered ever larger bodies. That would imply from (2) that $H_v > H$—a thermodynamic impossibility. In fact, there is a limiting length scale that keeps $g\alpha L/C_p < \sim 1$. This limit is only significant in much larger objects than the terrestrial planets, e.g. Sun and Jupiter.

on the same length scale L as that of the motions and of amplitude:

$$\delta T \sim \frac{10\bar{H}_{\text{v}}L^2}{\rho C_{\text{p}} K \text{Ra}^{1/2}} \tag{5}$$

where Ra is a Rayleigh number:

$$\text{Ra} = \frac{g\alpha L^5}{K^2 \eta C_{\text{p}}}(H - H_0)$$

Of course, regulation of η by the heat transfer process is ruled out in this case, since $\partial\eta/\partial T = 0$ and we have the problem of choosing appropriate values for η to put into this so called 'Boussinesq approximation' of the planetary heat transfer process that will cover a wide variety of circumstances. I have done this indirectly by choosing values of Ra in the range 10^6–10^7, which is typical of its values when self-regulation is present. With $L \sim 2 \times 10^8$ cm and typical chondritic values for the other parameters, we find from (4), (2), and (5) $\delta T \sim 3\,°\text{C}$—an unimportant effect. If $\partial\eta/\partial T$ is now made increasingly negative, the effective viscosity will be preferentially reduced where the rate of strain associated with the mass transfer (and hence H_{v}) has maxima. This reduction of viscosity increases the amplitude of the temperature perturbations roughly in proportion to a reduction in their 'width', since general thermodynamic principles expressed by (2) keep the total amount of energy to be dissipated by viscous heating about the same. Although we cannot make any reliable estimate of how the perturbations δT will vary as a function of $\partial\eta/\partial T$, we can obtain an instructive, if rather approximate, upper bound δT_{ub} to their value by allowing $\partial\eta/\partial T \to -\infty$. In this limiting case we can visualize the heat transfer motions as the movement of quasi rigid blocks of material past each other, so that all viscous dissipation is taking place on surfaces where the rate of shearing has a singularity. In this case one can show that the temperature fluctuations would be

$$\delta T_{\text{ub}} \sim \frac{10\bar{H}_{\text{v}}L^2}{\rho C_{\text{p}} K \text{Ra}^{1/4}} \tag{6}$$

which is $\text{Ra}^{1/4}$ times larger than the case $\partial\eta/\partial T = 0$ given by (5). With the same L value and the present radiogenic heat production used above we have $\delta T_{\text{ub}} \sim 200\,°\text{C}$. Since the temperature fluctuations vary as L^4 (remember $H_{\text{v}} \propto L^2$), this gives only a general idea of the magnitude of the heating; for an Earth sized object composed of chondritic material $\delta T_{\text{ub}} \sim 1000\,°\text{C}$ whereas for a lunar size object $\sim 100\,°\text{C}$. This difference mainly reflects the difference in gravitational acceleration in the two objects.† The actual temperature perturbations above the mean temperature that is associated with a viscosity $\sim 10^{21}$ poise may well be of the same order as δT_{ub}—particularly for the Earth size object, since in this general sub-solidus temperature regime the effective

†The length scale is probably not increasing quite in proportion to the radius (see (3)) because the effects of pressure in increasing T_{M} become important when $R > \sim 2000$ km.

viscosity is decreasing by a factor ~ 15 for each 100 °C rise in temperature and if the increase is ~ 500 °C the even more dramatic decline in viscosity that we call 'melting' will be encountered.

Although the difficulties of analysing convective flows of such high Rayleigh number make it impossible to create a precise predictive theory of the shear heating phenomenon, it is worth emphasizing that the effect of even a semi-quantitative analysis has been to introduce a coherence and simplicity into the picture of planetary evolution that was missing from the earlier attempts to base it on the theory of heat conduction. These made the mistake of oversimplifying the mathematics of heat transfer to such an extent that one could not consistently deal with the problems with which the theory was itself beset, e.g. the movement of heat sources, and it manufactured entirely new problems that do not enter the newer and more general theory. The most prominent of these was the very long time constant, i.e. the global memory (see above), of the heat conduction process in a planetary size object ($> 10^{11}$ years) compared with the half lives of the principle heat producing isotopes thought to be present. Not surprisingly, numerous versions of the theory, some absurdly detailed, made the point that for many billions of years after some initial state, the temperature distribution depended very directly on what one was prepared to assume about initial thermal conditions. In particular, the timing of any magma production was very uncertain, since the deep interior could act as an integrator of the heat production for very long periods of time. In contrast, we can now understand how virtually any 'initial' thermal state is regulated to a quasi-equilibrium situation of heat production and loss in only a few hundred million years. We have seen that this quasi steady mean state is very insensitive to the amount of radioactivity in the interior (see Figure 1 above) and to the changes that would have occurred in a chondritic-like heat production in a few billion years, but as far as magmatic activity is concerned, it is important to notice that the temperature perturbations δT produced by shear heating are directly proportional to $g(H - H_0)$—see equations (2) and (5). Therefore if we disregard a possible early period of adjustment ($\sim 2 \times 10^8$ years) following formation of a planet, we can predict that the level of magmatic activity, as measured by the volume rate of production of magma in localized zones of intense shearing declines steadily with the decay of a radiogenic heat production (see below). Since the possibility of differentiating such a well-mixed starting material as a chondrite hinges on whether the viscosity is lowered from $\sim 10^{20}$ poise to perhaps 10^6 poise or less, δT has to exceed a value ~ 500 °C, and there is consistently a second, even larger, threshold size than that mentioned for the onset of convection in chondritic bodies, below which the shear heating will be too small to induce differentiation. Obviously it will take a very long time for all the starting material to experience the low viscosity conditions necessary for a separation and in that time the radiogenic heat production may decay below that necessary to sustain the zones of differentiation. The larger the planet, the greater is the fraction of the starting material that can be 'processed' in such zones before this happens. It is interesting to note that although the important heat producing isotopes of uranium, thorium and potassium may be removed from the interior by the upward movement of crustal phases from these low viscosity zones, that specific differentiation process is self-limiting. It cannot deplete the interior of its radiogenic heat sources to less than will sustain a convection vigorous enough to produce magmatic conditions. This guarantees that the mean conditions we associate with a self-regulated state of heat transfer is a very persistent feature of a planet's history if its H_0 (see (2)) is small compared with its 'initial' heat source density.

Planetary Evolution away from a State of Material Homogeneity

Although the remarks of the previous paragraph indicate in a general way that large, 'initially' homogeneous objects will ultimately achieve a higher degree of differentiation than small ones, some additional factors that enter the problem once differentiation has been initiated by the heat transfer process require consideration before the degree of separation reached in some finite time can be estimated. In general, the rate of evolution of any chemical inhomogeneity has to be seen as a balance between factors controlling the rates at which material is separated and remixed. As long as the shear heating responsible for the low viscosity conditions that are necessary for a separation is effectively driven by a density difference produced by the thermal expansivity of the planetary material, one can probably rely on the much larger density contrasts between crust, mantle and core materials to prevent a significant rate of remixing. However, we shall see that thermally induced buoyancy does not remain in control of core separation in most terrestrial planets and while this ensures that core and shell material are not significantly remixed, it does have an important indirect effect on any prior separation of a crust and mantle from the silicate shell. Although the vigorous motions set up by the separation of a core would also produce the conditions necessary for a crust/mantle separation, remixing will prevent that process from making any progress until thermal buoyancy reasserts its control of the situation. Despite these complicated interactions between different differentiation processes, considerable insight can be gained by discussing the growth of chemical inhomogeneity as basically two processes (core/shell and crust/mantle) that each involve the separation of a two component, incompressible, mechanical mixture. For the moment I assume that the 'initial' thermal conditions in the homogeneous model planet are those which are fixed by the heat transfer process in carrying plausible amounts of radiogenic heat steadily to the outside.

We have seen that this assumption necessarily means that a very high average viscosity initially exists, but we now have to remember that the only reason separation occurs at all in some very localised zones of low viscosity associated with such an initial state is because it reduces the gravitational self energy of the object. In other words, differentiation is itself a source of heat energy and a degree of auto catalysis is always present. This is allowed for by writing relation (2) in the more general form:

$$\bar{H}_v \sim 0{\cdot}1 \frac{g\alpha L}{C_p}(H - H_0) + H_g \qquad (7)$$

where H_g is an effective average heat source density representing the rate at which gravitational self energy is being converted to heat by the material separation. As an approximation we write the volume rate of production of material with sufficiently low viscosity for the separation to occur as:

$$\frac{dV}{dt} = \frac{L^3 \bar{H}_v}{\rho C_p \Delta T} \qquad (8)$$

where ΔT is the temperature rise necessary to convert material from the average (10^{21} poise) state regulated by the heat transfer process into one in which the differential speed of the two components exceeds their mean speed. For any plausible model of the aggregate starting material ΔT is probably $\sim 500\,°C$ though rather larger in objects of several thousands kilometres radius, due to the effect of hydrostatic pressure in raising the melting point T_M of the planetary material (it can even be significantly less than this during times of accumulation—see the caption to Figure 3). For any object in which shear heating can produce such a large ΔT, we can approximate the length scale of the convection L by the radius of the planet. The reduction in the self-gravitational energy U of the object in separating the two phases constituting dV in our binary mixture model is given by:

$$dU \sim gR\Delta\rho f(1 - f) dV \qquad (9)$$

where f is the volume fraction occupied by one of the phases, which have a density difference $\Delta\rho$. This is converted entirely into heat and we have:

$$H_g \sim \frac{1}{R^3}\frac{dU}{dt}$$

Combining relations (8), and (9), and (4) we have:

$$H_g \sim \beta \bar{H}_v \qquad (10)$$

where

$$\beta = \frac{\pi G R^2 \Delta\rho f(1 - f)}{C_p \Delta T} \qquad (11)$$

Putting (10) into (7) and rearranging gives:

$$\bar{H}_v \sim \frac{(0{\cdot}1) g\alpha L(H - H_0)}{C_p(1 - \beta)} \qquad (12)$$

From this it can be seen that β is the parameter controlling the degree of autocatalysis present in a given differentiation process, and that if $\beta \geqslant 1$ the radiogenic heating rate will lose any control over the rate of differentiation until the separation is sufficiently complete i.e. f small enough, for β to take a value less than unity. Figure 3 shows the $\beta = 1$ condition for the cases $\Delta\rho = 4{\cdot}5\,\text{g/cm}^3$ and $\Delta\rho = 0{\cdot}3\,\text{g/cm}^3$, the first being an estimate of the effective $\Delta\rho$ value for a core/shell separation involving the stony and metallic phases of meteoritic material and the second for a separation of the stony fraction into 'crustal' and 'mantle' fractions. N.B.: The $\Delta\rho$'s are to be distinguished from say the zero pressure density differences of the regions which are separating, unless the differentiation has gone to completion. While the inherent instability

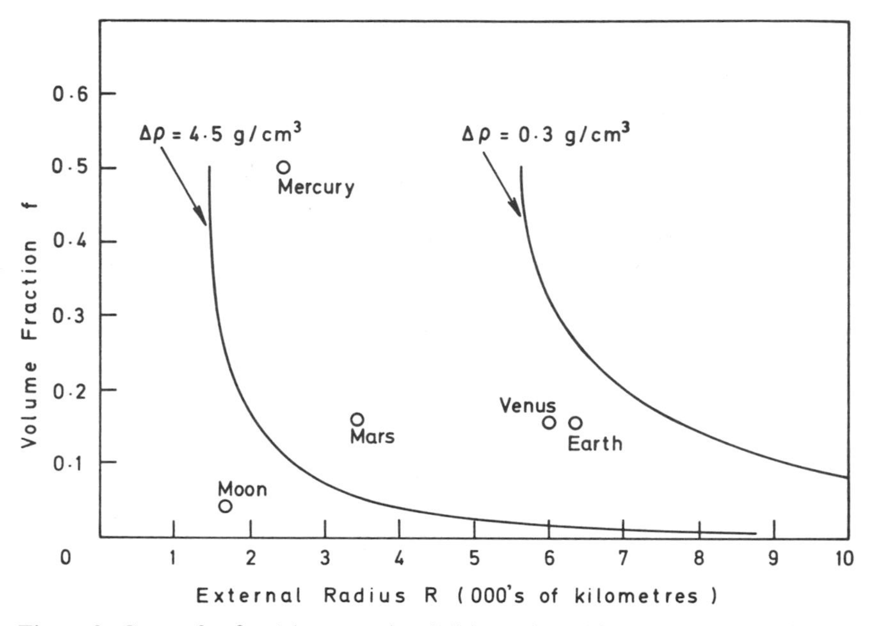

Figure 3. Curves for $\beta = 1$ (see equation (11) in text) marking the transition of core/shell and crust/mantle separations from induced to 'triggered' self-sustaining processes.

'f' is the volume fraction occupied by the less prevalent differentiating material, $\Delta\rho = 4{\cdot}5$ and $0{\cdot}3\,\text{g/cm}^3$ are estimates of the effective density differences for core/shell and mantle/crust separations respectively. ΔT, the temperature rise that is needed to make the speed of separation greater than that of remixing, has been put equal to 300 °C. This is somewhat less than if radiogenic heating had been controlling the average thermal conditions, and reflects the somewhat higher average temperatures that could have been produced by the input of the heat of accumulation over a 10^8 year period. The labelled points show the 'initial' state of homogeneous objects with the same properties as the planets *vis-à-vis* core/shell separations, and should be seen purely in relation to the left hand $\beta = 1$ curve. The Moon-like object is clearly distinguished from the others by the stability of its mixed state. Lunar core/shell separation, like crust/mantle separations in all the objects, has to be induced by shear heating associated with motions removing either the radiogenic or accumulation heat, or both of core/shell separations (see below) probably guarantees that the process of differentiating a core and shell has now ceased, this is not true of crust/mantle separations.

Despite the complications already referred to of having both crust/mantle and shell/core separations taking place in the same object, the great value of these simple binary models of differentiation is that they show that two qualitatively different types of separation process can exist, depending on the position occupied by a homogeneous mass of self-gravitating material on a diagram like Figure 3. For objects with properties that place them on the concave side of the relevant $\beta = 1$ curve, one has to regard the viscous dissipation of the motions of the heat transfer process in highly localized 'zones' as providing only the essential 'trigger' for a separation to take place. Even that role would be denied the object's own radiogenic heat sources if it was formed 'initially' in a sufficiently hot state for a separation to be immediately initiated. Completely chaotic conditions would very quickly ensue after any 'triggering' event with molten conditions spreading to most of the interior, and a transition to a stable state well on the other side of a $\beta = 1$ curve in times $\lesssim 1000$ years. This burst of heat energy—see Table 1—would immediately stimulate a very vigorous convective heat loss that would be trying to regulate the viscosity in the newly formed shell at a figure $\sim 10^{21}$ poise. However, one can see from Figure 2 above that this takes a few hundred million years to accomplish. As already stated there is little chance that any crust/mantle separation can be established during this period, and it is not surprising that the very oldest crustal rocks found on Earth are at least several hundred million years younger than the age assigned to the accumulation of the Earth.

In contrast, if the object's properties 'initially' place it on the convex side of all relevant $\beta = 1$ curves, the bodies' own heat sources are required to sustain any separation after the object has got rid of any heat resulting from its accumulation. Although we cannot confidently calculate the spatial form of the velocity field even in this less energetic situation, I think we can reasonably visualize the heat transfer motions as playing the dual role of bringing relatively

Table 1.

	Energy released by accumulation (joules/gm)[a]	Energy released by core/shell separation (joules/gm)
Moon	$1{\cdot}7 \times 10^3$	~ 5
Mercury	$5{\cdot}4 \times 10^3$	400
Mars	$7{\cdot}5 \times 10^3$	480
Venus	$3{\cdot}2 \times 10^4$	1500
Earth	$3{\cdot}8 \times 10^4$	1700

[a]Since the specific heat is ~ 1 joule/gm °C, the figures given in this table are approximately the general temperature rise that would occur if all the heat were stored in the planet.

cool, undifferentiated material into a number of highly localized hot regions which they have previously produced by shear heating.

On Figure 3 are marked states representing two component binary models of the terrestrial planets before differentiation. The f values for core formation are taken from estimates of the density structure in the planets as they exist today. From this one can reasonably infer that with the exception of Moon, all these bodies would have immediately undergone a catastrophic core-shell separation if they could ever have been produced in such a state (see below). In contrast if we take $f \sim 0{\cdot}15$ for a separation of stony meteorite material into crust and mantle, then it is clear that radiogenic heat production has always set an upper limit to the rate at which a crust will separate.

The Formation of the Terrestrial Planets

As a direct result of this demonstration that large, homogeneous bodies of a meteoritic-like material are prone to catastrophic core/shell separations, one is led to question whether such an unstable 'initial' state of the terrestrial planets (Moon excepted) could ever have arisen as the result of some earlier natural process, such as the accumulation of this material from a tenuous state. Clearly the key question here is the thermal state of the planetary matter throughout its accumulation.

Although there is an enormous and continuing flood of ideas about the history of planetary matter before the emergence of the familiar planetary objects, the desire to put discussion on a more rigorous footing has also led in the last few years to the acceptance of particular views of pre-planetary processes not because they have any firm foundations but simply because they have been extensively 'embroidered' with large numerical computations. Two of these which have to be mentioned because they have been very pervasive in fixing ideas about the chemical homogeneity and thermal state of planetary material during and immediately after its accumulation, are the so-called 'equilibrium condensation sequence' and what I shall call 'the embryo or asymmetric accumulation' hypothesis, i.e. the tacit assumption that planetary objects grew by the addition of bodies that were not themselves subject to such growth. I discuss the second of these ideas first because the significance one attaches to the idea of equilibrium condensation depends on one's view of the later accumulation process.

Dynamical justification for the picture that planets grew by the addition of small particles that liberate the kinetic energy of their impact essentially on the surface of a growing 'embryo' planet (Benfield, 1950), requires the demonstration of an interaction that 'picks out' for exclusive growth one particle necessarily existing among many millions at an early stage of accumulation. The only suggestion of such a 'runaway' accretion process has been based on the enhancement of the capture cross section by gravitational attraction but this is not a very selective influence, growing only slowly in importance if the escape velocity begins to exceed the relative velocities of the accumulating bodies. However, gravity is such a weak force that even this is not likely to happen until the bodies have reached very large dimensions, comparable in fact to a large asteroid unless one can justify extremely small relative velocities of the order of metres/sec or less throughout the period of accumulation. That is most unlikely if the bodies were moving in heliocentric orbits, but even if one tries to visualize accumulation in the absence of a sun, there are difficulties in the assumption of low velocities. It has been shown (Wetherill, 1976) that mutual perturbations prevent the escape velocity from ever becoming large compared with the relative velocities at any stage of the accumulation process, and no one body will achieve a dominant size until the exhaustive stage of accumulation has been reached.† The graphs that purport to show the growth of an 'embryo' radius, r, with time are better interpreted as showing how the radius corresponding to the peak of the mass distribution of accumulating bodies changes with time.

It is basically this belief that the relative velocities of the accumulating planetary material are of the order of several kilometres/sec that causes most writers on the dynamics of the accumulation process to agree that its time scale is in the range of 10^7-10^8 years (Levin, 1972), and there is now an independent line of evidence using the isotopes of xenon that indicates a 50–100 million year formation time for Earth (Pepin and Phinney, 1976). The much shorter accumulation times that one will frequently see quoted in the recent literature are entirely the product of trying to interpret certain lunar data with what I refer to as the asymmetric theory of planetary accumulation. It is a feature peculiar to this model (see, for example, Mizutani *et al.*, 1972) that because the gravitational energy of accumulation (Table 1 above) is converted to heat at the surface of the supposed embryo, the only way it can be prevented from radiating away is by more or less quick burial under later accumulating material. The temperature distribution that results is always one that rises from a minimum at the object's centre to a maximum somewhere near the external surface. This maximum becomes more definite the shorter the accumulation time or what amounts to the same thing, the smaller the relative kinetic energy of the accumulating bodies before acceleration in the embryo's gravitational field. If the accumulation of any planet took longer that 10^7 years, this theory predicts a negligible temperature maximum, e.g. only a few degrees Kelvin for a lunar-like object. Consequently when good evidence was found (Papanastassiou and Wasserburg, 1970, 1971) that the lunar highlands are a differentiate with a radiometric age only $\sim 2 \times 10^8$ years less than that of the meteorites, even short-lived radiogenic heating looked quite inadequate to supply in such a brief time the heat that would be missing if accumula-

†The word 'accretion' is frequently used to describe planetary formation, but its everyday use tends to convey the idea of small particles sticking to a large one. I, therefore, reserve this word for the terminal stage of planetary accumulation when a single body has emerged; I prefer 'agglomeration' as a description of the earlier phase of accumulation.

tion had occurred on the dynamically plausible time scale. It may be added that Moon is a very convenient sized object to test ideas about accumulation since its self-gravitational energy is only just sufficient to raise the temperature of its material to near melting point conditions if none of the energy is radiated away and its material starts at some plausible basal temperature (see below). However, instead of using this circumstance and the unanimity that existed about accumulation times exceeding 10^7 years to contest this view of gravitational energy release, the chosen solution, now hallowed by repetition, was to assert that accumulation of Moon had been accomplished in 10^4 years or less— a curious conclusion to draw when presented with such clear evidence that Moon had been bombarded with objects considerably more substantial than small particles throughout a subsequent history of $\sim 1{\cdot}5 \times 10^9$ years. Any disquiet about the enormous disparity in time scales was suppressed by introducing distinctions between the accumulation process and the flux of bodies responsible for the cratering events. These would not be necessary if one could reconcile the early lunar highland differentiation with the much longer time scale $\sim 10^8$ years for accumulation; the cratering events are then easily dismissed as the necessarily protracted terminal stage of this process.

Other voices than mine have expressed concern at the complexity of the conventional picture of Moon's accumulation (Gold, 1976) and felt unease about the inference of such a short time scale for its completion (Wetherill, 1976). Although some may feel that simple considerations of probability make it imperative to consider more symmetric accumulation theories in which the particle size distribution remains compact until the terminal stage is reached, I think it is worth mentioning observations that rather specifically point against the solution of the highland differentiation problem that is offered by the theory of asymmetric accumulation on a $\lesssim 10^4$ year time scale. The thermal evolution from an initial state in which the temperature has a maximum near the external surface is such as to promote cooling of the near surface material and radiogenic heating of the deep interior—a process that will clearly lead to tensional thermoelastic stresses in the surface material.† Although I do not personally believe that warming of the interior would go as far as is frequently suggested (e.g. Toksöz and Solomon, 1973)—essentially because convective heat loss regulates an average temperature distribution that is at the most 1000 °C (Tozer, 1972), it may be shown that a tensional strain $\sim 10^{-2}$ would now exist in Moon's surface rocks. Since this is more than ten times the elastic limit of any rock, one would have expected to see very obvious signs of tensional failure all over Moon's face. Another argument specifically against a pronounced temperature maximum in the outer part of a

†Solomon and Chaiken (1976) appear to infer that thermoelastic strain in surface material is solely linked to overall changes in the volume of the body. The observable consequences of thermal expansion arise from differential expansion and can exist even if total volume is unchanged. The appropriate theory of this thermoelastic strain has been given in several editions of *The Earth* by Sir Harold Jeffreys.

newly accumulated Moon follows from the observation of a definite hemispherical difference in the distribution of the dark mare material that is so prominent on the near side, and an offset of Moon's centre of mass from its centre of figure in the direction of Earth. These phenomena can both be attributed to a thicker layer of low density (highland type) rocks on the far side, which makes it more unlikely that the denser mare material would there be extruded onto the external surface. However, for this to be a valid explanation, the hemispherical contrast in crustal thickness must have been a very early feature of Moon's evolution. I have already explained why it is a lot more difficult to differentiate a lunar sized object than, say, Earth, with a plausible rate of radiogenic heating as the ultimate cause of shear heating, and this reinforces the inference that lunar crust formation was a product of much faster, non-radiogenic heating rates during the formation process. The only straightforward explanation of the hemispherical difference in crustal thickness is a vigorous two cell convection pattern involving nearly all lunar material in the terminal stage of accumulation. The enormous inward increase in the values of any plausible effective viscosity function, not to mention the convective stability that would be associated with the kind of temperature profiles predicted by fast asymmetric accumulation, would make any such velocity pattern quite impossible. Other observations pointing against the picture of surface heating of planetary 'embryos' are the survival of lunar mascons, very early lunar magnetism and the inference that a Mercurian core separated during the accumulation of that body (see below).

It is now appropriate to examine the case that has been made for believing that the observable chemical inhomogeneity, both in and between planets, is more a product of the accumulation process than of some process peculiar to large masses of an aggregated material. The evidence for compositional differences between planets—crudely represented in Figure 3 above—has not been conclusively explained but at least there are a number of plausible suggestions as to how chemically selective forces can act on particles while they are very small and tenuously distributed. Perhaps the essential message carried by the interplanetary chemical differences is that they show that planetary material was at one time in such a finely divided state that gravity was relatively unimportant, and that one need not be overly impressed by anyone claiming that any particular class of meteorite exactly represents the planetary material. The fundamental difficulty in explaining the chemical inhomogeneity of an individual planet as a product of an accumulation process is the non-selectivity of the gravitational force that must have inevitably controlled the dynamics of the process from a very early stage. The only known way in which chemical inhomogeneity on a much larger scale '*d*' than, say, centimetres might be introduced is for material with a systematically different chemical composition to be added to the population of accumulating bodies when either the 'original' or 'added' material has a particle size '*d*'. A very special version of this idea, popularly known as the 'inhomogeneous accretion hypothesis', assumes that the 'original' material has had time to reach the terminal stage of its accumulation, i.e. form a single dominant object that is the 'core' of a future planet, before the secondary or shell material arrives in much smaller pieces. In the usual version of the hypothesis, the secondary material is not visualised as coming from outside the region of primary accumulating bodies but of joining them by virtue of sudden changes in the composition of a condensate that is forming from a cooling gas phase that permeates the region. However, a great difficulty with this idea arises from the extremely low opacity of the gas phase to thermal radiation. This destroys the assumption of local thermodynamic equilibrium on which calculations of sudden changes in condensate composition are based (Grossman and Larimer, 1974). Very briefly, in a cloud with the dimensions and properties typically assigned to a preplanetary nebula, any condensate that has grown to the size of a few microns would become such a good radiator of its thermal energy to the 'cold' universe outside the nebula that its temperature would rapidly fall to a temperature perhaps only a tenth that of the surrounding gas. So effective is this as a process for cooling the gas phase, that one might well expect the rate of its cooling to accelerate for some time after the first appearance of a condensate. In fact it leads one to doubt that all the planetary material was ever simultaneously brought to the state of a 2000 K, optically thin planetary nebula. The disturbed isotopic ratios (Black, 1972; Lee *et al.*, 1976; Clayton *et al.*, 1973) rather definitely point to the quick formation of dust grains in the gas ejected by supernovae after nucleosynthesis has occurred.

The chemical significance of this very effective cooling of very small condensate grains is that virtually all metallic and silicate phases thought to be prominent in terrestrial planets will be simultaneously deposited on their surfaces. If there is any systematic spatial variation in mean grain composition it is only likely to be correlated with differences in mean orbital distance from a primitive sun. At the various distances of the terrestrial planets, a Sun that is not wildly different in luminosity from the present one will maintain grain surface temperatures somewhere in the range 200–800 K, and it is perhaps not surprising that suspected differences in the composition of the terrestrial planets mainly centre on components that would condense in this range. The degree of hydration of silicate phases is a case in point, and could well be invoked to explain the relative aridity of Venus and Mercury (see below).

The result of radiative losses of the gas phase's internal energy through the intermediary of condensate particles is the rapid collapse of the supposed planetary nebula in times that might be anything from the free fall time, estimated at 'less than 100 years' (Clark *et al.*, 1972) to $\sim 10^4$ years (Cameron, 1975). Any such period is far less than that needed for planetary accumulation to reach its terminal stage ($> 10^7$ years), and consequently one expects individual planets to be formed from a very well-mixed starting material.

The 'symmetrical' accretion theory sees this material being progressively concentrated in a progressively smaller number of bodies. As for the heating of the material by impact, it should be recalled that the amount of gravitational binding energy released per unit mass of a body in the making of it is proportional to the square of its radius so that most of the heating results from collisions among a relatively small number of large 'protoplanets' towards the end of the agglomerative stage of any planet's accumulation. In such collisions vast volumes of material, comparable to those of the colliding objects, would have suddenly been intensely deformed and buried; radiative losses from the surface and heat conduction from within would have been quite negligible as heat loss mechanisms even if these energetically important collisions were spread over 10^8 years or more.†

I therefore conclude that the interiors of the protoplanetary objects were very well-mixed and heated without significant losses of energy until a vigorous internal heat transfer by convection was achieved by the time they had reached roughly a lunar size. From that point onward the self-regulation of viscosity acts in a powerful way to ameliorate the temperature rise and homogenise its distribution. Note that the growth of a lunar sized object in 10^8 years represents a rate of gravitational energy release/unit mass that is ~ 50 times the radiogenic heat production rate, but were this maintained indefinitely it would not be sufficient to create generally molten conditions throughout the interior. Except for very small values of the iron/silicate ratio, it can be seen from Figure 3 that catastrophic core/shell separations would occur in the protoplanets that had reached about a lunar size, and since this is only a few per cent of the Earth's or Venus's mass, one very strongly suspects that the cores we now see in these objects evolved by the coalescence of several protoplanetary cores, rather than the kind of single body separation discussed above. Of the larger bodies inside the asteroid belt, only Moon would have avoided this phase of planetary formation, but of course the convection that would have occurred throughout its whole interior by the time the terminal state of its formation had been reached would have been vigorous enough to create low viscosity zones from which highland material was extracted.

The coalescence of protoplanetary cores would undoubtedly have been a very rapid process following their collision, allowing little time for any chemical equilibria to have been established. This latter point is important because it has to be remembered that at such an early stage in the formation of planets the size of Earth or Venus, the differentiated protoplanets would have been accompanied by a considerable amount of material in smaller objects that had not differentiated, but which was eventually to join the large objects. This may go far to explain the geochemical evidence (see, for example, Murthy, 1976) that Earth's core and shell materials were never intimately mixed. As Murthy

†As long as one was committed to a $\sim 10^3$ year time scale for accumulation, one had to assume that the initial kinetic energy of the accumulating bodies was negligible in order to get the large capture cross section (see above). The 'initial' kinetic energy associated with a 10^8 year accumulation time now helps to remove any worries that Moon's gravitational binding energy alone might not be sufficient to promote a differentiation.

remarks, the intense heating of the interior that occurs when cores form would have released a lot of the water and other volatiles that had originally condensed on the cold grains (see above). These volatiles would have oxidized *in situ* the dense metallic core material that was later to join the largest protoplanetary bodies. Incidentally, much of the expelled volatile material would have been able to escape entirely from lunar-sized protoplanetary bodies that had differentiated, so that the primitive atmosphere of Earth would never have been as massive as if core/shell separation had only occurred in the terminal stage of its accumulation.

There is independent geochemical evidence that indicates less than a 10^8 year difference in the times of Earth accumulation and its core/shell separation (Ulrych, 1967; Oversby and Ringwood, 1971), but even better, indirect evidence for the contemporaneity of the two processes is now available to us from the study of Mercury. As soon as the Mariner 10 probe had confirmed a $\sim 5{\cdot}5$ g/cm^3 mean density for the planet, and revealed a cratered surface with similar optical properties to that of Moon, it was immediately suggested that a core/shell separation had occurred in it, but that it must have predated the collisions responsible for the present appearance of its surface. The detection of a magnetic field (Ness *et al.*, 1974) was simply interpreted as evidence for the current existence of a Mercurian core, but I believe that this observation carries very clear implications about the timing and stimulus for core formation when viewed in the context of a *plausible* heat transfer theory for the planet. The key point at issue is whether the observed field should be interpreted as the effect of a currently active dynamo or as a remanence in the rocks that was produced by an ancient field. After 30 years' work we still know remarkably little about the conditions under which a self-exciting dynamo would function, but I think there is universal agreement that it definitely would not work if the core had solidified. There have been a number of discussions about the internal thermal state of Mercury (Siegfried and Solomon, 1974) and specifically dealing with the question of core liquidity (Fricker *et al.*, 1976), but they fail to come to any definite conclusion because no account is taken of the effect of solid state creep in regulating mean temperatures in planetary sized objects. If the shell of Mercury is made of any material that is remotely plausible to a geochemist, the temperature distribution in the interior would have very quickly dropped to temperatures below 1300 °C (Figures 1 and 2)—assuming that an accumulation process had been transiently capable of producing higher temperatures in the face of such a powerful regulatory process as convection. This is very substantially less than the melting point of iron/nickel alloys in a Mercurian core—see Figure 1 above. Furthermore, one cannot solve the problem of maintaining liquidity for billions of years, as it can be in the Earth's case, by adding a component, such as sulphur, to an iron/nickel alloy to reduce its melting point. If added in the quantities necessary to maintain a liquid Mercurian core, it reduces the density of the core so much that in keeping the mean density correct one has to reduce the amount of shell material to virtual non-existence. As explained above, a blanket of more refractory shell material is absolutely essential to prevent such a large mass of core material from quickly freezing and adopting an average viscosity $\gtrsim 10^{20}$ poise in a few hundred million years.† It was therefore a crucial test of modern ideas about the planetary heat transfer process that Mercury should not turn out to have such a large magnetic moment that only some form of dynamo or electrochemical theory involving a liquid core would be required to explain the requisite electric currents. Although it has been suggested (Runcorn, 1975) that remanence induced by an internal magnetic field source could not on its own give rise to an external magnetic field, Stephenson (1976) has since shown that Runcorn's result is only an approximation and that tenable theories of the Mercurian magnetic field ($\sim$ a few milligauss at its surface) can be based on the remanence of its near surface rocks, assuming that these have been cooled through their Curie point in a field of a few oersted. The attractiveness of Stephenson's theory is considerably enhanced by noting how the cooling of near surface rocks is correlated with the existence of an inducing field having its origin in a liquid core. The rapid convective loss of heat through a shell containing such a core guarantees that it would be very vigorously stirred right up to the onset of solidification. At just this time, the heat flow at the external surface would have been rapidly falling so that the thickness of a layer defined on its lower boundary by a Curie point isotherm would have been rapidly increasing—just the conditions for the planet to acquire a thermoremanent moment. Given effective viscosity functions that are typical of plausible silicate materials, one can quite confidently predict that an iron/nickel Mercurian core would have been frozen within a few tens of millions of years of the last coalescence of protoplanetary cores from which it formed (Figure 3). Like Strom and Cordell (1976), I interpret the 'lobate scarps' so profusely scattered on the Mercurian surface, and their complete absence on a body of similar size that has no comparable core—Moon, as evidence of the surface compression that would have accompanied a freezing core in the terminal stage of Mercury's accumulation. Peale (1976) has suggested a dynamical method of sensing the liquidity of a Mercurian core, but perhaps the most direct way of refuting my argument for its solidity would be to detect magnetic secular variation from orbiting spacecraft.

Conclusion

The surfeit of schemes about solar system development that do little but tentatively rationalize about a few known and basic parameters of its present state has prompted me to begin by explaining my different approach to the problem. For me it is not pictures about the past but comments and predictions

†It should be clear by now that the familiar inconclusive discussions about the radiogenic heat source distribution are quite incapable of solving this problem—Figure 1 shows the effect on the temperature of changing the heat source density inside such large objects.

about the observable present that give any evolutionary theory its scientific importance. I am much impressed by many demonstrations of man's present inability to predict the development of laboratory systems that appear to present well-posed problems in the dynamics of finite material deformation. This well-known propensity of fluid systems to amplify negligible perturbations from assumed 'initial' conditions tends to make any success in predicting the properties of the observable solar system from some supposedly simple 'initial' state look very unconvincing. If the word 'origin' has any meaning when applied to a natural system, I suggest that it be given a more subjective meaning, e.g. the point on an evolutionary time scale beyond which it becomes impossible to make any testable predictions about its present state. This has forced me to concentrate on evolutionary problems that are posed by the current existence of terrestrial planetary matter in the form of large self-gravitating objects, before contemplating the processes that might have affected it before the solar system was recognizable as such. Did the solar system then exist?

A new understanding of the importance of the heat transfer process in fixing the spatially averaged effective viscosity of matter at an enormous value when it constitutes such large objects as planets, has reopened the question of how individual terrestrial planets have evolved their present differentiated states. If these objects were 'initially' formed of a material that is at least well mixed down to a scale $\ll$ kilometres, it is shown that they will only appear significantly differentiated a few billion years later if:

(1) their initial thermal state was fixed by a process that could temporarily overwhelm the tendency of the heat transport process to maintain a spatially averaged material viscosity $\gtrsim 10^{20}$ poise; and/or
(2) some subsequent process has persistently but necessarily locally generated much lower viscosity material than the tightly regulated spatially averaged value.

Of the second type of process, shear heating associated with convective movements that are set up to carry the radiogenic heat in such large masses of material having a large negative temperature coefficient of viscosity, appears to have just the right characteristics to facilitate a phase separation over big distances, although a precise quantitative analysis of the problem is defeated by our known inability to calculate the form of any convective flows at such high Rayleigh numbers. If one takes the texture of primitive meteoritic material as indicative of the scale at which planetary material was well mixed, one has to explain how shear heating creates viscosities $> 10^{14}$ times less than the regulated average value. Consideration of the efficiency with which shear heat is produced by convective flows indicates that the necessary temperature rise (several hundred degrees) to produce such a large fall in viscosity will only be created in objects with radii greater than a definite, if hard to calculate, threshold value. For meteoritic objects radiogenically heated by chondritic amounts of uranium, thorium and potassium, this threshold size is probably about 1500 km.

It is shown that the picture of a localized separation of phases, whose rate is ultimately fixed by the rate of radiogenic heating, is valid for crust/mantle separations, but not for core/shell separations in the terrestrial planets (Moon excepted). The gravitational energy released by the latter processes is sufficient to create self-sustaining and catastrophic separations once the differentiation has been 'triggered' by shear heating.

These results again focus attention on the very early differentiation of the lunar highland material and the relevance of the 'initial' thermal state of planets to their present differentiated state. Evidence is given that leads one to reject the idea that these bodies are differentiated because of rapid changes in the composition of bodies arriving at the surface of a growing planetary embryo. Indeed, should one not be impressed by the sheer dynamical improbability of one body growing at the expense of an enormous number of much smaller ones, other evidence from Moon leads one to reject results obtained from the embryo model in favour of a more symmetrical view of planetary accumulation, i.e. one involving the parallel growth of roughly similar protoplanets of ever decreasing number. Not only does the amount of material involved in individual collisions towards the energetically important end of accumulation allow a more acceptable time scale ($\gtrsim 10^8$ years) for the process and still produce the thermal conditions necessary for a lunar highland differentiation, but the existence of protoplanets that are prone to catastrophic core/shell differentiations can probably resolve geochemical problems that have arisen in thinking of core/shell separation as a single body process. Cores in Earth and Venus and probably Mars and Mercury were largely formed by the rapid coalescence of cores that had previously separated in their respective protoplanets.

The cases of Mercury and Moon are particularly interesting in testing different views of differentiation, being roughly the same sized bodies but respectively with and without core material that could catastrophically differentiate. Arguments based on the morphology of Mercury's present external surface and magnetic moment strongly suggest core formation and accumulation were contemporaneous processes. This makes a very good case for believing that Earth material was separated into its core and shell phases before it formed the single body that gives it its name.

References

Benfield, A. E. (1950). *Trans. Am. Geophy. Union*, **31**, 53.
Black, D. C. (1972). *Geochim. Cosmochim Acta*, **31**, 1025.
Cameron, A. G. W. (1975). *Sci. Amer.*, **233** (**3**), 32.
Cathles, L. M. (1975). *The Viscosity of the Earth's Mantle*, Princeton.
Clark, S. P., Turekian, K. K., Grossman, L. (1972). *The Nature of the Solid Earth*, McGraw-Hill, 3.
Clayton, R. N., Grossman, L., Mayeda, T. K. (1973). *Science*, **182**, 485.
Fricker, P. E., Reynolds, R. T., Summers, A. L., Cassen, P. M. (1976). *Nature*, **259**, 293.

Gold, T. (1976). *Abstract 7th Lunar Science Conference*, 304.
Green, D. H. (1972). *Tectonophysics*, **13**, 47.
Grossman, L., Larimer, J. W. (1974). *Rev. Geoph. Space Phys.*, **12**, 71.
Gruntfest, I. J. (1963). *Trans. Soc. Rheology*, **7**, 195.
Jeffreys, H. (1976). *The Earth 6th Edition*, Cambridge, 430.
Lee, T., Papanastassiou, D. A., Wasserburg, G. J. (1976). *Geoph. Res. Letters*, **3**, 41.
Levin, B. J. (1972). *Tectonophysics*, **13**, 7.
Mizutani, H., Matsui, T., Takeuchi, H. (1972). *The Moon*, **4**, 658.
Murthy, V. Rama (1976). *The Early History of the Earth*, ed. B. Windley, Wiley, 21.
Ness, N. F., Behannon, K. W., Lepping, R. P., Whang, Y. C., Schatten, K. H. (1974). *Science*, **185**, 151.
Oversby, V. M., Ringwood, A. E. (1971). *Nature*, **234**, 462.
Pandit, B. I., Tozer, D. C. (1970). *Nature*, **236**, 335.
Papanastassiou, D. A., Wasserburg, G. J. (1970). *Earth Planet Sci. Lett.*, **8**, 269.
Papanastassiou, D. A., Wasserburg, G. J. (1971). *Earth Planet Sci. Lett.*, **11**, 37.
Peale, S. J. (1976). *Abstract 7th Lunar Science Conference*, 668.
Pepin, R. O., Phinney, D. (1976). *Abstract 7th Lunar Science Conference*, 682.
Runcorn, S. K. (1975). *Nature*, **253**, 701.
Siegfried, R. W., Solomon, S. C. (1974). *Icarus*, **23**, 192.
Solomon, S. C., Chaiken J. (1976). *Proc. 7th Lunar Science Conference*, 3229.
Stephenson, A. (1976). *Earth Planet Sci. Lett.*, **28**, 454.
Strom, R. G., Cordell, B. M. (1976). *Abstract from L. S. I. Conference on Comparisons of Mercury and Moon*, 7.
Tittman, R. B., Curnow, J. M., Housley, R. M. (1975). *Proc. 6th Lunar Sci. Conf.*, 3217.
Toksöz, M. N., Solomon, S. C. (1973). *The Moon*, **7**, 251.
Tozer, D. C. (1972a). *The Moon*, **5**, 356.
Tozer, D. C. (1972b). *Phy. Earth Planet Int.*, **6**, 182.
Ulrych, T. J. (1967). *Science*, **158**, 252.
Wetherill, G. W. (1976). *Proc. 7th Lunar Sci. Conference*, 3245.
Wyllie, P. J. (1971). *The Dynamic Earth*, Wiley, p. 175.

COMETS AND THE ORIGIN OF THE SOLAR SYSTEM

A. H. DELSEMME
Department of Physics and Astronomy, The University of Toledo, Ohio, U.S.A.

Introduction

The spectacular appearance of a comet in the sky is a transient phenomenon produced by the development of a head (coma) and a tail. But coma and tail appear only where the distance to the sun is short; they are made of gas and dust that are steadily lost to space. The single permanent feature, still present when coma and tail fade away, is the cometary nucleus. No cometary nucleus has ever been seen but as a pinpoint of light, reflecting the solar light just as the planets do. Whipple (1950) has convincingly argued that cometary nuclei are conglomerates of frozen gases and dust, in short, 'dirty snowballs' a few kilometres across. Although Lyttleton (1975) continues to claim that comets have no solid lump nucleus, he has never refuted any of the three major former criticisms against his loose 'sandbank' model, based on total mass loss, collapse of grain cluster, and gas drag deduced from observed gas production rates (Whipple, 1963; Delsemme, 1972–1973). A consensus in favour of the icy conglomerate has therefore been reached among the other astronomers.

Origin of Comets

About 600 different comets have actually been observed (Marsden, 1975). Approximately 100 are short-period comets (period from 3 years to one century) whose repeated returns can easily be predicted, and are often observed. They probably have been captured from the set of long-period comets by the combined perturbation of Jupiter and Saturn, and the solar heat makes them decay very fast, often in a few centuries or millenia. About 400 long-period comets have also been observed (period from one century to 10^5 years). Since they are also perturbed by the giant planets during each of their passages, this interaction will eventually lead either to their total capture and decay as short-period comets, or to their complete ejection out of the solar system, depending on the random walk of their orbital energy after each passage. Finally, there are almost 100 observed comets whose 'original' orbital energies (usually measured per unit mass by the reciprocal of their semi-major axes a) are concentrated in an extremely sharp interval of a^{-1}; the best determinations correspond to an aphelion distance in the range 70,000 AU $\pm$ 30,000 AU (Marsden and Sekanina, 1973). The sharp peak of these orbital energies has a half-width of some 2×10^{-5} in a^{-1} units. However, the random change of the orbital energy introduced by the giant planets in one single perihelion passage averages about 60×10^{-5} a^{-1} (Everhart and Raghavan, 1970), that is 30 times the half-width of the peak. We must therefore conclude that, as a group, these comets had never been through the solar system before because if they had, they would have diffused out of the peak. Following Oort (1950) we call them 'new' comets. They provide the steady supply by which the set of decaying periodic comets is being replenished. Since their aphelia extend roughly from 3×10^4 to 10^5 AU, Oort (1950) argues that the cause of their first passage must act most strongly between these distances. He has shown numerically that it can be explained by the perturbing influence of the nearby stars, that constantly reshuffle the slow velocities of a very large number of unobserved comets (10^{11}) that must therefore be present in a sphere surrounding the sun. Hence, a small number of them reappear every year in the inner solar system as 'new' comets, whereas the bulk of comets remain on very large orbits that are too far away to be influenced by the planets.

This view is still criticized by Lyttleton (1974), but his criticism is based on a misinterpretation of Oort's analysis, as already pointed out by Le Poole and Katgert (1968). At any rate, Lyttleton does not offer any alternate explanation of the surprising concentration of the binding energies of the 'new' comets in such a small energy interval. Although the evidence will remain circumstantial, because we obviously cannot observe comets much beyond 5 or 6 AU, the existence of Oort's cloud of comets is not much disputed any more by other astronomers, although this does not imply that we understand its origin or its age. The locus of the primeval belt of comets hypothesized by Oort (1950) was pushed first from the asteroid belt to the range of Uranus–Neptune, by Kuiper (1951) and Whipple (1964) and then to much larger distances by Cameron (1976). Whatever their origin, the fact that comets are permanent members of the solar system and must therefore be connected to its origin is generally accepted. However, Vsekhsvyatsky (1975) still tries to reverse the evolutionary process, claiming that short-period comets are still being made by violent eruptions from the surface of the giant planets, and that their orbits diffuse steadily into the long-period comets' and, eventually, into Oort's cloud. First envisaged by Lagrange (1814) this process is no more taken seriously by most astronomers; except as a possible although inefficient way to build Oort's cloud 4·6 billion years ago. However, numerical analyses of the steady-state of orbital diffusion are still order of magnitude estimates (Delsemme, 1973b) waiting for a final answer from numerical experiments of orbital diffusion (Everhart, 1973–1976).

The mere existence of the Oort's cloud implies however that comets are permanent members of the solar system. Although their actual place of origin is still in dispute, it sounds reasonable to assume that they have accreted from icy grains present somewhere in the solar nebula, or that they have condensed straight from its gases, or both (icy grains could have been condensation centres). At any rate, their chemistry must bring important clues on the primeval nebula and its fractionation processes.

Chemical Nature of Cometary Nuclei

Nothing is known from direct evidence. Qualitative observations of atoms, ions and radicals in the coma and tail, have been recently transformed into numerical estimates of production rates of parent molecules during the vaporization of the nucleus, mainly since Comet Kohoutek 1973 XII (Delsemme, 1976), but the uncertain history of the molecular fragments and ions (see Table 1) must first be reconstructed, since many charge-transfer reactions may reshuffle the components in a small (10^4 km) collision zone, while they are dissociated and/or ionized by the solar light and wind (Oppenheimer, 1975).

For instance, Delsemme and Combi (1976) argue that the origin of the oxygen atom in the ^1D state (source of the forbidden red line observed in comets) is the photodissociation: $CO_2 \rightarrow CO + O(^1D)$. This would imply that CO_2 was more abundant than water in Comet Bennett (1970 II), and would also imply that CO, recently detected by its ultraviolet bands, is not necessarily one of the stable primeval molecules. As a matter of fact, no stable molecules that could be primeval had been observed or deduced in any certain way until 1975. Since Comets Kohoutek (1975 XII) and Bradfield (1974b), three stable molecules: NCN, CH_3CN and H_2O have been detected by radio astronomers, but their production rates were assessed within an order of magnitude only. However, the author (Delsemme, 1976b) has mentioned that the brightness profiles of the resonance lines of H, C, N, O, observed at large distances from the nucleus can be used to deduce the total production rates of these atoms after all molecules have been dissociated by the solar light, and therefore yield a quantitative elementary analysis of the sublimating gas mixture. The resonance line of N has not yet been observed, but H, C and O have been observed in the vacuum ultraviolet for two recent comets. For Comet Kohoutek, Delsemme (1976b) quotes: $H/O = 1{\cdot}5 \pm 0{\cdot}7$ and $C/O = 0{\cdot}23 \pm 0{\cdot}1$; for Comet West, he gives

Table 1. Observed constituents in comas and tails[a]

Organic:	C, C_2, C_3, CH, CN, CO, CS, HCN, CH_3CN;
Inorganic:	H, NH, NH_2, O, OH, H_2O;
Metals:	Na, Ca, Cr, Co, Mn, Fe, Ni, Cu, V, Si;
Ions:	CH^+, CO^+, CO_2^+, CN^+, N_2^+, OH^+, H_2O^+;
Dust:	silicates (infrared reflection bands)

[a]This includes recent ultraviolet data from Comet West (Smith *et al.* 1976)

(1976c) C/O = 0·28 from Feldman and Brune's (1976) data. The low abundance of H clearly suggests that H was originally bound in molecular compounds, and that there is no free hydrogen in comets, contrarily to Öpik's (1973) belief.

Dust-to-Gas Ratio

The dust-to-gas mass-ratio can be deduced rather accurately from the momentum exchanged by the gas that drags the dust away (Finson and Probstein, 1968a). This ratio was 1·67 in Comet Arend–Roland 1957 III (Finson and Probstein, 1968b) and 0·50 in Comet Bennett 1970 II (Sekanina and Miller, 1973). The author submits that, by making the reasonable assumption that the dust is indeed silicate dust, as confirmed by its silicate signature in infrared (Ney, 1974), dust-to-gas ratios can be compared with models using solar abundances for C, N, O. The oxygen/metal (number) ratio of silicates goes from a low 4/3 for olivine to a high 9/5 for serpentine. Too-volatile He and Ne are assumed to be missing. The observed H/O ratio of 1·5 is used instead of the solar H/O = 1500, a depletion of 1000. For C, N, O, Cameron's solar abundances are used. If we assume that all metals and sulphur are dragged in the dust with the oxygen needed to make silicates, whereas the rest of the oxygen with H, C, N makes the bulk of the gas mass, the dust-to-gas ratio is predicted to be 0·41 (with silicates as serpentine) and 0·38 (with silicates as olivine).

It is remarkable to find that the production rates of both Comets Arend–Roland and Bennett should have only a depletion factor of 1·2 to 4 in CNO, in respect to this model based on solar abundances. At any rate, this depletion factor may be only applicable to the dust enrichment of the nuclear crust, and the proportions CNO/metals could still be exactly solar, a few metres deep.

Finally, the same model predicts a C/O ratio of 0·71 to 0·79 in the volatile fraction (depending on which silicates), to be compared with the observed 0·23 in Comet Kohoutek, and 0·28 in Comet West (Feldman and Brune, 1976) which would suggest that 4/5 of the primeval C and O was in CO, too volatile to be condensed. In conclusion, comets are undisputably very primitive and undifferentiated objects; much less differentiated than carbonaceous chondrites, since they are the only bodies without a sizeable gravitational field, that seem to have conserved their CNO in *almost* (but not quite) solar proportions.

Chemical Differences

Other spectra suggest chemical variations from comet to comet. The most conspicuous is the ratio CO^+/CN, with the extreme cases of Comets Humason 1962 VII and Morehouse 1908 III where it was 10 to 100 times the average; second, the ratio of the continuum to CN (the continuum measures the amount of light reflected by *fine* dust only); third, the ratio C_2/CN also may vary somehow. It is difficult at this stage to decide whether these variations imply chemical differences in the cometary nucleus, or can be entirely explainable by physical differences (the mere size of the comet changes its interaction with the solar wind; spectral variations with heliocentric distance are also well known). Giant comets with dust tails at very large heliocentric distances, may require more volatile gases, like methane or CO, to drag the dust away. Comet Schwassmann–Wachmann I, a very peculiar giant comet on an almost circular orbit beyond Jupiter's, also has irregular outbursts suggesting very volatile gases.

Thermal History of Comets

The four primeval molecules, H_2O, HCN, CH_3CN and CO_2 (probably) identified in the volatile fraction of comets, suggest that at least Comets Bennett and Kohoutek were condensed or accreted near 100 K or lower, and that they were never reheated since. In particular, the non-detection of large amounts of CH_4 as well as the interpretation of CO as coming from CO_2, suggests that their temperature was larger than 50 K during their formation or during a later processing (see Table 2).

It is however possible that, at places optically thin in infrared, silicate grains can radiate enough to cool down to 100 K in a hot nebula, and therefore condense the hot surrounding gases whose thermal equilibrium could be quenched very effectively. In order to produce HCN and CH_3CN in amounts that are not negligible in respect to water and CO_2, the temperature of the gas must be around 1000 K (see Table 3) which seems rather high; however, a large depletion of hydrogen would lower this temperature considerably by shifting the chemical equilibria.

The evidence is still inconclusive, because we do not know yet whether

Table 2.

	Sublimation Temperatures					
Pressure	H_2O	CH_3CN	HCN	CO_2	CH_4	CO
10^{-6} atm	190 K	170 K	140 K	105 K	48 K	38 K
10^{-8} atm	170 K	150 K	120 K	90 K	40 K	32 K

Table 3. Chemical reactions yielding H_2O, HCN, CH_3CN, and CO_2 in the primeval nebula

low-T side	$\Delta G = O$ at:	high-T side
$CH_4 + H_2O$	← 920 K →	$CO + 3H_2$
$CH_4 + NH_3$	← 1030 K →	$HCN + 3H_2$
$CO_2 + H_2$	← 1250 K →	$CO + H_2O$
$C_2H_2 + 2NH_3$	← 590 K →	$2HCN + 3H_2$
$C_2H_2 + NH_3$	← 650 K →	$CH_3CN + H_2$

charge-exchange reactions between molecules and ions could yield the same results in cooler surroundings. When we know better the cometary snows, we will probably be in a position to decide whether they condensed out of a gas phase, or if they accreted from interstellar grains.

References

Cameron, A. G. W. (1973). *Space Science Reviews*, **15**, 121.

Cameron, A. G. W. (1976). *The Origin of the Solar System*, NATO Advanced Study Institute, Newcastle upon Tyne (this book).

Delsemme, A. H. (1972). p. 305 in *On the Origin of the Solar System*, H. Reeves, ed., CNRS, Paris.

Delsemme, A. H. (1973a). *Space Science Reviews*, **15**, 89.

Delsemme, A. H. (1973b). *Astron. Astrophys.*, **29**, 377.

Delsemme, A. H. (1976a). *IAU Reports on Astronomy*, *XVIA*, 192 (Reidel).

Delsemme, A. H. (1976b). *Mem. Soc. Roy. Sci. Liège*, 6 série, **9**, 135.

Delsemme A. H. (1976c). *IAU Colloquium No 39*, Lyon, August 1976.

Delsemme, A. H., and Combi, M. (1976). *Astrophys. J. Letters*, **209**, L 149.

Everhart, E. (1973). *Astronom. J.*, **78**, 329.

Everhart, E. (1976). p. 51, in *IAU Colloquium No 39, Abstracts*, A. H. Delsemme, ed., University of Toledo.

Everhart, E., and Raghavan, N. (1970). *Astronom. J.*, **75**, 258.

Feldman P. D. and Brune W. H. (1976). *Astrophys. J. Letters*, **209**, L 45.

Finson, M. L., and Probstein, R. F. (1968a, b). *Astrophys. J.*, **154**, 327 and 353.

Kuiper, G. P. (1951). p. 357 in *Astrophysics: A Topical Symposium*, J. A. Hynek, ed. McGraw-Hill, publ.

Lagrange, J. L. (1814). addendum to *La Connaissance des Temps*, Paris.

Le Poole, R. S., and Catget, P. (1948). *Observatory*, **88**, L141.

Lyttleton, R. A. (1975). *Astrophys. Space Sci.*, **34**, 491.

Marsden, B. G. (1975). *Catalogue of Cometary Orbits*, IAU Central Bureau, Smithsonian Astrophys. Observatory, Cambridge, Mass.

Marsden, B. G., and Sekanina, Z. (1973). *Astronom. J.*, **78**, 1118.

Ney, E. P. (1974). *Astrophys. J. Letters*, **189**, L141.

Oort, J. H. (1950). *Bull. Astron. Inst. Netherlands*, **11**, 91.

Öpik, E. J. (1973). *Astrophys. Space Sci.*, **21**, 307.

Oppenheimer, M. (1975). *Astrophys. J.*, **196**, 251.

Sekanina, Z., and Miller, F. D. (1973). *Science*, **179**, 565.

Smith, A. M., Bohlin, R. C., Stecher, T. P. (1976). *IAU General Assembly Commission 15 meeting*, Grenoble, France.

Vsekhsvyatsky, S. K. (1975). *Problemi Kosmicheskoj Fiziki*, **10**, 96 (Kiev).

Whipple, F. L. (1950). *Astrophys. J.*, **111**, 375.

Whipple, F. L. (1963). p. 639 in *The Solar System*, Vol. 4, B. Middlehurst and G. Kuiper, ed., University of Chicago Press.

Whipple F. L. (1964). *Proc. Nat. Acad. Sci. U.S.A.*, **52**, 565.

Whipple, F. L. (1976). p. 622 in *The Study of Comets*, NASA-SP-393, Washington, D.C.

DYNAMICS OF PLANETESIMAL FORMATION AND PLANETARY ACCRETION

A. W. HARRIS
Jet Propulsion Laboratory, 4800 Oak Grove Drive, Pasadena, California, U.S.A.

Abstract

We review the dynamical processes by which condensed matter in the solar nebula accumulates into planets. The basic processes are: (1) gravitational instability; (2) conversion of the radial gradient of orbital motion into random motion between planetesimals; and (3) collisions, which damp the random motion and result in aggregation and/or fragmentation of planetesimals. Each of these processes is defined analytically and models, based on these results, of planetary growth, axial rotation rates, and formation of satellite systems are formulated.

Introduction

The accumulation of many small bodies into a few large planets is a general feature of essentially all theories of solar system origin. In this paper, we consider the dynamical processes involved in this transformation. We begin by defining the 'initial state' of the solar nebula inferred from the present distribution and composition of the solar system. This so-called 'minimum mass solar nebula' will be used for the purpose of numerical illustration of results. The processes described, however, are general and could equally be applied to other nebula models. (The term 'nebula' is used here, for lack of a better term, to denote the matter orbiting about the sun from which the planets are ultimately formed).

Initial State of the Solar Nebula

From the close parallel between solar composition and the bulk compositions of the earth, moon, and meteorites, one can infer that the matter from which the planets condensed was initially very close to solar composition, with the major fractionation due to failure of the very volatile constituents to condense before dissipation of the gaseous phase. The minimum mass of the nebula from which the planetary matter condensed can therefore be inferred by reconstituting the mass of the present planets to solar abundance. This exercise was done by Hoyle (1946) and has been repeated many times since. Table 1 lists the eight most abundant elements in the sun (Podolak and Cameron, 1974). Note that they fall in three distinct groups: H and He, which are gaseous in all regions of the nebula; C, N, and O, which with appropriate volumes of H form volatile compounds which are condensed at the temperatures of the outer planets ($\lesssim 150$ K) but are not condensed in the terrestrial zone; and finally Mg, Si, Fe, which with appropriate volumes of O are the major constituents of the terrestrial planets. In Table 2 are listed the present properties of the solar system along with reconstitution factors inferred from Table 1, and the characteristics of

Table 1. Elemental abundances in solar nebula

Element	Atomic fraction	Mass fraction		Mass fraction of compounds
H	0·930	0·762	Gas	0·984
He	0·068	0·224		
C	0·00033	0·0032	Volatiles	0·0121
N	0·00008	0·0009		
O	0·00055	0·0072		
Mg	0·000015	0·0003	Solids	0·0034
Si	0·000027	0·0006		
Fe	0·000027	0·0012		

Table 2. Solar nebula inferred from present planetary system

Planet	Orbit radius, cm	Mass gm	Re-constitution factor	Re-constituted mass, gm	Total surface density gm/cm^2	Surface density of solids gm/cm^2
Mercury	$5{\cdot}8 \times 10^{12}$	3×10^{26}	300	9×10^{28}	1000	3
Venus	$1{\cdot}1 \times 10^{13}$	5×10^{27}	300	$1{\cdot}5 \times 10^{30}$	4000	13
Earth	$1{\cdot}5 \times 10^{13}$	6×10^{27}	300	$1{\cdot}8 \times 10^{30}$	2500	8
Mars	$2{\cdot}3 \times 10^{13}$	6×10^{26}	300	2×10^{29}	200	0·7
Jupiter	$7{\cdot}8 \times 10^{13}$	2×10^{30}	3?	6×10^{30}	200	3
Saturn	$1{\cdot}4 \times 10^{14}$	6×10^{29}	5?	3×10^{30}	30	0·5
Uranus	$2{\cdot}9 \times 10^{14}$	9×10^{28}	30	3×10^{30}	10	0·15
Neptune	$4{\cdot}5 \times 10^{14}$	1×10^{29}	30	3×10^{30}	6	0·1

the solar nebula that result from spreading the reconstituted masses uniformly over each planet zone about the sun. For the terrestrial zone, it was assumed that all silicates were retained but no ices or gases; for Uranus and Neptune, it was assumed that ices and silicates make up about half the total masses of these planets (Cameron, 1975). The greatest uncertainty in reconstituting planetary masses is with Jupiter and Saturn. Since these bodies are mostly H and He, the bulk properties of the planet are rather insensitive to the small mass of the heavy elements. Podolak and Cameron (1974,1975) suggest a mass of ices and silicates as large as ~ 20 per cent of the total planet, while Stevenson (this volume) favours a much lower value, perhaps not much above solar abundance (~ 2 per cent). We adopt an intermediate value of 6 per cent which is unlikely to be wrong by more than a factor of 3. It is noteworthy that the total mass in each planetary zone is nearly constant. The surface densities listed in the last columns of Table 2 are the values which will be used for numerical results in the following sections.

Having deduced the above characteristics of the solar nebula, it is logical to ask how far back in the planetary growth process might these values be trusted. Cameron (this volume) has given evidence that the solar nebula was once much more massive than the mass obtained above. If this were so, by what point in the growth of the planets must the removal of excess mass be complete?

The most frequently suggested mechanism of removal of excess mass is by sweeping outward by mass ejected from the sun during a 'T-Tauri phase' stellar evolution. The observations of T-Tauri stars which motivate this hypothesis are reviewed by Herbig (this volume). These observations indicate that the sun may have ejected a mass ΔM of the order of 10^{-1} $M_{\odot}$ over a period $\tau \sim 10^5$–10^6 years at a velocity $v \sim 100$ km/sec. Because of the greater density of this wind over the present solar wind, the particles are unionized. In order to blow a particle out directly, it is necessary that the wind force on the particle exceed the solar gravitational force. For the values of ΔM, v, and τ given above, this will be the case only for particles of radius $\lesssim 0{\cdot}01$ cm. Hence outward ejection can only occur for gas or fine dust. On the other hand, inward spiralling analogous to the Poynting–Robertson effect will remove bodies up to a fraction of a kilometre in size, and would therefore be a more effective mechanism of modifying the early particle disk.

Since the velocity of the T-Tauri wind is at most only a few times the solar escape velocity in the planetary zone, it can sweep out with it a mass of only a few times its own mass. Even the hydrogen envelope about the nebula is very flattened. In the earth zone the thickness of the gas phase of the nebula is ~ 0·1 AU. For an isotropic flow out from the sun, only ~ 0·05 of the wind is intercepted by the cloud in the terrestrial zone. Thus only a mass of ~ 10^{-2} $M_{\odot}$ could be removed from the terrestrial zone. From Table 2, it is apparent that this is about the amount of mass which must be removed from the terrestrial planet zone under the minimum mass nebula assumption.

In conclusion, it appears plausible that only very fine dust and gas up to the order of the mass remaining in the planetary system could be removed from the solar nebula by a T-Tauri wind. This loss must have been completed before a substantial fraction of the condensed matter had reached metre to kilometre size. In the outer planetary zone, it is possible that a somewhat greater mass loss occurred due to ejection of planetesimals by growing planets and by Jeans escape of the hydrogen from the solar nebula. The latter effect may explain the depletion of hydrogen in Uranus and Neptune.

Formation of the Particle Disk

As a particle forms, starting from a height z_0 above the central plane, it settles through the gas until it arrives on the central plane. If all smaller particles that it overtakes stick to it, its rate of growth is given by:

$$\mathrm{d}m = \pi r^2 \rho_s \, \mathrm{d}z, \tag{1}$$

where ρ_s is the density of solid particles in the cloud. For a particle starting at the half thickness of the ring, $z_0 = \sigma_s/2\rho_s$, equation (1) can be integrated to obtain the final radius of the particle:

$$r \sim \tfrac{1}{8}\frac{\sigma_s}{\rho_p}, \tag{2}$$

where σ_s is the surface density of condensed material (from Table 2) and ρ_p the density of the condensate. For the earth zone, $r \sim 0{\cdot}5$ cm and for the Jupiter zone, $r \sim 1$ cm.

The rate of settling of particles to the central plane is obtained by equating the drag force as the particle moves through the gas with the normal component of the solar gravity field:

$$\tfrac{4}{3}\pi r^2 \rho_g \dot{z} v_g \approx m\Omega^2 z, \tag{3}$$

where ρ_g is the density of the gas, v_g is the mean thermal velocity of the molecules, and Ω the orbit frequency about the sun. The characteristic settling time is

$$\tau \sim \frac{z}{\dot{z}} \approx \frac{2\sigma_g}{\pi\Omega\rho_p r}, \tag{4}$$

where σ_g is the surface density of the gas, from Table 2, and is related to the thermal velocity and 'in plane density' as follows (Safronov, 1972, p. 25):

$$\sigma_g = \frac{\pi}{2\Omega}\rho_g v_g. \tag{5}$$

For particle sizes of the order of 1 cm, τ is ~ 30 years for the earth and ~ 100 years for Jupiter.

For a more massive nebula, the particles will grow larger before reaching the central plane. The time of fall is nearly constant, since r is proportional to σ_g.

Inefficient sticking of particles will result in slower growth, and also longer settling time. Goldreich and Ward (1973) discuss the process of particle growth and settling in much greater detail.

Gravitational Instability

A uniform flattened disk of gas and/or solids in Keplerian orbit about the sun will be subject to self-gravity, which will tend to cause clumping of the matter. This tendency will be counteracted by centrifugal force due to the rotational motion of the mass, and gas pressure. A clump of matter of density ρ and radius r in orbit about the sun at radius R has the following 'binding energy':

$$E = E_{\text{gravity}} + E_{\text{rotation}} + E_{\text{thermal}}$$

$$\approx -\frac{16\pi^2}{15}\rho^2 G r^5 + \tfrac{4}{15}\pi \rho r^5 \frac{GM_\odot}{R^3} + \tfrac{4}{3}\pi r^3 \rho \frac{KT}{m_{H_2}}. \tag{6}$$

If the above energy is negative, collapse will occur, if not, any perturbation in density will be damped out. If the energy of rotation is assumed small, the other two terms reduce to the Jean's criterion for gravitational collapse (Jeans, 1928, p. 347). If the gas pressure term is neglected, we obtain the following criterion:

$$\rho_{CR} = \frac{1}{4\pi}\frac{M_\odot}{R^3}, \tag{7}$$

where ρ_{CR} is the density at which the total energy is zero, i.e., the critical density above which collapse occurs.

The above dimensional argument is substantially correct, but neglects some details of Keplerian motion. We wish now to set up the equations of motion in detail. Consider a coordinate system with its origin a distance a from the sun and moving around the sun at the Kepler frequency. The x axis points straight away from the sun, y in the direction of orbit motion, and z normal to the orbit plane. The motion of a particle in this coordinate system can be described by the following equations:

$$\ddot{x} - 2\Omega\dot{y} - 3\Omega^2 x = R, \tag{8}$$

$$\ddot{y} + 2\Omega\dot{x} = T, \tag{9}$$

$$\ddot{z} + \Omega^2 z = N, \tag{10}$$

where R, T, and N are the radial, tangential, and normal components of any perturbing force acting on the particle. The displacements x, y, z are assumed small compared to a. For the case of no perturbations, $R = T = N = 0$, these equations yield solutions of elliptic motion. Consider the case of a radial perturbation only, $T = N = 0$. Equation (9) can be substituted into (8) to obtain

$$\ddot{x} + \Omega^2 x = R. \tag{11}$$

If the radial perturbation is due to a displacement wave, $x = x_0 \sin(ka + \omega t)$, the gas pressure term of R is given by

$$R_p = c^2 \frac{\partial^2 x}{\partial a^2} = -c^2 k^2 x, \tag{12}$$

where c is the sound speed. If the wavelength of the perturbation is much less than the nebula thickness, then the gravitational force can be estimated as that due to an infinite sheet of mass in the y–z plane, displaced to the side of the equilibrium point. For a displacement x, an excess of mass of ρx exists in the direction of x, and a deficiency of mass of ρx exists on the opposite side. Hence the gravitational attraction of this mass displacement is

$$R_g = 4\pi G \rho x. \tag{13}$$

If the wavelength of the disturbance is very long, then the nebula can be considered as a thin sheet of surface density σ. The perturbation in density is

$$\Delta\sigma = \sigma\left(1 - \frac{\partial x}{\partial a}\right),$$

$$= \sigma k x_0 \sin(ka + \omega t). \tag{14}$$

The gravitational perturbation due to this density difference is:

$$R_g \approx \int_{-\infty}^{\infty} \int_{-\infty}^{\infty} \frac{G\Delta\sigma(a' - a)}{[y^2 + (a' - a)^2]^{3/2}} da'\, dy, \approx 2\pi G \sigma k x. \tag{15}$$

We have assumed here that curvature is small over a wavelength; that is, the wavelength is still short compared to the orbit dimension. The radial equation of motion (11) becomes

$$\ddot{x} + (\Omega^2 + c^2k^2 - 2\pi G\sigma k)x = 0, \qquad \text{(thin disk)} \tag{16}$$

$$\ddot{x} + (\Omega^2 + c^2k^2 - 4\pi G\rho)x = 0. \qquad \text{(thick disk)} \tag{17}$$

Instability occurs when the frequency of the perturbation becomes imaginary, hence exponential with time rather than sinusoidal:

$$\sigma_{CR} = \frac{1}{2\pi G k}(\Omega^2 + c^2 k^2), \qquad \text{(thin disk)} \tag{18}$$

$$\rho_{CR} = \frac{1}{4\pi G}(\Omega^2 + c^2 k^2). \qquad \text{(thick disk)} \tag{19}$$

Note that the two approximations yield the same solution when the equivalent thickness, $H = \sigma/\rho$, is equal to $2/k$. This corresponds to a wavelength of πH.

It is clear from expression (18) that σ_{CR} will have a minimum at some value of k. Unfortunately, this occurs at a value of k which falls intermediate between the two solutions above. Safronov (1972, p. 47) shows that the critical wavelength is approximately $8H$, and that at that wavelength, the critical density is

$$\rho_{CR} \gtrsim 2{\cdot}1\, \rho^*, \tag{20}$$

where ρ^* is the density the sun would have if spread uniformly over a sphere of radius equal to the orbit radius:

$$\rho^* = \frac{3M_{\odot}}{4\pi a^3}. \tag{21}$$

The conditions of gravitational instability for the earth zone and the Jupiter zone are listed in Table 3. The velocity tabulated is that characteristic of particles or molecules in a disk of thickness H, and the temperature listed is that of hydrogen gas with thermal velocity v. It is clear that gas must condense into solid particles before gravitational instability can occur. In the earth zone, solid particles must have their random motions damped to less than a few cm/sec before instability will occur. In the Jupiter zone, v must be less than $\sim 1/2$ m/sec. If gravitational instability occurred in the solar system, the resulting planetesimal condensations would be of the order of a few km in size in the earth zone, up to $\sim$ 100 km in the Jupiter zone. As we have pointed out, such large objects could not have been subsequently removed from the solar system, so one cannot assume gravitational instability occurred in a more massive nebula.

Table 3. Gravitational instability in the solar nebula (all units c.g.s.)

	Earth zone		Jupiter zone	
	Total mass	Solids only	Total mass	Solids only
$\rho_{CR} \approx \frac{1}{2}\frac{M}{R^3}$	3×10^{-7}	3×10^{-7}	2×10^{-9}	2×10^{-9}
$H_{CR} = \sigma/\rho_{CR}$	2×10^{10}	3×10^{7}	5×10^{11}	5×10^{9}
$\lambda \approx 8H_{CR}$	10^{10}	2×10^{8}	4×10^{12}	4×10^{10}
$v = \frac{2}{\pi}H_{CR}\Omega$	2×10^{3}	4	5×10^{3}	50
$T = \frac{m_{H_2}v^2}{3k}$	0·04 K	—	0·3 K	—
$m \approx 64 H_{CR}^2 \sigma$	—	5×10^{17}	—	2×10^{22}
$r = \left(\frac{m}{\frac{4}{3}\pi\rho_p}\right)^{1/3}$	—	4×10^{5}	—	10^{7}

Growth of Random Motions in the Planetesimal Swarm

A swarm of planetesimals in nearly circular orbits about the sun will move relative to one another with velocities which have a random component due to eccentricities and inclinations of the individual orbits, plus a systematic shear velocity due to the differential orbit velocity with radial distance. Energy will be transferred from differential motion to random motion through gravitational interactions between particles, which act like elastic collisions. The energy transferred from directed motion into random motion, per encounter, is

$$\Delta E = \tfrac{1}{2}\overline{\Delta v^2} = \tfrac{1}{2}\overline{\Delta R^2}\left(\frac{\mathrm{d}v}{\mathrm{d}R}\right)^2, \tag{22}$$

where $\overline{\Delta R^2}$ is the mean squared radial excursion between encounters and $(\mathrm{d}v/\mathrm{d}R)$ is the radial velocity gradient. If the mean time between encounters is long compared to the orbit period, the mean squared radial travel is given by

$$\overline{\Delta R^2} \approx \frac{1}{2\pi}\int_0^{2\pi} e^2 R^2 \cos^2\phi \,\mathrm{d}\phi = \tfrac{1}{2}e^2 R^2. \tag{23}$$

The appropriate value of $(\mathrm{d}v/\mathrm{d}R)$ is not obvious : it is $\frac{1}{2}\Omega$ with respect to inertial space, but $\frac{3}{2}\Omega$ with respect to rigid body rotation. One would expect the effective value to lie within these limits. Safronov (1972, p. 69) gives the expression for $\overline{\Delta v^2}$ as follows:

$$\overline{\Delta v^2} = \tfrac{9}{8}\beta'\Omega^2 e^2 R^2, \tag{24}$$

where β' is a dimensionless parameter expected to have a value between 1/9 and 1.

It is possible to estimate the value of β' from a more detailed examination of the interactions of particles in orbit (Safronov, 1972, p. 77). Consider a planetesimal in orbit (a, e) interacting with a massive body in a circular coplanar orbit R. The velocity of encounter is

$$v^2 \approx \Omega^2 e^2 R^2 (1 - \tfrac{3}{4}\cos^2\phi), \tag{25}$$

where ϕ is the angle around from perihelion of the encounter. We assume there are many available bodies with which the planetesimal may interact, hence ϕ may have any value with equal probability. The mean squared value of v is thus

$$\overline{v^2} = \frac{\Omega^2 e^2 R^2}{2\pi}\int_0^{2\pi} (1 - \tfrac{3}{4}\cos^2\phi)\,\mathrm{d}\phi, = \tfrac{5}{8}\Omega^2 e^2 R^2. \tag{26}$$

Consider now a body scattered at an angle ψ with respect to the circular orbit motion and at velocity v with respect to that motion. For the moment, assume ψ lies in the plane of the circular orbit. The new orbit (a', e') of the planetesimal is defined by

$$\frac{R}{a'} = 1 - 2\frac{v}{\Omega R}\cos\psi - \left(\frac{v}{\Omega R}\right)^2, \tag{27}$$

$$1 - e'^2 = \frac{R}{a'}\left(1 + \frac{v}{\Omega R}\cos\psi\right)^2. \tag{28}$$

R/a' can be eliminated from (28) to give an expression for e'^2, which can be averaged over all scattering angles ψ:

$$e'^2 = 1 - \frac{1}{2\pi}\int_0^{2\pi}\left[1 - 2\frac{v}{\Omega R}\cos\psi - \left(\frac{v}{\Omega R}\right)^2\right]\left[1 + \frac{v}{\Omega R}\cos\psi\right]^2 \mathrm{d}\psi,$$

$$\approx \tfrac{5}{2}\left(\frac{v}{\Omega R}\right)^2. \tag{29}$$

When the mean squared value of v is substituted from (26), we see that the new eccentricity is on the average, larger than the original eccentricity. The increase in e, per encounter is:

$$\overline{\Delta e^2} = \overline{e'^2} - e^2 = \tfrac{9}{16}e^2 \tag{30}$$

since v is proportional to e, we also have the result that $\overline{\Delta v^2} = \frac{9}{16}\overline{v^2}$. From (26) we can thus obtain an expression for $\overline{\Delta v^2}$:

$$\overline{\Delta v^2} = \frac{45}{128}\Omega^2 e^2 R^2. \tag{31}$$

This expression implies a value of β' in equation (24) of 5/16. Since on the average 1/3 of the increase in random motion goes into out-of-plane motion, which was not considered, and which does not contribute to an increase in radial excursions, we can take a value of $\beta' \approx 0{\cdot}2$ for the three dimensional case (Safronov, 1972, p. 80).

The above result indicates that random motions will grow among particles without bound. This of course is not correct, since unlike molecules, planetesimals are not perfectly elastic when subjected to physical collisions rather than just close approaches. Since Δv^2 per interaction is of the order of v^2 itself, equilibrium will be reached when the probability of interaction is of the same order as the probability of collision. The condition for 'encounter' is that the relative velocity vector v be re-directed by a large amount. The radius of strong interaction is thus

$$r_{\mathrm{i}} \approx \frac{2v_{\mathrm{e}}^2}{v^2}r, \tag{32}$$

where r and v_{e} are the radius and surface escape velocity of the planetesimal.

On the other hand, the collision radius is ($v \ll v_e$):

$$r_c \approx \frac{v_e}{v} r \tag{33}$$

From these two relations, it is clear that the relative probability of collision increases as v increases, becoming comparable when $v \sim v_e$. Safronov (1972, p. 69) has derived expressions for the equilibrium value of v including such factors as unequal particle masses, gas drag, non-central collisions, and the cumulative effect of many distant interactions. He finds that v is well expressed in terms of the surface orbit velocity of the largest planetesimal in the distribution:

$$v^2 = \frac{Gm}{\theta r} \tag{34}$$

θ is found to have a value of ~ 2–5 for most situations in the planetary growth process.

Goldreich and Ward (1973) have shown that for thermodynamic reasons, gravitational instability does not happen in a single step as suggested in the previous section, but rather results in clusters of thousands of bodies of the order of hundreds of metres in size, rather than single bodies of several kilometres size. The equilibrium value of v attained by such first generation planetesimals, from equation (34), is $\sim$ 10 cm/sec. This velocity dispersion, while small, would be sufficient to halt further gravitational instability. Since v is at all times less than v_e for the largest bodies, gravitational sticking will result in accretion rather than fragmentation of the largest bodies. Thus gravitational instability is not necessary once a few bodies emerge with gravitational binding energies large compared to the energy of random motion.

Size Distribution of Planetesimals

As a planetesimal suffers collisions with other bodies, it may grow larger due to accretion of smaller bodies, smaller due to erosion or fragmentation, or even be consumed by collision with a larger body. The relative probability of these occurrences depends on the size of the body, its material properties, the velocity of collision, and the size distribution of the other bodies involved. The coagulation equation which describes these processes has the following form (Zvyagina *et al.*, 1973):

$$\frac{dn(m,t)}{dt} = \boxed{(\tfrac{1}{2}\text{ the number of particles of mass } m' \text{ and } m - m' \text{ involved in collisions})\cdot(\text{probability of aggregation})}$$

$$+ \boxed{\text{number of fragments of mass } m \text{ resulting from collisions of larger bodies}}$$

$$- \boxed{\text{number of particles of mass } m \text{ involved in collisions with particles of any other size}}$$

$$\frac{dn(m,t)}{dt} = \frac{1}{2}\int_0^m A(m', m-m')n(m',t)n(m-m',t)w(m',m-m')dm'$$
$$+\int_m^\infty n_1(m,m'')\int_0^{m''/2} A(m',m''-m)n(m',t)n(m''-m',t)[1-w(m',m''-$$
$$m')]dm'dm'' - n(m,t)\int_0^\infty A(m,m')n(m',t)dm'. \tag{35}$$

In the above expression, $A(m_1, m_2)$ is the 'coagulation coefficient', which describes the probability of a collision occurring which results in either aggregation or fragmentation; $w(m_1, m_2)$ is the probability of aggregation rather than fragmentation, $n_1(m_1, m_2)$ is the number distribution function of fragments resulting from a collision between m_1 and m_2, and $n(m, t)$ is the mass distribution function. Pechernikova (1975) has reviewed various recent attempts at solutions of the coagulation equation. It is usually assumed that $n(m,t)$ will tend to a steady state as $t \to \infty$, of the form

$$n(m) \propto m^{-q}. \tag{36}$$

One then attempts to find values of q for which $n(m,t)$ does indeed tend to a steady state. Dohnanyi (1969) has shown that for fragmentation only ($w(m_1, m_2) = 0$), q tends to a value of $\sim 11/6$. Zvyagina and Safronov (1972) show that in the absence of fragmentation ($w(m_1, m_2) = 1$) and with a coagulation coefficient $A(m_1, m_2) \propto m_1 + m_2$, q tends to a value of 3/2. Solutions including both fragmentation and aggregation (Zvyagina *et al.*, 1974) appear to yield solutions with intermediate values of $q \sim 5/3$, but are necessarily restricted by the larger number of simplifying assumptions required. One of the most severe is the assumption of steady state: the obvious end state of accretion is one single body containing all the mass. It is not at all obvious that a steady state is even approached before all the smaller bodies are swept up by the growing planet. Since only asymptotic solutions can be treated analytically (Pechernikova, 1975), numerical analysis is necessary to study the time dependent case, and must consider both the growth of mass and of relative velocities between bodies simultaneously. Several efforts to solve this problem are underway in the U.S.A. and U.S.S.R.

While the problem of the mass distribution of planetesimals is still not completely solved, it appears evident that the value of q in equation (36) lies in the range 3/2 to 2. It is important to note that for $q < 2$, most of the mass is in the largest bodies of the distribution and for $q < 5/3$, most of the surface area is in the largest bodies. The former condition is almost certainly satisfied, hence elimination of the smallest members of the distribution has negligible effect on the total mass of planetesimals. The latter condition is probably not satisfied, suggesting a large optical depth of the solar nebula particle layer throughout much of the accretion process.

Near the large end of the size distribution, the power law solutions discussed above become invalid, due to the increased collision cross-section and decreased tendency to fragment among bodies with significant gravitational fields (Zvyagina and Safronov, 1972). Consider the simplest case, where v is given by (34) and all colliding bodies stick. The rate of growth is proportional to the collision cross section :

$$\dot{m} \propto \pi r^2 \left(1 + \frac{v_e^2}{v^2} \right). \tag{37}$$

For the largest body, m_1, v_e^2/v^2 is equal to 2θ, according to (34). For a smaller body the rate of growth relative to the largest body is:

$$\frac{\dot{m}/m}{\dot{m}_1/m_1} = \frac{r_1}{r} \frac{1 + 2\theta \left(\frac{r}{r_1} \right)^2}{1 + 2\theta}. \tag{38}$$

When the above expression has a value < 1 $(r_1/2\theta < r < r_1)$, the relative sizes of the two bodies will diverge. If the expression has a value > 1 $(r < r_1/2\theta)$, the relative sizes will converge with time. Thus the asymptotic state is that of a single body maintaining a lead over all others by a factor of 2θ in radius (Safronov, 1972, p. 106) :

$$r_1 = 2\theta r. \tag{39}$$

For typical values of θ, the above expression suggests a mass ratio of $\sim 10^2$–10^3 between the largest and next largest body in a planetary zone. Again, it is not clear to what degree the asymptotic state is reached before accumulation is complete. Since the power law distribution (36) would suggest a much larger 'second largest' body, several authors have suggested that the largest planetesimals involved in the growth process may have been considerably larger than suggested by (39) (Wetherill, 1976; Hartmann and Davis, 1975). The numerical studies of the coagulation equation now in progress should improve knowledge of the actual distribution of large bodies near the conclusion of accretion.

Growth of Planets

If the random velocity between planetesimals is slow compared to the escape velocity of the largest body (the 'planet embryo'), then that body will accumulate essentially all of the matter which it encounters. The rate of growth of that body is given by (Safronov, 1972, p. 109).

$$\dot{m}_p = \pi r_p^2 \rho_0 (1 + 2\theta) v, \tag{40}$$

where ρ_0 is the mean density of matter in the planet zone and v is the mean speed of the random motion of the planetesimals relative to the planet. The volume density, ρ_0, is related to the surface density σ_0 and mean velocity v as follows

(Safronov, 1972, p. 25).

$$\rho_0 = \frac{4\sigma_0}{vP}, \tag{41}$$

where P is the planet's orbit period. The characteristic time scale of planetary growth, $\tau = m_p/\dot{m}_p$, is thus

$$\tau = \frac{r_p \rho_p P}{3\sigma_0(1+2\theta)}, \tag{42}$$

where ρ_p is the mean density of the planet. Detailed solutions to (40) have been worked out for various models of ρ_0 and v throughout accumulation (Safronov, 1972, p. 109, Weidenshilling, 1974). The general nature of the solution can be inferred from (42). In relative terms ($\dot{m}_p/m_p$), the growth is fastest in the early stages and slows down as r increases and σ_0 decreases. The absolute rate of growth ($\dot{m}_p$) peaks at an intermediate point in the growth process (Weidenshilling, 1974). For Mercury, Venus, and the Earth, equation (42) gives time scales of the order of 10^7 years. For Mars, the time is 5×10^8 years assuming no mass was lost from the zone. However it is probable that the Mars zone had much more mass in it initially (Weidenshilling, 1975a), and hence the growth time may have been much less. For the outer planets, τ as given by (42) becomes rather long, up to $\sim 10^{10}$ years for Neptune (Safronov, 1972, p. 136). The growth of the outer planets is complicated by the loss of mass through gravitational ejection from the solar system, the high compressibility of gas leading to large changes in ρ_p over the growth process, and possibly hydrodynamic effects in the nebular gas (i.e. turbulence). All of these effects tend to hasten the accumulation process. Detailed models of major planet growth are lacking and would be very valuable in understanding the evolutionary history of those planets and their satellite systems.

The time scales of formation of the planets obtained from equation (40) could be too short, especially in the concluding stage of accretion, due to the finite amount of time required to perturb neighbouring planetesimals into crossing orbits. As pointed out in the above discussion, the equilibrium random velocity in the outer planet zone is sufficient to cause planetesimals to cross the orbits of all of those bodies. In the inner planet zone, Venus and the earth produce a dispersion in planetesimal velocity great enough to cause overlapping accretion zones among all four inner planets, for a value of $\theta \lesssim 6$. According to the Safronov relations, then, the validity of equation (40) is guaranteed if the relaxation time to the equilibrium velocity distribution is short compared to the accretion time scale (42) at all times. Safronov (1972, p. 72) gives the relaxation time for velocity dispersion as

$$\tau_g \sim \frac{(\rho_p/G)^{1/2}}{8\rho_0 \theta^{3/2} \ln(1 + D_0/2\theta r_p)}, \tag{43}$$

where D_0 is the mean distance between bodies. τ_g becomes longer as growth proceeds, reaching a value of $\sim 10^5$ years near the conclusion of accretion, in both the earth and Jupiter zones. Hence the rate of accretion is dominated at all times by the rate at which the planet sweeps volume, not by the rate of input of matter into that volume.

Scattering of Planetesimals by the Outer Planets

For all of the outer planets, the equilibrium value of v obtained from (34) for $\theta \approx 3$–5 far exceeds the escape velocity from the solar system at those distances. Weidenshilling (1975a, 1975b) has shown that as v increases to the point where escape becomes possible, the probability of ejection quickly exceeds that of collision. In computing θ, one must therefore include the loss of high velocity planetesimals by ejection as well as by collision. The calculation has not been done in detail, however the value of v at which ejection becomes possible is given simply by:

$$v_{\mathrm{ej}} = (\sqrt{2} - 1)\left(\frac{GM_{\odot}}{R}\right)^{1/2}. \tag{44}$$

The corresponding value of θ is

$$\theta_{\mathrm{ej}} = \frac{1}{3 - 2\sqrt{2}} \frac{m_{\mathrm{p}}}{M_{\odot}} \frac{R}{r_{\mathrm{p}}}. \tag{45}$$

θ_{ej} has a value between 40 and 80 for all of the outer planets near the end of accumulation. Safronov (1972, p. 141) suggests that the value of θ characterizing the mean velocity should be 2 to 3 times larger, or $\sim$ 100–200.

Since planetsimals are preferentially ejected in the prograde direction, the planet experiences a recoil resulting in a decrease in its orbit radius. If one assumes ejection occurs exactly in the direction of motion at exactly escape velocity, the rate of orbit shrinking is given by

$$\frac{\mathrm{d}R}{R} = -2(\sqrt{2} - 1)\frac{\mathrm{d}m}{m_{\mathrm{p}}}, \tag{46}$$

If a planetesimal is scattered exactly opposite from the direction of orbital motion at a velocity v_{ej}, it will follow an orbit with a perihelion of $\sim 0{\cdot}21R$. Since the planetesimal swarms surrounding the growing outer planets probably had a sharp cutoff in the velocity distribution at v_{ej}, each planet scattered planetesimals inward to $\sim 0{\cdot}21R$ and outward to escape (Weidenshilling, 1975b).

The following conclusions may be drawn from the above results: (a) Ejection of planetesimals was more efficient than accumulation near the end of the growth of the outer planets. The comets may be icy planetesimals so ejected. (b) The total mass ejected is limited by the recoil on the planets to a mass comparable to that remaining. (c) The orbit evolution of Neptune caused by the ejection of many small planetesimals could account for the formation of the

Neptune/Pluto orbital resonance. (d) Planetesimals may have been injected in large numbers by Jupiter into the asteroid belt and the Mars zone, but the number reaching as far in as the Earth was probably much less. These planetesimals may have been responsible for the failure of a large planet to form in the asteroid belt and may have similarly limited the growth of Mars.

The Origin of Axial Rotation of the Planets

The rates and directions of axial rotation of the planets show significant regularity, but do appear to have a large dispersion from the mean. One can therefore not hope to obtain a precise result from a theory of planetary rotation, since the variance is of the same order as the mean.

Several analytical and numerical theories of rotation have been advanced. Safronov (1971; 1972, p. 113) has reviewed earlier work. The most satisfactory theory is that of Giuli (1968a, 1968b) in which intersecting planetesimal orbits were numerically integrated to collision with the growing planet and the contribution to angular momentum evaluated. Harris (1977) has derived an approximate analytical theory based on the results of Giuli (1968a). He notes that only planetesimals in certain orbits contribute significantly to the net angular momentum. These orbits are those with aphelia or perihelia within one collision radius of the planet's orbit radius. Figure 1 illustrates the collision geometry in sun-centred and planet-centred coordinates. If one imagines a stream of particles colliding with the planet from such a tangent orbit, it is clear that over the entire range of possible collisions more particles contribute prograde angular momentum than contribute retrograde momentum. One can therefore estimate the angular momentum acquired by the planet with a

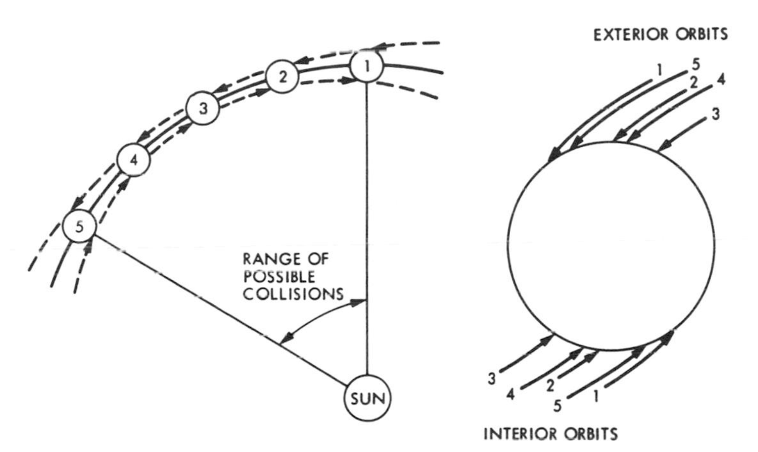

Figure 1. Collision geometry of streams of particles in exterior and interior grazing orbits in sun-centred and planet-centred coordinates. Note that both streams result in net prograde rotation. (Reproduced by permission of Academic Press, Inc.)

gain in mass of dm as follows:

$$\mathrm{d}h = \int_{V_{\min}}^{V_{\max}} n(v) f(v) \bar{l} v \,\mathrm{d}m \,\mathrm{d}v \tag{47}$$

where $n(v)$ is the relative density of planetesimals in the velocity range v to $v + \mathrm{d}v$, $f(v)$ is the fraction of planetesimals of velocity v which are in 'tangent' orbits, and $\bar{l}$ is the mean impact parameter averaged over the range of tangent orbits. For a Maxwellian distribution of planetesimal velocities, $n(v)$ is given by

$$n(v) \propto v \mathrm{e}^{-v^2/2v_0^2} \tag{48}$$

A factor $1/v$ arises from the effect of gravitational focusing. v_0^2 is the mean squared value of the velocity, without the effect of gravitational focusing.

The fraction of planetesimals arriving from 'tangent' orbits is obtained by evaluating the angle of intersection that a collision trajectory has when coming from an orbit with an aphelion or perihelion differing by one collision radius from the planet's orbit. If the velocity distribution is assumed isotropic, then $f(v)$ is found to be (two dimensional case):

$$f(v) \approx \frac{4}{\pi v} \sqrt{\Omega r v_{\mathrm{e}}} \tag{49}$$

where Ω is the planet orbit frequency, r the planet radius and v_{e} the surface escape velocity of the planet.

The mean value of the impact parameter can be estimated by assuming a uniform distribution of planetesimal orbits over the range of a necessary to span the range of 'tangent' orbits for a given value of eccentricity. The mean impact parameter is found to be a constant fraction of the collision radius, r_{c}, independent of v (Harris, 1977):

$$\bar{l} \approx 0{\cdot}143 r_{\mathrm{c}}. \tag{50}$$

Combining all of the above parameters into (47), and making use of the approximation to the collision radius $r_{\mathrm{c}} \approx r(v_{\mathrm{e}}/v)$, valid for $v \ll v_{\mathrm{e}}$, we obtain the following expression

$$\mathrm{d}h \approx 0{\cdot}143 \frac{4}{\pi} (\Omega r^3 v_{\mathrm{e}}^3)^{1/2} \mathrm{d}m \frac{\int_{v_{\min}}^{v_{\max}} \mathrm{e}^{-v^2/2v_0^2} \,\mathrm{d}v}{\int_{v_{\min}}^{v_{\max}} v \mathrm{e}^{-v^2/2v_0^2} \,\mathrm{d}v}. \tag{51}$$

The integral in the denominator is required for normalization since expression (48) is a proportionality rather than an equality. $v_{\max}$ is of the order of the escape velocity of a planetesimal from the solar system, and is in general $\gg v_0$. The upper limit of the integrals can therefore be taken to be ∞. The geometric pattern shown in Figure 1 above, and the approximate equations of motion used, are valid only if the radial excursion of the planetesimal in its orbit about the sun is large compared to the planet's collision radius:

$$ae > r_{\mathrm{c}}. \tag{52}$$

The eccentricity is related to the encounter velocity, for nearly tangential encounters:

$$v \approx \tfrac{1}{2} e\Omega R, \tag{53}$$

where R is the radius of the planet orbit. This condition implies a minimum value of the encounter velocity of

$$v_{\min} \sim \left(\tfrac{1}{2}\frac{m}{M_{\odot}}\frac{r}{R}\right)^{1/4} \Omega R \tag{54}$$

If we assume this velocity is small compared to v_0, then the lower limit of the integrals can be taken to be zero:

$$dh \approx 0{\cdot}143 \frac{4}{\sqrt{2\pi}} \frac{(\Omega r^3 v_e^3)^{1/2}}{v_0} dm. \tag{55}$$

If v_0 is taken to be proportional to the radius of the growing planet (34), and the mean density of the planet is assumed constant, then (55) can be integrated to obtain an expression for the rotation period, P:

$$P \approx 2{\cdot}2 \frac{I}{mr^2} \frac{v_0}{v_{\min}} T_0 \tag{56}$$

where I is the moment of inertia of the planet ($I = 0{\cdot}4\ mr^2$ for a homogeneous sphere) and T_0 is the surface orbit period about the planet. Since $v_{\min}$ is proportional to r, the ratio $v_0/v_{\min}$ should be approximately constant throughout the growth of the planet.

Note that the period of rotation is proportional to v_0. The slow rotations of Mars and Pluto may be the result of higher than normal values of v_0 induced by Jupiter and Neptune, respectively, in the planets' accretion zones.

Table 4 summarizes the results of this theory. The rotation rate for each planet is computed for a value of v_0 given by (34) with θ chosen appropriately

Table 4. Planetary rotation rates

	Period of rotation, days			
Planet	Observed	Calculated	(θ)	V_0/V_{Min}
Mercury	(direct)	8	(0·1?)	?
Venus	(retrograde)	2	(4)	?
Earth	~ 0·3	3	(4)	6
Mars	1·0	12	(0·5?)	18
Jupiter	0·4	2	(150)	4
Saturn	0·4	5	(100)	3
Uranus	0·4	8	(75)	4
Neptune	0·6	8	(150)	8
Pluto	6	120	(1?)	19

for each planet according to the estimates of Safronov (1972). The values of θ for Mercury and Pluto are controlled by the influence of neighbouring planets. For the large outer planets, θ is limited by escape from the solar system. Alternatively, the value of v_0/v_{min} is given for which equation (56) yields the correct value for P. It is noteworthy that v_0/v_{min} is remarkably constant for most planets, but differs by about an order of magnitude from the values suggested by Safronov's work. This discrepancy may be due to an error in the theory, a lower value of v_0 than given by Safronov (i.e., gas drag on planetesimals or gas accretion onto the outer planets), or by other sources of angular momentum (inward spiralling of satellite debris (discussed later in this chapter) is one such source). All of these possibilities should be considered further.

Planetary Axial Inclinations

As pointed out in the previous section, the rotation of planets is a phenomenon of small mean value and large variance. In that section, we computed the expected mean. If one assumes the observed variance ($\sim$ 50 per cent for most of the planets) is due to the finite number of discrete events which produced the rotational motion, it is possible to estimate the size of these events by the known variance. Safronov (1972, p. 129, or Safronov and Zvyagina, 1969) computes the 'largest body' impacting each of the major planets. This is a questionable computation since it requires the 'statistics of one' for a result. We shall here review his calculation, but only compute a single value of the 'second largest body' for all of the planets, based on the statistics of the six planets for which we know the initial spin state.

The mean squared increment of angular momentum acquired by the planet from a collision by a mass m is

$$\overline{h^2} = m^2 v^2 \overline{l^2}, \tag{57}$$

where v^2 is the mean squared velocity (34) and $\overline{l^2}$ is the mean squared impact parameter. For collisions uniformly distributed over the collision cross-section, $\overline{l^2} = \frac{1}{2} r_c^2 \approx \theta r^2$. Consider now the angular momentum acquired by all of the collisions making up an increment of mass dM where the individual masses follow a mass distribution given by (36) with a largest mass m_1:

$$n(m)\,\mathrm{d}m = \frac{2-q}{m_1^{2-q}}\,\mathrm{d}M\,m^{-q}\,\mathrm{d}m. \tag{58}$$

Since individual contributions to h add as their squares if they are randomly oriented, the squared increment of angular momentum is

$$\begin{aligned} \mathrm{d}\overline{H^2} &= \int_0^{m_1} \overline{h^2}\, n(m)\,\mathrm{d}m, \\ &= \frac{2-q}{3-q}\frac{m_1}{M}\omega_0^2 r^4 M\,\mathrm{d}M, \end{aligned} \tag{59}$$

where ω_0 is the surface orbit frequency about the planet. If we assume that as the planet grows, the sizes of the planetesimals grow in proportion, $m_1/M =$ constant, and that the mass law index q also remains constant, then the above relation can be integrated over the growth of the planet:

$$\overline{H^2} = \int_0^{m_p} \mathrm{d}\overline{H^2} = \frac{3}{10}\,\frac{2-q}{3-q}\,\frac{m_1}{M}\,\omega_0^2 r_p^4 m_p^2. \tag{60}$$

The out-of-plane component of a planet's rotation is presumably entirely due to the random component computed above. Since two of the three dimensions are out-of-plane, we have $\overline{H_{xy}^2} = \frac{2}{3}\overline{H^2}$. In terms of axial inclination, we have

$$\overline{\sin^2 i} = \frac{\overline{H_{xy}^2}}{H^2} = \frac{2}{3}\,\frac{\overline{H^2}}{I^2\omega_p^2}, \tag{61}$$

where I and ω_p are the moment of inertia and observed planetary rotation rate. Taking $\overline{H^2}$ from (60) and solving for m_1/M, we obtain

$$\frac{m_1}{M} = 5\,\frac{3-q}{2-q}\left(\frac{I}{m_p r_p^2}\,\frac{\omega_p \sin i}{\omega_0}\right)^2. \tag{62}$$

It should be noted again that ω_p sin i is an expectation value which can be estimated from a very few observations. One is tempted to consider each planet separately, since they may well have had differing values of m_1/M, yet one must be cautious of the statistical significance of a single observation point. Taken collectively, the inclinations of the planetary axes suggest a 'second largest body' of 10^{-2}–10^{-3} of the size of the planet for a value of $q \sim 3/2$–$5/6$. This is in good agreement with the prediction of (39) for a value of $\theta \sim 3$–5, but considerably larger than that obtained for a value of θ appropriate for the outer planets (45). This may be evidence that the asymptotic state described by (39) was not approached in the concluding stage of outer planet growth.

Origin of Satellite Systems

The common occurrence of satellites about planets requires a mode of origin of these bodies that is not only possible but probable, certainly in the case of the major planets. Ruskol (1960, 1963, 1972) has shown that in the course of accumulation of a planet from small planetesimals, a swarm of debris will accumulate in orbit about the planet, initiated by collisions between planetesimals within the sphere of influence of the planet and then accelerated by collisions of planetesimals with debris previously trapped in planetary orbit. In the course of the growth of the earth from bodies of hundred km size, she shows that it is reasonable to expect a mass as great as that of the moon to accumulate in orbit. Harris and Kaula (1975) have pointed out that due to the drag on orbiting particles of mass falling onto the planet, such a swarm would be continuously swept into the planet as it grew. They conclude that this was indeed the case in the outer planets, and hence the satellites of those planets constitute only a fraction of the total mass trapped which survived at the end of accretion. The earth, on the other hand, was able to preserve one of the satellites formed early in the accretion process through the counteraction of tidal friction. Nearly all of the trapped debris was therefore swept up by the moon embryo in the course of accretion of the earth, leading to a relatively large satellite-planet mass ratio.

The equations of growth and orbit evolution of a satellite about an accreting planet are (Harris and Kaula, 1975; Harris, 1978):

$$\frac{\mathrm{d}m_s}{\mathrm{d}m_p} \approx \frac{\left(1 + \frac{7}{3}\frac{\theta r_p}{R_s}\right)}{(1 + 2\theta)} \left(\frac{r_s}{r_p}\right)^2 + \beta\left[1 + \left(\frac{r_p}{r_p^*}\right)^2\right], \tag{63}$$

$$\frac{\mathrm{d}R_s}{\mathrm{d}m_p} \approx -\frac{8}{3}\frac{\left(1 + \frac{11}{5}\frac{\theta r_p}{R_s}\right)}{(1 + 2\theta)} \left(\frac{r_s}{r_p}\right)^2 \frac{R_s}{m_s} - \frac{R_s}{m_p}$$
$$+ \frac{3k_2}{Q}\left(\frac{G}{m_p}\right)^{1/2} \frac{m_s}{\dot{m}_p} \frac{r_p{}^5}{R_s^{11/2}}. \tag{64}$$

m_s and r_s are the mass and radius of the satellite, R_s is the satellite orbit radius, and r_p^* is the final planet radius at the conclusion of growth. Q and k_2 are the specific dissipation function and potential Love number of the planet respectively. The first term of (63) contains factors due to the relative effect of gravitational focusing at the satellite orbit radius and at the planet surface and the ratio of surface areas of satellite and planet. This term accounts for the rate of accretion of planetesimals directly by the satellite. The second term accounts for orbiting debris which spirals inward and is swept up by the growing satellite. The rate of trapping of matter is proportional to the square of the density of planetesimals, ρ_0, and to the volume of the planet's sphere of influence, which in turn is proportional to the planet's volume. The actual rate of accumulation is critically dependent on the size distribution of planetesimals. Ruskol (1960) derives the size distribution necessary to trap one lunar mass about the earth. While the result is reasonable, it is not possible to solve the problem in reverse since we do not know the actual size distribution of planetesimals well enough. The above proportionalities allow us to write the rate of accumulation in terms of the planet radius and a proportionality constant, β (Harris, 1978). We assume here that all of the trapped debris is swept up by the satellite. The three terms of equation (64) are due to (1) accretion drag, or mass gain by the satellite, (2) mass gain by the planet, and (3) tidal friction (Kaula, 1968, p. 202). Numerical integrations of these equations were attempted assuming an embryo satellite smaller than or roughly equal to the size of the present outer planet satellites, and Q/k_2 appropriate for the earth and for Jupiter (Goldreich and Soter, 1966). The results can be summarized as follows:

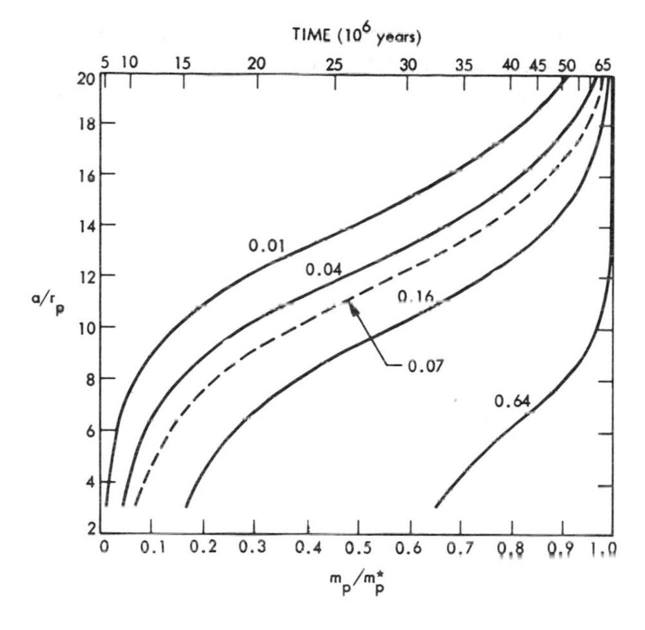

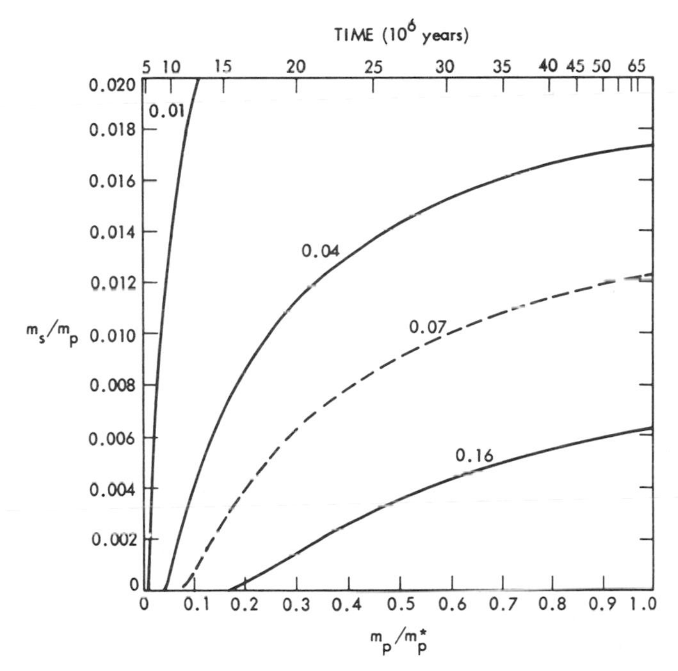

Figure 2. Satellite growth models computed from equations (63) and (64) with $\beta = 0$. The curves are identified by the mass of the earth at the time of introduction of the satellite embryo. The dashed curves (0·07) fit the actual moon/earth mass ratio

(1) For the outer planets, tidal friction was too weak to be of significance, even for an embryo the size of the present satellites. Without tidal friction, embryos quickly spiralled into the planet, indicating that the present satellite systems were formed only in the last few percent of the accumulation of their primaries.

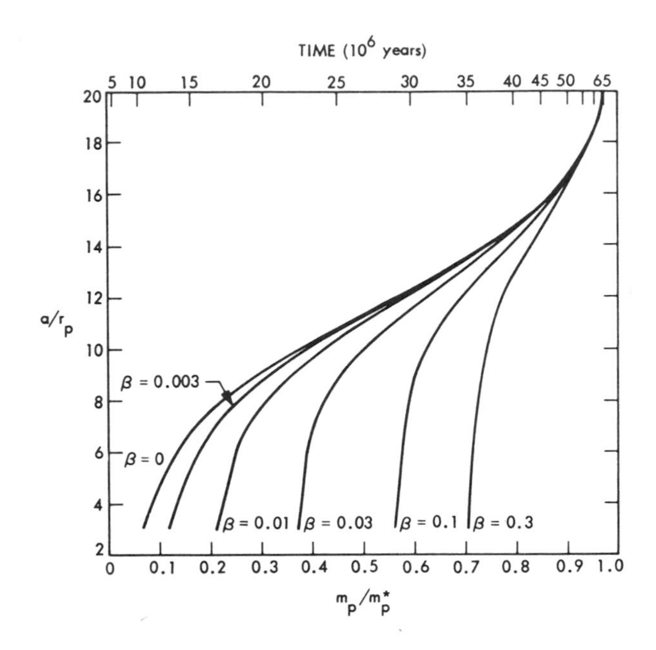

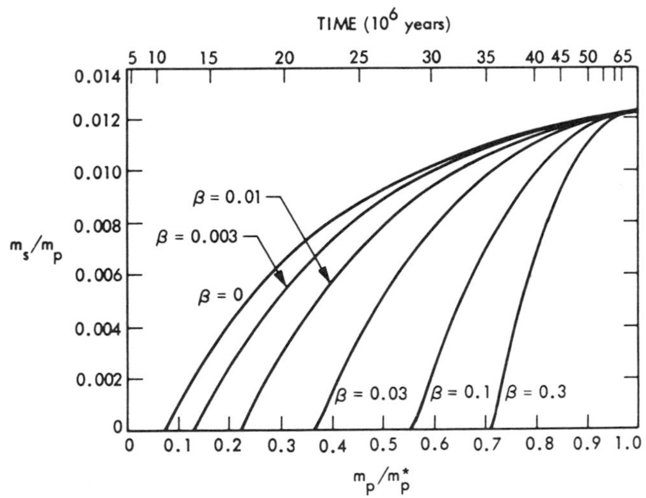

Figure 3. Satellite growth models including infall of debris from planetesimal–planetesimal collisions ($\beta \neq 0$). The value of β inferred from the outer planet satellite systems is $\sim 0{\cdot}03$

(2) For the Earth, tidal friction was sufficient to counteract accretion drag on an embryo as small as $\sim 10^{-5}\, m_p$, using a value of $Q/k_2 \sim 400$. Provided that a balance of these forces was achieved outside of the Roche limit, the subsequent growth history was only weakly affected by variations of Q/k_2, ρ_0, θ, or initial embryo size, by as much as an order of magnitude.

(3) A growth model of the Earth–Moon system assuming no contribution from ring debris ($\beta = 0$) would imply that the earth acquired the embryo moon when it was approximately Mars-sized (Figure 2). For non-zero values of β, the Moon could have started later (Figure 3). A value of $\beta \approx 0{\cdot}03$ is compatible with the amount of mass in the form of satellites orbiting about the outer planets (Harris, 1978). This value suggests that the lunar embryo began to grow when the Earth was about half accumulated.

Satellites of Mercury or Venus may have been removed by tidal friction (Burns, 1973, Ward and Reid, 1974), so the present lack of such satellites does not imply that none were formed in the course of accretion of those planets. Uncertainties in the time scale of accretion and the tidal effective Q/k_2 of Mars are such that one cannot be certain to what degree tidal friction should have been important in the growth of satellites about Mars. The lack of a large satellite about Mars is not inconsistent with present data.

Acknowledgement

This work is one phase of research carried out at the Jet Propulsion Laboratory, California Institute of Technology, under contract No. NAS 7–100, sponsored by the National Aeronautics and Space Administration.

References

Burns, J. A. (1973). 'Where are the satellites of the inner planets?' *Nature*, **242**, 23–25.

Cameron, A. G. W. (1975). 'Cosmogonical considerations regarding Uranus.' *Icarus*, **24**, 280–284.

Dohnanyi, J. S. (1969). 'Collisional model of asteroids and their debris.' *JGR*, **74**, 2531–2554.

Giuli, R. T. (1968a). 'On the rotation of the earth produced by gravitational accretion of particles.' *Icarus*, **8**, 301–323.

Giuli, R. T. (1968b). 'Gravitational accretion of small masses attracted from large distances as a mechanism for planetary rotation.' *Icarus*, **9**, 186–190.

Goldreich, P., and Soter, S. (1966). 'Q in the solar system.' *Icarus*, **5**, 375–389.

Goldreich, P., and Ward W. R. (1973). 'The formation of planetesimals.' *Ap. J.*, **183**, 1051–1061.

Harris, A. W. (1977). 'An analytical theory for the origin of planetary rotation.' *Icarus*, **31**, 168–174.

Harris, A. W. (1978). 'Satellite Formation II.' *Icarus*, in press.

Harris, A. W., and Kaula, W. M. (1975). 'A co-accretional model of satellite formation.' *Icarus*, **24**, 516–524.

Hartmann, W. K., and Davis, D. R. (1975). 'Satellite-sized planetesimals and lunar origin'. *Icarus*, **24**, 504–515.

Hoyle, F. (1946). 'On the condensation of the planets.' *Mon. Not. R. astr. Soc.*, **106**, 406–422.

Jeans, Sir James (1928). *Astronomy and Cosmogony*. Dover (1961). New York.

Kaula, W. M. (1968). *An Introduction to Planetary Physics: The Terrestrial Planets.* Wiley, New York.

Pechernikova, G. V. (1975). 'Mass distribution of protoplanetary bodies.' *Soviet Astron. A. J.*, **18**, 778–783.

Podolak, M., and Cameron, A. G. W. (1974). 'Models of the giant planets.' *Icarus*, **22**, 123–148.

Podolak, M., and Cameron, A. G. W. (1975). 'Further investigations of Jupiter models.' *Icarus*, **25**, 627–634.

Ruskol, E. L. (1960). 'The origin of the Moon I. Formation of a swarm of bodies around the earth.' *Sov. Astron. A. J.*, **4**, 690–702.

Ruskol, E. L. (1963). 'Origin of the Moon II. The growth of the moon in the circumterrestrial swarm of satellites.' *Sov. Astron. A. J.*, **7**, 221–227.

Ruskol, E. L. (1972). 'Origin of the Moon III. Some aspects of the dynamics of the circumterrestrial swarm.' *Sov. Astron. A. J.*, **15**, 646–654.

Safronov, V. S. (1971). 'Rotation of Giant Planets while accreting gas.' *Solar System Astronomy*, **5**, 139–144.

Safronov, V. S. (1972). 'Evolution of the protoplanetary cloud and formation of the earth and planets.' *Israel Program for Scientific Translations*, Jerusalem.

Safronov, V. S., and Zvyagina, E. V. (1969). 'Relative sizes of largest bodies during the accumulation of the planets.' *Icarus*, **10**, 109–115.

Ward, W. R., and Reid, M. J. (1973). 'Solar tidal friction and satellite loss., *Mon. Not. R. astr. Soc.*, **164**, 21–32.

Weidenshilling, S. J. (1974). 'A model for accretion of the terrestrial planets.' *Icarus*, **22**, 426–435.

Weidenshilling, S. J. (1975a). 'Mass loss from the region of Mars and the asteroid belt.' *Icarus*, **26**, 361–366.

Weidenshilling, S. J. (1975b). 'Close encounters of small bodies and planets.' *Astron. J.*, **80**, 145–153.

Wetherill, G. W. (1976). 'The role of large impacts in the formation of the earth and moon.' *Abstracts of the 7th Lunar Sci. Conference*, Lunar Science Institute, Houston, Texas.

Zvyagina, E. V., and Safronov, V. S. (1972). 'Mass distribution of protoplanetary bodies.' *Soviet Astron., A. J.*, **15**, 810–817.

Zvyagina, E. V., Pechernikova, G. V., and Safronov, V. S. (1974). 'Qualitative solution of the coagulation equation with allowance for fragmentation.' *Soviet Astron–A. J.*, **17**, 793–800.

ASPECTS OF ACCRETION IN THE EARLY SOLAR SYSTEM

J. F. KERRIDGE
Institute of Geophysics, University of California, Los Angeles, California, U.S.A.

Abstract

If formation of planetesimals took place by particle–particle interactions, rather than within gravitational instabilities, these interactions must have occurred at low relative velocities, leading to rates of growth which were two to three orders of magnitude slower than those characteristic of accretion within gravitational instabilities. Sticking efficiencies during particle–particle interactions are problematical, although early stages of growth were probably dominated by contact forces. Differences in accretion efficiency of metal and silicate particles probably led to the metal-silicate fractionation recorded in the composition of chondritic meteorites. It is likely that chondrules were made by particle–particle collisions prior to final accretion of planetesimals. This would have required a period of high relative velocities in an epoch otherwise characterized by low inter-particle velocities.

Introduction

Accretion is a term which can take on different meanings in different contexts. Here I define it as the growth of planetesimals by coalescence of smaller particles, initially those formed by direct condensation from the nebular gas. Equivalent terms in the literature include accumulation, aggregation, agglomeration, etc. This definition contains a number of implicit assumptions, most notably the concept of an initially gaseous nebula as a precursor to formation of the planetary system. Such a concept is by no means universally accepted. Cameron (1975 and this volume) has emphasized the role of clumping of interstellar grains prior to incorporation into the proto-solar system, and Alfvén and Arrhenius (1974) have proposed that gas and dust were added slowly to the circumsolar region over an extended period of time so that a nebula, as conventionally understood, never formed. These models, and many others, several of which are described in this volume, all involve coalescence of solid material into planetary objects but a review of the accretionary process under so many different, and frequently conflicting, circumstances is scarcely feasible. In addition, several chapters in this volume deal explicitly with specific versions of this question and, in particular, Harris gives a rigorous treatment of planetesimal formation, including the later stages involving gravitational interactions, not considered here.

The thrust of this chapter, therefore, is to focus on those aspects of the accretionary process, or processes, which tend to be neglected by modelists but which may nonetheless be instructive concerning the evolution of the early solar system. I have chosen the following topics for consideration. First, although formation of planetesimals was probably greatly accelerated by gravitational instabilities within the nebular dust (Goldreich and Ward, 1973), it is of interest to consider how rapidly planetesimals would have accreted in a 'classical' nebula, i.e. one in which such instabilities failed to develop, for example, because of turbulence. Second, a perennial problem in this area concerns sticking efficiencies during particle–particle interactions in the nebula, even though the effectiveness of gravitational instabilities in promoting accretion may have reduced the importance of such interactions. Third and fourth, at least two processes which apparently occurred at about the time that planetesimals were forming, are recorded in the meteorites. They are metal–silicate fractionation, also recorded by differences in planetary densities, and chondrule formation, and it is pertinent to ask what such observational evidence can tell us about conditions in the nebula during the accretion stage.

Accretion of Planetesimals: The Classical Approach

The role of gravitational instabilities in accelerating growth of protoplanetary objects within a dust-laden nebular disk, considered earlier by Kuiper (1951) and Urey (1966), was rigorously examined by Goldreich and Ward (1973) and shown to be effective in producing planetesimals with radii of a few kilometres a few thousand years after condensation began. It is of interest to compare this result with time scales calculated for growth of comparable objects by particle–particle interactions without the aid of gravitational instabilities. Such a situation might have pertained if turbulence inhibited development of local instabilities, as pointed out by Cameron and Harris during this meeting.

Following Hartmann (1970), consider a nebula containing about 0·2 $M_{\odot}$, emplaced in the circumsolar region during collapse of the protosolar fragment and heated to > 2000 K by release of gravitational energy during that collapse.

The mass of the nebula is calculated from the present masses of the planets with allowance for the volatile complements lost from most of the planets (Kuiper, 1956) and including an estimate of mass lost during a T-Tauri phase early in the evolution of the sun (Kuhi, 1964; Hartmann, 1970). For a nebula approximately 0·1 AU thick, this corresponds to a density of about 10^{-7} g/cm^3 (about 10^{-3} atm) in the plane of the ecliptic at about 1 AU from the sun. Such numbers should probably be regarded as illustrative rather than as firm estimates.

Condensation of the chemical elements from such a nebula has been considered by Lord (1965), Larimer (1967), and Grossman (1972, 1976). For our purposes three stages of condensation are of primary importance; first, formation of highly refractory compounds, typically containing calcium, aluminium, and titanium and comprising a small fraction of the nebular mass, above about 1500 K; second, condensation of the main mass of rocky material, as iron metal and magnesium silicate, at about 1400 K; third, formation of ices below about 300 K. Material from the first stage of condensation has apparently survived in some meteorites (Marvin *et al.*, 1970; Grossman, 1972). It is unlikely that the third stage operated to a significant extent in the vicinity of the earth's orbit (Laul *et al.*, 1972; Lewis, 1972 a, b). A quantitative treatment of nucleation and growth of these materials in the nebula requires knowledge of a number of factors which are imprecisely known at present. These include the number density of nucleation sites, whether nucleation was homogeneous or heterogeneous, the distribution of temperature as a function of time and space within the nebula, and the evolution of the inter-particle velocity spectrum.

Although the probable nucleation frequency is unknown, realistic limiting cases may be considered. Hartmann (1970) showed that if nucleation of iron grains occurred *via* formation of polyatomic molecules with a high (0·5) probability of sticking, a situation resulting in the maximum number density of nuclei, effectively all iron would have been in the form of grains within a few minutes. (It is generally agreed that iron condensed before magnesium silicate within the region of the terrestrial planets and was therefore the first major phase to condense.) The possible role of more refractory material, either earlier condensate (Grossman, 1972) or unvaporised, pre-solar system grains (Clayton *et al.*, 1973), in promoting metal or silicate condensation has not generally been considered.

A minimum density case was treated by Goldreich and Ward (1973) who calculated the maximum size that an iron grain could reach by growth from the vapour while falling under gravity from the edge of the nebular disc to the midplane. This led to a maximum radius of about 3 cm, achieved in about 10 years.

Whether magnesium silicate condensed onto earlier formed iron grains (heterogeneous nucleation) or whether metal and silicate formed separate populations (homogeneous nucleation) is not known. Hartmann (1970) assumed the former and suggested that iron grains rapidly acquired silicate coatings when they had achieved radii of about 10^{-5} cm. Larimer and Anders (1967), on the other hand, assumed homogeneous nucleation and were forced to postulate an *ad hoc* scenario for production of composite metal-plus-silicate grains in order to explain the subsequent introduction of ferrous iron into silicates. The observational evidence for non-igneous metal–silicate fractionation, discussed later, suggests that the two phases were distinct, i.e. that nucleation was homogeneous. Further support for this view may be provided by the refractory trace element abundances measured by Ganapathy and Grossman (1976) in apparently early condensate material surviving in meteorites. They found that siderophile elements correlated with each other, as did lithophile elements, but that the two groups did not inter-correlate, suggesting that the initial condensate was divided into metallic and non-metallic populations. However, the presence within some of these meteoritic inclusions of metal particles embedded within silicate grains (Wark and Lovering, 1976) argues against this simple picture. The possibility must also be borne in mind that at least part of this material may have been of pre-solar system origin.

As mentioned earlier, the classical approach assumes that condensation occurred during cooling of an initially hot nebula. Thus, Hartmann (1970) uses a cooling rate due to Cameron (1962) in which the temperature varied as $t^{-1/3}$. More recent calculations (e.g. Cameron and Pine, 1973) suggest that nebular cooling rates are highly uncertain, and may even be meaningless, in light of comments by Cameron at this meeting. Given, however, the assumption of a cooling nebula, heat would have been lost from the surface of the disc by radiation, resulting in a meridional temperature gradient and preferential condensation close to the surface of the disk. Such condensate particles would have experienced a gravitational acceleration towards the midplane (Goldreich and Ward, 1973), their resulting velocities being controlled by aerodynamic forces and by the fact that their masses were increasing with time as a result of accretion.

Lewis (1972b) has suggested that a radial temperature gradient may have led to a compositional gradient among condensed material by freezing in an equilibrium condensation sequence at different temperatures as a function of heliocentric distance. It should be noted that until accretion had led to grain sizes of centimetre to metre dimensions, the opacity of the nebula would have greatly modified the influence of solar radiation on the nebular temperature profile at planetary distances (Hartmann, 1970).

While residual gas remained in the vicinity of the condensate particles, its presence governed the magnitude of the inter-particle velocity spectrum and hence the frequency with which grains could interact with each other and coalesce. An upper limit to this velocity was established by a balance between external accelerating forces, such as gravitation, and gas drag, for which Stoke's Law was probably appropriate (Hartmann, 1970; Whipple, 1972). Hartmann (1970) showed that for a grain of radius r cm at 1 AU from the sun and falling radially under solar gravity, its Stoke's Law limiting velocity would have been given by

$$v_s = 4 \times 10^4 r^2 \text{ cm/sec}.$$

A lower limit to inter-particle velocities would have been set by their Brownian motion due to thermal velocities within the gas. This would have been given by

$$v_b = \sqrt{(9kT/4\pi\rho r^3)}$$

where ρ is the density of the condensed material. Evolution of these velocity

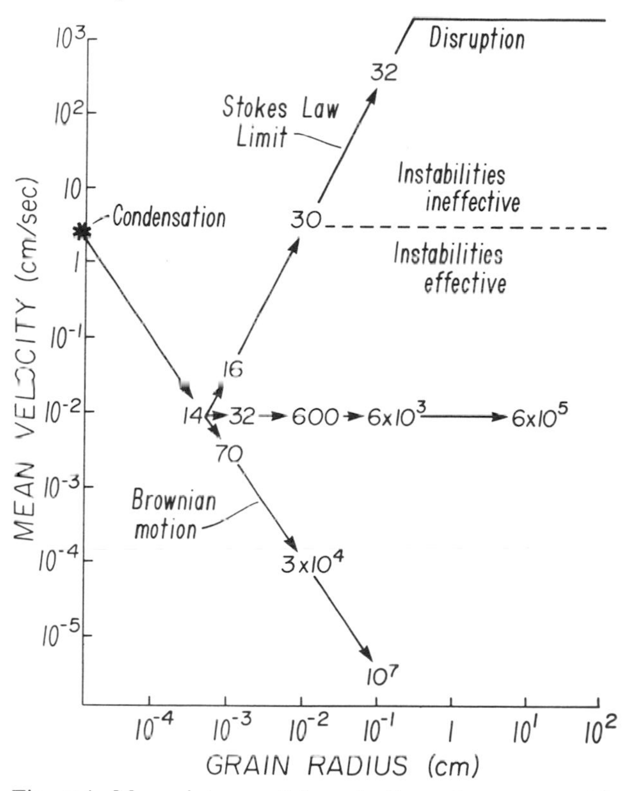

Figure 1. Mean inter-particle velocity and mean particle size in the nebula were inter-related, as shown here. The curve labelled 'Brownian motion' gives the r.m.s. velocity acquired by a grain from thermal motions in the gas. 'Stoke's Law Limit' gives the maximum velocity which a grain could have acquired relative to the gas under the influence of solar gravity. These curves represent limits to inter-particle velocities which in turn controlled rate of accretionary growth. Thus, evolutionary tracks may be drawn on this diagram, such as that corresponding to a constant velocity of 10^{-2} cm/sec, or the limiting curves themselves. Numbers at intervals along each track give the time in years taken to reach that point along the appropriate trajectory, starting with a particle radius of 10^{-3} cm, reached in about 14 years after condensation, assuming 0·5 sticking probability. If velocities lay below the dashed line, it is likely that gravitational instabilities dominated the accretion process

limits as functions of particle size is illustrated in Figure 1. The crucial role of inter-particle velocity in controlling rate of accretion may be demonstrated by the growth equation, given by Hartmann (1970), following work by Wickramasinghe (1967) and Greenberg (1968) as

$$\mathrm{d}r/\mathrm{d}t = f\rho_a v/\rho$$

where f is the sticking probability and ρ_a is the density of accretable material in the nebula.

Hartmann (1970) adopted a value of 1 cm/sec as an order of magnitude estimate for the effective average velocity during growth through the sub-kilometre size regime and showed that after about 200 years grains would have grown to about 2 mm radius and the temperature would have reached the condensation point for ice. This would have 'instantaneously' increased grain radii to about 3 mm, after which accretion would have continued until 10 km radii would have been achieved in about 10^7 years. This may be compared with a time scale of the order of 10^4 years for growth to the same size *via* gravitational instabilities (Goldreich and Ward, 1973). Figure 1 indicates the velocity regime within which such instabilities could have formed. If velocities were characteristically higher than this, accretion would probably have taken place *via* particle–particle interactions along the lines indicated above and on somewhat shorter time scales than estimated by Hartmann. However, the sticking probabilities at such velocities are problematical and, even assuming efficient coalescence, time scales still appear to be of the order of 10^6 years.

For as long as gas drag remained an important factor, the dependence of limiting velocity upon particle size would have resulted in large grains growing faster than smaller ones. Thus, any size distribution formed stochastically, for example as a result of the Maxwellian distribution of thermal velocities, would have been amplified during subsequent accretion. Thus, concepts such as average particle size may not be meaningful.

Sticking Efficiencies

Obviously, for particle–particle interactions to have been effective in accretion, such interactions must have resulted in net increase in particle size. This required some form of adhesion between particles with a strength adequate to overcome the disruptive effect of the particles' kinetic energy. It is not immediately obvious that condensate particles would have possessed such adhesive properties and numerous mechanisms have been advanced to account for the fact that sticking apparently did occur.

Welding

Although laboratory simulations employing microparticle accelerators have revealed that significant quantities of projectile material remain adhering to the target following impacts between metals in the velocity regime 0·5 to 13 km/sec (Neukum, 1968; Dietzel *et al.*, 1972), corresponding experiments using silicates showed no accretion at any velocity within experimental limits of 1·5 to 9·5 km/sec (Kerridge and Vedder, 1972). Thus, welding of particles during high velocity impact probably did not contribute to the main stream of planetary accretion, although preferential aggregation of metallic iron by this mechanism may possibly have resulted in some fractionation of metal from silicate.

As pointed out by Reeves (1972) in a perceptive review of this subject, cold welding has been the traditional *deus ex machina* of theoreticians anxious for their models of the early solar system to result in formation of planets. Unfortunately, any kind of empirical basis for this belief has been lacking so far, although possibly the mechanism suggested by Arnold and discussed below may qualify under this heading.

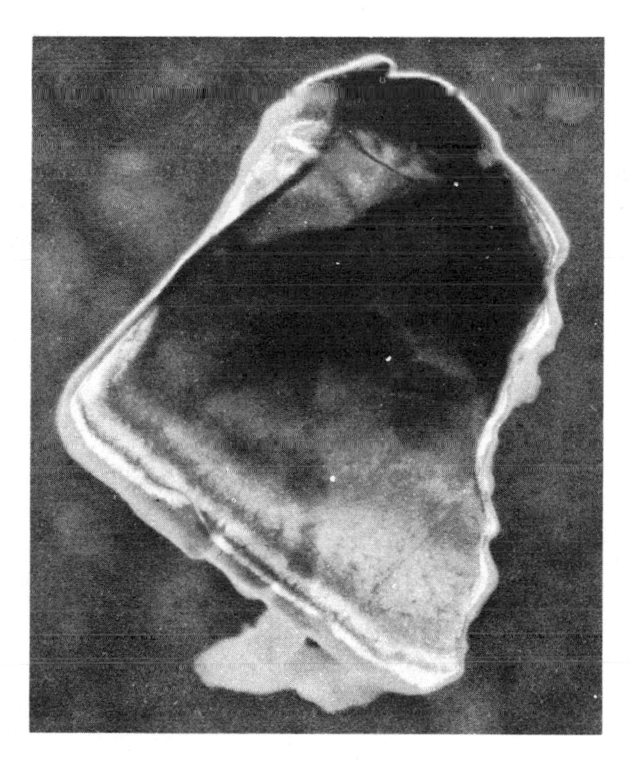

Figure 2. Dark-field electron micrograph of olivine grain from lunar soil 15031, showing amorphous rim formed by solar wind-induced radiation damage. Bibring and Maurette (1972) have demonstrated that such damage promotes adhesion between grains. Photographed by the author on a Philips EM-300 at 100 kV. Diameter of grain about 1 μm

Release of Stored Energy

Bibring and Maurette (1972) have presented experimental evidence which shows that lunar soil grains tend to bond more firmly to each other under the influence of moderate hydrostatic pressure than do mineralogically equivalent grains of terrestrial origin. They attributed this adhesion to release of stored energy from the amorphous rims which the same group (Dran *et al.*, 1970) had demonstrated were formed on lunar soil grains by irradiation damage produced by the solar wind, see Figure 2. Similar irradiation of condensate grains in the nebula, possibly by a T-Tauri solar wind, could have enhanced their tendency to coalesce. However, the opacity of the nebula would have been so great that only grains at the surface of the disc could have been influenced by a solar wind and their exposure time in such an environment would have been very brief.

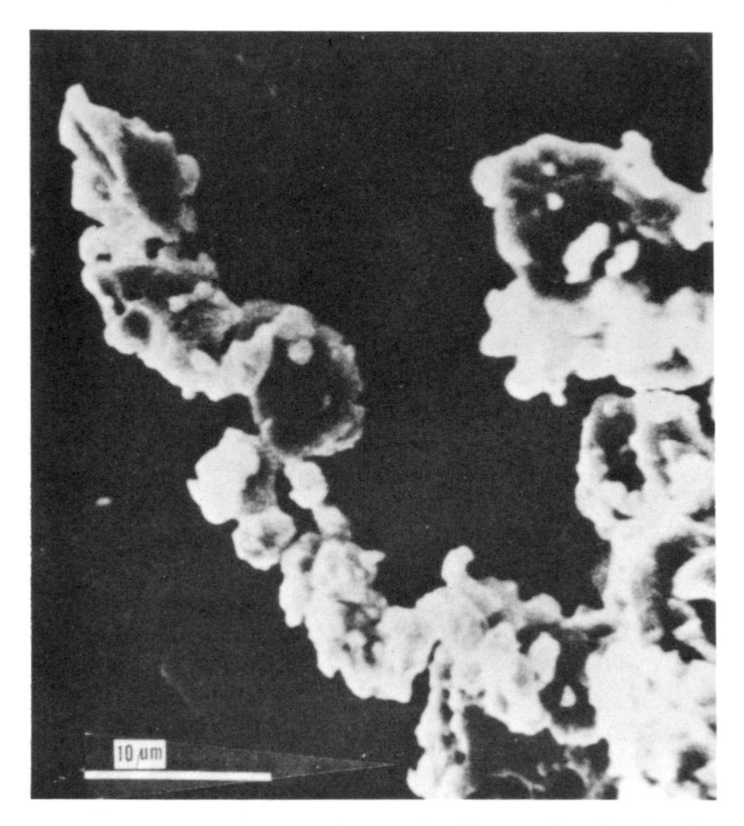

Figure 3. Scanning electron micrograph of lunar soil grains showing clustering to form a flexible chain extending about 40 μm from the surface of a glass sphere. The chain structure indicates that the individual microparticles are electric dipoles. (Reproduced with permission from G. Arrhenius and S. K. Asunmaa, 1973, *The Moon*, **8**, 368–391, copyrighted by D. Reidel Publishing Company, Holland)

Electrostatic Polarization

Arrhenius and Asunmaa (1973) have shown that many lunar soil grains retain a persistent dipolar electrostatic charge, apparently generated on the lunar surface, which can result in preferential aggregation of these grains, often into chains and other open structures, Figure 3. They showed that terrestrial mineral grains did not exhibit the same behaviour, Figure 4. They proposed that irradiation on the lunar surface was responsible for this electrostatic charge and that analogous irradiation in the early Solar System could have promoted accretion of grains. However, this mechanism suffers from the same drawback as that of Bibring and Maurette and it seems unlikely that the primitive condensate could have received sufficient irradiation for this effect to have been important.

Contact Forces

Arnold (1976) has considered the energy balance involved in interactions between micron-sized particles in the nebula. Similar considerations have been applied to terrestrial aerosols (see, for example, Fuchs, 1964). Arnold showed that the attractive energy due to contact forces such as Van der Waals or chemical bond formation would have been substantial if significant, but not unreasonable, proportions of their surface areas were in contact. The only energy

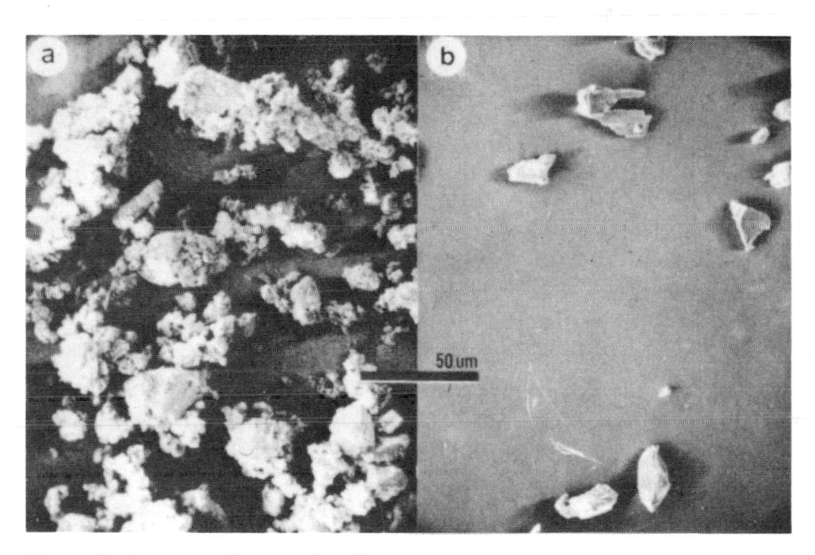

Figure 4. Lunar grains remain adhered on a fresh cleavage surface of phlogopite mica when the mica support is turned face down (left), whereas terrestrial pyroxene grains with a few exceptions fall off from the mica cleavage surface (right). Scanning electron micrograph, (Reproduced with permission from G. Arrhenius and S. K. Asunmaa, 1973, *The Moon*, **8**, 368–391, copyrighted by D. Reidel Publishing Company, Holland)

capable of inhibiting such contact forces was the kinetic energy of the particles and this would have been less than the attractive energy for relative velocities below about 1 m/sec. Velocities of this magnitude could not have been achieved by such small particles under gravitational forces because of gas drag, and sufficiently high electrostatic accelerations would have required unreasonable values for particle charge and/or electric field gradients. Electrostatic forces between the particles themselves, although capable of acting over much greater distances than contact forces and potentially as likely to be repulsive as attractive, would not have seriously inhibited accretion because the height of the electrostatic charge barrier would have been of order of kT, readily overcome during interactions. Also, particle charges would have fluctuated with time in both magnitude and sign and an aggregate of grains, once accreted, would have been very stable against breakup by electrostatic repulsive forces. Thus, accretion by contact forces would have been effectively irreversible and all forces were net attractive, either at one time or over the long run, so that the initial stages of accretion were probably efficient, albeit slow because of the low velocities involved.

Gravitational Instabilities

Before leaving the topic of sticking efficiency it should be noted that as pointed out by Goldreich and Ward (1973), accretion of planetesimals within gravitational instabilities would have rendered unnecessary any intrinsic stickiness of the grains themselves, in addition to accomplishing the task of accretion in less time.

Metal–Silicate Fractionation

The results of at least four chemical fractionation processes are recorded in the composition and mineralogy of chondritic meteorites. Three of these appear to be based upon differences in volatility, i.e. to result from separation of gaseous and condensed phases from each other. Although instructive concerning conditions in the early solar system, we shall not consider them here but will focus upon a fractionation which apparently occurred between two condensed phases, namely metallic iron and magnesium silicate. It is likely that this separation took place during accretion and that the causative mechanism was probably based upon the nature of the accretion process itself.

An inter-relationship between metallic and oxidized iron in chondrites was recognized by Prior (1916) and shown by Urey and Craig (1953) to be due to variation in both total iron content and its state of oxidation. From distributions of siderophile elements, such as nickel and cobalt, Urey and Craig concluded that variations in total iron were due to gain or loss of metal and they showed further that these properties were discontinuous so that bulk analyses of ordinary chondrites fell into two distinct compositional fields, termed High Iron (H) and Low Iron (L). A third group, Low Iron, Low Metal (LL), was added later by Keil and Fredriksson (1964). These groups are illustrated in Figure 5 in which the metal contents of chondrites are plotted against their contents of oxidized iron, after the manner of Urey and Craig. The correlation between iron and nickel, normalized to magnesium content, is shown in Figure 6, after Larimer and Anders (1970) who suggested that the different chondrite groups were related by fractionation of metal, containing 6·6 wt per cent nickel, with respect to silicate, characterized by an iron/magnesium ratio of 0·14. This fractionation probably reflects loss of metal, as the data in Figures 5 and 6 correspond to proportions of iron which are less than or equal to the generally accepted solar system value (Anders, 1964, 1971). The starting material is identified with material chemically equivalent to the most primitive carbonaceous chondrites, though in a more reduced state than presently observed (Larimer and Anders, 1970).

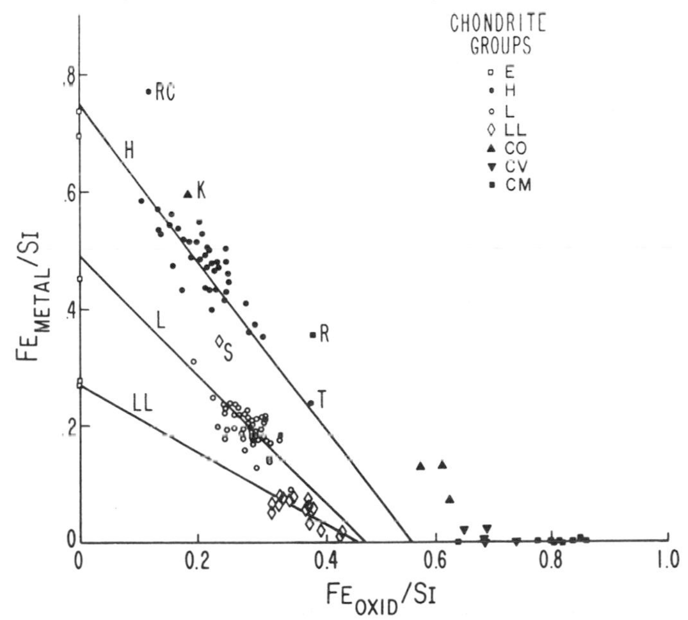

Figure 5. Bulk analyses of ordinary chondrites fall into three distinct compositional groups when plotted in terms of metallic iron against oxidised iron, both parameters normalized to silicon content. These groups can only be inter-related by gain or loss of iron in addition to change in oxidation state, as shown by Urey and Craig (1953), who also demonstrated that this was due to metal–silicate fractionation. Least squares regression lines are drawn through the data for each group of ordinary chondrites. Letters beside some symbols identify anomalous meteorites. (Reproduced with permission from J. F. Kerridge, 1977, *Space Science Review* **20**, 3–68, copyrighted by D. Reidel Publishing Company, Holland)

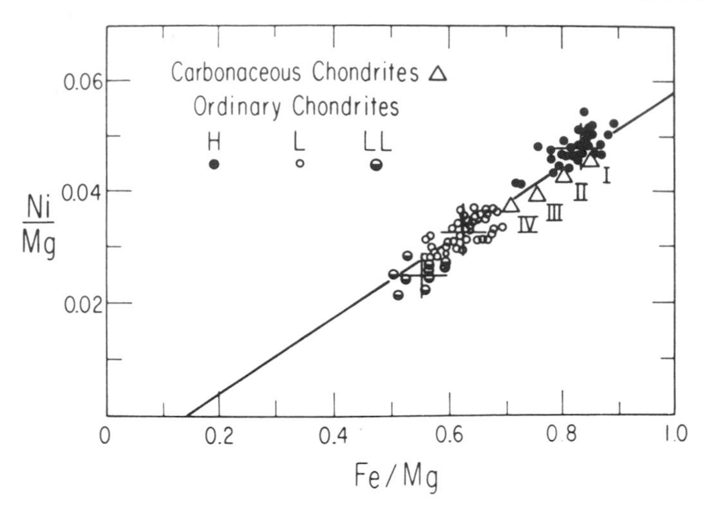

Figure 6. Metal–silicate fractionation in ordinary and carbonacious chondrites. Only the average compositions of carbonaceous chondrites have been plotted to avoid crowding. Loss or gain of metal would displace a point along a straight line joining the present composition of the meteorite with that of its parent material. The slope and intercept of this line give (Fe/Ni) of the metal and (Fe/Mg) of the oxidized species during fractionation, respectively. (Reproduced with permission from J. W. Larimer and E. Anders, 1970, *Geochim. Cosmochim. Acta*, **37**, 1603–1623, copyrighted by Pergamon Press)

The metal–silicate fractionation recorded in chondrites probably took place in the nebula rather than on the meteorite parent body because segregation of phases in even a weak gravitational field would have resulted in all-or-nothing separation of metal from silicate, whereas the relatively metal-deficient chondrites still contain significant amounts of metal (Anders, 1964; Wasson, 1972). Although separation of gas from dust constituted the most efficient kind of fractionation mechanism operating in the nebula (Suess, 1965; Wasson, 1972), it seems unlikely that this could have been responsible for the observed metal-silicate fractionation. The condensation temperatures calculated for iron metal and magnesium silicate, Figure 7 are very close together over the pressure range 10^{-5} to 10^{-3} atm, believed to correspond to the formation location of the chondrites (Larimer, 1973). Thus, a very short interval of time would have been available during which gas–solid fractionation could have led to significant separation of metal from silicate and such an effective segregation of gas and solid should have perturbed the overall composition of chondrites further away from the cosmic proportions than is observed. However, on the basis of systematic differences in the chondritic abundances of nickel, gold and iridium, Wasson (1972) has suggested that chondritic metal–silicate fractionation involved just such a gas–solid process.

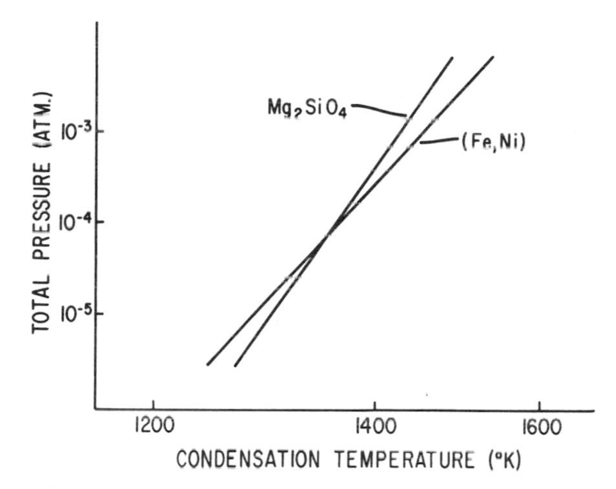

Figure 7. Condensation temperatures of magnesium silicate and metallic nickel–iron as functions of total pressure in the primitive nebula. Metal condensed at higher temperatures than silicate for pressures greater than about $5{\cdot}5 \times 10^{-5}$ atm., the reverse being true at lower pressures, but the differences were not great over the pressure range believed to be characteristic of the formation location of the chondrites and terrestrial planets. (Data from Grossman, 1976)

From the inter-relations between metal contents and abundances of moderately volatile siderophile elements in chondrites, Larimer and Anders (1970) gave an upper limit of about 1050 K to the temperature at which metal–silicate fractionation occurred. By a series of arguments based upon the chondritic distribution of sulphur, they inferred that fractionation took place before condensation of sulphur, i.e. above 680 K. In fact this conclusion may be incorrect, as the intercept in Figure 6, giving the amount of non-metallic iron during fractionation, corresponds to a silicate composition which would not have been stable in the nebula until below the condensation point of sulphur (Grossman, 1972). It follows that the intercept probably denotes the amount of iron tied up as sulphide plus silicate during fractionation (actually almost entirely as sulphide) and that fractionation therefore took place below 680 K (Kerridge, 1977).

Segregation of two condensed phases most likely resulted from differences in some physical property affecting the efficiency with which each phase accreted. Differences in density, mechanical strength, volatility or ease of magnetization have been advanced by various workers as responsible for the observed effects. These properties invariably appear to favour metal over silicate during accretion, an effect observed also in high velocity impact experiments (Kerridge and Vedder, 1972; Dietzel *et al.*, 1972), whereas the chondritic pattern was apparently caused by loss of metal relative to silicate. This suggests that chondrites may represent residues following accretion of some earlier metal-rich material. Clearly more information is required concerning the conditions in which metal–silicate fractionation occurred in meteorites and also among the terrestrial planets (Urey, 1952; Lewis, 1972b; Turekian and Clark, 1969) and to establish the role which the accretion process played in effecting these fractionations.

Chondrule Formation

The petrographic structure of chondrites is dominated by the presence of submillimetre, spheroidal particles, generally polymineralic but occasionally monomineralic, known as chondrules, Figure 8. These were almost certainly formed by the solidification and crystallization of formerly molten droplets but the mechanism which generated the melt has been the subject of debate, too extensive to summarise adequately here. Those theories based upon impact melting during high velocity collisions (Urey, 1952; Fredriksson, 1963; Kurat, 1967) received considerable support from the study of lunar soil and breccia samples, which showed that impacts into the surface of the moon have produced

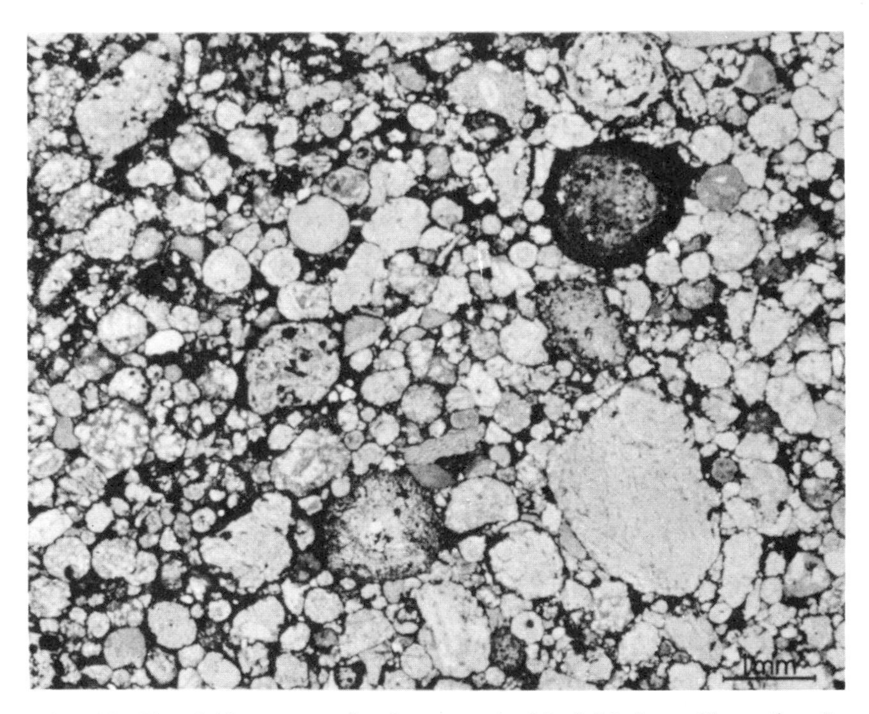

Figure 8. Chondritic structure in the meteorite Mezö Madaras, illustrating the spheroidal nature and abundant proportion of chondrules. (Micrograph by W. R. Van Schmus in J. T. Wasson, 1972, *Rev. Geophys. Space Phys.*, **10**, 711–759, reproduced by permission)

glass spherules, some of which have crystallized into textures resembling chondrules (Fredriksson *et al.*, 1970). However, impact into lunar or other planetary surfaces produces a variety of characteristic glasses in addition to spherules and the proportion of spherule to non-spherule types is very much greater in chondrites than in lunar samples, suggesting that chondrule production took place by impacts in a different kind of environment from that of a planetary surface (Kieffer, 1975; Kerridge and Kieffer, 1975). Specifically, Kieffer (1975) has shown that collisions between centimetre-sized objects at velocities greater than about 3 km/sec would have led to production of abundant spherules, in the size range observed in chondrites, *via* a phenomenon known as jetting. Similar conclusions regarding the nature of the chondrule-forming process were reached by Wasson (1972), using geochemical arguments, and Whipple (1972) and Cameron (1973), from dynamical considerations. It seems likely, therefore, that chondrules were produced at an intermediate stage in the accretion process.

This conclusion is not inconsistent with the scanty evidence relating to the time when chondrules formed. Dating by $^{39}Ar/^{40}Ar$, $K/^{40}Ar$ and (U,Th)/He techniques has given ages around $4{\cdot}5 \times 10^9$ years (Podosek, 1971; Fireman *et al.*, 1970) while retention of radiogenic ^{129}Xe began in chondrules at about the same time as in chondrite matrix and whole-rock samples (Podosek, 1970). Larimer and Anders (1970) pointed out that the presence of troilite, FeS, in some chondrules showed that their production occurred after the nebula had cooled below 680 K and they also argued that chondrule formation probably took place after metal–silicate fractionation because chondrules reflect the bulk chemistry of the chondrite in which they occur regardless of the proportion of chondrules to matrix. In addition, the chondritic structure would not have survived a process capable of segregating metal from silicate.

Depletion of moderately volatile elements, such as copper, tin, and sulphur, in chondrites, approximately reflects the proportion of chondrules to matrix and indicates that these elements were fairly efficiently vaporized during chondrule formation (Anders, 1972). The fact that these elements did not significantly recondense following vaporization has been interpreted by Wasson (1972) as showing that chondrule formation took place after condensation had effectively ceased.

An interesting consequence of the collisional model for chondrule formation is that it requires a high velocity 'spike' during an epoch characterized in general by velocities which were low enough to permit accretion to occur (Kieffer, 1975). More precise dating techniques should allow this sequence of events to be defined in detail.

Acknowledgements

I thank J. R. Arnold, W. V. Boynton, A. W. Harris, I. R. Kaplan, S. W. Kieffer, and J. T. Wasson. Support by NASA through grants NGL–05–00–005 Supplement 6, NGR 05–007–289 and NSG 7121 is gratefully acknowledged.

References

Alfvén, H., and Arrhenius, G. (1974). 'Structure and evolutionary history of the Solar System, IV., *Astrophys. Space Sci.*, **29**, 63–159.

Anders, E. (1964). 'Origin, age and composition of meteorites.' *Space Sci. Rev.*, **3**, 583–714.

Anders, E. (1971). 'How well do we know "cosmic" abundances?' *Geochim. Cosmochim. Acta*, **35**, 516–522.

Anders, E. (1972). 'Physico-chemical processes in the solar nebula, as inferred from meteorites.' In *Origin of the Solar System*, CNRS, Paris, pp. 179–195.

Arnold, J. R. (1976). Paper presented at *IAU Colloq. No. 39*, Lyons, France.

Arrhenius, G., and Asunmaa, S. K. (1973). 'Aggregation of grains in space.' *The Moon*, **8**, 368–391.

Bibring, J. P., and Maurette, M. (1972). 'Stellar wind radiation damage in cosmic dust grains: implications for the history of early accretion in the solar nebula.' In *Origin of the Solar System*, CNRS, Paris, pp. 284–291.

Cameron, A. G. W. (1962). 'The formation of the sun and planets.' *Icarus*, **1**, 13–69.

Cameron, A. G. W. (1973). 'Accumulation processes in the solar nebula., *Icarus*, **18**, 407–450.

Cameron, A. G. W. (1975). 'Clumping of interstellar grains during formation of the primitive solar nebula.' *Icarus*, **24**, 128–133.

Cameron, A. G. W., and Pine, M. R. (1973). 'Numerical models of the primitive solar nebula.' *Icarus*, **18**, 377–406.

Clayton, R. N., Grossman, L., and Mayeda, T. K. (1973). 'A component of primitive nuclear composition in carbonaceous meteorites.' *Science*, **182**, 485–488.

Dietzel, H., Neukum, G., and Rauser, P. (1972). 'Micrometeoroid simulation studies on metal targets.' *J. Geophys. Res.*, **77**, 1375–1395.

Dran, J. C., Durrieu, L., Jouret, C., and Maurette, M. (1970). 'Habit and texture studies of lunar and meteoritic materials with a 1 MeV electron microscope.' *Earth Planet. Sci. Lett.*, **9**, 391–400.

Fireman, E. L., DeFelice, J., and Norton, E. (1970). 'Ages of the Allende meteorite.' *Geochim. Cosmochim. Acta*, **34**, 873–881.

Fredriksson, K. (1963). 'Chondrules and the meteorite parent bodies.' *Trans. N.Y. Acad. Sci.*, **25**, 756–769.

Fredriksson, K., Nelen J., and Melson, W. G. (1970). 'Petrography and origin of lunar breccias and glasses.' *Proc. Apollo 11 Lunar Sci. Conf.*, pp. 419–432.

Fuchs, N. (1964). 'The mechanics of aerosols.' Macmillan, New York.

Ganapathy, R., and Grossman, L. (1976). 'Chemical characteristics of high-temperature condensates.' In *Lunar Science VII*, Lunar Science Institute, Houston, pp. 278–280.

Goldreich, P., and Ward, W. R. (1973). 'The formation of planetesimals.' *Ap. J.*, **183**, 1051–1061.

Greenberg, J. M. (1968). 'Interstellar grains.' In *Nebulae and Interstellar Matter*, Univ. Chicago Press, Chicago, pp. 221–361.

Grossman, L. (1972). 'Condensation in the primitive solar nebula.' *Geochim. Cosmochim. Acta*, **36**, 597–619.

Grossman, L. (1976). 'Chemical fractionation in the solar nebula.' *Proc. Soviet–American Conf. Cosmochemistry of the Moon and Planets*, in the press.

Hartmann, W. K. (1970). 'Growth of planetesimals in nebulae surrounding young stars.' *Liège Collec. 8°, 5th Ser.*, **19**, 215–227.

Keil, K., and Fredriksson, K. (1964). 'The iron, magnesium and calcium distribution in coexisting olivines and rhombic pyroxenes of chondrites.' *J. Geophys. Res.*, **69**, 3487–3515.

Kerridge, J. F. (1977). 'Iron: whence it came, where it went.' *Space Sci. Rev.* **20**, 3–68.

Kerridge, J. F., and Kieffer, S. W. (1975). 'Agglutinates as indicators of regolith environments and their significance in theories of chondrule formation.' *Trans. Amer. Geophys. U.*, **56**, 1016.

Kerridge, J. F., and Vedder, J. F. (1972). 'Accretionary processes in the early Solar System: an experimental approach.' *Science*, **177**, 161–163.

Kieffer, S. W. (1975). 'Droplet chondrules.' *Science*, **189**, 333–340.

Kuhi, L. V. (1964). 'Mass loss from T-Tauri stars.' *Ap. J.*, **140**, 1409–1433.

Kuiper, G. P. (1951). 'On the origin of the Solar System.' *Proc. Nat. Acad. Sci. USA*, **37**, 1–14

Kuiper, G. P. (1956). 'The formation of the planets.' *J. Roy. Astr. Soc. Canada*, **50**, 57.

Kurat, G. (1967). 'Formation of chondrules.' *Geochim. Cosmochim. Acta*, **31**, 491–502.

Larimer, J. W. (1967). 'Chemical fractionations in meteorites, I. Condensation of the elements.' *Geochim. Cosmochim. Acta*, **31**, 1215–1238.

Larimer, J. W. (1973). 'Chemical fractionations in meteorites, VIII. Cosmothermometry and cosmobarometry.' *Geochim. Cosmochim. Acta*, **37**, 1603–1623.

Larimer, J. W., and Anders, E. (1967). 'Chemical fractionations in meteorites, II. Abundance patterns and their interpretation.' *Geochim. Cosmochim. Acta*, **31**, 1239–1270.

Larimer, J. W., and Anders, E. (1970). 'Chemical fractionations in meteorites, III. Major element fractionations in chondrites.' *Geochim. Cosmochim. Acta*, **34**, 367–387.

Laul, J. C., Keays, R. R., Ganapathy, R., Anders, E. and Morgan, J. W. (1972). 'Chemical fractionations in meteorites, V. Volatile and siderophile elements in achondrites and ocean ridge basalts.' *Geochim. Cosmochim. Acta*, **36**, 329–345.

Lewis J. S. (1972a). 'Low temperature condensation from the solar nebula.' *Icarus*, **16**, 241–252.

Lewis J. S. (1972b). 'Metal/silicate fractionation in the solar system.' *Earth Planet. Sci. Lett.*, **15**, 286–290.

Lord H. C. (1965). 'Molecular equilibria and condensation in a solar nebula and cool stellar atmospheres.' *Icarus*, **4**, 279–285.

Marvin U. B., Wood J. A. and Dickey J. S. (1970). 'Ca–Al-rich phases in the Allende meteorite.' *Earth Planet. Sci. Lett.*, **7**, 346–350.

Neukum G. (1968). 'Investigations of projectile material in microcraters.' *Thesis*, University of Heidelberg.

Podosek F. A. (1970). 'Dating of meteorites by the high-temperature release of iodine-correlated Xe^{129}.' *Geochim. Cosmochim. Acta*, **34**, 341–365.

Podosek F. A. (1971). 'Neutron-activation potassium–argon dating of meteorites.' *Geochim. Cosmochim. Acta*, **35**, 157–173.

Prior, G. T. (1916). 'On the genetic relationship and classification of meteorites.' *Mineral. Mag.*, **18**, 26–44.

Reeves, H. (1972). 'Presentation of the models.' In *Origin of the Solar System*, CNRS, Paris, pp. 28–55.

Suess, H. E. (1965). 'Chemical evidence bearing on the origin of the Solar System.' *Ann. Rev. Astron. Astrophys.*, **3**, 217–234.

Turekian, K. K., and Clark, S. P. (1969). 'Inhomogeneous accumulation of the earth from the primitive solar nebula.' *Earth Planet. Sci. Lett.*, **6**, 346–348.

Urey, H. C. (1952). 'The planets–their origin and development.' Yale University Press, New Haven.

Urey, H. C. (1966). 'Chemical evidence relative to the origin of the Solar System.' *Mon. Not. Roy. astr. Soc.*, **131**, 199–223.

Urey, H. C., and Craig, H. (1953). 'The composition of stone meteorites and the origin of the meteorites.' *Geochim. Cosmochim. Acta*, **4**, 36–82.

Wark, D. A., and Lovering, J. E. (1976). 'Refractory platinum metal grains in Allende calcium–aluminium-rich clasts (CARCs): possible exotic presolar material?' In *Lunar*

Science VII, The Lunar Science Institute, Houston, pp. 912–914.

Wasson, J. T. (1972). 'Formation of ordinary chondrites.' *Rev. Geophys. Space Phys.*, **10**, 711–759.

Whipple, F. L. (1972). 'On certain aerodynamic processes for asteroids and comets.' In *From Plasma to Planet*, Wiley, New York, 211–232.

Wickramasinghe, N. C. (1967). *Interstellar Grains*. Chapman and Hall, London.

ON THE INTERACTION BETWEEN NEBULAR DRAG, PLANETARY PERTURBATIONS, AND ACCRETION

T. A. HEPPENHEIMER
Center for Space Science, 11040 Blue Allium Avenue, Fountain Valley, California, U.S.A.

Theories of accretion, such as those of Hartmann (1968), Weidenschilling (1974), or Safronov (1972), typically seek to treat the physics of planetesimal infall and of planetary growth. However, in at least the earliest stages of planet formation, the planetesimals were immersed in the solar nebula. Moreover, their orbital motions may well have been subject to perturbations from Jupiter, which must have formed in the brief interval prior to nebular dissipation. Thus, one is led to inquire into the dynamical effects of nebular drag as well as of Jupiter perturbations, insofar as they may have influenced conditions permitting accretion.

The essential observation is that planetesimal collision velocities must be small, e.g. $\lesssim 100$ m/sec, for accretion to proceed. Higher values lead to fragmentation. The collision velocity $v_c = ev_0$; $e =$ orbit eccentricity, $V_0 =$ orbit velocity. For the asteroids, then, one requires $e \lesssim 0{\cdot}005$ for accretion to proceed. e tends to be increased by planetary perturbations; e is damped or decreased by nebular drag. These simple observations, it turns out, lead to fairly interesting consequences, some of which are described here.

Planetary perturbations are studied using the Lagrange planetary equations together with an appropriate disturbing function. This topic is treated by Brouwer and Clemence (1961a) as well as by Heppenheimer (1976). In what follows, results from perturbation theory will be cited when required. Nebular drag, however, requires further discussion.

A solar nebula does not rotate with locally Keplerian velocity but rather has local velocity which is reduced from the Keplerian by

$$\Delta V = -c^2/2V_0 \tag{1}$$

(Goldreich and Ward, 1973). The reason is that the nebula receives partial support from its pressure gradient. In equation (1), $c =$ sound speed, approximately $0{\cdot}1\ V_0$. In what follows we use normalized units: unit mass, length, and time are respectively taken as one solar mass, Jupiter's semimajor axis, and $1/2\pi$ times Jupiter's orbit period.

Equation (1) implies there are two regimes of motion. For $e \leqslant c^2/2V_0^2$, the motion always has velocity greater than that of the nebula. Then the decay law for e is found to be, very nearly,

$$de/dt = -(\rho_g \beta c^2/2V_0)e \tag{2}$$

where ρ_g = nebular density, β = ballistic coefficient = $AC_D/2m$; A = planetesimal cross-sectional area, C_D = drag coefficient $\sim 0{\cdot}5$, m = mass. For $e > c^2/2V_0^2$, the motion at aphelion has velocity less than that of the nebula, so a planetesimal alternately is faster and slower than the nebula. Then the decay law is

$$de/dt = -(2/\pi)\rho_g \beta V_0 e^2 \tag{3}$$

The derivations of equations (2) and (3) are given elsewhere (Heppenheimer, 1978a).

The presence in the equations of the model-dependent terms ρ_g, A, m may appear to limit the usefulness of equations (2) and (3). The problem of model-dependence can be largely avoided, however, by considering Goldreich–Ward (1973) planetesimals. Thus, one notes that for circular bodies, $\beta = 3C_D/4\rho_p D$; ρ_p = material density ~ 3 g/cm^3, D = diameter. Goldreich–Ward planetesimals form by two stages of gravitational instability, leading respectively to bodies of diameter D_1, D_2:

$$D_2 = \frac{4\pi\alpha\rho_g c}{n^2}\left[\frac{3G^2}{2\rho_p n}\right]^{1/3};\ D_1 = D_2\left[\frac{3}{2\pi^3}\frac{n^2}{G\rho_p}\right]^{1/6} \tag{4}$$

α = fraction of condensibles in nebula $\sim 0{\cdot}005$; n = orbital mean motion; G = constant of gravitation = unity, in normalized units. Hence equations (2) and (3) are rendered independent of ρ_g, D. There is a weak dependence on ρ_g through c, since $c \propto T^{1/2}$ (T = temperature) and a nebular adiabat will give $T \propto \rho_g^{2/3}$. But it is found that to a good approximation, $c \propto V_0$. Then, in normalized units,

$$(\rho_g\beta)_1 = 5{\cdot}5 \times 10^{-3} n^{5/3};\ (\rho_g\beta)_2 = 1{\cdot}1 \times 10^{-4} n^2 \tag{5}$$

respectively for first and second generation Goldreich–Ward planetesimals. It now is possible to proceed to address some problems of accretion.

Jupiter's Influence on the Formation of the Asteroids

It has often been proposed that Jupiter perturbations disrupted the planet-forming process in the zone of the asteroids, preventing their coalescence into a single planet. This proposal can be studied. The perturbations of interest are the so-called secular perturbations (Brouwer and Clemence, 1961a). For a body initially at near-zero eccentricity, the subsequent variation in e is given by

$$e = \tfrac{5}{2} ae' \left| \sin \tfrac{1}{8} \frac{\mu}{na} b_{3/2}^{(1)} t \right| \tag{6}$$

a = semimajor axis (Jupiter = 1), e' = Jupiter's eccentricity, μ = Jupiter mass (Sun = 1), t = normalized time. Also, $b_{3/2}^{(1)}$ is a Laplace coefficient:

$$b_{3/2}^{(1)} = 3a\left[1 + \tfrac{15}{8} a^2 + \tfrac{175}{64} a^4 + \dots\right] \tag{7}$$

Thus, at Jupiter's present mass ($\mu = 10^{-3}$) and eccentricity ($e' = 0{\cdot}05$), a planetesimal at $a = 0{\cdot}5$ has e varying up to 0·06, with period of variation $\sim$ 5000 time units or 10,000 years. This is confirmed by more detailed computations (Brouwer, 1951).

Where $a = 0{\cdot}5 = 2{\cdot}6$ AU, in the middle of the asteroids, $n \sim 3$, $V_0 \sim 1{\cdot}5$. Then equations (2), (3), and (5) serve to give the time scale τ for damping of e due to drag. For first-generation planetesimals, $\tau \sim 10^3$ time units so that damping may have been an important effect. But for second-generation planetesimals, $\tau \sim 10^5$ time units so that damping is quite negligible and secular perturbations may proceed unhindered. Hence, if Jupiter existed as it does today, when the asteroids were forming, then asteroids could not have grown by accretion. Typical nebular models give $D_1 \sim 1$ km, $D_2 \sim$ a few tens of km. Hence one cannot propose that asteroids are identical with Goldreich–Ward planetesimals, since asteroids exist with D = several hundred km.

One way out of this difficulty is to consider that Jupiter may have formed slowly, so that it would have been of small size, giving perturbations of long period, during the time of asteroid formation. This is consistent with initial formation of a rocky core of Jupiter, by accretion, which subsequently swept up gas from the solar nebula. From equations (6) and (7), a suitable Jupiter mass is some 0·01 times its present mass.

However, it is difficult to see how such a proposal can be supported. Jupiter clearly formed while the solar nebula still existed; but this appears to have involved times of the order of 10^6 years, possibly much less (Cameron and Pines, 1973). By contrast, accretion time scales may be of the order of 10^7 years, possibly much more (Weidenschilling, 1974). Thus, while the concept of a slowly-growing Jupiter indeed appears adequate to explain the limited sizes of asteroids, it appears difficult to reconcile with cosmogonic considerations.

A simple explanation (Heppenheimer, 1978b) lies in the fact that in the presence of a solar nebula, equation (6) must be modified to reflect the effects due not only to the drag of the nebula but also its gravitational effects. Gravitationally, the nebula is similar to an ensemble of perturbing planets on circular coplanar orbits. The modification is approximated by writing $5\mu/2\mu_N$ for 5/2 in equation (6), where μ_N = fractional mass of the nebula. Since even low-mass solar nebula models give $10 < \mu_N/\mu < 100$, it follows that asteroidal accretion could readily proceed in the presence of the nebula, even with Jupiter also present. The dissipation of the nebula, however, would lead to equation (6) applying without modification. Brouwer and Clemence (1961b) have shown that at least in the geologically recent past, e' has varied between 0·027 and 0·062, so that nebular dissipation could have immediately produced termination of asteroidal accretion.

A variety of other mechanisms also may have influenced this. Kaula and Bigeleisen (1975) note that proto-Jupiter could have scattered planetesimals at high velocity throughout the asteroid zone; Weidenschilling (1975) points out that this flux could have reached Mars but not Earth. The lectures by Alan Harris (this volume) indicate how asteroidal close approaches could have perturbed their eccentricities. But if this effect was important in limiting asteroid growth, it would also have limited terrestrial planets to asteroidal size. Harris has also noted how a T-Tauri solar wind could have removed planetesimals up to a fraction of kilometre size, thus possibly depleting the mass reservoir from which asteroids would have grown. However, none of these mechanisms can readily account for the limited asteroidal sizes.

The key to these limited sizes may lie in the asteroids' unusually high eccentricities and inclinations, e.g. $e \sim 0{\cdot}15$, $i \sim 10°$. Whatever process led to these e's and i's would certainly have brought asteroid growth to an immediate halt. This process is at present unknown.

Another explanation, which may also be adequate, again invokes Weidenschilling's (1974) theory of accretion, in noting that asteroidal accretion time scales are systematically longer than those for the inner planets. If the dissipation of the solar nebula brought about a crisis for the accretion process, the asteroids may simply have been interrupted before they could accrete into bodies large enough to continue to grow in the absence of the nebula. This explanation may also shed light on the fact that the asteroids together aggregate much less mass than any planet, a fact which suggests that there has been large-scale mass loss from the asteroids. Initially, the asteroids' mass would of course have been distributed in planetesimals, of sizes given by equation (4), which would readily be destroyed by collisions in the present dynamical environment. It is a characteristic of the accretion process that it leads to growth of large bodies in which the mass is predominantly concentrated. But it is possible that asteroidal accretion may have been interrupted at an intermediate stage wherein, despite the growth of some bodies, most mass was still distributed in small bodies of sizes comparable to those of planetesimals. Safronov (1972) has given a time-variable mass distribution for an accreting population of planetesimals, in which precisely such an intermediate stage is evident.

The Kirkwood Gaps and the Galilean Satellites

If one plots the number density of asteroids as a function of semimajor axis, it is found that this density is zero near orbits which are in resonance with Jupiter. The gaps are at the 2 : 1, 3 : 1, and 5 : 2 resonances, corresponding to orbits with period 2·0, 3·0, and 2·5 times Jupiter's. Some authors also refer to a gap at the 7 : 3 resonance, but there the number density is not zero.

An explanation of the gaps has proven difficult. Purely dynamical theories, e.g. perturbational ejection of asteroids, appear to be ruled out (Weisel, 1974). Thus, in recent years investigators have sought physical effects which could produce gaps.

The principal such effect, on which to build a theory, is that asteroids in resonance tend to have markedly higher eccentricities, and more rapid variations in e, than would be predicted from equations (6) and (7). At the 2:1 gap, e varies on a time scale not of μ^{-1} but of $\mu^{-1/2}$, with amplitude $\sim 0{\cdot}1$, even for $e' = 0$. At the 3:1 and 5:2 gaps, the time scales are similar and the amplitudes are reduced but still may be $\sim 0{\cdot}01$ for $e' = 0$, rather larger for $e' \sim 0{\cdot}05$.

A number of authors have proposed that these increased e's would have been associated with increased collision probabilities, so that asteroids would have preferentially been destroyed within the gaps. Heppenheimer (1975) has pointed out numerous difficulties with this approach. The main problem is that the resonance-induced eccentricities are too small to give any major increase in collision probability.

But from the foregoing material of this paper, one may propose that the Kirkwood gaps are primordial, representing regions where asteroids failed to form. The nebular-induced reduction of eccentricity (by the factor μ/μ_N) does not occur at a resonance. The magnitudes and time scales of resonance-induced eccentricity variation then are quite adequate to prevent accretion. A development of this concept has been given elsewhere (Heppenheimer, 1978a).

Such a theory then represents the Kirkwood gaps as giving a constraint on the problem of asteroid formation. The gaps, if filled in, could not be cleaned out, either by collisions or by dynamical effects. Hence the asteroids must have acquired their e's and i's in a manner whereby they would have avoided changes in semimajor axis. The only evident way this could have happened is through secular perturbations, which indeed can change e and i but not a. One is then faced with the difficult question: How could there have been a planet with the required elements, $e \sim 0{\cdot}2$, $i \sim 20°$, as required to achieve the asteroidal e's and i's?

The primordial-origin theory of the Kirkwood gaps gains credibility inasmuch as it throws light upon the curious fact that in the asteroids, resonances are associated with gaps, but in the Galilean satellites, resonances are associated with the presence of large bodies.

From the Goldreich–Ward theory, one finds that the number density of planetesimals increases rapidly as one proceeds inward. Hence it is plausible that Io formed first. We then enquire as to the values of $\rho_g \beta$ when our normalizing parameters are taken as one Jupiter mass, Io's semimajor axis, and 1/2 times Io's period. The result is, $\rho_g \beta \sim 1$. Since for Io, $\mu \sim 10^{-4}$ it follows that drag effects in a Jupiter nebula will completely dominate effects due to secular perturbations.

But at the 1:2 resonance (twice Io's period), one finds a major effect due to resonant perturbations. This is a marked increase in orbit decay rate, da/dt, such that planetesimals will rapidly cross this resonance region. Hence there is a large decrease there, in planetesimal number density. When this decrease is superimposed over the general gradient in number density, one finds a local maximum in number density just outside Io's 1:2 resonance. This is precisely the location of Europa. If Europa formed there, by the same process it could give rise to Ganymede, just outside Europa's 1:2 resonance—as observed.

To conclude, then, the results of this paper provide new constraints on any theoretical origin of the asteroids. They also appear to shed light on the origin of some of the structural features of the solar system. Certainly, they offer the prospect of a rich variety of a new results, associated with the interaction between planetary or nebular perturbations and nebular drag.

References

Brouwer, D. (1951). 'Secular variations of the orbital elements of minor planets.' *Astron. J.*, **56**, 9–32.

Brouwer, D. and Clemence, G. M. (1961a). *Methods of Celestial Mechanics*, Academic Press, New York.

Brouwer, D. and Clemence, G. M. (1961b). 'Orbits and masses of planets and satellites.' In *Planets and Satellites*, (G. P. Kuiper and B. Middlehurst, eds.), pp. 42–54. University of Chicago Press, Chicago.

Cameron, A. G. W. and Pine, M. R. (1973). 'Numerical models of the primitive solar nebula.' *Icarus*, **18**, 377–406.

Goldreich, P. and Ward, W. R. (1973). 'The formation of planetesimals.' *Astrophys. J.*, **183**, 1051–1061.

Hartmann, W. K. (1968). 'Growth of asteroids and planetesimals by accretion.' *Astrophys. J.*, **152**, 337–342.

Heppenheimer, T. A. (1975). 'On the alleged collisional origin of the Kirkwood gaps.' *Icarus*, **26**, 367–376.

Heppenheimer, T. A. (1976). 'Introduction to the restricted Jupiter orbiter problem.' *Celestial Mechanics*, **14**, 175–200.

Heppenheimer, T. A. (1978a). 'On the origin of the Galilean satellites and Kirkwood gaps.' *Astronomy and Astrophysics*, under review.

Heppenheimer, T. A. (1978b). 'On the early history of the asteroids.' *Astronomy and Astrophysics*, under review.

Kaula, W. and Bigeleisen, P. (1975). 'Early scattering by Jupiter and its collision effects in the terrestrial zone.' *Icarus*, **25**, 18–33.

Safronov, V. S. (1972). 'Accumulation of the planets.' In *On the Origin of the Solar System* (H. Reeves, ed.) pp. 89–113. Centre National de la Recherche Scientifique, Paris.

Weidenschilling, S. J. (1974). 'A model for accretion of the terrestrial planets.' *Icarus*, **22**, 426–435.

Weidenschilling, S. J. (1975). 'Mass loss from the region of Mars and the asteroid belt.' *Icarus*, **26**, 361–366.

Wiesel, W. E. (1974). 'A statistical theory of the Kirkwood gaps.' *Center for Astrophysics Preprint Series*, Nos. 191 and 204, Cambridge, Massachusetts.

CHEMICAL SEGREGATION IN PROTOPLANETARY THEORIES

I. P. WILLIAMS
Queen Mary College, Mile End Road, London, U.K.

It is first essential to clearly define what is meant by a protoplanet as opposed to an embryoplanet. An embryoplanet is a small object which will grow into a planet, presumably by the accretion of suitable material. A protoplanet is a much larger object which evolves into a planet by contraction coupled with possible mass loss. The first protoplanetary theory was therefore that of Buffon (1745) who envisaged material being ejected from the sun after a collision with a comet. This material then fragmented into planetary mass objects and the planets simply formed by the contraction of these fragments. The classical tidal theories, made famous by Jeans (1916) and Jeffreys (1918), are also early protoplanetary theories.

We know now that planet formation cannot be such a simple process for the planets have widely different masses and, more significantly, perhaps, widely different chemical compositions. The primary composition of the four terrestrial planets is iron silicates, Jupiter and possibly Saturn are similar to the solar composition while Uranus and Neptune are generally thought to be primarily composed of the materials based on the elements C, N and O (Stevenson (1978) discusses these aspects in this volume.) In Table 1 we show the basic planetary masses, their chemical compositions and the cosmic abundance of this material (all in round numbers). From this table, a rather remarkable fact emerges. If we take as a standard initial protoplanet an object of 300 $M_{\oplus}$ of cosmic material, it is obviously identical in the sense under discussion to Jupiter. The CNO content of such an object is about 10 Earth masses, thus similar to the outer planets while the iron-silicate content is similar to a terrestrial planet. Thus, if this type of hypothetical protoplanet is to evolve into an outer planet it must lose hydrogen and helium and leave only the CNO based compounds (ammonia, methane, water) while if the object is to become a terrestrial planet it must lose everything but the iron-silicate fraction and in both cases that is all that is

Table 1

Planetary type	Mass ($M_{\oplus}$)	Composition	Cosmic abundance (per cent)
Terrestrial	1	Iron silicate	0·3
Uranus/Neptune	15	CNO	2
Jupiter/Saturn	300	Cosmic	100

necessary. Confirmation of this picture comes from the work described by Stevenson (1978) who shows that it is necessary for both Jupiter and Saturn to have an icy–rocky core of order 10–20 $M_{\oplus}$ in agreement with the CNO content predicted. In fact this suggests that the original protoplanets may be somewhat larger than Jupiter, with Jupiter and Saturn losing some hydrogen and helium. It is therefore interesting to discuss ways by which the iron silicates, or the CNO could have become segregated in only some of the protoplanets.

Before doing this it is pertinent to remember that the theory of planetary formation described by McCrea (1960) (also described in this book) postulates the existence of protoplanets similar to those described above with a mean density of the order of 10^{-9} g cm^{-3}. Woolfson (1964) has similar objects resulting from his capture theory while Cameron (1978) in this volume now also advocates the existence of such objects.

Clearly no mechanism can exist that can work efficiently in segregating gaseous hydrogen and helium from gaseous iron silicates or gaseous CNO. It is much easier to segregate solids (or liquids if the physics allows for such a phase to exist) from the gas. Thus either we must have a cooling phase in the protoplanets evolution during which time the iron silicates and the CNO in certain planets condense out or the protoplanets must be originally at such a low temperature that iron silicates and, if necessary, the CNO existed as solid grains. Since the tendency for a contracting object is to heat up the latter scheme would appear to be the most likely and this agrees well with the McCrea model where the initial temperatures are taken to be 50 K. Cooling by CO molecules could even reduce the temperatures of protoplanets in isolation to below this value. Thus we have the situation where the iron silicates would most certainly exist predominantly in the form of grains possibly even being the original interstellar grains that were present in the solar nebula when it first formed out of a high-density molecular cloud.

In 1965 McCrea and Williams (1965) suggested that these grains would fall through the ambient gas in response to the gravitational field of the protoplanet and form a central core, a very simple suggestion which becomes obvious to everybody once it has been made. McCrea and Williams showed that normal interstellar grains would take far too long to segregate for the mechanism to be of any practical use when applied to protoplanets. Now the gravitational field varies as the cube of the grain radius but viscous drag only as the radius squared, and so increasing the size of the grains decreases the segregation time and McCrea and Williams showed that if the grains were assumed to coagulate with each other on collision (the mechanism now known as cold welding) then the segregation time became acceptably short. Experimental work by Kerridge and Vedder (1972) indicated that such cold welding could occur while Orowan (1969) suggested that iron grains might adhere easier than silicates, thus falling faster and offering an explanation for the iron core of the Earth. Numerical work by Williams and Crampin (1971) took account of the complexities of expressions for the resistive forces (discussed by Baines *et al.*, 1965) and growth rates (discussed by Baines and Williams, 1965) caused by the grains having speeds close to the sound speed in the gas. This showed that the McCrea and Williams estimate for the time of segregation was reasonable. Williams and Handbury (1974) showed that the results were still valid even if the protoplanet was taken to be highly centrally condensed. Hence segregation is certainly possible if iron silicate grains coalesce on collision. It should perhaps be pointed out that the work of McCrea and Williams remains valid if the growth of the grains occurs by condensation onto them rather than by coagulation. The reason that McCrea and Williams assumed grain–grain type growth was because of a general belief prevalent at the time that all the iron silicates found in interstellar space were present in grain form. The fairly recent discovery of numerous complex molecules might alter this picture and indeed if CO molecules actually cause the protoplanets to cool, condensation may be as realistic a growth mechanism. It is to be noted that large grains, however formed, will segregate and it may be of interest to note that Rowan-Robinson (1975) claims that substantially larger than average grains are detectable in the Orion nebula, a region of considerable interest because star formation is suspected to be occurring there.

So far we have discussed only the formation of an iron silicate core and, because of the proximity of the Sun which keeps these at a reasonable temperature, it is only iron silicates that would form grains in the terrestrial planet region. Handbury and Williams (1975) have suggested that in the outer planet region methane and ammonia could also condense, thus leading to the formation of a CNO type ice core in the outer protoplanets. They suggest further that the gravitational energy released by the formation of such a core could be responsible for driving off the gaseous (i.e. hydrogen–helium) envelope, thus accounting for the mass and composition of Uranus and Neptune. One could speculate further by suggesting that in Saturn the least volatile of the CNO compounds condensed and segregated leading to some energy release and loss of hydrogen, while Jupiter being a little closer to the Sun lost very little.

This brings us back to the terrestrial planet region and the remaining problem here is the removal of all the gases (hydrogen, helium and CNO) to leave the core as a terrestrial planet. Donnison and Williams (1975) have shown that the tidal effect of the Sun can be very efficient at disrupting a protoplanet in this region and it is tempting to suggest that as the original elliptic orbits of these protoplanets became rounded and reduced, they moved inside the sphere of influence of the tidal effect which duly dispersed all the free gas. Donnison and Williams (1976) have investigated this rounding phenomena. It is interesting to note that for protoplanets of the density postulated by McCrea, this sphere of influence extends out to the asteroidal belt so that Jupiter and the planets beyond would remain unscathed. Another possible reason for the loss of gas from the terrestrial planet region is the proximity of the Sun itself with its heating effect and possible enhanced primordial solar wind (see Handbury, 1975).

In conclusion, proplanetary theories envisage the initial existence of a number of protoplanets, identical in mass and composition and comparable to or perhaps somewhat larger than Jupiter. In all of these the condensable material forms grains which then fall, under the influence of gravity, into the central regions of the protoplanet. For the protoplanets nearer the Sun only the iron silicate material is condensable but in the outer regions ammonia and methane may also have condensed giving rise to the much more massive icy cores. It may be that the formation of such a core releases enough energy to evaporate the residual hydrogen–helium outer layers leaving an Uranus/Neptune type of planet. In the terrestrial planet region solar effects of one form or another also cause the loss of all gaseous material again leaving only the core which this time is Earth-like as ammonia and methane are gaseous. The remaining two protoplanets condense in bulk in much the manner described by Donnison and Williams (1974) to become a Jupiter-like planet though some hydrogen loss can occur. In this way planets are formed having a mass and chemical composition closely resembling the existing planets and grains play a vital role in this process, though it is a matter of speculation as to whether these grains are original interstellar grains or later condensates out of the nebula.

References

Baines, M. J. and Williams, I. P. (1965). *Nature*, **205**, 59.
Baines, M. J., Williams, I. P. and Asebiomo, A. S. (1965). *Mon. Not. R. astr. Soc.*, **130**, 63.
Buffon, G. L. L. (1745). *De La Formation des Planets*, Paris.
Cameron, A. G. W. (1978). this volume.
Donnison, J. R. and Williams, I. P. (1974). *Astrophys. Space Sci.*, **29**, 397.
Donnison, J. R. and Williams, I. P. (1975). *Mon. Not. R. astr. Soc.*, **172**, 257.
Donnison, J. R. and Williams, I. P. (1976). *Mon. Not. R. astr. Soc.*, (submitted).
Handbury, M. J. (1975). *Ph.D. Thesis*, London University.
Handbury, M. J. and Williams, I. P. (1975). *Astrophys. Space Sci.*, **38**, 29.
Jeans, J. (1916). *Mon. Not. R. astr. Soc.*, **77**, 84.
Jeffreys, H. (1918). *Mon. Not. R. astr. Soc.*, **78**, 424.
Kerridge, J. F. and Vedder, J. K. (1972). *Nice Symposium on the Origin of the Solar System*, (ed. Reeves, H.) C.N.R.S., 282.
McCrea, W. H. (1960). *Proc. Roy. Soc.*, **A256**, 245.
McCrea, W. H. and Williams, I. P. (1965). *Proc. Roy. Soc.*, **A287**, 143.
Orowan, E. (1969). *Nature*, **222**, 867.
Rowan-Robinson, M. M. (1975). *Mon. Not. R. astr. Soc.*, **127**, 109.
Stevenson, D. (1978). this volume.
Williams, I. P. and Crampin, D. J. (1971). *Mon. Not. R. astr. Soc.*, **152**, 261.
Williams, I. P. and Handbury, M. J. (1974). *Astrophys. Space Sci.*, **30**, 215.
Woolfson, M. M. (1964). *Proc. Roy. Soc.*, **A282**, 485.

CHEMICAL ASPECTS OF THE FORMATION OF THE SOLAR SYSTEM

G. ARRHENIUS

Scripps Institution of Oceanography and Institute for Pure and Applied Physical Sciences, University of California, San Diego, California, U.S.A.

Introduction

All scientific theories ultimately rest on some basic postulates that by definition cannot be proved or disproved by the scientific method. The reasons for choosing such a set of postulates are consequently non-scientific; they are mostly aesthetic, religious, or political (Frank, 1954). A commonly and often productively used aesthetic argument is simplicity of the resulting theory. Its obvious limitations are due to the fact that apparent simplicity in theories for complex natural processes is a reflection of the state of the art and of the specific training and background of the scientists who develop the theories. Thus purists such as Mach and Ostwald well into the 20th century, and in the face of overwhelming experimental evidence, chose to reject the atomistic theory as a complicated and hence undesirable alternative to a 'simple' continuum description of Nature by the laws of thermodynamics. Similarly Lord Kelvin for fifty years successfully imposed on the scientific world the notion of stars and planets forming from hot balls of gas. This permitted simple and elegant calculations to the effect that the Earth would be very young; protests from geologists, relying on a network of complicated facts, were kept underground.

In fundamental fields, where theories can be convincingly substantiated by experiment, the realistic concepts are relatively quickly sorted out; few scientists doubt any longer that atoms exist. In complex natural sciences such as astrophysics and space chemistry where the objects of study are remote, obsolete concepts have a life of their own. A rigorous new branch of chemical physics is clarifying the actual state of matter in interstellar clouds and in the solar system by the methods of molecular quantum chemistry. Meanwhile, however, most chemists interested in the processes in the early solar system continue to discuss these in terms of thermally insulated neutral gas systems from which solids would nucleate in thermodynamic equilibrium. The purpose of the present paper is to draw attention to some of the ways in which modern knowledge of processes in space can be used as a basis for reconstruction of the formation of the solar system.

Strategy of Approach

In discussing such an exceedingly complex problem as the chemical evolution of the early solar system, simplicity, although useful as a book-keeping device, cannot be used as a primary guide to the speculative selection of postulates. With the large number of branching points in a purely theoretical treatment one is bound to go astray, possibly already in the very first hypothetical assumptions, and to carry out increasingly refined calculations on developments that may never have taken place. Any realistic strategy instead must place major emphasis on verifying by experiments in space and in the laboratory, which processes are likely to have been of importance in the formative era. Any chemical process suggested must also be in concordance with physical laws and with the wealth of information gleaned from modern space research on the structure of the solar system and on the state of matter in space.

In order to give strength and validity to any chemical arguments they must consequently be coupled to such present-day physical solar system theory, that relies on experiment and observation for realism. There are several important contributions that discuss specific details of the origin of the solar system in terms of modern space physics. However, to the author's knowledge the only comprehensive theory for the formation of the solar system which attempts to explain in some quantitative detail the entire body of observations, and which seriously takes into account the actual behaviour of particles and fields in space, arises from a series of early contributions by Hannes Alfvén and his collaborators (for example, Alfvén, 1942–1946, 1954; Alfvén and Fälthammar, 1963; and references in these works). Because of the enormous scope of the subject, the subordinate criterion of simplicity still has to be used to sketch many of the details in this framework of theories, where coming space experiments will doubtlessly reveal a richer and still more complex reality on many points. Nonetheless, the experimental criteria that characterize Alfvén's approach appear to provide testable guidelines for a realistic picture of the chemical evolution in the formative era.

It has been a privilege to join in an effort to integrate the large amount of new information, arising from space exploration and from associated modern theory of particles and fields into a series of papers (Alfvén and Arrhenius, 1970–1974) and two recent books (Alfvén and Arrhenius, 1975, 1976). In the discussion that follows below I shall make extensive reference to material developed in detail in these books.

In the approach, introduced by Alfvén, and involving only known processes, the present day properties of the regular satellite systems and of the planetary system are found to depend uniquely on two properties of the central bodies; their spin and mass.

Constraints from Chemical Composition of Space Materials

The chemical information is in general less restrictive than the physical observational constraints. This is due to the vagueness of our present knowledge about the chemical composition of the bodies in the solar system, extrapolated in most cases from physical parameters. A wealth of chemical information has been derived from meteorites—the difficulty in applying this to the problem of the formation and early history of the solar system is partly due to the fact that we do not know with certainty where the meteorites come from, and still less where the material originally condensed, which later aggregated into the meteorite parent bodies.

Nonetheless, some important constraints are given by chemical evidence (section on *Composition of the Source Medium*). Also the fact that there cannot be two evolutions—one for chemists and one for physicists—places severe restrictions on otherwise more expansive chemical interpretation. For example, the need for angular momentum transfer (by known processes) to the secondary bodies (including the satellites) from their primaries places an upper limit on number densities in local regions of the solar nebula in the range 10^{10}–10^{13} cm^{-3}; average number densities must have been lower by orders of magnitude. The characteristic physical phenomena associated with the release of large amounts of gravitational energy at emplacement and with the motion of vast bodies of 'gas' with high velocities through stellar, and planetary magnetic fields, also help to focus allowable chemical speculation.

In the following the evidence is first summarized for the chemical properties of the space medium and the chemical consequences of the postulated physical differentiation processes are outlined. Finally the range of interpretations is discussed which are based on structure and composition of the material found in meteorites.

Particles, Fields, and Chemical Reactions in Dark Interstellar Clouds

The chemical processes in dark interstellar clouds are of great interest since they can be observed in considerable detail and since it is reasonable to regard them as source clouds for stellar and planetary nebulae. The term 'planetary nebulae' is used here for plasma nebulae around magnetized planets, assumed to give rise to the regular satellite systems, in contrast to the common astrophysical use referring to a large luminous diffuse object in interstellar space. These are likely to be much denser local regions—an upper limit for local densities of the order of 10^{13} cm^{-3} is imposed by the need for angular momentum transfer in our early solar system. However, the comparatively well known chemical state of the dark clouds is a far better base for extrapolation toward primordial conditions in circumsolar space than the also well known, but in this case inapplicable chemistry of hot gases at atmospheric pressure from which most cosmochemical speculation has arisen in the past.

Average number densities in the dark clouds are of the order 10^4 to 10^6 cm^{-3}. They are often somewhat misleadingly referred to as 'neutral' (HI) clouds because the most abundant species are neutrals such as H_2, CO, and He. Yet the degree of ionization is sufficiently high ($\gtrsim 10^{-17}$ per H atom s^{-1}) so that the chemical reactions are practically entirely determined by ion–molecule interactions. The energetic ions H^+ and He^+ rapidly transfer charge to a multitude of neutral species, giving rise to ionic reactants such as C^+, N^+ and O^+ and polyatomic species such as HCO^+, HN_2^+, H_3O^+ and H_2CO^+. The hundred most important reaction mechanisms have been analysed by Herbst and Klemperer (1973), and are found to be in consonance with the rotational molecular spectra observed in emission from the dark clouds in the microwave region. Because of the low collision rates and low kinetic temperatures only binary collisions and exothermic reactions without activation energy are of importance. This limits polyatomic molecule synthesis in the dark clouds to ion molecule reactions (Herbst and Klemperer, 1973).

As the source of ionization, energetic cosmic rays have been invoked (photoionization and photoactivation are negligible except in near-stellar regions because of opacity). The high degree of ionization, the observed magnetization of the clouds (Beichman and Chaisson, 1974; Chaisson and Beichman, 1975) and general magnetohydrodynamic considerations make it necessary, however, to conclude that large currents exist (see, for example, Alfvén, this volume) which are likely to be the main source of the ionization. Such hydromagnetically induced currents would sustain ionization also in the densest clouds which may be practically opaque to cosmic rays $\lesssim 100$ MeV.

Like in other space plasmas, which have been studied in more detail, electric currents in the dark clouds must also cause large scale heterogeneities which make homogeneous models inapplicable; local number densities orders of magnitude above the average would be expected in plasma filaments and sheets, and strongly enhanced electric fields would arise in electric double layers and at current disruption.

Such strong local enhancements in number density and excitation above the smoothed-out average indicated by astronomical observation are expected to affect locally the kinetics of chemical reactions in the dark clouds and in the solar and planetary nebulae.

Nucleation and Growth of Grains

Chemical treatment of the interstellar medium as a homogeneous, neutral gas made it difficult to understand the observed chemical reactions, and to explain the formation of dust in interstellar space. The long collision times, the low kinetic temperatures and the high activation energies for reaction would effectively preclude practically any reaction. For this reason it has been common to assume that interstellar grains are formed elsewhere, for example, in the atmospheres of cool stars, and transported into their present location. If this were entirely the case, the redistribution of grains through the dust clouds would pose great difficulties in view of the long time needed (Salpeter, 1974 a, b). Homogeneous nucleation would require the extremely improbable event of more than five to six atoms by chance to find themselves in contact with each other, and for this unstable cluster to survive until the arrival of yet other atoms, necessary to stabilize it. Classical homogeneous nucleation theory thus has the distinction of a record gap between prediction and observation; a factor of about 10^4. Nonetheless, it continues in the literature to be used as a basis for discussion of condensation even in the dilute space medium.

Laboratory investigations of ion–molecule reactions, together with studies of plasma reactions in the Earth's exosphere and in the interstellar clouds (Massey, 1972) cast new light on nucleation and condensation in space. Applicable theory relies on quantum chemical consideration of molecular bond formation and of the characteristic thermal regime in the space medium.

The relatively low abundance of species other than those containing carbon, hydrogen, oxygen, nitrogen, and sulphur presently limit relevant observation in space to molecules in the system C–H–O–N–S. Detailed kinetic models have focused on reactions involving ions and molecules of H, He, C, O, and N.

Returning to the dense interstellar clouds as an observational base, the very short lifetime of excited states (mostly $\lesssim 10^{-3}$ s) compared to the long collision times (hours or days), preclude any reactions that require activation energy, and leave ion–molecule reactions as the only process of practical significance for the formation of diatomic and polyatomic molecules and molecular clusters (Herbst and Klemperer, 1973). Taking into account only the five most abundant elements, these authors have identified the one hundred kinetically most important reactions, all in pronounced thermodynamic disequilibrium, and most proceeding practically only in one direction. The resulting reaction paths form a network, connecting the reservoirs of reactant ions and molecules; the size of each individual reservoir is determined by the excess of rates of formation by various processes, over the sum of the rates of destruction, mostly through other channels than the buildup processes. The oxygen containing segment of this network is shown in Figure 1.

Within a quasi-steady state domain of the plasma (which is certainly a very small part of an interstellar cloud, and probably also a small fraction of the individual clouds in the solar nebula) the ions and molecules, particularly those with low dissociation energies of the corresponding solids, are continuously cycled through the reservoirs in the reaction network. New material may be added from adjacent reservoirs, such as stars, including supernovae, embedded in the interstellar clouds, and in stellar nebulae by gravitational trapping of neutrals falling in from the surrounding source cloud.

Exit channels are presumably provided by growth of small molecules into macromolecules and eventually into grains. Solid grains, after sufficient equalization of kinetic energy, accrete into larger bodies. Such accretion mechanisms, which are applicable both to planet and satellite formation, are reviewed in Alfvén and Arrhenius, 1976 (Section 11 and 12).

The mechanisms by which the formation of polynuclear carbon compounds take place in interstellar clouds from this and other molecules have not yet been clarified. Herbst and Klemperer (1973) have pointed out that C^+ as the most abundant carbon containing ion is likely to play a dominant role.

The polymeric nature of the carbon and silicon compounds in carbonaceous meteorites, together with the demonstrated ion–molecule reaction control in dense clouds suggests that ion and radical polymerization reactions are important for the growth of molecules to grains. Such self perpetuating polymerization

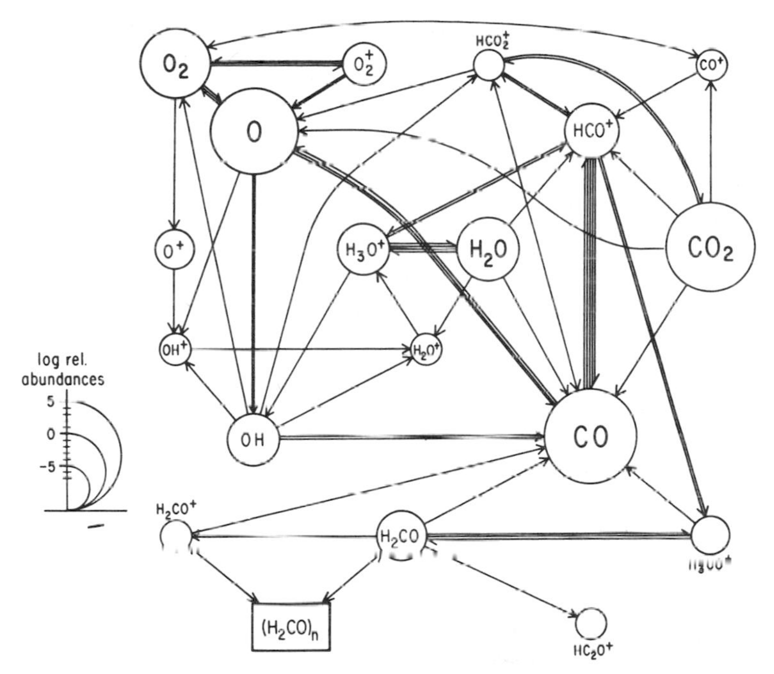

Figure 1. Reaction paths of oxygen containing species in the system H–C–O in dense interstellar clouds.

The reservoir magnitudes are approximated by the diameter of the circles on the scale in the figure. The reactions shown assume a number density of molecular hydrogen of 10^6. In the solar nebula, number densities, in local filamentary regions, could have been as high as the order of 10^{13} cm^{-3}, and the ionization rates strongly enhanced by the solar magnetic field and by ionization at the critical velocity limit. These effects, together with chemical differentiation between the clouds A–D (section on *Chemical Differentiation in the Emplacement Process*) would strongly change the reaction kinetics in the present figure, which is included here to give an example based on direct observation.

Lines between the reservoirs indicate reaction paths, and the arrows their directions. (Data from Herbst and Klemperer, 1973)

reactions, involving transfer of a double or triple bond to a colliding ion, are typified schematically by

$$Y^+ \; CH_2 = \underset{\substack{|\\G_m}}{CH} \rightarrow \left(Y - CH_2 - \underset{\substack{|\\G_m}}{C^+H}\right)^* \rightarrow Y - CH_2 - \underset{\substack{|\\G_m}}{C^+H} + h\nu$$

$$Y - CH_2 - \underset{\substack{|\\G_m}}{C^+H} \quad CH_2 = \underset{\substack{|\\G_n}}{CH} \rightarrow \left(Y - CH_2 - \underset{\substack{|\\G_m}}{CH} - CH_2 - \underset{\substack{|\\G_n}}{CH}\right)^* \quad \text{etc.} \qquad (1)$$

where Y^+ is a chain initiating ion, and G any functional group.

The rates at which different molecular bond donors contribute to polymer growth would, as a first approximation, be proportional to their number densities. Hence molecules such as $H_2C=CH_2$ and $HC\equiv CH$ would appear to be likely neutral participants. Their symmetry rules out observation by rotational emission but there are reasons to assume that they are among the more abundant molecules.

Vanysek and Wickramasinghe (1975) and Mendis and Wickramasinghe (1976) have proposed that polyoxymethylene $n{\cdot}CH_2O$ (Figure 1) forms a significant fraction of the volatile component of interstellar and cometary dust. Model experiments (for example, Miller, 1955; Ponnamperuma *et al.*, 1976) (cf. Figure 6) as well as the material in meteorites, demonstrate the formation of a range of different molecules. The residual material, left in meteorites, consists of complex polymers, crosslinked into refractory solids.

Excited collision complexes such as in equation (1) stabilize either by infrared emission, or by transfer of the excess energy at a second collision; the former process would dominate at densities characteristic of the dark clouds (Herbst and Klemperer, 1973). In calculating the rate constant for the formation of H_3CO^+, the precursor of H_2CO, these authors point at the controlling influence of dissociation, which keeps the forward reaction rate eight orders of magnitude below the Langevin value. Increase in size of the molecule, however, leads to a rapid decrease in the dissociation rate. This is due to the increase in the rate of spontaneous emission stabilizing the complex against dissociation. The combined effect of this increase on the forward and backward reaction rates are such that the net formation rate would increase by several orders of magnitude at the first addition of a mononuclear carbon complex. Hence the rate limiting step in molecular growth is the destruction of the simplest species, while already dinuclear and trinuclear complexes would have strongly enhanced stabilities.

Radical or ion polymerization reactions could control also the formation of inorganic polymers such as silicates, oxides and sulphides in dark clouds and in solar and planetary nebulae. For example,

$$\begin{aligned}
&Ca^+ + Si=O \rightarrow Ca^+\ Si=O \\
&Ca^+\quad Si=O \rightarrow (CaOSi^+)^* \\
&(CaOSi^+)^* \rightarrow (CaOSi^+) + h\nu \\
&(CaOSi^+)\ O=O \rightarrow (CaO_3Si^+)^* \\
&(CaO_3Si^+)^* \rightarrow (CaO_3Si^+) + h\nu \\
&(CaO_3Si^+)\quad Si=O \rightarrow (CaO_3SiOSi^+)^*
\end{aligned}$$

$$\text{etc.} \qquad (2)$$

Polyatomic molecules with rapidly increasing number of modes of vibration can effectively cool themselves by radiation, and when immersed in a thin, transparent medium, maintain a lower temperature than the medium that they condense from. Polymeric grains growing at small optical depth in a cloud in space will consequently be more stable than if they were in thermodynamic equilibrium with the temperature modes of the source plasma. This is further discussed in the sections below on *Temperature Differential* and *Thermal Equilibrium* between grains and the medium.

Condensation in space can consequently in principle be understood as a self-sustained ion or radical polymerization reaction which at low rate and at the small optical depth, where condensation is possible, proceeds in pronounced thermodynamic disequilibrium. The polymeric condensate particles are efficient radiators, but do not efficiently absorb photons until they reach a size comparable to the photon wavelength. Such properties of the initial condensate were predicted by Platt (1956); the presence of a substantial amount of matter in this size range is suggested by interstellar absorption in the far ultraviolet (Morton *et al.*, 1975).

Regardless of the assumptions made about internal pressure or other properties of the solar nebula, condensation is likely to have taken place in ion–molecule and radical–molecule interactions, possibly of the type discussed above, at shallow optical depth (section on *Arbitrary Assumptions about Pressure*). This process has little to do with the often made assumption of condensation by homogeneous nucleation of solids from a supersaturated neutral gas, cooling from high temperature under conditions similar to those in the laboratory where temperature equilibrium is maintained between the growing solid nuclei and the surrounding gas by means of heated furnace walls.

Chemical Differentiation in the Interstellar Medium

In consonance with the obsolete picture of the interstellar medium as a homogeneous neutral gas, efforts have been made in the past to explain the observable universe as also chemically homogeneous with specific 'cosmic' abundances of the elements (approximated by evaluation of solar photospheric spectra and by chemical analysis of carbonaceous meteorites, and idealized by the Suess–Urey (1956) nuclear–systematic criteria). In the meantime marked deviations from 'cosmic abundances' were observed in many stars, and Herbig (1968) noted a large underabundance of calcium in the interstellar medium compared to the Sun. Recent satellite measurements in the far ultraviolet (Rogerson, Spitzer *et al.*, 1973; Morton *et al.*, 1974, 1975) have made it possible to evaluate the composition of the interstellar medium in greater detail. The resulting element distribution indicates, in comparison with the solar photosphere, marked underabundances, especially of those elements whose oxygen compounds have high negative free energies of formation, and which thus are particularly stable against dissociation.

The 'underabundances' in the interstellar medium range from a factor 3 to 4 for elements such as carbon and oxygen to a factor 3000–4000 for calcium and aluminium. These observations suggest that for elements capable of forming refractory compounds, the steady state distribution (rather than equilibrium) between monomeric components on one hand and molecular aggregates on the other is displaced far in the direction to the latter in the interstellar medium, and that such molecular aggregates, including grains, contain about one half of the store of oxygen (Morton, 1974).

Chemical Differentiation around Magnetized Bodies

The manner in which densification of matter takes place to form stars within dark interstellar clouds is not known from observational evidence. Numerous hypothetical studies of such evolution have been made using methods of gas dynamics. Since we now know that the development takes place from a magnetized medium it is clear that any treatment that ignores the magneto-hydrodynamic aspects of the problem is about as realistic as a gas dynamic treatment of cometary plasmas or of the planetary magnetospheres. Rather than speculating *ab initio* and without observational basis about the origin of the Sun, we attempt to place boundary conditions on the state of the early Sun and the protoplanets from the observed structuring of matter around these central bodies (Alfvén and Arrhenius, 1976; Ch. 25).

In the present context we are limiting ourselves to the conclusions about the state of the primary bodies in the solar system which can be obtained from analysis of the distribution and properties of their secondaries. As for the processes responsible for the observed distribution we rely on the results of space probe and satellite observations of the actual behaviour of particles and fields in interplanetary space, on observations from interstellar clouds, discussed above and on the result of laboratory experiments.

Striking illustrations of the differentiation processes in the circumstellar environment are provided by the strong fluctuations in the composition of the solar wind. This phenomenon has been analysed by Geiss *et al.* (1969, 1971) in terms of dynamic friction, thermal diffusion and gravitational effects in the solar plasma.

An element separation process, possibly of quite general cosmic significance has been discussed by Lehnert (1969) and more extensively by Lehnert (1976) and by Bonnevier (1976). In the boundary regions between magnetized plasmas with widely different degrees of ionization, diffusive equilibrium would lead to elemental abundance differences in the two media. In permeable plasmas the differences are shown by Bonnevier (1976) to be controlled by the difference in ionization probability, in impermeable plasmas mainly by resonant charge exchange collisions.

In the regions surrounding the early protosun and later the protoplanets, gravitational acceleration of gas toward these bodies would seem to have been a major phenomenon responsible for the accumulation of matter in the solar system. The associated hydromagnetic effects would be expected to cause the large scale separation of matter, reflected by the preserved structure of the planet and satellite systems. The experimentally observed phenomenon assumed foremost of importance in this respect is the ionization at the critical velocity

$$V_{\text{crit}} = \sqrt{\frac{2eV_{\text{i}}}{m}} \tag{3}$$

at which neutral gas molecules, falling through a thin plasma in a magnetic field, become ionized and are then stopped by the field (Alfvén and Arrhenius, 1976; Section 21.8). The critical velocity for any given monatomic gas depends, as in equation (3), on the ionization energy eV_{i}, where V_{i} is the ionization potential and m the atomic mass. The critical velocities and ionization potentials of some of the most common elements are indicated in Figure 2.

Besides in the original thermonuclear experiment and later in many other types of laboratory plasma experiments, the critical velocity phenomenon is believed to be observed in the interaction of neutral gas with solar wind on

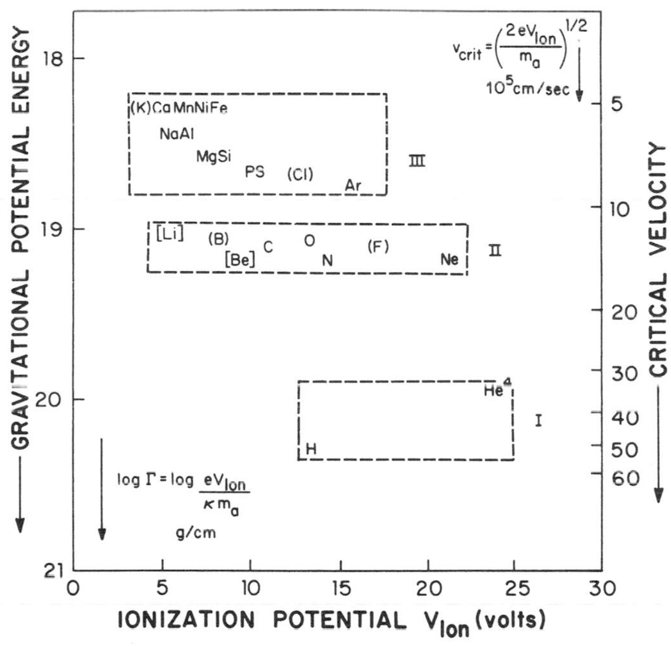

Figure 2. Critical velocity and ionization potential of the most abundant elements.

The left hand ordinate scale shows gravitational potential energy Γ, and the right hand scale the critical velocities of the controlling elements (and, in parentheses, some of the minor elements) of the A, B, C, and D clouds. (Reproduced with permission from H. Alfvén and G. Arrhenius, *Structure and Evolutionary History of the Solar System*, copyrighted by D. Reidel Publishing Co., Holland, 1975) the Moon (Manka *et al.*, 1972; Srnka, 1976). Theoretical analyses of the critical velocity phenomenon have been given by Petschek (1960), Lin (1961), Drobyshevskii (1964), Hassan (1966), Lehnert (1966,1967), Sockol (1968), and most recently and extensively by Sherman (1972, 1973) and by Raadu (1975). A comprehensive review and discussion of the various types of plasma experiments, which demonstrate the critical velocity phenomena, has been given by Danielsson and Sherman (1973).

Origin of Band Structure in the Planetary and Satellite Systems

A notable aspect of the Alfvén critical velocity effect is that it was originally predicted from the systematic distribution of the secondary bodies around their primaries in the solar system (Alfvén, 1942) long before it was verified by the laboratory experiments referred to above. The characteristic feature which prompted the prediction of this effect was the occupancy of specific potential energy bands by the planets and regular satellites, and the separation of these bands by energy gaps, practically devoid of matter (Figure 3).

Alfvén suggested that this band structure of matter in the present day solar system is a direct effect of selective ionization of groups of elements during the infall of neutral gas toward the developing Sun, and subsequently around the magnetized protoplanets in the late stages of their formation. The discovery of the critical velocity effect placed this suggestion on an experimental basis, and related theory made it possible to predict and evaluate the details of the associated phenomena.

Alfvén and Arrhenius (1975 and 1976) give a detailed discussion of these processes and of the subsequent evolution of the planet and satellite systems, including transfer of angular momentum to the plasma clouds, emplaced over a time period of the order of 10^8 years, their state of partial corotation with the central body, the orbital evolution of condensed and captured grains, the accretion of protoplanets, and the formation of satellite plasma clouds and satellite systems in a manner analogous with their counterparts in the planetary system. In the present context we will limit the discussion to the chemical implications of this proposed reconstruction.

Chemical Differentiation in the Emplacement Process

From the summary discussion above and from the detailed derivation of the dynamics of the emplacement process in Alfvén and Arrhenius (1975 and 1976) it is apparent that a major chemical differentiation would be expected between the plasma clouds from which the secondary bodies in the solar system are assumed to have formed. As in most natural processes, the differentiation would not be expected to lead to perfect separation of components; hence it is important to consider, and experimentally verify the spread and overlap, and the influence of minor elements on the zoning and differentiation phenomena.

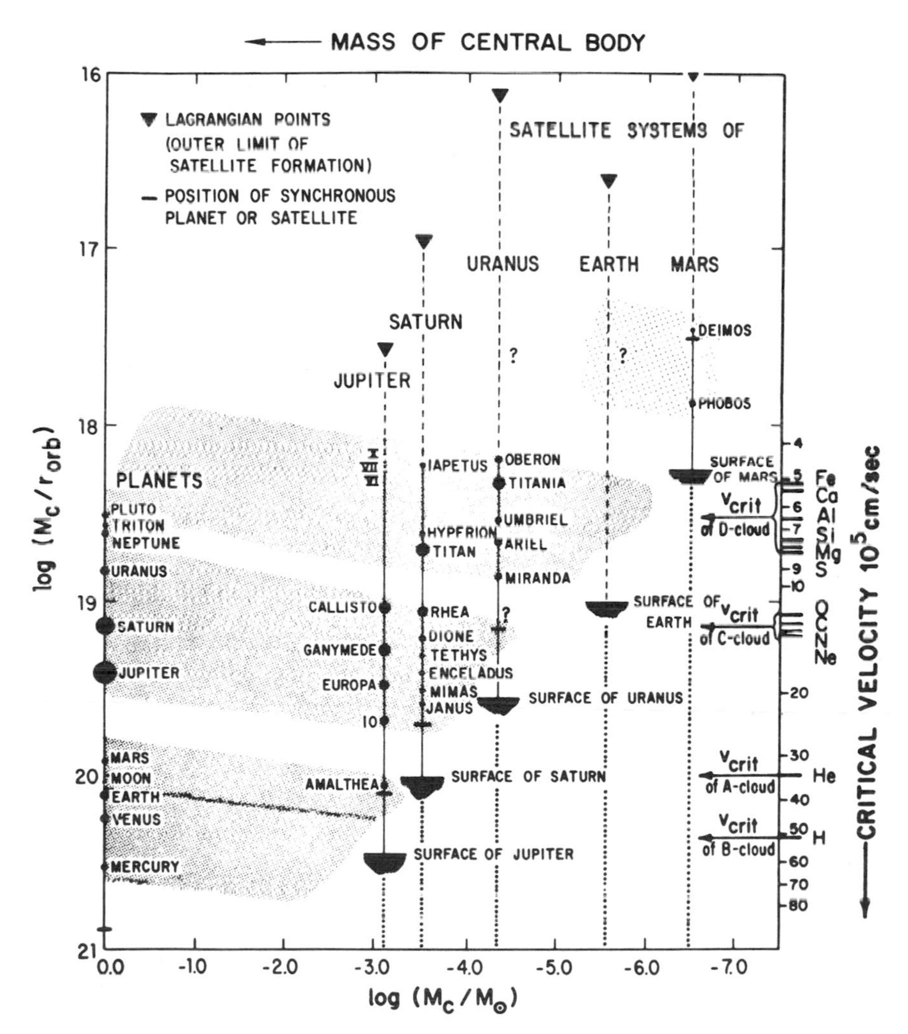

Figure 3. Structure of the planetary and regular satellite systems.

The individual systems are spaced along the horizontal axis with mass of the central body decreasing to the right; the Earth, which may have had a regular satellite system, and the rudimentary Martian system, are included for comparison.

The vertical scale to the left indicates the potential energy of the secondary bodies (M_c denotes the mass of the central body; r_{arb} the orbital radius of the secondary). The critical velocities of the major elements from Figure 2 are shown on the right. The band structure, proposed to result from the critical velocity zoning of infalling neutral gas, is indicated by shading. The details of the band structure including the slope of the bands, and the distribution of mass within them, and also the absences of regular satellite systems around Mercury, Venus, Earth and Neptune, are discussed in Alfvén and Arrhenius (1975, 1976). (Reproduced with permission from H. Alfvén and G. Arrhenius, *Structure and Evolutionary History of the Solar System*, copyrighted by D. Reidel Publishing Co., Holland, 1975)

Contamination of already emplaced plasma at subsequent infall events could also be of importance, and the capture of infalling dust in the chemically discrete plasma clouds would tend to modify their original composition. The major elements regulating plasma emplacement in the A (He), B (H), and C(C, CH, O, N, Ne) clouds are only to a small extent condensable. The solid residues condensed from or trapped in the clouds, and now accreted into planets and satellites by mechanisms outlined in detail in Alfvén and Arrhenius (1975 and 1976), would thus consist of minor contaminants of the primordial plasma clouds. The physical effects of such minor components is discussed in the section below on *The Effect of Minor Components on the Critical Velocity*.

The four major plasma clouds, stopped at different gravitational potential energy levels around the Sun and the magnetized planets (Figure 3), would thus function as regulators of the distribution of primordial matter in the solar system, derived from infalling neutral gas and dust.

A and B Clouds

Since the regulatory, major components of the A and B clouds, hydrogen and helium are volatile and since furthermore the terrestrial planets forming from these clouds are not massive enough to capture the light gases gravitationally, these have not been retained but must have diffused inward into the Sun and outward into the region of the C cloud and the giant planets during the time span of the formative era, of the order of 10^8 years. The condensable material in the region of the A and B clouds (terrestrial planets around Sun; Amalthea around Jupiter; Figure 3) would thus be expected to consist of impurities carried in the infalling A–B gases. These, as well as the other clouds, would furthermore be expected to electromagnetically capture and bring into corotation infalling interplanetary dust with particles in the size range up to 10^{-4} cm (Alfvén and Arrhenius, 1976; Section 5).

Larger grains, grain aggregates and transplanetary accreted bodies of large size (perhaps to the order of 10^3 m) in parabolic orbits would, in analogy with present day meteors in the terrestrial atmosphere (or, with a chemically more close analogy, the Jovian atmosphere) undergo frictional ablation in the tenuous but extensive A and B clouds. This contaminant ablation plasma would add to the condensable components in the A and B clouds. It is possible that the difference in volatilization of magnesium silicates and iron in the reductively inert A cloud (He) and the reducing B cloud (H) are responsible for the higher density and hence probably higher iron content of the inner terrestrial planets, formed in the B cloud, inside the overlap zone with the A cloud in the terrestrial region.

C Cloud

Ionization at the critical velocity limit would stop infalling gas jets with carbon, nitrogen, oxygen and neon species as main components, in the gravita-

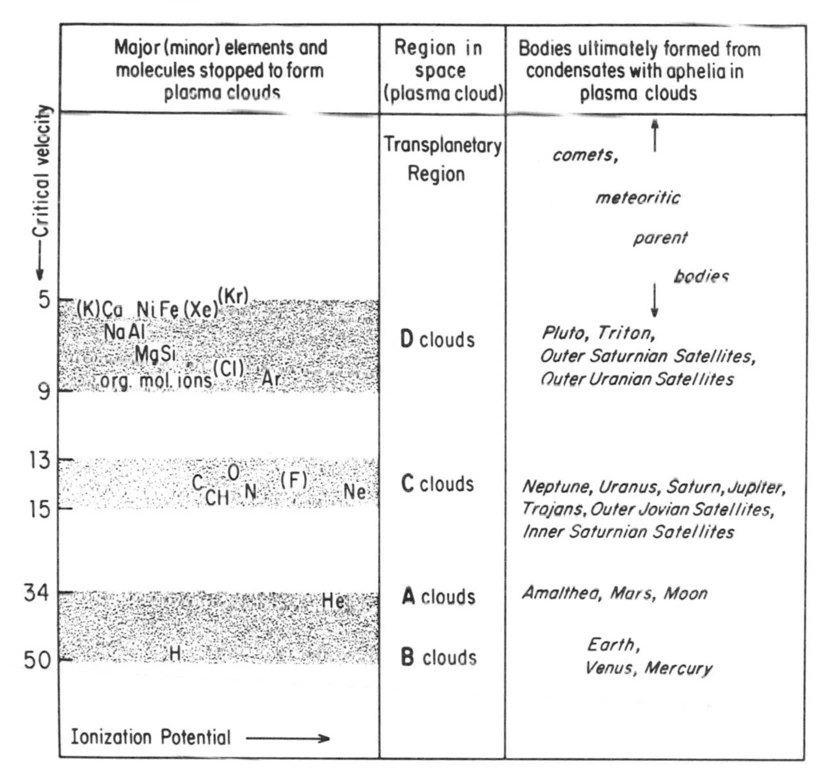

Major (minor) elements and molecules stopped to form plasma clouds
Region in space (plasma cloud)
Bodies ultimately formed from condensates with aphelia in plasma clouds
Critical velocity
5
9
13
15
34
50
(K)Ca Ni Fe (Xe)(Kr)
Na Al
MgSi
org. mol. ions (Cl) Ar
C O CH N (F) Ne
He
H
Ionization Potential
Transplanetary Region
D clouds
C clouds
A clouds
B clouds
comets, meteoritic parent bodies
Pluto, Triton, Outer Saturnian Satellites, Outer Uranian Satellites
Neptune, Uranus, Saturn, Jupiter, Trojans, Outer Jovian Satellites, Inner Saturnian Satellites
Amalthea, Mars, Moon
Earth, Venus, Mercury

Figure 4. Critical velocity and main components of plasma clouds formed from infalling neutral gas around the Sun and the magnetized planets. In the right hand part of the diagram are indicated the bodies formed from the respective clouds

tional potential energy region of the giant planets, and of the outer Jovian and the inner Saturnian satellite systems (Figures 3 and 4). This energy region is widely separated from the A–B region, containing the terrestrial planets and the Jovian satellite Amalthea. The intervening energy gap is practically devoid of matter (Figure 5), although the small amount present in the form of asteroids is of great physical and chemical diagnostic importance since it is close to us.

Condensable molecules in the C cloud would be likely to include the polymers which characteristically form in ion–molecule reactions in the C–H–O system (Figure 6) and which are the dominant carbon compounds in carbonaceous meteorites.

Extraneous impurities in the C cloud would, for the same reasons as in the A–B clouds, probably include electromagnetically captured transplanetary dust. A large contribution of hydrogen, normally stopped in the B cloud, would also be expected to accumulate in the C cloud from several sources. One of these comprises molecules of CH-species which have critical velocities in the C-cloud range and hence would be ionized and stopped there (Figure 4).

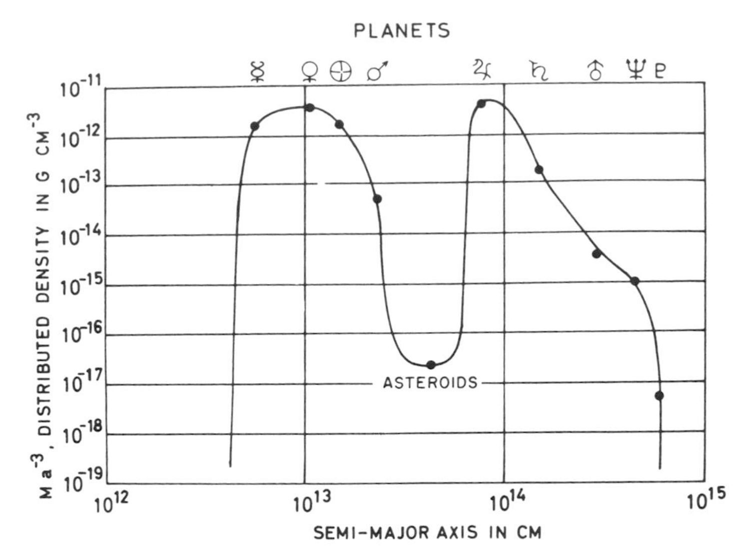

Figure 5. Distributed density versus semi-major axis for the planets, showing the density gap between the A–B region (terrestrial planets: cf. Figure 3) and the C region (Jupiter, Saturn, Uranus, Neptune) (Reproduced with permission from H. Alfvén and G. Arrhenius, *Structure and Evolutionary History of the Solar System*, copyrighted by D. Reidel Publishing Co., Holland, 1975)

C-cloud hydrogen would be contributed also as an impurity in the infalling gas controlled by major species such as C, N, O, CH and Ne. Neutral hydrogen and helium, falling toward the B and A regions would furthermore be expected to interact with any plasma intercepted in the C and D clouds, and be partially ionized and trapped there. As demonstrated by Axnäs (1976) (see also section on *The Effect of Minor Components on the Critical Velocity*), binary oxygen–

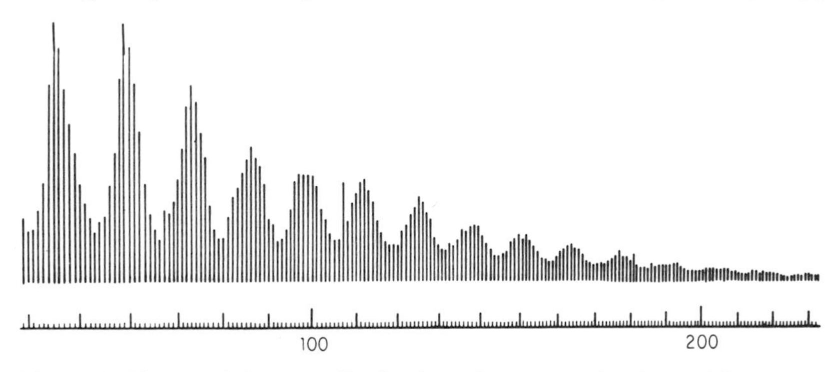

Figure 6. Characteristic mass distribution of parent molecules and fragments in the range 40–230 *m*/*e* in a plasma synthesis experiment.
Source composition $CO/H_2 = 0{\cdot}20$; ionization > 0·5 per cent kinetic temperature 77 K

hydrogen mixtures, containing as much as 50–60 per cent H have critical velocities similar to that of oxygen (within 20 per cent of the critical velocity difference between H and O, Figure 7). This critical velocity entrapment of disproportionately large amounts of hydrogen by heavier gas species is observed also for Ar, less pronouncedly for N and not at all for Ne.

A fourth source of hydrogen and helium in the C cloud is probably the dissipation of these non-condensable gases from the A–B clouds, which are emplaced closer to the central bodies (in the region of the terrestrial planets in the planetary systems; in the region of Amalthea in the Jovian system; see Figure 3 above). At such a stage in the planetesimal accretion in the C cloud ($t \sim 10^6$ years) when the protoplanets have grown to sufficiently large size (a few Earth masses) to efficiently retain hydrogen, a fraction of the hydrogen and helium, diffusing outward in the ecliptic plane, would possibly be swept up by these growing bodies with rapidly increasing efficiency.

In conclusion, although the bodies forming in the C cloud, particularly the

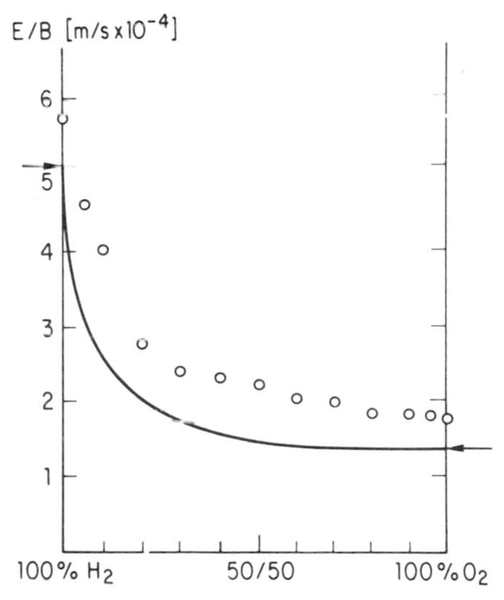

Figure 7. Critical velocity (E/B) as a function of mixing ratio of hydrogen in oxygen. Circles indicate experimental data; curve theoretical relationship.

The experiments demonstrate that substantial amounts of hydrogen in a hydrogen–oxygen mixture of neutral gas will attain critical velocity and become ionized at or near the critical velocity limit of oxygen. (Reproduced with permission from I. Axnäs, *Report No. TRITA-EPP-76-02, Department of Electron and Plasma Physics*, copyrighted by Royal Institute of Technology, Stockholm, 1976) giant planets, would be expected to contain considerable amounts of hydrogen, their compositions are likely to be dissimilar from the solar surface composition, frequently assumed in homogeneous Laplacian models. Due to the complexity of the several contributing processes it is impossible to predict the compositions accurately; direct observational evidence is necessary. Since we will never be able to obtain direct information on the interior of the giant planets, and since model calculations become highly uncertain when simplifying assumptions are abandoned (Alfvén and Arrhenius, 1976; Section 20.5), sampling of the small bodies in the C region, including the Trojans, which possibly represent a residue of Jovian source material, is a particularly important aim for understanding the origin of the solar system.

D Cloud and Transplanetary Space

The outermost of the well-defined clouds, the D cloud, would be controlled by the elements with the lowest critical velocity and would have provided the source material for Pluto and Triton in the planetary system, and the outer Jovian and Uranian satellites in the satellite systems. Triton, nowadays orbiting Neptune, has characteristics strongly suggesting that it has been captured by Neptune from a planetary orbit (McCord, 1966); see also discussion in (Alfvén and Arrhenius, 1976; Section 24.4).

The controlling elements in the D cloud are largely condensable, the most abundant being iron, magnesium, and silicon. It would be expected, however, that the enhanced plasma densities in the cloud would lead to collision interaction and capture of A, B, and particularly C material, falling through this region. Substantial amounts of carbon, oxygen, and nitrogen would thus be expected as contaminants. The noble gases in the D cloud would be enriched in the heavier species and correspondingly depleted in helium (A cloud) and neon (C cloud). Chlorine (and the heavier halogens) would also be enriched in the D cloud, relative to fluorine.

Observations that indicate the presence of higher density material in the outermost region of the solar system are the increasing density from Saturn to Uranus and further to Neptune, and the suggested enhanced densities of Triton and Pluto. In the latter two cases the uncertainties in the measurements are very large, and no definite conclusion can be made. The increased densities of Uranus over Saturn, and of Neptune over Uranus are, however, beyond observational error, and cannot be due to higher internal compression, since Neptune's radius is comparable to but somewhat smaller than that of Uranus.

The D cloud would be expected to have a diffuse delimitation outward. Beyond the Lagrangian points L_5 of the central bodies, their gravitational control of the emplacement process would be perturbed. Also the decreasing strength of the solar magnetic field with increasing radial distance would, beyond the outer region of the planetary system, curtail the transfer of angular momentum to the infalling matter. Material of D cloud type condensing in transplanetary space would consequently assume highly eccentric Kepler orbits such as those typical of comets.

It is commonly assumed on the basis of dynamic arguments (Öpik, 1968; Wetherill, 1971) that most meteorites, particularly carbonaceous meteorites, are of cometary origin. If we assume that comets derive from the D cloud or the transplanetary region of the solar system, the most direct and detailed evidence of the chemical composition of extraterrestrial material, unmodified by planetary differentiation, relates to these outermost regions of the solar system and beyond. The chemical separation effects due to the critical velocity phenomenon are perhaps most suggestively demonstrated by the noble gases. In the condensed grains, later accreted to form the parent bodies of carbonaceous meteorites, these occluded gases occur in proportions (Signer and Suess, 1963; see also Black, this volume) which characteristically differ from those in the outer regions of the Sun. The high relative concentration of the heavier noble gases in carbonaceous meteorites (Weber *et al.*, 1971, 1976) may be the result of the critical velocity separation effect in the cloud plasma from which the meteorite material formed (cf. section on *State of Excitation of the Medium in the Solar and Planetary Nebulae*). Whether this condensation took place in the D cloud of our solar system or in the interstellar source cloud from which the material fell in, remains conjectural.

The Effect of Minor Components on the Critical Velocity

As indicated above, the discrete zoning of the plasma clouds around the Sun and around the magnetized planets would be regulated by the critical velocity of the major infalling neutral gas species. The presence of other minor components is of obvious importance, since in the inner solar and Jovian systems, minor condensable components form the only residue now remaining of the primordial nebulae transformed into planets and satellites.

Experimental studies of these effects have been initiated by Axnäs (1976) who measured the critical velocity effect of addition of a second gas component. The results indicate that minor gas components (in the range of a few per cent) do not become ionized at their own characteristic critical velocity, but follow the major controlling components. The critical velocity of the major component is modified to an extent which is as a first approximation proportional to the fraction of the minor component. Important deviations are, however, found for hydrogen as a second component in oxygen (Figure 7 above) or in argon; the critical velocities of these two gases is only little affected by addition of hydrogen in amounts as high as 50 per cent.

Thus one would expect impurities in the infalling neutral gas jets to follow the distribution of the major components. The best place to study the actual perturbations of the sharp, ideal critical velocity effects would be in the widest cloud gap, which occurs between the C and A clouds (Figure 3 ; Figure 5). In the solar system this void region contains a small mass ($\sim 10^{24}$ g) now in the form of the asteroids. Systematic differences in composition between asteroids in the outer and inner part of the belt could provide indications of the difference between the C and A cloud composition of condensables, and of the impurity levels expected in order to perturb the small amounts of C type material to penetrate inside the region of the giant planets and small amounts of A material to reach the critical velocity limit outside of the region of the outer terrestrial planets. These chemical considerations add to the importance of controlled sampling and analysis of the asteroids by space missions to them (Alfvén and Arrhenius, 1972; Arrhenius *et al.*, 1973). Suggestive indications of chemical differences are provided by optical reflection spectra from asteroids (Chapman and Salisbury, 1973; Johnson and Fanale, 1973) but the distinctive resolution in these measurements is marginal, and the information comes only from a thin surface layer.

Chemical Differentiation at Condensation

The discussion above of the fractionation associated with emplacement of gas in the circumsolar and circumplanetary regions is largely based on observation of the present day physical properties of the solar system and on indications from physical laws governing processes in space. Direct chemical information, available now, is limited to observations, which although suggestive, provide no definite proof. This situation cannot be improved until chemical exploration of the solar system has proceeded further; we need samples of undifferentiated solid material of unquestionable primary origin in the C and D clouds. In contrast, attempts to solve the problems by speculation in the framework of terrestrial equilibrium processes are unencumbered by such needs. In these views (see, for example, Larimer and Anders, 1970; Anders, 1971, 1972; Lewis, 1974; Grossman and Larimer, 1974), the chemical evolutionary processes can be deduced in minute detail from the composition of meteorites reaching the surface of Earth, by assuming that the location of their origin is known, that they formed from a space medium consisting of neutral gas of solar photospheric composition, and that the condensing material did not obey the radiation laws (Alfvén and Arrhenius, 1976, Section 1; and discussion below).

A realistic reconstruction of the condensation processes, and of the differentiation at condensation necessarily has a different character. Tangible samples of space condensates are indeed available which can be analysed in the laboratory with highly discriminating and sensitive techniques. A wealth of data from such research is available, and the quality and detail of the analyses surpass most investigations of terrestrial materials. However, the source of this material is not well known, and without adequate coupling of the compositional information to the known laws of behaviour of matter in space the interpretation is bound to be unrelated to reality. In order to clarify all aspects of the complex differentiation and condensation processes, additional chemical and structural measurements are needed, and they must be placed in context with relevant observationally based physical theory.

Quantum phenomena in energy regions favourable for observation make it possible to identify individual molecular species and excitation states even in regions remote from our solar system, and thus to obtain realistic insights into the processes of extreme thermodynamic disequilibrium that give rise to solid matter in space (sections above on *Space Materials*, *Dark Interstellar Clouds*, *Grains*, and *Interstellar Medium*). Comets provide nearby demonstrations of the actual chemistry of space plasmas and their interaction with solid dust (Mendis, 1976; Mendis and Ip, 1976; Mendis and Wickramasinghe, 1976). Experiments on the Moon and on earth-orbiting space stations make it possible to study condensation phenomena and particle behaviour in the actual space environment. Model experiments in terrestrial laboratories offer control over individual parameters, and make it possible to draw realistic conclusions about the corresponding behaviour of complex chemical systems in space (Meyer, 1971; Arrhenius and Alfvén, 1971; Brecher, 1972; Kristoferson, 1975; Kristoferson and Fredga, 1976, Do and Arrhenius, 1978)

All of these favourable circumstances are of use for establishing boundary conditions for condensation in the early solar system and to eliminate parochial inferences from the bottom of the atmosphere. However, when analytical data from condensates agglomerated in meteorites have to be interpreted in terms of specific properties of specific regions of the solar nebula, additional difficulties arise. They can be grouped in three different categories.

Uncertainty in Location

We have no reliable information yet where the condensates formed, which are observed in carbonaceous meteorites. They could have formed in the source cloud of the solar nebula outside the planetary region of our solar system. Or they could have condensed in the outer planetary region from the partially corotating nebula.

If the latter were the case, we could learn something about condensation in the outer planetary regions of the nebula from the material in the carbonaceous meteorites. The understandable emotional appeal of this possibility has been so great that in most discussions today this rather arbitrary assumption is taken for granted and hardly subjected to debate.

Obsolete Models of the Solar Nebula

Cosmochemists have embraced a concept of the solar nebula as a hot luminous mass of gas with kinetic temperature in the range of a few thousand kelvins, collapsing on itself in disregard of modern knowledge of hydromagnetics. This concept is rooted, ultimately, in the Laplacian disk notion, which was not taken very seriously by Laplace himself, but which became firmly based on scientific necessity in the nineteenth century, when a source had to be sought for the thermal energy of the stars. Although a celestial object of this nature has never been observed, the ability of the Helmholtz–Kelvin contraction process to explain why stars are hot made this lack of observational basis a minor matter. A more likely source of stellar energy was discovered only a few decades later, however, practically all cosmochemical discussion of condensation in the solar nebula continues to rely on the Helmholtz–Kelvin concept (see, for example, Wood (1963) and numerous papers by Larimer, Anders, Grossman, and other outstanding experimentalists, using this nineteenth century notion in the interpretation of their data).

Although we still do not know much about the detailed properties of the solar and planetary nebulae, modern theory based on observation of the properties of diffuse matter in space (e.g. Alfvén, 1954; Spitzer, 1968; Alfvén and Fälthammar, 1963; Freeman, this volume) places drastically different boundary conditions on the discussion of condensation in the circumsolar region. Such constraints are imposed not only by observationally based theories which take hydromagnetic processes into detailed account, but also by models analysing the gas dynamic development in the absence of, or after eliminating, magnetohydrodynamic effects. The most elaborate modern study of this kind is probably that of Cameron (1976, this volume).

Arbitrary Assumptions about Pressure

Deductions or *ad hoc* assumptions about pressure in the solar and planetary nebulae vary over more than ten orders of magnitude. The actual pressure relations are of importance from the point of view of allowable dynamics of formation of planets and satellites, particularly for the necessary angular momentum transfer from the central bodies to their secondaries.

The question of the condensation of solids from the nebulae must of course ultimately be compatible also with these physical constraints. However, the problem of the possible range of pressures at condensation appears largely independent of the question of internal pressures and pressure distributions in the nebulae.

The obvious basis for this is that condensation can occur at a high relative rate only in the outermost regions of any imagined massive nebulae. This has to do with the fact (see section below on *Temperature Differential between Condensing Grains and the Surrounding Medium*) that grains, in order to cool by radiation, have to be located in an optically thin region. Hence the maximum pressure that could be of any importance in this context is the pressure at about one optical depth or less in the nebulae. Whether pressures inside any Laplacian type of nebula is conjectured at 10^{-4} or 10^{2} bar would consequently appear irrelevant from the point of view of the discussion of condensation by radiative cooling.

The important question then is how far below the radiation imposed limit the actual pressures at condensation would have been, introducing other important constraints. Assessment of this pressure problem in terms of magnetohydrodynamic theory stresses the notion that the formation interval of the solar system must have had a considerable duration. During this interval, estimated to be of the order of 10^7–10^8 years, the actual density in any given region of the nebula would depend on the balance between rate of infall and emplacement of matter, rate of removal into larger bodies by accretion, and rate of diffusion of non-condensable gases. As a result the pressure in the medium would always remain low, on the basis of this argument probably below 10^{-5} bar at any given time, also in the densest local regions in the emplacement clouds.

Another critical constraint on the average pressure (number density) is placed by the need to transfer angular momentum from the Sun to the planetary precursor plasma and from the magnetized planets to the satellite source plasma. As far as I am aware, the only process proposed to quantitatively satisfy this fundamental need of any solar system theory, including numerous critical detail features such as the inclination of the Uranian satellite system, is hydromagnetic angular momentum transfer (Alfvén 1942–46; Alfvén and Arrhenius, 1976, Section 16; Alfvén, this volume). This process requires average number densities not to exceed the order of 10^7–10^9 cm^3. In view of the extreme inhomogeneity associated with currents and fields in space, local number densities in filaments and superprominences could probably rise two to three orders of magnitude above the average. In the temperature range of 5000 to 15000 K this would correspond to local gas pressures of the order of 10^{-8} to 10^{-6} bar.

Within the framework of consistent partial models for the formation of the solar system outlined in Alfvén and Arrhenius (1975, 1976) and earlier work, local plasma and gas densities in this range would consequently be considered an upper limit but the uncertainty about actual local and average pressures ranges several orders of magnitude downward from this approximate limit.

Temperature–Pressure Path of Condensation

The locking of cosmochemical interpretation onto the pre-nuclear Helmholtz Kelvin nebular model, and the lack of recognition of the constraints on radiative cooling discussed in section on *Arbitrary Assumptions about Pressure* and *Temperature Differential between Grains and Medium*, has limited practically all present discussion of condensation by meteoriticists to cooling at a practically constant and arbitrarily chosen high pressure from neutral gas, originally at an assumed temperature of about 2000 K. Occasionally, suitable but more or less isolated features have been borrowed from physical models developed in the 20th century, most commonly from those of Cameron and collaborators in the era before this meeting; the new developments by Cameron (1976; this volume) introduce concepts which are clearly distinguished from those invoked by meteoriticists in recent years.

It is in fact possible to fit the evidence from meteorites into several entirely different interpretations some of which have the further merit of relying on observationally based physical models. It is desirable that all such chemically and at the same time, physically acceptable possibilities be discussed, in order to break the currently prevailing circularizing arguments.

It is, for example, important to realize that the minerals and mineral associations as presently described from the extensively investigated Allende meteorite and other meteorites of similar type, must not necessarily have formed at isobaric cooling of a medium from high temperature (and are for reasons discussed below, unlikely to have grown at the imagined high grain temperatures). Purely from the point of view of thermodynamic phase stability the minerals, to the detail at which they are now known, could equally well have formed at isothermal compression of a lower temperature 'gas' originally of interstellar density. Much in the same way as along an assumed cooling path, a mineral formation sequence would be expected where phases with large negative free energy of formation would form first, followed by increasingly weakly bonded phases. Phases with extreme dissociation energies include aluminium oxide, calcium aluminium silicates and some of the platinum metals. These are now commonly and misleadingly referred to as 'high temperature minerals', although they are perfectly stable at low temperature. The current misuse of terms such as 'high temperature minerals' has thus created a false, but widely held belief that a high temperature origin of such minerals (often specified with numerical temperature values to the nearest degree) must be taken for granted.

From a thermodynamic point of view, the term 'high temperature mineral' is in fact well-defined and refers to a phase, which at a particular pressure is stable above a specific temperature, but which upon cooling below this temperature converts to a different structure or exists only metastably. An example of a correctly termed high temperature phase, occurring in meteorites, is γ iron with face-centred cubic structure, which (in ideally pure form) at 1183 K and a pressure of one bar becomes unstable against transformation into the low-temperature α-modification with body centred cubic structure.

Among the meteorite minerals misleadingly labelled 'high temperature phases' in an increasingly voluminous literature are corundum (α-Al_2O_3), perovskite ($CaTiO_3$), gehlenite ($Ca_2Al[SiAlO_7]$) and spinel ($MgAl_2O_4$). All of these are, as far as is known, perfectly stable at low as well as at high temperature, and their presence in itself does not give any indication of their temperature of formation. Several minerals such as corundum are grown commercially in thermodynamic equilibrium at temperatures of the order of 500 K. They are equally stable at room temperature, but kinetic economics dictates the use of comparatively high growth temperatures.

The obvious reason for current use of thermodynamically inappropriate terminology by *a priori* classifying any strongly bonded and hence in most cases refractory compound as a 'high temperature phase' is the tacit assumption of a series of processes and circumstances in the environment of formation, some of which we claim to be highly unlikely, others to be incompatible with fundamental physical laws. The refractory phases would be the first ones to condense, and at a relatively high temperature, *if* they crystallized solely as a result of cooling of a medium at high temperature. From the fact that such 'high temperature phases' are observed, the conclusion is drawn that the model is correct. The argument is perfectly circular; however, attempts to draw attention to this and related conceptions are considered as eccentric (Metz, 1975; Chou *et al.*, 1976). (In an apparent misunderstanding of the discussion in articles by Alfvén, Arrhenius and author combinations, Chou *et al.* (1976) claim that these authors, unlike the rest of the scientific community, do not believe in the fact that refractory compounds are refractory.)

It is the purpose of our investigations to develop model independent objective criteria for determination of the grain temperature at which meteorite condensate crystals formed, and for inferring the characteristic temperatures of the medium from which they condensed. Some such criteria are discussed below.

The intention here is not to arbitrarily propose just another of the many condensation paths compatible with the observed phases in meteorites. Since the mineral properties discussed so far in the literature do not define any specific path we simply do not know from this evidence if cooling was more important than compression, or in which combination the two processes could have operated.

Instead we prefer to be guided by unequivocal thermometric observations to the extent that they exist in the minerals, and by the boundary conditions imposed by the present day structure of the solar system, and by the actual properties of matter in space, established by space and laboratory experiments.

Such a framework of limiting conditions, discussed in detail in Alfvén and Arrhenius (1975, 1976) suggests, in agreement with most other literature dealing with the actual space medium, that condensation takes place in regions of circumstellar and interstellar space where the gas pressure is increasing due to gravitational densification. This process is necessarily associated with localized conversion of kinetic energy into electronic excitation and ionization, resulting in regions with drastically increased excitation temperatures, both in the inferred critical velocity bands around magnetized bodies and in the observed magnetized source clouds in transplanetary space. The translational temperatures in clouds of the solar nebula are more uncertain; in the observed source clouds they are, on the average, low.

Another important mode of condensation in primordial circumsolar space must have been from plasmas generated at the collision of solid particles with sufficiently high relative velocities to result in vaporization. It is possible that, as is generally thought, the meteoritic chondrules result from such collision melting and vaporization. Collisional interaction of solid grains is, regardless of the chondrule evidence, a necessary stage in all theories of evolution of grain orbits and planetesimal accretion (Alfvén and Arrhenius, 1976, Sections 12, 13).

Furthermore, condensation from expanding plasma shells around stars may possibly have contributed to interstellar solids eventually accreted into meteorites (see, e.g., Salpeter 1974 a, b).

Temperature Differential between Condensing Grains and the Surrounding Medium

The lack of realism of the Laplacian nebular concept in terms of modern knowledge of processes in space has been discussed above. In addition to, and entirely independent of this, another important misconception is a part of the foundation of the 'equilibrium cooling hypothesis' as formulated by Larimer and Anders (1970) and Grossman and Larimer (1974) to rationalize the numerous important chemical observations in meteorites that they and many others have assembled.

Under the conditions visualized in the nebular models used as basis for this hypothesis, solid nuclei in order to grow by condensation, need to cool together with the source medium from which they are thought to form. For radiative cooling of an element of gas and grains to be of importance it needs to be at shallow optical depth in the cloud. Since solid grains radiate much more efficiently than gases, particularly homonuclear diatomic gases such as H_2, or monatomic He, the grains will necessarily assume a lower temperature than the source medium, and actually provide the by far most efficient coolant for the medium.

The rate of condensation increases with the magnitude of the temperature difference between the grains and the source medium, and with the flux of atomic particles to the grain surface. The grain–gas temperature differential decreases exponentially with increasing optical depth while the particle flux increases exponentially. As a result, a maximum in the rate of condensation, dominating the production of solids, would be expected at shallow optical depth in the surface layer of an imagined Laplacian disk. At this production maximum a large temperature difference would prevail between grains and the surrounding medium.

Condensation in thermodynamic equilibrium between grains and gas under the physical conditions assumed by the proponents of this condensation concept consequently appears unrealistic (entirely aside from the fact that these conditions themselves are in conflict with observation and theory as discussed above and in detail in Alfvén and Arrhenius (1975, 1976)).

Straightforward and quantitative accounts of these fundamental considerations have been published (Arrhenius and De, 1973; Arrhenius, 1972; De, 1973; Alfvén and Arrhenius, 1975, 1976), and presented at numerous public discussions involving the community of meteoriticists. The controlling importance of the thermal disequilibrium between solids and surrounding 'gas' in space is well-known to astrophysicists and experimental space chemists since the fundamental work by Planck, Jeans, Lindblad and Spitzer, decades ago. The conclusion of strong thermal disequilibrium at primordial condensation was consequently self evident to Urey (1952) in his classical discussion of the chemical aspects of the origin of the solar system.

In spite of all this, the lure of apparent simplicity and the gratification from arithmetic answers calculable with great precision appears so overwhelming to most meteoritic theorists, that the hypothesis of equilibrium condensation on cooling of a Laplacian nebula and a myriad of ramifications have become articles of faith.

Needless to say, there are several situations realizable in actual space, where thermal equilibrium between solids and the space medium can be approached. One is at very low temperature, here, however, large differences between kinetic and vibrational temperatures are bound to occur (section on *Kinetic Isotope Effects in Space*). Another is at slow gravitational compression, such as discussed above. Strong excitational singularities, such as ionization at the critical velocity limit, are however associated with the gravitational densification phenomena, and affect the approach to thermal equilibrium.

Finally, condensation at cooling during expansion includes identical temperature of grains and expanding medium as a fortuitous transient. This cannot, however, be characteristic of the process as a whole.

Progress in this field obviously depends on evaluation of all possible conditions for condensation within a physically realistic framework, and on the analysis of parameters of the solids which bear a unique relationship to their temperature at condensation (sections below on *Temperature Parameters*, *Volatile Components*, and *Evidence*).

In contrast to the temperatures of the condensing grains, the temperatures of the source medium is more difficult to determine uniquely from the presently observed properties of the solid grains. However, some general conclusions can be drawn and are discussed below.

Only if and when the various temperature parameters of the vapour phase are defined (electron, ion, and atom temperatures, and kinetic as well as excitational temperatures of molecules and molecular ions) can the properties of solid condensates provide information about the number densities in the vapour medium; we are still far from this possibility. In the meantime we can only rely on boundary conditions established from other considerations such as discussed above in Alfvén and Arrhenius (1976) and in the literature referred to there.

Evidence for the Sequence of Condensation

As pointed out above, the thermodynamic properties alone of the characteristic refractory minerals in C3 chondrites convey no specific information about the temperature at which they were formed. Nor does the presence of isolated grains aggregated in meteorites give any indication about a sequence of formation; they may have formed in different regions and at different times.

In contrast, from the way in which minerals are intergrown, and show interdiffusion of components, suggestions can be made about the time sequence, and about possible reactions between the phases or with the medium from which they grew.

The inclusions in the Allende meteorite have attracted considerable attention for this reason. A wide variety of suggestions have been made to explain the observed interrelation between different minerals in the aggregates. Some of the features are very likely the result of reaction between the solid phases. Others are highly suggestive of change of growth conditions during the formation of aggregates, or individual crystals. In some observed cases, where the changes happen to be in the right direction, these features have been invoked to support the notion of an equilibrium reaction series at cooling from a high temperature gas in which the radiating grains are in an unspecified manner kept at the same temperature as the gas.

In numerous other cases, however, a variety of other interpretations are forced by the observations. Among them are suggestions of condensation as subcooled liquids (Blander and Fuchs, 1975; Nagasawa *et al.*, 1976); formation as residue after vaporization (Kurat, 1970), or accidental arrangement of originally free grains of 'high temperature phases' to surround 'low temperature phases' resulting in a sequence reverse to the one dictated by a hypothetical reaction series (Grossman and Larimer, 1974). However, in spite of the fact that the observed textures demand various conflicting explanations of this kind, the general discussion among meteoriticists, often even by the same authors, continues to fall back on a concept of reaction equilibrium between the already condensed phases and the residual components in the vacuum of the nebula at decreasing temperature and constant pressure. This concept is derived from the misleading assumption of equilibrium cooling condensation discussed above.

Indeed, the number of such textural and mineralogical observations, conflicting with this assumption and its consequences, have been mounting to the extent that they are no longer entirely ignored. A survey of the measurements indicating various disequilibria at initial formation as well as secondary partial or complete equilibration after formation of the aggregates is given by Nagasawa *et al.* (1976) and Wark (1977).

In conclusion, neither the relationship between minerals in mutual contact in C3 meteorite aggregates, nor the compositional variations within single crystals have given any systematic indications of effects of monotonic cooling in the environment in which they were formed. One of the few things that can be concluded is that they do not consistently show the record calculated from cooling in chemical equilibrium with a gas of solar composition in a container with opaque, heated walls, nor is it possible for physical reasons to expect such a history.

One possibility, suggested by textural evidence, is that some of the material in the C3 inclusion aggregates condensed at grain temperatures below the glass transition temperature (Arrhenius and Alfvén, 1971) of most of the silicates and oxides present, and with compositional differences, reflecting changes in composition, number density and thermal parameters of the source medium during their formation. The aggregates might subsequently have been heated by friction in parabolic orbits (Alfvén and Arrhenius, 1976, Sections 11.8 and 21.12), by short-lived radioactive nuclides (Lee *et al.*, 1976), and by release of the large latent heat of crystallization, stored in the refractory glasses (for example, 14·7 kJ/mol for $MgSiO_3$; 13·4 kJ/mol for $CaSiO_3$; 105 kJ/mol for Y_2O_3).

The result is a crystalline aggregate occasionally with remnants of glass (Marvin *et al.*, 1970), and with exsolved or occluded sulphide and refractory metal grains (Wark and Lovering, 1976; Palme and Wlotzka, 1976). Low temperature phases (in the thermodynamically correct sense of the word) such as grossularite ($Ca_3Al_2[Si_3O_{12}]$) and in some instances, alkali halide silicates (section on *Volatile Components in Lattices with High Dissociation Energy*), occur with fine-grained textures suggestive of devitrification.

Such a course of condensation is in principle the same as the one concluded by Blander and Fuchs (1975) and Nagasawa *et al.* (1976) from textural and chemical evidence, namely that condensation took place in thermodynamic disequilibrium such that the condensate formed a 'metastable subcooled liquid'. This is equivalent to the suggestion that the liquid was a glass and condensation took place at a temperature of the solid below the glass–crystal transformation temperature.

The actual grain temperature, and other conditions at which the meteorite components formed, and their subsequent thermal histories are at the present time poorly defined. Textural and crystallographic evidence, studied with ultrahigh resolution techniques such as electron diffraction and electron transmission microscopy, could give information on these questions.

Individual crystals and inclusions within crystals are in general more reliable indicators (Arrhenius *et al.*, 1974) of the evolution of the solid than aggregates, since there is often uncertainty if the aggregate components formed at the same time and location, and in contact with each other. In special cases experiments have been designed that resolve some of these questions. Examples are the reconstruction of the time sequence of aggregation by accelerated particle and fission track techniques (Macdougall and Price, 1974; Price *et al.*, 1975; Alfvén and Arrhenius, 1976, Section 22.9, and references therein), and evidence from the preservation of ^{26}Mg at the site of decay from ^{26}Al (Lee *et al.*, 1976).

Ultrahigh resolution transmission techniques are also capable of giving evidence of solid state reactions between crystals in contact with each other. Such studies could also, in principle, reveal the structural effects of phase transformations by reaction of solid grains with a surrounding gas phase. Reactions of this kind are a basic speculative ingredient of hypotheses that infer condensation of meteorite solids at cooling of grains and gas, held at the same temperature and maintaining chemical equilibrium between each other. In view of the various objections against such schemes it is not surprising that uniquely distinctive structural evidence for such grain–gas reactions have never been reported in meteorite crystals.

Estimates of Temperature Parameters

For a meaningful discussion of condensation in an open, radiating system with the magnetohydrodynamic parameter $L \gg 1$ (Alfvén and Arrhenius, 1976, Section 15.1) such as in space, it is necessary to specify the temperature of electrons, ions, neutral atoms, and of molecules ranging into the size of what is commonly referred to as grains. At increasing number densities, ion and electron temperatures approach each other. Asymmetric diatomic molecules, polyatomic molecules and molecular aggregates (grains) are capable of vibrational cooling and assume lower temperatures than surrounding atomic and homonuclear molecular gases. Under the conditions where these are transparent (as discussed in sections on *Temperature Differential Thermal Equilibrium* and *between Grains and Medium*) and hence permit the growth of grains, the temperature differential is a necessary condition for condensation in a non-expanding system (section on *Temperature Differential*). A detailed discussion of the material and heat balance of condensing grains is given by Lehnert (1970).

The large grain–gas temperature differential that would occur at condensation at high grain temperature in a medium where the grains can radiate much more efficiently than the gas make it difficult to deduce the gas temperature solely from the grain temperatures that may be recorded in the solids.

From analysis of solid condensates the temperatures of the grains at their formation have most promise of being derivable, at least semiquantitatively, or in combination with other variables such as pressure or time. Since grains may have been heated (not necessarily 'reheated') after condensation, it may only be possible, however, to specify upper grain temperature limits.

There are at least four types of phenomena, largely unexplored, which promise to provide direct and model independent information on grain temperature or on specific temperature–pressure combinations. One is polymorphic phase transformation between high- and low-temperature phases (in the correct thermodynamic sense of the word; see section above on *Temperature–Pressure Path of Condensation* and related crystalline ordering phenomena). The work by Ogilvie, Wood, Goldstein, Short, and others on the diffusive transformation of the high temperature FeNi γ-phase in meteorites at monotonic cooling or at oscillatory heating (Alfvén and Arrhenius, 1976, Section 11.8) is a classical example. It is possible and likely that analysis of the structure and ordering in phases such as plagioclase, melilite, spinel, and polymorphic alumina structures can provide definite or suggestive informaton on actual temperature relationships. The possibility of formation of metastable phases and the sluggishness of displacive phase transformations must be taken into account.

Diffusion gradients provide criteria for the length of time that given temperatures can have been sustained by grains. The annealing of charged particle tracks in lunar minerals, for example, has in some cases been calibrated in terms of time–temperature relationships (e.g. Macdougall *et al.*, 1973). The lack of annealing of cosmic ray tracks in many types of chondrites (together with many other thermally sensitive features) shows, for example, that they cannot have participated in thermally disastrous events such as the explosion of a parent body of lunar size as is often implied in the meteoritical literature.

Gradients from internal and peripheral regions of high iron concentration in olivine ($[Mg, Fe]_2SiO_4$) crystals such as in Figure 9 in a similar way place limits on the time–temperature relationships when considered together with the activation energies of diffusion of Fe^{2+} and Mg^{2+} in the olivine structure.

A third diffusion controlled phenomenon potentially useful for determination of the actual thermal effect received by silicate and oxide grains and grain aggregates is the state of dispersion within them of metal droplets, described in particular detail by Wark and Lovering (1976) (Figure 8). Empirical data on surface diffusion, sintering and alloying of platinum metals and of iron suggest that a nickel–iron grain, in contact with an osmium–ruthenium alloy

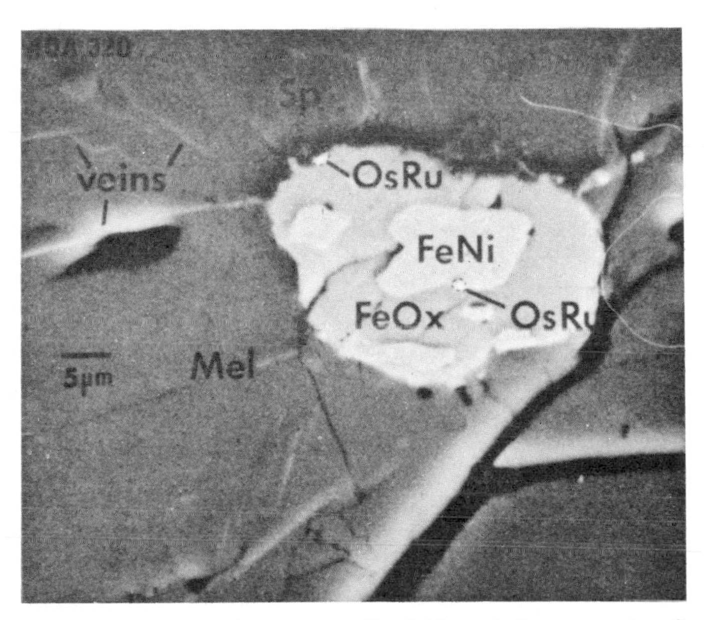

Figure 8. Metal grains ('nuggets') of Type A in aggregate of refractory minerals from the Allende carbonaceous meteorite. The absence of interdiffusion between the Os–Ru grain and the Fe–Ni alloy places an upper limit on the temperature at which these grains can have been in contact for any specific length of time. (Reproduced with permission from D. Wark and J. Lovering, 1976, *Lunar Science VII*, 912–914 copyrighted by Lyndon B. Johnson Space Center, Texas)

such as in Figure 8 would interdiffuse and react in a few hours or less in a hydrogen atmosphere if the temperature were above 1300 K. Since the grains remain intact, the aggregate in which they occur cannot have been in this time–temperature range in the presence of the metal grains.

Other textural-chemical clues to the temperature of the condensate at formation includes cases where there is evidence for formation of the solid below the glass transformation temperature (Section 14). The main features of the present aggregate textures could in these cases conceivably arise from devitrification at heating by short lived radioactive species (Lee and Papanastassiou, 1976; Lee *et al.*, 1976) and by gas friction around perihelion (Alfvén and Arrhenius, 1976, Section 11.8), both accompanied by the release of the large latent heat of crystallization.

Volatile Components in Lattices with High Dissociation Energy

The occlusion of volatiles at condensation is another process which is highly dependent on the temperature of the growing solid and hence useful for estimates of the upper limits of grain temperature (Arrhenius and Alfvén, 1971). The noble gases, primarily occluded in the carbon compounds of meteorites (Phinney *et al.*, 1976) bear promise in this respect particularly since any attendant fractionation between heavier and lighter noble gases is also temperature sensitive, and to be effective, can only take place at very low grain temperatures (Fanale and Cannon, 1972).

From the point of view of the thermal history of the Ca–Al rich inclusions in carbonaceous meteorites the presence of volatile components such as potassium, zinc, hydroxyl, and halogens is of significance. The fact that the halogens comprise also ^{129}I, the radioactive precursor of ^{129}Xe places time limits on the incorporation of the halogens. They occur in the form of sodalite (equation (4)) which could in principle have formed directly from the source medium, or by devitrification of a precursor low temperature glass phase.

A proposition sometimes made, but for reasons discussed below considered unlikely, is the formation of sodalite by a reaction such as

$$6(Na, K)[AlSiO_4] + 2(Na, K)(Cl, Br, I) \rightleftarrows (Na, K)_8[(Cl, Br, I)_2|(AlSiO_4)_6]$$

nepheline salt vapour sodalite (4)

X-ray diffraction analysis of the products of this reaction with NaCl and NaI as vapour phase have shown the iodide aluminosilicate to be isostructural with the chloride mineral (sodalite) and a continuous solid solution to exist between these two end members.

This is of interest in connection with the finding by Fireman *et al.* (1970) and Marti (1973) that the uniquely high contents of ^{129}Xe, presumably derived from radioactive decay of ^{129}I originally incorporated in the solid, is correlated with the strongly varying alkali content in the C3 meteorite inclusions. The conclusion is that the primordial iodine 129 is likely to have been incorporated together with chlorine (and bromine) in the sodalite which is a characteristic component in some C3 inclusions. The uniquely high content of ^{129}Xe found by Marti suggests that the incorporation of alkali halides took place early in the evolution of the aggregate minerals.

The inclusions, characterized also by their high content of refractory minerals, divide into two groups, one with high and one with low content of potassium and chlorine (Arrhenius, 1973). Marti (1973) also observed such a grouping on the basis of the concentrations of potassium and ^{129}Xe.

The aggregates which are 'depleted' of alkali halides (in the form of sodalite) are mostly rounded bodies with characteristics that have led to the suggestion by several authors that they have been melted after formation. The sodalite containing aggregates in contrast, are frequently irregular. Microcrystals of sodalite also occur in the dark, fine grained material of the meteorite often referred to as 'matrix'. No chlorine bearing mineral other than sodalite has been identified in the 'matrix' or in the refractory aggregates, a fact of importance in the discussion of the chemical history of the aggregates.

In some of the white aggregates the sodalite crystals are concentrated near the rim, however, in others they show a patchy distribution in the interior of the aggregate.

Since the alkali halide silicates are stable only below about 1000 K under the conditions invoked, their presence is yet another embarrassment for the hypothesis that the refractory aggregates condensed at exceptionally high temperatures. The explanation has therefore been proposed that the alkali halide silicates would be due to secondary reactions of a primary 'high temperature' phase with an (unspecified) chlorine containing phase occurring in the 'matrix' of the meteorite such as in equation (4). This reaction would thus have taken place at a time when individual grains had already accreted to form the meteorite parent bodies.

Such suggestions conflict with the facts discussed above. One is that the halogen minerals have uniquely high concentrations of ^{129}Xe and thus are likely to have formed in an early stage of condensation. Furthermore, the sodalite occurs also in the interior of aggregates, occasionally in textures that suggest devitrification (Mason, 1972), in other instances as large individual crystals (Nagasawa *et al.*, 1976). None of these cases show the characteristics of vapour diffusion and reactive phase transformation. Reaction of alkali halide vapour with nepheline would be likely to result in rims of sodalite on parent nepheline crystals, such as observed in laboratory diffusion experiments, but not in meteorite crystals.

The texture, and preservation of thermally sensitive features of the minerals in the 'matrix' surrounding the aggregates in the meteorites (see, for example, Alfvén and Arrhenius, 1976, Figure 2.1) precludes that the aggregated meteorite material could have provided sufficient thermal power to drive the proposed secondary diffusion reaction. Analysis of the remanent magnetization of the

Allende meteorite specifically indicates that it cannot have been heated near the Curie temperature of magnetite, 858 K, at any time after accretion (Brecher and Arrhenius, 1974).

The 'matrix' of the meteorite indeed contains microscopic sodalite containing aggregates of refractory minerals of the same kind as in the macroscopic inclusions, but there is no evidence of halide source minerals, which could have provided a chemical potential to drive a nepheline–sodalite transformation reaction.

Present evidence thus indicates that the formation of relatively low temperature minerals (in the thermodynamically correct sense of the term) such as grossularite ($Ca_3Al_2[Si_3O_{12}]$) and the alkali halide silicates, incorporating also ^{129}I, took place concurrent with the formation of the refractory components of the white aggregates, which later accreted together with 'matrix' material to form the C3 chondrite parent bodies. The large negative free energy of formation of most of the minerals in the inclusions certainly speak for the notion that they represent an early condensation of material, from the source cloud of the solar nebula, or from the nebula itself. Fluctuations in compositional or other parameters of the source medium would give rise to the observed inhomogeneities in the aggregate composition with patches or rims higher, sometimes, in more volatile components, at other occasions in more refractory components (Arrhenius *et al.*, 1974). Uncomplicated evidence of such fluctuations in the source medium is provided by the record in single crystals discussed in the section on *Evidence from the Formation of Ferromagnesian Silicates* (cf. Figure 9 below).

The grain temperature of the aggregate material at condensation is difficult to judge on the basis of present evidence. An upper limit is set by the presence of the volatile components of minerals such as sodalite, and zinc in spinel (Nagasawa *et al.*, 1976) and by the state of coalescence and interdiffusion of refractory metal grains in the interior of silicate aggregates. The former criterion is difficult to translate into grain temperature in the absence of information on composition, number density and temperature parameters of the source plasma. The latter requires model experiments, and estimates of growth rates of the silicate material if it occluded the metal at condensation. It is not always clear either if the individual crystals that now constitute the aggregates were in contact with each other at the time when they formed.

The presence of monomorphic refractory minerals such as gehlenite, spinel or perovskite does obviously not carry any temperature information; they (or precursor glass) could form at lower or higher grain temperature at incipient densification of the solar nebula as visualized in most physically based models of the early solar system regardless of the way in which hydromagnetic effects are taken into account (see, for example, Cameron, this volume).

Investigations of the fine structure of Ca–Al–Na-rich inclusions and other meteorite solids, coupled with model experiments and appropriate theoretical considerations are likely to yield information that can place narrower limits on the formation conditions than is now possible.

Evidence from the Formation of Ferromagnesian Silicates

As discussed above, the solid grains contain information which, if properly interpreted, may give at least upper limits for the temperatures which they have experienced at their formation or in later heating events. The corresponding thermal characteristics and state of excitation of the medium from which they formed is by necessity more difficult to assess, since any effect on the preserved record is indirect. Some important conclusions can, however, be drawn from the composition of the common iron–magnesium silicates, olivine ($[Fe, Mg]_2SiO_4$) and pyroxene ($[Fe, Mg]SiO_3$) in meteorites, provided that one can show the crystals to be primordial condensates.

The variable and often large (up to 70 mole per cent) concentration of Fe^{2+} in these minerals in carbonaceous meteorites shows that these crystals cannot have formed at a combination of high grain temperature, high temperature of the source vapour, and high activity of hydrogen species in the source vapour.

This is of interest as one of the many observational facts, which (in addition to fundamental physical principle discussed in other sections) demonstrate that the assumption of primordial condensation by cooling in thermodynamic equilibrium with a gas of solar photospheric composition is an unrealistic conjecture. According to this, iron metal and practically pure magnesium silicates would form by condensation. To account for the observed reality, the suggestion has been made that at a low temperature the metal grains would become oxidized, somehow the oxidized iron would appear in a vapour state and, together with oxygen, attack the magnesium silicate grains. The oxygen and iron (presumably as Fe^{2+}) would in this reaction diffuse into the magnesium silicate grains and convert them to the observed iron–magnesium silicates, as indicated in equation (5) below.

For reasons presented in the literature and reviewed below, this hypothesis is not only unsupported by observation but is also wrong in principle. This has been obscured by the fact that the equilibrium cooling hypothesis, which has derived its appeal from quantitative calculations becomes vague and non-specific on this and related reactions of major significance.

Olivine–Iron Reaction

Iron cannot simply 'diffuse' into a structure like Mg_2SiO_4 (forsterite olivine) thereby converting it to isostructural $(Mg_xFe_{1-x})_2SiO_4$; $x < 1$ (fayalitic olivine). For such a reaction to be at least hypothetically possible, iron as a divalent ion has to enter the structure, and an equivalent amount of magnesium ion has to be removed from the crystal. No mechanism has been suggested that could achieve this exchange in space, and there is no observable indication of such processes in those olivine crystals in meteorites which can be presumed to be vapour condensates.

Distribution of Fe^{2+} in Olivine Crystals

Diffusion reactions of the kind surmised would lead to unique compositional gradients in the crystals with iron contents falling toward the centre. Compositional gradients in euhedral olivine crystals, presumed to be vapour condensates (Grossman and Olsen, 1975) are indeed common (Arrhenius *et al.*, 1974). However the iron content in some cases decreases, in other cases increases in the rim of the crystal, and in some cases a distribution is found which indicates oscillatory zoning or a patchy distribution of iron and magnesium (Arrhenius *et al.*, 1974; Kerridge and Macdougall, 1976; see also Figure 9). These features indicate variations in the Fe^{2+}/Mg^{2+} activity ratio in the medium from which the crystals grew rather than a secondary introduction of iron from a vapour phase with concomitant removal and escape of magnesium ions into this vapour medium.

Conversion of Pyroxene to Olivine

It is also presumed in the cooling equilibrium hypothesis that the grains of the other major magnesium silicate mineral $MgSiO_3$ (enstatite pyroxene), orbiting

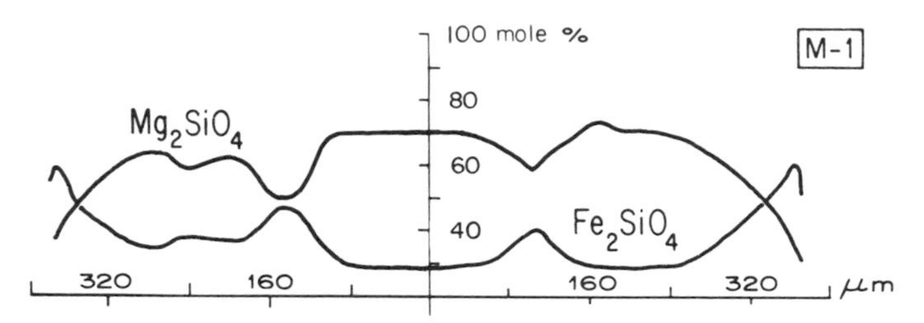

Figure 9. Electron microprobe cross-section from edge to edge through center of zoned olivine crystal from the Murchison meteorite. The section shows alternating regions of high and low iron content. This distribution cannot be achieved by conversion of an original magnesium silicate crystal by introduction of iron from the surroundings since this, aside from other difficulties (see the first three parts of the section *Evidence from the Formation of Ferromagnesian Silicates*) would lead to a monotonic diffusion gradient. The observed distribution must derive from differences in iron and magnesium content established at the surface of the crystal during its growth. The original gradients may have been smoothed by thermally activated diffusion of Fe^{2+} and Mg^{2+}. Upper limits for various temperature-time combinations may be obtained by assuming original, sharply bounded regions of pure Mg_2SiO_4 and Fe_2SiO_4. (Reproduced with permission from G. Arrhenius *et al.*, 1974, *Meteoritics*, **9(4)**, 313–314)

in space, were converted to the actually observed iron–magnesium solid solution pyroxene crystals $Mg_yFe_{1-y}SiO_3$; $y < 1$ by a similar diffusion process taking place at low temperature. This proposal meets the same conceptual and observational difficulties as the presumed olivine transformation in the last subsection, or a different mechanism and end result has to be postulated. This alternative would imply the reaction

$$\underset{\text{'vapour'}}{Fe^{2+} + O^{2-}} + \underset{\text{enstatite}}{MgSiO_3} \rightleftarrows \underset{\text{fayalitic olivine}}{(Mg, Fe)_2SiO_4} \tag{5}$$

i.e. simultaneous diffusion of ferrous ion and oxygen into the enstatite crystals, converting them to an iron containing a variety of olivines with rearrangement also of the silicate polymer framework (Arrhenius, 1975). The characteristic crystallographic properties of such pseudomorphic replacement structures are in principle well-known. The olivine crystals in carbonaceous chondrites do not show such properties, and it can, also on this basis, be concluded that reaction (5) is not the source of those crystals.

The question of the source and mode of transport and selective deposition of iron on such an immense scale as to provide all of the ferromagnesian minerals, a major component of meteorites, poses yet other difficulties in this kind of hypothesis.

Source and Transport of Iron in the Magnesium Silicate Conversion Conjecture

The proposition in the cooling equilibrium hypothesis is, as discussed above, that the metallic iron condensate grains formed at high temperature, and became oxidized at lower temperature. The oxidized iron would then somehow be transported through space and, as in reaction (5), impregnate the grains of magnesium silicate together with oxygen.

As one of the few iron compounds with sufficiently high vapour pressure, $FeCl_2$ has been suggested as a vehicle for transport of iron. The orbiting iron metal grains would become corroded by chlorine or hydrogen chloride, and by some undisclosed process the iron chloride would select the surface of the independently orbiting pyroxene grains, react there with oxygen, presumably in the form of water vapour, and diffuse together with oxygen into the pyroxene, converting it into iron containing olivine.

This process would, from a quantitative point of view, be of paramount importance, since it would have to produce all of the ferromagnesian minerals, now observed as the main components of carbonaceous meteorites. An interrelated problem of similar dimensions is the unknown fate of the doubtlessly large amounts of partially corroded and oxidized nickel iron grains (see also, Herndon, 1976; this volume).

One attempted solution to some of the practical aspects of this dilemma (but not resolving the theoretical reaction problems) is that all of the evidence would have been removed by reaction of the iron–magnesium silicates and the corroded iron metal with water to form the equally ubiquitous magnesium hydroxysilicates and magnetite ($FeO \cdot Fe_2O_3$). Composition and structural features of these minerals (Arrhenius and Alfvén, 1971) suggest that they may be primary rather than alteration products; synthesis experiments (Meyer, 1970; Brecher, 1972) demonstrate that they can form by direct condensation from partially ionized gas. All olivine, pyroxene, and nickel–iron metal crystals found in carbonaceous meteorites would then have to be regarded as products of secondary processes involving vaporization and recondensation or remelting of the primordial material. This would include also the iron containing olivine in the Allende inclusions.

Such an assumed reprocessing would in fact mean that no primordial silicate or metal condensates remain to be inspected in meteorites, and that the observational basis for the anyhow physically untenable iron impregnation hypothesis has vanished.

The information conveyed by the composition and structure of the ferromagnesian silicates in carbonaceous meteorites is that they certainly did not form by the metamorphic processes invoked in the cooling equilibrium hypothesis. It is also clear that they formed from a medium that contained species of magnesium and iron as primary components. If this medium was a melt, a low temperature glass condensate, or a vapour phase is open to question, but can probably be established by high resolution structural investigations (see, for example, Grossman and Olsen, 1975; Grossman *et al.*, 1975; Kerridge and Macdougall, 1976; Arrhenius *et al.*, 1974).

Thermal Equilibrium between Grains and the Condensing Medium

If any of the solids in carbonaceous meteorites can reasonably be assumed to be primary condensates, unaltered by later transformations, how can we from their structure and composition tell anything about the nature of the condensing medium in the solar nebula or in its source cloud in interstellar space?

If it is assumed that the isolated crystals or aggregates with ferromagnesian olivine resulted from primary condensation processes, then it is necessary to assume that the medium had comparable number densities of both Fe and Mg species. This requires that the source medium in which the growing grains were immersed, had a kinetic temperature well above solid–vapour equilibrium at the prevailing pressure. The grain temperature, on the other hand, must have been sufficiently low to permit growth to the observed size during the (unknown) time available.

The actual grain temperature is determined by the balance between heat loss by radiation and the heat received from different energy sources. The heat and material balance of grains in condensation equilibrium has been analysed by

Lehnert (1970a). The rate of heat loss for a grain radiating as a blackbody is

$$L = 4\pi a^2 \sigma T_g^4 \tag{6}$$

where T_g is the bulk temperature of the grain, a its radius and σ the Stefan–Boltzmann constant. Equation (6) transforms to the appropriate quantum mechanical expression for radiative emission, when the molecular properties of the grain are taken into account.

The heat flux to the grains derives from the heating by collisions with ions, electrons and neutrals, by neutralization, recombination and bond formation of ions and atoms on the grain surface, and by any photon radiation field. In a simplified form these terms can be expressed:

heating power of ion–electron component:

$$P_{i,e} = 4\pi a^2 n_i v_p [\tfrac{3}{2} c_p k(T_e + T_i - 2T_g) + c_d w_d + c_i w_i] \tag{7}$$

where a is the grain radius, n the ion or electron density, v_p the effective velocity of the charged particles impinging on the grain, c_p, c_d, and c_i transfer efficiency factors, k the Boltzmann constant, T_e, T_i, and T_g the electron-ion- and grain-temperatures respectively, w_d the association and bond energies and w_i the neutralization energy.

The heating power of neutral component

$$P_n = 4\pi a^2 n_n v_n [\tfrac{3}{2} c_n k(T_n - T_g) + c_d w_d] \tag{8}$$

where variables with the subscript 'n' are the neutral gas analogues of the plasma variables in equation (7).

The velocities v_n and v_p are unequal due to the differences in the kinetic temperatures T_n, T_e, and T_i and because of the acceleration of ions toward the surface, negatively charged because of the higher electron mobility. Hence

$$v_n = k_n (kT_n/2\pi m_n)^{1/2} \tag{9}$$

and

$$v_p = (k/m_i)^{1/2} [k_e T_e^{1/2} + k_i (T_i/2\pi)^{1/2}] \tag{10}$$

The first term in equation (10) is due to the ion acceleration mentioned above; k_n, k_e and k_i are dimensionless constants near unity; m_i and m_n are the masses of the ions and neutrals.

The heating power of the radiation field

$$P_r = 4\pi a^2 \sigma T_r^4 \cdot f(\lambda/a) \tag{11}$$

where T_r is the temperature of the radiation field at the location of the grain and λ the effective wavelength of the heating radiation. The function $f(\lambda/a)$ represents the decoupling of the particle from heating by the radiation field when the particle radius decreases below the effective wavelength.

In thermal equilibrium (which obviously does not imply equal temperature between the grain and the plasma components)

$$P_{i,e} + P_n + P_r = L \tag{12}$$

Assuming a radiation field temperature around 100 K, a plasma pressure of the order of 10^{-6} b, and a degree of ionization of 10^{-3} (hydrogen and helium as neutral species but major condensable elements completely ionized), the condensing grains would assume a temperature of about 300 K at a temperature of the neutral gas component of the order of 5,000 K.

If one could (which is unlikely) demonstrate by reliable criteria that some refractory meteorite crystals actually formed at as high a grain temperature as their stability against dissociation and evaporation permits ($\sim 2 \times 10^3$ K), then the kinetic temperature of ions and neutrals in the surrounding medium would exceed 10^4 K at radiation field temperatures of the order of magnitude of a few hundred degrees.

Composition of the Source Medium

From the densities and, in a few cases, also from other properties of planets and satellites we have some rather unspecific indications of their bulk chemical compositions, and hence indirectly some suggestions about the variation in chemical composition of the source media for the solid material in different regions of the solar system. In the above sections on *Magnetized Bodies*, *Band Structure*, *Chemical Differentiation*, and *Critical Velocity*, the corresponding inferences from the band structure have been reviewed.

Before we can learn much more about the present and primeval chemical structure of the solar system, new results from space exploration are needed, most importantly from the outer regions in gravitational potential energy space (Figure 3 above). The Jupiter–Saturn–Uranus probe, to be launched in 1977, will be of unique importance in this respect.

Although we do not know with any certainty where the material condensed which is now aggregated into meteorites, attempts at reconstruction of the source medium from which these materials formed are obviously of great general interest. If we make such an assessment on the broadest possible basis, and in a manner consistent with the extensive modern knowledge of physical–chemical processes in space, it is possible that the understanding of the manner in which this material formed can be brought to the same advanced level as the present understanding of the time sequence of formation.

One of the conclusions from the composition of meteorites is that those materials which are considered to have formed by primary condensation, appear to derive from media with chemical composition different from the composition observed in the solar photosphere. Considering the marked fractionation that is suggested from other structural features in the solar system and the considerable length of time during which the formative processes now are known to have been active, it would indeed be surprising if chemical homogeneity could have been maintained.

Some of the evidence for the extent to which the source medium in one region was different from the solar photospheric comparison material comes from the oxidation state of iron, discussed in the section on *Evidence from the Formation of Ferromagnesian Silicates*. The fact that carbonaceous chondrite crystals have been formed with oxidized iron (Fe^{2+} and Fe^{3+}) as primary components places limits on the activity ratio of oxidizing/reducing species in the medium, such as some of the oxygen containing ions and neutrals shown in Figure 1, relative to species such as H^+, H, and H_2. Because of the very high activity factors of ionic, monatomic, and radical species (Urey and Donn, 1956) these influence the redox balance more strongly (section on *Particles, Fields, and Chemical Reactions in Dark Interstellar Clouds*) than would be indicated by mass action of molecular hydrogen alone.

The redox activity limit on a condensation environment for ferrous olivine and pyroxene crystals, ferroferric spinel, and ferric garnet, characteristic of carbonaceous meteorites, most likely requires an elemental composition of the source plasma, different from, for example, solar plasma. Specifically it indicates a lower partial number density of hydrogen species. This is the situation characteristic of the C and D clouds, and the transplanetary region of the solar system source cloud (sections on *Band Structure* and *Chemical Differentiation* and Figure 4). It is in these regions that the meteorite source material also for independent orbital dynamic reasons are thought to have formed (Alfvén and Arrhenius, 1976; Section 22.2).

Variations in the distribution of iron between divalent states and more oxidized states such as reflected in the formation of $Fe^{2+}O{\cdot}Fe_2^{3+}O_3$ give rise to olivines $(Fe^{2+}, Mg^{2+})_2SiO_4$ with a wide range of iron contents within single carbonaceous meteorites, and hence probably within single parent jet streams or in streams, adjacent in velocity space. Similar variations are found within single olivine crystals (Figure 9 above). This indicates marked variations in the redox conditions during the formation of the material, and on a time scale, short compared to the growth time of a single crystal. Such marked, highly localized and changing conditions are characteristic also of plasma phenomena observed in space today (see, for example, Mendis and Ip, 1976).

There are numerous other indications of temporal or spatial differences in the composition of the source medium, including those components which affect the redox state impressed on the condensate. Among such indications are those arising from the variations in europium and ytterbium content, unrelated to crystal field effects (section below on *Rare Earth Elements*). Another is the growth of wollastonite ($CaSiO_3$) over, and hence subsequent to, the formation of diopside ($MgCaSi_2O_6$) (Fuchs, 1971). Variations in the content of volatile components in refractory aggregates (section above *Volatile Components in Lattices*) may also be caused by differences in the speciation or composition of source medium. Other effects of compositional variations in the source media are indicated by the highly reduced phases observed in enstatite meteorites.

The examples discussed above indicate what kind of boundary conditions can be placed on the properties of the source medium on the basis of the structure and composition of preserved records in solid condensates. The boundaries can doubtlessly be narrowed by more detailed investigations guided by realistic assumptions. One of the most crucial questions to be answered is obviously how convincing the arguments can be made for a direct condensation origin of the solids in carbonaceous meteorites. The return of solid samples from comets also from this point of view appears as one of the most fundamental tasks of the space program.

State of Excitation of the Medium in the Solar and Planetary Nebulae

At gravitational accumulation of interstellar gas and dust into protostars, stars, and stellar envelopes, the motion of matter and the release of gravitational energy would necessarily generate corresponding quantities of hydromagnetic energy, as discussed in the companion paper by Alfvén. Partial ionization of the source cloud and of the nebular medium (by electric currents and by the critical velocity effect rather than mainly by photoionization) would thus appear to be an unavoidable stage in the formation and evolution of solar and planetary nebulae, regardless of optical opacity. The degree of observed and inferred ionization covers a wide range, from practically total ionization close to massive magnetized bodies such as the Sun and the magnetized planets (HII regions in astrophysical terminology) to degrees of ionization of the order of 10^{-3} and many orders of magnitude lower (HI regions).

The HI regions are probably identifiable with the regions outside of Mars in the large-scale pattern of the solar nebula, and with the corresponding region in the Jovian, Saturnian, and Uranian planetary nebulae (C and D clouds and transplanetary region; section on *Origin of Band Structure* above). HI regions are in astrophysical parlance referred to as 'neutral' because the major components of the plasma, hydrogen, and helium, are ionized only to a low (but chemically important degree. In such 'neutral' plasma regions with a degree of ionization as low as 10^{-3}, all of the elements are still ionized that are major components of condensable materials. At exceedingly 'low' degrees of ionization around 10 parts per million, the important rock forming alkali metals remain practically fully ionized.

These conditions of excitation, that are characteristic for dilute matter in space, control the chemical speciation and the reactions that are observed to occur. Any matter that ultimately densifies to solids and to atmospheric gases must by necessity pass through a stage where this control is exerted. Postulation, for example, of a nebula with atmospheric densities in the interior provides no exemption from this evolution.

If, as is likely, many of the original features have been modified in the condensates which form the precursors of meteorite material, what are the chances for deriving information from these materials about the energy state of the medium from which they formed? It would appear that there are a number of possibilities to estimate, at least qualitatively, some of the critical parameters. One of the most promising is the characteristic isotope fractionation in the solid materials, probably reflecting the kinetic effects of concurrent ion–molecule and ion–solid reactions (the concluding six sections of this chapter). Another is the chemical composition of crystalline solid solutions such as the ferromagnesian minerals discussed above. If, for example, the ferrous olivine crystals formed from a vapour phase, or from a low temperature glass formed from a vapour phase, then it is clear that this vapour must have had such ion–atom and molecule temperatures and such number densities of the various chemical species that iron and magnesium could enter simultaneously and in comparable amounts in the growing solid.

To unravel the variables is a difficult task requiring measurement of many parameters in the presumed space condensates, and most importantly, verification by laboratory experiments. The total effort needed to bring order and fundamental understanding into the field of petrology of space plasmas would appear to be at least as large as the corresponding effort in petrology of terrestrial melts. This latter field is now, after sixty years of intense efforts by a number of specialized research groups, understood to the extent necessary for realistic application to the complex natural systems.

The variables of major importance to be experimentally controlled include total number density, kinetic temperatures of electrons, ions, atoms, and molecules, the state of excitation, and ionization of different components, including isotopes, the relative proportions of the different chemical species, and the temperature of the solids forming from the plasma.

The degree of ionization is important not so much because of any easily distinguishable effect on the composition of condensing solids—such effects would with certain exceptions be expected to be subtle—but because of the profound effect of this parameter on the physical behaviour and distribution of the medium, and on the kinetics of the reactions leading to the formation of solids by polymerization. The control of the state of ionization on these reactions is evident from the chemical processes now observed in space, for example, in dark clouds (Herbst and Klemperer, 1973), in comets where some of the fundamental insights on real space chemistry were first obtained (Biermann, 1951; for further discoveries, see, for example, Mendis and Ip, 1976; Oppenheimer, 1974; Shimuzu, 1975), and in the upper atmosphere of the Earth (see, for example, Narcisi and Roth, 1969).

In view of the large amount of information already accumulated for organic ion–molecule reaction mechanisms, including polymerization reactions, analysis of meteoritic polymers, as well as corresponding plasma synthesis experiments, promise fertile ground for early advances in this field. The compounds of diagnostic interest in meteorites and in synthesis experiments include also carbonates and oxalates.

For reconstruction of excitation parameters from the record in meteorites, evaluation of the isotope disequilibrium effects in light element compounds appears promising. Furthermore, elemental fractionation caused or influenced by excitation effects would be most easily identifiable among elements characterized by large differences in degree of ionization or critical velocity, but as similar as possible in other respects. Such element pairs or groups are, for example, mercury–thallium (Arrhenius, 1972), the noble gases (Jokipii, 1964; Weber *et al.*, 1971, 1976) and chlorine–fluorine (Figure 2 above). In the case of the noble gases, the advantage lies in their relative chemical inertness in the neutral ground state and, on the other hand, the high reactivities of their ions, and the large fractionation effects possible by several of the mechanisms discussed in the sections on *Electrodynamic Fractionation*, *Kinetic Isotope Fractionation* and *Kinetic Isotope Effects in Space.* As pointed out by Weber *et al.* (1976), the extreme fractionation of the noble gases observed in the ureilite meteorites requires condensation of the host solids at kinetic temperatures in the range $10–10^2$ K or partial ionization of the source medium. For occlusion and solid growth reactions to proceed at sufficient rate in this low temperature range, excitation of the source medium is required. Hence the two processes discussed by Weber *et al.* would be likely to operate in conjunction.

Rare Earth Elements as Indicators of Fractionation in the Source Medium

The large free energies of rare earth–oxygen compounds cause concentration of these elements into the Ca–Al rich inclusions in carbonaceous meteorites. The similarity in electronic properties of the rare earth elements due to the to joint $5d\,6s^2$ structure of the neutral atoms, modified by their systematic differences in the internal $4f$ levels, make them of interest in tracing the details of the separation processes that they have gone through. For these reasons an increasing number of accurate measurements are being made of their distribution. The most detailed analyses of coexisting phases in carbonaceous meteorites are those by Nagasawa *et al.* (1976), who also review and interpret the results of earlier measurements.

At the present time, the major fractionation effects observed are: (1) systematic changes in relative concentration with atomic number, (2) marked singularities in relative concentration of the rare earth elements, related to corresponding singularities in their ionization dynamics, (3) differences in total concentration. These properties have been studied in different minerals occurring together in single inclusions and in different inclusions with similar or different mineral composition.

The results, schematically summarized in Table 1, show a number of permutations where, until recently, interpretations were based only on the distribution in Group 2. Most importantly, these new combinations of characteristic disequilibrium features are not systematically associated with any specific mineral assemblages. Hence Nagasawa *et al.*, conclude that the characteristics of the rare earth element distributions are primarily related to properties of the medium from which the solids formed, rather than controlled, by the crystal field properties of the solid phases, which also give recognizable effects. From the differences in distribution between coexisting minerals in the different types

Table 1. Characteristics of rare earth element distributions in Allende inclusions, generalized from investigations by Nagasawa *et al.* (1976) and earlier work reviewed by them

Group number	Concentration of La relative to chondrite standard	Concentration trend with increasing atomic number	Eu anomaly	Yb anomaly	Mineral assemblage
1	15	Invariant	Small positive	None	Melilite, clinopyroxene, anorthite, spinel, grossularite
2	20	Decreasing	Negative	Positive	Sodalite, nepheline, spinel, clinopyroxene
3	100	Invariant	Strong negative	Strong negative	Spinel
4	1·5	Invariant	Positive	Positive	Sodalite, nepheline, clinopyroxene, spinel

of inclusions the authors also draw important conclusions about the highly varying degree of equilibration between these minerals, and hence about the thermal histories of the inclusions.

On the basis of these investigations, Nagasawa *et al.*, draw the conclusion, obvious also for fundamental physical reasons (see, for example, the section on *Temperature–Pressure Path*, *Temperature Differential*, and *Thermal Equilibrium*) that the solids analysed cannot have formed by condensation in thermodynamic equilibrium with the source medium. The results indicate condensation at a grain temperature, low compared to the Tammann reaction range.

To clarify the processes responsible for the different fractionation effects in the source medium, it would be necessary to measure in the laboratory the parameters that determine the speciation and separation of the rare earth elements under realistically modelled space conditions.

Of the factors that determine the degree of ionization due to electron collision, neither the ionization energies, and still less the oscillator strengths are now known with sufficient accuracy to predict the necessary details of speciation patterns, analogous for example, to those in Figure 10. However, it is clear from the low ionization energies of this element group as a whole that it will carry a fraction of the total charge, which is high in proportion to its abundance in regions of low H–He ionization (HI regions; C and D clouds). Rare earth ions would for this reason be expected to be kinetically effective in the reactions controlling exchange between the reservoirs (section on

Particles, Fields, and Chemical Reaction in Dark Interstellar Clouds) and they would, like elements such as aluminium, the alkalis, alkaline earths, and actinides flow at relatively high rates also through plasma-solid exit reaction channels. Unlike the case of elements such as sodium and potassium, it is difficult to distinguish these effects from those caused by the high formation energy of their solid oxygen compounds, which also would be expected to strongly influence the observed removal of elements such as calcium and aluminium from the space medium. The persistent occurrence of sodalite and potassium bearing nepheline in the refractory mineral associations possibly indicates the isolated effect of the ionization of the medium.

The characteristic anomalies in the relative abundances of Eu and Yb are doubtlessly linked to the stabilization of the $4f^7$ and $4f^{14}$ states respectively by the $5d$ electron at ionization involving $6s^2$. This well-known stabilization of these two ions in divalent state by the half-filled respectively filled $4f$ group is reflected by a marked elevation of the third ionization potential of Eu and Yb (Figure 10). This does not necessarily imply that the two elements during the separation process in space, responsible for the anomalies in the rare earth element distribution in the condensates, prevailed as free divalent ions. The electronic stabilization effect is affecting all chemical properties of these elements, including the thermochemical properties of molecular ions, complexes and other compounds. Boynton (1975) has pointed out the corresponding stability properties of oxide species in the gas phase, but appears to assume that these properties somehow are independent of, rather than fundamentally induced by, electronic effects. Rare earth element distribution features calculated from thermochemical data of the kind now available thus cannot be used to determine in principle if the reactants were ionized or not. For reasons men-

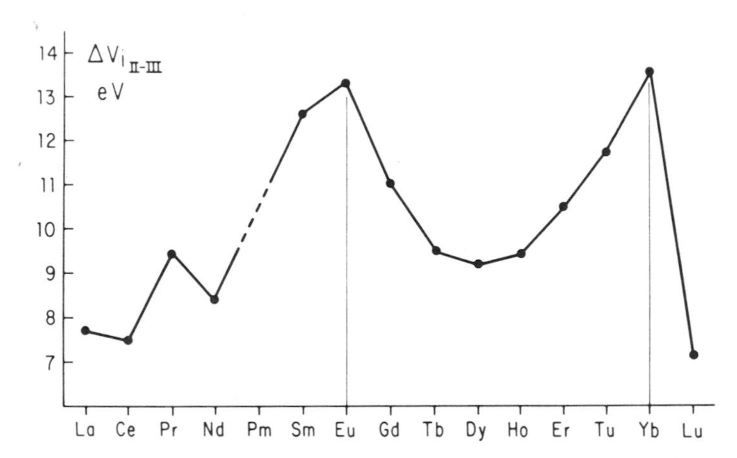

Figure 10. Ionization potential for rare earth element transitions from second to third ionization reflecting the stabilization of Eu^{2+} in the $4f^7$ and Yb^{2+} in the $4f^{14}$ (filled $4f$ shell) configurations

tioned above, the basis for thermochemical calculations, including charged species of importance, is at the present time too uncertain to treat the problem in a rigorous way in the manner attempted by Boynton for neutral species in their ground states.

Boynton's paper has the merit of drawing attention to the likelihood that the rare earth elements were fractionated in the medium, from which the Ca–Al–Na-rich inclusions formed. The attempt to make deductions of the properties of the source medium, developed by Boynton on the basis of a pink portion of one single inclusion (Type 2) stands to be modified, however, not only because of the inherent uncertainties mentioned above, but also by the new data, particularly by Nagasawa *et al.* (1976), demonstrating anomalies in various differnt combinations, summarized in Table 1 above. Boynton's suggestion of a space medium, unlike any known to exist, and mitigating against Planck's radiation laws, appears partly to arise from some misunderstandings of the physically acceptable alternatives discussed in the papers quoted by him.

The chemical coherence of the rare earth elements make them an interesting group with distribution anomalies potentially useful as tracers of the formative processes in space. The range of their first ionization potentials is, however, narrow, the remaining uncertainties in these are considerable, and the oscillator strengths approximate only within an order of magnitude. This makes the speciation under space conditions difficult to evaluate in detail from existing spectroscopic data alone; extensive additional measurements and model experiments are necessary. Space observations, except of artificially injected plasmas, will hardly give any direct information on the rare earth chemistry in space because of the low abundance of these elements. Their behaviour as a group can, however, be approximated from observations of elements with similar electronic and thermochemical properties such as calcium, aluminium, and potassium, which are strongly depleted by condensation in the interstellar medium.

More immediately useful information on the state of excitation of the source medium, possibly preserved in meteorites, should be sought from elements with similarities in low temperature behaviour but with large differences in ionization properties such as discussed in section on *The State of Excitation of the Medium in the Solar and Planetary Nebulae.* Such elements are, for example, Hg–Tl, F–Cl, and the noble gases. The latter also offer the advantage that their oscillator strengths are more accurately known and theoretically predictable than those of the transition elements, particularly the electronically complex actinides and rare earth elements.

The most crucial information, however, comes from direct observation of the speciation, state of excitation and reaction kinetics in dense regions in space, and from model experiments. Conclusions drawn from systems with chemical and physical characteristics vastly different from actual space media may be perfectly valid in their own realm, but they may also have little to do with condensation in space, now or in the past.

Isotope Fractionation Effects

Under conditions known to prevail in space, isotope fractionation effects may occur, which are more pronounced than in systems which are in thermal or in chemical and thermodynamic equilibrium. The latter types of isotope fractionation effects are mostly the only ones considered in geochemical and planetary processes. If the space environment could be regarded as an equilibrium system of neutral gas, such a treatment would be applicable also in cosmochemistry. As is well-known from space observations and modern theory, this is not the case.

Disequilibrium fractionation effects in space plasmas, would tend to decay unless they are maintained in gravitational fields, such as in planetary atmospheres or if they are frozen into solid phases, capable of sustaining the disequilibrium. Until space missions with direct sampling of the small bodies in the solar system have been realized, carbonaceous meteorites provide the only solid material that, at least partially, appears to represent relatively unaltered condensates. Since the solids in these meteorites probably condensed in the transplanetary region (Alfvén and Arrhenius, 1976, Section 19), they may not provide very direct evidence on the emplacement and excitation of source medium condensing to solids in the planetary region. Nonetheless the record in the carbonaceous meteorites would seem to be of prime importance as a link between the chemical processes observed in the solar system and those taking place in dark interstellar clouds; the meteorite record also connects to laboratory experiments simulating space processes under controlled conditions.

Electrodynamic Fractionation

Several isotope fractionation processes have been proposed to be of importance in space plasmas on the basis of laboratory experiments. Among these are primarily the centrifugal effects on ions spiralling in magnetic fields (Bonnevier, 1970, 1972; James and Simpson, 1976). The isotope effect observed in critical velocity ionization (Alfvén and Arrhenius, 1976, Section 21.8) may also be considerable in the space environment. In contrast it has been pointed out by Bonnevier (1976) that the element separation effects due to ion–ion collision discussed by him and by Lehnert (1976) are not likely to be accompanied by isotope separation.

For any fractionation to have lasting effects, the separated components need to form reservoirs, physically or chemically isolated sufficiently well and for sufficiently long time for solids to form and to aggregate, preserving the integrity of their fractionation record.

Kinetic Isotope Fractionation at Chemical Reaction

Chemical isotope fractionation effects are probably of particular importance in space because of their magnitude under disequilibrium conditions, demonstrated in laboratory experiments, and because of the fact that ion and molecule reservoirs can maintain their identities chemically, although they are not necessarily separated physically.

The abundance of chemical species in space plasmas (like in the biosphere) is largely controlled by reaction rates, not by chemical equilibria. The most abundant ionic species in dense clouds for example is C^+ which is four times as abundant as H^+ at a number density and ionization rate of H_2 of 10^8 cm^{-3} and 10^{-17} cm^{-3} sec^{-1} respectively (Herbst and Klemperer, 1973). C^+ is produced by the reaction

$$CO + He^+ \rightarrow C^+ + O + He \qquad k = 2{\cdot}0 \times 10^{-9}\ \text{cm}^2\ \text{sec}^{-1} \tag{13}$$

where He^+ is probably mainly produced by electron collision ionization in electric currents in the magnetized dark clouds, and also by cosmic rays and UV photons in regions transparent to them.

C^+ under these conditions is destroyed mainly by reaction with polyatomic neutrals such as O_2, OH, NH, H_2O, H_2CO, *etc.* However, the abundances of these molecules, likewise kinetically controlled, are low and C^+ does not react effectively with the most abundant molecular species, H_2 and CO, or with electrons. As a result of the high rate of formation and the low rate of destruction, C^+ builds up to form the largest ion reservoir. Such balances between the reaction rates in the major production and destruction channels control the abundances of most ions and neutral molecules in space (Massey, 1972; Herbst and Klemperer, 1973); thermodynamic equilibrium considerations are obviously irrelevant. This includes particularly the gravitational densification of gas around massive objects such as protostars, stars and planets, where the release of potential energy alone suffices to ionize each atom several times (Alfvén, this volume) and where, like in other dense regions in space, hydromagnetic processes are likely to provide the dominant driving force for chemical reactions.

The kinetic control of chemical reactions extends also to any isotope fractionation resulting from these reactions. Such fractionation effects which are of magnitudes unparalleled in equilibrium systems, are frozen into the reaction products preserved in meteorites. The effects are particularly notable among such elements which, due to their formation of volatile compounds, are likely to have been extensively cycled through the fractionation reactions. Documented cases include the noble gases (see, for example, Black, this volume), carbon (Clayton, 1963; Krouse and Modzeleski, 1970), oxygen (Clayton *et al.*, 1973, 1976), magnesium 24/25 (Lee *et al.*, 1976), and among the heavy elements, mercury (Reed and Jovanovich, 1967). The $^{12}C/^{13}C$ fractionation in meteorite materials typically spans over a range much larger than the range resulting from geochemical equilibrium processes. As is discussed below, the 'large' effects recorded in meteorites are modest compared to kinetic isotope effects obtained in the laboratory under simulated space conditions, and to those directly observed in space (see concluding section).

The isotope distribution in light nuclides with three stable species has been studied so far only for oxygen, leading to the important observation (Clayton *et al.*, 1973, 1976) that the isotope ratio $^{16}O/^{17}O$ is highly variable compared to the ratio $^{17}O/^{18}O$ in meteorites, and that the isotope distribution differs markedly from the ratios found in equilibrium reactions. Apparently unaware of the large kinetic isotope fractionation effects predicted and observed in cold plasmas, and influenced by the equilibrium cooling condensation conjecture, meteoriticists attempting to explain these phenomena have resorted to assumptions such as the inclusion in meteorite crystals of 'extrasolar' particles with the remarkable property of having practically pure ^{16}O. Clayton *et al.* (1976) have convincingly demonstrated the difficulties in identifying any actual inclusion particles that have the physical properties required in this interpretation. Such particles would have to be sufficiently numerous to give the large isotope effects observed, and refractory enough to avoid equilibration, yet small enough to be invisible.

Mechanistic Basis for Kinetic Isotope Effects in Space

The discussion in the above section about *Particles, Fields, and Chemical Reactions in Dark Interstellar Clouds* suggests that isotope fractionation at condensation in space is probably entirely a kinetically controlled phenomenon. The observed non-radiogenic isotope distributions in space materials under these conditions would provide potential information on the state of excitation and reaction paths in the source media from which these materials condensed.

One of the conditions of major importance for chemical isotope fractionation in space materials is the strong disequilibrium between excitational and kinetic energy, which characterizes the reactants in most regions in space. In cold plasmas, with relatively low translational temperatures, high reaction rates are maintained due to the fact that the predominant reaction channels are ion–molecule interactions. These lack activation barriers since one of the reactants carries sufficient energy for a transition state to form exothermically and decay into the reaction product.

The effect of low translational temperature on the magnitude of the isotope fractionation, even in thermodynamic equilibrium, is illustrated by the simplified form of the general Biegeleisen–Mayer (1947) relationship in the low temperature limit, referred to as the zero order approximation to primary kinetic isotope effects (Biegeleisen, 1964).

Neglecting in this context symmetry factors and transmission coefficients, the ratio of the reaction rates K_1 and K_2 of the two analogous reactions, transferring the isotopes in (1) and in (2) from reactant to product form, can for the general case be written (van Hook, 1970)

$$\frac{K_1}{K_2} = \mathrm{MMI} \cdot \mathrm{EXC} \cdot \mathrm{ZPE} \tag{14}$$

where MMI is the term including mass and moment of inertia, EXC the excitation factor term and ZPE the zero point energy term.

With the assumptions made, EXC has the form

$$\mathrm{EXC} = \frac{\prod_{i}^{3n-6} \frac{1 - \mathrm{e}^{-u_{i1}}}{1 - \mathrm{e}^{-u_{i2}}}}{\prod_{i}^{3n-7} \frac{1 - \mathrm{e}^{-u_{i1}^{\ddagger}}}{1 - \mathrm{e}^{-u_{i2}^{\ddagger}}}} \tag{15}$$

where $u_i = hv_i/kT = hc\omega_i/kT$ and v_1 and ω_i are the vibrational frequencies and wavenumbers (assuming harmonic oscillation). The subscripts refer to the transition states.

In the low temperature limit hv/kT becomes large and $(1 - \mathrm{e}^{-hv/kT}) \rightarrow 1$; hence the factor EXC also approaches 1 and

$$\frac{K_1}{K_2} = \mathrm{MMI} \cdot \mathrm{ZPE} \tag{16}$$

Assuming only one critical isotopic frequency, equation (16) with the terms written out, can be reasonably approximated (van Hook, 1970) by

$$\ln \frac{K_1}{K_2} = \frac{hc}{2kT}\left(\omega_{1i}\left[1 - \left(\frac{\mu_2}{\mu_1}\right)^{1/2}\right]\right) \tag{17}$$

where μ_1 and μ_2 are the reduced masses of the vibrations in question.

Numerical evaluation of equation (17) for the case where the fractionation reaction proceeds by dissociation in the transition state of one of the reactant molecules into a massive fragment and a smaller fragment m_f, gives for $m_f = 1, 2$ (protium–deuterium) and $m_f = 10, 11$ the isotopic fractionation factors in Table 2.

Although equation (17) is a very rough approximation, the evaluation demonstrates in principle the large isotope fractionation due to the effect of low translational temperature.

As is also seen from Table 2, temperatures low enough to give rise to

Table 2. Kinetic isotope fractionation effects approximated in the low temperature limit (van Hook, 1970)

Vibrational frequency of mass 1, mass 10, respectively (cm^{-1})	K_H/K_D		K_{10}/K_{11}	
	50 K	298 K	50 K	298 K
1000	73	2·1	2·0	1·12
2000	5×10^3	4·2	3·9	1·26
3000	4×10^5	8·4	7·7	1·41
4000	3×10^7	17·7	15·3	1·58
5000	2×10^9	36·1	30·6	1·78

large kinetic isotope effects, compared to those expected in equilibrium reactions (see, for example, Urey, 1947), are of the order of 300 K for the hydrogen isotopes. Substantial kinetic deuterium and tritium effects are thus well-known to investigators of organic reaction mechanisms (Swain *et al.*, 1958; Melander, 1960; Collins and Bowman, 1970).

The effect of low temperature alone on the isotopic reaction rate ratios would obviously be of no practical consequence under equilibrium conditions, except for the lightest nuclides, since the absolute magnitude of the reaction rates decreases exponentially with temperature. In the space environment, however, the equipartition of kinetic and hydromagnetic energy inevitably generates magnetic fields and electric currents, which strongly amplify the background ionization from cosmic rays and radioactivity (Alfvén and Arrhenius, 1975, 1976; Alfvén, this volume). Excitation and ionization by electron collision in such electric currents may raise the vibrational temperature out of equilibrium with the translational temperature. This further increases ratio $h\nu/kT$ and also reduces or removes the activation barriers so that reaction rates become comparable to the collision frequencies.

In regions where neutral atoms and molecules are accelerated gravitationally by massive bodies (such as presumably in the solar and planetary nebulae of our early solar system) the critical velocity effect provides another important mode for conversion of translational energy into ionization, thus abruptly increasing $h\nu/kT$.

The various processes enhancing the vibrational excitation of reactants at low kinetic temperature would further increase the maximum effects of kinetic isotope fractionation above the values estimated from equation (17) in Table 2 for the equilibrium case in the low temperature limit.

Several different reaction mechanisms may be of controlling importance under different circumstances in achieving the isotope fractionation effects observed in laboratory plasmas and space materials. One such mechanism proposed by Basov *et al.* (1974), arises from the population inversion in vibrational states that has been demonstrated theoretically by Treanor *et al.* (1968) to occur as an anharmonicity effect at a combination of low translational temperature ($T \lesssim 10^3$ K) with high vibrational energy. Basov *et al.* (1974) demonstrated the proposed fractionation effect experimentally on the nitrogen isotope ratio $^{14}N/^{15}N$ using oxidation reactions of nitrogen in a plasma around 80 K.

Manuccia and Clark (1976) while criticizing the experimental interpretation by Basov and his collaborators, confirmed the effect and measured an enrichment of ^{15}N by 20 per cent over the mean terrestrial isotope ratio, using the same reactions and similar excitation conditions.

Considering the similarity in principle (but difference in scale) between these conditions and those in the regions of interest in space, kinetic isotope fractionation effects of this magnitude would be expected to arise in single reaction steps and in reaction loops (Figure 1 above) in the space environment. If ions and molecules are selectively removed from such isotopically fractionated reservoirs by polymerization and growth into solids, these may preserve the fractionation pattern. This would seem to be the reason for the isotopic anomalies observed in meteorites.

Fractionation in Three-isotope Systems

Fractionation involving more than two isotopes of an element is of particular interest since the nonlinearities that occur in kinetic fractionation effects potentially provide additional information on the reaction history.

In thermodynamic equilibrium the relative fractionations $c_{i(1)}/c_{i(n)}$ of three isotopes $i(1)$, $i(2)$ and $i(3)$

$$\frac{c_{i(1)}}{c_{i(3)}} = \left(\frac{c_{i(1)}}{c_{i(2)}}\right)^{\beta} \tag{18}$$

are in a fixed relationship, with β determined by the masses of the isotopes.

Under the conditions prevailing in space, discussed above, where the characteristic energies of the internal modes of reactant molecules and molecular ions are normally different from the translational energies, which again differ between electrons, ions and neutrals, the relationship between the kinetic isotope effects, determined by the rate constants K_1, K_2, and K_3, is a variable. In the isotope fractionation mechanisms considered by Basov *et al.* (1974)

$$\beta = \frac{h\nu_1 r_1\left(\frac{1}{T} - \frac{1}{T_1}\right) - h\nu_3 r_3\left(\frac{1}{T} - \frac{1}{T_3}\right) - \frac{h\nu_{r_1} - h\nu_{r_3}}{T}}{h\nu_1 r_1\left(\frac{1}{T} - \frac{1}{T_1}\right) - h\nu_2 r_2\left(\frac{1}{T} - \frac{1}{T_2}\right) - \frac{h\nu_{r_1} - h\nu_{r_2}}{T}} \tag{19}$$

where $h\nu_1$, $h\nu_2$ and $h\nu_3$ are the first vibrational levels in the reactant molecules substituted with the isotopes (1), (2) and (3), T_1, T_2, and T_3 are their vibrational temperatures, and T the translational temperature; r_1, r_2, and r_3 are the level numbers and $h\nu_{r_1}$, $h\nu_{r_2}$ and $h\nu_{r_3}$ the corresponding energies in the transition states.

It is likely that yet other mechanisms contribute significantly to the isotope fractionation in space condensates, and possibly are more important than the vibrational relaxation mechanism, particularly in ion–molecule reactions. Manuccia and Clark (1976), in measuring the nitrogen and hydrogen isotope fractionation at reaction in cold plasmas, hypothesize that the observed, up to 20 per cent enrichment of ^{15}N is a cumulative effect of isotope fractionation in the excitation, transfer, reaction and deactivation steps involved in the plasma reaction.

Extreme selective effects (from a modest geocentric point of view) of the type discussed above have been found to be characteristic of the fractionation pattern of the three isotopes of oxygen, ^{16}O, ^{17}O, and ^{18}O in meteorites (Clayton *et al.*, 1973, 1976) and have been discussed in the section on *Kinetic Isotope*

Fractionation above). It has been stated by these and other authors that such effects cannot be the result of chemical fractionation, a belief that seems to be rooted in the concept of the space medium as a gas with components in chemical equilibrium. As a solution to this apparent dilemma, it is commonly suggested that the anomalous excesses of ^{16}O would be caused by admixture of material from an exotic nucleosynthetic process, yielding practically pure ^{16}O. We propose instead that the observed oxygen isotope distributions may be the result of kinetic isotope fractionation processes characteristic of the known physical and chemical state of matter in space, understood in principle, and verified experimentally for two-isotope systems. Further experimental and theoretical investigations of these processes in polyisotopic systems would appear to be the proper course for understanding of the record preserved in space materials. It is interesting to note, for example, that fractionation by the vibrational relaxation mechanism cannot take place in plasma shock fronts such as postulated around active stars, while the effect would be expected to be notable in rapid expansion flows of initially excited diatomic or polyatomic gases (Treanor *et al.*, 1963), and in cold plasmas. Such flow probably results, for example, from hypervelocity collision of solid grains in space, a phenomenon which must have been extensive in the early solar system, controlling the orbital and accretional evolution of grains, embryos, and planetesimals (Alfvén and Arrhenius, 1976, Section 12).

Direct Observations of Isotope Distribution in Space and the Non-Radiogenic Isotope Anomalies in Meteorites

Although the most accurate measurements of the fractionation effects can be made on solids retrieved from space, possibilities also exist for establishing effects of the order of a factor two or larger (which are enormous in comparison with effects observed in meteorites) from optical measurements (Bortolot and Thaddeus, 1972) and from the isotope frequency shift in the rotational radio emission spectra. Isotopes of hydrogen, carbon, nitrogen, oxygen and sulphur have been distinguished in this way. The accuracy with which the observed line intensities can be interpreted in terms of isotopic abundances is at the present time only sufficient to indicate that the fractionation of ^{16}O relative to ^{18}O and of ^{12}C relative to ^{13}C are less than a factor of two of the range of terrestrial values (Penzias *et al.*, 1972).

In contrast the deuterium/hydrogen ratio in hydrogen cyanide in the Orion cloud was discovered by Jefferts *et al.* (1973) and Wilson *et al.* (1973) to be 100 times the ratio in methane in the Jovian atmosphere and 40 times the terrestrial value. HCN and HCN^+ are produced mainly by the reactions (Herbst and Klemperer, 1973):

$$C^+ + NH_2 \rightarrow HCN + H^+$$

$$H_2CN^+ + e \rightarrow HCN + H$$

$$H_3^+ + CN \rightarrow HCN^+ + H_2$$

and destroyed by

$$HCN^+ + H_2 \rightarrow H_2CN^+ + H$$

$$HCN + H_3^+ \rightarrow H_2CN^+ + H_2$$

$$HCN + C^+ \rightarrow C_2N^+ + H$$

Reaction loops that may lead to cumulative oxygen isotope effects, at least at the modest level observed in meteorites, and that contain exit channels into solids, which could preserve these effects include

$$H_2CO + HCO^+ \rightarrow H_3CO^+ + CO \tag{20}$$

$$H_3CO^+ + e \rightarrow H_2CO^* + H \tag{21}$$

Other potentially important reactions which can be conjectured to preserve a large fractionation of ^{16}O include the oxidation of Si^+, Mg^+ and Al^+, possibly by atomic oxygen, and polymerization of the oxidation products to silicates and oxides of the types common in meteorites. It appears reasonable to assume that the isotopically highly fractionated magnesium aluminosilicates and spinel ($MgAl_2O_4$) in carbonaceous meteorites (Clayton *et al.*, 1976) are products of this type of process.

Acknowledgements

This work was sponsored by Grant NGL05–009–002 from NASA's Lunar and Planetary Program Office. Helpful comments were obtained from Drs. H. Alfvén, J. Axnäs, B. Bonnevier, B. De, J. Isaacs, W. Klemperer, B. Kothari, B. Lehnert, K. Marti, A. Mendis, and D. Wark. Encouragement and support from the organizers of the Newcastle Symposium on *The Origin of the Solar System*, particularly Dr. S. F. Dermott, is also gratefully acknowledged.

References

Alfvén, H. (1942). 'On the cosmogony of the solar system.' *Stockholms Observatorium Ann.*, **14** (2), 3–33.
Alfvén, H. (1943). 'On the cosmogony of the solar system.' *Stockholms Observatorium Ann.*, **14**(5), 3–32.
Alfvén, H. (1946). 'On the cosmogony of the solar system.' *Stockholms Observatorium Ann.*, **14**(9), 3–29.
Alfvén, H. (1954). On the *Origin of the Solar System*, Oxford University Press, London.
Alfvén, H. (1976). 'The band-structure of the solar system.' This volume.
Alfvén, H., and Arrhenius, G. (1970)–(1974). 'Structure and evolutionary history of the solar system, Parts I–IV.' *Astrophys. Space Sci.*, **8**, 338–421; **9**, 3–33; **21**, 117–176; **29**, 63–159.
Alfvén, H., and Arrhenius, G. (1972). 'Exploring the origin of the solar system by space missions to asteroids.' *Naturwiss.*, **59**, 183–187.
Alfvén, H., and Arrhenius, G. (1975). *Structure and Evolutionary History of the Solar System*. D. Reidel, Dordrecht, Holland. 276 pp.

Alfvén, H., and Arrhenius, G. (1976). *Evolution of the Solar System*, NASA SP-345, U.S. Government Printing Office, Washington, D. C. 599 pp.

Alfvén, H., Arrhenius, G., and Mendis, A. (1976). 'The role of plasma in the primeval nebula.' *IAU Colloquium 39*, Lyon, France.

Alfvén, H. and Fälthammar, C. -G. (1963). *Cosmical Electrodynamics, Fundamental Principles*, 2nd Ed., Oxford University Press, London.

Anders, E. (1971). 'Reasons for not having an early asteroid mission.' In *Physical Studies of Minor Planets*, T. Gehrels, ed., NASA SP–267, U.S. Government Printing Office, Washington, D.C. 20402.

Anders, E. (1972). 'Conditions in the early solar system, as inferred from meteorites.' In *From Plasma to Planet*, Proc. Nobel Symp. 21, A. Elvius, ed., Wiley, New York, pp. 117–132.

Arrhenius, G. (1971). 'Low temperature inclusions in C3–4 carbonaceous chondrites. *Meteoritics*, **6**, 248.

Arrhenius, G. (1972). 'Reconstruction of the history of the solar system.' In *Origin of the Solar System*, H. Reeves, ed., Centre Nationale de la Recherche Scientifique, Paris, France, pp. 80–88.

Arrhenius, G. (1975). 'Iron in space condensates.' *Meteoritics*, **10**, 354.

Arrhenius, G., and Alfvén, H. (1971). 'Fractionation and condensation in space.' *Earth Planet. Sci. Lett.*, **10**, 253–267.

Arrhenius, G., Alfvén, H., and Fitzgerald, R. (1973). *Asteroid and Comet Exploration*, NASA CR-2291, U.S. Government Printing Office, Washington, D. C.

Arrhenius, G., and Asunmaa, S. K. (1974). 'Adhesion and clustering of dielectric particles in the space environment. l. Electric dipole character of lunar soil grains.' In *Lunar Science V* (Lunar Science Institute, Houston), pp. 22–24.

Arrhenius, G., Asunmaa, S. K., Fitzgerald, R. W., Kothari, B. J., and Macdougall, D. (1974). 'Record of primordial growth environment in olivine crystals from carbonaceous chondrites.' *Meteoritics*, **9**(4), 313–314.

Arrhenius, G., and De, B. R. (1973). 'Equilibrium condensation in a solar nebula.' *Meteoritics*, **8**, 297–313.

Axnäs, I. (1976). 'Velocity limitations in coaxial plasma gun experiments with gas mixtures.' *Report No. TRITA-EPP-76-02* (Department of Electron and Plasma Physics, Royal Institute of Technology, Stockholm).

Basov, N. B., Belenov, E. M., Gavilina, L. K., Isakov, V. A., Markin, E. P., Oraevskii, A. N., Romanenko, V. I., and Ferapontov, N. B. (1974). 'Isotope separation in chemical reactions occurring under thermodynamic nonequilibrium conditions.' *Sov. Phys. JETP*, **19**, 190.

Beichman, C. A., and Chaisson, E. J. (1974). 'Possible evidence for a large magnetic field in the Orion infrared nebula.' *Ap. J.*, **190**, L21–L24.

Biegeleisen, J. (1964). 'Correlation of kinetic isotope effects with chemical bonding in three-centre reactions.' *Pure Appl. Chem.*, **8**, 217.

Biegeleisen, J., and Goeppert-Mayer, M. (1947). 'Calculation of equilibrium constants for isotopic exchange.' *J. Chem. Phys.*, **15**, 261.

Biermann, L. (1951). 'Kometenschweife und solare Korpuskularstrahlung.' *Z. Astrophys.*, **29**, 274.

Blander, M., and Fuchs, L. H. (1975). 'Calcium–aluminum-rich inclusions in the Allende meteorite: Evidence for a liquid origin.' *Geochim. Cosmochim. Acta*, **39**, 1605–1619.

Bonnevier, B. (1970). 'Experimental evidence of element and isotope separation in a rotating plasma.' *Plasma Phys.*, **13**, 763–774.

Bonnevier, B. (1972). 'Experiment and observation of isotope and element separation in a plasma with cosmical applications.' In *Origin of the Solar System*, H. Reeves, ed., Centre Nationale de la Recherche Scientifique, Paris, France.

Bonnevier, B. (1976). 'Ion–ion collisions and element separation effects in the boundary

region of a plasma surrounded by neutral gas.' *Astrophys. Space Sci.*, **40**, 231–240.
Bortolot, V. J. and Thaddeus, P. (1972). *Ap. J.*, **175**, L17.
Brecher, A. (1972). 'The paleomagnetic record in carbonaceous chondrites.' *Ph.D. Thesis*, University of California, San Diego.
Brecher, A., and Arrhenius, G. (1971). 'Hydrogen recombination by non-activated chemisorption on metallic grains.' *Nature*, **230**, 107–109.
Brecher, A., and Arrhenius, G. (1974). 'The paleomagnetic record in carbonaceous chondrites: natural remanence and magnetic properties.' *J. Geophys. Res.*, **79**, 2081–2106.
Cameron, A. G. W. (1976). 'The primitive solar accretion disk and the formation of the planets.' This volume.
Chaisson, E. J., and Beichman, C. A. (1975). 'Further evidence for magnetism in the Orion region.' *Ap. J.*, **199**, L39–L42.
Chapman, C. R., and Salisbury, I. W. (1973). 'Comparison of meteorite and asteroid spectral reflectivities.' *Icarus*, **19**, 507.
Chou, C., Baedecker, P., and Wasson, J. (1976). 'Allende inclusions: volatile-element distribution and evidence for incomplete volatilization of presolar solids.' *Geochim. Cosmochim. Acta*, **40**, 85.
Clayton, R. (1963). 'Carbon isotope abundance in meteoritic carbonates.' *Science*, **140**, 192–193.
Clayton, R. N., Grossman, L., and Mayeda, T. K. (1973). 'A component of primitive nuclear composition in carbonaceous meteorites.' *Science*, **182**, 485–488.
Clayton, R., Onuma, N., Grossman, L., Mayeda, T. (1976). 'Distribution of presolar component in Allende and other carbonaceous meteorites.' *Earth Planet. Sci. Lett.* (in the press).
Collins, C. J., and Bowman, N. S., eds. (1969). *Isotope Effects in Chemical Reactions*. ACS Monograph. Van Nostrand Reinhold Co., New York.
Conrad, R. L., Schmitt, R. A., and Boynton, W. V. (1975). 'Rare earth and other elemental abundances in Allende inclusions.' *Meteoritics*, **10**, 384.
Danielsson, L. (1973). 'Review of the critical velocity of gas–plasma interaction.' *Astrophys. Space Sci.*, **24**, 459.
De, B. (1973). 'Some astrophysical problems involving plasmas and plasma–solid systems.' *Ph.D. Thesis*, University of California, San Diego.
De, B., Alfvén, H., and Arrhenius, G. (1976). 'Experiments, observations, and theory relating to the formation of the regular satellites.' *IAU Colloquium 28*, Planetary Satellites, ed. J. A. Burns, University of Arizona Press, Tucson.
De, B. and Arrhenius, G. (1978). 'Colloidal plasmas in space'. In *Advances in Colloid and Interface Science*.
Drobyshevskii, E. M. (1964). 'The volt–ampere characteristics of a homopolar cell.' *Soviet Physics–Technical Physics*, **8**, 903.
Fanale, F., and Cannon, W. (1972). 'Origin of planetary primordial rare gas: the possible role of adsorption.' *Geochim. Cosmochim. Acta*, **36**, 319.
Fireman, E., DeFelice, J., and Norton, E. (1970). 'Ages of the Allende meteorite.' *Geochim. Cosmochim. Acta*, **34**, 873.
Frank, P. G. (1954). 'The variety of reasons for the acceptance of scientific theories.' *Scientific Monthly*, **79**, 139.
Fuchs, L. H. (1971). 'Occurrence of wollastonite, rhönite, and andradite in the Allende meteorite.' *Amer. Mineral.*, **56**, 2053–2068.
Fuchs, L. H., and Blander, M. (1973). 'Calcium–aluminum rich inclusions in the Allende meteorite: Textural and mineralogical evidence for a liquid origin (abstract).' *EOS Trans. AGU*, **54**, 345.
Geiss, J., Hirt, P., and Leutwyler, H. (1970). *Solar Physics*, **12**, 458.
Geiss, J. (1972). 'Elemental and isotopic abundances in the solar wind.' *Proc. Asilomar Conference, NASA SP 308*, pp. 559–581.

Grossman, L., and E. Olsen, (1974). 'Origin of the high-temperature fraction of C2 chondrites.' *Geochim. Cosmochim. Acta*, **38**, 173–187.

Grossman, L. (1972). 'Condensation in the primitive solar nebula.' *Geochim. Cosmochim. Acta*, **36**, 597–619.

Grossman, L., Fruland, R., and McKay, D. (1975). 'Scanning electron microscopy of a pink inclusion from the Allende meteorite.' *Geophys. Res.*, **2**, 37.

Grossman, L., and Larimer, J. W. (1974). 'Early chemical history of the solar system.' *Rev. Geophys. Space Phys.*, **12**, 71–101.

Hassan, H. A. (1966). 'Characteristics of a rotating plasma.' *Phys. Fluids*, **9**, 2077–2078.

Herbig, G. H. (1968). 'The interstellar line spectrum of Zeta Ophiuchi.' *Zeit. Astrophys.*, **68**, 243–277.

Herbig, G. H. (1976). 'Some aspects of early stellar evolution that may be relevant to the origin of the solar system'. This volume.

Herbst, E., and W. Klemperer. (1973). 'The formation and depletion of molecules in dense interstellar clouds'. *Ap. J.*, **185**, 505–533.

Herndon, J. (1976). 'Chondrites and chondrules—macroscopic chemical aspects of their origins.' This volume.

Hollenbach, D. J., Werner, M. W., and Salpeter, E. E. (1971). 'Molecular hydrogen in HI regions.' *Ap. J.*, **163**, 165–180.

Hollenbach, D. J., and Salpeter, E. E. (1971). 'Surface recombination of hydrogen molecules'. *Ap. J.*, **163**, 155–164.

James, B. W., and Simpson, S. W. (1976). 'Isotope separation in the plasma centrifuge.' *Plasma Phys.*, **18**, 289–300.

Jefferts, K. B., Penzias, A. A., and Wilson, R. W. (1973). 'Deuterium in the Orion nebula'. *Ap. J.*, **179**, L57.

Johnson, T. V., and Fanale, F. P. (1973). 'Optical properties of carbonaceous chondrites and their relationship to asteroids.' *J. Geophys. Res.*, **78**, 8507.

Jokipii, J. R., (1964). 'The distribution of gases in the protoplanetary nebula'. *Icarus*, **3**, 248.

Jovanovic, S., and Reed, G. (1976). '^{196}Hg and ^{202}Hg isotopic ratios in chondrites: revisited.' *Earth Planet. Sci. Lett.*, **31**, 95–100.

Kerridge, J., and Macdougall, D. (1976). 'Mafic silicates in the Orgueil carbonaceous meteorite.' *Earth Planet. Sci. Lett.*, **29**, 341.

Kristoferson, L. (1975). 'The interaction of a plasma stream with solid bodies'. *Div. of Electron and Plasma Physics*, Royal Institute of Technology, Stockholm, Report 75–12.

Kristoferson, L., and Fredga, K. (1976). 'Laboratory simulation of erosion by space plasma'. *Div. of Electron and Plasma Physics*, Royal Inst. of Technology, Stockholm, Report 76–06.

Krouse, H. R., and Modzeleski, V. E. (1970). 'C^{13}/C^{12} abundances in components of carbonaceous chondrites and terrestrial samples'. *Geochim. Cosmochim. Acta*, **34**, 459–474.

Kurat, G. (1970). 'Zur Genese der Ca–Al-reichen Einschlüsse im Chondriten von Lancé'. *Earth Planet. Sci. Lett.*, **9**, 225–231.

Larimer, J. W., and Anders, E. (1970). 'Chemical fractionations in meteorites. III. Major element fractionations in the chondrites.' *Geochim. Cosmochim. Acta*, **34**, 367–387.

Lee, T., Papanastassiou, D. A., and Wasserburg, G. J. (1976). 'Demonstration of ^{25}Mg excess in Allende and evidence for ^{26}Al.' *Geophys. Res. Lett.*, **3**, 109.

Lehnert, B. (1967). 'Space charge effects by non-thermal ions in a magnetized plasma', *Phys. Fluids*, **10**, 2216.

Lehnert, B. (1970a). 'On the condition for cosmic grain formation', *Cosmic Electrodyn.*, **1**, 219–232.

Lehnert, B. (1970b). 'Minimum temperature and power effect of cosmical plasmas interacting with neutral gas', *Cosmic Electrodyn.*, **1**, 397–410.

Lehnert, B. (1971). 'Rotating plasmas', *Nuclear Fusion*, **11**, 485.
Lehnert, B. (1976). 'Element separation effects in the boundary region of a plasma surrounded by neutral gas', *Astrophys. Space Sci.*, **40**, 225–230.
Lewis, J. (1974). 'The temperature gradient in the solar nebula', *Science*, **186**, 440.
Lin, S. -C. (1961). 'Limiting velocity for a rotating plasma', *Phys. Fluids*, **4**, 1277–1288.
Lindblad, B. (1935). 'A condensation theory of meteoritic matter and its cosmological significance', *Nature*, **135**, 133–135.
Macdougall, D., Rajan, R., Hutcheon, I., and Price, P. (1973). 'Irradiation history and accretionary processes in lunar and meteoritic breccias', *Proc. 4th Lunar Science Conf.*, **3**, 2319.
Macdougall, D., and Price, P. (1974). 'Low-energy particle irradiation and possible age indicators for components of carbonaceous chondrites', *Meteoritics*, **9**, 370.
Manka, R. H., Michel, F. C., Freeman, J. W., Dyal, P., Parkin, C. W., Colburn, D. S., and Sonett, C. P. (1972). 'Evidence for acceleration of lunar ions'. In *Lunar Science—III*, C. Watkins, Ed. (The Lunar Science Institute, Houston, Texas), pp. 504–506.
Manuccia, T. J., and Clark, M. D. (1976). 'Enrichment of N^{15} by chemical reactions in a glow discharge at 77 K'. *Appl. Phys. Lett.*, **28**, 372.
Marti, K. (1973). 'Ages of the Allende chondrules and inclusions', *Meteoritics*, **8**, 55.
Martin, P. M. and Mason, B. (1974). 'Major and trace elements in the Allende meteorite'. *Nature*, **259**, 333–387.
Marvin, U. B., Wood, J. A., and Dickey, J. S., Jr. (1970). 'Ca–Al-rich phases in the Allende meteorite.' *Earth Planet. Sci. Lett.*, **7**, 346–350.
Massey, H. (1972). 'Atomic and molecular reactions in space'. In *From Plasma to Planet*, Proc. Nobel Symp. 21, A. Elvius, ed. (Wiley, New York), pp. 17–37.
McCord, T. B. (1966). 'Dynamical evolution of the Neptunian system'. *Astron. J.*, **71**, 585–590.
Melander, L. (1960). *Isotope Effects on Reaction Rates*, Ronald Press, New York.
Mendis, A. (1976). 'The cometary atmosphere'. *IAU Colloquium 15: Transitory Phenomena in Comets*, Grenoble, France.
Mendis, A. and Ip, W. -H. (1976a). 'The ionospheres and plasma tails of comets.' *Space Science Reviews* (in the press).
Mendis, A. and Ip, W. -H. (1976b). 'On the synthesis of complex organic molecules in cometary electric discharges'. *IAU Colloquium 39*, Lyon, France.
Mendis, A. and Wickramasinghe, N. C. (1975). 'On the composition of cometary dust'. *Astrophys. Space Sci.*, (in the press).
Mendis, A. and Wickramasinghe, N. C. (1976a). 'On the acceleration of interstellar grains'. *Astrophys. Space Sci.*, (in the press).
Mendis, A. and Wickramasinghe, N. C. (1976b). 'Cometary dust: The case against silicates'. *Astrophys. Space Sci.*, **37**, L13.
Metz, W. D. (1974). 'Exploring the solar system (IV)'. *Science*, **186**, 814.
Meyer, C., Jr. (1971). 'An experimental approach to circumstellar condensation'. *Geochim. Cosmochim. Acta*, **35**, 551–556.
Miller, S. (1953). 'A production of amino acids under possible primitive earth conditions'. *Science*, **117**, 3046.
Morton, D. (1974). 'Interstellar abundances toward Zeta Ophiuchi'. *Ap. J.*, **193**, L35.
Morton, D. (1975). 'Interstellar absorption lines in the spectrum of Zeta Ophiuchi'. *Ap. J.*, **197**, 85.
Nagasawa, N., Blanchard, D., Jacobs, J., Brannon, J., Philpotts, J., and Onuma, N. (1976). 'Trace element distribution in mineral separates of the Allende inclusions and their genetic implications'. *Lunar Science Institute Contribution*, No. XXX.
Narcisi, R., and Roth, W. (1970). 'The formation of cluster ions in laboratory sources and in the ionosphere'. *Adv. Electronics and Elec. Physics*, **29**, 79.
Öpik, E. J. (1968). 'The cometary origin of meteorites', *Irish Astron. J.*, **8**, 185.

Oppenheimer, M. (1974). 'The gas phase chemistry in comets'. *Ap. J.*, **196**, 251.
Palme, H. and Wlotzka, F. (1976). 'A metal particle from the meteorite Allende, and the condensation of refractory siderophile elements.' *Earth Planet. Sci. Lett.* (in the press).
Penzias, A. A., Jefferts, K. B., Wilson, R. W., Liszt, H. S., and Solomon, P. M. (1972). 'Interstellar deuterium'. *Ap. J.*, **178**, L35.
Petshek, H. E. (1960). Comment following Alfvén, H., 'Collision between a nonionized gas and a magnetized plasma.' *Rev. Mod. Phys.*, **32**, 710–712.
Phinney, D., Frick, U., and Reynolds, J. (1976). 'Rare-gas-rich separates from carbonaceous chondrites'. *Lunar Science VII*, 690–697. Lunar Science Institute, Houston, Texas.
Platt, J. R. (1956). 'On the optical properties of interstellar dust'. *Ap. J.*, **123**, 486.
Ponnamperuma, C., Woeller, F., Flores, J., Romiez, M., and Allen, W. (1969). 'Synthesis of organic compounds by the action of electric discharge in simulated primitive atmospheres.' In *Chemical Reactions in Electrical Discharges*, American Chemical Society, Washington, D.C., p. 280.
Price, P., Hutcheon, I., Braddy, D., and Macdougall, D. (1975). 'Track studies bearing on solar-system regoliths.' *Proc. 6th Lunar Sci. Conf.*, 3449.
Raadu, M. A. (1975). 'Critical ionization velocity and electrostatic instabilities.' *Report No. TRITA-EPP-75-28*, Division of Electron and Plasma Physics, Royal Institute of Technology, Stockholm.
Reed, G. W., and Jovanovic, J. (1969). '^{196}Hg and ^{202}Hg isotopic ratios in chondrites.' *J. Inorg. Nucl. Chem.*, **31**, 3783–3788.
Rogerson, J., Spitzer, L., Drake, J., Dreissler, K., Jenkins, E., Morton, D., and York, D. (1973). 'Spectrophotometric results from the Copernicus satellite. I. Instrumentation and performance.' *Ap. J.*, **181**, L97.
Salpeter, E. (1974a). 'Nucleation and growth of dust grains.' *Ap. J.*, **193**, 579.
Salpeter, E. (1974b). 'Dying stars and reborn dust.' *Rev. Mod. Phys.*, **45**, 3.
Sherman, J. C. (1973). 'Review of the critical velocity gas–plasma interaction. II. Theory.' *Astrophys. Space Sci.*, **24**, 487–510.
Shimuzu, M. (1975). 'Ion chemistry in the cometary atmosphere.' *Astrophys. Space Sci.*, **36**, 353.
Signer, P. and H. E. Suess, (1963). 'Rare gases in the Sun, in the atmosphere, and in meteorites.' In *Earth Science and Meteoritics*, dedicated to F. G. Houtermans; Eds. J. Geiss and E. D. Goldberg, North-Holland Publ. Co., Amsterdam, pp. 241–272.
Sockol, P. M. (1968). 'Analysis of a rotating plasma experiment.' *Phys. Fluids*, **11**, 637–645.
Spitzer, L. (1968). *Diffuse Matter in Space*. Interscience, New York.
Srnka, L. (1976). 'Critical velocity phenomena as LTP source excitation mechanism.' *Physics of the Earth and Planetary Interiors* (in press).
Suess, H. E., and Urey, H. C. (1956). 'Abundances of the elements.' *Rev. Mod. Phys.*, **28**, 53–74.
Swain, C. G., Stivers, E. C., Reuwer, J. F., and Schaad, L. J. (1958). 'Use of hydrogen isotope effects to identify the attacking nucleophile in the enolization of ketones catalyzed by acetic acid.' *J. Am. Chem. Soc.*, **80**, 5885.
Treanor, C. W., Rich, J. W., and Rehm, R. G. (1967). 'Vibrational relaxation of anharmonic oscillators with exchange-dominated collisions.' *J. Chem. Phys.*, **48**, 1798.
Urey, H. C. (1947). 'The thermodynamic properties of isotopic substances.' *J. Chem. Soc.*, p. 562.
Urey, H. C. (1952). *The Planets: Their Origin and Development* (New Haven, Connecticut: Yale University Press).
Urey, H. C., and Donn, B. (1956). 'Chemical heating for meteorites.' *Ap. J.*, **124**, 307.
Van Hook, W. A. (1969). 'Kinetic isotope effects: introduction and discussion of the theory.' In *Isotope Effects in Chemical Reactions*, C. J. Collins and N. S. Bowman, eds., ACS Monograph. Van Nostrand Reinhold Co., New York.

Vanysek, K., and Wickramasinghe, N. (1975). 'Formaldehyde polymers in comets.' *Astrophys. Space Sci.*, **33**, L19.

Wark, D., and Lovering, J. (1976). 'Refractory/platinum metal grains in Allende carcs.' *Lunar Science VII*, 912–914. Lyndon B. Johnson Space Center, Houston, Texas.

Wark, D. and Lovering, J. (1977). 'Marker events in the early evolution of the solar system—evidence from rims on calcium–aluminum-rich inclusions in carbonaceous chondrites.' Unpublished manuscript. Abstract in *Lunar Science*, **8**, Lyndon B. Johnson Space Center, Houston, Texas.

Weber, H. W., Hintenberger, H., and Begemann, F. (1971). 'Noble gases in the Haverö ureilite.' *Earth Planet. Sci. Lett.*, **13**, 205

Weber, H. W., Begemann, F., and Hintenberger, H. (1976). 'Primordial gases in graphite–diamond–kamacite inclusions from the Haverö urelite.' *Earth Planet. Sci. Lett.*, **29**, 81–90.

Wetherill, G. W. (1971). 'Cometary versus asteroidal origin of chondritic meteorites.' In *Physical Studies of Minor Planets*, NASA SP–267, T. Gehrels, ed. (U.S. Government Printing Office, Washington, D.C.), pp. 447–460.

Wilson, R. W., Penzias, A. A., Jefferts, K. B., and Solomon, P. M. (1973). 'Interstellar deuterium: the hyperfine structure of DCN.' *Ap. J.*, **179**, L107.

Wood, J. A. (1963). 'On the origin of chondrules and chondrites.' *Icarus*, **2**, 152–180.

ISOTOPIC ANOMALIES IN SOLAR SYSTEM MATERIAL—WHAT CAN THEY TELL US?

DAVID C. BLACK
Space Science Division, NASA–Ames Research Center, Moffett Field, California, U.S.A.

Introduction

Speculation concerning the origin of the solar system is not a recent phenomenon. Although primitive man did not possess any knowledge that the solar system as we know it existed, recorded history gives ample evidence that our forebears were vitally concerned with the question of how their perceivable universe, particularly heavenly objects, were created and behaved. These early thinkers could do little beyond speculate as to the workings of their surroundings.

As with almost all studies of those aspects of nature which are amenable to abstraction and quantification, the study of the formation and evolution of the solar system is characterized by certain critical developments which afford a vast increase in the extent to which one can quantitatively specify, and hence constrain, the behaviour of the system under study. One such watershed in the study of the solar system was the advent of the telescope in the early 1600's and the subsequent increase in man's knowledge of the structure of the solar system. This increased observational knowledge, coupled with theoretical advances concerning the motion of bodies under the influence of gravity, led immediately to more elaborate speculation by people such as Laplace and Kant concerning the formation of the solar system, and indeed this era is generally regarded as the beginning of 'modern' cosmogony.

It is not surprising to present-day students of our subject that in spite of the aforementioned advances, there was little fundamental advancement in cosmogonic hypotheses from the time of these early 'modern' notions until the 1960's. There were many fascinating alternative hypotheses advanced, but it is beyond the scope of this communication to go into the history of this subject in detail. I recommend that interested readers consult a paper by ter Haar and Cameron (1963) for a succinct review of 'modern' cosmogonic hypotheses.

What happened in the late 1950's and early 1960's to create the beginnings of another watershed in the history of this subject? In my view there were two fundamental events. The first, and certainly best known, was the advent of the space age brought about by a small Russian device called Sputnik which did little more than 'beep' as it orbited the Earth. The device was of no scientific merit, but it precipitated major efforts to *explore* our solar system, the results of which are well known. We now have much more detailed *quantitative* knowledge about the inner five planets in the solar system than was available as little as a decade ago. The second, and less well-known event, was the discovery by John Reynolds that the Richardton meteorite contained unambiguous evidence for the existence of an extinct radioactive isotope, ^{129}I. Meteorites had been studied for some time before Reynold's discovery, and were known to be extraterrestrial objects. The significance of Reynold's work was that it confirmed that these objects retained traces of very ancient events in the history of the solar system, and it thereby catalysed a marked increase in the level of examination of these objects. As a consequence of this increased study, we are now more able to *quantitatively* constrain conditions and events in the formative stages of the solar system.

I have taken this slight introductory excursion from my main topic for what is, in my opinion, an important message. Namely, as a result of these watersheds, hypotheses on the origin of the solar system must now be judged in a new light. The key word which characterizes this new era in our subject is 'quantitative'. Prior to 1960, it was sufficient for a cosmogonic hypothesis to account in a fairly vague manner for the basic structure of the solar system. However, we must now *require* all serious cosmogonic hypotheses to be sufficiently detailed to at least afford a comparison with our increased quantitative knowledge of the solar system. All cosmogonic hypotheses are necessarily speculative, but those which do not provide sufficiently detailed predictions, to be tested by what we know, can no longer be considered serious contenders.

Before moving on to the aspect of this new era that is the main theme of this lecture, namely isotopic anomalies, there is one final philosophical point I would make. The increased quantitative awareness of our solar system can be a two-edged device. If we are ever to understand the *origin* of our solar system, we must assume that it is *not* a unique phenomenon. A practical implication of this assumption is that we must sift our quantitative data with an eye toward general principles. The beauty of our problem is its highly interdisciplinary nature, involving subjects ranging from geology to astronomy and astrophysics. This interdisciplinary character of the subject, however, presents a potential problem. There is a tendency for researchers in a given discipline to view the problem of the origin of the solar system from their own vantage point. In moderation, that is necessary and valuable. However, as more data become available, we must constantly guard against too narrow a view. It is similar to the parable of the blind men trying to describe an elephant. Each has his 'area' of expertise, and provides a description of the beast in keeping only with his knowledge of the problem. The blind men will only reach a correct description of the animal if they take full cognizance of what each has learned. As with our blind friends, we must constantly be aware of knowledge gained in other special fields, so as not to push our description of our 'elephant' beyond what is consistent with other information.

The subject we will be concerned with in this lecture is that of isotopic anomalies in solar system material and what they might be telling us. We shall concentrate our attention on three relatively recently discovered anomalies, specifically an anomaly known as Neon E, an oxygen anomaly and a magnesium anomaly.

The Role of Isotopic Research

Prior to discussing specific isotopic anomalies, it is perhaps worthwhile to briefly indicate the type of information one can obtain from isotopic data. It is generally assumed that the solar system was isotopically uniform in composition at some early epoch in its history, and that variations in isotopic composition among meteorites, lunar, and terrestrial samples arise as a consequence of chemical and physical phenomena which altered the initial 'master composition' of the solar system. The goal of isotopic research is to understand what phenomena are responsible for the observed isotopic anomalies, and thereby to determine the initial solar system isotopic composition. The search for this

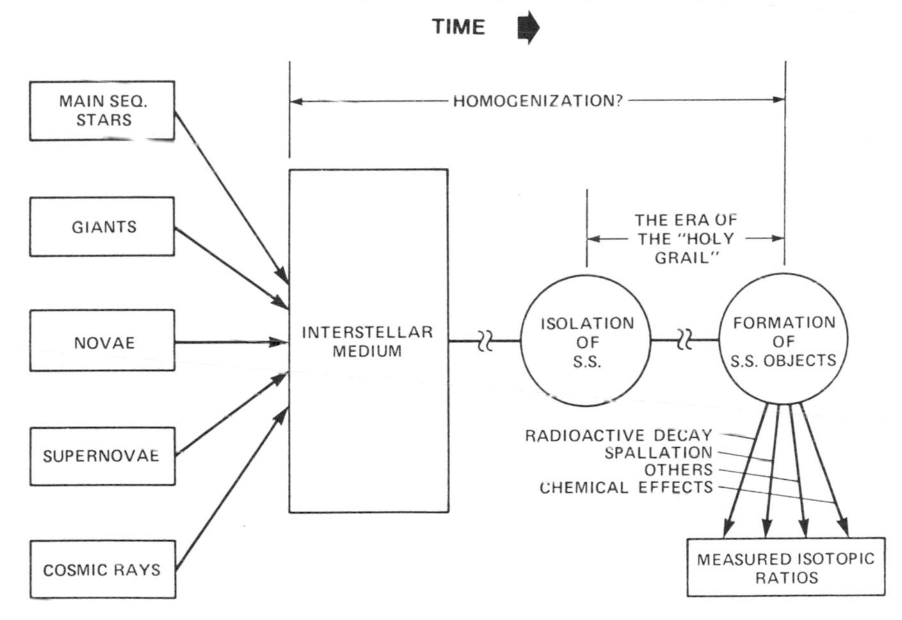

Figure 1. A schematic representation of the phenomena and objects which contributed to the history of elements and isotopes found in present-day solar system material. Evolution proceeds generally from left to right in the figure.

initial master isotopic composition has been likened to a search for the Holy Grail. Shown in Figure 1 is a schematic representation of the history of the elements found in solar system material. The arrow of time may be thought of as moving from left to right in the figure. The boxes in the left-hand portion of the figure indicate objects or processes which supply elements to the interstellar medium, out of which the solar system formed. Sometime after the solar system isolated itself in the sense of elemental and isotopic exchange with its surroundings, and after isotopic homogenization of the solar system, solid bodies began to form. Those systems evolved, perhaps influenced by any one or all of the indicated processes, until the present time when the relevant isotopic ratios are measured.

Until recently there has been no compelling evidence from isotopic research to indicate that the assumption of isotopic homogeneity in the primitive solar system was invalid. Stated in a more practical manner, until recently there had been no measured isotopic ratios which could not be understood in the context of reasonable physical or chemical processes acting to alter a uniform initial solar isotopic composition. This situation changed is 1969 with the discovery by Black and Pepin of an unusual neon isotopic composition in certain carbonaceous meteorites. This unusual composition was later designated as Neon E, and attributed to the existence of a non-solar isotopic composition contained in interstellar dust grains which survived the formation of the solar system (Black, 1972b). Four years after the discovery of Neon E, an unusual or anomalous oxygen isotopic composition was discovered (Clayton *et al.*, 1973). More recently, a magnesium isotopic anomaly has been discovered by two research groups (Gray and Compston, 1974; Lee and Papanastassiou, 1974). The remainder of this lecture will deal with these three isotopic anomalies in more detail, emphasizing particularly the experimental evidence, possible causes for the observed isotopic composition and finally, possible implications of these anomalies with regard to models of solar system formation and evolution.

Isotopic Anomalies in Neon

Rare gases, more so than other elements, provide sensitive isotopic monitors of physical phenomena throughout the history of a sample. There are several reasons for this sensitivity. First, under most conditions rare gases do not participate in chemical interactions; their abundance in solid material is dictated by physical phenomena. As a consequence of their high volatility rare gases are relatively strongly depleted in solid material, as compared to their relative abundance in the Sun. This strongly depleted character means that physical phenomena which can produce easily recognized isotopic patterns in the rare gases can be more readily discerned. Effects of spallation (production of nuclei from heavier nuclei through the interaction of high energy particles with the heavier nuclei), fission (e.g. $U \longrightarrow {}^{136,134,132,131}Xe$) and radioactive decay (e.g. $^{129}I \longrightarrow {}^{129}Xe$, see Kirsten, 1978) are not overwhelmed by the 'natural' isotopic patterns of the rare gases.

Historically, much of the early research on the isotopic composition of rare gases was concerned with the two heavier rare gases, krypton, and xenon. The reason for this interest was simply that of the twenty-three stable rare gas isotopes, fifteen are found in xenon and krypton. This richness of isotopes offers a corresponding richness in terms of detecting effects due to a variety of phenomena. It was generally thought that the lighter rare gases, helium, neon, and argon contained isotopic effects which although important were not as varied and complex as those found in the heavier gases. Work by Manuel (1967) and Pepin (1967) provided the first comprehensive analysis of the isotopic composition of neon in bulk samples of a variety of meteoritic material. Their work emphasized the findings of earlier research, namely that the isotopic composition of neon in gas-rich meteorites is characterized by a $^{20}Ne/^{22}Ne$ ratio of about 12·5, whereas the corresponding ratio for carbonaceous meteorites is typically ~ 8·2. There was no ready explanation for the wide range in isotopic composition (the largest range in isotopic composition known at the time). It was, however, generally thought that the difference in isotopic composition could be due to fractionation of the neon isotopic composition from diffusion of neon from the meteoritic material. The lighter isotope would be lost preferentially, thereby lowering the $^{20}Ne/^{22}Ne$ ratio as more neon was lost. Simple diffusion theory predicts that there will be roughly a ten per cent decrease in the $^{20}Ne/^{22}Ne$ ratio for every order of magnitude reduction in the abundance of neon. The difference between 12·5 and 8·2 for a $^{20}Ne/^{22}Ne$ ratio requires a significant depletion (99·99 per cent), but not one that is so large as to be unreasonable.

Shown in Figure 2 is a graphical representation of the isotopic composition of neon. This type of plot is known as a 'three-isotope plot' and is used frequently in isotopic studies. It is a property of such a plot that a mixture of two distinct isotopic compositions will fall on a straight line joining the two parent compositions. The major neon compositions known prior to 1969 are designated as components A, B and S. Also shown is the composition of neon in the Earth's atmosphere. Component B is found in gas-rich meteorites, lunar soils, and lunar breccias, and is now known to be due to solar wind rare ions directly implanted in the samples (Bühler *et al.*, 1969), confirming earlier suggestions (Signer, 1964; Wänke, 1965) as to the origin of component B. Component A is found in certain carbonaceous meteorites, but is not present in less primitive or more metamorphosed meteorites. Its origins are still a subject of debate. Component S represents the isotopic composition of spallation neon. The composition of S can vary significantly depending on the target nuclei available for spallation, the energy spectrum of the spalling particles and the nature of the spalling particle (mainly protons and/or neutrons). The components labelled C and D were first studied and discussed in detail by Black (1972a). Although they are of interest in the general context of understanding pheno-

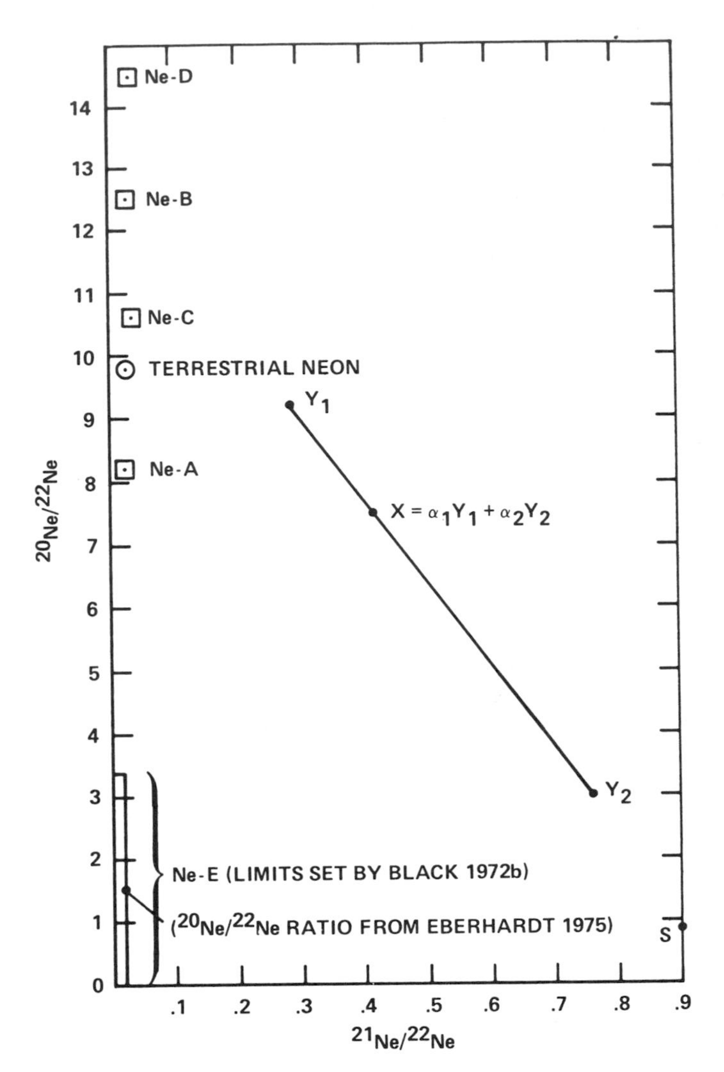

Figure 2. A three-isotope plot for neon, showing $^{20}Ne/^{22}Ne$ *vs.* $^{21}Ne/^{22}Ne$. The compositions of the principal components found to date in solar system material are shown. Also shown is a mixing line between two hypothetical neon compositions, Y_1 and Y_2. The composition designated as X is derived from mixing Y_1 and Y_2 such that the fraction of ^{22}Ne in X due to Y_1 is given by the ratio $[(^{20}Ne/^{22}Ne)_X - (^{20}Ne/^{22}Ne)_{Y_2}]/[(^{20}Ne/^{22}Ne)_{Y_1} - (^{20}Ne/^{22}Ne)_{Y_2}]$. All mixtures of Y_1 and Y_2 must lie on the line joining Y_1 and Y_2

mena involved in the evolution of solar system material, they will not be discussed here.

Component E was first discovered in stepwise heating experiments (Black and Pepin, 1969) where an upper limit on the $^{20}Ne/^{22}Ne$ ratio of this component was set at 3·4. Subsequently, Eberhardt (1975) isolated portions of a carbonaceous meteorite which were enriched in Neon E. These samples were characterized by $^{20}Ne/^{22}Ne$ ratios of 1·5. The abundance of Neon E in meteorites is somewhat uncertain, but the available evidence suggests that it is characterized by ^{22}Ne abundance of 10^{-9}–10^{-8} ccSTP/gm. This is the present state of experimental knowledge concerning this unusual neon isotopic composition. Several laboratories are presently attempting to further isolate the host phase for this composition with the hope of determining the composition of Neon E (at present, Eberhardt's value of 1·5 can only be considered as an upper limit to the $^{20}Ne/^{22}Ne$ ratio), and how this component was sited in the meteoritic material.

As can be seen from Figure 2 above, the composition of Neon E is significantly different than that of Neon B. The latter is known to represent the composition of neon in the solar wind, which is assumed to reasonably well approximate the neon composition of the bulk Sun, which in turn is assumed to represent the neon isotopic composition of the primative solar nebula. If the above chain of assumptions is correct, and indeed the last two links have *not* been subjected to exhaustive testing, one is forced to the conclusion that Neon E is very non-solar in character.

Can Neon E be produced from Neon B by diffusive fractionation of the type discussed earlier in connection with Neon A? The answer is almost certainly no. In order to reduce the $^{20}Ne/^{22}Ne$ ratio from 12·5 to 1·5 by diffusive fractionation, the loss of neon must be complete to approximately one part in 10^{20}! Knowing the amount of gas present in the meteorites which contain Neon E, and multiplying that amount by 10^{20} leads to the unlikely scenario that the required initial neon abundance in the sample would have exceeded by many orders of magnitude the abundance of all other elements in the meteorite; in fact the initial sample would have to have been a pure neon drop contaminated with traces of magnesium, silicon and so on.

As noted earlier, the composition of spallation neon is sensitive to a number of parameters. The question has been raised in connection with Neon E, and the other anomalies that we will discuss later in the lecture, as to whether these anomalies might not be a consequence of a particle irradiation of some kind. Black and Pepin first considered this suggestion as a possible explanation for Neon E. The proposed mechanism involves the formation of ^{22}Na which decays by way of positron emission to form ^{22}Ne. This process could build a reservoir of pure ^{22}Ne in a sample, and such a reservoir is consistent with what is known about Neon E. Subsequent to this suggestion and motivated by discoveries of anomalies in oxygen and magnesium, other more detailed irradiation models have been advanced (cf. Heymann and Dziczkaniec, 1976; Audouze *et al.*, 1976;

Clayton, Dwek, and Woosley, 1977). Those interested in the many fascinating and important subtle effects which arise from irradiation of gas or particulate matter should read these three references. To summarize the general results of these studies, one can say that there is no particular difficulty in producing sufficient amounts of neon by an irradiation process; the problems arise in producing the correct isotopic ratios ($^{20}Ne/^{22}Ne \leqslant 1{\cdot}5$, $^{21}Ne/^{22}Ne \leqslant 0{\cdot}02$). It is, I believe, a fair statement that the only way one could hope to produce Neon E by any realistic irradiation mechanism is as Black and Pepin had originally suggested, namely the production of ^{22}Na. At present, such a model is consistent with the available data. There are, however, difficulties with that model in terms of the details of how a grain which traps the sodium parent isotope is able to retain the highly mobile neon daughter isotope. An important test of the sodium hypothesis will be forthcoming when the host phase(s) of Neon E are isolated and their composition determined. If they are composed of material which is geochemically unrelated to sodium, it will be very difficult to defend the sodium hypothesis; conversely, if the material is geochemically similar to sodium, the hypothesis will receive empirical support.

A third possibility is that originally suggested by this author in 1972, namely that Neon E is a result of a stellar nucleosynthetic process, and that Neon E was implanted in dust grains which eventually found their way into the primitive solar system. It is hypothesized that these grains were not heated sufficiently to degas the neon contained in them, and have thereby preserved an isotopic remembrance of events which predate the solar system formation process. It should be noted that this view can also be made consistent with an irradiation hypothesis of the type discussed above. The key question would be where the irradiation took place. Was it in the young solar system, or was it at some site outside of the solar system?

These points of view represent the major contenders as explanations of the origins of Neon E. To summarize to this point, it does not seem at all possible to produce Neon E from solar system neon (by diffusive fractionation or any other mechanism), but it may be possible to produce Neon E in the solar system by means of some variant of a particle irradiation scheme; it is also possible that Neon E does not represent any phenomenon in the solar system, but rather is the peculiar isotopic signature of some more general astrophysical phenomenon which has been preserved throughout the history of the solar system. It should be clear that whatever the correct hypothesis might be, the explanation it affords of Neon E will tell us a great deal about events in the early history of the solar system, and perhaps even more. Let us turn now to the other two isotopic anomalies of interest in this lecture.

Oxygen Isotopic Anomalies

As remarked earlier, Neon E was the only example (prior to 1973) of an isotopic anomaly which was possibly inconsistent with the concept of the 'Holy

Grail'. All other known isotopic variations could be readily understood within the framework of the assumption of a master solar system isotopic composition for all elements, with that composition altered and/or added to by solar system phenomena. I recall a conversation I had with John Reynolds shortly after I suggested that Neon E was a consequence of some presolar event which had been preserved in solar system material until the present. He argued, prophetically as it turned out, that if presolar grains did exist in meteoritic material, one ought to see evidence of isotopic anomalies in major elements, and there was no evidence for such anomalies. He remarked that oxygen provided a critical test. Subsequent to that conversation, Clayton and coworkers at the University of Chicago announced the discovery of such an oxygen isotopic anomaly (Clayton *et al.*, 1973). We turn now to a brief review of the current experimental understanding of this important oxygen isotopic anomaly.

Oxygen, like neon, has three stable isotopes. The most frequently used graphical representation of the isotopic composition of oxygen is the three-isotope plot discussed earlier. There is one small difference between this plot as used to represent oxygen data and the plot as used to represent neon data, namely the use of the *per mil* or δ notation to express the isotopic ratios. Utilizing this notation, one defines δ^{17} and δ^{18} as follows:

$$\delta^{17} = (R^{17}/R^{17}_{std} - 1) \times 1000\,; \ \delta^{18} = (R^{18}/R^{18}_{std} - 1) \times 1000, \qquad (1)$$

where $R^{17} = {}^{17}O/{}^{16}O$, $R^{18} = {}^{18}O/{}^{16}O$ and the subscript 'std' refers to an arbitrary oxygen isotopic standard, such as SMOW (Standard Mean Ocean Water). Shown in Figure 3 is a schematic representation of the oxygen isotopic composition in a variety of meteoritic material, lunar samples and terrestrial rocks. The δ-values shown are with respect to the so-called carbonaceous chondrite reference standard (CCRS; see Clayton *et al.*, 1973). The line in the Figure designated as line (a) is defined by the equation

$$\delta^{18} = 2\delta^{17}, \qquad (2)$$

while the line designated as line (b) is defined by the equation

$$\delta^{18} = \delta^{17}. \qquad (3)$$

The physical significance of these two lines will be discussed later. Note that most of the data, representing meteoritic, lunar, and terrestrial material, tends to lie either along line (a), or along a line parallel to line (a). There are, however, a number of samples whose oxygen isotopic composition tends to fall along line (b), and it is those samples which most clearly indicate the presence of the anomalous composition.

We turn now to a discussion of possible interpretations of the behaviour evidenced in Figure 3 above. It will be recalled from the earlier discussion in connection with Figure 2 that mixtures of two distinct isotopic compositions produce compositions which fall along a line joining the two parent compositions. Is there any other way in which a straight line data array could be genera-

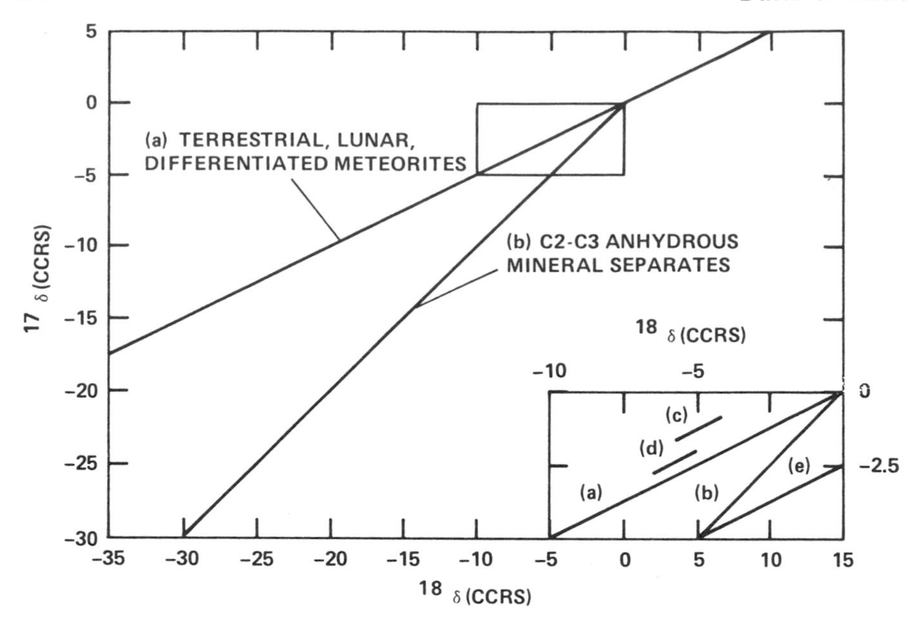

Figure 3. A three-isotope plot for oxygen expressed in permil or δ-notation (see text). Line (a) is the trend line predicted for compositions produced through isotopic fractionation during equilibrium chemical reactions. Line (b) is the trend line discovered by Clayton *et al.* (1973). Lines (c), (d) and (e) in the inset are respectively, trend lines defined by whole-rock analyses of L-group and H-group chondrites, and matrix material from C2 carbonaceous meteorites

ted in a plot such as Figure 3? As one might expect, there are a number of processes which produce straight line behaviour in such a plot, at least over a restricted range in δ-values. One such process is fractionation of isotopic ratios during a chemical reaction.

Chemical reactions are, to a good approximation, dependent only on the elements involved—the reaction of H plus OH to form H_2O is basically the same whether the oxygen involved is ^{16}O, ^{17}O or ^{18}O. However, if the reactions in question take place at temperatures T such that kT (where k is Boltzmann's constant) is small compared to the relevant free energies in the reaction, minor differences in free energy due to differences in isotopic mass can lead to significant changes in isotopic composition of an element as measured in the pre- and post-reaction states. We will not go into the physics of this phenomena here, but interested readers may consult any text on chemical kinetics (cf. Van Zeggeren and Storey, 1970) for details. A consequence of this chemical fractionation is that changes in a given isotopic ratio which are *small* compared to the value of the isotopic ratio are proportional to the mass difference between the isotopes involved. In the case of oxygen, one would expect that the isotopic shift in the $^{18}O/^{16}O$ ratio would be twice the shift in the $^{17}O/^{16}O$ ratio. This is the condition governed by equation (2) and represented by line (a). We see then how the oxygen isotopic data prior to Clayton *et al.*'s discovery was consistent with the Holy Grail concept—the master oxygen isotopic composition altered only by straightforward isotopic fractionation in chemical reactions. We also see how the data which define line (b) in Figure 3 are *not* consistent with isotopic changes during chemical reactions. Another way to generate a straight line in Figure 3 is by mixing two distinct compositions. If line (b) is due to a mixture, one of the endpoint compositions is at the lower end of line (b) and may be at the origin ($^{17}O/^{16}O = {}^{18}O/^{16}O = 0$) of the plot. Such an explanation has been advanced by Clayton *et al.* (1973); they suggested that a pure ^{16}O component was preserved in interstellar grains which survived the origin of the solar system.

It should be noted for the sake of completeness that the oxygen anomaly is not easily generated by means of a particle irradiation process of the type discussed in connection with Neon E. The difficulties are two-fold; the isotopic composition (pure ^{16}O) and more importantly, the high abundance of oxygen, which necessitates such a high particle fluence that one would expect to observe irradiation produced anomalies in a number of elements, and such effects are not present.

Although the oxygen isotopic anomaly is most likely to derive from a nuclear effect outside of the solar system, there are still questions relating to the type of nuclear phenomena, the astrophysical site where the anomalous composition was created, and finally the means by which the non-solar oxygen was introduced into the solar system (in the gas phase, or in pre-existing particulate matter). We will say something more on these questions after reviewing the third anomaly of interest.

Magnesium Isotopic Anomalies

Magnesium isotopic anomalies, specifically an anomaly characterized by an excess of ^{26}Mg, have been the subject of many searches. The major impetus for these searches stems from an earlier suggestion (Urey, 1955) that the radioactive isotope ^{26}Al may have been an important heat source in the early solar system. The importance of ^{26}Al as a heat source lies in the fact that it has a relatively short half-life ($\sim 7{\cdot}2 \times 10^5$ years), or a high specific activity which makes it an ideal candidate as a heat source for small bodies (several to hundreds of kilometres in radius) which are virtually uneffected by the longer lived radioactivities (e.g. ^{40}K) that successfully heat the larger (and hence less able to cool) bodies in the solar system. The very short half-life means that the isotope would now be extinct in meteoritic or other types of samples; one must look for an excess in the stable daughter product of ^{26}Al, namely ^{26}Mg.

Discovery of isotopic effects in elements such as magnesium, which are not usually strongly depleted in solid material, requires very precise measurements. The type of sample best suited for the purpose of revealing a ^{26}Mg excess due to ^{26}Al is one which has a high Al/Mg ratio. The first serious claim to have detected a ^{26}Mg anomaly came in 1970 (Clarke *et al.*), however, subsequent higher precision analyses (Schramm *et al.*, 1970) failed to confirm the reported anomaly. The work by Schramm *et al.* also set strong limits on the magnitude of ^{26}Mg variations ($\lesssim 0{\cdot}03$ per cent) in a variety of materials.

The discovery by Gray and Compston (1974), and by Lee and Papanastassiou (1974) of marginal ($\lesssim 0{\cdot}2$–$0{\cdot}4$ per cent) magnesium isotopic shifts in certain Ca–Al rich inclusions from the Allende carbonaceous meteorite, rekindled interest in ^{26}Al. One of the difficulties with the aforementioned discoveries was that the effects were comparable in magnitude to instrumental effects and as such were ambiguous in the sense of revealing which isotope, or isotopes (^{24,25,26}Mg) were causing the measured variations. More recent work (Lee *et al.*, 1976) has found larger effects ($\sim 1{\cdot}3$ per cent) which show conclusively that the positive shifts in $\delta\,^{26}$Mg are due solely to an *excess* of ^{26}Mg. As Lee *et al.* point out, the presence of a ^{26}Mg excess in a Ca–Al rich inclusion is highly suggestive that the effect is due to now-extinct ^{26}Al. At the time of this Institute, additional experiments are under way which will undoubtedly further elucidate this fascinating and important story.

Given the experimental evidence reviewed above, the next question is what are the possible processes which could produce the required effect? There is, in my opinion, little doubt that the positive ^{26}Mg anomaly is due to the decay of now extinct ^{26}Al. The question thus centres on the origin of the ^{26}Al—did it derive from some solar system process, or did it derive from a nucleosynthetic process outside of the solar system? Also, if the latter origin is correct, how and when did the ^{26}Al enter the solar system?

The only nuclear process within the confines of the solar system which could have given rise to ^{26}Al is some type of particle irradiation of circumsolar material. The Sun is incapable of significant nuclear processing of almost all elements other than hydrogen. The particle fluence needed to produce the amount of ^{26}Al implied by the data of Lee *et al.*, is $\sim 10^{20}$ cm^{-2}, far more than required to yield Neon E. If the abundance of ^{26}Al found by Lee *et al.* (^{26}Al/^{27}Al $\sim 6 \times 10^{-5}$) represents the average ^{26}Al/^{27}Al ratio for solar system material, we can draw some interesting conclusions. First, the irradiation necessary to produce this level of ^{26}Al should have left other distinct isotopic traces of its existence. Second, this abundance of ^{26}Al would have been sufficient to melt a body $\geqslant 6$ km in radius if it were of chondritic composition (thermal diffusivity $\kappa \simeq 0{\cdot}007$ cm^2/s), or to melt a body $\geqslant 0{\cdot}6$ km in radius if it were of a structure similar to the lunar regolith ($\kappa = 0{\cdot}00013$ cm^2/s) (Lee *et al.*, 1976).

An alternative to a solar system irradiation mechanism for producing ^{26}Al is the pre-solar hypothesis, which takes the view that the ^{26}Al was produced outside of the solar system and subsequently mixed into the primitive solar system. There are a variety of astrophysical sites which could produce sufficient ^{26}Al. For example, most models of supernova nucleosynthesis indicate that the production ratio of ^{26}Al to ^{27}Al would be $\sim 10^{-3}$ (cf. Schramm, 1971). Allowing for free-decay of the ^{26}Al according to the law $N(t) = N(0)e^{-\lambda t}$, with

$\lambda = 0{\cdot}693/\tau_{1/2}$, and using an initial $^{26}Al/^{27}Al$ ratio of 10^{-3} in conjunction with the value of 6×10^{-5} for $^{26}Al/^{27}Al$ at time t, one finds that $t \lesssim 2{\cdot}9 \times 10^6$ years. Further, if the ratio of 6×10^{-5} was characteristic of the *entire* solar system at some early epoch, the ^{26}Al producing event also provided *at least* 5–6 per cent of the total solar system inventory of ^{27}Al. In other words, this nucleosynthetic event was a significant contributor to the total abundance of a number of elements, and more importantly, specific isotopes.

A variant on the pre-solar hypothesis is the notion that the ^{26}Al was *not* introduced into the primitive solar system, but rather it was trapped in interstellar grains where it decayed to form an excess of ^{26}Mg, and the grains with excessive ^{26}Mg were mixed with 'normal' solar magnesium. In the context of this model, there is no limit on the time of the nucleosynthetic event prior to the formation of the solar system.

Not surprisingly, both versions of the presolar hypothesis have good points. The sample studied by Lee *et al.*, is a white spherical inclusion about 2 mm in diameter. It is argued that it is unlikely that such an inclusion could have formed outside of the solar system, and hence the ^{26}Al must have been present when the solar system formed. On the other hand, cosmochronologies based on ^{129}I and ^{244}Pu (cf. Schramm and Wasserburg, 1970) indicate that the last nucleosynthetic contribution to the solar system inventory of these radioactive (or fissionable) nuclei occurred $\sim 10^8$ years before the solar system formed. The event which produced ^{26}Al (if stellar in nature) should also have produced significant amounts of ^{129}I and ^{244}Pu—but, if we take the view that ^{26}Al was present at the time the solar system formed, this event occurred less than 3×10^6 years earlier, not 10^8 years as indicated by the ^{129}I and ^{244}Pu chronologies. These then are some of the puzzles. Many solutions are, and will be advanced (cf. Cameron, 1976). Much more data are needed, particularly with regard to the siting of the ^{26}Al–^{26}Mg anomaly and its frequency of occurrence in solar system material (it has only been measured in samples from one meteorite).

Concluding Remarks

The discovery of Neon E, followed by the subsequent discoveries of oxygen and magnesium isotopic anomalies has ushered in a new era in solar system research. Just what understanding that new era holds for us remains to be seen. At present, there are two models which are advanced to explain one or all of the anomalies. The first model would preserve the basic concept of the Holy Grail, and produce the isotopic anomalies in the primitive solar system by means of a particle irradiation of solar system material. The strengths and weaknesses of this type of model for each of the anomalies was discussed earlier in the lecture. The second model accounts for the isotopic anomalies by having them be the products of nucleosynthetic formation outside of the solar system. In keeping with the spirit of this new era in solar system research, these anomalies provide further constraints on models of the formation and evolution of the solar system.

References

Audouze, J., Bibring, J. P., Dran, J. C., Maurette, M., and Walker, R. M. (1976). *Ap. J. (Letters)*, **206**, L185.
Black, D. C., and Pepin, R. O. (1969). *Earth Planet. Sci. Lett.*, **6**, 345.
Black, D. C. (1972a). *Geochim. Cosmochim. Acta*, **36**, 347.
Black, D. C. (1972b). *Geochim. Cosmochim. Acta*, **36**, 377.
Bühler, F., Eberhardt, P., Geiss, J., Meister, J., and Signer, P. (1969). *Science*, **166**, 1502.
Cameron, A. G. W. (1976). This volume.
Clarke, W. B., DeLaeter, J. R., Schwarcz, H. P. and Shane, K. C. (1970). *J. Geophys. Res.*, **75**, 448.
Clayton, R. N., Grossman, L., and Mayeda, T. K. (1973). *Science*, **182**, 485.
Clayton, D. D., Dwek, E., and Woosley, S. E. (1976). Preprint.
Eberhardt, P. (1975). *Meteoritics*, **10**, 401.
Gray, C. M., and Compston, W. (1974). *Nature*, **251**, 495.
Heymann, D., and Dziczkaniec, M. (1976). *Science*, **191**, 79.
Kirsten, T. (1976). This volume.
Lee, T., and Papanastassiou, D. A. (1974). *Geophys. Res. Letters*, **1**, 227.
Lee, T., Papanastassiou, D. A., and Wasserburg, G. J. (1976). *Geophys. Res. Letters*, **3**, 109.
Manuel, O. K. (1967). *Geochim. Cosmochim. Acta*, **31**, 2413.
Pepin, R. O. (1967). *Earth Planet Sci. Lett.*, **2**, 13.
Schramm, D. N., Tera, F., and Wasserburg, G. J. (1970). *Earth Plan. Sci. Lett.*, **10**, 44.
Schramm, D. N. (1971). *Astrophys. and Space Sci.*, **13**, 249.
Signer, P. (1964). In *Origin and Evolution of Atmospheres and Oceans*, (eds. P. S. Brancazio and A. G. W. Cameron), John Wiley.
ter Haar, D., and Cameron, A. G. W. (1963). In *Origin of the Solar System*, eds. R. Jastrow and A. G. W. Cameron), Academic Press, New York.
Urey, H. C. (1955). *Proc. Natl. Acad. Sci. U.S.*, **41**, 127.
Van Zeggren, F., and Storey, S. H. (1970). *The Computation of Chemical Equilibria*, Cambridge University Press.
Wänke, H. (1965). *Z. Naturforsch*, **20a**, 946.

CHEMICAL EVIDENCE FOR THE ORIGIN, MELTING, AND DIFFERENTIATION OF THE MOON

G. MALCOLM BROWN
Department of Geological Sciences, University of Durham, U.K.

Abstract

The composition and origin of the aluminous lunar crust and of the mantle-derived lunar basalts are discussed in relation to two major melting events. The first involved large-scale melting and fractionation into crust and mantle at −4·6 to −4·5 Gy and the second, partial melting of the uppermost mantle at −3·8 to −3·2 Gy. Using recent lunar heatflow evidence for a bulk-Moon uranium content of 30 p.p.b., a model is presented which favours early melting of at least 90 per cent by volume, and a development of the high U-contents of the crust and upper mantle by intense crystal–liquid fractionation.

A downward revision of the bulk U-content of the Moon results in down-scaling of the other refractory lithophile elements by analogy with the solar-nebula condensation models. This means that the bulk Moon is fairly close in composition to that of the Earth's mantle, including its iron content but excluding the volatile elements which are strongly depleted in the Moon. Low contents of siderophile and chalcophile elements, and high contents of lithophile refractory elements in the lunar basalts are attributable to the large-scale fractionation into a core, mantle and crust.

The hypothesis of an origin for the Moon by fission from a proto-Earth is revived. Early collision with a large body may have caused fission and formation of a proto-Moon from the Earth's iron-poor proto-mantle, with loss of volatiles. Early melting of most of the proto-Moon led to strong fractionation such that the crust and mantle-derived basalts appear to have more extreme compositions, relative to Earth basalts, than is indicated by the likely bulk composition of the Moon.

Subdivision of the Moon

Layered Structure

Seismic data for the Moon (e.g. Toksöz *et al.*, 1973) have provided evidence for a major discontinuity at about 60 km depth, between a crust ($V_p = 7$ km/sec) and an upper mantle ($V_P \geqslant 8$ km/sec). Subsequent information from deep moonquakes and attenuation of transmitted shear waves has led to the definition of an 'asthenosphere' zone below the thick, rigid lithosphere. The attenuation of the S-waves could be attributed either to a small degree (*ca.* 1 per cent) of partial melting or to a material with slightly higher volatile contents than the lithosphere. The model presented by Dainty *et al.* (1975) gives an 'asthenosphere' extending for about 1000 km radius from the Moon centre, a 60 km crust, and thus a rigid intervening lithospheric mantle of about 680 km thickness. Nakamura *et al.* (1974) divide the rigid mantle into two zones (upper 250 km, middle 500 km), call the 'asthenosphere' a lower mantle (*ca.* 600 km), and subdivide from the latter a core of 170–360 km radius with low *P* velocity. Thus within the Moon radius (1738 km) there is clear evidence for a layered structure, broadly analogous to that of Earth except for a smaller-volume core, and a much thicker rigid lithosphere which has supported large positive gravity anomalies (mascons) since at least the period of mare basin filling ($-$ 3·8 Gy, where 1 Gy = 10^9 years).

Taylor and Jakeš (1974) proposed a model in which the asthenosphere or lower mantle is undifferentiated, primitive lunar material. Such material would be an attractive possibility because it would have retained its original content of radioactive heat-producing elements and (if they were not lost prior to accretion) of volatile elements. Hence either effect could account for the present-day attenuation of shear waves in this inner zone, since the first could result in partial melting through 4·6 Gy of radioactive-decay heating. However, this model would require that a major differentiation involving the crust and upper-middle mantle did not extend deeper than about 700–1000 km. In other words, that the Moon was never totally melted. This could cause difficulties in explaining the generation of a core by sinking of an Fe–Ni–S liquid to the central region (discussed later).

Crustal Highlands Rocks

The pale-coloured, densely cratered highlands consist chiefly of breccias covered by a veneer of soil. The breccias consist of rock and glass fragments, produced by impact explosions and usually welded. The fragments consist chiefly of aluminous material rich in the calcic plagioclase–feldspar component ($CaAl_2Si_2O_8$). The average composition of such material is of anorthositic gabbro (feldspar $\geqslant$ pyroxene). It is customary, therefore, to refer to the primitive crust as being a 60 km-thick layer of anorthositic gabbro, but this is a risky oversimplification. Other fragments in the breccias include coarse-grained, plutonic rocks consisting of feldspar, orthopyroxene (chiefly $Mg_2Si_2O_6$), olivine (chiefly Mg_2SiO_4) and spinel (chiefly $MgAl_2O_4$) as well as rare granitic fragments. Also, there is a type of basalt very rich in K, REE, and P ('KREEP' basalt) that has been melted and mixed with the other crustal materials during certain major impact events or, possibly, only during the Imbrian impact event. This material, rich also in Zr, U, and Th, has a problematical history but is almost certainly produced originally by extensive crystal liquid fractionation that included the evolution of the whole crust and a major part of the mantle.

Crustal Mare Basalts

This, the second major unit of the crust, is less complex than the highlands but is by no means internally uniform. The crystallization ages of the basalt lavas indicate flooding of the ringed impact basins between about − 3·8 and − 3·2 Gy, so a period of volcanism spanning about 600 my is involved and hence would be expected to produce some chemical variation in the products. There is a division into the following groups if other localized, minor variations are discounted:

(1) *High-Ti basalts*
   (a) Apollo 11 (high K).
   (b) Apollo 11 and Apollo 17 (low K).

(2) *Low-Ti basalts*
   (a) Apollo 12, Apollo 15, Luna 16 (olivine basalts and quartz-normative basalts).
   (b) Apollo 12, Luna 16 (aluminous mare basalts).
   (c) KREEP aluminous basalts (pre-mare material, highlands-derived fragments especially from Apollo 14 'Fra Mauro' formation).

There is considerable chemical variation between and within these groups. Average compositions (e.g. Rose *et al.*, 1975) indicate, between groups, Al_2O_3 variation from 8·6 to 21·4 per cent, TiO_2 from 1·2 to 13·0 per cent, and K_2O (as a trace) from 0·06 to 0·48 per cent, for example. The most significant chemical grouping is related to the abundances of the refractory, large-ion lithophile (LIL) elements. Especially, REE contents relative to chondrites (Taylor, 1975) are enriched by factors of about 20 (Low Ti, quartz and olivine normative basalts), 50 (High Ti basalts) and 200 (KREEP basalts).

Chemical Affinities between Lunar Crust and Mantle

The highly aluminous, refractory nature of the lunar crust, its enrichment in heat-producing elements such as U and Th, and its 'primitive' age compared with the mare basalts, led to early views that it was accreted from material chemically distinct from that of the sub-crustal part of the Moon. Element fractionation in the solar nebula would give a condensation sequence, related to temperature and pressure distribution, such that the early condensates would consist chiefly of Ca, Al, Ti and other refractory lithophile elements (e.g. Grossman and Larimer, 1974). Although such condensates have accreted within the solar nebula (as evidenced by the white inclusions in the Allende carbonaceous chondrite meteorite), they would not be expected to accrete during the latest stages of formation of a planetary body. Also, whereas the lunar crust is rich in Ca and Al, the enrichment in Ti resides in the mantle-derived mare basalts, and the high REE, U, Th and other LIL elements in the KREEP basalts (believed to originate at the crust/mantle interface). For these reasons alone, the crust cannot be viewed as having formed by heterogeneous accretion of the proto-Moon. The hypothesis advanced early by Wood *et al.* (1970) and Smith *et al.* (1970), that the aluminous crust is chiefly a result of low-density plagioclase-crystal fractionation from a molten or partly molten, homogeneously accreted Moon, is the one that has since received extensive petrological and geochemical support. Since the mare basalts are believed to have originated by later partial melting of a sub-crustal zone, their composition must be associated with the chemical differentiation into primitive crust and upper mantle. Chemical differences and affinities between crustal rocks and mare basalts are hence a reflection of much of that primitive differentiation event.

Taylor (1975) has made an extensive summary of the evidence for chemical affinities between the early highlands crust and the mantle-derived mare basalts. The most significant factors are:

(1) Plots of volatile *vs.* involatile elements show similar ratios in both highlands and mare rocks (e.g. K/Zr, K/U, K/La, Rb/Ba, Tl/U) except for the rare cases of volatile-element contamination from meteoritic components in highlands rocks.

(2) The 'europium anomaly' shows a complementary relationship between highlands and mare rocks. The entry of Eu^{2+} preferentially into plagioclase gives mostly a positive anomaly for the anorthosites, anorthositic gabbros, troctolites and norites. The fact that the mare basalts show a strong negative anomaly means that they are derived from a source from which the plagioclase-rich rocks were earlier subtracted. The KREEP basalts show the strongest negative Eu anomaly and are thus related to a subcrustal source, probably at the crust/mantle interface.

The mare basalts thus provide evidence for a chemical relationship with the highlands crust. They also possess chemical characteristics that relate to their upper mantle source. Certain chemical properties common to all the mare basalts are to be noted insofar as they may relate to overall lunar chemistry. Compared with terrestrial primary basalts and solar-system abundances, the lunar basalts are enriched in the refractory lithophile elements, and depleted in the chalcophile, siderophile, and volatile elements (e.g. Taylor, 1975; Ganapathy and Anders, 1974). Enrichment in the LIL elements is characteristic also of the highlands crust, and hence is likely to be characteristic of the bulk Moon. Volatile/involatile element ratios, similar in both highlands and mare rocks, indicate also a bulk-Moon depletion in the volatile elements (rather than local surface loss from the low gravity field). Depletion in chalcophile and siderophile elements (neglecting local meteoritic contamination of highlands rocks) could be attributed to concentration in an Fe–Ni–S core although terrestrial basalts do not show this depletion (see later).

The mare basalts are derived from shallow depths in the mantle, probably not deeper than about 250 km (e.g. experimental studies by Walker *et al.*, 1976), and from mafic material that, according to the Eu-anomaly and other chemical data, has participated in an early chemical fractionation process involving the feldspathic crust. Hence the chemistry of the mare basalts alone cannot reflect derivation from a 'bulk-Moon' source. However, grouping the crust and mare basalts together, and assuming an underlying refractory residuum of upper mantle material, rich in olivine and clinopyroxene (from the experimental data), such an outer 250-km zone of melted Moon ought to reflect the significant elemental enrichments and depletions of the bulk Moon.

Dated Events in Lunar Evolution

The isotopic age data indicate that there were two major igneous events in the Moon's history. The first took place shortly after accretion (– 4·6 to – 4·5 Gy) and the second began about 800 my later and lasted for about 600 my (– 3·8 to – 3·2 Gy). A summary based on Rb–Sr internal isochrons is given by Papanastassiou and Wasserburg (1976, Figure 2) and a range of age data is tabulated by Taylor (1975, p. 181).

The crystallization ages (T_X) of the mare basalts (internal isochrons) fall into two groups. The high-Ti basalts (Apollo 11 and 17) crystallized at about – 3·8 to – 3·6 Gy, and the low-Ti basalts (Apollo 12 and 15, and Luna 16) at about – 3·5 to – 3·2 Gy. Both groups represent the second major igneous episode.

Model ages (T_{BABI}) based on whole-rock isochrons (e.g. soil analyses) show, in contrast, a majority with ages of about – 4·6 to – 4·5 Gy. The only exceptions are the Apollo 11, high-K basalts with model ages of – 3·9 to – 3·8 Gy. The high model ages represent the first major igneous episode. The data suggest that at about – 4·6 Gy, when the Earth and Moon accreted (and other solar system planetary bodies such as the meteorite sources), the Moon then melted and crystallized, at least throughout the outer 200 km or so, over a very short period of time (10^6 years has been suggested). The major Rb–Sr fractionation occurred then, associated with gravity differentiation into a feldspathic crust and a mafic (Mg–Fe pyroxene and olivine rich) upper mantle.

The heat-producing elements were concentrated in the crust and KREEP layer, but the underlying mafic-layer source for the mare basalts contained enough for radioactive decay, over 800 my, to generate partial melting and the first eruption of the high-Ti mare basalts at – 3·8 Gy. Deeper mafic layers were more depleted in heat-producing elements and generation, there, of low-Ti mare basalts did not begin until – 3·5 Gy. Even deeper layers were probably so depleted that no partial melting could occur, and mare basalt generation apparently ceased at about –3·2 Gy. This model differs in one vital respect from models of basalt generation from the Earth's mantle. The partial melting of the lunar mantle must have been non-equilibrium melting, such that the liquid did not equilibrate with a Sr-bearing residual phase. Hence the model ages (*ca.* 4·6 Gy) continued to reflect the early Rb–Sr fractionation and no subsequent Rb–Sr fractionation was superimposed except for the Apollo 11 high-K basalts which were presumably generated in equilibrium with plagioclase near the base of the crust. Albee and Gancarz (1974), who explain the Rb–Sr systematics in the context of the model ages, suggest that lack of water on the Moon may have inhibited equilibrium (i.e. diffusion between partial melt and residual crystals).

Between the first and second of the igneous events, lies 800 my of indistinct and highly complex lunar history. Intense meteoritic bombardment of the feldspathic crust occurred, and culminated in the 'cataclysmic' production of the huge ringed basins such as Imbrium (– 3·9 Gy). The KREEP basalts, that formed a sub-crustal layer during the first igneous event (from model ages), were excavated and recrystallized by the Imbrian impact and possibly in part by earlier impacts around – 4·0 Gy (e.g. Nectaris). Working back in time from then, anorthositic fragments are revealing Ar^{39}/Ar^{40} ages of about – 4·2 Gy (Eberhardt *et al.*, 1976). The most dramatic age results of all, however, relate to crystallization (T_X) ages of – 4·55 to – 4·6 Gy for a troctolite (olivine + feldspar) and a dunite (olivine) fragment, with the most primitive (BABI) of solar-system initial Sr^{87}/Sr^{86} ratios. Papanastassiou and Wasserburg (1976) give these data and the internal isochron for the troctolite.

Source of Mare Basalts

If the mare basalts were derived simply by partial melting of primitive lunar mantle material, it would be necessary to place their source deeper than the thick, outer shell that gave rise to the low density (*ca.* 2·9 g cm^{-3}) plagioclase-rich crust and, inevitably, to an underlying complementary layer of dense (*ca.* 3·4 g cm^{-3}) mafic cumulates (crystal accumulations rich in magnesium–iron silicates). It would also be necessary to provide an alternative explanation for the basaltic negative Eu anomaly that is complementary to the positive Eu anomaly of the highlands rocks. In the absence of any compelling reason for seeking a source in an undifferentiated layer of the mantle, the most likely source would be a mafic cumulate layer beneath the feldspathic crust.

The Taylor–Jakeš model (summarized by Taylor, 1975) has many attractive features. It requires melting of the outer 1000 km of the Moon during the – 4·6 Gy event, the inner 'asthenosphere' being possibly undifferentiated primary material. It then provides for the deposition of a series of mafic cumulus minerals, later joined by plagioclase which floated to form the crust. The upper mafic cumulate layers would thus be depleted in Eu^{2+} and be the source of the mare basalts. Such a mafic cumulate sequence would, by analogy with terrestrial layered intrusions (Wager and Brown, 1968), show upward enrichment in the LIL refractory elements. Hence the mare basalts richest in those elements (the high-Ti basalts) would be derived by partial melting of mafic cumulate layers overlying the source of the low-Ti basalts. As noted earlier, this also correlates with the eruption ages which are presumably related to the relative concentration of heat-producing elements in the mafic cumulates, being highest in the source of the earliest-erupted mare basalts.

The Taylor–Jakeš model was not related to a layered-intrusion model in detail, because of neglect of the role of intercumulus material and other more subtle aspects of igneous cumulate formation. The writer has discussed some of those features (Brown, 1976) although a more rigorous approach is still needed. Igneous cumulates consist of crystals concentrated under the influence of gravity (cumulus phases), and the products of the intercumulus liquid which accounts for about 40–50 per cent by volume under simple packing conditions. The intercumulus liquid may be trapped (to crystallize later as lower-temperature phases), usually because of rapid crystal accumulation, the final rock being termed an 'orthocumulate'. Slower deposition rates can, however, allow element diffusion between intercumulus and overlying (main bulk) liquid such that the pore spaces are filled by growth of the high-temperature cumulus crystals. The resultant rock, from which the contemporary liquid products have therefore been expelled, is termed an 'adcumulate' (Wager *et al.*, 1960). Adcumulate layers within, say, the lunar mantle would thus be refractory and would be responsible for concentrating the LIL trace elements in an overlying zone more effectively than would layers of orthocumulates (analagous to the 'zone refining' process of Hubbard and Minear, 1976). Conversely, orthocumulates would, by partial melting of their less refractory pore material, yield liquids enriched in those trace elements. To account for the high Ti contents of certain mare basalts, for example, melting of pore material enriched in Ti would therefore be more feasible than melting of an assemblage of Ti-rich cumulus minerals. The latter would require near-total melting of the layer, and such a layer would occur at a higher level in the fractionated layered sequence than the Ti-rich pore material. The next point is that to obtain, say, Ti-rich pore material would require, from a melted Moon material of moderate Ti content, extensive earlier removal of Ti-poor phases. This could be best effected if those phases were cemented not by trapped Ti-bearing pore material (contemporary liquid) but by adcumulus-growth material. Clearly, therefore, any consideration of large-scale melting of the Moon, followed by crystal–liquid fractionation, leads to the recognition of variables that cannot easily be quantified, but which could be used to explain certain anomalies.

Uranium Distribution and Melting History

The distribution of an 'incompatible' element that is rejected by the major crystallizing phases can be used to illustrate likely constraints, as noted above.

Uranium has been chosen because there are independent means for calculating its abundance in the bulk Moon, and plenty of data on its abundance in the observed rocks and its distribution across the lunar surface from γ-ray spectrometer data (Trombka *et al.*, 1974). The calculations are summarized in Table 1. Initially, a mean U-value for the Moon was taken to be 50 p.p.b. (Brown, 1976) but a drastic revision downward of the heatflow values (Langseth *et al.*, 1976) now indicates that a value closer to about 30 p.p.b. is more likely.

In order to obtain the postulated, average crustal abundances of about 450 p.p.b. uranium, a very large volume of the Moon must be 'reworked'. The 60 km crust accounts for about 10 per cent of the Moon's volume, and if homogeneous the crust alone would therefore contribute 45 p.p.b. uranium to a Moon assumed to have only 30 p.p.b. uranium! In order to make any type of meaningful calculation, it is therefore proposed that because the KREEP-rich materials are erratically distributed on the lunar surface and would contribute to high average, near-surface, U-contents out of proportion to their volume in the whole crust, an average crustal U-value of about 250 p.p.b. is preferable by subtracting about 6 per cent of KREEP-rich material (with 3500 p.p.b.).

Table 1. Uranium distribution estimated for layers of the Moon (Less significant values rounded-off for a generalized model)

Unit (Subtraction stages in parenthesis)	Shell thickness (radial km)	Volume (% of Moon)	U content (p.p.b.)	U contribution (p.p.b.) to total volume
1. Total Moon (see text)	1738	100	30	30
2. Undifferentiated part	738	8	30	2
3. Differentiated part (1–2)	1000	92	30	28
4. L. mantle adcumulates (3–5)	800	62	0	0
5. Crust + U: mantle orthocumulates	200	30[b]	94	28
6. Average feldspathic crust	60	10	250[a]	25
7. U: mantle orthocumulates including low-volume KREEP layer (5–6)	140	20	13[c]	3

Notes:

[a]From γ-ray orbital values (Trombka *et al.*, 1974) of 450 p.p.b. for average crust, less about 6 per cent surface-distributed KREEP material (3500 p.p.b.) that distorts upward the γ-ray spectrometer estimate of average crustal value. Values for various crustal rock-types summarized by Brown (1976).

[b]Volume (and hence shell thickness) calculated to best-fit the need to generate the crustal U-content (6). A lower volume would contribute too little, and a higher volume too much U for the required 27 p.p.b. of the differentiated part (3). A lower volume for the undifferentiated part would have only a slight effect on these U figures and an appreciably higher volume (i.e. less initial melting) could not be tolerated if the values in (6) are to be explained.

[c]A maximum value. Any increase would reduce the crustal contribution which is based on a probable minimum[a] of about 250 p.p.b. As this uranium (13 p.p.b.) would reside in the orthocumulate pore material, 5 per cent partial melting could generate basalts with 260 p.p.b. uranium, as observed in the average low-K mare basalts (240 p.p.b.).

As shown in Table 1 above, even with a 1000 km-zone of melting and the assumed U-value for the average crust, much of the early melt must have crystallized as deep-level adcumulates (lower mantle) so that elements such as uranium could be concentrated in an outer zone of mafic orthocumulates (upper mantle, basalt source) and feldspathic cumulates (crust). Total melting would not affect the balance significantly, since the postulated asthenosphere is only about 8 per cent of the lunar volume. As a result of this calculation, the upper mantle plus crust is only 30 per cent of the lunar volume. Hence mare basalt generation would be confined, according to this model, to a zone extending no deeper than about 200 km. Deeper zones would be ultramafic adcumulates and thus devoid of heat-producing elements and also of basaltic components.

At such a late stage of major fractionation, whereby the highlands crust and the mare basalt source material have evolved in the proportion, to early adcumulates, of 1 : 2, major elements such as Fe and Ti, as well as the incompatible trace elements, would have risen appreciably over lunar bulk abundances. Further fractionation of plagioclase and mafic minerals to form the crust and upper mantle cumulates would extend this fractionation to give material such as the KREEP-basalt layer, highly enriched in REE and other LIL elements and probably located as the final unit of a 'sandwich' between feldspathic crust and underlying mafic cumulates (cf. Hubbard and Minear, 1976). Partial melting of the mafic cumulates would concentrate elements such as Fe, Ti, REE, etc. in the mare basalt liquids, beyond the levels reached by the early, major fractionation process. Before the 200 km-thick 'sandwich' had become stabilized, opportunities would arise for magma to invade the proto-crust (to form fractionated plutons), at a stage when the crustal skin was only a few kilometres thick and hence the 'upper mantle' magma was little advanced and could precipitate fairly magnesian olivines and pyroxenes (as in the 4·6 Gy-old troctolites and dunites).

Ringwood (1975) has pointed out some chemical problems in the Taylor–Jakeš model and a more complex and problematical, hybrid history has since been proposed (Kesson and Ringwood, 1976).

Bulk Composition and Origin of the Moon

The aluminous nature of the lunar crust led to an early conclusion that the Moon has a refractory composition compared with the Earth and solar-system elemental abundances. This view was substantiated by comparisons with carbonaceous chondrites in terms of the lunar basalt contents of U, Th, REE and other refractory lithophile elements. Since the Moon is clearly a strongly fractionated body, however, the crustal and upper-mantle derived basalt compositions cannot give more than a qualitative estimate of bulk composition.

Consideration of condensation processes in a cooling solar nebula (e.g. Grossman, 1972) led to the assumption that the inner planets formed by the processes responsible for chondrite formation. Ganapathy and Anders (1974) present a model based on these processes for the formation of the Moon and the Earth. The equilibrium condensation sequence of a solar gas (Larimer, 1967) would give dust condensates in the sequence (a) early refractory condensate, (b) metallic nickel–iron, and (c) magnesium silicates, there being subsequent reactions with volatiles to produce FeS, for example, and FeO for the silicate phases.

The refractory component of a planet (i.e. proportion of early condensate) is estimated from the bulk uranium content, taken to be 18 p.p.b. for the Earth (Larimer, 1971) and 59 p.p.b. for the Moon, from heatflow data (Ganapathy and Anders, 1974). Other established ratios (e.g. Th/U, K/U, Ca/Al) are then used to calculate abundances for a planet relative to cosmic (solar system) abundances. Unfortunately this model has always produced anomalies in regard to the postulated Ca and Al contents versus the petrological evidence for the composition of the Moon's mantle. Ringwood (e.g. 1975) has been concerned for some time that the estimated Ca and Al contents for the Moon's mantle are far too high in relation to the basalt petrology, as evidenced by the pyroxene compositions.

The recent discovery (Langseth *et al.*, 1976) that the Moon's heatflow is much less than estimated from earlier measurements, and based on more reliable determinations over $3\frac{1}{2}$ years (Apollo 15 station) and 2 years (Apollo 17 station), requires a reappraisal of the compositional estimate. The revised heatflow values indicate a bulk U-content closer to about 30 p.p.b. than the 59 p.p.b. used in the Ganapathy and Anders model (1974). In that case, all values for the refractory lithophile elements need to be scaled downward by a factor of 2. They become $1\frac{1}{2}$ times rather than 3 times cosmic abundances. This brings them much closer to abundances for the Earth (*ca.* 0·7 times cosmic).

This less refractory composition for the Moon is particularly significant in terms of its thermal history. As shown earlier (Table 1), large-scale melting is now even more necessary in order to concentrate the U into the high crustal values, from the lower initial bulk content now indicated. It also means that the models for bulk-Moon composition that show high Al and Ca contents, and which were unacceptable in details from the experimental petrology data (e.g. Kushiro and Hodges, 1974; Delano, 1975), are no longer necessary. Instead, the model composition proposed by Binder (1974) is more attractive. This composition approaches closely the composition of the Earth's mantle, either in terms of a pyrolite composition (Ringwood, 1966) or by subtraction of the Fe + FeS core from the solar-nebula condensation model for the Earth (Ganapathy and Anders, 1974).

The bulk density of the Moon (3·34 g cm^{-3}) is close to that of uncompressed, low-temperature Earth mantle material. In terms of the refractory lithophile element abundances, it now seems likely that the bulk Moon composition is not far removed from that of the Earth's mantle. The bulk Earth contains about 35 per cent Fe (Mason, 1966) whereas the Moon contains about 9 per cent Fe

(± 4 per cent) based on magnetometer data (Parkin *et al.*, 1973). However, the Earth's mantle probably contains about 7 per cent Fe (Mason, 1966), similar again to that of the bulk Moon.

The bulk Moon is depleted in chalcophile, siderophile and volatile elements relative to cosmic abundances (Ganapathy and Anders, 1974, Figure 3). Whereas the latter depletion must be a bulk characteristic, the other two depleted groups could be due to concentration of those elements in a metallic core. Their abundances are related to Fe abundances and Fe–S reactions, and to depletion in the lunar basalts. If, however, large-scale melting of the Moon has occurred, then the basalt source would be depleted in siderophile and chalcophile elements relative to the core. (If the central 'asthenosphere' of the Moon is hot undifferentiated material, as discussed earlier, the metal-rich phase could have concentrated at the interface between this and the molten shell and later diffused to a core region.)

It is noteworthy that the Earth's mantle is not strongly depleted in siderophile elements such as nickel, suggesting that the mantle was not at any time in chemical equilibrium with the iron-rich core. In contrast, the Moon's mantle is so strongly depleted in nickel that the element could not be accommodated in the core (restricted in volume by bulk density and moment of inertia) if the Moon had accreted a cosmic abundance of nickel. Nickel could be so accommodated, however, if the Moon had begun with a partly-depleted, Earth-mantle nickel abundance, and large-scale melting had 'scavenged' the siderophile elements to the core, more effectively than on Earth.

This raises the question of a possible difference in the evolution of the Earth and Moon. If chemical equilibrium were not attained by the Earth's core and mantle, heterogeneous accretion could have been responsible for a layering in the sequence metal → silicates → volatile-rich veneer, and large-scale initial melting, although likely, would not be necessary. The volatile-rich outer layer would contribute to the generation of a granitic crust. Shortly after accretion, collision with another body (Cameron, 1976, suggested a 'Mars-sized object') could have caused fission and the production of a proto-Moon from the proto-mantle of the Earth. The proto-Moon would need to have been heated extensively at this stage, such that the volatile-rich material was lost and the bulk of the body melted. Loss of volatiles would impart to the proto-Moon its slightly more refractory composition relative to that of the Earth's proto-mantle. Large-scale fractionation then concentrated the siderophile and chalcophile elements towards the core region, and the lithophile refractory elements towards the crust region.

The hypothesis of a fission-origin for the Moon has not been popular for dynamic reasons, but on chemical grounds there is no doubt that the Moon shows much in common with the Earth's mantle. The oxygen isotope data (Clayton *et al.*, 1976) are particularly compelling in that respect. Dynamical problems need to be resolved for the Earth–Moon system, as do the problems concerning the heat-source for large-scale melting of the proto-Moon. Insofar as we do not yet have a satisfactory hypothesis for the origin of the Moon, however, there is a case for reconsidering the fission hypothesis.

References

Albee, A. L., and Gancarz, A. J. (1974). 'Petrogenesis of lunar rocks: Rb-Sr constraints and lack of H_2O.' *Proc. Soviet–Amer. Conf. Cosmochem. Moon and Planets (Moscow)*, Lunar Sci. Inst. Preprint, **201**, 1–19.

Binder, A. (1975). 'On the petrology and structure of gravitationally differentiated Moon of fission origin.' *Lunar Science VI*, Lunar Sci. Inst., Part 1, 54–56.

Brown, G. M. (1977). 'Two-stage generation of lunar mare basalts.' *Phil. Trans. Roy. Soc., Lond.*, in the press.

Cameron, A. G. W. (1976). 'The primitive solar accretion disk and the formation of the planets.' *The Origin of the Solar System* (NATO Adv. Study Inst., Newcastle), elsewhere in this volume.

Clayton, R. N., Onuma, N., and Mayeda, T. K. (1976). 'A classification of meteorites based on oxygen isotopes.' *Earth Planet. Sci. Lett.*, **30**, 10–18.

Dainty, A. M., Goins, N. R., and Toksöz, M. N. (1975). 'The structure of the Moon as determined from natural lunar seismic events.' *Lunar Science VI*, Lunar Sci. Inst., Part 1, 175–177.

Delano, J. W. (1976). 'Experimental petrology of a refractory whole Moon composition.' *Lunar Science VII*, Lunar Sci. Inst., Part 1, 190–192.

Eberhardt, P., Geiss, J., Grögler, N., Maurer, P., Stettler, A., Peckett, A., Brown, G. M., and Krähenbühl, U. (1976). 'Young and old ages in the Descartes Region.' *Lunar Science VII*, Lunar Sci. Inst., Part 1, 233–235.

Ganapathy, R., and Anders, E. (1974). 'Bulk compositions of the moon and earth, estimated from meteorites.' *Proc. 5th Lunar Sci. Conf.*, **2**, 1181–1206.

Grossman, L. (1972). 'Condensation in the primitive solar nebula.' *Geochim. Cosmochim. Acta*, **36**, 597–619.

Grossman, L., and Larimer, J. W. (1974). 'Early chemical history of the solar system.' *Rev. Geophys. Space Phys.*, **12**, 71–101.

Hubbard, N. J., and Minear, J. W. (1976). 'Hybridization: an answer to the heterogeneous source materials.' *Lunar Science VII*, Lunar Sci. Inst., Part 1, 393–395.

Kesson, S. E. and Ringwood, A. E. (1976). 'Mare basalt petrogenesis in a dynamic Moon.' *Lunar Science VII*, Lunar Sci. Inst., Part 1, 448–450.

Kushiro, I. and Hodges, F. N. (1974). 'Differentiation of the Model Moon.' *Carnegie Inst. Yb.*, **73**, 454–457.

Langseth, M. G., Keihm, S. J., and Peters, K. 'The revised lunar heat flow values.' *Lunar Science VII*, Lunar Sci. Inst., Part 1, 474–475.

Larimer, J. W. (1967). 'Chemical fractionations in meteorites—I. Condensation of the elements.' *Geochim. Cosmochim. Acta*, **3**, 1215–1238.

Larimer, J. W. (1971). 'Composition of the Earth: Chondritic or achondritic?' *Geochim. Cosmochim. Acta*, **35**, 769–786.

Mason, B. (1966). *Principles of Geochemistry* (3rd edn.). Wiley and Sons, New York, 329 pp.

Nakamura, Y., Latham, G., Lammlein, D., Ewing, M., Duennebier, F., and Dorman, J. (1974). 'Deep lunar interior inferred from recent seismic data.' *Geophys. Sci. Lett.*, **1**, 137–140.

Papanastassiou, D. A., and Wasserburg, G. J. (1976). 'Early lunar differentiates and lunar initial $^{87}Sr/^{86}Sr$.' *Lunar Science VII*, Lunar Sci. Inst., Part 2, 665–667.

Parkin, C. W., Dyal, P., and Daily, W. D. (1973). 'Iron abundance in the moon from magnetometer measurements.' *Proc. 4th. Lunar Sci. Conf.*, **3**, 2947–2961.

Ringwood, A. E. (1966). 'Chemical evolution of the terrestrial planets.' *Geochim. Cosmochim. Acta*, **30**, 41–104.

Ringwood, A. E. (1975). 'Limits on the bulk composition of the Moon.' *Publ. Res. School Earth Sciences*, A.N.U., No. 1160, 1–36.

Rose, H. J. Jr., Baedecker, P. A., Berman, S., Christian, R. P., Dwornik, E. J., Finkelman, R. B., and Schnepfe, M. M. (1975). 'Chemical composition of rocks and soils returned by the Apollo 15, 16 and 17 missions.' *Proc. 6th Lunar Sci. Conf.*, **2**, 1363–1373.

Smith, J. V., Anderson, A. T., Newton, R. C., Olsen, E. J., and Wyllie, P. J. (1970). 'A petrologic model for the Moon based on petrogenesis, experimental petrology, and physical properties.' *J. Geol.*, **78**, 381–405.

Taylor, S. R. (1975). *Lunar Science: A Post-Apollo View*. Pergamon, New York, 372 pp.

Taylor, S. R., and Jakeš, P. (1974). 'The geochemical evolution of the Moon.' *Proc. 5th Lunar Sci. Conf.*, **2**, 1287–1305.

Toksöz, M. N., Dainty, A. M., Solomon, S. C., and Anderson, K. R. (1973). 'Velocity structure and evolution of the moon. *Proc. 4th Lunar Sci. Conf.*, **3**, 2529–2547.

Trombka, J. I., Arnold, J. R., Adler, I., Metzger, A. E., and Reedy, R. C. (1974). 'Lunar elemental analysis obtained from the Apollo gamma-ray and X-ray remote sensing experiment.' *Proc. Soviet-Amer. Conf. Cosmochem. Moon and Planets (Moscow)*, Lunar Sci. Preprint **224**, 1–50.

Wager, L. R., Brown, G. M., and Wadsworth, W. J. (1960). 'Types of igneous cumulates.' *J. Petrol.*, **1**, 73–85.

Wager, L. R., and Brown, G. M. (1968). *Layered Igneous Rocks*. Oliver and Boyd, Edinburgh, 584 pp.

Walker, D., Longhi, J., and Hays, J. F. (1976). 'Heterogeneity in titaniferous lunar basalts.' *Earth Planet. Sci. Lett.*, **30**, 27–36.

Wood, J. A., Dickey, J. S., Jr., Marvin, U. B., and Powell, B. N. (1970). 'Lunar anorthosites and a geophysical model of the Moon.' *Proc. Apollo 11 Lunar Sci. Conf.*, **1**, 965–988.

THE CHEMICAL EQUILIBRIUM MODEL FOR CONDENSATION IN THE SOLAR NEBULA: ASSUMPTIONS, IMPLICATIONS, AND LIMITATIONS

K. A. GOETTEL[a] AND S. S. BARSHAY[b]

[a]*Department of Earth and Planetary Sciences, McDonnell Center for the Space Sciences, Box 1169, Washington University, St. Louis, Missouri, U.S.A.*

[b]*Department of Chemistry and Department of Earth and Planetary Sciences, Room 54-1220, Massachusetts Institute of Technology, Cambridge, Massachusetts, U.S.A.*

Abstract

The chemical and physical evolution of the solar nebula is of considerable importance, because it is widely accepted that the planets and other solar system objects were formed directly or indirectly from the solar nebula. A critical review of the assumptions, implications, and limitations inherent in (1) calculation of the composition of condensed material as a function of temperature and pressure in the solar nebula, and(2) application of these results to obtain quantitative models for the compositions of the terrestrial planets, suggests that the chemical equilibrium model is the most viable model.

Introduction

In the last decade, considerable progress has been made in understanding the chemistry of condensation in the primitive solar nebula (Lord, 1965; Larimer, 1967, 1973; Grossman, 1972a, 1972b; Grossman and Larimer, 1974;

Lewis, 1972a, 1972b, 1973, 1974a; Barshay and Lewis, 1975, 1976). Recently, results of these condensation calculations have been applied to obtain quantitative models for the compositions of planets (e.g. Lewis, 1972a, 1972b, 1973, 1974a) and meteorites (e.g. Larimer and Anders, 1967, 1970; Grossman and Clark, 1973; Larimer, 1973; Grossman and Olsen, 1974; Grossman and Larimer, 1974). The purpose of the present review is to examine the basic assumptions, implications and limitations inherent in (1) calculation of the composition of condensed material as a function of temperature and pressure in the solar nebula, and (2) application of these results to the present compositions of planets and other solar system objects. Particular emphasis is placed on the condensation model which assumes chemical equilibrium in the solar nebula, because a variety of evidence suggests that equilibrium was probably approached in the nebula. The chemical equilibrium model thus appears to be the most viable, albeit oversimplified, working model for condensation in the solar nebula.

Composition of Condensed Material as a Function of Temperature and Pressure in the Solar Nebula

Assumptions

Several basic assumptions must be made prior to calculating the composition of condensed or gaseous material in the solar nebula. A zeroth-order assumption is that the nebula (i.e. a neutrally-charged, low-density cloud of gas and dust) existed. The nebular hypothesis for the origin of the solar system is widely accepted (e.g. Safronov, 1972; Cameron, 1975; Ward, 1976) and is supported by observations of such clouds in other regions of the galaxy (Herbig, 1975, 1976) and by dynamic collapse calculations which suggest that evolving protostars go through a nebular phase (Larson, 1974). However, the nebular hypothesis is not universally accepted; major competing hypotheses include the near stellar collision model of Woolfson (1976), and the plasma-physics model of Alfvén and Arrhenius (1975, 1976). Woolfson's model suffers from probability considerations; Alfvén and Arrhenius have not had notable success in predicting the observed chemistry of the solar system. Thus, the basic assumption of a nebular origin for the solar system, while not yet universally accepted, does appear to be sound.

Knowledge of the elemental composition of the solar nebula is a necessary prerequisite to any condensation calculations. It is commonly assumed that the composition of the primitive solar nebula was nearly the same as the present composition of the Sun, because solar nuclear reactions have effected only the H/He ratio and a few low stability nuclides such as the Li isotopes and ^2H. The composition of solar material has been determined by direct observation of the Sun in various spectral regions, by study of solar cosmic rays, and by analysis of meteorites (particularly carbonaceous chondrites). In addition, for some elements for which data are sparse, nucleosynthetic theory may be used to estimate elemental or isotopic abundances. Cameron (1973) has presented a compilation of recent data on the composition of solar material. Trimble (1975) has presented a more general review of the present state of knowledge of the abundances of the elements.

It has also been generally assumed that turbulent mixing in the solar nebula (Cameron and Pine, 1973) would result in virtually complete homogenization of elemental and isotopic abundances. However, the recent discovery of isotopic anomalies in meteoritic neon (Black, 1972), oxygen (Clayton *et al.*, 1973), magnesium (Lee and Papanastassiou, 1974), and xenon (Drozd and Podosek, 1976) apparently indicates that complete elemental and isotopic homogenization of the nebula did not occur. Without in any way discounting the significance of these isotopic anomalies, the dynamic arguments for turbulent mixing in the nebula and the data indicating that Cl carbonaceous chondrites have non-volatile elemental compositions virtually identical to solar composition, suggest that the simplifying assumption of chemical homogeneity in the solar nebula is not likely to be grossly in error, at least as far as major element compositions are concerned.

Once the elemental composition of the solar nebula is known, it is a relatively straightforward thermodynamic exercise to calculate the molecular composition of the gas phase and the composition of the solids condensed from the gas phase, as a function of temperature and pressure in the nebula. Grossman (1972b, pp. 598–603) has outlined the general computational techniques of such calculations. However, it is essential to note that the deduced compositions depend critically on the extent to which chemical equilibrium is assumed to be attained. If equilibrium is not attained, the composition of both the gas and condensed solids will be substantially different than would be the case if equilibrium were attained. A discussion of the evidence for equilibrium having been approached in the solar nebula is presented below in the *Limitations* part of this section.

Implications

With the above assumptions, it becomes possible to calculate directly the composition of solids condensed from the solar nebula as a function of temperature and pressure. The condensation temperatures of important elements and compounds, *assuming chemical equilibrium* in the nebula, are presented in Figure 1 (after Barshay and Lewis, 1976). Figure 1 was calculated assuming that complete equilibrium was attained in the nebula: equilibrium in the gas phase, equilibrium between the gas and condensed solids, and equilibrium between condensed solids. The temperature at which a compound or element 'condenses' is the highest temperature, for a given pressure, at which the condensate is stable with respect to evaporation in a system of solar composition. Some of the 'condensation' curves shown in Figure 1, for example, the FeS curve, are more precisely 'reaction temperature' curves, since they represent the

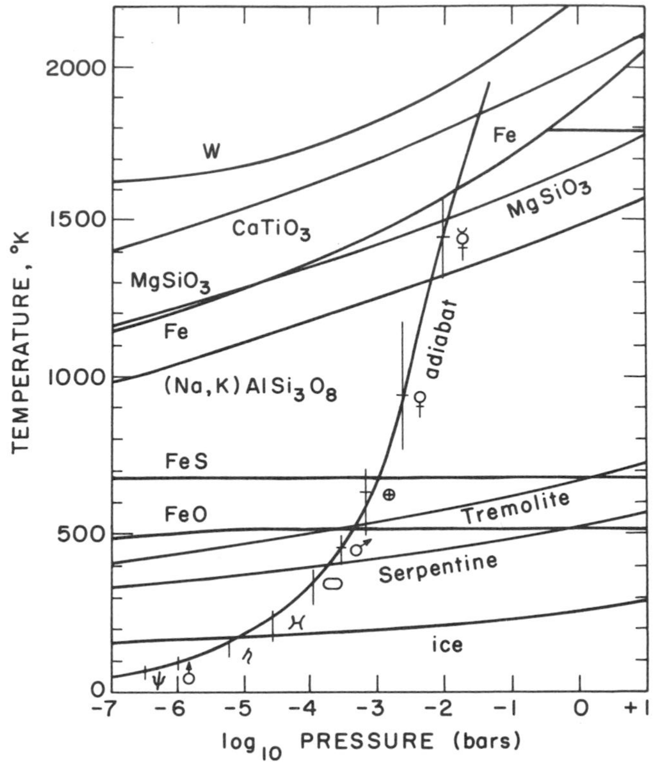

Figure 1. The equilibrium chemistry of solar composition material as a function of temperature and pressure (Reproduced from S. S. Barshay and J. S. Lewis, 1976, *Ann. Rev. Astron. Astrophys.*, **14**, 81–94, by permission of Annual Reviews, Inc.)

reaction of a previously condensed solid with the gas to form a new compound. Nevertheless, for convenience, these temperature curves, as well as those in which condensates are precipitated from the gas, will be referred to as 'condensation temperatures'.

Because the condensation sequence shown in Figure 1 is an *equilibrium* chemical process, the composition of both the gas and condensed phases are fixed by temperature and pressure, and are independent of path. Therefore, the results at any temperature and pressure do *not* depend, for example, on the previous temperature and pressure history of the system. In particular, the results do *not* change if (1) the material constituting the nebula was completely vaporized and subsequently partially condensed upon cooling, or (2) the material was never completely vaporized. The latter case may in fact have occurred in the solar nebula.

In Figure 1, the nebula consists only of a gas phase at temperatures above the curve labeled 'W'. The 'W' curve represents the condensation temperature interval for the most refractory metals and oxides, including W, Os, Sc_2O_3, and

ZrO_2. The '$CaTiO_3$' curve represents the condensation temperature interval for a second group of refractory species, including oxides of Ca, Ti, Al, V, the rare earths, U and Th. Details of the condensation sequence for these refractory materials have been given by Grossman (1972a,b). It should be noted, however, that the total of all of these refractory condensates constitutes only about 5 per cent of the total amount of rock-forming material in a system of solar composition.

Metallic iron (actually an alloy containing Ni and other siderophile elements) and magnesium silicates (enstatite, $MgSiO_3$, and forsterite, Mg_2SiO_4) are the most stable of the abundant condensates. In Figure 1, iron alloy is simply labelled 'Fe', and magnesium silicates are labelled '$MgSiO_3$', since enstatite is the major magnesium silicate phase when the magnesium silicates are fully condensed. Fe, Si, and Mg have nearly equal abundances in the solar nebula, and are ten times more abundant than any other rock-forming element except sulphur. Therefore, iron alloy and magnesium silicates, which condense at nearly the same temperature, constitute the majority of the total rock-forming material in the solar nebula, and thus may be expected to be the major constituents of rocky bodies formed from the solar nebula.

At lower temperatures, several other important species condense and several important gas–solid reactions occur. Na and K (important as a major radiogenic heat source) condense as alkali feldspar by reaction of Na and K with previously condensed Al and Si compounds. FeS (troilite) forms by reaction of H_2S gas with previously condensed iron alloy; the supply of H_2S is exhausted before all the Fe has been oxidized to FeS. The curve labelled 'FeO' represents the end-point of oxidation of the remaining Fe metal to FeO, which is incorporated into silicates. Oxidation of Fe to FeO, which is complete by about 500 K, is a gradual process, which is, for example, about one per cent complete at 750 K. Water is first retained in condensates as water of hydration in minerals such as tremolite and serpentine. At lower temperatures, water ice becomes stable. Finally, at very low temperatures, a variety of other volatiles condense as ices; details of the low-temperature condensation sequence have been presented by Lewis (1972b).

The condensation temperature curves shown in Figure 1 above are simplified in one respect; condensation of an element or compound actually occurs, at a given pressure, over a range of temperature, ΔT, typically 6 to 10 per cent of T. Condensation of most major rock-forming minerals is 80–90 per cent complete within about 50 K. Thus, representing the condensation interval of a given substance by the temperature at which condensation is essentially complete (as in Figure 1) is an incomplete description *only* in the narrow temperature interval when the substance is partially condensed.

Limitations

The validity of the condensation sequence presented in Figure 1 (i.e. how well the calculated condensation sequence corresponds to the actual solar nebula) depends, obviously, on the validity of the assumptions inherent in the calculations. Uncertainties in our knowledge of the composition of solar material, errors in the thermodynamic data, and the lack of certain thermodynamic data all limit the accuracy of the calculated condensation sequence. Most importantly, the degree to which equilibrium is attained in the nebula has a great effect on the composition of condensed material.

The present uncertainty in the relative abundances of major elements in solar composition material is of the order of $\pm$ 20 per cent (Anders, 1971). Uncertainties of this magnitude do not appreciably alter the condensation sequence as presented in Figure 1. For example, changing the sulphur abundance by 20 per cent would change the condensation temperature of FeS by only about 12 K. Therefore, uncertainties in the abundances of major elements are unlikely to have a major impact on the condensation sequence. Likewise, minor elemental and/or isotopic inhomogeneities in the nebula are not likely to affect significantly the condensation temperatures of major species. However, uncertainties in the relative abundances of elements do affect the relative proportions of condensates. For example, changing the Fe/Si ratio would affect the proportions of metal and silicate phases present, and changing the Mg/Si ratio would affect the proportions of $MgSiO_3$ and Mg_2SiO_4.

Results of condensation sequence calculations are also limited by the availability and accuracy of the requisite thermodynamic data. In most cases, data for major condensates are sufficiently accurate so that resulting uncertainties in calculated condensation temperatures are only a few degrees K (Grossman, 1972b, p. 602). However, in a few cases, errors in the thermodynamic data may be more significant. As shown in Figure 1, Fe-alloy and magnesium silicates condense only a few degrees apart. Thus, small errors in the data may significantly affect the pressure and temperature range in which Fe-alloy is more completely condensed than magnesium silicates. Errors thus introduced may be very significant in calculating the composition of condensed material within this narrow temperature interval. However, this uncertainty does *not* affect the calculated composition of condensed material at lower temperatures, once Fe-alloy and magnesium silicates have both fully condensed.

Another area where uncertainties in the thermodynamic data may be significant is in the calculation of the amount of FeO present in magnesium silicates as a function of temperature. Uncertainties in activity coefficients introduce large errors into the calculated FeO content of the silicates at temperatures above about 500 K, when the oxidation of Fe to FeO is completed. For example, at 600 K Grossman (1972b) estimates the range of equilibrium fayalite (Fe_2SiO_4) content in olivine (Fe_2SiO_4–Mg_2SiO_4 solid solution) which is allowed by the uncertainty in the data to be 3·3 to 15 mol per cent.

Accurate calculation of the behaviour of minor and trace elements during the condensation process is limited by the paucity of free energy of formation data, and by the absence of activity coefficient data for most of the complex solid solutions involved. A few minor and trace elements (e.g., W., ZrO_2)

probably condense as separate phases, but most probably condense as solid solutions in major condensates. The lack of the requisite thermodynamic data may be partially surmounted by making reasonable estimates of the relevant thermodynamic parameters (e.g. Boynton, 1975), or by utilizing the considerable amount of analytic data in meteoritic and other geochemically relevant systems. Partition coefficient data and major element–trace element ratios may be used to estimate trace element behaviour during condensation by estimating in which major phases the various trace elements are most likely to be concentrated. The condensation behaviour of minor and trace elements is a prime area for future research.

The assumption which has the greatest effect on the calculated condensation sequence is the assumption that chemical equilibrium is attained. Thus, the validity of the calculated equilibrium condensation sequence is limited largely by the validity of the assumption of chemical equilibrium. Blander and Katz (1967) suggested that kinetic barriers to the homogeneous nucleation of condensed phases would result in supersaturation of the gas phase, and thus actual condensation temperatures which are lower than the calculated equilibrium temperatures. The meteoritic evidence for supersaturation in the solar nebula is, however, somewhat ambiguous. Incomplete vaporization of interstellar grains or the presence of refractory condensates may have provided sufficient nuclei to initiate heterogeneous nucleation of the major condensates, and thus have bypassed the homogeneous nucleation barrier. In the actual solar nebula, supersaturation of the gas phase may have occurred to a limited extent, but probably not to such an extent as to reduce significantly the calculated equilibrium condensation temperatures of major condensates.

A more difficult barrier to attainment of equilibrium may be attainment of gas solid equilibrium, which requires that the grain size of condensates be small enough to allow diffusion to maintain equilibrium between grain interiors and the gas. If gas–solid equilibrium were not attained, then several important differences would be observed in the condensation sequence. Fe-alloy would not be homogeneous, as in the equilibrium case, because the first Fe-alloy to condense is greatly enriched in Ni and Co, relative to the solar proportions of these elements (Grossman and Larimer, 1974). Thus, in the disequilibrium case, Fe-alloys would occur either as zoned crystals with Ni- and Co-rich cores and Fe- and Cr-rich rims, or as a mixture of grains with early condensing grains rich in Ni and Co, and later condensing grains richer in Fe and Cr. $MgSiO_3$ would occur in the disequilibrium case in greatly reduced quantities because the initial major Mg-bearing condensate, Mg_2SiO_4, would be precluded from reaction with the gas to form $MgSiO_3$ at lower temperatures. Thus, the gas remaining after Mg_2SiO_4 condensation would be markedly enriched in Si, relative to solar proportions, and SiO_2 would condense at a lower temperature. In addition, if gas/solid equilibrium were not attained in the nebula, the following important steps in the equilibrium condensation sequence would not occur: (1) reaction of Na and K with aluminosilicates to form alkali feldspar, (2) reaction of H_2S with Fe to produce FeS, (3) reaction of H_2O with Fe to produce FeO, and (4) hydration of magnesium silicates. Differences in the composition of condensed material in the equilibrium and disequilibrium cases have recently been discussed in detail by Barshay and Lewis (1975, 1976).

Equilibrium and disequilibrium between gas and solids are actually only the end-member cases. Partial equilibrium (i.e. incomplete reaction of the gas with previously condensed solids) may also occur, and is probably the most likely case for the actual solar nebula. However, a variety of evidence suggests that equilibrium was probably approached in the solar nebula, at least at the temperatures and pressures corresponding to the source regions of the terrestrial planets and meteorites. The mineral troilite (FeS) is ubiquitous in meteorites of every type, and there are good geochemical arguments for the Earth having retained roughly the solar proportion of FeS mainly in the Fe–FeS core (Murthy and Hall, 1970). Formation of FeS in the solar nebula requires that gas–solid equilibrium be at least approached to temperatures as low as 680 K, the pressure-independent temperature at which FeS is formed. The presence of substantial amounts of FeO (in magnesium silicates) in the Earth and in ordinary chondrites suggests that equilibrium was approached at temperatures not far above 500 K. Also, models for the compositions of the terrestrial planets (e.g. Lewis, 1972a) and models for the composition of Allende meteorite inclusions (e.g. Grossman and Clark, 1973), which are based on the assumption of chemical equilibrium in the nebula, are in good general agreement with observational data, as discussed in the next section of this chapter, suggesting that equilibrium was in fact approached in the nebula.

Attainment of gas/solid equilibrium in the nebula at low temperatures will be more difficult, since diffusion may be kinetically inhibited. Thus, disequilibrium effects may be very important in considering the low-temperature icy condensates important in the outer solar system (Lewis, 1972b). Disequilibrium effects are probably also important in the formation of the low-temperature carbonaceous matrix in carbonaceous chondrites (Anders *et al.*, 1973). However, the evidence to date supports the hypothesis that equilibrium, while not completely attained, was probably approached in that portion of the solar nebula which corresponds to the terrestrial planets.

Application to Planets and Other Solar System Objects

Assumptions

The primary result obtained from chemical equilibrium condensation sequence calculations, as discussed in the last section, is the composition of condensed material as a function of temperature and pressure in the solar nebula. Application of these results to the compositions of the outer planets and satellites has been discussed by Lewis (1972b, 1974a). Application of condensation calculations to meteorites, particularly to Allende inclusions, has been discussed by Grossman and Clark (1973), Larimer (1973), Grossman and Olsen (1974), Grossman and Larimer (1974), and by several other researchers. The following discussion focuses on the terrestrial planets, with the model proposed by Lewis (1972a, 1973, 1974a, 1974b) taken as the starting point for the discussion.

Lewis' model for the compositions of the terrestrial planets is based directly on the equilibrium condensation sequence. Although in many respects this model is very simple, application of the equilibrium condensation sequence to the compositions of the terrestrial planets does necessitate making several additional assumptions. First, Lewis assumed an adiabatic temperature–pressure profile in the solar nebula. The adiabat chosen by Lewis, and the positions along the adiabat assigned to each of the planets, are shown in Figure 1 above. An adiabatic model for the nebula is in agreement with the nebular models of Cameron and Pine (1973), and is further supported by the fact that an adiabat provides a good fit to the observed or inferred chemical compositions of solar system objects (Lewis, 1974b). The results of equilibrium calculations are however, only slightly dependent on pressure; therefore, choosing a different adiabat at somewhat higher or lower nebular pressures would make little difference in the condensation sequence. Likewise, moderate departures from an adiabatic gradient in the nebula would not appreciably affect the calculated condensation sequence.

Second, Lewis assumed that each planet accreted material from a narrow, well-defined region of the nebula. Thus, each planet accreted material with a relatively narrow range of compositions, with the compositional range determined by the temperature and pressure region from which the planet accreted material. The approximate ranges in temperature and pressure from which each planet accreted material are shown in Figure 1 above. By assuming that the composition of each planet is determined by temperature and pressure in the nebula, Lewis assumed that physical fractionation processes (e.g. magnetic effects, surface effects, density or size separations), which could significantly modify the composition of material condensed directly from the solar nebula, were not important. In Lewis' model, the formation temperature of each planet is the temperature at which the material constituting the planet ceased to react further with the nebular gas. This temperature may represent the conditions when the grain size of the condensates became too large to permit further reactions, or when the nebular gas was removed by an intense solar wind or by some other process.

Third, by assuming that the equilibrium condensation sequence is applicable to the compositions of the terrestrial planets, Lewis requires that accretion of planets occurred *after* the composition of most of the material constituting each planet had been fixed by equilibration in the nebula. Thus, condensates must remain small enough to maintain equilibrium with the gas, and not accrete into planetesimals immediately upon condensation. A contrary hypothesis has been advocated by Turekian and Clark (1969), who suggested that planets accreted *during* condensation. In their inhomogeneous accretion model, planets accreted in layers: a proto-core of refractory condensates, an Fe-alloy layer, a magnesium silicate layer, and subsequent layers of lower temperature condensates, with each condensate accreted as soon as it condensed. In Lewis' model, however, accretion of planets is homogeneous, with the differentiation of planets into mantles and cores occurring after accretion.

Fourth, by postulating that the composition of each of the terrestrial planets is the same as the composition of condensed material in the nebula at the temperature and pressure range corresponding to each planet, Lewis requires that accretion of the planets must have been 'slow', rather than 'rapid'. In the present context, 'slow' accretion means that the rate of accretion was sufficiently slow to allow most of the gravitational potential energy of the accreting material to be lost via radiation from the surface, and thus to limit heating during accretion. 'Rapid' accretion would imply that sufficient gravitational potential energy was retained to produce substantial heating. In Lewis' model, accretion must have been slow enough to avoid chemical fractionation during accretion by volatilization and subsequent loss of the more volatile fractions of the material constituting each planet. A contrary hypothesis has been advocated by Ringwood (1975, Chapter 16), who suggested that heating and volatilization during accretion greatly modified the compositions of the terrestrial planets, particularly the Earth.

It is not possible to evaluate directly the simplifying assumptions made by Lewis. Present knowledge of the evolution of the solar nebula, of the extent to which mixing and various physical fractionation processes may have occurred, and of the process of planetary accretion, is not sufficient to support or refute these assumptions. Rather, the validity of these assumptions must be tested by comparing the calculated compositions of the terrestrial planets, based on these assumptions, with present data on the terrestrial planets. The consequences of relaxing some of these simple assumptions are discussed below in the *Limitations* part of this section. First, however, we briefly summarize the results of the chemical equilibrium model for the terrestrial planets.

Implications

The results of the chemical equilibrium model for the terrestrial planets were presented initially by Lewis (1972a), and have been subsequently discussed by Lewis (1973, 1974a, 1974b) and by Barshay and Lewis (1975, 1976). Therefore, we limit our discussion to a brief summary of these results.

Mercury, as a direct result of its high formation temperature, is postulated to be composed of material in which refractories (CaO, Al_2O_3, U, Th, etc.) were fully condensed, Fe–Ni alloy nearly fully condensed, and magnesium silicates only about 15 per cent condensed. Mercury is expected to contain virtually no alkali metals, FeO, FeS or H_2O. Thus, Mercury is predicted to have a massive Fe–Ni core, and a small magnesium silicate mantle enriched in refractories. If Mercury has fully differentiated, the crust should be very thick, and be greatly enriched in refractories including U and Th. Lewis suggested that the mantle of Mercury would be dominantly $MgSiO_3$, which is the major magnesium silicate phase once the magnesium silicates are fully condensed. However, as noted by Goettel (1976a), the magnesium silicate component of the Mercurian mantle is enriched in Mg_2SiO_4 because, in the chemical equilibrium model, Mg_2SiO_4 is the major magnesium silicate in the initial stages of condensation of magnesium silicates, and is largely converted to $MgSiO_3$ at somewhat lower temperatures. Thus, Mercury is predicted to have Fe/Si and Mg/Si ratios substantially greater than the solar ratios.

Venus is postulated to be composed of material in which refractories, Fe–Ni alloy, magnesium silicates, and alkali metals (as feldspar) are all fully condensed. Venus is expected to contain virtually no FeO, FeS, or H_2O. Thus, Venus is predicted to have a large Fe–Ni core, and a large magnesium silicate mantle. If Venus is fully differentiated, the crust should be similar to the Earth's crust (i.e., enriched in silica, refractories, and alkali metals), except for the probable absence of FeO. Venus, as well as Earth and Mars, is predicted to have the solar proportions (relative to Fe or Si) of all three major heat-producing elements: U, Th, and K.

Earth, because of its lower formation temperature, is postulated to contain, in addition to the constituents of Venus, significant amounts of S (as FeS) and FeO (in magnesium silicates), and small (but important) amounts of H_2O. Thus, Earth is predicted to have a large Fe–Ni–S core, and a large magnesium silicate mantle (containing about 8 wt per cent FeO).

Mars, as well as Earth and Venus, is postulated to have solar proportions of Fe, Mg, and Si. Mars' low density results from the iron in Mars being fully, or nearly fully, oxidized to FeS or FeO, because of its postulated low formation temperature. Thus, Mars is predicted to have an FeS core, and a large, dense mantle rich in FeO. Mars is expected to contain more total water than does the Earth, because Mars' formation temperature is within the stability fields of several hydrated phases in the solar nebula.

The Earth's Moon, which was briefly discussed by Lewis (1974a), was recently discussed by Goettel (1976a) in the context of the chemical equilibrium model for the terrestrial planets. There are two temperature ranges where the density of condensed material in the solar nebula equals or approaches the mean density of the Moon. In the high-temperature case, the formation temperature of the Moon would be even higher than the formation temperature of Mercury, and the Moon would be composed almost entirely of refractory condensates, plus a small amount of Fe–Ni alloy. This model matches the lunar density, but is inconsistent with lunar petrologic data. In the low-temperature case, the formation temperature would be similar to or slightly lower than the formation temperature of Mars, and the Moon would be composed of material in which Fe was fully oxidized to FeS and FeO. This model could be consistent with lunar density and petrology *if* the outer portion of the Moon were very strongly heated (perhaps *via* solar wind induction or tidal dissipation), resulting in substantial loss of FeO, FeS, alkali metals, and water. This low-temperature chemical equilibrium model for the Moon may be viable, and warrants more detailed consideration. Alternatively, the Moon may not be composed of material condensed directly from the solar nebula, but rather may be composed of material which has undergone extensive mixing and/or fractionation.

A full discussion of the evidence suggesting that the chemical equilibrium model is a reasonable model for the compositions of the terrestrial planets is beyond the scope of the present review. However, as noted by Lewis (1972a, 1973, 1974a, 1974b) and by Barshay and Lewis (1975, 1976), the chemical equilibrium model is in good agreement with the mean densities of the terrestrial planets, and with present knowledge of their chemical compositions, as deduced from moment of inertia data, Earth-based spectral data, and data from fly-by, orbiter, and lander space missions. Goettel (1976b) also recently discussed the evidence supporting the chemical equilibrium model, with particular emphasis on the Earth.

Limitations

The chemical equilibrium model for the compositions of the terrestrial planets is subject to the assumptions and limitations inherent in the calculations of the composition of condensed material as a function of temperature and pressure in the solar nebula. In particular, the assumption that complete equilibrium was attained in the nebula may not be entirely correct, although the evidence discussed above and in the last section on *Composition* suggests that equilibrium was probably closely approached. The most significant effects of non-attainment of complete equilibrium are that the following reactions would not be as complete, at a given temperature, as is assumed in the chemical equilibrium model: (a) reaction of H_2O with Fe to form FeO, (b) reaction of H_2S with Fe to form FeS, and (c) hydration of magnesium silicates. As a result, Earth and Mars could contain somewhat less FeO, FeS, and H_2O than predicted on the basis of their postulated formation temperatures. However, the good agreement between the calculated mean densities of Earth and Mars, based on the chemical equilibrium model, with the observed densities (Lewis, 1972a) suggests that departures from equilibrium were probably fairly minor.

As discussed in the last section on *Composition*, uncertainties in the relative abundances of elements in the solar nebula of about ± 20 per cent have little effect on condensation temperatures, but do affect the relative proportions of condensates present at a given temperature. Thus, this uncertainty affects directly the model compositions calculated for the terrestrial planets in the chemical equilibrium model. However, the good fit between calculated and observed mean densities and compositions of the terrestrial planets suggests that uncertainties in our knowledge of the elemental composition of the nebula probably do not introduce large errors into calculated planetary compositions.

Application of the results of equilibrium condensation calculations to obtain models for the compositions of the terrestrial planets involves, as discussed in the *Assumptions* part of this section, several simplifving assumptions. In particular, the effects of mixing materials from different heliocentric distances, the effects of various possible fractionation mechanisms, and the effects of fractionation during the accretion process were not considered in the simple chemical equilibrium model.

The simplest model assumes that no mixing of material from different heliocentric distances occurred in the early solar system. However, the fact that meteorites of many different compositions, which presumably originated at varied heliocentric distances, intersect the Earth's orbit, and the dynamic calculations presented recently by Hartmann (1976) suggest strongly that some mixing must have occurred. Hartmann's calculations suggest that sizeable fractions of the final few per cent of material accreting onto each of the terrestrial planets may have originated in orbits nearer to other planets. In Hartmann's simple model, Mercury, Venus, Earth, and Mars received 47, 45, 37, and 52 per cent respectively, of their late-accreted mass (*not* total mass) from material formed closer to other planets.

Hartmann's calculations suggest that 60 per cent of the final material accreting onto the moon may have originated near Venus. Interestingly, this result suggests that the isotopic composition of Venus (e.g. oxygen) may be the same as the Earth and Moon, since the Moon falls on the same chemical fractionation oxygen isotope line as the Earth (Clayton *et al.*, 1973). Thus, it appears that the portion of the solar nebula which corresponds to the inner solar system, or at least the Earth–Moon–Venus region, may have been isotopically homogenized.

Many mechanisms have been proposed for mixing and/or fractionating material in the solar nebula with respect to grain size, density or other physical properties. Proposed mechanisms have included magnetic forces, surface effects (e.g., stickiness), the Poynting–Robertson effect and many others. Recently, Peterson (1976) has suggested that the Yarkovsky effect, which arises from the asymmetric reradiation emitted by an illuminated rotating object, may be important.

Another mechanism for mixing materials from different heliocentric distances has been proposed recently by Weidenschilling (1976). He noted that as a result of aerodynamic drag, solids present in the solar nebula will move with less than Keplerian velocities, and thus will spiral inward toward the sun. Radial velocities as high as 100 m/sec may result, with the rate of inward spiral being a complex function of grain size and density. Thus, this mechanism could have resulted in both (a) mixing of materials from different heliocentric distances, and (b) fractionation of metal and silicate condensates. For grain sizes of 1 cm or larger, Weidenschilling's calculated radial velocities are substantially greater at the radius of Mercury than at the radii of the other planets. Thus, this mechanism could have a greater effect on the composition of Mercury than on the compositions of the other terrestrial planets.

In Lewis' simple model, the high density of Mercury results from Mercury accreting much of its mass from the rather narrow region of the nebula in which Fe-alloy is more completely condensed than magnesium silicates. However, the relative proportions of Fe-alloy and magnesium silicates condensed varies sharply with temperature, and a number of factors could influence the fraction of Mercury's mass actually accreted from this region of the nebula. One alternate model for the composition of Mercury is that much of the silicate fraction of Mercury actually condensed further out in the solar nebula, and was subsequently mixed with material condensed near Mercury. If mixing/fractionation processes have significantly affected the composition of Mercury, Goettel (1976a) noted that the following changes are most likely, in decreasing order of probable importance: (1) replacement of Mg_2SiO_4 by $MgSiO_3$, (2) addition of Na and K (as feldspar), (3) addition of FeS, and (4) addition of FeO. Chemical analysis of a sample of the Mercurian crust might be indicative of the extent to which the composition of Mercury may have been affected by mixing and fractionation processes.

Volatiles such as N_2, CO_2, or H_2O constitute a virtually negligible fraction of the total mass of each of the terrestrial planets. Thus, the volatile content of a planet is particularly susceptible to modification by mixing/fractionation processes. Addition of small amounts of volatile-rich material to a planet could greatly change the volatile content and/or composition without appreciably affecting the bulk composition of the planet. As noted by Lewis (1974c), the small amount of S present in the atmosphere of Venus may be the result of infall of small amounts of volatile-rich material. However, in general it is very difficult to evaluate the importance of mixing–fractionation processes in governing the volatile content of the terrestrial planets.

Some mixing and fractionation of materials from different heliocentric distances must have occurred in the early solar system. However, the observed fact that the terrestrial planets retain distinct differences in mean density is strong evidence that compositional differences produced by differing formation temperatures were not obliterated by mixing processes. Thus, the simple chemical equilibrium model for the compositions of the terrestrial planets still appears to be a viable, first-order model. Quantitative evaluation of the effects of mixing and/or fractionation processes on the compositions of the terrestrial planets is, however, a prime area for more detailed consideration.

Summary and Conclusions

The chemical and physical evolution of the solar nebula is of considerable interest because it is widely accepted that planets, satellites, meteorites, and other solar system objects were formed directly or indirectly from the nebula. If one assumes that (a) the nebular hypothesis for the origin of the solar system is valid, (b) the elemental composition of the nebula was equal to present solar composition and thus is known, and (c) the nebula was homogeneous in elemental abundances, then sufficient thermodynamic data exist to calculate directly, as a function of temperature and pressure in the nebula, the composition of solids condensed from the nebula and the composition of the gaseous portion of the nebula. The composition of condensed solids is generally of greater interest, since most solar system objects, with the exception of the outer planets, are composed of condensed material. The accuracy of the calculated composition of condensed material in the solar nebula is limited somewhat by uncertainties in the knowledge of the composition of the nebula, and by errors and gaps in the requisite thermodynamic data. However, the calculated composition of condensed material depends principally on the extent to which chemical equilibrium is assumed to be attained.

A variety of evidence, particularly the ubiquitous presence of FeS in meteorites of every class, and the presence of substantial amounts of FeO in ordinary chondrites and in the Earth, strongly suggests that chemical equilibrium was probably closely approached in the nebula, at least in that portion of the nebula which corresponds to the inner solar system. Therefore, the present review has focused on the chemical equilibrium model for condensation in the solar nebula. The composition of condensed material, as a function of temperature and pressure, was presented in Figure 1 above for the chemical equilibrium case.

Application of the results of chemical equilibrium condensation calculations to obtain quantitative models for the compositions of the terrestrial planets necessitates making several additional assumptions, most important of which is that the composition of each planet is governed by its formation temperature, and thus by its heliocentric distance. In the simple chemical equilibrium model, mixing and/or fractionation processes are not considered. The simple chemical equilibrium model predicts densities and chemical compositions for the terrestrial planets which are in good agreement with the observed densities and with present data on planetary compositions. Therefore, the simple model does appear to be a viable working model. However, some mixing and/or fractionation must have occurred to at least a limited extent, in the early solar system, as suggested by the dynamic calculations presented by Hartmann (1976) or the effects of aerodynamic drag as discussed by Weidenschilling (1976).

Thus, while the chemical equilibrium model appears to be a good first-order model for the chemical evolution of the solar nebula and for the compositions of the terrestrial planets, it is certainly not completely accurate. Detailed consideration of the modifications which must be made to account for modest departures from attainment of equilibrium and more quantitative consideration of the effects of mixing and/or fractionation processes are fruitful research topics for planetary scientists interested in the origin and evolution of the solar system.

It is important to note that the technique of calculating the composition of condensed material in the solar nebula is a very powerful and very general technique which could be applied equally fruitfully to systems of different composition or with differing assumptions about the degree of equilibrium attained. Therefore, planetary scientists who disagree with the assumptions inherent in the chemical equilibrium model for condensation in the solar nebula are encouraged, in fact obligated, to perform similar calculations under a set of assumptions they consider more relevant to the actual evolution of the solar nebula.

Acknowledgements

We thank J. S. Lewis for many helpful discussions and for constructive comments on this manuscript. This work was supported in part by NASA Grant NGL–22–009–521.

References

Alfvén, H., and Arrhenius, G. (1975). *Structure and Evolutionary History of the Solar System*, Reidel Press, Boston, 276 pp.

Alfvén, H., and Arrhenius, G. (1976). 'Evolution of the Solar System.' *NASA*, SP–345.

Anders, E. (1971). 'How well do we know cosmic abundances?' *Geochim. Cosmochim. Acta*, **35**, 516–522.

Anders, E., Hayatsu, R., and Studier, M. H. (1973). 'Organic compounds in meteorites.' *Science*, **182**, 781–790.

Barshay, S. S., and Lewis, J. S. (1975). 'Chemistry of solar material.' in *The Dusty Universe*, eds. G. B. Field and A. G. W. Cameron, Neale Watson Academic, New York, pp. 34–46.

Barshay, S. S., and Lewis, J. S. (1976). 'Chemistry of primitive solar material.' *Ann. Rev. Astron. Astrophys.*, **14**, 81–94.

Black, D. C. (1972). 'On the origins of trapped helium, neon and argon isotopic variations in meteorites, II. Carbonaceous meteorites.' *Geochim. Cosmochim. Acta*, **36**, 377–394.

Blander, M., and Katz, J. L. (1967). 'Condensation of primordial dust.' *Geochim. Cosmochim. Acta*, **31**, 1025–1034.

Boynton, W. V. (1975). 'Fractionation in the solar nebula: Condensation of yttrium and the rare earth elements.' *Geochem. Cosmochim. Acta*, **39**, 569–584.

Cameron, A. G. W. (1973). 'Abundances of elements in the solar system.' *Space Sci. Rev.*, **15**, 121–146.

Cameron, A. G. W. (1975). 'The origin and evolution of the solar system. *Sci. Am.*, **233(3)**, 32–41.

Cameron, A. G. W., and Pine, M. R. (1973). 'Numerical models of the primitive solar nebula.' *Icarus*, **18**, 377–406.

Clayton, R. N., Grossman, L., and Mayeda, T. K. (1973). 'A component of primitive nuclear composition in carbonaceous meteorites.' *Science*, **182**, 485–488.

Drozd, R. J., and Podosek, F. A. (1976). 'Primordial ^{129}Xe in meteorites.' *Earth Planet. Sci. Lett.*, **31**, 15–30.

Goettel, K. A. (1976a). 'Cosmochemical constraints on the composition of Mercury and the Moon.' *Paper presented at Conference on Comparisons of Mercury and the Moon*, Lunar Science Institute, Houston, Texas, 15–17 November, 1976. Conference proceedings to be published.

Goettel, K. A. (1976b). 'Models for the origin and composition of the Earth, and the hypothesis of potassium in the Earth's core.' *Geophys. Surv.*, **2**, 369–397.

Grossman, L. (1972a). 'Condensation, chondrites, and planets.' *Ph.D. Thesis*, Yale University, New Haven, Connecticut. 97 pp.

Grossman, L. (1972b). 'Condensation in the primitive solar nebula.' *Geochim. Cosmochim. Acta*, **36**, 597–619.

Grossman, L., and Clark, S. P. (1973). 'High-temperature condensates in chondrites and the environment in which they formed.' *Geochim. Cosmochim. Acta*, **37**, 635–649.

Grossman, L., and Larimer, J. W. (1974). 'Early chemical history of the solar system.' *Rev. Geophys. Space Phys.*, **12**, 71–101.

Grossman, L., and Olsen, E. (1974). 'Origin of the high-temperature fraction of C2 chondrites.' *Geochim. Cosmochim. Acta*, **38**, 173–187.

Hartmann, W. K. (1976). 'Planet formation' Compositional mixing and lunar compositional anomalies.' *Icarus*, **27**, 553–559.

Herbig, G. H. (1970). 'Early stellar evolution at intermediate masses.' in *Spectroscopic Astrophysics*, ed. G. H. Herbig, University of California Press, Berkeley, California.

Herbig, G. H. (1976). Papers presented at NATO Advanced Study Institute, *The Origin of the Solar System*, 29 March–9 April, 1976, Newcastle upon Tyne, England. Elsewhere in this volume.

Larimer, J. W. (1967). 'Chemical fractionations in meteorites, 1, Condensation of the elements.' *Geochim. Cosmochim. Acta*, **31**, 1215–1238.

Larimer, J. W. (1973). 'Chemistry of the solar nebula.' *Space Sci. Rev.*, **15**, 103–119.

Larimer, J. W., and Anders, E. (1967). 'Chemical fractionations in meteorites, 2, Abundance patterns and their interpretation.' *Geochim. Cosmochim. Acta*, **31**, 1239–1270.

Larimer, J. W., and Anders, E. (1970). 'Chemical fractionations in meteorites, 3, Major element fractionations in chondrites.' *Geochim. Cosmochim. Acta*, **34**, 367–387.

Larson, R. B. (1974). 'The evolution of protostars—theory.' *Fund. Cosmic. Phys.*, **1**, 1–70.

Lee, T., and Papanastassiou, D. A. (1974). 'Mg isotopic anomalies in the Allende meteorite and correlation with O and Sr effects.' *Geophys. Res. Lett.*, **1**, 225–228.

Lewis, J. S. (1972a). 'Metal/silicate fractionation in the solar system.' *Earth Planet. Sci. Lett.*, **15**, 286–290.

Lewis, J. S. (1972b). 'Low temperature condensation from the solar nebula.' *Icarus*, **16**, 241–252.

Lewis, J. S. (1973). 'Chemistry of the planets.' *Ann. Rev. Phys. Chem.*, **24**, 339–351.

Lewis, J. S. (1974a). 'The chemistry of the solar system.' *Sci. Am.*, **230(3)**, 50–65.

Lewis, J. S. (1974b). 'The temperature gradient in the solar nebula.' *Science*, **186**, 440–443.

Lewis, J. S. (1974c). 'Volatile element influx on Venus from cometary impacts.' *Earth Planet. Sci. Lett.*, **22**, 239–244.

Lord, H. C. (1965). 'Molecular equilibria and condensation in a solar nebula and cool stellar atmospheres.' *Icarus*, **4**, 279–288.

Murthy, V. R., and Hall, H. T. (1970). 'The chemical composition of the Earth's core: Possibility of sulphur in the core.' *Phys. Earth Planet. Interiors*, **2**, 276–282.

Peterson, C. (1976). 'A source mechanism for meteorites controlled by the Yarkovsky effect.' *Icarus*, **29**, 91–111.

Ringwood, A. E. (1975). *Composition and Petrology of the Earth's Mantle*, McGraw-Hill, New York, 618 pp.

Safronov, V. S. (1972). *Evolution of the Protoplanetary Cloud and Formation of the Earth and Planets*, IPST, Jerusalem, 207 pp.

Trimble, V. (1975). 'The origin and abundances of the chemical elements.' *Rev. Mod. Phys.*, **47**, 877–976.

Turekian, K. K., and Clark, S. P. (1969). 'Inhomogeneous accumulation of the earth from the primitive solar nebula.' *Earth Planet. Sci. Lett.*, **6**, 346–348.

Ward, W. R. (1976). 'The formation of the solar system' in *Frontiers of Astrophysics*, ed. E. H. Avrett, Harvard University Press, Cambridge, Massachusetts, pp. 1–40.

Weidenschilling, S. J. (1976). 'Aerodynamics of solid bodies in the solar nebula.' *Mon. Not. R. astron. Soc.*, submitted.

Woolfson, M. M. (1976). Papers presented at NATO Advanced Study Institute, *The Origin of the Solar System*, 29 March–9 April, Newcastle upon Tyne, England. Elsewhere in this volume.

CHONDRITES AND CHONDRULES—MACROSCOPIC CHEMICAL ASPECTS OF THEIR ORIGINS

J. M. HERNDON
Department of Chemistry B-017, University of California, San Diego, La Jolla, California, U.S.A.

Abstract

The presence in chondrules of iron in chemical states other than metal indicates that chondrule formation might have occurred after hydrogen was fractionated from solar matter by a factor of 10^2–10^3. Such an assumption is contrary to current chemical models of meteorite formation which assume that hydrogen was present at the time mineral compositions were established.

It is not yet possible to state with certainty many aspects and parameters fundamental to the question of the origin of the solar system. Most theories, hypotheses, or interpretations of planetary and meteoritic data bearing on this question can be replaced by alternate, equally plausible, ones.

Long ago, authorities as eminent as Urey and Suess recognized that meteorites formed as a consequence of complex processes. They also realized that the conditions which prevailed during meteorite formation resulted in remarkably uniform chemical features among many different types of meteorites (Urey and Craig, 1953; Suess, 1964, 1965; Urey, 1966). Meteorites are indeed complex as far as their detailed structure is concerned (Ramdohr, 1973). However, broad generalizations can be made by considering the smallest number of components which are necessary and sufficient to characterize macroscopic chemical aspects of their origins. Such generalizations provide an interesting contrast to current chemical models of meteorite formation.

The great majority of meteorites which fall to earth are termed chondrites. About 1600 of these relatively undifferentiated meteorites are presently known. There are conspicuous physical and chemical features common to practically all chondrites—important exceptions are the very rare Cl and enstatite chondrites.

Chondrites are so named because they contain chondrules, typically spherical structures which have an average diameter of ~ 1mm. Many, although not all, chondrules bear the unmistakable signs of once having been molten droplets that obtained their shapes from the surface tension of the melt which presumably solidified rapidly from a supercooled liquid (Nelson *et al.*, 1972). In a given meteorite, the mineral constitutions and chemical compositions of the chondrules are essentially identical to those of the matrix minerals, consisting primarily of:

Olivine	$(Mg, Fe)_2 SiO_4$
Pyroxene	$(Mg, Fe) SiO_3$
Metal	Fe
Troilite	FeS

These minerals all contain iron in three different chemical states: as oxide (FeO) in silicates, sulphide (FeS) and metal (Fe).

Individual chondrules typically contain one of these chemical states of iron. Occasionally, two and even three states co-exist in a single chondrule (Ramdohr, 1973).

At temperatures at which silicates are molten, chemical reactions between condensate and ambient gas phase progress at rapid rates particularly for chondrule-sized droplets, and equilibrium is approached. (The term 'condensate' is used in this connection to emphasize that at such high temperatures condensation–evaporation equilibria cannot be completely ignored). Thermodynamic calculations, assuming equilibrium, give information about the composition of the ambient gas phase.

A boundary condition that determines the existence of FeS in equilibrium with an ambient gas is given by

$$Fe(c) + H_2S(g) \rightleftarrows FeS(c) + H_2(g)$$

The direction of this reaction is determined from the well known relationship

$$\Delta G = -2{\cdot}3\,RT \log K = -2{\cdot}3\,RT \log \frac{P_{H_2}}{P_{H_2S}}$$

The essentially pressure independent, equilibrium P_{H_2}/P_{H_2S} ratio is shown in Figure 1 as a function of temperature. At elevated temperatures FeS can only be in stable equilibrium with a gas phase, without being reduced to metal, provided the P_{H_2}/P_{H_2S} ratio is a factor of 100 to 1000 smaller than that in solar matter.

Similar, although slightly more complicated, calculations on the thermodynamic stability of FeO in silicates lead to an analogous conclusion: oxidized iron in the silicate minerals, olivine and pyroxene, can only exist in equilibrium with a gas phase at near melt temperatures provided the P_{H_2}/P_{H_2O} ratio is a factor of 100 to 1000 smaller than that in solar matter.

In current chemical models of meteorite formation, it is either assumed that oxidized iron and iron sulphide formed at low temperatures, $\lesssim 700$ K, (Wood, 1962; Larimer and Anders, 1967; Anders, 1968, 1971, 1972, a,b; Larimer, 1973; Grossman and Larimer, 1974) or it is assumed that disequilibrium effects were responsible for establishing the observed mineral compositions (Blander and

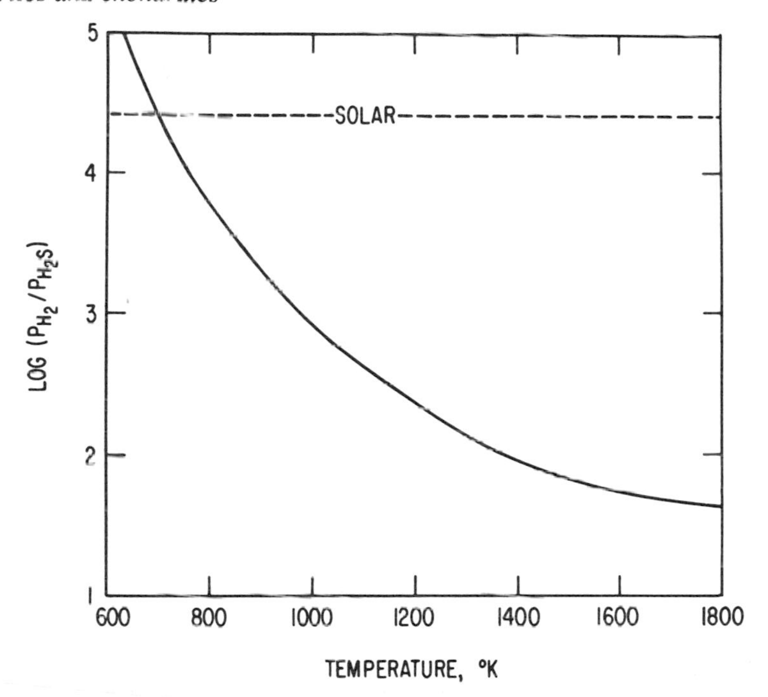

Figure 1. Equilibrium P_{H_2}/P_{H_2S} ratio as a function of temperature at the Fe–FeS boundary. FeS is thermodynamically stable below the solid curve. The unfractionated solar matter ratio is indicated by the broken curve

Katz, 1967; Blander and Abdel-Gawad, 1969; Blander, 1971; Arrhenius and Alfvén, 1971). These models share a common assumption—that the P_{H_2}/P_{H_2O} and P_{H_2}/P_{H_2S} ratios in the gas phase during mineral formation were typical of solar matter, $\sim 1{\cdot}4 \times 10^3$ and $\sim 2{\cdot}6 \times 10^4$, respectively. In such an atmosphere, iron oxide and iron sulphide in equilibrium are only stable as solids at comparatively low temperatures. However, loss of hydrogen by a factor of 10^2–10^3 allows iron oxide and iron sulphide to be thermodynamically stable as liquids at higher temperatures.

At temperatures near the melting points of silicates, gas-condensate interaction becomes rapid, particularly for chondrule-sized droplets. The resulting chemical reactions will, of course, depend not only on temperature, pressure and on the vapour pressures of the individual minerals, but also on the chemical composition of the ambient gas. Herndon and Suess (1976) have shown that the chondrules could have obtained their chemical compositions in liquid–vapour equilibrium, and, hence, by condensation from solar gas which is depleted in hydrogen by a factor of 10^2–10^3.

The similarity in chemical compositions of the minerals in the chondrules and those in the bulk meteorite suggests that these compositions were established in chemically similar environments. Wood (1962) attempted to explain the

FeO content of chondrules by iron diffusion in the solid state, assuming that the oxidized iron condensed at a much lower temperature than the melting points of the chondrules. Since that time, it has been shown by many authors [for example, Fredriksson *et al.* (1974)] that such diffusion could not have taken place.

It has to be concluded, therefore, that when chondrules were molten the gas phase, at that time, was depleted in hydrogen by a factor of 10^2–10^3 relative to solar matter.

There are several mechanisms that could possibly account for the loss of hydrogen during the formation of the solar system. At the present time, it is not possible to state with certainty which is correct. This will be learned, however, not by sophisticated mathematical treatment of arbitrary chemical models, but by tedious efforts to place many seemingly independent observations in a logical sequence so that causal relationships become evident.

References

Anders, E. (1968). 'Chemical processes in the early solar system, as inferred from meteorites.' *Accounts Chem. Res.*, **1**, 289–298.

Anders, E. (1971). 'Meteorites and the early solar system.' *Ann. Rev. Astron. Astrophys.*, **9**, 1–34.

Anders, E. (1972a). 'Conditions in the early solar system, as inferred from meteorites.' In *From Plasma to Planet*, edited by A. Elvius. Almqvist and Wiksell, Stockholm, pp. 133–150.

Anders, E. (1972b). 'Physico–chemical processes in the solar nebula as inferred from meteorites.' In *Symposium on the Origin of the Solar System*, edited by H. Reeves. Centre National de la Recherche Scientifique, Paris, pp. 179–195.

Arrhenius, G., and Alfvén, H. (1971). 'Fractionation and condensation in space.' *Earth Planet. Sci. Lett.*, **10**, 253–267.

Blander, M. (1971). 'The constrained equilibrium theory: Sulfide phases in meteorites.' *Geochim. Cosmochim. Acta*, **35**, 61–76.

Blander, M., and Abdel-Gawad, M. (1969). 'The origin of meteorites and the constrained equilibrium condensation theory.' *Geochim. Cosmochim. Acta*, **33**, 701–716.

Blander, M., and Katz, J. L. (1967). 'Condensation of primordial dust.' *Geochim. Cosmochim. Acta*, **31**, 1025–1034.

Fredriksson, K., Dube, A., Jarosewich, E., Nelson, J. A., and Noonan, A. F. (1974). 'The Pulsora anomaly. A case against metamorphic equilibration in chondrites.' Smithson, Contrib. *Earth Sci.*, **14**, 41–55.

Grossman, L., and Larimer, J. W. (1974). 'Early chemical history of the solar system.' *Revs. Geophys. Space Phys.*, **12**, 71–101.

Herndon, J. M., and Suess, H. E. (1976). 'Can the ordinary chondrites have condensed from a gas phase.' *Geochim. Cosmochim. Acta*, (in press).

Larimer, J. W. (1973). 'Chemistry for the solar nebula.' *Space Sci. Rev.*, **15**, 103–119.

Larimer, J. W., and Anders, E. (1967). 'Chemical fractionations in meteorites-II. Abundance patterns and their interpretations.' *Geochim. Cosmochim. Acta*, **31**, 1239–1270.

Nelson, L. S., Blander, M., Skaggs, S. R., and Keil, K. (1972). 'Use of a CO_2 laser to prepare chondrule-like spherules from super-cooled molten oxide and silicate droplets.' *Earth Planet. Sci. Lett.*, **14**, 338–344.

Ramdohr, P. (1973). *The Opaque Minerals in Stony Meteorites*. Elsevier.
Suess, H. E. (1964). 'The Urey-Craig groups of chondrites and their state of oxidation.' Chapter 25 in *Isotopic and Cosmic Chemistry*, ed. H. Craig, S. L. Miller and G. J. Wasserburg, North-Holland Publishing Co.
Suess, H. E. (1965). 'Chemical evidence bearing on the origin of the solar system.' *Ann. Rev. Astron. Astrophys.*, **3**, 217–234.
Urey, H. C. (1966). 'Chemical evidence relative to the origin of the solar system.' *Mon. Not. Roy. astron. Soc.*, **131**, 199–223.
Urey, H. C., and Craig, H. (1953). 'The composition of the stone meteorites and the origin of the meteorites.' *Geochim. Cosmochim. Acta*, **4**, 36–82.
Wood, J. A. (1962). 'Metamorphism in chondrites.' *Geochim. Cosmochim. Acta*, **26**, 739–749.

THE PRIMORDIAL SOLAR MAGNETIC FIELD

JOHN W. FREEMAN
Department of Space Physics and Astronomy, Rice University, Houston, Texas, 77001 U.S.A.

Abstract

We envision an epoch of quasi-dipolar field for the pre-main sequence Sun with a magnetic moment in the range 1×10^{34} to 5×10^{36} gauss cm^3. By analogy with the Jovian magnetosphere we calculate equatorial field values at 1 AU possibly as high as about 10^{-2} gauss. Higher field values may be possible during the earlier epoch of a less developed Sun. Various magnetospheric phenomena suggest themselves during this quasi-dipolar period.

Introduction

A variety of evidence suggests the presence of a substantial magnetic field associated with the young Sun or the solar nebula. This includes the ubiquitous nature of planetary magnetism (Runcorn, this volume), the remanent magnetism of lunar rocks and carbonaceous chondrites and the observation of an interstellar magnetic field. A 'seed' field is necessary for the initiation of a planetary dynamo. The ancient magnetic field at the lunar surface deduced from laboratory measurements of the remanent magnetism of lunar rocks ranges from 0·02 to 1·2 oersteds and may result from cooling in the presence of an internal field such as a dynamo generated field (Runcorn, this volume), a permanently magnetized core (Strangway, Sharpe, and Peltier, 1976) or, as has been argued by Banerjee and Mellema (1976), cooling in the presence of an external solar field. The existence of a general interstellar magnetic field of the order of several microgauss (1 gauss $= 10^{-4}$ tesla) guarantees a magnetic field in the solar nebula of this or higher value since some of the interstellar field will be trapped in the collapsing cloud.

The present solar field is of the order of 5×10^{-5} gauss at 1 AU and the surface field ranges from a few gauss up to several hundred gauss near solar flares (Ness, 1968).

In this paper we review a range of possible values for the magnetic moment of the pre-main sequence sun, $\mu'_{\odot}$, and model the external field to obtain the radial dependence of the equatorial magnetic field.

The Magnetic Moment

Reeves (1972) estimated that if the magnetic field of the interstellar gas cloud from which the sun was formed was compressed without slippage of the field through the cloud the resulting magnetic moment, $\mu'_{\odot}$, would be of the order of 10^{41} gauss cm^3. Since some loss of the field through the gas must have occurred due to the finite conductivity of the gas, this value represents an extreme upper limit.

Alfvén and Arrhenius (1973) require that $\mu'_{\odot} \simeq 5 \times 10^{38}$ gauss cm^3 in order to support the plasma upon field lines in their model of the solar nebula.

To establish a lower limit to the solar primordial dipole moment, we assume the magnetic moment of a rotating convective body to be proportional to some power, α, of the product of the spin rate, ω, and volume, V, of the body:

$$\mu'_{\odot} = k(\omega V)^{\alpha} \tag{1}$$

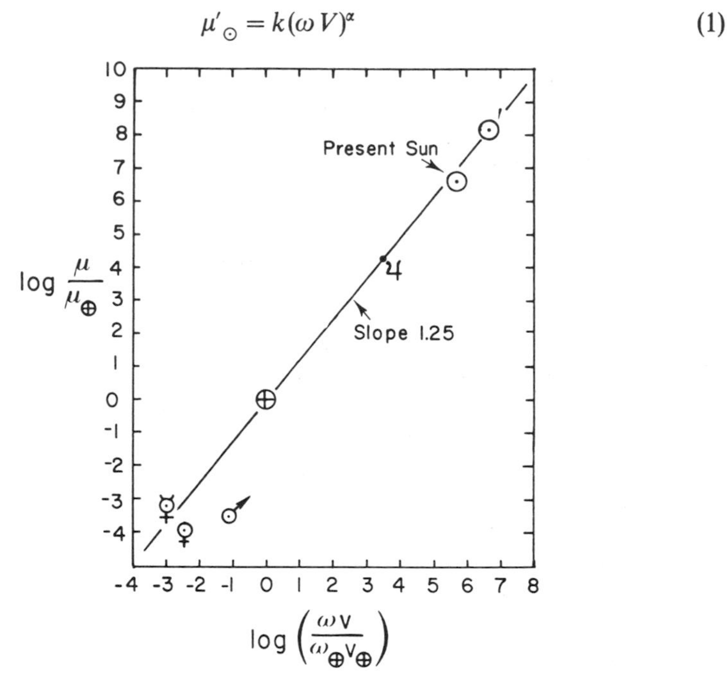

Figure 1. Logarithms of the solar and planetary magnetic moments (relative to earth) *vs.* the spin rate–volume products (relative to earth). The planetary moments are as summarized by Russell (1976). The point marked 'present sun' assumes the ambient surface solar field to arise from a fictional (pseudo) dipole moment

Table 1. B in gauss (1 gauss = 10^{-4} tesla)

$\mu'_\odot$ (gauss cm^3)	B (surf.)[b]	B (0·39 AU)	B (1 AU)	B (10 AU)
10^{41}	4×10^7	1×10^3	4×10^2	40
5×10^{38}	2×10^5	5	2	0·2
6×10^{36}	2×10^3	6×10^{-2}	2×10^{-2}	2×10^{-3}
1×10^{34}[a]	3	1×10^{-4}	4×10^{-5}	4×10^{-6}

[a]Based on $R'_\odot = 2R_\odot$ and $\omega'_\odot = 10\omega_\odot$
[b]$R'_\odot = 2R_\odot$

where k is a constant of proportionality. In a dynamo the magnetic moment should be proportional to the total convective energy within the object. This is in turn proportional to the volume and spin rate.

Figure 1 shows a log plot of the magnetic moments of the planets relative to earth *vs.* the log of the spin rate-volume product relative to earth. The linear nature of the plot confirms our hypothesis of a power law dependence. The exponent (or slope) is about 1·25. For reference, we also show the present Sun.

We next assume that the protosun was twice its present radius and rotated 10 times as fast and that the moment lies along the slope 1·25 line. We then read the magnetic moment off the ordinate. This point is marked as $\odot'$ in Figure 1. The resulting magnetic moment, $\mu'_\odot$, is about 1×10^{34} gauss cm^3. The surface field from such a moment is several gauss.

Table 1 shows the foregoing three dipole moments and their resulting surface flux densities. The intermediate $\mu'_\odot$ shown in Table 1 will be discussed below.

The Distant Field

We next address the question of the shape of the external field. The currents within the Sun may be highly irregular since they arise from field lines drawn into the collapsing protostar by the finite conductivity of the nebular gas. However, some distance above the surface the dipole term will dominate. We wish to consider the effect of external currents on this dipole field. We assume that some nebular gas remains to interact with the magnetic field tending to inflate it by plasma pressure (or currents). In a search for a way in which to model this field we turn to the Jovian magnetosphere as an example of a quasi-dipole field in which plasma currents are important. Here equatorial ring currents inflate or distend the field. Hill *et al.* (1974) have shown that to within 25 per cent the Jovian field can be represented by a dipole out to 30 R_J. Beyond this the field falls off approximately as R^{-1} out to the magnetopause. We then take for the functional form of the primordial field

$$B_0(R) = \mu'_\odot \left[\frac{1}{R^3} + \frac{1}{(30R_\odot)^2 R} \right] \tag{2}$$

where B_0 is the field value at the equator and R the distance from the centre of the Sun.

Table 1 gives the result of applying equation (2) to the previous three estimates of the primordial solar dipole moments. Field values are given for the solar surface, the orbit of Mercury and 1 and 10 AU. In addition, we have shown the dipole moment and resulting field values necessary to obtain 2×10^{-2} gauss at 1 AU. This 6×10^{36} gauss cm^3 magnetic moment represents the solar field necessary if the *weaker* lunar rocks were magnetized by the solar field directly. It is interesting to note that the surface field for such a moment ($\sim 10^3$ gauss) is close to that observed for many magnetic stars (cf. Lang, 1974). In addition to the foregoing, we expect substantial azimuthal skewing due to the high co-rotation velocities.

Plausible Values

Because the surface fields required by the first two magnetic moments listed in Table 1 above greatly exceed those reported for other magnetic stars, we consider them implausible. The remaining two moments, 1×10^{34} and 6×10^{36} gauss cm^3, both result in fields throughout the solar system above the interstellar magnetic field value. These fields would seem to be adequate to provide the 'seed' field necessary for planetary dynamos. We suggest that these moments and their specified equatorial field values represent the range of probable values appropriate to the sun and solar system *before the solar wind turned on.*

The Primordial Solar Magnetosphere

Next we examine the gas pressures (or energy densities) for which this magnetic field could influence the nebula dynamics. The magnetic pressure *vs.* gas pressure is compared by examining

$$\frac{B^2}{8\pi} \gtrless nkT + n\bar{m}v^2 + \ldots \tag{3}$$

The magnetic pressure is on the lefthand side and the particle pressures, thermal and bulk flow respectively are on the righthand side. n is the particle number density (charged particles if the mean free path is long, neutral otherwise), T is the temperature, k Boltzman's constant, $\bar{m}$ the mean atomic or molecular mass and v the bulk flow velocity. Only if the lefthand side is larger than the right can the magnetic field control the gas or plasma motion.

For the 6×10^{36} gauss cm^3 moment the magnetic pressure at 1 AU is only about 10^{-5} dynes/cm^2 or about 10^{-11} atmospheres. For an ideal gas at 500 K the equivalent hydrogen number density is about 10^8 atoms/cm^3. These pressures and densities are much lower than those in most nebular models (cf. Cameron, this volume). Hence this magnetic field cannot be dynamically important until the solar system is well cleared of gas.

Suppose we hypothesize an epoch of low gas density before the solar wind has turned on. What role might a quasi-dipolar magnetic field play in the solar system? Such a configuration will be confined on the outside by the interstellar magnetic field and gas (the heliopause). Inside this heliopause, individual charged particle motion will be determined by the electric and magnetic field configuration. This is essentially a magnetosphere. We may then look at other known examples of magnetospheres and see which, if any, familiar phenomena might be expected to occur in this primordial solar magnetosphere.

One of the most striking phenomena of the terrestrial and Jovian magnetospheres are the Van Allen radiation belts. One important source for the inner Van Allen belt of the terrestrial magnetosphere is protons from the radioactive decay of neutrons produced in the earth's atmosphere by cosmic rays. A flux of the order of 10^6 protons/cm^2 sec of energy > 5 MeV is seen near the heart of the inner zone (cf. Freeman, 1965). One would expect an analogous process from cosmic-ray-produced or indigenous neutrons from the solar atmosphere. It is interesting to consider the possibility that such a radiation source might explain isotopic anomalies such as the neon E (Black, this volume) or the ^{26}Mg anomalies in the Allende meteorite (Heymann and Dziczhaniec, 1976). To provide the observed isotopic fractionations the irradiation must occur in the gas phase whereas most nebular models have surface densities large compared to the range of MeV protons. One might consider irradiation of the surface of the nebula. The detailed examination of this process is beyond the scope of this paper. We hope to investigate this and other magnetospheric processes in the future.

The Subsequent Field Configuration

Following this epoch of quasi-dipolar field we expect the solar wind to turn on and further stretch out the field. A T-Tauri solar wind could easily overwhelm the field calculated from the 6×10^{36} gauss cm^3 magnetic moment. For example, at 1 AU a solar wind corresponding to a mass loss of $10^{-7}\ M_{\odot}$/year (Herbig, this volume) would yield a particle pressure equivalent to a magnetic field of about 1 gauss compared with our predicted value of 2×10^{-2} gauss. Fields of the order of 1 gauss at 1 AU are possible only if the field falls off much more slowly than indicated by equation (2).

As the solar wind begins to blow, the field lines will be drawn out and angular momentum will be transferred from the sun. As the sun slows, it also approaches a less convective state and, in turn, the magnetic moment and attendant field will weaken and approach the present day configuration.

Summary

We have examined the range of possible values for the primordial solar dipole moment. We find likely values of 1×10^{34} to about 5×10^{36} gauss cm^3. Using the naive approach of a direct scaling with the Jovian magnetosphere we obtain values for the distant solar field of no more than [6, 2, and 0·2] $\times 10^{-2}$

gauss at the orbits of Mercury and Earth and Saturn respectively. These substantial fields are probably sufficient to provide the 'seed' fields for planetary dynamos. They would, however, fall short of providing an adequate field to magnetize *some* of the lunar samples. The quasi-dipole nature of the solar field during this epoch may give rise to interesting magnetospheric-type phenomena. Subsequent solar wind interaction and solar spin down will reduce this field to the present-day configuration.

References

Alfvén, H., and Arrhenius, G. (1973). 'Structure and evolutionary history of the solar system, III., *Astrophys. Space Sci.*, **21**, 117.

Banerjee, S. K., and Mellema, J. O. (1976). 'A solar origin for the large lunar magnetic field at $4{\cdot}0 \times 10^9$ years ago?' Abstracts of papers submitted to the *Seventh Lunar Science Conference*, March 1976.

Freeman, J. W. (1965). 'The geomagnetically trapped radiation, protection against radiation in space.' *NASA*, SP–71.

Heymann, Dieter, and Dziczkaniec, Marlene (1976). 'Early irradiation of matter in the solar system: Magnesium (proton, neutron) scheme.' *Science*, **191**, 79.

Hill, T. W., Dessler, A. J., and Michel, F. C. (1974). 'Configuration of the Jovian magnetosphere.' *Geophys. Res. Lett.*, **1**, 3.

Lang, Kenneth R. (1974). *Astrophysical Formulae*, Springer–Verlag.

Ness, Norman F. (1968). *The Interplanetary Medium, Introduction to Space Science*, Gordon and Breach.

Reeves, H. (1972). 'Seven questions relating to the solar system and the answers given in various models,' *Symposium on the Origin of the Solar System*, (H. Reeves, ed.), C.N.R.S., Nice.

Russell, Christopher T. (1976). 'The magnetic moment of Venus: Venera-4 measurements reinterpreted.' *Geophys. Res. Lett.*, **3**, 125.

Strangway, D. W., Sharpe, H. N., and Peltier, W. H. (1976). 'Planetary magnetism—primordial or dynamo.' *Seventh Lunar Science Conference Abstracts*.

EVIDENCE FROM THE EARLY MOON

S. K. RUNCORN
Institute of Lunar and Planetary Sciences, School of Physics, The University, Newcastle upon Tyne, U.K.

Important evidence concerning the processes of formation of the solar system is that of the primeval melting and differentiation processes in its solid bodies. If shortlived radioactive isotopes have played a role in this, important conclusions follow about the time of formation. The evidence for such melting is widespread: for example, the iron meteorites were melted and cooled gradually as is evident from their Widmanstätten structures. In the Earth, the iron core and the sialic continent material also differentiated early, but we only know that this occurred far back in the geological record of the Pre-Cambrian. This dates the differentiation earlier than $3{\cdot}6 \times 10^9$ y. Reasoning from the lead isotopes, core formation has now been dated very soon after the Earth's formation. Mars, Mercury and probably Venus possess evidence for varying degrees of differentiation but again the date of this process is conjectural. Evidence of present magnetic fields in the terrestrial planets suggest weak dynamo processes in fluid electrically conducting cores. Thus the record of such planetary melting and differentiation is obscure particularly in its timing, so that alternative mechanisms of generating the required heat have been proposed. For example, Sonett *et al.* (1975) have suggested that in an early T-Tauri phase of the Sun, intense solar-wind magnetic fields could have heated planetary bodies by induction. The ferromagnetic silicates which they contain are semiconductors so that weak induced currents heat the bodies and the electrical conductivity increases: the currents then increase further and thus the magnetic energy dissipates in the bodies and rapidly heats them. Such electromagnetic heating seems particularly applicable for small bodies like the meteorites. With large bodies, like the Earth and Venus, it may simply heat the exterior and might not cause core formation. In these planets, however, except in the early stages of accretion, substantial amounts of gravitational potential energy are released which, provided that the heat is not radiated into space but retained by opaque dust clouds around the accreting object, is easily adequate to raise the temperature to the melting point. However, we are not sure that it was immediately after accretion that this process of differentiation and core formation took place. Core formation itself releases in the Earth twice the amount of energy released through the whole of the Earth's life by U, Th, and K^{40}.

In discussion of the nature of these early melting processes, evidence from the Moon is critical. This is so firstly because the Moon is too small an object for the gravitational energy released by accretion to be so large that early melting presents no problem. After fragments of anorthosite were found in the regolith and identified as originating in the highlands, it became clear that the Moon had differentiated early to form a highland shell of anorthosite gabbroic composition. Geochemists (e.g. Taylor, 1975) generally interpreted this as requiring that an outer shell of the primitive Moon, perhaps a few hundred km thick, had been melted to yield the observed thickness of the anorthosite shell. As this differentiation could be shown to be very early, it was natural to suggest as a cause the gravitational energy released in the accretion process. That the energy would be released in sufficient quantity to raise silicates to the melting point only when the accreting Moon began to approach its present size seemed to be in accord with the observed phenomena. However, there are strong reasons for suggesting that the Moon was originally melted to the centre to form a core and a second shell of basalt-rich material a few hundred million years after the origin.

The second reason why evidence from the Moon is so critical is that the record of lunar events between say 4·4 and $3{\cdot}2 \times 10^9$ y is now remarkably well documented. This is in marked contrast with the Earth where knowledge of this period is restricted to fragmentary outcrops of rocks dating back only to about $3{\cdot}6 \times 10^9$ y. One of the striking results of the Apollo project was that although the lunar basalts have ages of crystallization ranging between 3·2 and $3{\cdot}9 \times 10^9$ y, their model ages all cluster around $4{\cdot}4 \times 10^9$ y. This result is interpreted by geochemists in terms of closed-system melting, which means that the original differentiation producing the source materials of the mare basalts occurred about $4{\cdot}4 \times 10^9$ y and that at the later melting of these sources to provide the magma which filled the basins, no isotopic exchange with the surrounding Moon took place. This geochemical concept has been interpreted geophysically (Runcorn, 1976): below the anorthosite shell there is a shell rich in basalt and below this is the olivine mantle. There are three geophysical data which suggest the existence of this basalt–olivine shell. The variation of electrical conductivity with depth, the seismic data suggesting a shell of properties different from the mantle between 60 km and 250 km deep and the moment of inertia factor requiring some differentiation all point to the existence of this basalt shell but they are not conclusive. Recently, the date of this differentiation has been further determined by the Samarium–Nyodynium decay scheme as about $4{\cdot}4 \times 10^9$ y. The seismological and electrical conductivity data suggest that this basalt-rich-shell extends down to 250 km depth. Thus, the melting of the Moon which yielded this differentiated shell must have extended to a much greater depth.

Thirdly and perhaps this is most critical, evidence for a completely molten early Moon is provided by the lunar gravitational and magnetic data. There has been up to recent times, no compelling evidence for an iron core in the moon: the lunar mean density is in fact compatible with the absence of one. An iron core was first suggested in a theory of the non-hydrostatic shape of the moon (Runcorn, 1967). It was supposed that this is dynamically maintained and is not a consequence of an early distortion of the Moon retained by finite strength. The two-cell pattern of convection necessary to explain the observations seemed compatible only with the existence of a core of radius between 1/10 and 1/3 of the lunar radius. Recently more compelling evidence for the existence of a core has been obtained from the extensive remanent magnetization present in the lunar surface rocks. Most of the returned samples of basalt and high grade breccia possess natural remanent magnetization and the magnetic fields observed at the surface, local anomalies observed in various experiments, and the anomalies seen in the lunar magnetic maps obtained from orbiting subsatellites launched by Apollo 15 and 16 are all consistent with these laboratory findings. The distribution of directions of the magnetization of the lunar crust must be such as to produce no dipole field outside the Moon at the present time, for an exceedingly small limit to the external dipole field has recently been set from the analysis of the subsatellite magnetometer experiments on Apollo 15 and 16. It has been shown that this distribution implies that the lunar shell was magnetized by a field of internal origin (Runcorn, 1975). Although other mechanisms for generating such a field have been suggested, the dynamo mechanism in which fluid motions in a liquid iron core are responsible for the field seems to be the only tenable one. Some geophysical evidence for the existence of an iron core of 400 km radius has been obtained but is not conclusive.

The magnetic maps of the far side of the Moon show large anomalies over the deep basins and it seems likely that parts of the highland anorthosite shell possess magnetizations stronger by an order of magnitude than the mare basaltic flows. This suggests that the lunar magnetic field was present a few hundred million years after the lunar origin and persisted until at least $3{\cdot}2 \times 10^9$ y. Recent studies of the palaeointensity of this field (Stephenson *et al.*, 1974) have further suggested that it was about 1 G at 4×10^9 y ago and had diminished by one to two orders of magnitude by $3{\cdot}2 \times 10^9$ y ago: subsequently the dynamo action ceased. While this data is still only based on rather few samples the evidence suggests that the dynamo process was more vigorous in the early history of the Moon than later. Simple thermodynamic arguments concerning the flux of convective heat in this core 4×10^9 y ago indicate that heat sources were present that were many times more powerful than those inferred from the present concentration of U, Th, K^{40} in the Moon, assuming that the Moon's interior has the composition of chondritic meteorites (Runcorn, 1977).

For these reasons it must be concluded that radioactive elements, now extinct or virtually so, may have played a critical role in the early physical and chemical process of the Moon (Runcorn *et al.*, 1977). The discovery in the Allende meteorite of Mg^{26} has provided strong evidence for the existence of Al^{26} in the early accreted matter of the solar system (Lee *et al.*, 1976). However, if it was present in the abundance needed to melt the Moon in its first few hundred million years, it would be necessary to suppose that the time interval between the end of nucleogenesis and the condensation from the solar system nebula and dust cloud of small solar system bodies was within a few times the half life of Al^{26}, which is 700,000 years. Recent studies of the dynamics of evolution of this dust cloud have shown that the time from the condensation into planetesimals until the formation of Moon-sized bodies, is of the order of 10^8 to 10^9 years. Thus, Al^{26} cannot be the cause of the early heating of the Moon, unless solar system condensation processes occur much more rapidly as has been suggested by Mizutani *et al.* (1972). Radioactive elements of longer half lives are required. But those with half lives much in excess of 10^9 years would heat the Moon too slowly, for when temperatures above about half the melting temperature are reached, solid-state creep becomes important and the much more efficient thermal convection replaces conduction. Thus unless the heat sources deliver their supply quickly, convective transport prevents the internal temperatures rising to the melting point. The recent report (Gentry *et al.*, 1976) that atoms of superheavy elements are present in monazite inclusions from giant pleichroic haloes suggests that these may have half lives of the right order to melt the Moon early without requiring an unacceptably short period between the end of nucleogenesis and the formation of the Moon. The reported discovery remains unproven but has served to stimulate a search for evidence of superheavy elements in the early history of the solar system. Should such elements be proved not to have existed in the necessary abundance or to have too short a half life, then the presence of the abundant Al^{26} will have to be assumed and this will set a most stringent limit on the time of formation of the solar system.

References

Gentry, R. V., Cahill, T. A., Fletcher, N. R., Kaufmann, H. C., Medsker, L. R., Nelson, J. W., and Flocehini, R. G. (1976). *Phys. Rev. Letts.*, **37**, 11–14.

Lee, T., Papanastassiou, D. A., and Wasserburg, G. J. (1976). *Geophys. Res. Lett.*, **3**, 109–112.

Mizutani, H., Matsui, J., and Takeuchi, H. (1972). *The Moon*, **4**, 476–489.

Runcorn, S. K. (1967). *Proc. Roy. Soc.*, **A296**, 270–284.

Runcorn, S. K. (1975). *Phys. Earth and Planet. Int.*, **10**, 327–335.

Runcorn, S. K. (1976). *Proceedings of the 7th Lunar Science Conference.*

Runcorn, S. K. (1977). *Science.* In the press.

Runcorn, S. K., Libby, L. M., and Libby, W. F. (1977). *Nature.* In the press.

Sonett, C. P., Colborne, D. S., and Schwartz, K. (1975). *Icarus*, **24**, 231–255.

Stephenson, A., Collinson, D. W., and Runcorn, S. K. (1974). *Proc. 5th Lunar Conf., Geochim. et Cosmochim. Acta*, Suppl. 5, **3**, 2859–2871.

Taylor, S. R. (1975). *Lunar Science: A post-Apollo View*. Pergamon Press.

Recent General References Summarizing Recent Work on the Moon

International Astronomical Union, *Symposium, No. 47, The Moon.* 1971. (Ed. S. K. Runcorn and H. C. Urey). D. Reidel.

Proceedings of the Apollo 11 Lunar Science Conference. 1970. Vols. 1 to 3 (Ed. A. A. Levinson). Pergamon Press: *Geochimica et Cosmochimica Acta*, Suppl. 1.

Proceedings of the 2nd Lunar Science Conference. 1971. Vols. 1 to 3 (Ed. A. A. Levinson). The MIT Press: *Geochimica et Cosmochimica Acta*, Suppl. 2.

Proceedings of the 3rd, 4th, 5th, 6th, 7th, 8th. Lunar Science Conferences. 1972. Vols. 1 to 3. The MIT Press: *Geochimica et Cosmochimica Acta*, Suppl. 3.

Lunar Geophysics (Ed. Z. Kopal and D. Strangway). *The Moon*, **4**, 3–249, 271–504; **5**, 3–160.

'The Moon: a recent appraisal from space missions and laboratory analysis.' 1977. (Ed. G. M. Brown et al). *Phil. Trans. R. Soc.*, **A285**, 1–600.

AUTHOR INDEX

A contributor's references to his own publications are not repeated here. References to publications are enclosed in parentheses.

SUBJECT INDEX